D1230994

Energy and Finite Element Methods in Structural Mechanics

Energy and Finite Element Methods in Structural Mechanics

SI Units Edition

IRVING H. SHAMES
State University of New York at Buffalo

CLIVE L. DYM
University of Massachusetts, Amherst

PUBLISHING FOR ONE WORLD

NEW AGE INTERNATIONAL (P) LIMITED, PUBLISHERS
New Delhi • Bangalore • Chennai • Cochin • Guwahati • Hyderabad
Jalandhar • Kolkata • Lucknow • Mumbai • Ranchi
Visit us at **www.newagepublishers.com**

Copyright © 1991, New Age International (P) Ltd., Publishers
Published by New Age International (P) Ltd., Publishers
First Edition: 1991
Reprint: 2009

All rights reserved.
No part of this book may be reproduced in any form, by photostat, microfilm, xerography, or any other means, or incorporated into any information retrieval system, electronic or mechanical, without the written permission of the copyright owner.

Branches:

- 36, Malikarjuna Temple Street, Opp. ICWA, Basavanagudi, **Bangalore**. ✆ (080) 26677815
- 26, Damodaran Street, T. Nagar, **Chennai**. ✆ (044) 24353401
- Hemsen Complex, Mohd. Shah Road, Paltan Bazar, Near Starline Hotel, **Guwahati**. ✆ (0361) 2543669
- No. 105, 1st Floor, Madhiray Kaveri Tower, 3-2-19, Azam Jahi Road, Nimboliadda, **Hyderabad**. ✆ (040) 24652456
- RDB Chambers (Formerly Lotus Cinema) 106A, Ist Floor, S.N. Banerjee Road, **Kolkata**. ✆ (033) 22275247
- 18, Madan Mohan Malviya Marg, **Lucknow**. ✆ (0522) 2209578
- 142C, Victor House, Ground Floor, N.M. Joshi Marg, Lower Parel, **Mumbai**. ✆ (022) 24927869
- 22, Golden House, Daryaganj, **New Delhi**. ✆ (011) 23262370, 23262368

ISBN (10): 81-224-0749-8
ISBN (13): 978-81-224-0749-5

Rs. 265.00

C-09-01-3305

Printed in India at Glorious Printers, Delhi.

PUBLISHING FOR ONE WORLD
NEW AGE INTERNATIONAL (P) LIMITED, PUBLISHERS
4835/24, Ansari Road, Daryaganj, New Delhi-110002
Visit us at **www.newagepublishers.com**

CONTENTS

*Indicates that section may be bypassed.
**Indicates that section is in fine print.

PREFACE TO SI UNITS EDITION

In this edition we have used SI units rather than USCA units. We wish to thank Mr. Shenyao Yang, a doctoral student in mechanical engineering at State University of New York at Buffalo, for his tireless work in changing units and checking the entire text.

This text is the outgrowth of material used in senior and graduate courses for students in civil, mechanical, and aeronautical engineering. To meet the needs of this varied audience, we have labored to make this text flexible as to possible use. Consequently, the book is divided into three distinct parts of approximately equal size. Part I is entitled "Foundations of Solid Mechanics and Variational Methods." Part II is entitled "Structural Mechanics." These two parts are taken (with some deletions, additions, and alterations) from the authors' text *Solid Mechanics: A Variational Approach,* published by McGraw-Hill in 1973. Part III is entitled "Finite Elements." Depending on the background of the students and the aims of the course, the instructor can use selected portions from some or all of the three parts of the text to form the basis of an individual course. To aid the instructor in this regard, we have made liberal use of asterisks and somewhat smaller printer, respectively, to delineate text material that one can bypass without loss of continuity and material that is of a decidedly more advanced level.* The asterisk is also used to identify problems that are more difficult or longer than usual. Following this preface we present two course outlines based on our teaching experiences with this text.

Our purpose is to afford the student who goes primarily through Part I and Part III of the text a sound foundation in variational calculus and energy methods before

*To ensure that the reader notices the somewhat finer print used for advanced material, there is a small horizontal line just before and just after all such material.

delving into finite elements. The goal is to make finite elements more understandable in terms of fundamentals and also to provide the student with the background needed to extrapolate the finite element method to areas of study other than solid mechanics. Students who go primarily through Parts I and II, on the other hand, will be able to derive working equations of structural mechanics through variational principles, so that they will have a good understanding of the limits of validity of these equations. Furthermore, they will know that the resulting equations and associated kinematic and natural boundary conditions stemming from the variational process form properly posed boundary-value problems. In addition, a number of approximation techniques will be available to them, using the quadratic functional for a boundary-value problem. These approximations will be continuous functions over the entire domain of the problem rather than the discretized results obtained from finite elements. Finally, it is our purpose to give students who go through the entire text a balanced and connected exposure to certain key aspects of modern structural and solid mechanics. With these goals in mind, we briefly describe the contents of this text.

In Part I we formulate certain key principles of elasticity (Chapter 1). In Chapter 2 we present a self-contained treatment of the calculus of variations. We have chosen in this study to use the single- or multiple-parameter family approach as well as the delta-operator approach and, in doing so, have made a great effort to clearly relate these approaches. In Parts A and B of Chapter 3 we carefully derive the various energy methods of solid mechanics and relate these methods to each other. We use the energy methods to solve various aspects of truss problems. The energy principles are presented from a continuum viewpoint and then are specialized for specific structures. In Parts A and B of this chapter we start with a functional and then derive Euler-Lagrange equations as well as valid kinematic and natural boundary conditions. In Part C of the chapter we go the opposite way, starting with certain classes of boundary-value problems and then formulating the appropriate quadratic functional. We then show how a boundary-value problem may be approximately solved by extremizing the quadratic functional, using the full domain of the problem. Thus the Ritz method is presented, as is the closely related nonvariational Galerkin method. Part C of Chapter 3 forms invaluable background material for finite elements in Part III.

In Part II, on structural mechanics, we derive the working equations for beams; torsion members; plates; the dynamics of beams and plates; and the elastic stability of columns, beam-columns, and plates. Our general procedure is to extremize the total potential energy functional to derive a properly posed boundary-value problem. In each case, after deriving a set of equations and boundary conditions, we investigate in a variety of ways the validity of the resulting formulation. Specifically, we consider how well the equations portray the behavior of the actual structural member. This is necessary since simplifications of geometry, of constitutive laws, and of mode of behavior will have been built into the total potential energy functional out of which the working equations have evolved. The limitations of the resulting working equations are thus pointed out. Also, specific approximation methods are presented for these structural elements by working with the quadratic functional of the boundary-value problem. Thus the methods of Ritz, Rayleigh, Rayleigh-Ritz, Kantorovich, and Trefftz are presented for structural problems at various points in Part II. Exact solutions for various boundary-value problems are worked out so as to afford means of comparison with the results of the approximation techniques.

We are now ready to proceed to finite elements in Part III. We assume that the student has no background in this area, and so in Part A of Chapter 9 we come back to plane trusses. Here we consider the members to be the finite elements and the joints to be nodal points. We solve for nodal deflections of trusses in a simple, natural way, yet using the format and notation of finite elements. We are thus able to see the finite element process in a most familiar setting. After this introduction, we proceed in Part B of Chapter 9 to set forth certain preliminary basic considerations of finite element methodology, developing key equations and notation to be used many times over later in the text. In Chapter 10 we go to another simple structural element—the beam—using small lengths of the beam as finite elements. We now apply the formulations of Part B of the previous chapter to this simple structure to solve for deflections and section forces at the nodes. Thus in Chapters 9 and 10 we have presented a careful, deliberate means of easing the student into the finite element methodology.

In Chapter 11 we can now discuss the kinds of elements that may be used and the degrees of freedom for such elements. The requirements for convergence of the finite element process with decreasing mesh size are discussed. Various interpolation formulas are presented, including Lagrange and Hermite polynomials. Such vital topics as completeness, geometric isotropy, levels of continuity, natural coordinates, and isoparametric elements are discussed. After Chapter 11 we apply this material to plane stress, plates, torsion, elastic stability, and dynamics in succeeding chapters. The examples and homework problems are done with five or six elements, with much of the initial work performed with hand calculators. The digital computer is used to solve the resulting simultaneous equations. Results from complete computer programs are used for fine meshes to study convergence properties. For a term project, a student or group of students might put together a complete program. As a guide, we present a full program for plates in Appendix IV. Also, we include in Appendix III a gridding program that is very useful, albeit one that cannot reasonably be expected to be developed by students as part of the course. Hence, in Chapters 9 to 16 of Part III of the text, we cover the same topics that were covered in Part II but from a finite element point of view.

We present in Chapter 17 a reexamination of the variational principles underlying the energy methods—primarily as these principles are related to the finite element process. We believe this to be very important in two respects. First, it will give the student a better understanding of the finite element process and will enable him or her to do complex problems. Second, and of no less importance, it will enable the student to use finite elements in other areas of study with minimal additional considerations. In Chapters 9 to 16 of Part III, we use the total potential energy principle and the total complementary potential energy principle as the vehicles for developing the finite element equations. We see in Chapter 17 that, in fact, we have been using the Ritz method to minimize the aforementioned functionals in small domains. We next point out that what we have been doing is only a special case of a more general undertaking. Specifically, we see that for cases where a quadratic functional exists for a specific boundary-value problem, we can use the finite element approach by minimizing the quadratic functional in each of the small domains comprising the mesh inside the boundary of the problem. Thus the study of quadratic functionals in Part C of Chapter 3 now provides the means for extending finite elements to many areas of study. (Indeed, many recent texts on finite elements consider solid mechanics as only one of many areas of application.) We then apply the finite element approach briefly to ordinary differential equations, steady-state heat transfer, and two-dimensional potential flow. Thus in the last chapter we return to Chapter 3 of Part I to close the loop, as it were, and at the same time open up some additional horizons. Finally, we point out that when quadratic functionals cannot be found, we use for finite elements the method of weighted residuals and the associated Galerkin method presented earlier in Chapter 3.

In the Instructor's Manual, in addition to problem solutions, we present some useful material that the instructor may wish to duplicate and make available to the students. This includes a treatment of matrices and a discussion of programming for finite elements. We have put this material in the manual rather than in the text itself to avoid making the text excessively large. In a separate Instructor's Computer Manual we have all the programs used in the text in both SI and British units as well as data file instructions for these programs. Furthermore, all this material is on an enclosed floppy disk.

At this point, we wish to acknowledge the help we received for the earlier version of this book, and for this purpose we quote from the earlier preface:

> We wish to thank Prof. T. A. Cruse of Carnegie-Mellon University for reading the entire manuscript and giving us a number of helpful comments. Also our thanks go to Prof. J. T. Oden of the University of Alabama for his valuable suggestions. Dr. A. Baker and Dr. A. Frankus helped out on calculations and we thank them for their valuable assistance. One of the authors (C.L.D.) wishes to pay tribute to three former teachers–Prof. J. Kempner of Brooklyn Polytechnic Institute and Profs. N. J. Hoff and J. Mayers of Stanford University–who inspired his interest in variational methods. The other author (I.H.S.) wishes to thank his colleague Prof. R. Kaul of State University of New York at Buffalo for many stimulating and useful conversations concerning several topics in this text.

For the present volume, we wish to thank Prof. G. Manolis of the Civil Engineering Department at the State University of New York (SUNY) at Buffalo for his interest in this undertaking and his numerous helpful suggestions. Dr. Manolis used the voluminous notes for this text for this graduate course in finite elements for several years. We are particularly grateful for the help of Dr. R. Gu, who, while a doctoral student in engineering science at SUNY, Buffalo, did much of the programming for Part III of the text. He also carefully checked and rechecked our work over parts of the book. Mr. P. Wipasuramonton, a doctoral student in engineering science at SUNY, Buffalo, developed a number of computer programs for this text and tirelessly checked the manuscript. We thank him sincerely. Dr. M. Elattari, Dr. J. Pitarresi, Mr. B. J. Wang, and Mr. C. Ng at SUNY, Buffalo, worked on certain problems for us, and we appreciate their very competent efforts. We were fortunate in having had helpful reviews of our manuscript by Prof. J. Tuma of Arizona State University, Tempe, and Prof. S. Batterman of the University of Pennsylvania. Finally, we wish to thank Ms. Barbara Willgens for her excellent typing.

Irving H. Shames
Clive L. Dym

OUTLINE
COURSE I (Senior–Graduate Level)*
FINITE ELEMENTS
(3 credit hours)

Chapter 1	Linear Elasticity 1.1–1.6 to Eq. (1.24); 1.10–1.12, 1.17, 1.18. Delete fine-print material except for examples	4 hr
Chapter 2	Calculus of Variations 2.1–2.6, 2.9, 2.10	5 hr
Chapter 3	Variational Principles of Elasticity. Delete 3.6; delete material in 3.8 after Example 3.6; delete fine-print material except for examples and Case 1 of 3.11	8 hr
Chapter 6	Classical Theory of Plates 6.1–6.4	3 hr
Chapter 9	Introduction to Finite Elements	3 hr
Chapter 10	Finite Elements for Beams	3 hr
Chapter 11	Elements and Interpolation Functions	4 hr
Chapter 12	Finite Elements for Plane Stress	3 hr
Chapter 13	Finite Elements for Plates	4 hr
Chapter 17	Broader Aspects of Finite Elements 17.1–17.3; 17.8 on weighted residuals; 17.9–17.11 on potential flow of fluids	5 hr
Appendix I	Cartesian Tensors and Integral Theorems	3 hr
		45 hr

*See Instructor's Manual for a detailed discussion of this proposed course.

OUTLINE

COURSE II (Senior–Graduate Level)

STRUCTURAL MECHANICS

(3 credit hours)

Chapter 1	Linear Elasticity 1.1–1.6 to Eq. (1.24); 1.10–1.12; 1.17, 1.18. Delete fine-print material except for examples	4 hr
Chapter 2	Calculus of Variations	6 hr
Chapter 3	Variational Principles of Elasticity. Delete 3.6; delete material in 3.8 after Example 3.6. Delete fine-print material except for examples.	8 hr
Chapter 4	Beams. Delete 4.5 and 4.8	3 hr
Chapter 5	Torsion. Delete 5.5	4 hr
Chapter 6	Classical Theory of Plates, up to 6.10	5 hr
Chapter 7	Dynamics of Beams and Plates, 7.1–7.6; 7.8–7.11	5 hr
Chapter 8	Elastic Stability. Delete fine-print material except for examples. Delete 8.11, 8.12, and 8.16	7 hr
Appendix I	Cartesian Tensors and Integral Theorems	3 hr
		45 hr

I

FOUNDATIONS OF SOLID MECHANICS AND VARIATIONAL METHODS

We begin by presenting key features of elasticity. Readers unfamiliar with index notation should first study Appendix I. We then set forth in Chapter 2 a self-contained treatment of the calculus of variations. In this chapter we are immediately made aware of one of the remarkably large number of applications stemming directly and indirectly from the calculus of variations–that is, optimization theory. Other applications occur in Chapter 3. There the calculus of variations is used to develop the vital energy methods of solid mechanics. Here we begin the study of structures by applying these methods to the solution of truss problems. In addition, in Chapter 3 we are able to present an introduction to the theory of quadratic functionals. This study then permits the derivation of properly posed boundary-value problems as well as the formulation of useful approximation techniques. These techniques yield a continuous function over the entire domain of the problem rather than a set of discretized results. Finally, we point out that quadratic functional theory forms the basis for the introduction of finite element methodology in Part III of the text.

Although the applications are directed toward solid mechanics, the key ideas of the variational process pervade many other areas of study where, instead of dealing with the differential equation and the boundary conditions, one works with the associated quadratic functional.

Before embarking, we wish to point out that Chapter 1 on elasticity will more than serve our needs throughout the entire text. Accordingly, depending on the goals and level of the course, the instructor may wish to cover only selected portions of Chapter 1. More is said about this point in the Instructor's Manual.

1

THEORY OF LINEAR ELASTICITY

1.1 INTRODUCTION

In this text we shall be concerned with the study of elastic bodies. Accordingly, we shall now present a brief treatment of the theory of elasticity. In developing the theory we shall set forth many concepts that are needed for understanding the variational techniques soon to be presented.

We will not attempt solutions of problems directly with the full theory of elasticity; indeed, very few such analytical solutions are available. Generally we will work with special simplifications of the theory in which a priori assumptions are made about

1. The stress field (plane stress problems for example),
2. The strain field (plane strain problems for example), and
3. The deformation field (structural mechanics of beams, plates, and shells).

The significance of these simplications and the limitations of their use can best be understood in terms of the general theory.

Specifically, in this chapter we will examine the concepts of stress, strain, constitutive relations, and various forms of energy. This will permit us to present the equations of linear elasticity and to consider the question of uniqueness of solutions to these equations. The plane stress simplification will be considered in this chapter; examples of simplification 3 will be presented in subsequent chapters.

Part A
Stress

1.2 FORCE DISTRIBUTIONS

In the study of continuous media, we are concerned with the manner in which forces are transmitted through a medium. At this time, we set forth two classes of forces that will concern us. The first is the so-called *body force distribution*, distinguished by the fact that it acts directly on the distribution of matter in the domain of specification. Accordingly, it is represented as a function of position and time and will be denoted as $\mathbf{B}(x, y, z, t)$ or, in index notation, as $B_i(x_1, x_2, x_3, t)$. The body force distribution is an intensity function and is generally evaluated per unit mass or per unit volume of the material acted on. (In the study of fluids the basis of measure is usually per unit mass, while in the study of solids the basis of measure is usually per unit volume.)

In discussing a continuum there may be some apparent physical boundary that encloses the domain of interest, such as the outer surface of a beam. On the other hand, we may elect to specify a domain of interest and thereby generate a "mathematical" boundary. In either case, we will be concerned with the force distribution that is applied to such boundaries directly from material outside the domain of interest. We call such force distributions *surface tractions* and denote them as $\mathbf{T}(x, y, z, t)$ or $T_i(x_1, x_2, x_3, t)$. The surface traction is again an intensity, given on the basis of per unit area.

Now consider an infinitesimal area element on a boundary (see Fig. 1.1) over which we have a surface traction distribution $\mathbf{T}$ at some time t. The force df_i transmitted across this area element can then be given as follows:

$$df_i = T_i \, dA$$

Note that $\mathbf{T}$ need not be normal to the area element and so this vector and the unit

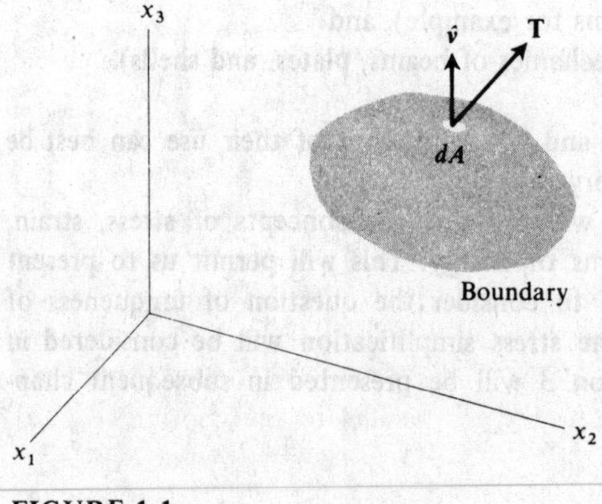

FIGURE 1.1
Traction force intensity on the boundary.

outward normal vector $\boldsymbol{\nu}$ may have any orientation whatever relative to each other. We have not brought the unit normal $\boldsymbol{\nu}$ into consideration thus far, but we will find it useful, as we proceed, to build into the notation for the surface traction a superscript referring to the direction of the area element at the point of application of the surface traction. Thus we will give the traction vector as

$$\mathbf{T}^{(\nu)}(x, y, z, t)$$

or as

$$T_i^{(\nu)}(x_1, x_2, x_3, t)$$

where (ν) is not to be considered as a power. If the area element has the unit normal in the x_j direction, then we would express the traction vector on this element as $\mathbf{T}^{(j)}$ or as $T_i^{(j)}$. In the following section we shall show how we can use the superscript to good advantage.

1.3 STRESS

Consider now a vanishingly small rectangular parallelepiped taken at some time t from a continuum. Choose reference x_1, x_2, x_3 to be parallel to the edges of this rectangular parallelepiped, as shown in Fig. 1.2. We have shown surface tractions on three rectangular boundary surfaces of the body. Note that we have employed the superscript to identify the surfaces. The Cartesian components of the vector $\mathbf{T}^{(1)}$ are then $T_1^{(1)}$, $T_2^{(1)}$, $T_3^{(1)}$. We shall now represent these components by employing τ in place of T and moving the superscript down to be the first subscript while deleting the enclosing parentheses. The components of $\mathbf{T}^{(1)}$ are then given as τ_{11}, τ_{12}, τ_{13}. In general, we have

$$T_i^{(1)} \equiv \tau_{11}, \tau_{12}, \tau_{13}$$

$$T_i^{(2)} \equiv \tau_{21}, \tau_{22}, \tau_{23}$$

$$T_i^{(3)} \equiv \tau_{31}, \tau_{32}, \tau_{33}$$

In a more compact manner we have

$$T_j^{(i)} \equiv \tau_{ij}$$

where the nine quantities comprising τ_{ij} are called *stresses* and are forces per unit area, the first subscript gives the coordinate direction of the normal of the area element, and the second subscript gives the direction of the force intensity itself. These nine force intensities are shown in Fig. 1.3 on three orthogonal faces of an infinitesimal rectangular parallelepiped with faces parallel to the coordinate planes of x_1, x_2, x_3.

Representing the set τ_{ij} as an array, we have

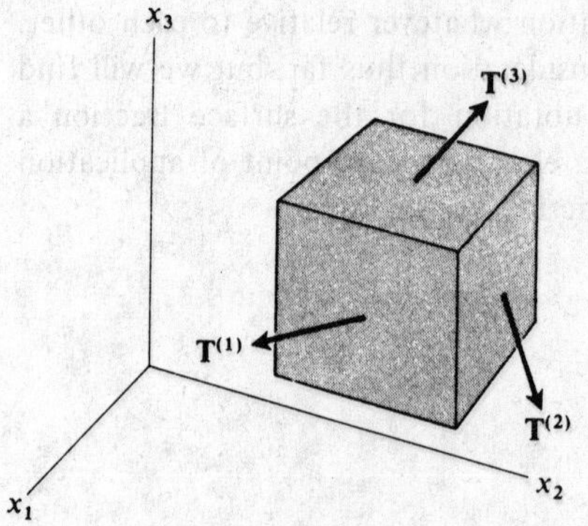

FIGURE 1.2
Traction forces on orthogonal faces.

$$\begin{pmatrix} \tau_{11} & \tau_{12} & \tau_{13} \\ \tau_{21} & \tau_{22} & \tau_{23} \\ \tau_{31} & \tau_{32} & \tau_{33} \end{pmatrix}$$

where the terms in the main diagonal are called normal stresses, since the force intensities corresponding to these stresses are normal to the surface, while the off-diagonal terms are the shear stresses.

We shall employ the following sign convention for stresses. A normal stress directed outward from the interface is termed a tensile stress and is taken by definition as positive. A normal stress directed toward the surface is called a compressive stress (it is exactly the same as pressure on the surface) and is, by definition, negative. For shear stresses we employ the following convention:

> A shear stress is positive if (1) both the stress itself and the unit normal point in positive coordinate directions (the coordinates need not be the same), or (2) both point in the negative coordinate directions (again, the coordinates need not be the same). A mixture of signs for coordinate directions corresponding to shear stress and the unit normal indicates a negative value for this shear stress.†

Note that all the stresses shown in Fig. 1.3 are positive stresses according to this sign convention.

Knowing τ_{ij} for a set of axes, i.e., for three orthogonal interfaces at a point, we can determine a stress vector $\mathbf{T}^{(\nu)}$ for an interface at the point having *any* direction whatever. We shall now demonstrate this. Consider any point P in a continuum (Fig. 1.4*a*). Form as a free body a tetrahedron with P as a corner and with three orthogonal faces parallel to the reference planes, as shown enlarged in Fig. 1.4*b*. The legs of the tetrahedron are given as Δx_1, Δx_2, and Δx_3, as indicated in the diagram, and the inclined face $\overline{ABC}$ has a normal vector $\boldsymbol{\nu}$. The stresses have been shown for the orthogonal faces, as have the stress vector $\mathbf{T}^{(\nu)}$ and its components for the inclined face. We denote by h the perpendicular distance from

† Note that normal stress follows this same convention.

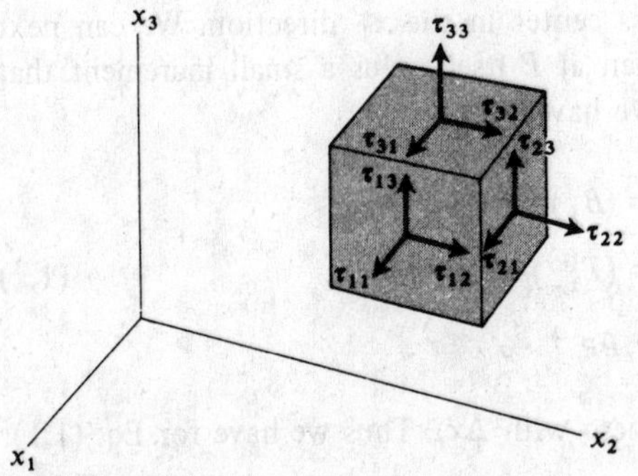

FIGURE 1.3
Stresses on orthogonal faces.

P to the inclined face (the "altitude" of the tetrahedron). The stresses and traction components given on the faces of the tetrahedron are average values over the surfaces on which they act. Also, the body force vector $(\mathbf{B})_{av}$ (not shown) represents the average intensity over the tetrahedron. Using these average values as well as an average density ρ_{av}, we can express Newton's law for the mass center of the tetrahedron in the x_1 direction as follows:

$$
\begin{aligned}
-(\tau_{11})_{av}\overline{PCB} - (\tau_{21})_{av}\overline{ACP} - (\tau_{31})_{av}\overline{APB} + (B_1)_{av}\rho_{av}\tfrac{1}{3}\overline{ABC}h \\
+ (T_1^{(\nu)})_{av}\overline{ABC} = \rho_{av}\tfrac{1}{3}\overline{ABC}ha_1
\end{aligned}
\tag{1.1}
$$

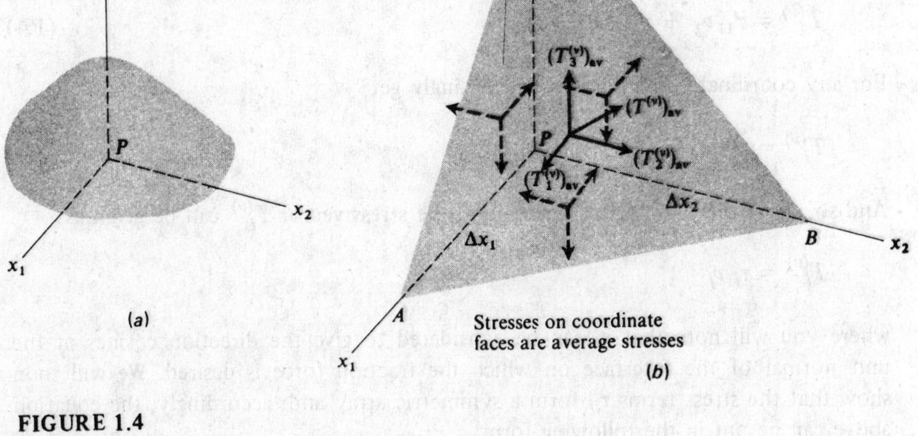

FIGURE 1.4
Tetrahedron at point P.

where a_1 is the acceleration of the mass center in the x_1 direction. We can next replace the average values by values taken at P itself, plus a small increment that goes to zero as Δx_i goes to zero.† Thus we have

$$\begin{aligned} (\tau_{11})_{av} &= (\tau_{11})_P + e_1 & (B_1)_{av} &= (B_1)_P + e_B \\ (\tau_{21})_{av} &= (\tau_{21})_P + e_2 & (T_1^{(\nu)})_{av} &= (T_1^{(\nu)})_P + e_T \\ (\tau_{31})_{av} &= (\tau_{31})_P + e_3 & \rho_{av} &= \rho_P + e_\rho \end{aligned} \tag{1.2}$$

where e_1, e_2, e_3, e_B, e_T, and e_ρ go to zero with Δx_i. Thus we have for Eq. (1.1):

$$\begin{aligned} -[(\tau_{11})_P + e_1]\,\overline{PCB} - [(\tau_{21})_P + e_2]\,\overline{ACP} - [(\tau_{31})_P + e_3]\,\overline{APB} \\ + [(B_1)_P + e_B](\rho_P + e_\rho)\tfrac{1}{3}\overline{ABC}h + [(T_1^{(\nu)})_P + e_T]\,\overline{ABC} \\ = \tfrac{1}{3}(\rho_P + e_\rho)\overline{ABC}ha_1 \end{aligned} \tag{1.3}$$

Next, dividing through by area $\overline{ABC}$ and noting that $\overline{PCB}/\overline{ABC} = \nu_1$, $\overline{ACP}/\overline{ABC} = \nu_2$, etc., we have

$$-[(\tau_{11})_P + e_1]\,\nu_1 - [(\tau_{21})_P + e_2]\,\nu_2 - [(\tau_{31})_P + e_3]\,\nu_3 +$$

$$[(B_1)_P + e_B](\rho_P + e_\rho)\frac{h}{3} + [(T_1^{(\nu)})_P + e_T] = (\rho_P + e_\rho)\frac{h}{3}a_1$$

Now go to the limit as $\Delta x_i \to 0$ in such a manner as to keep ν_i constant. Clearly, the e's disappear and $h \to 0$. We then get in the limit:

$$-(\tau_{11})_P\nu_1 - (\tau_{21})_P\nu_2 - (\tau_{31})_P\nu_3 + (T_1^{(\nu)})_P = 0$$

Since P is any point of the domain, we need no longer carry along the subscript. Rearranging terms, we get

$$T_1^{(\nu)} = \tau_{11}\nu_1 + \tau_{21}\nu_2 + \tau_{31}\nu_3 \tag{1.4}$$

For any coordinate direction i we accordingly get

$$T_i^{(\nu)} = \tau_{1i}\nu_1 + \tau_{2i}\nu_2 + \tau_{3i}\nu_3$$

And so, using the summation convention, the stress vector $T_i^{(\nu)}$ can be given as

$$T_i^{(\nu)} = \tau_{ji}\nu_j$$

where you will note that ν_j can be considered to give the direction cosines of the unit normal of the interface on which the traction force is desired. We will soon show that the stress terms τ_{ij} form a symmetric array and, accordingly, the equation above can be put in the following form:

†We are thus tacitly assuming that the quantities in Eq. (1.1) vary in a continuous manner.

$$\boxed{T_i^{(\nu)} = \tau_{ij}\nu_j} \tag{1.5}$$

Thus, knowing τ_{ij} we can get the traction vector for any interface at the point. The equation above is called *Cauchy's formula* and may be used to relate tractions on the boundary to stresses directly next to the boundary.

1.4 EQUATIONS OF MOTION

Consider an element of the body of mass dm at any point P. Newton's law requires that

$$d\mathbf{f} = dm\ \dot{\mathbf{V}}$$

where $d\mathbf{f}$ is the sum of the total traction force on the element and the total body force on the element. Integrating this equation over some arbitrary spatial domain having a volume D and a boundary surface S, we note that, as a result of Newton's third law, only tractions on the bounding surface do not cancel out, so that we have in index notation:

$$\oiint_S T_i^{(\nu)}\,dA + \iiint_D B_i\ dv = \iiint_D \dot{V}_i\rho\ dv \tag{1.6}$$

Now employ Eq. (1.5) to replace the stress vector $T_i^{(\nu)}$ by stresses. Thus:

$$\oiint_S \tau_{ij}\nu_j\,dA + \iiint_D B_i\ dv = \iiint_D \dot{V}_i\rho\ dv$$

Next employ Gauss' theorem for the first integral and collect terms under one integral. We obtain:

$$\iiint_D (\tau_{ij,j} + B_i - \dot{V}_i\rho)\,dv = 0$$

Since the domain D is arbitrary we conclude that at any point the following must hold:

$$\boxed{\tau_{ij,j} + B_i = \rho\dot{V}_i} \tag{1.7}$$

This is the desired equation of motion.

Suppose that next we consider an integral form of the moment of momentum equation derivable from Newton's law, i.e., $\mathbf{M} = \dot{\mathbf{H}}$. Thus we may say (considering Fig. 1.5):

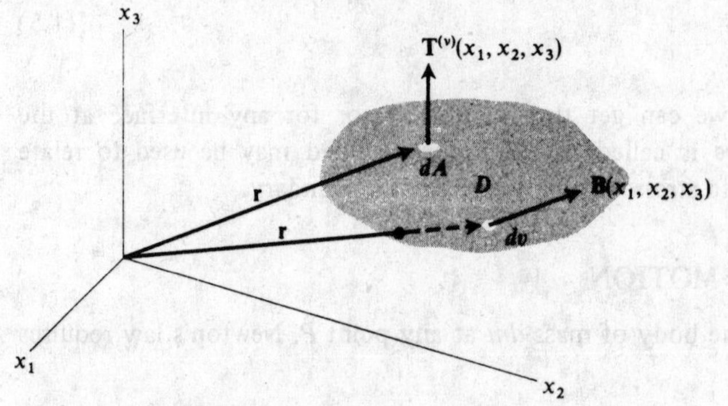

FIGURE 1.5
Body force and surface traction distributions.

$$\oint\!\!\oint_S \mathbf{r} \times \mathbf{T}^{(\nu)}\, dA + \iiint_D \mathbf{r} \times \mathbf{B}\, dv = \iiint_D \mathbf{r} \times \frac{d}{dt}(\mathbf{V}\, dm)$$

Considering the continuum to be composed of elements whose mass dm is constant but whose shape may be changing, we can express the integrand of the last expression as $\mathbf{r} \times \dot{\mathbf{V}}\, dm$. Thus we have for the above equation in tensor notation:

$$\oint\!\!\oint_S \epsilon_{ijk} x_j T_k^{(\nu)}\, dA + \iiint_D \epsilon_{ijk} x_j B_k\, dv - \iiint_D \epsilon_{ijk} x_j \dot{V}_k \rho\, dv = 0$$

where we replace dm by $\rho\, dv$. Now replace $T_k^{(\nu)}$ by $\tau_{lk}\nu_l$ and employ Gauss' theorem. Thus we may write the equation as:

$$\iiint_D \epsilon_{ijk}\left[(x_j \tau_{lk})_{,l} + x_j B_k - \rho x_j \dot{V}_k\right] dv = 0$$

Since this is true for any domain D, we can set the integrand equal to zero. Carrying out differentiation of the first expression in brackets and collecting terms, we then obtain:

$$\epsilon_{ijk} x_j(\tau_{lk,l} + B_k - \rho \dot{V}_k) + \epsilon_{ijk} x_{j,l} \tau_{lk} = 0$$

Because of Eq. (1.7) we can set the first expression equal to zero and so we get

$$\epsilon_{ijk} x_{j,l} \tau_{lk} = 0$$

Noting that $x_{j,l} = \delta_{jl}$, we have

$$\epsilon_{ijk}\delta_{jl}\tau_{kl} = \epsilon_{ijk}\tau_{kj} = 0 \tag{1.8}$$

It should be clear by considering these equations for each value of the free index i that the stresses with reversed indices with respect to each other are equal. That is,

$$\tau_{kj} = \tau_{jk} \tag{1.9}$$

The stress components form a symmetric array.†

1.5 TRANSFORMATION EQUATIONS FOR STRESS

Thus far, we have shown how a stress vector $\mathbf{T}^{(\nu)}$ on an arbitrarily oriented interface at a point P in a continuum can be related to the set of nine stress components on a set of orthogonal interfaces at the point. Furthermore, we have found through consideration of Newton's law that the stress terms form a symmetric array. We shall now show that the stress components for a set of Cartesian axes (i.e., for three orthogonal interfaces at a point) transform as a second-order tensor.

For this purpose consider Fig. 1.6, showing axes $x_1x_2x_3$ and $x_1'x_2'x_3'$ rotated arbitrarily relative to each other. Suppose we know the set of stresses for the unprimed reference, i.e., for a set of orthogonal interfaces having $x_1x_2x_3$ as edges, and wish to determine stresses for reference $x'y'z'$. Accordingly, in Fig. 1.7 we have shown a vanishingly small rectangular parallelepiped having $x_1'x_2'x_3'$ as edges

†We have considered only body force distributions here. If one assumes *body couple* distributions, which may occur as a result of magnetic or electric fields on certain kinds of dielectric and magnetic materials, we will have in Eq. (1.8) the additional integral $\iiint \mathbf{M}\, dv$, where $\mathbf{M}$ is the couple moment vector per unit volume. The result is that Eq. (1.8) becomes $\epsilon_{ijk}\tau_{kj} + M_i = 0$. The stress tensor is now no longer a symmetric tensor. Such cases are beyond the scope of this text.

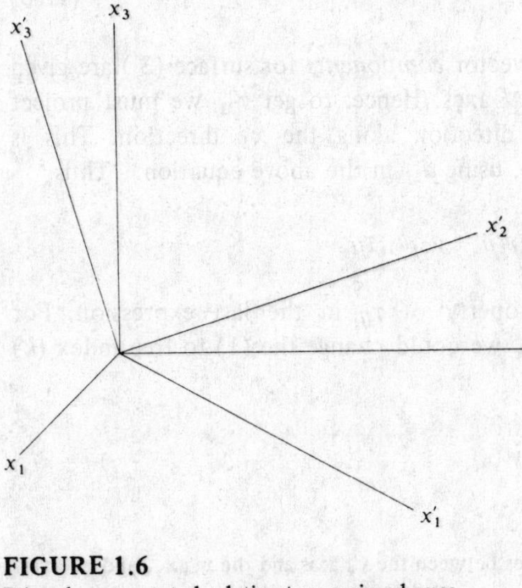

FIGURE 1.6
Primed axes rotated relative to unprimed axes.

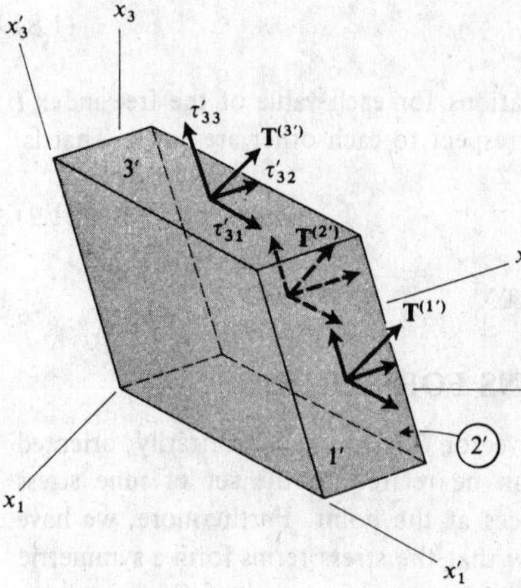

FIGURE 1.7
Traction forces and stresses.

to present the set of orthogonal interfaces for the desired stresses. The stress vectors for these interfaces have been shown as well as the corresponding stresses.

Suppose we wish to evaluate the stress component τ_{31}'. We can do this by first computing the stress vector on the interface corresponding to this stress [this interface is denoted as (3')]. The direction cosines ν_1', ν_2', and ν_3' for this interface can be given in our familiar notation as a_{31}, a_{32}, and a_{33}, where the first subscript identifies the normal (x_3') to the interface and the second subscript identifies the proper *unprimed axis*. Thus we can say, using Eq. (1.5),

$$T_i^{(3')} = \tau_{ij} a_{3j} \tag{1.10}$$

It must be remembered that the stress vector *components* for surface (3') are given above as components along the *unprimed* axes. Hence, to get τ_{31}' we must project each of the above components in a direction along the x_1' direction. This is accomplished by taking an inner product, using a_{1i} in the above equation.† Thus

$$\tau_{31}' = T_i^{(3')} a_{1i} = (\tau_{ij} a_{3j}) a_{1i} = a_{3j} a_{1i} \tau_{ij} = a_{3j} a_{1i} \tau_{ji}$$

where we have used the symmetry property of τ_{ij} in the last expression. For another component of stress such as τ_{3k}' we could change the (1) to free index (k) in the above equation and get

$$\tau_{3k}' = a_{3j} a_{ki} \tau_{ji}$$

† a_{1i} represents the set of direction cosines between the x_1' axis and the x_1, x_2, and x_3 axes. It is therefore a *unit vector* in the x_1' direction.

To get a corresponding result on *any other* interface we need only change the subscript (3) to free index p. Thus we can say for any stresses τ'_{pk}.

$$\boxed{\tau'_{pk} = a_{pj}a_{ki}\tau_{ji}} \tag{1.11}$$

We have thus shown that stress components transform as a second-order tensor.

1.6 PRINCIPAL STRESSES

We have shown that, given a system of stresses for an orthogonal set of interfaces at a point, we can associate a stress vector for interfaces having any direction in space according to the formulation

$$T_i^{(\nu)} = \tau_{ij}\nu_j$$

We now ask, is there a direction ν such that the stress vector is *collinear* with ν? That is, is there an interface having a normal such that there is only one nonzero stress–the normal stress? We shall call such a stress, if it exists, a *principal* stress and we shall denote it as σ (see Fig. 1.8). Thus we can express the above equation for this case as follows:

$$\sigma\nu_i = \tau_{ij}\nu_j \tag{1.12}$$

Now replace $\sigma\nu_i$ by $(\sigma\nu_j)(\delta_{ij})$ and rearrange the above equation to form the relation:

$$(\tau_{ij} - \sigma\delta_{ij})\nu_j = 0 \tag{1.13}$$

We have here three simultaneous equations with unknowns σ, ν_1, ν_2, and ν_3. However, we do have a fourth equation involving these direction cosines, namely:

$$\nu_i\nu_i = 1 \tag{1.14}$$

A nontrivial solution to the set of Eqs. (1.13) requires

$$|\tau_{ij} - \sigma\delta_{ij}| = 0 \tag{1.15}$$

Expanding out, we get

$$\begin{vmatrix} \tau_{11} - \sigma & \tau_{12} & \tau_{13} \\ \tau_{21} & \tau_{22} - \sigma & \tau_{23} \\ \tau_{31} & \tau_{32} & \tau_{33} - \sigma \end{vmatrix} = 0$$

Hence, we have

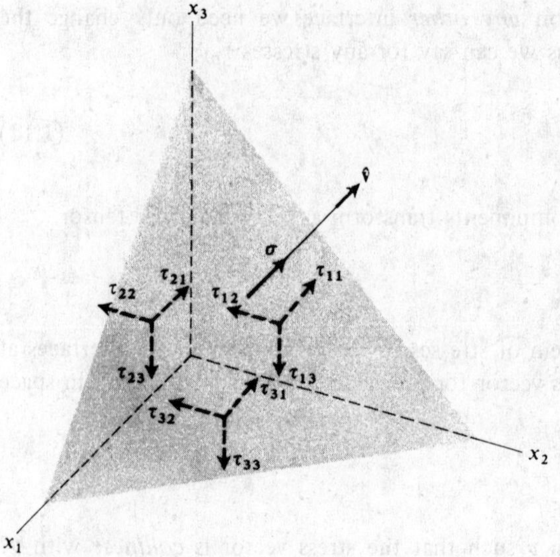

FIGURE 1.8
Tetrahedron with three orthogonal interfaces.

$$\sigma^3 - (\tau_{11} + \tau_{22} + \tau_{33})\sigma^2 + (\tau_{11}\tau_{22} + \tau_{22}\tau_{33} + \tau_{11}\tau_{33} - \tau_{12}^2 - \tau_{13}^2 - \tau_{23}^2)\sigma$$
$$- (\tau_{11}\tau_{22}\tau_{33} - \tau_{11}\tau_{23}^2 - \tau_{22}\tau_{13}^2 - \tau_{33}\tau_{12}^2 + 2\tau_{12}\tau_{13}\tau_{23}) = 0 \qquad (1.16)$$

We have three roots to this equation, which are the principal stresses and which we denote as σ_1, σ_2, and σ_3, respectively. We can associate for any root σ_α three direction cosines corresponding to the normal direction of the interface on which σ_α acts. We denote these direction cosines as $\overset{\alpha}{\nu}_1$, $\overset{\alpha}{\nu}_2$, and $\overset{\alpha}{\nu}_3$. To determine these direction cosines, substitute the value σ_α into Eq. (1.13). In attempting to solve for the ν's from these equations you will find that only two (any two) of the three equations are independent. A third independent equation to use for solving for the ν's is Eq. (1.14). In this way we may generate three sets of direction cosines $\overset{\alpha}{\nu}_i$ associated with the three roots σ_α. Thus:

$$\sigma_1 : \overset{1}{\nu}_1, \overset{1}{\nu}_2, \overset{1}{\nu}_3$$

$$\sigma_2 : \overset{2}{\nu}_1, \overset{2}{\nu}_2, \overset{2}{\nu}_3$$

$$\sigma_3 : \overset{3}{\nu}_1, \overset{3}{\nu}_2, \overset{3}{\nu}_3$$

We will first show that when the three stress roots are not equal, the principal directions $\overset{1}{\nu}_i$, $\overset{2}{\nu}_i$, and $\overset{3}{\nu}_i$ are *mutually orthogonal*. We can do this by examining σ_1 and σ_2. Equation (1.12) for these stresses becomes

$$\sigma_1 \overset{1}{\nu}_i = \tau_{ij}\overset{1}{\nu}_j \qquad (1.17a)$$

$$\sigma_2 \overset{2}{\nu}_i = \tau_{ij}\overset{2}{\nu}_j \qquad (1.17b)$$

Now take an inner product of Eq. (1.17*a*) with $\overset{2}{\nu}_i$ and of Eq. (1.17*b*) with $\overset{1}{\nu}_i$. Thus we get

$$\sigma_1 \overset{1}{\nu}_i \overset{2}{\nu}_i = \tau_{ij} \overset{1}{\nu}_j \overset{2}{\nu}_i \tag{1.18a}$$

$$\sigma_2 \overset{2}{\nu}_i \overset{1}{\nu}_i = \tau_{ij} \overset{2}{\nu}_j \overset{1}{\nu}_i \tag{1.18b}$$

We can rewrite the right side of Eq. (1.18*b*) as follows:

$$\tau_{ij} \overset{2}{\nu}_j \overset{1}{\nu}_i = \tau_{ji} \overset{2}{\nu}_i \overset{1}{\nu}_j = \tau_{ij} \overset{2}{\nu}_i \overset{1}{\nu}_j$$

where we have first changed the repeated indices and then, because of the symmetry of the stress tensor, have interchanged indices for τ in the last term. Note that it is now identical with the expression on the right side of Eq. (1.18*a*). Accordingly, when we subtract Eq. (1.18*b*) from Eq. (1.18*a*) we get

$$\overset{1}{\nu}_i \overset{2}{\nu}_i (\sigma_1 - \sigma_2) = 0$$

Since we have assumed that the principal stresses are different, we can conclude that:

$$\overset{1}{\nu}_i \overset{2}{\nu}_i = 0 \tag{1.19}$$

Thus the vectors $\overset{1}{\nu}_i$ and $\overset{2}{\nu}_i$ are at right angles to each other. We can draw the same conclusion about any other pair of principal axes and so we can conclude that for nonequal principal stresses, the principal axes must be mutually orthogonal.

We can also show that the principal stresses must be real. Since the components of τ_{ij} are all real, it is clear that the coefficients of the cubic equation (1.16) are all real and so we know that there must be at least one real root—say σ_1. This means there is a reference x_1', x_2', x_3' such that $\tau_{11}' = \sigma_1$, $\tau_{12}' = 0$, and $\tau_{13}' = 0$. For such a reference, Eq. (1.15) becomes

$$\begin{vmatrix} \sigma_1 - \sigma & 0 & 0 \\ 0 & \tau_{22}' - \sigma & \tau_{23}' \\ 0 & \tau_{32}' & \tau_{33}' - \sigma \end{vmatrix} = 0$$

Expanding out the determinant, we get

$$(\sigma_1 - \sigma)[\sigma^2 - \sigma(\tau_{22}' + \tau_{33}') - (\tau_{23}')^2 + \tau_{22}'\tau_{33}'] = 0$$

$\sigma = \sigma_1$ clearly is a root and so for the other two roots we have the quadratic equation:

$$\sigma^2 - (\tau_{22}' + \tau_{33}')\sigma + \tau_{22}'\tau_{33}' - (\tau_{23}')^2 = 0$$

We get

$$\sigma_{2,3} = \frac{(\tau_{22}' + \tau_{33}') \pm \sqrt{(\tau_{22}' + \tau_{33}')^2 + 4(\tau_{23}')^2 - 4\tau_{22}'\tau_{33}'}}{2}$$

$$= \frac{(\tau_{22}' + \tau_{33}') \pm \sqrt{(\tau_{22}' - \tau_{33}')^2 + 4(\tau_{23}')^2}}{2} \tag{1.20}$$

Clearly, the discriminant is not a negative term and so we must have real roots for the principal stresses.

We have thus found three planes on which the shear stresses are zero. These are the so-called *principal planes* corresponding to the principal stresses. We have computed the principal stresses in terms of the stress components τ_{ij} for reference $x_1x_2x_3$ and can get the corresponding sets of direction cosines $\overset{1}{\nu}_i, \overset{2}{\nu}_i, \overset{3}{\nu}_i$ relative to this reference by employing Eq. (1.13), as explained earlier. But the stresses on any interface are uniquely established, via the transformation equation (1.11). Accordingly, the principal stresses and planes as computed for τ_{ij} are unique for the point and can be considered as characteristics of the state of stress at the point. Thus, if one started with any other reference $x_1'x_2'x_3'$ at the point under discussion, where we have τ_{ij}', we would find the same set of principal stresses at the point. Thus we should always get the same cubic equation [(1.16)] at a point no matter what set of axes we may choose to employ at the point. Accordingly, we can conclude that the coefficients of the cubic equation are *independent of the reference* at a point. That is, these expressions have values that are independent of a rotation of axes at a point and they are called the *first, second,* and *third tensor invariants.* Thus

$$\boxed{\begin{aligned} \tau_{11} + \tau_{22} + \tau_{33} &= \mathrm{I}_\tau \\ \tau_{11}\tau_{22} + \tau_{11}\tau_{33} + \tau_{22}\tau_{33} - \tau_{12}^2 - \tau_{23}^2 - \tau_{31}^2 &= \mathrm{II}_\tau \\ \tau_{11}\tau_{22}\tau_{33} - \tau_{11}\tau_{23}^2 - \tau_{22}\tau_{13}^2 - \tau_{33}\tau_{12}^2 + 2\tau_{12}\tau_{23}\tau_{13} &= \mathrm{III}_\tau \end{aligned}} \tag{1.21}$$

The *first tensor invariant* is simply the sum of the terms of the left-to-right diagonal from τ_{11} to τ_{33}—the so-called *principal diagonal*. This sum is called the *trace* of the tensor and can be given in tensor notation as

$$\mathrm{I}_\tau = \tau_{ii} \tag{1.22}$$

The *second tensor invariant* is the sum of three subdeterminants formed from the matrix representation of the stress tensor. These subdeterminants form the minors of terms of the principal diagonal, and so II_τ is the sum of the principal minors. Thus we have:

$$\mathrm{II}_\tau = \begin{vmatrix} \tau_{11} & \tau_{12} \\ \tau_{21} & \tau_{22} \end{vmatrix} + \begin{vmatrix} \tau_{22} & \tau_{23} \\ \tau_{32} & \tau_{33} \end{vmatrix} + \begin{vmatrix} \tau_{11} & \tau_{13} \\ \tau_{31} & \tau_{33} \end{vmatrix} \tag{1.23}$$

Finally, the *third tensor invariant* can be seen to be simply the determinant of the tensor itself. Thus:

$$\mathrm{III}_\tau = \begin{vmatrix} \tau_{11} & \tau_{12} & \tau_{13} \\ \tau_{21} & \tau_{22} & \tau_{23} \\ \tau_{31} & \tau_{32} & \tau_{33} \end{vmatrix} \tag{1.24}$$

These tensor invariants characterize a tensor to a great extent just as the invariant length $V_i V_i$ of a vector characterizes a vector to some extent.

As a next step in this section, we shall show that the principal axes correspond to the directions for which the normal stress forms an *extremum* when compared to normal stresses in neighboring directions. Consider an arbitrary interface having a normal $\boldsymbol{\nu}$ as shown in Fig. 1.9. The surface traction vector $\mathbf{T}^{(\nu)}$ is shown for the interface with its components parallel to the reference $x_1 x_2 x_3$. We can give the traction vector in terms of stresses associated with reference $x_1 x_2 x_3$ as follows, using Cauchy's formula:

$$T_i^{(\nu)} = \tau_{ij} \nu_j \tag{1.25}$$

To get the traction component τ_{nn} normal to the interface, we have

$$\tau_{nn} = T_i^{(\nu)} \nu_i = \tau_{ij} \nu_j \nu_i \tag{1.26}$$

We now ask what direction $\boldsymbol{\nu}$ will extremize τ_{nn}. A necessary condition for an extremum would be that the partials of the quantity $(\tau_{ij}\nu_i\nu_j)$ be zero with respect to ν_1, ν_2, and ν_3 if these were independent variables. However, the variables ν_1, ν_2, and ν_3 are not independent, since there is the constraining equation

$$\nu_i \nu_i = 1$$

that must always be satisfied. Accordingly, using λ as a Lagrange multiplier, we must extremize the function $[\tau_{ij}\nu_i\nu_j + \lambda(\nu_i\nu_i - 1)]$ Thus

$$\frac{\partial}{\partial \nu_s} [\tau_{ij}\nu_i\nu_j - \lambda(\nu_i\nu_i - 1)] = 0$$

We have here three equations. Noting that τ_{ij} is constant with respect to ν_s:

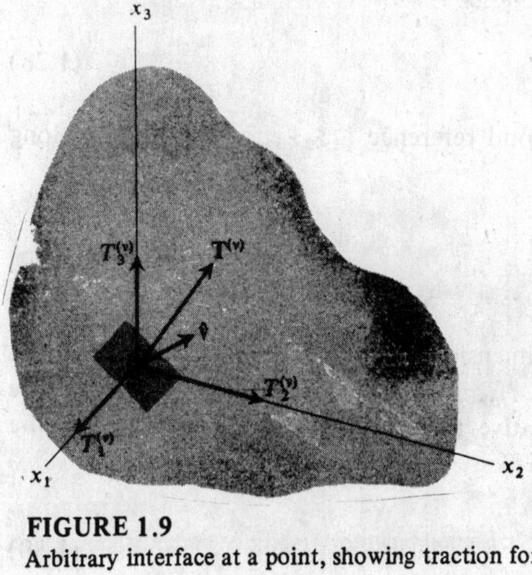

FIGURE 1.9
Arbitrary interface at a point, showing traction forces.

$$\tau_{ij}\nu_i \frac{\partial \nu_j}{\partial \nu_s} + \tau_{ij} \frac{\partial \nu_i}{\partial \nu_s} \nu_j + 2\lambda\nu_i \frac{\partial \nu_i}{\partial \nu_s} = 0 \qquad s = 1, 2, 3$$

But

$$\frac{\partial \nu_j}{\partial \nu_s} = \delta_{js} \qquad \frac{\partial \nu_i}{\partial \nu_s} = \delta_{is}$$

We get, on making these substitutions,

$$\tau_{ij}\nu_i\delta_{js} + \tau_{ij}\delta_{is}\nu_j + 2\lambda\nu_i\delta_{is} = 0$$

This becomes

$$\tau_{is}\nu_i + \tau_{sj}\nu_j + 2\lambda\nu_i\delta_{is} = 0$$

Using the symmetry of the stress tensor in the second term and changing repeated indices j to i, we can put the above equation into the form:

$$(\tau_{is} + \lambda\delta_{is})\nu_i = 0 \tag{1.27}$$

Note that λ, the Lagrange multiplier, has the dimensions of stress and accordingly represents a stress quantity. Note further that Eq. (1.27) is of identically the same form as Eq. (1.13) for determining the principal stresses σ_α. Accordingly, the three values of λ that are obtained on carrying through the above calculations, $(\lambda_1, \lambda_2, \lambda_3)$, and their corresponding sets of direction cosines $(\overset{1}{\nu}_1, \overset{1}{\nu}_2, \overset{1}{\nu}_3)$, $(\overset{2}{\nu}_1, \overset{2}{\nu}_2, \overset{2}{\nu}_3)$, $(\overset{3}{\nu}_1, \overset{3}{\nu}_2, \overset{3}{\nu}_3)$ will be respectively identical to σ_1, σ_2, σ_3 and the corresponding sets of direction cosines. Thus the directions that emerge for extremizing the normal stress are the directions corresponding to the principal stresses.

Now consider the variations of normal stress τ_{nn} as the direction $\boldsymbol{\nu}$ is varied at a point to cover all directions (Fig. 1.9). Using Cauchy's formula:

$$T_i^{(\nu)}\nu_i = \tau_{nn} = \tau_{ij}\nu_i\nu_j \tag{1.28}$$

For convenience we lay off, using a second reference $\xi_1\xi_2\xi_3$, a distance $\overline{OA}$ along direction $\boldsymbol{\nu}$ (see Fig. 1.10) such that

$$(\overline{OA})^2 = \frac{\pm d^2}{\tau_{nn}} \tag{1.29}$$

where d is an arbitrary constant with dimensions to render $\overline{OA}$ dimensionless. The plus sign obviously must be used when τ_{nn} is positive (i.e., tensile stress) and the negative sign is used when τ_{nn} is negative (compression). We can then give the direction cosines ν_i as follows:

$$\nu_i = \frac{\xi_i}{\overline{OA}} = \frac{\xi_i}{\sqrt{\pm d^2/\tau_{nn}}} \tag{1.30}$$

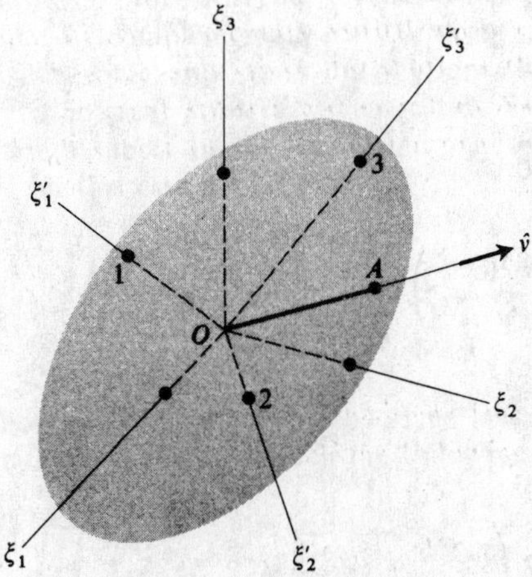

FIGURE 1.10
Stress ellipsoid.

Hence Eq. (1.28) can be given as

$$\tau_{nn} = \tau_{ij}\left(\frac{\xi_i \xi_j}{\pm d^2/\tau_{nn}}\right) \tag{1.31}$$

Canceling τ_{nn}, we get the following result on rearranging terms in the above equation:

$$\xi_i \xi_j \tau_{ij} = \pm d^2 \tag{1.32}$$

For a particular choice of sign of d^2, this represents a real second-order surface in a particular region of $\xi_1\xi_2\xi_3$. We call this surface the *stress quadric*. The distance from the origin to this surface in some direction $\boldsymbol{\nu}$ is inversely proportional to the square root of the normal stress τ_{nn} [see Eq. (1.29)] for the same direction $\boldsymbol{\nu}$ in physical space (reference x_1, x_2, x_3). For instance, with τ_{11}, τ_{22}, and τ_{33} positive, $+d^2$ in Eq. (1.32) gives the only real surface–that of an ellipsoid, as shown in Fig. 1.10. It is apparent that for the three symmetrical semiaxes (ξ'_1, ξ'_2, ξ'_3) the distances from origin to surface are local extrema, and so these directions correspond to the principal axes in accordance with the previous remarks. Since one of these distances (O3) is a maximum for the ellipsoid, the corresponding principal stress must be the minimum normal stress at the point in the body, and since the distance (O1) is a minimum for the ellipsoid, the corresponding normal stress must be the maximum normal stress at the point in the body. The third principal stress corresponding to direction (O2) must then have some intermediate value such that the sum of the principal stresses gives the proper first tensor invariant at the point in the body. Other kinds of second-order surfaces are possible if the signs of τ_{11}, τ_{22}, and τ_{33} are not all positive or all negative. Thus for the particular case where $\sigma_1 \geqslant \sigma_2 > 0$, $\sigma_3 < 0$ we get the surfaces shown in Fig. 1.11, where $+d^2$ is needed in Eq. (1.32) to generate surface (*c*) as a real surface and $-d^2$ is needed in Eq. (1.32) to generate (*a*) and (*b*) as real surfaces. The stress for any direction is found by

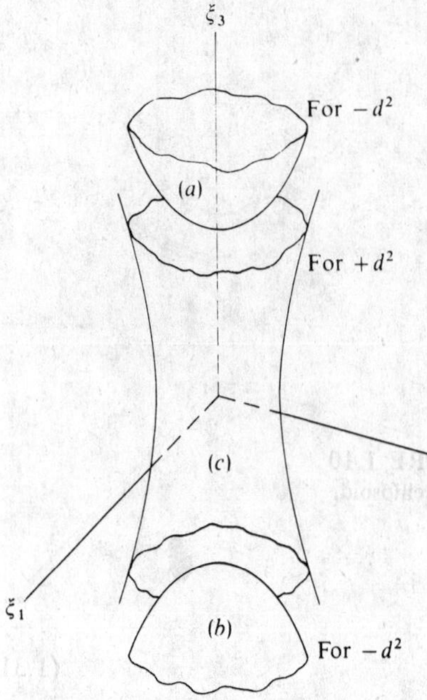

FIGURE 1.11
Stress quadric surfaces.

measuring $\overline{OA}$ for that direction and employing Eq. (1.29), using the same sign with d^2 as is associated with that part of the surface intercepted by $\overline{OA}$. Thus the proper sign of τ_{nn} is then determined for use in Eq. (1.29) for that direction. The earlier conclusion that the largest (algebraically) normal stress and the smallest (algebraically) normal stress at a point are principal stresses still holds. We thus can see why principal stresses are so important in engineering work.

We will not formally use the so-called stress quadric. However, it does serve as a graphic representation of stress (or any other second-order symmetric tensor) just as an arrow is a graphic representation of a vector.

It should be clearly understood that the main ingredient in arriving at the conclusions in this section was the fact that stress transforms according to the formula

$$\tau'_{ij} = a_{ik}a_{jl}\tau_{kl}$$

and the fact that $\tau_{ij} = \tau_{ji}$. Thus, all the conclusions made concerning principal stresses, tensor invariants, etc., apply to any *second-order symmetric tensor*. We shall make ample use of these results in studies to follow.

Part B
Strain

1.7 STRAIN COMPONENTS

We shall now propose means of expressing the deformation of a body. Accordingly, we consider an undeformed body, as shown in Fig. 1.12. If this body is given a

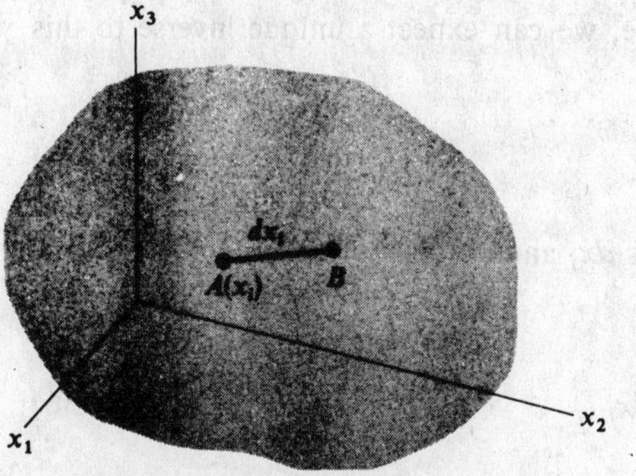

FIGURE 1.12
Line segment in the undeformed geometry.

rigid-body motion we know that each line segment in the body undergoes no change in length. Accordingly, the change in length of line segments in the body, or, more appropriately, the distance between points in the body can serve as a measure of the deformation (change of shape and size) of the body. We therefore consider two points A and B in the body at positions x_i and $x_i + dx_i$, as shown in Fig. 1.12. The distance between these points is given by

$$(\overline{AB})^2 = (ds)^2 = dx_i \, dx_i \tag{1.33}$$

When external forces are applied, the body will deform so that points A and B move to points A^* and B^*, respectively. It is convenient now to consider that the x_i reference is labeled the ξ_i reference when considering the deformed state, as shown in Fig. 1.13. That is, we may consider the deformation as depicted by a mapping of each point from coordinate x_i to coordinate ξ_i. We can then say for a deformation

$$\xi_i = \xi_i(x_1, x_2, x_3) \tag{1.34}$$

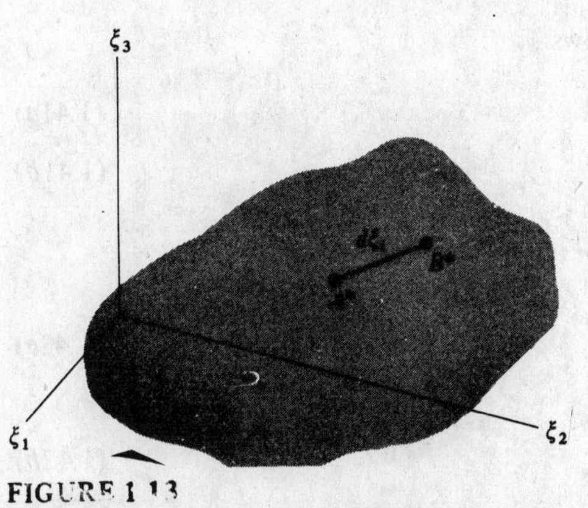

FIGURE 1.13
Line segment mapped to deformed geometry.

And since the mapping must be one-to-one, we can expect a unique inverse to this formulation in the form:

$$x_i = x_i(\xi_1, \xi_2, \xi_3) \tag{1.35}$$

We can accordingly express the differentials dx_i and $d\xi_i$ as follows by making use of the above relations:

$$dx_i = \left(\frac{\partial x_i}{\partial \xi_j}\right) d\xi_j \qquad d\xi_i = \left(\frac{\partial \xi_i}{\partial x_j}\right) dx_j \tag{1.36}$$

This permits us to express $(ds)^2$ in Eq. (1.33) in the following way:

$$(ds)^2 = dx_i\, dx_i = \frac{\partial x_i}{\partial \xi_m}\frac{\partial x_i}{\partial \xi_k}\, d\xi_m\, d\xi_k \tag{1.37}$$

It should then be clear that in the defomed state we have for the line segment $\overline{A^*B^*}$:

$$\overline{A^*B^*}^2 = (ds^*)^2 = d\xi_i\, d\xi_i = \frac{\partial \xi_i}{\partial x_k}\frac{\partial \xi_i}{\partial x_l}\, dx_k\, dx_l \tag{1.38}$$

Now we will examine the change in length of the segment, that is, we investigate the measure of deformation mentioned earlier. We may do this in either of two ways, by using either formulation (1.37) or (1.38). Thus:

$$(ds^*)^2 - (ds)^2 = \left(\frac{\partial \xi_k}{\partial x_i}\frac{\partial \xi_k}{\partial x_j} - \delta_{ij}\right) dx_i\, dx_j \tag{1.39}$$

$$(ds^*)^2 - (ds)^2 = \left(\delta_{ij} - \frac{\partial x_k}{\partial \xi_i}\frac{\partial x_k}{\partial \xi_j}\right) d\xi_i\, d\xi_j \tag{1.40}$$

We now rewrite these equations as follows:

$$(ds^*)^2 - (ds)^2 = 2\epsilon_{ij}\, dx_i\, dx_j \tag{1.41a}$$

$$(ds^*)^2 - (ds)^2 = 2\eta_{ij}\, d\xi_i\, d\xi_j \tag{1.41b}$$

where we have introduced so-called *strain* terms:

$$\epsilon_{ij} = \frac{1}{2}\left(\frac{\partial \xi_k}{\partial x_i}\frac{\partial \xi_k}{\partial x_j} - \delta_{ij}\right) \tag{1.42a}$$

$$\eta_{ij} = \frac{1}{2}\left(\delta_{ij} - \frac{\partial x_k}{\partial \xi_i}\frac{\partial x_k}{\partial \xi_j}\right) \tag{1.42b}$$

The set of terms ϵ_{ij} contains the implicit assumption that they are expressed as functions of the coordinates in the undeformed state–i.e., the so-called *Lagrange coordinates*. The set of terms ϵ_{ij}, formulated by Green and St. Venant, is called Green's strain tensor. The second set, η_{ij}, is formulated as a function of the coordinate for the deformed state–the so-called *Eulerian coordinates*. This form was introduced by Cauchy for infinitesimal strain (soon to be discussed) and by Almansi and Hamel for finite strains. It is often called the Almansi measure of strain.

We now introduce the *displacement field* u_i defined such that

$$u_i = \xi_i - x_i \tag{1.43}$$

Thus u_i gives the displacement of each point in the body from the initial undeformed configuration to the deformed configuration, as shown in Fig. 1.14. We may express u_i as a function of the Lagrangian coordinates x_i, in which case it expresses the displacement from the position x_i in the undeformed state to the deformed position ξ_i. On the other hand, u_i can equally well be expressed in terms of ξ_i, the Eulerian coordinates, in which case it expresses the displacement that must have taken place to get to the position ξ_i from some undeformed configuration. The following relations can then be written from Eq. (1.43):

$$\frac{\partial x_i}{\partial \xi_j} = \delta_{ij} - \frac{\partial u_i}{\partial \xi_j} \tag{1.44a}$$

$$\frac{\partial \xi_i}{\partial x_j} = \frac{\partial u_i}{\partial x_j} + \delta_{ij} \tag{1.44b}$$

Substituting these results into Eq. (1.42), we obtain the following:

$$\epsilon_{ij} = \frac{1}{2}\left(\frac{\partial u_i}{\partial x_j} + \frac{\partial u_j}{\partial x_i} + \frac{\partial u_k}{\partial x_i}\frac{\partial u_k}{\partial x_j}\right) \tag{1.45a}$$

$$\eta_{ij} = \frac{1}{2}\left(\frac{\partial u_i}{\partial \xi_j} + \frac{\partial u_j}{\partial \xi_i} - \frac{\partial u_k}{\partial \xi_i}\frac{\partial u_k}{\partial \xi_j}\right) \tag{1.45b}$$

The Green strains ϵ_{ij} are thus referred to the initial undeformed geometry and indicate what must occur during a given deformation; the other strain terms η_{ij} are referred to the deformed or instantaneous geometry of the body and indicate what must have occurred to reach this geometry from an earlier undeformed state.

Up to this point in the discussion we have made no restriction on the magnitude of deformations for which Eq. (1.45) is valid. We now restrict ourselves to what is commonly called *infinitesimal strain*, wherein the derivatives of the displacement components are small compared to unity. Thus:

$$\frac{\partial u_i}{\partial x_j} \ll 1 \qquad \frac{\partial u_i}{\partial \xi_j} \ll 1$$

With this in mind, consider the following operator acting on an arbitrary function $J(x_i)$:

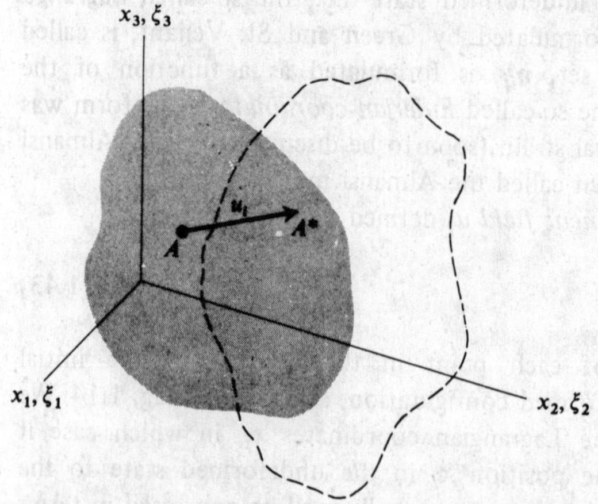

FIGURE 1.14
Displacement field vector.

$$\frac{\partial J}{\partial \xi_i} = \frac{\partial J}{\partial x_j}\left(\frac{\partial x_j}{\partial \xi_i}\right) = \frac{\partial J}{\partial x_j}\left[\frac{\partial}{\partial \xi_i}(\xi_j - u_j)\right] = \left(\delta_{ij} - \frac{\partial u_j}{\partial \xi_i}\right)\frac{\partial J}{\partial x_j}$$

For infinitesimal strain we may drop the term $\partial u_j/\partial \xi_i$ above to reach the result†

$$\frac{\partial}{\partial \xi_i} = \frac{\partial}{\partial x_i}$$

and we conclude that we need no longer distinguish between Eulerian and Lagrangian coordinates in expressing strain. A further simplification occurs when we note that the product of the derivatives of displacement components in Eq. (1.45) can now be considered negligible compared to the linear terms. We then get the following formulation for strain‡:

†This assumes that $\partial J/\partial x_i$, $i = 1, 2, 3$, are all of the same order of magnitude. Thus consider

$$\frac{\partial J}{\partial \xi_1} = \left(1 - \frac{\partial u_1}{\partial \xi_1}\right)\frac{\partial J}{\partial x_1} - \frac{\partial u_2}{\partial \xi_1}\frac{\partial J}{\partial x_2} - \frac{\partial u_3}{\partial \xi_1}\frac{\partial J}{\partial x_3}$$

It is clear we can neglect $\partial u_1/\partial \xi_1$ compared to unity. To neglect the last two terms means that

$$\frac{\partial u_2}{\partial \xi_1}\frac{\partial J}{\partial x_2} \quad \text{and} \quad \frac{\partial u_3}{\partial \xi_1}\frac{\partial J}{\partial x_3}$$

must be small compared to $\partial J/\partial x_1$. For this to be true the $\partial J/\partial x_i$ must be of the same order of magnitude.

‡You will recall that in developing the equations of equilibrium we employed only a single reference, and the tacit assumption was that the geometry employed for the equation was the *deformed* geometry. For infinitesimal deformation we may use the undeformed geometry rather than the deformed geometry for expressing equations of equilibrium.

$$\epsilon_{ij} = \eta_{ij} = \frac{1}{2}\left(\frac{\partial u_i}{\partial x_j} + \frac{\partial u_j}{\partial x_i}\right) = \frac{1}{2}(u_{i,j} + u_{j,i}) \tag{1.46}$$

Accordingly, we shall now use only ϵ_{ij} for infinitesimal strain. In unabridged notation we have

$$\begin{aligned}
\epsilon_{xx} &= \frac{\partial u_x}{\partial x} & \epsilon_{xy} &= \frac{1}{2}\left(\frac{\partial u_x}{\partial y} + \frac{\partial u_y}{\partial x}\right) = \frac{1}{2}\gamma_{xy} \\
\epsilon_{yy} &= \frac{\partial u_y}{\partial y} & \epsilon_{yz} &= \frac{1}{2}\left(\frac{\partial u_y}{\partial z} + \frac{\partial u_z}{\partial y}\right) = \frac{1}{2}\gamma_{yz} \\
\epsilon_{zz} &= \frac{\partial u_z}{\partial z} & \epsilon_{xz} &= \frac{1}{2}\left(\frac{\partial u_x}{\partial z} + \frac{\partial u_z}{\partial x}\right) = \frac{1}{2}\gamma_{xz}
\end{aligned} \tag{1.47}$$

where the γ_{ij} are called the *engineering shear strains*.

1.8 PHYSICAL INTERPRETATION OF STRAIN TERMS

In the previous section we introduced the strain terms ϵ_{ij} by considering changes in length between points separated by infinitesimal distances ("adjacent points"). Clearly, ϵ_{ij} also affords us directly some measure of the deformation of each vanishingly small element of the body and thus gives us a means of describing the deformation of the body as a whole.

We shall examine this local deformation in this section, and we will find it helpful to employ for this purpose a small rectangular parallelepiped at point P in the body, as shown in Fig. 1.15. Notice that we have placed a Cartesian reference at P. Let us imagine next that the body has some deformation and let us focus on line $\overline{PQ} = \Delta y$. As shown in Fig. 1.16, point P moves to P^* and point Q moves to Q^* as a result of the deformation. The projection of $\overline{P^*Q^*}$ in the y direction, which we

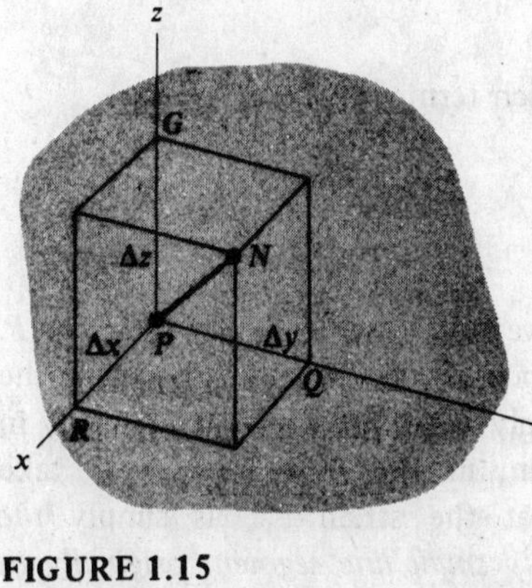

FIGURE 1.15
Rectangular parallelepiped for adjacent points.

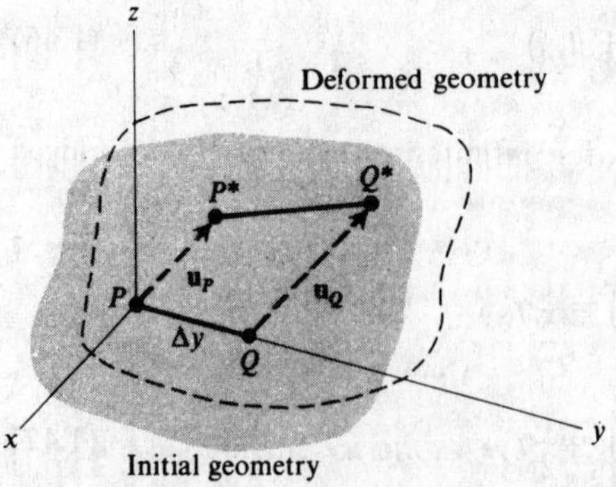

FIGURE 1.16
Line segment in the initial and deformed geometries.

denote as $(\overline{P^*Q^*})_y$, is computed in terms of the original length Δy and in terms of the displacement of points P and Q in the following way:

$$(\overline{P^*Q^*})_y = \Delta y + (u_y)_Q - (u_y)_P$$

Now express $(u_y)_Q$ as a Taylor series in terms of $(u_y)_P$ in the above equation. We then get

$$\begin{aligned}(\overline{P^*Q^*})_y &= \Delta y + \left[(u_y)_P + \left(\frac{\partial u_y}{\partial y}\right)_P \Delta y + \cdots\right] - (u_y)_P \\ &= \Delta y + \left(\frac{\partial u_y}{\partial y}\right)_P \Delta y + \cdots \end{aligned} \tag{1.48}$$

The net y component of elongation of the segment $\overline{PQ}$ can then be given as

$$(\overline{P^*Q^*})_y - \Delta y = \left(\frac{\partial u_y}{\partial y}\right)_P \Delta y + \cdots$$

Dividing through by Δy and taking the limit of each term as $\Delta y \rightarrow 0$, we get

$$\frac{(\overline{P^*Q^*})_y - \Delta y}{\Delta y} = \frac{\partial u_y}{\partial y} = \epsilon_{yy}$$

where with the coalescence of points P and Q we may now drop the subscript P. We can conclude that the normal strain ϵ_{yy} at a point is the change in length in the y direction per unit original length of a vanishingly small line segment originally in the y direction. For small deformations we can, in the above formation, take $(\overline{P^*Q^*}) = (\overline{P^*Q^*})_y$. This permits us to say that the strain ϵ_{yy} is simply *the elongation per unit original length of a vanishingly small line segment originally in* *the* y *direction*. We can make corresponding interpretation for ϵ_{xx} and ϵ_{zz} or, given a direction p, for ϵ_{pp}.

Now let us consider respectively line segments $\overline{PR}$ of length Δx and $\overline{PQ}$ of length Δy in Fig. 1.17. In the deformed state P, Q, and R move to P^*, Q^*, and R^*, respectively, as shown in Fig. 1.17. We shall be interested in the projection of $\overline{P^*R^*}$ and $\overline{P^*Q^*}$ onto the xy plane–i.e., onto the plane of these line segments in the undeformed state. We show this projection in Fig. 1.18. Note that θ is the angle between the projection of $\overline{P^*Q^*}$ and the y axis while β is the angle between the projection of $\overline{P^*R^*}$ and the x axis. The displacement component of point P in the x direction has been shown simply as $(u_x)_P$ and the displacement component of point Q in the x direction has been given with the aid of a Taylor series expansion in the form

$$(u_x)_P + \left(\frac{\partial u_x}{\partial y}\right)_P \Delta y + \cdots$$

Finally, the component of the projection of $\overline{P^*Q^*}$ taken in the y direction has been given as $\Delta y + \delta_y^2$, where δ_y^2 is a second-order increment for a small deformation. Now we can give $\tan\theta$ as follows:

$$\tan\theta = \frac{(\partial u_x/\partial y)_P\, \Delta y + \cdots}{\Delta y + \delta_y^2}$$

Take the limit as $\Delta y \to 0$. Higher-order terms in the numerator vanish. Also we delete the second-order increment δ_y^2. We then have

$$\tan\theta = \theta = \frac{\partial u_x}{\partial y}$$

at any point P. Similarly, we have for β:

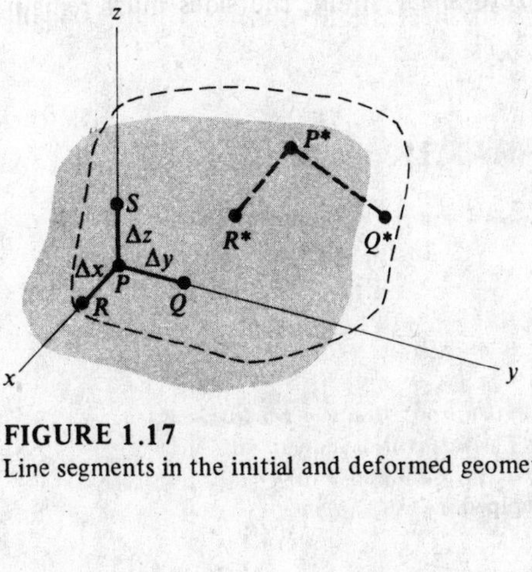

FIGURE 1.17
Line segments in the initial and deformed geometries.

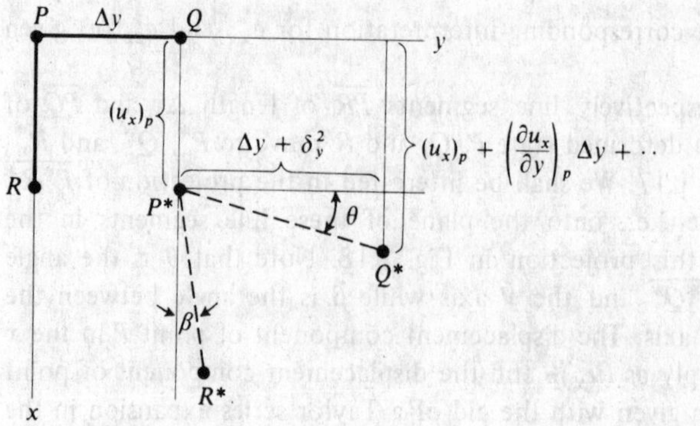

FIGURE 1.18
Projections of line segments in the deformed geometry back onto the xy plane.

$$\beta = \frac{\partial u_y}{\partial x}$$

The sum $\theta + \beta$ can then be directly related to the shear strain as follows:

$$(\theta + \beta) = \frac{\partial u_x}{\partial y} + \frac{\partial u_y}{\partial x} = 2\epsilon_{xy} = \gamma_{xy}$$

The sum of the angles $\theta + \beta$, and hence the engineering shear γ_{xy}, is the decrease from a right angle of a pair of infinitesimal line segments originally in the x and y directions at P when we project the deformed pair onto the xy plane in the undeformed geometry. Because of the infinitesimal deformation restriction, however, the change in right angle of the line segments themselves can be taken as equal to that of the projections of these line segments. In general, then, *γ_{ij} gives the change from a right angle of vanishingly small line segments originally in the* i *and* j *directions at a point.*

Now consider the effects of strain on an infinitesimal rectangular parallelepiped in the undeformed geometry. With zero shear strain, the sides must remain

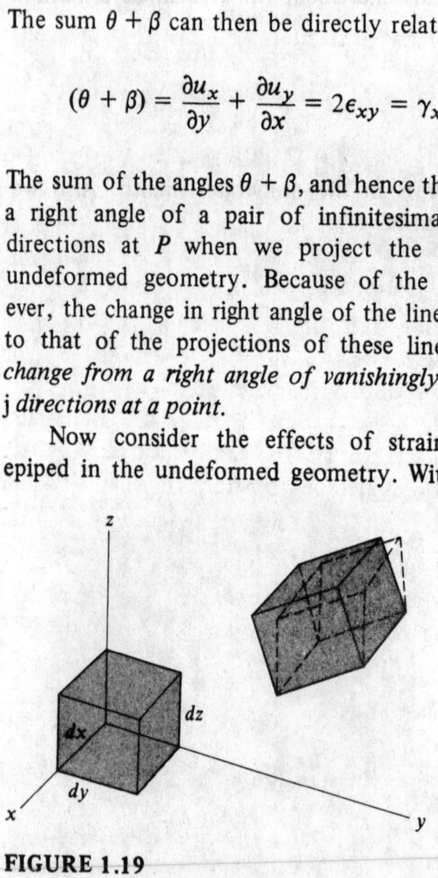

FIGURE 1.19
Deformation of an infinitesimal rectangular parallelepiped.

orthogonal on deformation. However, the position and orientation of the element may change, as may the length of sides and volume. This has been shown with full lines in Fig. 1.19. The shear strain can now be applied [the effects of normal strain and shear strain superpose in an uncoupled manner in accordance with Eq. (1.47)]. The result is that the sides may lose their mutual perpendicularity, so that we have parallelograms instead of rectangles for the sides (see dashed lines in Fig. 1.19). In short, we can say that *the size of the rectangular parallelepiped is changed by normal strain while the basic shape is changed by shear strain*. In fact, you are asked to show (in the exercises) that for infinitesimal deformation the volume change per unit volume is simply ϵ_{ii}.

1.9 THE ROTATION TENSOR

In the previous sections we considered stretching of a line element to generate the strain tensor ϵ_{ij} and then used the deformation of a vanishingly small rectangular parallelepiped to give physical interpretations to the components of the strain tensor.

We shall now go through a similar discussion to introduce the rotation tensor. This time, rather than considering just the stretch of a vanishingly small line element, we consider the *complete* mutual relative motion of the endpoints of the line element (thus we include rotation as well as stretching of the element). For this purpose consider line element $\overline{PN}$ in Fig. 1.15. The relative movement of the endpoints can be given by using the displacement field as follows:

$$\mathbf{u}_N - \mathbf{u}_P = \left[\mathbf{u}_P + \left(\frac{\partial \mathbf{u}}{\partial x_j}\right)_P \Delta x_j + \cdots\right] - \mathbf{u}_P \tag{1.49}$$

where we have expanded $\mathbf{u}_N$ as a Taylor series about point P. In the limit $\Delta x_j \to 0$ we have from the above equation:

$$d\mathbf{u} = \frac{\partial \mathbf{u}}{\partial x_j}\, dx_j$$

In index notation,

$$du_i = \frac{\partial u_i}{\partial x_j}\, dx_j \tag{1.50}$$

Thus the relative movement du_i between the two adjacent points dx_i apart is determined by the tensor $u_{i,j}$. But this tensor obviously can be written as

$$u_{i,j} = \tfrac{1}{2}(u_{i,j} + u_{j,i}) + \tfrac{1}{2}(u_{i,j} - u_{j,i}) \tag{1.51}$$

The first expression on the right side of the equation is the strain tensor ϵ_{ij}. The second expression is denoted as ω_{ij} and is called the *rotation tensor*, for reasons soon to be made clear. Note that the rotation tensor ω_{ij} is a *skew-symmetric tensor*. We can then express the above equation as

$$u_{i,j} = \epsilon_{ij} + \omega_{ij} \tag{1.52}$$

where

$$\omega_{ij} = \tfrac{1}{2}(u_{i,j} - u_{j,i}) \tag{1.53}$$

Thus the general relative movement between adjacent points is a result of the strain tensor plus a second effect, which we will now interpret by examining again an infinitesimal parallelepiped having $\overline{PN}$ as a diagonal, as shown in Fig. 1.15. For now we assume that there is no deformation of the body–only rigid-body motion. Then line segments $\overline{PQ}$ and $\overline{PG}$ of the rectangular parallelepiped each undergo the same rotation $\delta\phi_x$ about the x axis. We have shown these lines (Fig. 1.20), after rigid-body motion, projected onto the original plane yz. We can then give $\sin \delta\phi_x$ as follows:

$$\sin \delta\phi_x = \frac{[(u_z)_P + (\partial u_z/\partial y)_P\, \Delta y + \cdots] - (u_z)_P}{\Delta y'}$$

where $\Delta y'$ is the projection of $\overline{P^*Q^*}$ onto plane yz. Now take the limit as $\Delta y \to 0$. Noting that for small rotations we can replace the sine of the angle by the angle itself and that $\Delta y'$ may be replaced by Δy, we get for this equation in the limit

$$\delta\phi_x = \frac{\partial u_z}{\partial y}$$

at any point P. Similarly, considering line element PG, we get for $\sin \delta\phi_x$

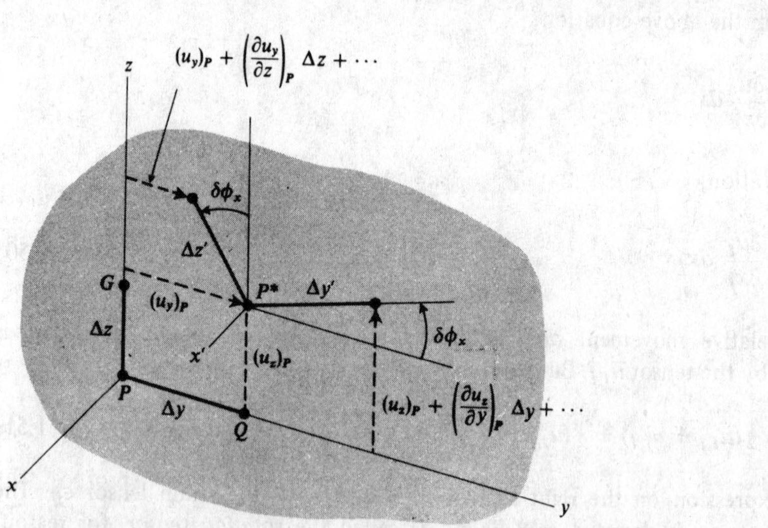

FIGURE 1.20
Line segments Δy and Δz undergo rigid-body displacement. $\delta\phi_x$ is rotation about x axis.

$$\sin \delta\phi_x = \frac{(u_y)_P - [(u_y)_P + (\partial u_y/\partial z)_P \, \Delta z + \cdots]}{\Delta z'}$$

In the limit of small deformations, we get

$$\delta\phi_x = -\frac{\partial u_y}{\partial z}$$

Thus we may express $\delta\phi_x$ in the following manner:

$$\delta\phi_x = \frac{1}{2}\left(\frac{\partial u_z}{\partial y} - \frac{\partial u_y}{\partial z}\right) \tag{1.54}$$

For the other two components of rotation we have, on permuting subscripts,

$$\begin{aligned} \delta\phi_y &= \frac{1}{2}\left(\frac{\partial u_x}{\partial z} - \frac{\partial u_z}{\partial x}\right) \\ \delta\phi_z &= \frac{1}{2}\left(\frac{\partial u_y}{\partial x} - \frac{\partial u_x}{\partial y}\right) \end{aligned} \tag{1.55}$$

The expressions on the right sides of Eqs. (1.54) and (1.55) are clearly the off-diagonal terms of the rotation tensor ω_{ij}. Thus we can say

$$\begin{aligned} \delta\phi_x &= \omega_{32} = -\omega_{23} \\ \delta\phi_y &= \omega_{13} = -\omega_{31} \\ \delta\phi_z &= \omega_{21} = -\omega_{12} \end{aligned} \tag{1.56}$$

We see that for a rigid-body movement of the element the nonzero components of the rotation tensor give the infinitesimal rotation components of the element. What does ω_{ij} represent when the rectangular parallelepiped is undergoing a movement including deformation of the element and not just rigid-body rotation? For this case each line segment in the rectangular volume has its own angle of rotation and we can show that ω_{ij} for such a situation gives the *average rotation* components of all the line segments in the body. However, we shall term the components of ω_{ij} the rigid-body rotation components.

In summary, in considering the general relative motion between adjacent points and then going to the infinitesimal rectangular parallelepiped, we have by the first step introduced ω_{ij} and by the second step given a physical interpretation for ω_{ij}. Now going back to the adjacent point approach, we would expect that the strain tensor should be directly related to the force intensities (i.e., the stress tensor) at a point. (This is to be expected from studies in atomic physics, where interatomic forces were considered as a function of the separation of atoms.) Indeed, we know from experiment that it is the ϵ_{ij} portion of Eq. (1.52) that is related to the stress τ_{ij} at a point. We shall discuss constitutive laws shortly.

1.10 TRANSFORMATION EQUATIONS FOR STRAIN

Since $u_{i,j}$ must be a second-order tensor (see the conclusion of Section I.5 of Appendix I on taking the partial derivative of a tensor), clearly ϵ_{ij} must be a second-order tensor. However, we shall show this formally here since this is a simple instructive procedure. We shall first consider the normal strain at a point P in the direction α (see Fig. 1.21). For this purpose a line segment $\Delta\alpha$ has been shown connecting points P and Q in the undeformed geometry. The positions of P and Q in the deformed geometry have been shown as P^* and Q^*, respectively. We express the displacement component of point Q in the α direction in terms of the displacement component of point P in the α direction as follows:

$$(u_\alpha)_Q = (u_\alpha)_P + \left(\frac{\partial u_\alpha}{\partial x_i}\right)_P \Delta x_i + \frac{1}{2!}\left(\frac{\partial^2 u_\alpha}{\partial x_i\,\partial x_j}\right)_P \Delta x_i\,\Delta x_j + \cdots$$

Therefore

$$\frac{(u_\alpha)_Q - (u_\alpha)_P}{\Delta\alpha} = \left(\frac{\partial u_\alpha}{\partial x_i}\right)_P \left(\frac{\Delta x_i}{\Delta\alpha}\right) + \frac{1}{2!}\left(\frac{\partial^2 u_\alpha}{\partial x_i\,\partial x_j}\right)_P \frac{\Delta x_i}{\Delta\alpha}\,\Delta x_j + \cdots$$

Take the limit as Δx_i and $\Delta\alpha$ go to zero. The left side clearly is the normal strain $\epsilon_{\alpha\alpha}$. On the right side only the first term remains, so that we have

$$\epsilon_{\alpha\alpha} = \left(\frac{\partial u_\alpha}{\partial x_i}\right)\left(\frac{dx_i}{d\alpha}\right) = \left(\frac{\partial u_\alpha}{\partial x_i}\right) a_{\alpha i} \qquad (1.57)$$

where $a_{\alpha i}$ is the direction cosine between the α axis and the x_i axis. Now express u_α in terms of u_i as follows by projecting components u_i in the direction α:

$$u_\alpha = a_{\alpha j} u_j$$

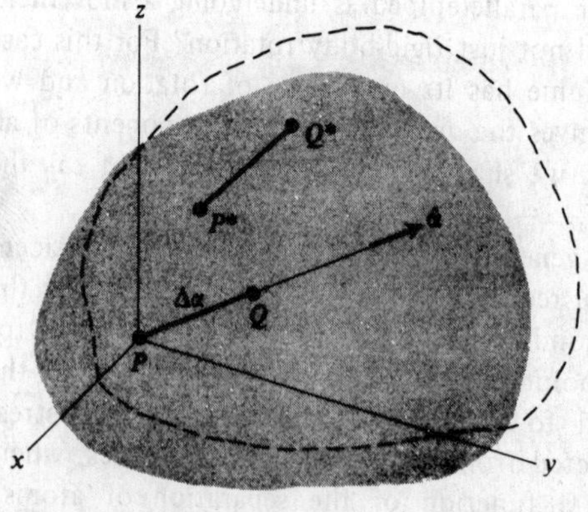

FIGURE 1.21
Mapping of line segment *PQ*.

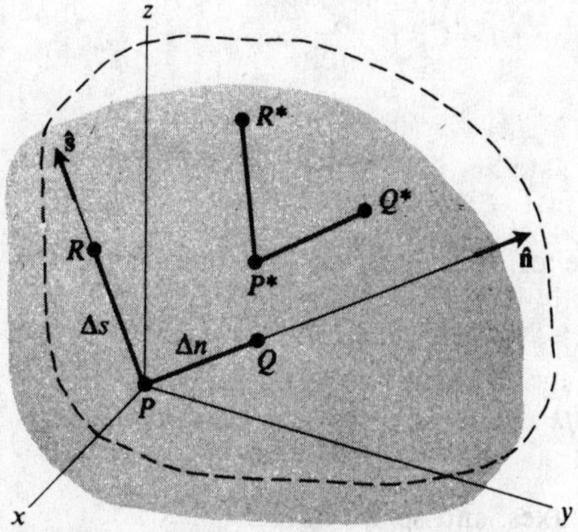

FIGURE 1.22
Mapping of line segments PQ and PR.

Substituting into Eq. (1.57) we get

$$\epsilon_{\alpha\alpha} = u_{j,i}a_{\alpha i}a_{\alpha j} = \tfrac{1}{2}(u_{j,i} + u_{i,j})a_{\alpha i}a_{\alpha j} = \epsilon_{ij}\, a_{\alpha i}a_{\alpha j} = a_{\alpha i}a_{\alpha j}\,\epsilon_{ij}$$

where we have interchanged dummy indices i and j in expanding $u_{j,i}$ above. We can imagine that the α direction coincides with the axes of a primed reference x'_i. Accordingly, we may then say

$$\epsilon'_{\alpha\alpha} = a_{\alpha i}a_{\alpha j}\,\epsilon_{ij} \tag{1.58}$$

Now consider two segments $\overline{PQ}$ and $\overline{PR}$ at right angles to each other in the directions **n** and **s** in the undeformed geometry, as shown in Fig. 1.22. The points P^*, R^*, and Q^* correspond to the deformed geometry. If n and s are to be coordinate axes, the shear strain ϵ_{ns} for these axes is given as

$$\epsilon_{ns} = \frac{1}{2}\left(\frac{\partial u_n}{\partial s} + \frac{\partial u_s}{\partial n}\right)$$

Expressing u_n and u_s as follows, using vector projections,

$$u_n = a_{nj}u_j \qquad u_s = a_{sj}u_j \tag{1.59}$$

we can write ϵ_{ns} in the following way:

$$\epsilon_{ns} = \frac{1}{2}\left(\frac{\partial u_j}{\partial s}\, a_{nj} + \frac{\partial u_j}{\partial n}\, a_{sj}\right) \tag{1.60}$$

Next, using the chain rule, we note that

$$\frac{\partial u_j}{\partial s} = \left(\frac{\partial u_j}{\partial x_k}\right)\left(\frac{\partial x_k}{\partial s}\right) = \frac{\partial u_j}{\partial x_k}\, a_{sk} \qquad \frac{\partial u_j}{\partial n} = \left(\frac{\partial u_j}{\partial x_k}\right)\left(\frac{\partial x_k}{\partial n}\right) = \frac{\partial u_j}{\partial x_k}\, a_{nk} \tag{1.61}$$

Substituting Eqs. (1.61) into Eq. (1.60) we get

$$\epsilon_{ns} = \frac{1}{2}\left(\frac{\partial u_j}{\partial x_k} a_{sk} a_{nj} + \frac{\partial u_j}{\partial x_k} a_{nk} a_{sj}\right)$$

In the second expression on the right side of this equation we interchange the dummy symbols j and k. We thus get

$$\epsilon_{ns} = \frac{1}{2}\left(\frac{\partial u_j}{\partial x_k} + \frac{\partial u_k}{\partial x_j}\right) a_{sk} a_{nj} = a_{nj} a_{sk} \epsilon_{jk}$$

We may think of n and s as a pair of primed axes, and so we have

$$\epsilon'_{ns} = a_{nj} a_{sk} \epsilon_{jk} \tag{1.62}$$

And if $n = s = \alpha$ in the above formulation, we get back to Eq. (1.58). Thus we have reached the transformation equation of strain at a point from an unprimed reference to a primed, rotated reference, and the transformation equation is that of a second-order tensor.

The various properties set forth for the stress tensor stemming from the transformation equations then apply to the strain tensor.

1.11 COMPATIBILITY EQUATIONS

Let us consider further the strain-displacement relations

$$\epsilon_{ij} = \tfrac{1}{2}(u_{i,j} + u_{j,i}) \tag{1.63}$$

If the displacement field is given, we can readily compute the strain tensor field by substituting u_i into the above equations. The inverse problem of finding the displacement field from a strain field is not so simple. Here the displacement field, composed of *three* functions u_i, must be determined by integration of *six* partial differential equations given by Eq. (1.63). In order to ensure *single-valued, continuous* solutions u_i, we must impose certain restrictions on ϵ_{ij}. That is, we cannot set forth any strain field ϵ_{ij} and expect it automatically to be associated with a single-valued, continuous displacement field. But actual deformations must have single-valued displacement fields. Furthermore, the deformations of interest to us will be those having continuous displacement fields. Hence the restrictions we will reach in rendering u_i single-valued and continuous apply to our formulations. The resulting equations are called the *compatibility equations.*

We shall first set forth necessary conditions on ϵ_{ij} for single-valuedness and continuity of u_i. The use of Eq. (1.63) to give ϵ_{ij} requires that u_i have the aforestated properties. Accordingly, we shall ensure these properties for u_i by working directly with these equations. We now form the following derivatives from these equations:

$$\epsilon_{ij,kl} = \tfrac{1}{2}(u_{i,jkl} + u_{j,ikl})$$
$$\epsilon_{kl,ij} = \tfrac{1}{2}(u_{k,lij} + u_{l,kij})$$
$$\epsilon_{lj,ki} = \tfrac{1}{2}(u_{j,lik} + u_{l,jik})$$
$$\epsilon_{ki,lj} = \tfrac{1}{2}(u_{k,ijl} + u_{i,kjl})$$

By adding the first two equations and then subtracting the last two equations we may eliminate the u_i components and thus arrive at a set of relations involving only strains. That is,

$$\boxed{\epsilon_{ij,kl} + \epsilon_{kl,ij} - \epsilon_{lj,ki} - \epsilon_{ki,lj} = 0} \tag{1.64}$$

These form a set of 81 equations known as the *compatibility equations* that a strain field must satisfy if it is to be related to u_i via Eq. (1.63), which in turn means that u_i is single-valued and continuous. These equations are thus *necessary requirements*.

We shall now consider the *sufficiency* of these equations for generating a single-valued continuous displacement field for the case of simply connected regions.† We first consider Eq. (1.64) valid.

Let $P^0(x_1, x_2, x_3)$ be a point at which displacement u_i^0 and rotation terms ω_{ij}^0 are known. Then the displacement of *any other point* P^* in the domain is representable by a line integral along a continuous curve C from P^0 to P^* in the following way:

$$u_i^* = u_i^0 + \int_{P^0}^{P^*} du_i = u_i^0 + \int_{P^0}^{P^*} u_{i,j}\, dx_j$$

We now employ Eq. (1.52) in the last expression to give

$$u_i^* = u_i^0 + \int_{P^0}^{P^*} \epsilon_{ij}\, dx_j + \int_{P^0}^{P^*} \omega_{ij}\, dx_j \tag{1.65}$$

Since x_j^* is a given fixed point, as far as integration is concerned, we can employ $d(x_j - x_j^*)$ in the last line integral. Integrating this expression by parts we then may say

$$\int_{P^0}^{P^*} \omega_{ij}\, d(x_j - x_j^*) = \omega_{ij}(x_j - x_j^*)\Big|_{P^0}^{P^*} - \int_{P^0}^{P^*} (x_j - x_j^*)\, d\omega_{ij}$$

$$= -\omega_{ij}^0(x_j^0 - x_j^*) - \int_{P^0}^{P^*} (x_j - x_j^*)\omega_{ij,k}\, dx_k \tag{1.66}$$

†A simply connected region is one for which each and every closed curve can be shrunk to a point without cutting a boundary.

Now the expression $\omega_{ij,k}$ can be written as follows:

$$\omega_{ij,k} = \tfrac{1}{2}(u_{i,jk} - u_{j,ik}) = \tfrac{1}{2}(u_{i,jk} + u_{k,ij}) - \tfrac{1}{2}(u_{j,ik} + u_{k,ij})$$

where we have added and subtracted $\frac{1}{2}u_{k,ij}$. From this we conclude:

$$\omega_{ij,k} = \tfrac{1}{2}(u_{i,kj} + u_{k,ij}) - \tfrac{1}{2}(u_{j,ki} + u_{k,ji}) = (\epsilon_{ik,j} - \epsilon_{jk,i}) \tag{1.67}$$

Substituting from Eqs. (1.67) and (1.66) into Eq. (1.65) we get

$$u_i^* = u_i^0 - \omega_{ij}^0(x_j^0 - x_j^*) + \int_{P^0}^{P^*} U_{ik}\, dx_k \tag{1.68a}$$

where

$$U_{ik} = \epsilon_{ik} - (x_j - x_j^*)(\epsilon_{ik,j} - \epsilon_{jk,i}) \tag{1.68b}$$

If u_i^* is to be single-valued and continuous, the integral of Eq. (1.68a) must be independent of the path. This in turn means that the integrand $U_{ik}\, dx_k$ must be an exact differential. In a simply connected domain the necessary and sufficient condition for $U_{ik}\, dx_k$ to be an exact differential is that

$$U_{ik,l} - U_{il,k} = 0 \tag{1.69}$$

Now using Eq. (1.68b) in the above equation and noting that $x_{j,l} = \delta_{jl}$, we find that

$$\begin{aligned}\epsilon_{ik,l} - \delta_{jl}(\epsilon_{ik,j} - \epsilon_{jk,i}) - (x_j - x_j^*)(\epsilon_{ik,jl} - \epsilon_{jk,il}) - \epsilon_{il,k} \\ + \delta_{jk}(\epsilon_{il,j} - \epsilon_{jl,i}) + (x_j - x_j^*)(\epsilon_{il,jk} - \epsilon_{jl,ik}) = 0\end{aligned}$$

Rearranging the equation and simplifying, we get

$$\begin{aligned}\epsilon_{ik,l} - \epsilon_{il,k} - (\epsilon_{ik,l} - \epsilon_{lk,i}) + (\epsilon_{il,k} - \epsilon_{kl,i}) \\ + (x_j - x_j^*)(\epsilon_{il,jk} + \epsilon_{jk,il} - \epsilon_{ik,jl} - \epsilon_{jl,ik}) = 0\end{aligned}$$

The first six terms cancel each other, while the expression multiplied by $(x_j - x_j^*)$ has been assumed equal to zero at the outset, thus ensuring compatibility. We have thus proved the sufficiency requirements for the compatibility equations.

Actually, only 6 of the 81 equations of compatibility are independent; the rest are either identities or repetitions due to the symmetry of ϵ_{ij}. The 6 independent equations of compatibility are given as follows in unabridged notation:

$$\frac{\partial^2 \epsilon_{xx}}{\partial y^2} + \frac{\partial^2 \epsilon_{yy}}{\partial x^2} = \frac{\partial^2 \gamma_{xy}}{\partial x\, \partial y} \tag{1.70a}$$

$$\frac{\partial^2 \epsilon_{yy}}{\partial z^2} + \frac{\partial^2 \epsilon_{zz}}{\partial y^2} = \frac{\partial^2 \gamma_{yz}}{\partial y\, \partial z} \tag{1.70b}$$

$$\frac{\partial^2 \epsilon_{zz}}{\partial x^2} + \frac{\partial^2 \epsilon_{xx}}{\partial z^2} = \frac{\partial^2 \gamma_{zx}}{\partial z\, \partial x} \tag{1.70c}$$

$$2\frac{\partial^2 \epsilon_{xx}}{\partial y\, \partial z} = \frac{\partial}{\partial x}\left(-\frac{\partial \gamma_{yz}}{\partial x} + \frac{\partial \gamma_{xz}}{\partial y} + \frac{\partial \gamma_{xy}}{\partial z}\right) \tag{1.70d}$$

$$2\frac{\partial^2 \epsilon_{yy}}{\partial z\, \partial x} = \frac{\partial}{\partial y}\left(-\frac{\partial \gamma_{zx}}{\partial y} + \frac{\partial \gamma_{yx}}{\partial z} + \frac{\partial \gamma_{yz}}{\partial x}\right) \tag{1.70e}$$

$$2\frac{\partial^2 \epsilon_{zz}}{\partial x\, \partial y} = \frac{\partial}{\partial z}\left(-\frac{\partial \gamma_{xy}}{\partial z} + \frac{\partial \gamma_{zy}}{\partial x} + \frac{\partial \gamma_{zx}}{\partial y}\right) \tag{1.70f}$$

Part C
General Considerations

1.12 ENERGY CONSIDERATIONS

We have described the stress tensor arising from equilibrium considerations and the strain tensor arising from kinematic considerations. These tensors are related to each other, as noted earlier, by laws that are called *constitutive laws*. In general, such relations include temperature and time as other variables. In addition, they often require knowledge of the history of the deformations leading to the instantaneous condition of interest in order to properly relate stress and strain. In this text we shall consider that the constitutive laws relate stress and strain directly and uniquely. That is,

$$\tau_{ij} = \tau_{ij}(\epsilon_{11}, \epsilon_{12}, \ldots, \epsilon_{33}) \tag{1.71}$$

We will discuss specific constitutive laws later in the text. Now it will be of interest to us to consider such constitutive laws that render the following integral

$$\mathfrak{U} = \int_0^{\epsilon_{ij}} \tau_{ij}\, d\epsilon_{ij} \tag{1.72}$$

a point function of the upper limit ϵ_{ij}. For $\mathfrak{U}$ to be a point function, the integral must be independent of the path, and this in turn means that $\tau_{ij}\, d\epsilon_{ij}$ must be a perfect differential. When that is the case we can say

$$d\mathfrak{U} = \tau_{ij}\, d\epsilon_{ij} \tag{1.73a}$$

Therefore

$$\frac{\partial \mathfrak{U}}{\partial \epsilon_{ij}} = \tau_{ij} \tag{1.73b}$$

The function $\mathcal{U}$ under such circumstances is called the *strain energy density function*. We now pose two queries: What physical attributes can be ascribed to this function, and when does it exist? As to the first query, consider an infinitesimal rectangular parallelepiped under the action of normal stresses, as shown in Fig. 1.23. The displacements on faces 1 and 2 in the x direction are given as u_x and $u_x + (\partial u_x/\partial x)\,dx$, respectively, so that the increment of mechanical work done by the stresses on the element during deformation is

$$-\tau_{xx}\,du_x\,dy\,dz + \left(\tau_{xx} + \frac{\partial \tau_{xx}}{\partial x}\,dx\right) d\left(u_x + \frac{\partial u_x}{\partial x}\,dx\right) dy\,dz$$
$$+ B_x\,dx\,dy\,dz\,d\left(u_x + \kappa\,\frac{\partial u_x}{\partial x}\,dx\right)$$

where κ is some factor between 0 and 1. Collecting terms and deleting higher-order expressions, we then get for the above expressions:

$$\left[\tau_{xx}\,d\left(\frac{\partial u_x}{\partial x}\right) + \left(\frac{\partial \tau_{xx}}{\partial x} + B_x\right) du_x\right] dx\,dy\,dz$$

We may employ equilibrium considerations to delete the second expression in the brackets, leaving the following expression for the increment of mechanical work:

$$\tau_{xx}\,d\left(\frac{\partial u_x}{\partial x}\right) dx\,dy\,dz = \tau_{xx}\,d\epsilon_{xx}\,dv$$

By considering normal stress and strain in the y and z directions, we may form similar expressions for the element. Thus, for normal stresses on an element, the increment of mechanical work for isotropic materials is

$$(\tau_{xx}\,d\epsilon_{xx} + \tau_{yy}\,d\epsilon_{yy} + \tau_{zz}\,d\epsilon_{zz})\,dv$$

We shall denote w as the mechanical work per unit volume. From the above expression we see that for normal stresses only we have

$$dw = \tau_{xx}\,d\epsilon_{xx} + \tau_{yy}\,d\epsilon_{yy} + \tau_{zz}\,d\epsilon_{zz}$$

Next consider a case of pure shear strain such as that shown in Fig. 1.24. The mechanical increment of work can be given here as follows:

$$\left[\left(\tau_{xy} + \frac{\partial \tau_{xy}}{\partial y}\,dy\right) dz\,dx\right]\left\{d\left[\gamma_{xy} + \beta\left(\frac{\partial \gamma_{xy}}{\partial x}\right) dx\right] dy\right\}$$
$$+ B_x\,dx\,dy\,dz\,d\left(\gamma_{xy} + \eta\,\frac{\partial \gamma_{xy}}{\partial x}\,dx\right)(\kappa\,dy)$$

where β, η, and κ are factors between 0 and 1. Now carrying out arithmetical

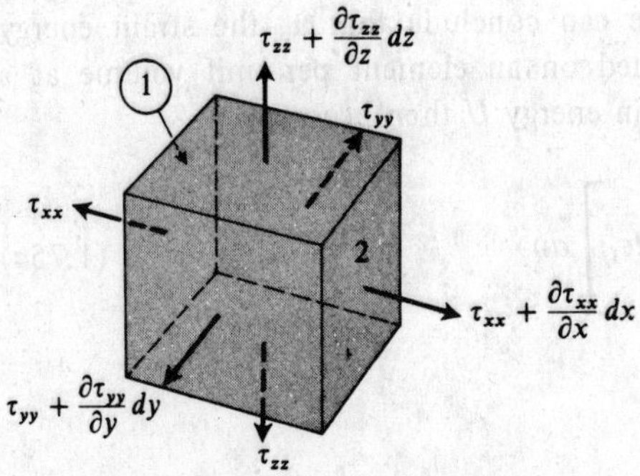

FIGURE 1.23
Infinitesimal rectangular parallelepiped with normal stresses only.

operations and dropping higher-order terms, we get for the above expression:

$$\tau_{xy}\, d\gamma_{xy}\, dx\, dy\, dz = 2\tau_{xy}\, d\epsilon_{xy}\, dv$$

Thus for pure shear stresses on all faces we get the following result for the increments of mechanical work:

$$2(\tau_{xy}\, d\epsilon_{xy} + \tau_{xz}\, d\epsilon_{xz} + \tau_{yz}\, d\epsilon_{yz})\, dv$$

Accordingly, the mechanical work increment per unit volume at a point for a *general* state of stress is then given as

$$dw = \tau_{ij}\, d\epsilon_{ij} \tag{1.74}$$

It should be clear that this result is valid only for infinitesimal deformations. And now integrating from 0 to some strain level ϵ_{ij} we get

$$w = \int_0^{\epsilon_{ij}} \tau_{ij}\, d\epsilon_{ij} = \mathfrak{U}$$

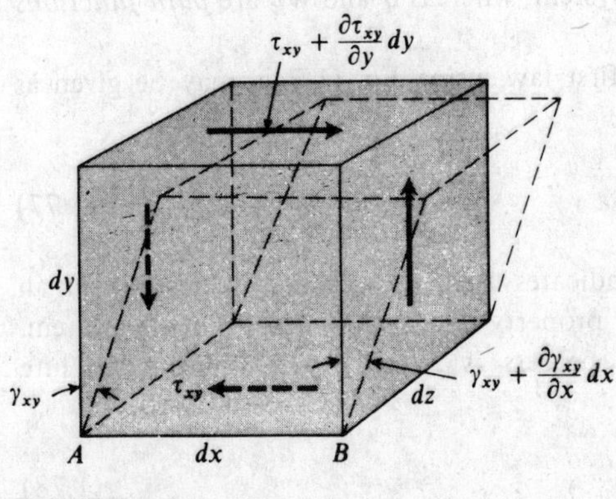

FIGURE 1.24
Shear stresses only act on infinitesimal rectangular parallelepiped.

in accordance with Eq. (1.72). Thus we can conclude that $\mathcal{U}$, the strain energy density, is the mechanical work performed on an element per unit volume at a point during a deformation. The total strain energy U then becomes

$$U = \iiint_V \mathcal{U}\, dv = \iiint_V \left[\int_0^{\epsilon_{ij}} \tau_{ij}\, d\epsilon_{ij}\right] dv \tag{1.75a}$$

where

$$\mathcal{U} = \int_0^{\epsilon_{ij}} \tau_{ij}\, d\epsilon_{ij} \tag{1.75b}$$

As to the second query (when does $\mathcal{U}$ exist?), we may proceed by using two different arguments. First, we may rely on a strictly mathematical approach whereby it is known from calculus that for $\tau_{ij}\, d\epsilon_{ij}$ to be a perfect differential (and thus to permit existence for $\mathcal{U}$) it is necessary and sufficient that

$$\frac{\partial \tau_{ij}}{\partial \epsilon_{kl}} = \frac{\partial \tau_{kl}}{\partial \epsilon_{ij}} \tag{1.76}$$

On the other hand, we may resort to arguments from *thermodynamics*. The *first law of thermodynamics* states that

$$du = d'q + d'w_k$$

where u is the internal energy per unit volume (specific internal energy), q is the heat transfer per unit volume, and w_k is the work done on the system per unit volume by the surroundings. The primes associated with the differentials of q and w_k are to indicate that these quantities are not *perfect differentials*–they simply represent vanishingly small increments of these quantities. In other words, u is a *point function* depending on the state of the system, whereas q and w_k are *path functions* depending on the process.

We can say for a solid that the first law, using Eq. (1.74), may be given as follows:

$$du = d'q + \tau_{ij}\, d\epsilon_{ij} \tag{1.77}$$

The second law of thermodynamics indicates that no process is possible in an isolated system which would decrease a property called the *entropy* of the system. The specific entropy s in a reversible process is related to q and the absolute temperature T according to the equation:

$$d'q = T\, ds \tag{1.78}$$

Combining Eqs. (1.77) and (1.78) to eliminate $d'q$ we get a *combined* form of the *first and second laws*:

$$du = T\,ds + \tau_{ij}\,d\epsilon_{ij} \tag{1.79}$$

Let us now examine a reversible *adiabatic process* (no heat transfer). For such a process there can be no change in entropy and the above equation becomes:

$$du = \tau_{ij}\,d\epsilon_{ij}$$

Since u is a point function we can conclude for this process that $\tau_{ij}\,d\epsilon_{ij}$ is a perfect differential. Indeed, noting Eq. (1.73*a*), we can conclude that the strain energy density $\mathcal{U}$ is simply the specific internal energy u for such processes (isentropic processes).

Let us now consider another function used in thermodynamics called the *Helmholtz function, F*, defined as a specific value in the following way:

$$F = u - Ts$$

Since u, T, and s are point functions, F must be a point function. Now consider a reversible *isothermal process*. For such a process, the above equation in differential form becomes

$$dF = du - T\,ds \tag{1.80}$$

Now replacing du in Eq. (1.79) by using Eq. (1.80), we arrive at the result:

$$dF = \tau_{ij}\,d\epsilon_{ij} \tag{1.81}$$

Once again, $\tau_{ij}\,d\epsilon_{ij}$ is a perfect differential. Indeed, for such processes we conclude from Eq. (1.73*a*) that the strain energy density function equals the Helmholtz function.†

We thus see that the energy density function $\mathcal{U}$ exists for certain reversible processes. Because of the reversibility requirement we see that the existence of $\mathcal{U}$ calls for *elastic behavior* of the bodies, in that a body must return identically to its original condition when the loads are released. In most situations in elasticity, furthermore, the process is somewhere between isothermal and adiabatic. Since in solid mechanics, unlike fluid mechanics, the difference between adiabatic and isothermal processes is usually not great, the strain energy function $\mathcal{U}$ is assumed to exist for most processes involving elastic behavior.

Furthermore, it can be shown from thermodynamic considerations that the

†From our discussion thus far, we can say that the Helmholtz function represents the energy that can be converted to mechanical work in a reversible isothermal process. This is often called *free energy*. This is one reason for its importance in thermodynamics.

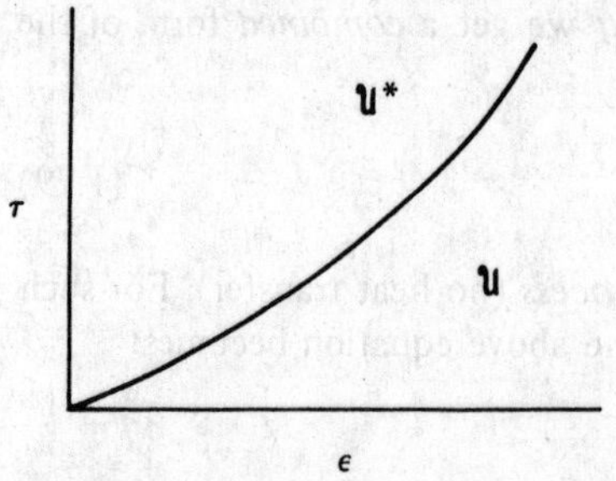

FIGURE 1.25
Stress-strain curve showing $\mathfrak{u}$ and $\mathfrak{u}^*$.

strain energy density functions are *positive definite* functions† of the strains for small deformation. We will ask you to demonstrate this in Problem 1.18 for the case of linear elastic materials.

Let us next consider a stress-strain diagram for a tensile test in a case where the energy density function exists (Fig. 1.25). For τ to be a single-valued function of ϵ the curve τ versus ϵ must be monotonically increasing, as shown in the diagram. Furthermore, the value of $\mathfrak{u}$ is simply the area under the curve for a given strain.

Now just as the curve generates a single-valued function $\mathfrak{u}$ as the area between the curve and the ϵ axis, it also generates a single-valued function, giving the area between the curve and the τ axis (shown darkened in the diagram). This function is called the *complementary energy density function* and is denoted as $\mathfrak{u}^*$. For the one-dimensional case we have

$$d\mathfrak{u}^* = \epsilon\, d\tau$$

Therefore

$$\mathfrak{u}^* = \int_0^{\tau} \epsilon\, d\tau$$

For a general state of stress at a point we may generalize the above result as follows:

$$d\mathfrak{u}^* = \epsilon_{ij}\, d\tau_{ij} \tag{1.82a}$$

Therefore

$$\mathfrak{u}^* = \int_0^{\tau_{ij}} \epsilon_{ij}\, d\tau_{ij} \tag{1.82b}$$

†A positive definite function has the property of never being less than zero for the range of the variables involved and is zero only when the variables are zero. This means that $\mathfrak{u} = 0$ at a point when the strains are zero at the point.

It is apparent from Eq. (1.82*a*) that:

$$\frac{\partial \mathfrak{U}^*}{\partial \tau_{ij}} = \epsilon_{ij} \tag{1.83}$$

For linear elastic behavior (to be discussed in the next section) the one-dimensional stress-strain curve is that of a straight line and the strain energy density function (the area between the curve and the ϵ axis) equals the complementary strain energy density function (the area between the curve and the τ axis). It is similarly true, for a general state of stress, that for linear elastic behavior:

$$\mathfrak{U} = \mathfrak{U}^* \tag{1.84}$$

We shall find that both the functions $\mathfrak{U}$ and $\mathfrak{U}^*$ will be of considerable use in this text.

1.13 HOOKE'S LAW

In the previous section we pointed out and used the fact that the stress tensor and the strain tensor are related. These relations depend on the nature of the material and are called constitutive laws. We shall be concerned in most of this text with *linear elastic behavior*, wherein each stress component is linearly related, in the general case, to all the strains by equations of the form

$$\tau_{ij} = C_{ijkl}\epsilon_{kl} \tag{1.85}$$

where the C's are at most functions only of position. This law is called the *generalized Hooke's law*. Because ϵ_{ij} and τ_{ij} are second-order tensor fields, one may show that C_{ijkl} is a fourth-order tensor field. Since τ_{ij} is symmetric it should be clear that C_{ijkl} must also be symmetric in i,j. That is,

$$C_{ijkl} = C_{jikl} \tag{1.86}$$

And since ϵ_{kl} is symmetric we can always express the components C_{ijkl} in a form symmetric in kl without violating Eq. (1.85). That is, we will stipulate that

$$C_{ijkl} = C_{ijlk} \tag{1.87}$$

We next assume the existence of a strain energy function $\mathfrak{U}$, as discussed in the previous section. Now in order to satisfy Eq. (1.73) and Hooke's law simultaneously this function must take the form

$$\mathfrak{U} = \tfrac{1}{2}C_{ijkl}\epsilon_{ij}\epsilon_{kl} \tag{1.88}$$

where the tensor C_{ijkl} must also have the symmetry property†

$$C_{ijkl} = C_{klij} \tag{1.89}$$

Thus starting with 81 terms for C_{ijkl}, we may show, using the three aforementioned symmetry relations for C_{ijkl}, that only 21 of these terms are independent. We will assume now that the material is homogeneous‡ so we may consider C_{ijkl} to be a set of *constants* for a given reference.

We next consider the property of *isotropy* for the mechanical behavior of the body. This property requires that the mechanical properties of a material at a point are not dependent on direction. Thus a stress such as τ_{xx} must be related to all the strains ϵ_{ij} for reference xyz exactly as the stress $\tau_{x'x'}$ is related to all the strains ϵ'_{ij} for a reference $x'y'z'$ rotated relative to xyz. Accordingly, C_{ijkl} must have the *same components* for all references. A tensor such as C_{ijkl} whose components are invariant with respect to a rotation of axes is called an *isotropic* tensor. For an isotropic second-order tensor:

$$A_{ij} = A'_{ij} \tag{1.90}$$

Transforming the right side of this equation according to the rules of second-order tensors, we have

$$A_{ij} = a_{il}a_{jk}A_{lk} \tag{1.91}$$

You may demonstrate (Problem 1.15) that the following set satisfies the above requirements and is the most general second-order isotropic tensor:

$$A_{ij} = \alpha\delta_{ij} \tag{1.92}$$

where α is a scalar constant. One can show that this is the only such second-order tensor. As for a fourth-order isotropic tensor, we require that:

$$D_{mnop} = D'_{mnop} \tag{1.93}$$

Using the fourth-order transformation formula on the right side of this equation we get

$$D_{mnop} = a_{mi}a_{nj}a_{ok}a_{pl}D_{ijkl}$$

†To understand why this is so, consider the computation of τ_{12} from $\mathcal{U}$. That is:

$$\frac{\partial \mathcal{U}}{\partial \epsilon_{12}} = \frac{\partial}{\partial \epsilon_{12}}[\tfrac{1}{2}C_{ijkl}\epsilon_{ij}\epsilon_{kl}] = \tfrac{1}{2}C_{12kl}\epsilon_{kl} + \tfrac{1}{2}C_{ij12}\,\epsilon_{ij} = C_{12kl}\epsilon_{kl} = \tau_{12}$$

Thus by using Eq. (1.88) for $\mathcal{U}$ in conjunction with Eq. (1.89) we are able to compute stresses from $\mathcal{U}$ properly and still have the basic relation for Hooke's law intact.

‡A homogeneous material has the same composition throughout.

You may demonstrate (Problem 1.16) that the following set is a fourth-order isotropic tensor:

$$D_{ijkl} = \lambda\delta_{ij}\delta_{kl} + \beta\delta_{ik}\delta_{jl} + \gamma\delta_{il}\delta_{jk} \tag{1.94}$$

where λ, β, and γ are scalar constants. It is convenient to express the constants β and γ as follows:

$$\beta = G + B \qquad \gamma = G - B$$

where G and B are constants determined by the above equations. Then we can give Eq. (1.94) as

$$\begin{aligned} D_{ijkl} &= \lambda\delta_{ij}\delta_{kl} + (G + B)(\delta_{ik}\delta_{jl}) + (G - B)(\delta_{il}\delta_{jk}) \\ &= \lambda\delta_{ij}\delta_{kl} + G(\delta_{ik}\delta_{jl} + \delta_{il}\delta_{jk}) + B(\delta_{ik}\delta_{jl} - \delta_{il}\delta_{jk}) \end{aligned} \tag{1.95}$$

Clearly, the tensor associated with G is symmetric in the indices i,j while the tensor associated with B is skew-symmetric in these indices. For the tensor D_{ijkl} to be used for isotropic elastic behavior it should be clear that the constant B must be zero [see Eq. (1.86)]. Setting $B = 0$, we can consider the above tensor as the most general isotropic tensor that may be used for Hooke's law. Thus we can say

$$\tau_{ij} = [\lambda\delta_{ij}\delta_{kl} + G(\delta_{ik}\delta_{jl} + \delta_{il}\delta_{jk})]\,\epsilon_{kl}$$

Carrying out the operations with the Kronecker delta terms, we get

$$\tau_{ij} = \lambda\delta_{ij}\epsilon_{ll} + 2G\epsilon_{ij} \tag{1.96}$$

This is the general form of Hooke's law giving stress components in terms of strain components for isotropic materials. The constants λ and G are the so-called *Lamé constants*. We see that as a result of isotropy the number of independent elastic moduli has been reduced from 21 to 2. The inverse form of Hooke's law yielding strain components in terms of stress components may be given as follows:

$$\epsilon_{ij} = \frac{1}{2G}\tau_{ij} - \left[\frac{\lambda}{2G(3\lambda + 2G)}\tau_{kk}\right]\delta_{ij} \tag{1.97}$$

In terms of the commonly used constants E and ν–respectively, Young's modulus and the Poisson ratio stemming from one-dimensional test data–you may readily show (see Problem 1.21) that[†]

$$G = \frac{E}{2(1 + \nu)} \tag{1.98a}$$

[†] G is the shear modulus and is often represented by the letter μ in the literature.

$$\lambda = \frac{E\nu}{(1+\nu)(1-2\nu)} \tag{1.98b}$$

The inverse forms of these results are given as

$$E = \frac{G(3\lambda + 2G)}{\lambda + G}$$
$$\nu = \frac{\lambda}{2(\lambda + G)} \tag{1.99}$$

Using Eqs. (1.98), we express Eq. (1.97) as follows:

$$\epsilon_{ij} = \frac{1+\nu}{E}\tau_{ij} - \frac{\nu}{E}\tau_{kk}\delta_{ij} \tag{1.100}$$

In unabridged notation we then have the following familiar relations:

$$\epsilon_{xx} = \frac{1}{E}[\tau_{xx} - \nu(\tau_{yy} + \tau_{zz})]$$
$$\epsilon_{yy} = \frac{1}{E}[\tau_{yy} - \nu(\tau_{xx} + \tau_{zz})]$$
$$\epsilon_{zz} = \frac{1}{E}[\tau_{zz} - \nu(\tau_{xx} + \tau_{yy})]$$
$$\epsilon_{xy} = \frac{1+\nu}{E}\tau_{xy} = \frac{1}{2G}\tau_{xy} \tag{1.101}$$
$$\epsilon_{yz} = \frac{1+\nu}{E}\tau_{yz} = \frac{1}{2G}\tau_{yz}$$
$$\epsilon_{xz} = \frac{1+\nu}{E}\tau_{xz} = \frac{1}{2G}\tau_{xz}$$

In closing this section it is well to remember that a material may be both isotropic and inhomogeneous or, conversely, anisotropic and homogeneous. These two characteristics are independent.

1.14 BOUNDARY–VALUE PROBLEMS FOR LINEAR ELASTICITY

The complete system of equations of linear elasticity for homogeneous, isotropic solids includes the equilibrium equations:

$$\tau_{ij,j} + B_i = 0 \tag{1.102}$$

the stress-strain law:

$$\tau_{ij} = \lambda \epsilon_{ll} \delta_{ij} + 2G\epsilon_{ij} \tag{1.103}$$

and the strain-displacement relations:

$$\epsilon_{ij} = \tfrac{1}{2}(u_{i,j} + u_{j,i}) \tag{1.104}$$

We have here a complete system of 15 equations for 15 unknowns. When explicit use of the displacement field is not made, we must be sure that the compatibility equations are satisfied. It must also be understood that B_i and $T_i^{(\nu)}$ have resultants that satisfy equilibrium relations for the body as dictated by rigid-body mechanics. We shall say in this regard that B_i and $T_i^{(\nu)}$ must be *statically compatible.*

We may pose three classes of boundary-value problems.

1. Determine the distribution of stresses and displacements in the interior of the body under a given body force distribution and a *given surface traction over the boundary*. This is called a boundary-value problem of the *first kind.*
2. Determine the distribution of stresses and displacements in the interior of the body under the action of a given body force distribution and a *prescribed displacement distribution over the entire boundary*. This is called a boundary-value problem of the *second kind.*
3. Determine the distribution of stresses and displacements in the interior of a body under the action of a given body force distribution with a *given traction distribution over part of the boundary*, denoted as S_1, and a *prescribed displacement distribution over the remaining part of the boundary*, S_2. This is called a *mixed boundary-value problem*.

It should be noted that on the surfaces where the $T_i^{(\nu)}$ are prescribed, Cauchy's formula $T_i^{(\nu)} = \tau_{ij}\nu_j$ must apply.

For boundary-value problems of the *first kind*, we find it convenient to express the basic equations in terms of stresses. To do this we substitute for ϵ_{ij}, using Eq. (1.100) in the compatibility equation (1.64). Using the equilibrium equation we can then arrive (see Problem 1.24) at the following system of equations:

$$\nabla^2 \tau_{ij} + \frac{1}{1+\nu}\kappa_{,ij} - \frac{\nu}{1+\nu}\delta_{ij}\nabla^2\kappa = -(B_{i,j} + B_{j,i}) \tag{1.105}$$

where $\kappa = \tau_{kk}$. These are the *Beltrami-Michell equations.* The solution of these equations, subject to the satisfaction of Cauchy's formula on the boundary for simply connected domains, will lead to a set of stress components that both satisfy the equilibrium equations and are derivable from a single-valued continuous displacement field.

As for boundary-value problems of the *second kind*, we employ Eqs. (1.103) and (1.104) in the equilibrium equation (1.102) to yield differential equations with the displacement field as the dependent variable. By straightforward substitutions we then get the well-known *Navier equations of elasticity*:

$$G\nabla^2 u_i + (\lambda + G)u_{j,ji} + B_i = 0 \tag{1.106}$$

For dynamic conditions we need only employ Eq. (1.7) in place of the equilibrium equations. The result is the addition of the term $\rho\ddot{u}_i$ on the right side of the above equation. If the above equation can be solved in conjunction with the prescribed displacements on the surface and if the resulting solution is single-valued and continuous the problem may be considered solved.

Solutions for *mixed boundary-value* problems will be investigated throughout the text with emphasis on techniques stemming from the variational approach.

1.15 ST. VENANT'S PRINCIPLE

Sometimes it is advantageous to simplify the specification of surface tractions in the boundary-value problems mentioned in the previous section. Thus as in rigid-body mechanics, where point forces are used to replace a force distribution, we have the analogous situation in elasticity where, as a result of St. Venant's principle, a surface traction distribution over a comparatively small part of a boundary may be replaced by a statically equivalent† system without altering the stress distribution at points sufficiently far away from the surface traction. We can say that the effects of surface tractions over a part of the boundary that are felt relatively far into the interior of an elastic solid are dependent only on the rigid-body resultant of the applied tractions over this part of the boundary. By this principle we can (Fig. 1.26) replace the complex supporting force distribution exerted by the wall on the cantilever beam by a single force and couple, as shown in the diagram, for the purpose of simplifying the computation of stress and strain in the domain to the right of the support.

Although we shall not present them here, it is pointed out that mathematical justifications have been advanced for St. Venant's principle (Goodier, 1937; Hoff, 1945; Fung, 1965).

†Having the same rigid-body resultant.

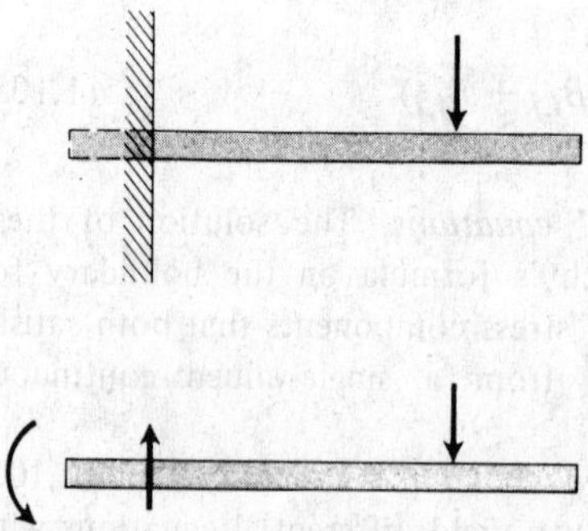

FIGURE 1.26
Cantilever beam with statically equivalent support.

1.16 UNIQUENESS

We shall present a proof, due to the German mathematician Franz Neumann, that the solution to the mixed boundary-value problem of classical elasticity is unique—that is, all the stresses and strains can be found without ambiguity. We shall consider the linearized *dynamic* case where the boundary conditions are expressed as functions of time as follows:

$$\text{On } S_1: \quad T_i^{(\nu)}(x, y, z, t) = f_i(x, y, z, t) \qquad \text{for } t \geqslant 0$$
$$\text{On } S_2: \quad u_i(x, y, z, t) = g_i(x, y, z, t) \qquad \text{for } t \geqslant 0$$

where f_i and g_i are known functions. Furthermore, we must also express *initial conditions* here of the form:

$$u_i(x, y, z, 0) = h_i(x, y, z) \qquad \text{in } V$$

$$\dot{u}_i(x, y, z, 0) = k_i(x, y, z) \qquad \text{in } V$$

Let us assume that there are two solutions u_i' and u_i'' for the displacement field of a given mixed boundary-value problem. The corresponding stress and strain fields are given respectively as τ_{ij}', ϵ_{ij}', and τ_{ij}'', ϵ_{ij}''. Now form a new displacement field given as

$$u_i = u_i' - u_i'' \tag{1.107}$$

Because of the *linearity* of the basic equations, this new function must be a solution to these equations for the situation where we have zero body forces, zero surface tractions on S_1, and zero displacements u_i on S_2. Furthermore, the initial values of u_i and $\dot{u}_i$ are zero. The equation of motion for such a case is then

$$\tau_{ij,j} - \rho \frac{\partial^2 u_i}{\partial t^2} = 0$$

Using the strain energy density function $\mathcal{U}$, this becomes

$$\frac{\partial}{\partial x_j}\left(\frac{\partial \mathcal{U}}{\partial \epsilon_{ij}}\right) - \rho \frac{\partial^2 u_i}{\partial t^2} = 0$$

Now multiply by $\partial u_i/\partial t$ and integrate first over the entire volume and then with respect to time from t_0 to t, where for t_0 we take $u_i = \dot{u}_i = 0$:

$$\int_{t_0}^{t} dt \iiint_V \left[\frac{\partial}{\partial x_j}\left(\frac{\partial \mathcal{U}}{\partial \epsilon_{ij}}\right) - \rho \frac{\partial^2 u_i}{\partial t^2}\right] \frac{\partial u_i}{\partial t}\, dv = 0 \tag{1.108}$$

Consider the second expression in the integrand. We may express it as follows:

$$\int_{t_0}^{t} dt \iiint_V \rho \frac{\partial^2 u_i}{\partial t^2} \frac{\partial u_i}{\partial t}\, dv = \iiint_V \left[\int_{t_0}^{t} \left(\rho \frac{\partial^2 u_i}{\partial t^2} \frac{\partial u_i}{\partial t}\right) dt\right] dv \tag{1.109}$$

$$= \iiint_V \left[\int_{t_0}^{t} \frac{1}{2} \rho \frac{\partial}{\partial t} \left(\frac{\partial u_i}{\partial t} \right)^2 dt \right] dv$$

$$= \frac{1}{2} \iiint_V \rho \left(\frac{\partial u_i}{\partial t} \right)^2 \Bigg|_{t_0}^{t} dv = \frac{1}{2} \iiint_V \rho \left(\frac{\partial u_i}{\partial t} \right)^2 dv \qquad \text{(1.109) (Cont.)}$$

where we have noted that $\dot{u}_i(t_0) = 0$. Now consider the first expression in the integrand of Eq. (1.108). We may express this as follows:

$$\int_{t_0}^{t} dt \iiint_V \left[\frac{\partial}{\partial x_j} \left(\frac{\partial \mathcal{U}}{\partial \epsilon_{ij}} \right) \right] \frac{\partial u_i}{\partial t} dv = \int_{t_0}^{t} dt \iiint_V \frac{\partial}{\partial x_j} \left(\frac{\partial \mathcal{U}}{\partial \epsilon_{ij}} \frac{\partial u_i}{\partial t} \right) dv - \int_{t_0}^{t} dt \iiint_V \frac{\partial \mathcal{U}}{\partial \epsilon_{ij}} \frac{\partial^2 u_i}{\partial x_j \partial t} dv \qquad (1.110)$$

Examine next the expression

$$\frac{\partial \mathcal{U}}{\partial \epsilon_{ij}} \frac{\partial^2 u_i}{\partial x_j \partial t}$$

in the second integral on the right side of the above equation. We may express it as

$$\frac{\partial \mathcal{U}}{\partial \epsilon_{ij}} \frac{\partial^2 u_i}{\partial x_j \partial t} = \frac{\partial \mathcal{U}}{\partial \epsilon_{ij}} \frac{\partial}{\partial t} (u_{i,j}) = \frac{1}{2} \frac{\partial \mathcal{U}}{\partial \epsilon_{ij}} \frac{\partial}{\partial t} u_{i,j} + \frac{1}{2} \frac{\partial \mathcal{U}}{\partial \epsilon_{ji}} \frac{\partial}{\partial t} (u_{j,i})$$

wherein we have simply decomposed the expression into two equal parts while interchanging the dummy indices. Now, making use of the fact that ϵ_{ij} is symmetric and employing Eq. (1.46), we get

$$\frac{\partial \mathcal{U}}{\partial \epsilon_{ij}} \frac{\partial^2 u_i}{\partial x_j \partial t} = \frac{\partial \mathcal{U}}{\partial \epsilon_{ij}} \frac{\partial \epsilon_{ij}}{\partial t} = \frac{\partial \mathcal{U}}{\partial t} \qquad (1.111)$$

Employing this result and using Gauss' theorem for the first integral on the right side, Eq. (1.110) becomes:

$$\int_{t_0}^{t} dt \iiint_V \left[\frac{\partial}{\partial x_j} \left(\frac{\partial \mathcal{U}}{\partial \epsilon_{ij}} \right) \right] \frac{\partial u_i}{\partial t} dv = \int_{t_0}^{t} dt \oiint_S \frac{\partial \mathcal{U}}{\partial \epsilon_{ij}} \frac{\partial u_i}{\partial t} \nu_j \, dA - \int_{t_0}^{t} dt \iiint_V \frac{\partial \mathcal{U}}{\partial t} dv$$

Noting that $\partial \mathcal{U}/\partial \epsilon_{ij} = \tau_{ij}$ is zero on S_1 and that $\partial u_i/\partial t$ is zero on S_2, we conclude that the surface integral vanishes. After interchanging the order of integration in the last integral, we get for that expression, on noting that we have taken $\mathcal{U}(t_0) = 0$,

$$-\iiint_V \mathcal{U}\, dv$$

Now substitute this result and that of Eq. (1.109) into Eq. (1.108). We get:

$$\iiint_V \left[\mathcal{U} + \frac{1}{2}\rho\left(\frac{\partial u_i}{\partial t}\right)^2\right] dv = 0 \tag{1.112}$$

Since both expressions in the integrand are positive definite, we must conclude that each must vanish everywhere in the domain. This means that $\mathcal{U} = 0$ and hence the strain field is zero throughout the body. Accordingly, except possibly for rigid-body motions, which we may ignore, the displacement fields u_i' and u_i'' are identical, proving uniqueness of the solution.

It is well to note that this proof hinged on the *linearity* of the governing equations. No such uniqueness theorem is possible in nonlinear elasticity. Indeed, we shall see in our discussion of stability later that a multiplicity of solutions is possible when nonlinearities are involved.

Part D
Plane Stress

1.17 INTRODUCTION

The rigorous solutions of three-dimensional problems of elasticity are few. Consequently, there is a need to simplify problems so that we can obtain a mathematical solution that is reasonably close to representing the actual physical problem. As an illustration of such an approach we now present briefly the plane stress problem. Other ways of simplifying the problem by using variational approaches will be presented later; the present undertaking will provide a basis for comparison.

1.18 EQUATIONS FOR PLANE STRESS

We define plane stress in the z plane as a stress distribution where

$$\tau_{xz} = \tau_{yz} = \tau_{zz} = 0 \tag{1.113}$$

Thin plates acted on by loads lying in the plane of symmetry of a plate (Fig. 1.27*a*) can often be considered to be in a state of plane stress with the z direction taken normal to the plate. Clearly, with no loads normal to lateral surfaces of the plate, τ_{zz} must be zero there, and since the plate is thin we consider τ_{zz} to be zero inside. If the loads are distributed so as to have a constant intensity over the thickness of the plate, then the shearing stresses τ_{zx} and τ_{zy}, in addition to being zero on the faces of the plate, will be zero throughout the thickness. If, however, the peripheral load is symmetrically distributed over the thickness (see Fig. 1.27*b*), then the shear

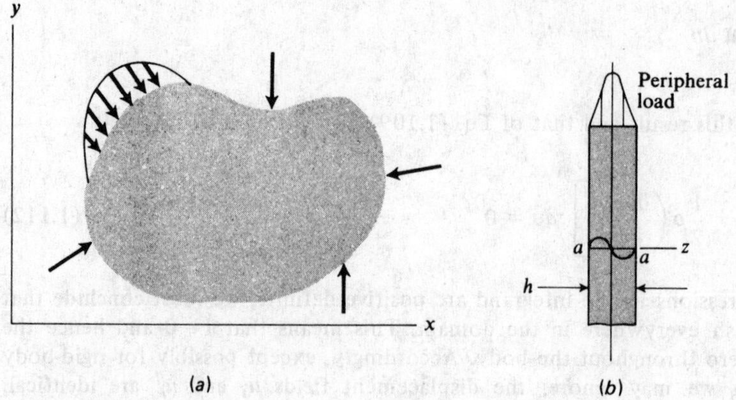

FIGURE 1.27
Symmetrically loaded plane stress.

stresses τ_{zx} and τ_{zy} may have considerable magnitudes at points along the thickness of the plate, as shown along *a–a* in the diagram. This is possible in spite of the fact that these shear stresses must be zero at the lateral faces of the plate. However, the shear stresses must be antisymmetric over the thickness with a net area of zero for the curve τ versus z. Thus these shear stresses have an average value of zero over the thickness of the plate. This motivates the theory of *generalized plane stress*, where the stress tensor is replaced by the average values of stress over the plate thickness. We shall consider for simplicity that the surface loads are applied so as to have uniform intensity over the plate thickness, and so we can assume that the shear stress τ_{xz} and τ_{yz} are zero everywhere. The equations of equilibrium then become

$$\frac{\partial \tau_{xx}}{\partial x} + \frac{\partial \tau_{xy}}{\partial y} = 0 \qquad \frac{\partial \tau_{yx}}{\partial x} + \frac{\partial \tau_{yy}}{\partial y} = 0 \tag{1.114}$$

We may permanently satisfy these equations by expressing the stresses in terms of a function Φ, called the *Airy stress function*, as follows:

$$\tau_{xx} = \frac{\partial^2 \Phi}{\partial y^2} \qquad \tau_{yy} = \frac{\partial^2 \Phi}{\partial x^2} \qquad \tau_{xy} = -\frac{\partial^2 \Phi}{\partial x\, \partial y} \tag{1.115}$$

A substitution for stresses in Eq. (1.114) with the Airy stress function renders these equations trivial identities. Next, Hooke's law for strain gives us†

$$\epsilon_{xx} = \frac{1}{E}(\tau_{xx} - \nu\tau_{yy}) \qquad \gamma_{xy} = \frac{1}{G}\tau_{xy} \tag{1.116}$$

†Hooke's law for plane stress giving stress in terms of strain will be used later in the text and can be found from Eq. (1.116) by straightforward algebraic steps to be:

$$\tau_{xx} = \frac{E}{1-\nu^2}(\epsilon_{xx} + \nu\epsilon_{yy}) \qquad \tau_{yy} = \frac{E}{1-\nu^2}(\epsilon_{yy} + \nu\epsilon_{xx}) \qquad \tau_{xy} = 2G\epsilon_{xy} \tag{1.117}$$

$$\epsilon_{yy} = \frac{1}{E}(\tau_{yy} - \nu\tau_{xx}) \qquad \gamma_{xz} = 0$$
$$\epsilon_{zz} = -\frac{\nu}{E}(\tau_{xx} + \tau_{yy}) \qquad \gamma_{yz} = 0 \tag{1.116 (Cont.)}$$

Now replace the stresses with Airy's function in the above equations. We get

$$\epsilon_{xx} = \frac{1}{E}\left(\frac{\partial^2 \Phi}{\partial y^2} - \nu \frac{\partial^2 \Phi}{\partial x^2}\right) \qquad \gamma_{xy} = -\frac{1}{G}\frac{\partial^2 \Phi}{\partial x\, \partial y}$$
$$\epsilon_{yy} = \frac{1}{E}\left(\frac{\partial^2 \Phi}{\partial x^2} - \nu \frac{\partial^2 \Phi}{\partial y^2}\right) \qquad \gamma_{yz} = \gamma_{xz} = 0 \tag{1.118}$$
$$\epsilon_{zz} = -\frac{\nu}{E}\left(\frac{\partial^2 \Phi}{\partial x^2} + \frac{\partial^2 \Phi}{\partial y^2}\right)$$

Because we will be working with stress quantities (and not displacements) we shall have to satisfy the compatibility equations. Accordingly, we now turn to the compatibility equations (1.70). Examine the first of these. Substituting from the preceding equations, we get

$$\frac{1}{E}\left(\frac{\partial^4 \Phi}{\partial y^4} - \nu \frac{\partial^4 \Phi}{\partial y^2\, \partial x^2} + \frac{\partial^4 \Phi}{\partial x^4} - \nu \frac{\partial^4 \Phi}{\partial x^2\, \partial y^2}\right) = -\frac{1}{G}\left(\frac{\partial^4 \Phi}{\partial y^2\, \partial x^2}\right)$$

Multiply through by E and replace E/G by $2(1 + \nu)$ in accordance with Eq. (1.98*a*). We then get

$$\frac{\partial^4 \Phi}{\partial x^4} + 2\frac{\partial^4 \Phi}{\partial y^2\, \partial x^2} + \frac{\partial^4 \Phi}{\partial y^4} = \nabla^4 \Phi = 0 \tag{1.119}$$

We see that Φ must satisfy the so-called *biharmonic* equation and is thus a biharmonic function as a result of compatibility. As for the other compatibility equations, we see on inspection that Eqs. (1.70*d*) and (1.70*e*) are satisfied identically. The remaining equations are not satisfied but involve only the strain ϵ_{zz}, which is of no interest to us here. We shall accordingly disregard these latter equations.† Equation (1.119) incorporates equations of equilibrium, Hooke's law, and satisfies the aspects of compatibility that are of concern to us. It thus forms the key equation for plane stress problems. The function Φ desired must be a biharmonic function that satisfies, within the limits of St. Venant's principle, the given surface tractions of the problem. We now consider the boundary conditions that we must impose on Φ. We confine ourselves here to the case where surface tractions are prescribed. Applying Cauchy's formula to plane stress [Eq. (1.5)], we get

†One can show that the error incurred by using Eq. (1.119) to determine Φ leads to results that are very close to the correct result for thin plates (Timoshenko and Goodier, 1951).

$$T_x^{(\nu)} = \tau_{xx} \cos(\nu, x) + \tau_{xy} \cos(\nu, y)$$
$$T_y^{(\nu)} = \tau_{yx} \cos(\nu, x) + \tau_{yy} \cos(\nu, y) \tag{1.120}$$

Now, employing the Airy stress function, we have

$$T_x^{(\nu)} = \frac{\partial^2 \Phi}{\partial y^2} \cos(\nu, x) - \frac{\partial^2 \Phi}{\partial x\, \partial y} \cos(\nu, y)$$
$$T_y^{(\nu)} = -\frac{\partial^2 \Phi}{\partial x\, \partial y} \cos(\nu, x) + \frac{\partial^2 \Phi}{\partial x^2} \cos(\nu, y) \tag{1.121}$$

Next observe Fig. 1.28, showing an arbitrary boundary where element ds of the boundary is depicted with coordinate elements dx and dy. We can say, on inspecting this diagram, that

$$\cos(\nu, x) = \cos\alpha = \frac{dy}{ds}$$
$$\cos(\nu, y) = \sin\alpha = -\frac{dx}{ds} \tag{1.122}$$

Substituting these results into Eqs. (1.121), we get

$$T_x^{(\nu)} = \frac{\partial^2 \Phi}{\partial y^2} \frac{dy}{ds} + \frac{\partial^2 \Phi}{\partial x\, \partial y} \frac{dx}{ds} = \frac{d}{ds}\left(\frac{\partial \Phi}{\partial y}\right)$$
$$T_y^{(\nu)} = -\frac{\partial^2 \Phi}{\partial x\, \partial y} \frac{dy}{ds} - \frac{\partial^2 \Phi}{\partial x^2} \frac{dx}{ds} = -\frac{d}{ds}\left(\frac{\partial \Phi}{\partial x}\right) \tag{1.123}$$

where we use the chain rule of differentiation to reach the rightmost expression in the equations. These equations are the usual form employed for the boundary conditions on Φ. We have thus posed the boundary-value problem for plane stress.

We now illustrate the entire procedure in solving a plane stress problem.

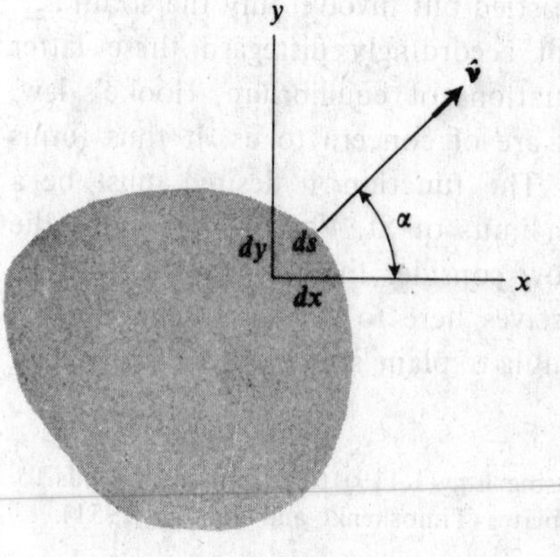

FIGURE 1.28
Segment ds on boundary.

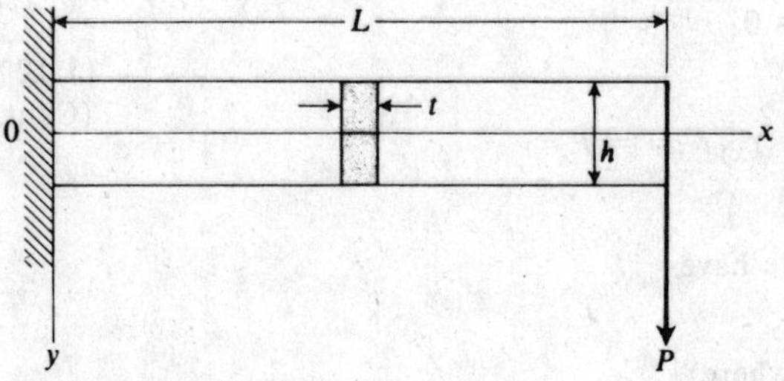

FIGURE 1.29
Tip-loaded cantilever beam.

1.19 PROBLEM OF THE CANTILEVER BEAM

Consider now a cantilever beam (Fig. 1.29) loaded at the tip by a force P. The cross section of the beam is that of a rectangle with height h and thickness t. The length of the beam is L. For t small compared to both h and L, we may use the plane stress analysis for determining deflections in the midplane of the beam.

We shall postulate the form of the stress function Φ for this problem by examining the results of elementary strength of materials for the beam. You will recall that

$$\tau_{xx} = -\frac{My}{I_{zz}} = -\frac{P(L-x)}{I_{zz}}\,y \tag{1.124a}$$

$$\tau_{yy} = 0 \tag{1.124b}$$

$$\tau_{xy} = \frac{VQ}{I_{zz}t} = \frac{P(h^2/4 - y^2)}{2I_{zz}} \tag{1.124c}$$

where V is the shear force and Q is the first moment about the z axis of the section area above the elevation y. Thus we can see that the stresses are of the form

$$\begin{aligned} \tau_{xx} &= C_1 y + C_2 xy \\ \tau_{yy} &= 0 \\ \tau_{xy} &= C_3 + C_4 y^2 \end{aligned} \tag{1.125}$$

You may demonstrate directly that the following function Φ yields a stress distribution of the desired form given above:

$$\Phi = Ay^3 + Bxy^3 + Dxy \tag{1.126}$$

It is also readily demonstrated that the above function is a biharmonic function, and so this represents a plane stress problem. Accordingly, we must investigate to see whether we can satisfy the boundary conditions of the problem by adjusting the constants A, B, and D. Thus we require

$$\text{For } y = \pm\frac{h}{2}: \quad (1)\ \tau_{yy} = 0 \qquad (2)\ \tau_{xy} = 0 \tag{1.127}$$

For $x = L$: (3) $\tau_{xx} = 0$

Everywhere: (4) $\displaystyle\int_{-h/2}^{h/2} \tau_{xy} t\, dy = P$ (1.127) (Cont.)

Applying these conditions to Φ, we have

$$(1)\quad \tau_{yy} = \frac{\partial^2 \Phi}{\partial x^2} = 0 \quad \text{(everywhere)}$$

$$(2)\quad (\tau_{xy})_{y=\pm h/2} = -\left(\frac{\partial^2 \Phi}{\partial x\, \partial y}\right)_{y=\pm h/2} = (3By^2 + D)_{y=\pm h/2} = 0$$

$$\text{Therefore } D = -\tfrac{3}{4}Bh^2 \tag{a}$$

$$(3)\quad (\tau_{xx})_{x=L} = \left(\frac{\partial^2 \Phi}{\partial y^2}\right)_{x=L} = (6Ay + 6Bxy)_{x=L} = 0$$

$$\text{Therefore } A = -LB \tag{b}$$

$$(4)\quad \int_{-h/2}^{h/2} \tau_{xy} t\, dy = \int_{-h/2}^{h/2} -\left(\frac{\partial^2 \Phi}{\partial x\, \partial y}\right) t\, dy = \int_{-h/2}^{h/2} -(3By^2 + D)$$

$$\text{Therefore } -Bt\,\frac{h^3}{4} - Dth = P \tag{c}$$

Solving Eqs. (*a*), (*b*), and (*c*) simultaneously, we get for the constants B, D, and A:

$$B = 2\,\frac{P}{th^3}$$

$$D = -\frac{3}{2}\,\frac{P}{th}$$

$$A = -2\,\frac{PL}{th^3}$$

Thus the following form of Φ satisfies the stipulated boundary conditions:

$$\Phi = -2\,\frac{PL}{th^3}\,y^3 + \frac{2P}{th^3}\,xy^3 - \frac{3}{2}\,\frac{P}{th}\,xy \tag{1.128}$$

If we now compute the stresses using the above function we find results identical to those given by Eq. (1.124). Thus the stress distribution from strength of materials is identical to that from the theory of elasticity. Indeed, if the load P has the parabolic distribution stipulated in Eq. (1.124*c*) for τ_{xy} we have an "exact" solution of the problem for regions to the right of the base support.† If this is not

†One can show that when the bending moment varies linearly with x the flexure formula is exact. See Borg (1962).

the case, then the solution is valid away from the support and away from the right end.

Let us next consider the displacement field for the problem. We have, using Hooke's law and the notation $u_x = u$ and $u_y = v$:

$$\epsilon_{xx} = \frac{\partial u}{\partial x} = \frac{1}{E}\tau_{xx} = -\frac{P(L-x)}{EI_{zz}}y \tag{1.129a}$$

$$\epsilon_{yy} = \frac{\partial v}{\partial y} = -\frac{1}{E}(\nu\tau_{xx}) = \frac{P\nu}{EI_{zz}}(L-x)y \tag{1.129b}$$

$$\gamma_{xy} = \left(\frac{\partial u}{\partial y} + \frac{\partial v}{\partial x}\right) = \frac{P}{2GI_{zz}}\left(\frac{h^2}{4} - y^2\right) \tag{1.129c}$$

We integrate Eqs. (1.129a) and (1.129b) as follows:

$$\begin{aligned} u &= -\frac{P}{EI_{zz}}\left(Lyx - \frac{x^2}{2}y\right) + g_1(y)\frac{P}{EI_{zz}} \\ v &= \frac{P\nu}{EI_{zz}}\left(\frac{Ly^2}{2} - \frac{xy^2}{2}\right) + g_2(x)\frac{P}{EI_{zz}} \end{aligned} \tag{1.130}$$

where g_1 and g_2 are arbitrary and P/El_{zz} is attached for later convenience. Now substitute these results into Eq. (1.129c).

$$-\frac{P}{EI_{zz}}\left(Lx - \frac{x^2}{2}\right) + g_1'\frac{P}{EI_{zz}} + \frac{P\nu}{EI_{zz}}\left(-\frac{y^2}{2}\right) + g_2'\frac{P}{EI_{zz}} = \frac{P}{2GI_{zz}}\left(\frac{h^2}{4} - y^2\right)$$

Replacing G by $E/2(1+\nu)$ [see Eq. (1.98a)], we may cancel out the term P/EI_{zz}. We then have, on rearranging terms,

$$-\left(Lx - \frac{x^2}{2}\right) + g_2'(x) = \frac{\nu y^2}{2} + (1+\nu)\left(\frac{h^2}{4} - y^2\right) - g_1'(y)$$

Note that each side is a function of a different variable and hence each side must equal a constant, which we denote as K. Thus:

$$-\left(Lx - \frac{x^2}{2}\right) + g_2' = K$$

$$\frac{\nu y^2}{2} + (1+\nu)\left(\frac{h^2}{4} - y^2\right) - g_1' = K$$

We may integrate these equations now to get g_1 and g_2 as follows:

$$g_2 = Kx + \frac{Lx^2}{2} - \frac{x^3}{6} + C_1$$

$$g_1 = -Ky + \frac{\nu y^3}{6} + (1+\nu)\left(\frac{h^2 y}{4} - \frac{y^3}{3}\right) + C_2$$

We then have for u and v:

$$u = -\frac{P}{EI_{zz}}\left[Lxy - \frac{x^2y}{2} - Ky + \frac{\nu y^3}{6} + (1+\nu)\left(\frac{h^2y}{4} - \frac{y^3}{3}\right) + C_2\right] \quad (1.131a)$$

$$v = \frac{P}{EI_{zz}}\left[\nu\left(\frac{Ly^2}{2} - \frac{xy^2}{2}\right) + Kx + \frac{Lx^2}{2} - \frac{x^3}{6} + C_1\right] \quad (1.131b)$$

We have three arbitrary constants C_1, C_2, and K that must now be determined. We shall find these constants by fixing the support in some way. Let us say that point 0 (see Fig. 1.29) is stationary. Then $u = v = 0$ when $x = y = 0$. It is clear that the constants C_1 and C_2 must then be zero. Next let us say that the centerline of the beam at 0 remains horizontal. That is, $\partial v/\partial x = 0$ when $x = y = 0$. We see on inspection of Eq. (1.131b) that K is also zero. The displacement field for this case is then

$$u = -\frac{P}{EI_{zz}}\left[Lxy - \frac{x^2y}{2} + \frac{\nu y^3}{6} + (1+\nu)\left(\frac{h^2y}{4} - \frac{y^3}{3}\right)\right] \quad (1.132a)$$

$$v = \frac{P}{EI_{zz}}\left[\nu\left(\frac{Ly^2}{2} - \frac{xy^2}{2}\right) + \frac{Lx^2}{2} - \frac{x^3}{6}\right] \quad (1.132b)$$

In the one-dimensional study of beams to be soon undertaken, we center attention on the vertical deflection of the centerline of the beam. We may get this result from above by setting $y = 0$ in the formulation of v. Thus:

$$(v)_{y=0} = \frac{P}{2EI_{zz}}\left(Lx^2 - \frac{x^3}{3}\right) \quad (1.133)$$

This coincides with the result from strength of materials.†

1.20 CLOSURE

In this chapter we have presented a self-contained treatment of classical linear elasticity that will serve as background for the ensuing chapters of this text. We have essentially restricted ourselves to small deformation. In Chapter 8 we shall very briefly consider finite deformation in anticipation of our study of elastic stability.

In the next chapter we continue to lay the foundation for the rest of this text by considering certain salient features of the elegant calculus of variations.

REFERENCES

Borg, S. R., "Matrix-Tensor Methods in Continuum Mechanics," sect. 5.5, Van Nostrand, New York, 1962.

Fung, Y. C., "Foundations of Solid Mechanics," p. 300, Prentice-Hall, Englewood Cliffs, N.J., 1965.

†It is to be pointed out that this solution does not adequately account for the shear deformation in the beam. This is because it is really more appropriate to use the boundary condition $\partial u/\partial y = 0$. See Timoshenko and Goodier (1951); pp. 45–46.

Goodier, J. N., "A General Proof of St. Venant's Principle," *Philos. Mag.* 7, no. 23, 637 (1973).
Hoff, N. J., "The Applicability of Saint Venant's Principle to Airplane Structures," *J. Aero. Sci.* 12, 455 (1945).
Timoshenko, S., and Goodier, J. N., "Theory of Elasticity," McGraw-Hill, New York, 1951.

READING

Boresi, A. P., "Theory of Elasticity," Prentice-Hall, Englewood Cliffs, N.J., 1965.
Filonenko-Boroditch, M., "Theory of Elasticity," Dover, New York, 1965.
Fung, Y. C., Foundations of Solid Mechanics," Prentice-Hall, Englewood Cliffs, N.J., 1965.
Hodge, P. G., "Continuum Mechanics," McGraw-Hill, New York, 1970.
Shames, I. H., "Mechanics of Deformable Solids," Krieger Publishers, Melbourne, Florida, 1964.
Sokolnikoff, I. S., "Mathematical Theory of Elasticity," McGraw-Hill, New York, 1956.
Timoshenko, S., and Goodier, J. N., "Theory of Elasticity," McGraw-Hill, New York, 1970.

PROBLEMS

1.1 Label the stresses shown on the infinitesimal rectangular parallelepiped in Fig. 1.30 and give their correct signs.

1.2 The stress components at orthogonal interfaces parallel to $x_1x_2x_3$ at a point are known:

$$\tau_{11} = 7 \times 10^6 \text{ Pa} \qquad \tau_{12} = \tau_{21} = 1.4 \times 10^6 \text{ Pa}$$

$$\tau_{22} = -42 \times 10^6 \text{ Pa} \qquad \tau_{13} = \tau_{31} = 0$$

$$\tau_{33} = 0 \qquad \tau_{23} = \tau_{32} = -2.8 \times 10^6 \text{ Pa}$$

(*a*) Find the components of the surface traction vector for an interface whose normal vector is

$$\boldsymbol{\nu} = 0.11\mathbf{i} + 0.35\mathbf{j} + 0.93\mathbf{k}$$

(*b*) What is the component of this force intensity in the $\boldsymbol{\epsilon}$ direction?

$$\boldsymbol{\epsilon} = 0.33\mathbf{i} + 0.90\mathbf{j} + 0.284\mathbf{k}$$

*1.3 Suppose you had a body couple distribution given as M N-m/m^3. What is the relation between the stresses and M at a point? In particular, if

$$M_1 = 1.4 \times 10^6 \text{ N-m/m}^3 \qquad \tau_{11} = 7 \times 10^5 \text{ Pa} \qquad \tau_{22} = -1.4 \times 10^6 \text{ Pa}$$

$$\tau_{33} = 0 \text{ Pa} \qquad \tau_{13} = 3.5 \times 10^6 \text{ Pa} \qquad \tau_{31} = 7 \times 10^5 \text{ Pa}$$

$$\tau_{23} = 7 \times 10^5 \text{ Pa} \qquad \tau_{21} = 3.5 \times 10^5 \text{ Pa} \qquad \tau_{12} = 5.6 \times 10^5 \text{ Pa}$$

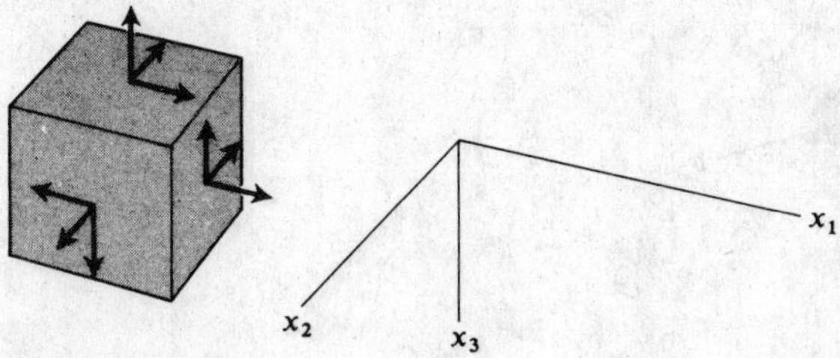

FIGURE 1.30

What is the stress τ_{32}? What are the body couple components M_2 and M_3?

1.4 The state of stress at a point in a reference $(x_1 x_2 x_3)$ is given by

$$\begin{pmatrix} 14 & 7 & 0 \\ 7 & 0 & 0 \\ 0 & 0 & 35 \end{pmatrix} \times 10^5 \text{ Pa}$$

What is the array of stress terms for a new set of axes $(x'_1 x'_2 x'_3)$ formed by rotating $(x_1 x_2 x_3)$ 60° about the x_3 axis?

1.5 Given the following stress tensor:

$$\tau_{ij} = \begin{pmatrix} -7 & 7 & 0 \\ 7 & 7 & 0 \\ 0 & 0 & 0 \end{pmatrix} \times 10^6 \text{ Pa}$$

Determine:

(*a*) The principal stresses σ_1, σ_2, σ_3.

(*b*) The three stress tensor invariants. Show that they are indeed invariant by calculating them from the given τ_{ij} and from the array of principal values.

(*c*) The principal directions $\overset{1}{v}_j$, $\overset{2}{v}_j$, $\overset{3}{v}_j$; check them for orthogonality.

1.6 Show that the stress quadric becomes:

(*a*) For *uniaxial tension* a pair of plane surfaces.

(*b*) For *plane stress* (i.e., $\tau_{zz} = \tau_{yz} = \tau_{xz} = 0$) a cylinder.

(*c*) For *simple shear* a set of four surfaces forming as traces rectangular hyperbolas on a plane normal to the surfaces.

*1.7 The *octahedral shear stress* is the maximum shear stress on a plane equally inclined toward the principal axes at a point (Fig. 1.31). Using the transformation of stress equations, show that the normal stress on the octahedral plane is related to the principal stresses σ_i as follows:

$$(\tau_{nn})_{\text{oct}} = \tfrac{1}{3}(\sigma_1 + \sigma_2 + \sigma_3)$$

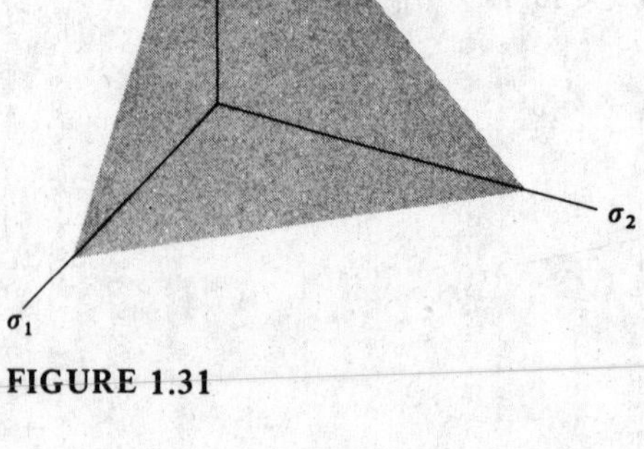

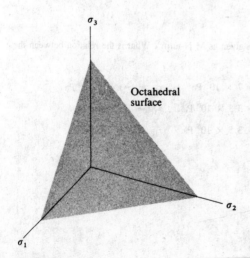

FIGURE 1.31

Now, using Fig. 1.31, show from Newton's law that

$$\tau_{oct}^2 = \tfrac{1}{9}[(\sigma_1 - \sigma_2)^2 + (\sigma_2 - \sigma_3)^2 + (\sigma_1 - \sigma_3)^2]$$

The octahedral shear stress is often considered the stress to observe to predict *yielding* in a general state of stress at a point. If the yield stress from a one-dimensional test is Y, what is then the octahedral stress for yielding under general conditions according to the preceding test? This test for yielding is the so-called *Mises-Hencky criterion.*

*1.8 Show that

$$9\tau_{oct}^2 = 2(\text{I}_\tau)^2 - 6(\text{II}_\tau)$$

Thus the tensor invariants are shown to be simply related to a physically meaningful stress measure. Use the results of Problem 1.7.

1.9 Given the descriptions below of various types of deformation, determine the various Green tensor representations [Eq. (1.42*a*)]:

(*a*) Simple dilatation: $\xi_1 = \lambda x_1$, $\xi_2 = x_2$, $\xi_3 = x_3$.
(*b*) Pure deformation: $\xi_1 = \lambda_1 x_1$, $\xi_2 = \lambda_2 x_2$, $\xi_3 = \lambda_3 x_3$.
(*c*) Cubic dilatation: $\lambda_1 = \lambda_2 = \lambda_3 = \lambda$ in (*b*).
(*d*) Simple shear: $\xi_1 = x_1 + \Gamma x_2$, $\xi_2 = x_2$, $\xi_3 = x_3$.

1.10 Determine the displacements ($u_i = \xi_i - x_i$) corresponding to the deformations of Problem 1.9 and from them compute the infinitesimal strain components. Can any physical meanings be deduced for the parameters λ and Γ involved?

1.11 For the case of pure deformation (see Problem 1.9) show that the volume V of an element $dx_1 dx_2 dx_3$ in the deformed state is represented as $V^* = \lambda_1 \lambda_2 \lambda_3 \, dx_1 dx_2 dx_3$. Further, show that if the strains are infinitesimal, $V^*/V = 1 + \epsilon_{ii}$.
Hint: Consider ds^{*2}/ds^2 for an element such as dx_1. Use Eq. (1.41*a*) and the fact that $(ds^{*2}/dx_1^2) = d\xi_1^2/dx_1^2$.

*1.12 Working in rectangular coordinates, show that integration of the equations $\epsilon_{ij} = 0$ (in terms of displacements) leads to rigid-body rotations. Consider infinitesimal strain.

1.13 The strain tensor ϵ_{ij} can be written as follows:

$$\epsilon_{ij} = \bar{\epsilon}_{ij} + \frac{\epsilon_{kk}}{3}\delta_{ij} = \bar{\epsilon}_{ij} + \frac{e}{3}\delta_{ij}$$

where

$$\bar{\epsilon}_{ij} = \begin{pmatrix} \epsilon_{11} - e/3 & \epsilon_{12} & \epsilon_{13} \\ \epsilon_{21} & \epsilon_{22} - e/3 & \epsilon_{23} \\ \epsilon_{31} & \epsilon_{32} & \epsilon_{33} - e/3 \end{pmatrix}$$

The tensor $\bar{\epsilon}_{ij}$ is called the strain *deviator* tensor. Using the results of Problem 1.11, explain why $\bar{\epsilon}_{ij}$ characterizes the distortion of an element with no change in volume. (This is the shearing distortion.) Show that the first, second, and third tensor invariants for the strain deviator are given in terms of principal strains as follows:

$$\text{I}_{\bar{\epsilon}} = 0$$

$$\text{II}_{\bar{\epsilon}} = -\tfrac{1}{6}[(\epsilon_1 - \epsilon_2)^2 + (\epsilon_2 - \epsilon_3)^2 + (\epsilon_1 - \epsilon_3)^2]$$

$$\text{III}_{\bar{\epsilon}} = \left(\epsilon_1 - \frac{e}{3}\right)\left(\epsilon_2 - \frac{e}{3}\right)\left(\epsilon_3 - \frac{e}{3}\right)$$

The strain deviator invariants (as well as those for stress) play an important role in the theory of plasticity, since for the plastic state we have primarily shear distortion deformation.

1.14 Show that $\lambda\delta_{ij}$ satisfies the equation for an isotropic second-order tensor.

1.15 Show that $\lambda\delta_{ij}$ is the most general isotropic second-order tensor by starting with a second-order tensor A_{ij} and considering the following rotation of axes while employing the isotropic condition [Eq. (1.91)]. Thus with a rotation of 180° about the z axis show that:

$$A_{23} = A_{32} = 0 \qquad A_{13} = A_{31} = 0$$

For a rotation about the x axis of 180° show that $A_{12} = A_{21} = 0$. Thus the nondiagonal terms must be zero. Next, by considering successively rotations of 90° about two axes (see Fig. 1.32), show that the diagonal terms are equal. But this means that $A_{ij} = A\delta_{ij}$, and so the proposed tensor is the most general isotropic second-order tensor.

1.16 Show that D_{ijkl} as given by Eq. (1.94) is an isotropic fourth-order tensor.

*1.17 Consider u_r to be the displacement component in the radial direction and u_θ to be the displacement component in the transverse (θ) direction. Show that:

$$\epsilon_{\theta\theta} = \frac{1}{r}\frac{\partial u_\theta}{\partial \theta} + \frac{u_r}{r}$$

As for $\gamma_{r\theta}$, observe that the shaded element (Fig. 1.33) is in the undeformed geometry while the unshaded element is in the deformed geometry. What is the change in right angles at corner A in terms of angles γ, β, and α? Show that:

$$\gamma_{r\theta} = \frac{\partial u_\theta}{\partial r} - \frac{u_\theta}{r} + \frac{1}{r}\frac{\partial u_r}{\partial \theta}$$

What is ϵ_{rr}?

1.18 We want to find the conditions on the elastic constants to make the strain energy density $\mathcal{U}$ *positive definite*. Thus integrate Eq. (1.75*b*), using the stress-strain law [Eq. (1.96)], to obtain

$$\mathcal{U} = \frac{\lambda}{2}e^2 + G\epsilon_{ij}\epsilon_{ij} \qquad e = \epsilon_{kk}$$

Then, introducing the strain deviator (Problem 1.13), one can write

$$\mathcal{U} = \tfrac{1}{2}Ke^2 + G\bar{\epsilon}_{ij}\bar{\epsilon}_{ij} \qquad K = \lambda + \tfrac{2}{3}G$$

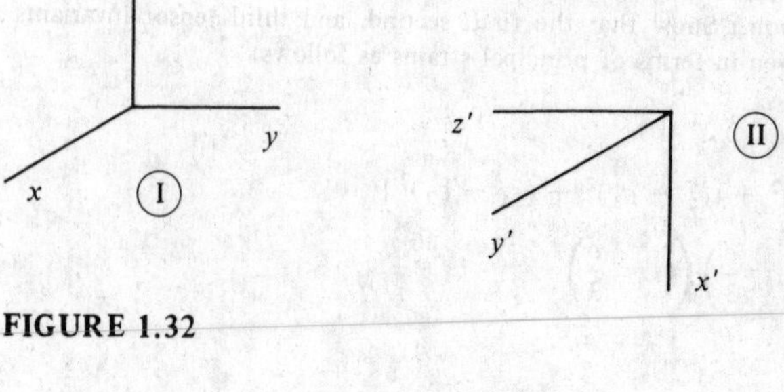

FIGURE 1.32

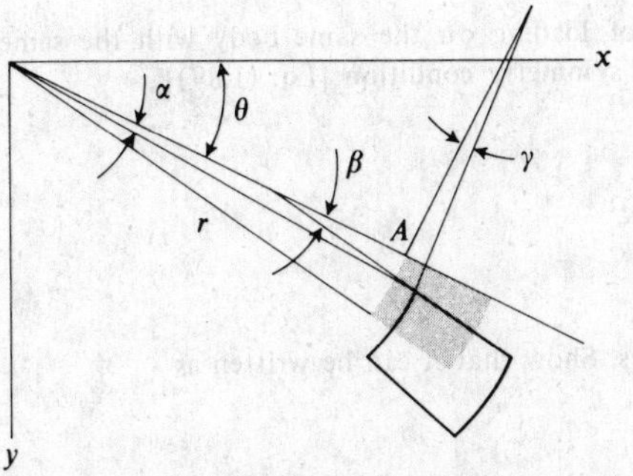

FIGURE 1.33

From this result, deduce appropriate conditions on E, for $\mathcal{U} > 0$. Also notice that $e \equiv \mathrm{I}_\epsilon$ and $\bar{\epsilon}_{ij}\bar{\epsilon}_{ij} = 2\mathrm{II}_{\bar{\epsilon}}$ (Problem 1.13 extended). Hence $\mathcal{U}$ is an invariant!

*1.19 The stress-strain law for an isotropic elastic body undergoing a temperature change ΔT is

$$\tau_{ij} = \lambda\epsilon_{kk}\delta_{ij} + 2G\epsilon_{ij} - (3\lambda + 2G)\alpha\,\Delta T\delta_{ij}$$

where α is the *thermal expansion coefficient*. Show for *plane stress* ($\tau_{yz} = \tau_{xz} = \tau_{zz} = 0$) that:

$$\tau_{xx} = \frac{E}{1-\nu^2}(\epsilon_{xx} + \nu\epsilon_{yy}) - \frac{E}{1+\nu}\alpha\,\Delta T$$

$$\tau_{yy} = \frac{E}{1-\nu^2}(\nu\epsilon_{xx} + \epsilon_{yy}) - \frac{E}{1+\nu}\alpha\,\Delta T$$

$$\tau_{xy} = 2G\epsilon_{xy}$$

Next construct the strain energy density function for this case [see Eq. (1.72)]. Is it equal to the simple expression $\frac{1}{2}\tau_{ij}\epsilon_{ij}$? Explain why you should expect your conclusion when you consider the basic forms of constitutive law for the isothermal and non-isothermal cases.

1.20 For a two-dimensional (plane stress) *orthotropic* continuum, the constitutive law is usually given as

$$\tau_{xx} = C_{11}\epsilon_{xx} + C_{12}\epsilon_{yy} \qquad \tau_{yy} = C_{21}\epsilon_{xx} + C_{22}\epsilon_{yy} \qquad \tau_{xy} = G_{12}\epsilon_{xy}$$

Construct an appropriate strain energy density. Also, verify that we must make $C_{12} = C_{21}$. Check your constitutive law by computing stresses τ_{xx}, τ_{yy}, and τ_{xy}.

1.21 Using the stress-strain law [Eq. (1.96)] consider a one-dimensional (uniaxial) tensile test in the z direction (i.e., $\tau_{zz} \neq 0$, $\tau_{xx} = \tau_{yy} = 0$). Relate ϵ_{xx} and ϵ_{yy} to ϵ_{zz}, and τ_{zz} to ϵ_{zz}. Next, compare these "analytical" results with the following "experimental data":

$$\epsilon_{xx} = \epsilon_{yy} = -\nu\epsilon_{zz} \qquad \tau_{zz} = E\epsilon_{zz}$$

and hence derive Eqs. (1.98).

1.22 Using only the linearity of the constitutive (stress-strain) law, derive the *Betti reciprocal theorem*:

$$\iiint_V \tau_{ij}^{(1)}\epsilon_{ij}^{(2)}\,dv = \iiint_V \tau_{ij}^{(2)}\epsilon_{ij}^{(1)}\,dv$$

The superscripts represent different states of loading on the same body with the same kinematic constraints. Hint: make use of the symmetry condition [Eq. (1.89)].

*1.23 Consider the expression

$$\iiint_V \tau_{ij}^{(1)} \epsilon_{ij}^{(2)}\, dv$$

where the superscripts refer to different states. Show that it can be written as

$$\iiint_V [(\tau_{ij}^{(1)} u_i^{(2)})_{,j} - \tau_{ij,j}^{(1)} u_i^{(2)}]\, dv$$

Now, using the divergence theorem, the equations of equilibrium, and Cauchy's formula, as well as the Betti reciprocal theorem (Problem 1.22), show that

$$\iint_S T_i^{(\nu)(1)} u_i^{(2)}\, dS + \iiint_V B_i^{(1)} u_i^{(2)}\, dv = \iint_S T_i^{(\nu)(2)} u_i^{(1)}\, dS + \iiint_V B_i^{(2)} u_i^{(1)}\, dv$$

1.24 To develop the *Beltrami-Michell equation* substitute ϵ_{ij} in Eq. (1.64), using Eq. (1.100) with τ_{kk} replaced by κ. Set indices k and l equal to a dummy variable. Use Eq. (1.102) to replace terms of the form $\tau_{ij,j}$ by body forces $-B_i$. Show that the result is that given by Eq. (1.105).

1.25 (*a*) In the example of the cantilever beam show that plane sections do not, in fact, remain plane but distort slightly as a third-order curve.

(*b*) Find the deflection curve for the case where we fix the end so that $\partial u/\partial y = 0$. For a steel ($E = 2.1 \times 10^{11}$ Pa) beam of length $L = 3$ m, $h = 150$ mm, $t = 25$ mm, and $F = 4.5 \times 10^4$ N, what is the difference between the deflection at the end for this case and for the case where $\partial v/\partial x = 0$ at the base?

1.26 Show that the following function is the proper *Airy function* for the simply supported rectangular beam of unit width problem carrying a uniform load as shown in Fig. 1.34.

$$\Phi = \frac{Az^3}{6} + \frac{Bx^2z^3}{6} - \frac{C}{2}x^2z - \frac{qx^2}{4} - \frac{B}{30}z^5$$

where

$$B = -\frac{6q}{h^3} \qquad C = -\frac{3}{2}\frac{q}{h} \qquad A = \frac{12}{h^3}\left(\frac{qL^2}{8} - \frac{qh^2}{20}\right)$$

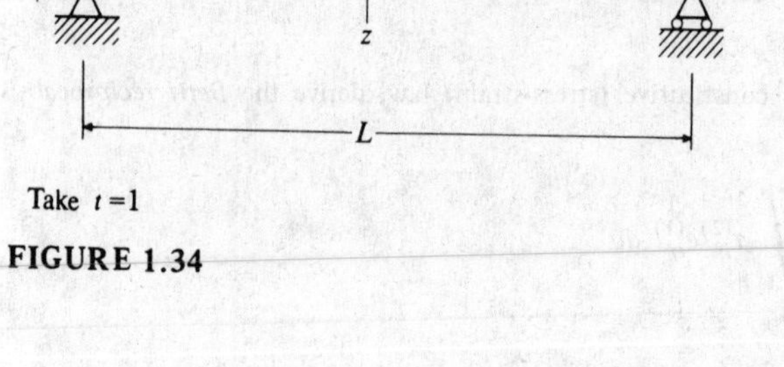

Take $t = 1$

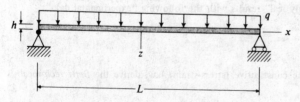

FIGURE 1.34

Check the boundary conditions. Show that:

$$\tau_{xx} = -\frac{q}{2I}\left[x^2 - \left(\frac{L}{2}\right)^2\right]z + \frac{q}{2I}\left(\frac{2}{3}z^3 - \frac{h^2}{10}z\right)$$

$$\tau_{zz} = -\frac{q}{2I}\left(\frac{1}{3}z^3 - \frac{h^2}{4}z + \frac{h^3}{12}\right)$$

$$\tau_{xz} = -\frac{q}{2I}\left(\frac{h^2}{4} - z^2\right)x$$

1.27 Show for the beam of the previous problem that the deflection of the centerline is given as

$$w(x, 0) = \frac{qL^4}{24EI}\left[\frac{5}{16} - \frac{3}{2}\left(\frac{x}{L}\right)^2 + \left(\frac{x}{L}\right)^4\right]$$

$$+ \frac{qL^4}{24EI}\left(\frac{h}{L}\right)^2\left[\left(\frac{12}{5} + \frac{3\nu}{2}\right)\left(\frac{1}{4} - \left(\frac{x}{L}\right)^2\right)\right]$$

2

INTRODUCTION TO THE CALCULUS OF VARIATIONS

2.1 INTRODUCTION

In dealing with a function of a single variable, $y = f(x)$, in the ordinary calculus, we often find it useful to determine the values of x for which the function y is a *local maximum* or a *local minimum*. By a local maximum at position x_1, we mean that f at position x in the neighborhood of x_1 is less than $f(x_1)$ (see Fig. 2.1). Similarly, for a local minimum of f to exist at position x_2 (see Fig. 2.1), we require that $f(x)$ be larger than $f(x_2)$ for all values of x in the neighborhood of x_2. The values of x in the neighborhood of x_1 or x_2 may be called the *admissible values* of x relative to which x_1 or x_2 is a maximum or minimum position.

To establish the condition for a local extremum (maximum or minimum), let us expand the function f as a Taylor series about a position $x = a$. Thus assuming that $f(x)$ has continuous derivatives at position $x = a$, we have

$$f(x) = f(a) + \left(\frac{df}{dx}\right)_{x=a} (x - a) + \frac{1}{2!}\left(\frac{d^2 f}{dx^2}\right)_{x=a} (x - a)^2 + \frac{1}{3!}\left(\frac{d^3 f}{dx^3}\right)_{x=a} (x - a)^3 + \cdots$$

We next rearrange the series and rewrite it in the following more compact form:

$$f(x) - f(a) = f'(a)(x - a) + \frac{1}{2!} f''(a)(x - a)^2 + \frac{1}{3!} f'''(a)(x - a)^3 + \cdots \tag{2.1}$$

For $f(a)$ to be a minimum it is necessary that $f(x) - f(a)$ be a positive number for

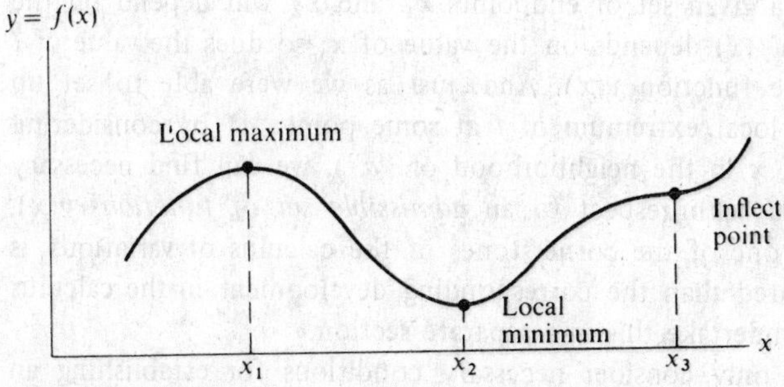

FIGURE 2.1
Local extrema.

all values of x in the neighborhood of "a." Since $x - a$ can be positive or negative for the admissible values of x, then clearly the term $f'(a)$ must be zero to prevent the dominant term in the series from yielding positive and negative values for the admissible values of x. That is, a necessary condition for a local minimum at "a" is that $f'(a) = 0$. By similar reasoning we can conclude that the same condition prevails for a local maximum at "a". Considering the next term in the series, we see that there will be a constancy in sign for admissible values of x and so the sign of $f''(a)$ will determine whether we have a local minimum or a local maximum at position "a". Thus with $f'(a) = 0$, the sign of $f''(a)$ [assuming $f''(a) \neq 0$] supplies sufficient information for establishing a local minimum or a local maximum at position "a".

Suppose next that both $f'(a)$ and $f''(a)$ are zero but that $f'''(a)$ does not equal zero. Then the third term of the series of Eq. (2.1) becomes the dominant term, and for admissible values of x there must be a change in sign of $f(x) - f(a)$ as we move across point "a". Such a point is called an *inflection point* and is shown in Fig. 2.1 at position x_3.

Thus we see that point "a", for which $f'(a) = 0$, may correspond to a local minimum point, to a local maximum point, or to an inflection point. Such points as a group are often of much physical interest (Courant, 1947), and they are called *extremal* positions of the function.

We have presented a view of elements of the theory of local extrema in order to set the stage for the introduction of the calculus of variations,† which will be of considerable use in the ensuing studies of elastic structures. In place of the function of the preceding discussion we shall be concerned now with *functionals*, which are, plainly speaking, functions of functions. Specifically, a functional is an expression that takes on a particular value that is dependent on the function used in the functional. A form of functional that is employed in many areas of applied mathematics is the integral of $F(x, y, y')$ between two points (x_1, y_1) and (x_2, y_2) in two-dimensional space. Denoting this functional as I, we have

$$I = \int_{x_1}^{x_2} F(x, y, y')\, dx \tag{2.2}$$

†For a rigorous study of this subject refer to Gelfand and Fomin (1963) or Fox (1954).

Clearly, the value of I for a given set of endpoints x_1 and x_2 will depend on the function $y(x)$. Thus, just as $f(x)$ depends on the value of x, so does the value of I depend on the form of the function $y(x)$. And, just as we were able to set up necessary conditions for a local extremum of f at some point "a" by considering *admissible values* of x (i.e., x in the neighborhood of "a"), we can find necessary conditions for extremizing I with respect to an *admissible set of functions* $y(x)$. Such a procedure, forming one of the cornerstones of the calculus of variations, is considerably more complicated than the corresponding development in the calculus of functions, and we shall undertake this in a separate section.

In this text we shall only consider necessary conditions for establishing an extremum. We usually know a priori whether this extremum is a maximum or a minimum by physical arguments. Accordingly, the complex arguments† needed in the calculus of variations for giving sufficiency conditions for maximum or minimum states of I will be omitted.

We may generalize the functional I in the following ways:

1. The functional may have many independent variables other than just x.
2. The functional may have many functions (dependent variables) of these independent variables other than just $y(x)$.
3. The functional may have higher-order derivatives.

We shall examine such generalizations in subsequent sections.

In the next section we set forth some very simple functionals.

2.2 EXAMPLES OF SIMPLE FUNCTIONALS

Historically, the calculus of variations became an independent discipline of mathematics at the beginning of the 18th century. Much of the formulation of this mathematics was developed by the Swiss mathematician Leonhard Euler (1707–1783). It is instructive here to consider three of the classic problems that led to the growth of the calculus of variations.

(a) The Brachistochrone

In 1696 Johann Bernoulli posed the following problem. Suppose one were to design a frictionless chute between two points (1) and (2) in a vertical plane such that a body sliding under the action of its own weight goes from (1) to (2) in the *shortest* interval of time. The time for the descent from (1) to (2) we denote as I, and it is given as follows:

$$I = \int_{(1)}^{(2)} \frac{ds}{V} = \int_{(1)}^{(2)} \frac{\sqrt{dx^2 + dy^2}}{V} = \int_{x_1}^{x_2} \frac{\sqrt{1 + (y')^2}}{V}\, dx$$

where V is the speed of the body and s is the distance along the chute. Now

†See references cited earlier.

employ conservation of energy for the body. If V_1 is the initial speed of the body, we have at any position y measured downward:

$$\frac{mV_1^2}{2} - mgy_1 = \frac{mV^2}{2} - mgy$$

Therefore

$$V = [V_1^2 - 2g(y_1 - y)]^{1/2}$$

We can then give I as follows:

$$I = \int_{x_1}^{x_2} \frac{\sqrt{1 + (y')^2}}{[V_1^2 - 2g(y_1 - y)]^{1/2}}\, dx \tag{2.3}$$

We shall later show that the chute [i.e., $y(x)$] should take the shape of a *cycloid*.†

(b) Geodesic Problem

The problem here is to determine the curve on a given surface $g(x, y, z) = 0$ having the *shortest length* between two points (1) and (2) on this surface. Such curves are called *geodesics*. (For a spherical surface the geodesics are segments of the so-called great circles.) The solution to this problem lies in determining the extreme values of the functional:

$$I = \int_{x_1}^{x_2} ds = \int_{x_1}^{x_2} \sqrt{1 + (y')^2 + (z')^2}\, dx \tag{2.4}$$

We have here an example where we have two functions y and z in the functional, although only one independent variable, i.e., x. However, y and z are not independent of each other but must have values satisfying the equation:

$$g(x, y, z) = 0 \tag{2.5}$$

The extremizing process here is analogous to the *constrained* maxima or minima problems of the calculus of functions, and indeed Eq. (2.5) is called a constraining equation in connection with the extremization problem.

If we are able to solve for z in terms of x and y or for y in terms of z and x in Eq. (2.5), we can reduce Eq. (2.4) so as to have only one function of x rather than two. The extremization process with the one function of x is then no longer constrained.

†This problem has been solved by both Johann and Jacob Bernoulli, Sir Isaac Newton, and the French mathematician L'Hôpital.

(c) Isoperimetric Problem

The original isoperimetric problem is given as follows. Of all the closed nonintersecting plane curves having a given *fixed length* L, which curve encloses the *greatest* area A?

The area A is given from the calculus by the following line integral:

$$A = \tfrac{1}{2}\oint (x\,dy - y\,dx)$$

Now suppose that we can express the variables x and y parametrically in terms of τ. Then we can give the above integral as follows:

$$A = I = \frac{1}{2}\int_{\tau_1}^{\tau_2} \left(x\frac{dy}{d\tau} - y\frac{dx}{d\tau} \right) d\tau \tag{2.6}$$

where τ_1 and τ_2 correspond to the beginning and end of the closed loop. The constraint on x and y is now given as follows:

$$L = \oint ds = \oint (dx^2 + dy^2)^{1/2}$$

Introducing the parametric representation we have:

$$L = \int_{\tau_1}^{\tau_2} \left[\left(\frac{dx}{d\tau}\right)^2 + \left(\frac{dy}{d\tau}\right)^2 \right]^{1/2} d\tau \tag{2.7}$$

We have thus one independent variable, τ, and two functions of τ, which are x and y, constrained this time by the *integral relationship* (2.7).

In the historical problems set forth here we have shown how functionals of the form given by Eq. (2.2) may enter directly into problems of interest. Actually, the extremization of such functionals or their more generalized forms is equivalent to solving certain corresponding differential equations, and so this approach may give alternative viewpoints of various areas of mathematical physics. Thus in optics the extremization of the time required for a beam of light to go from one point to another in a vacuum relative to an admissible family of light paths is equivalent to satisyfing Maxwell's equations for the radiation paths of light. This is the famous Fermat principle. In the study of particle mechanics the extremization of the difference between the kinetic energy and the potential energy (i.e., the Lagrangian) integrated between two points over an admissible family of paths yields the correct path as determined by Newton's law. This is Hamilton's principle.† In the theory of elasticity, which will be of primary concern to us in this text, we shall among other things extremize the so-called total potential energy of a body with respect to an admissible family of displacement fields to satisfy the equations of equilibrium

†We shall consider Hamilton's principle in detail in Chapter 7.

for the body. We can note similar dualities in other areas of mathematical physics and engineering science, notably electromagnetic theory and thermodynamics. Thus we conclude that the extremization of functionals of the form of Eq. (2.2) or their generalizations affords us a different view of many fields of study. We shall have ample opportunity in this text to see this new viewpoint as it pertains to solid mechanics. One important benefit derived by recasting the approach as a result of variational considerations is that some very powerful *approximate procedures* will be made available to us for the solution of problems of engineering interest. Such considerations will form a significant part of this text.

It is to be further noted that the series of problems presented required, respectively: a *fastest* time of descent, a *shortest* distance between two points on a surface, and a *greatest* area to be enclosed by a given length. These problems are examples of what are called *optimization* problems.†

We now examine the extremization process for functionals of the type described in this section.

2.3 THE FIRST VARIATION

Consider a functional of the form

$$I = \int_{x_1}^{x_2} F(x, y, y')\, dx \tag{2.8}$$

where F is a known function, twice differentiable for the variables x, y, and y'. As discussed earlier, the value of I between points (x_1, y_1) and (x_2, y_2) will depend on the path chosen between these points, i.e., it will depend on the function $y(x)$ used. We shall assume the existence of a path, which we shall henceforth denote as $y(x)$ (see Fig. 2.2), having the property of extremizing I with respect to other

†In seeking an *optimal solution* in a problem we strive to attain, subject to certain given constraints, that solution, among other possible solutions, which satisfies or comes closest to satisfying a certain criterion or certain criteria. Such a solution is then said to be optimal relative to this criterion or criteria, and the process of arriving at this solution is called optimization.

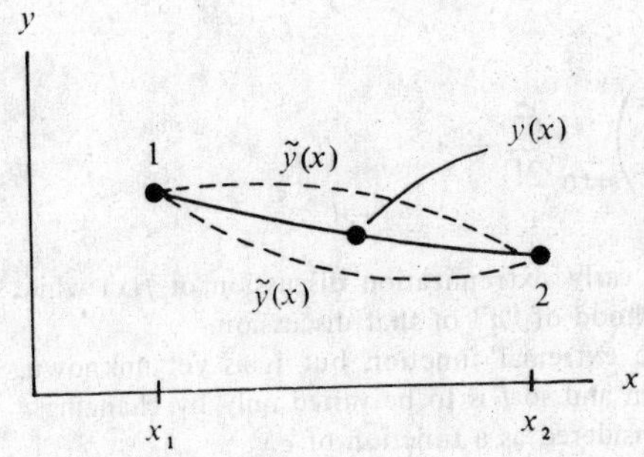

FIGURE 2.2
Extremal path and varied paths.

neighboring paths, which we now denote collectively as $\tilde{y}(x)$.† We assume further that $y(x)$ is twice differentiable. We shall for simplicity refer henceforth to $y(x)$ as the *extremizing path* or the *extremal function*, and to $\tilde{y}(x)$ as the *varied paths.*

We will now introduce a single-parameter family of varied paths as follows:

$$\tilde{y}(x) = y(x) + \epsilon\eta(x) \tag{2.9}$$

where ϵ is a small parameter and $\eta(x)$ is a differentiable function having the requirement that

$$\eta(x_1) = \eta(x_2) = 0$$

We see that an infinity of varied paths can be generated for a given function $\eta(x)$ by adjusting the parameter ϵ. All these paths pass through points (x_1, y_1) and (x_2, y_2). Furthermore, for any $\eta(x)$ the varied path becomes coincident with the extremizing path when we set $\epsilon = 0$.

With the agreement to denote $y(x)$ as the extremal function, then I in Eq. (2.8) becomes the extreme value of the functional

$$\int_{x_1}^{x_2} F(x, \tilde{y}, \tilde{y}')\, dx$$

We can then say

$$\tilde{I} = \int_{x_1}^{x_2} F(x, \tilde{y}, \tilde{y}')\, dx = \int_{x_1}^{x_2} F(x, y + \epsilon\eta, y' + \epsilon\eta')\, dx \tag{2.10}$$

By having employed $y + \epsilon\eta$ as the admissible functions, we are able to use the extremization criteria of simple function theory as presented earlier since $\tilde{I}$ is now, for the desired extremal $y(x)$, a function of the parameter ϵ and thus it can be expanded as a power series in terms of this parameter.‡ Thus

$$\tilde{I} = (\tilde{I})_{\epsilon=0} + \left(\frac{d\tilde{I}}{d\epsilon}\right)_{\epsilon=0} \epsilon + \left(\frac{d^2\tilde{I}}{d\epsilon^2}\right)_{\epsilon=0} \frac{\epsilon^2}{2!} + \cdots$$

Hence:

$$\tilde{I} - \tilde{I}_{\epsilon=0} = \tilde{I} - I = \left(\frac{d\tilde{I}}{d\epsilon}\right)_{\epsilon=0} \epsilon + \left(\frac{d^2\tilde{I}}{d\epsilon^2}\right)_{\epsilon=0} \frac{\epsilon^2}{2!} + \cdots$$

†Thus $y(x)$ will correspond to "a" of the early extremization discussion of $f(x)$ while $\tilde{y}(x)$ corresponds to the values of x in the neighborhood of "a" of that discussion.

‡Note that $y(x)$ is assumed to exist as the extremal function but is as yet unknown. Therefore, $y(x)$ is a particular function yet unknown and so $\tilde{I}$ is to be varied only by changing ϵ for any particular function $\eta(x)$. Thus $\tilde{I}$ is to be considered as a function of ϵ.

For $\tilde{I}$ to be extreme when $\epsilon = 0$ [i.e., for the extremal function $y(x)$] we know from our earlier discussion that

$$\left(\frac{d\tilde{I}}{d\epsilon}\right)_{\epsilon=0} = 0$$

is a *necessary condition*. This, in turn, means that

$$\left[\int_{x_1}^{x_2} \left(\frac{\partial F}{\partial \tilde{y}}\frac{d\tilde{y}}{d\epsilon} + \frac{\partial F}{\partial \tilde{y}'}\frac{d\tilde{y}'}{d\epsilon}\right) dx\right]_{\epsilon=0} = 0$$

Noting that $d\tilde{y}/d\epsilon = \eta$ and that $d\tilde{y}'/d\epsilon = \eta'$, and realizing that deleting the tilde for $\tilde{y}$ and $\tilde{y}'$ in the derivatives of F is the same as setting $\epsilon = 0$ as required above, we may rewrite the above equation as follows:

$$\int_{x_1}^{x_2} \left(\frac{\partial F}{\partial y}\eta + \frac{\partial F}{\partial y'}\eta'\right) dx = 0 \tag{2.11}$$

where henceforth $y(x)$ must be the extremal function. We now integrate the second term by parts as follows:

$$\int_{x_1}^{x_2} \frac{\partial F}{\partial y'}\eta'\, dx = \frac{\partial F}{\partial y'}\eta\,\Bigg|_{x_1}^{x_2} - \int_{x_1}^{x_2} \left[\frac{d}{dx}\left(\frac{\partial F}{\partial y'}\right)\right]\eta\, dx$$

Noting that $\eta = 0$ at the endpoints, we see that the first expression on the right side of the above equation vanishes. We then get, on substituting the above result into Eq. (2.11),

$$\int_{x_1}^{x_2} \left[\frac{\partial F}{\partial y} - \frac{d}{dx}\left(\frac{\partial F}{\partial y'}\right)\right]\eta\, dx = 0 \tag{2.12}$$

With $\eta(x)$ arbitrary between the endpoints, a basic lemma of the calculus of variations† indicates that the bracketed expression in the above integrand is zero. Thus:

$$\boxed{\frac{\partial F}{\partial y} - \frac{d}{dx}\left(\frac{\partial F}{\partial y'}\right) = 0} \tag{2.13}$$

†For a particular function $\phi(x)$ continuous in the interval (x_1, x_2), if $\int_{x_1}^{x_2} \phi(x)\eta(x)\, dx = 0$ for every continuously differentiable function $\eta(x)$ for which $\eta(x_1) = \eta(x_2) = 0$, then $\phi \equiv 0$ for $x_1 \leqslant x \leqslant x_2$.

This is the famous *Euler–Lagrange equation*. It is the condition required for $y(x)$ in the role we have assigned it of being the extremal function. Substitution of $F(x, y, y')$ will result in a second-order ordinary differential equation for the unknown function $y(x)$. In short, the variational procedure has resulted in an ordinary differential equation for getting the function $y(x)$ which we have "tagged" and handled as the extremizing function.

We shall now illustrate the use of the Euler-Lagrange equation by considering the brachistochrone problem presented earlier.

EXAMPLE 2.1 BRACHISTOCHRONE PROBLEM We have from the earlier study of the *brachistochrone problem* of Section 2.2 the requirement to extremize

$$I = \int_1^2 \frac{\sqrt{1 + (y')^2}}{[V_1^2 - 2g(y_1 - y)]^{1/2}}\, dx \tag{a}$$

If we take the special case where the body is released from rest at the origin, the above functional becomes

$$I = \frac{1}{\sqrt{2g}} \int_1^2 \sqrt{\frac{1 + (y')^2}{y}}\, dx \tag{b}$$

The function F can be identified as $\{[1 + (y')^2]/y\}^{1/2}$. We go directly to the Euler-Lagrange equation to substitute for F. After some algebraic manipulation we obtain

$$y'' = -\frac{1 + (y')^2}{2y} \tag{c}$$

Now make the substitution $u = y'$. Then we can say

$$u\frac{du}{dy} = -\frac{1 + u^2}{2y}$$

Separating variables and integrating, we have

$$y(1 + u^2) = C_1$$

Therefore

$$y[1 + (y')^2] = C_1$$

We may arrange for another separation of variables and perform another quadrature as follows:

$$x = \int \frac{\sqrt{y}}{\sqrt{C_1 - y}}\, dy + x_0$$

Next make the substitution

$$y = C_1 \left(\sin^2 \frac{t}{2} \right) \tag{c}$$

We then have

$$x = C_1 \int \sin^2 \frac{t}{2} \, dt + x_0 = C_1 \left(\frac{t - \sin t}{2} \right) + x_0$$

Since at time $t = 0$ we have $x = y = 0$, then $x_0 = 0$ in the above equation. We then have as results

$$x = \frac{C_1}{2} (t - \sin t) \tag{d}$$

$$y = \frac{C_1}{2} (1 - \cos t) \tag{e}$$

wherein we have used the double-angle formula to arrive at Eq. (*e*) from Eq. (*c*). These equations represent a cycloid, which is a curve generated by the motion of a point fixed to the circumference of a rolling wheel. The radius of the wheel here is $C_1/2$.†

2.4 THE DELTA OPERATOR

We now introduce an operator δ, termed the *delta operator*, in order to give a certain formalism to the procedure of obtaining the first variation. We define $\delta[y(x)]$ as follows:

$$\delta[y(x)] = \tilde{y}(x) - y(x) \tag{2.14}$$

Notice that the delta operator represents a small arbitrary change in the dependent variable y for a *fixed* value of the independent variable x. Thus in Fig. 2.3 we have shown the extremizing path $y(x)$ and some varied paths $\tilde{y}(x)$. At the indicated position x any of the increments *a–b, a–c,* or *a–d* may be considered as δy–i.e., as a variation of y. Most important, note that we *do not associate a* δx with each δy. This is in contrast to the differentiation process, where a dy is associated with a given dx. We can thus say that δy is simply the vertical distance between points on different curves at the same value of x, whereas dy is the vertical distance between points on the *same* curve at positions dx apart. This is illustrated in Fig. 2.4.

We may generalize the delta operator to represent a small (usually infinitesimal) change of any function wherein the independent variable is kept fixed. Thus we may take the variation of the function dy/dx. We shall agree here,

†Note that we have represented x and y parametrically in terms of t.

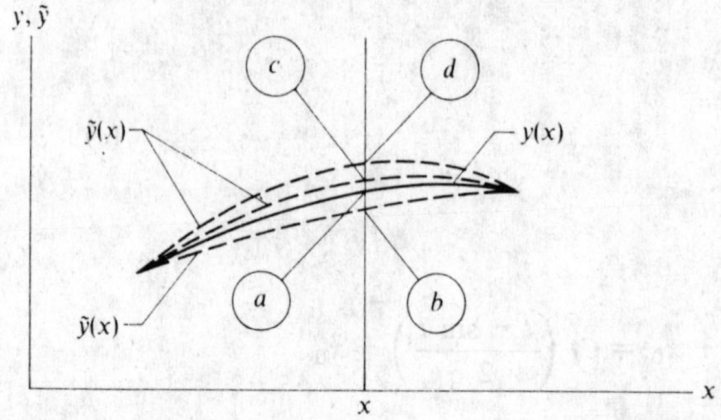

FIGURE 2.3
Extremal path and varied paths.

however, to use as the varied function the derivatives $d\tilde{y}/dx$, where the $\tilde{y}$ are varied paths for y. We can then say

$$\delta\left(\frac{dy}{dx}\right)=\left(\frac{d\tilde{y}}{dx}\right)-\left(\frac{dy}{dx}\right)=\frac{d}{dx}(\tilde{y}-y)=\frac{d(\delta y)}{dx} \tag{2.15}$$

As a consequence of this arrangement, we conclude that the δ operator is *commutative* with the derivative operator. In a similar way, by agreeing that $\int \tilde{y}(x)\,dx$ is a varied function for $\int y(x)\,dx$, we can conclude that the variation operator commutes with the integral operator.

Henceforth we shall make ample use of the δ operator and its associated notation. It will encourage the development of mechanical skill in carrying out the variational process and will aid in developing "physical" intuition in the handling of problems.

It will be well to go back now to Eq. (2.10) and reexamine the first variation, using the delta-operator notation. Considering $y(x)$ to be again the extremal function, we have

$$\delta y=\tilde{y}-y \tag{2.16a}$$

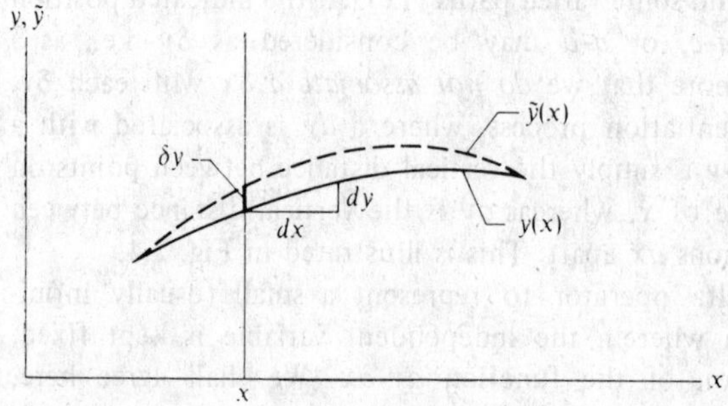

FIGURE 2.4
Difference between δy, dx, and dy.

and

$$\delta y' = \tilde{y}' - y' \tag{2.16b}$$

Accordingly, we can give F along a varied path as follows, using the δ notation and noting from Eqs. (2.16) that $\tilde{y} = y + \delta y$ and $\tilde{y}' = y' + \delta y'$:

$$F(x, y + \delta y, y' + \delta y')$$

Assuming the existence of a definite extremal function $y(x)$ and its derivative $y'(x)$, we see that, at position x, F depends on the increments δy and $\delta y'$. And so we expand F as a Taylor series about the extremal path and its derivative, using increments δy and $\delta y'$. That is,

$$F(x, y + \delta y, y' + \delta y') = F(x, y, y') + \left(\frac{\partial F}{\partial y}\,\delta y + \frac{\partial F}{\partial y'}\,\delta y'\right) + O(\delta^2)$$

and so

$$F(x, y + \delta y, y' + \delta y') - F(x, y, y') = \left(\frac{\partial F}{\partial y}\,\delta y + \frac{\partial F}{\partial y'}\,\delta y'\right) + O(\delta^2) \tag{2.17}$$

where $O(\delta^2)$ refers to terms containing $(\delta y)^2$, $(\delta y')^2$, $(\delta y)^3$, etc., which are of negligibly higher order. The left side of the equation is the *total variation* of F and we shall denote it as $\delta^{(T)}F$. The bracketed expression on the right side of the equation with δ's to the *first power* we call the *first variation*, $\delta^{(1)}F$. Thus we define

$$\delta^{(T)}F = F(x, y + \delta y, y' + \delta y') - F(x, y, y') \tag{2.18a}$$

$$\delta^{(1)}F = \left(\frac{\partial F}{\partial y}\,\delta y + \frac{\partial F}{\partial y'}\,\delta y'\right) \tag{2.18b}$$

On integrating Eq. (2.17) from x_1 to x_2 we get

$$\int_{x_1}^{x_2} F(x, y + \delta y, y' + \delta y')\,dx - \int_{x_1}^{x_2} F(x, y, y')\,dx$$

$$= \int_{x_1}^{x_2} \left(\frac{\partial F}{\partial y}\,\delta y + \frac{\partial F}{\partial y'}\delta y'\right) dx + \int_{x_1}^{x_2} O(\delta^2)$$

This may be written as

$$\tilde{I} - I = \int_{x_1}^{x_2} \left(\frac{\partial F}{\partial y}\,\delta y + \frac{\partial F}{\partial y'}\,\delta y'\right) dx + O(\delta^2)$$

Note that I represents the extreme value of the functional occurring at $\tilde{y} \to y$. We shall call $\tilde{I} - I$ the total variation of I and denote it as $\delta^{(T)}I$, while, as expected, the first expression on the right side of the equation is the first variation of I, $\delta^{(1)}I$. Hence we define

$$\delta^{(T)}I = \tilde{I} - I$$

$$\delta^{(1)}I = \int_{x_1}^{x_2} \left(\frac{\partial F}{\partial y} \delta y + \frac{\partial F}{\partial y'} \delta y' \right) dx$$

Integrating the second expression in the above integral by parts after rewriting $\delta y'$ as $d(\delta y)/dx$, we get for $\delta^{(1)}I$:

$$\delta^{(1)}I = \int_{x_1}^{x_2} \left(\frac{\partial F}{\partial y} - \frac{d}{dx} \frac{\partial F}{\partial y'} \right) \delta y \, dx + \left(\frac{\partial F}{\partial y'} \delta y \right) \Bigg|_{x_1}^{x_2}$$

We shall consider that all $\tilde{y}(x)$ take the specific values corresponding to those of $y(x)$ at x_1 and x_2. Clearly the variation in $y(x)$ must then be zero at these points–i.e., $\delta y = 0$ at x_1 and x_2. Thus we have for $\delta^{(1)}I$:

$$\delta^{(1)}I = \int_{x_1}^{x_2} \left(\frac{\partial F}{\partial y} - \frac{d}{dx} \frac{\partial F}{\partial y'} \right) \delta y \, dx \tag{2.19}$$

We can then say for $\delta^{(T)}I$:

$$\delta^{(T)}I = \delta^{(1)}I + O(\delta^2) + \cdots$$

$$= \int_{x_1}^{x_2} \left(\frac{\partial F}{\partial y} - \frac{d}{dx} \frac{\partial F}{\partial y'} \right) \delta y \, dx + O(\delta^2) + \cdots$$

In order for I to be a maximum or a minimum $\delta^{(T)}I$ must retain the same sign for all possible variations δy over the interval. (Note that δy at any position x could be $\pm K$, where K is a small number.) For this to be possible the expression in parentheses in the integrand on the right side of the equation has to be zero, which in turn leads to the familiar Euler-Lagrange equations. We may thus conclude that

$$\boxed{\delta^{(1)}I = 0} \tag{2.20}$$

Accordingly, the requirement for extremization of I is that its *first variation be zero*. Suppose now that $F = F(e_{ij}, x, y, z)$. Then for independent variables x, y, z we have for the first variation of I, on extrapolating from Eq. (2.18*b*),

$$\delta^{(1)}I = \iiint_V \delta^{(1)}F\,dx\,dy\,dz = \iiint_V \frac{\partial F}{\partial \epsilon_{ij}}\,\delta\epsilon_{ij}\,dx\,dy\,dz$$

where we observe the summation convention of the repeated indices. If F is a function of ϵ as a result of using a one-parameter family approach, then one might suppose that

$$\delta^{(1)}I = \int_{x_1}^{x_2} \frac{\partial F}{\partial \epsilon}\,\delta\epsilon\,dx$$

But the dependent variable represents the extremal function in the development and for the one-parameter development $\epsilon = 0$ corresponds to this condition. Hence we must compute $\partial F/\partial \epsilon$ at $\epsilon = 0$ in the above formulation and $\delta\epsilon$ can be taken as ϵ itself. Hence we get

$$\delta^{(1)}I = \int_{x_1}^{x_2} \left(\frac{\partial \tilde{F}}{\partial \epsilon}\right)_{\epsilon=0} \epsilon\,dx \tag{2.21}$$

If we use a two-parameter family (as we soon shall) then it should be clear for ϵ_1 and ϵ_2 as parameters that

$$\delta^{(1)}I = \int_{x_1}^{x_2} \left[\left(\frac{\partial \tilde{F}}{\partial \epsilon_1}\right)_{\substack{\epsilon_1=0\\ \epsilon_2=0}} \epsilon_1 + \left(\frac{\partial \tilde{F}}{\partial \epsilon_2}\right)_{\substack{\epsilon_1=0\\ \epsilon_2=0}} \epsilon_2\right] dx \tag{2.22}$$

Now going back to Eq. (2.12) and its development, we can say

$$\left(\frac{d\tilde{I}}{d\epsilon}\right)_{\epsilon=0} = \int_{x_1}^{x_2} \left[\frac{\partial F}{\partial y} - \frac{d}{dx}\left(\frac{\partial F}{\partial y'}\right)\right] \eta\,dx \tag{2.23}$$

Next, rewriting Eq. (2.19), we have

$$\delta^{(1)}I = \int_{x_1}^{x_2} \left[\frac{\partial F}{\partial y} - \frac{d}{dx}\left(\frac{\partial F}{\partial y'}\right)\right] \delta y\,dx$$

Noting that $\delta y = \epsilon\eta$ for a single-parameter family approach, we see by comparing the right sides of the above equations that

$$\delta^{(1)}I \equiv \left(\frac{d\tilde{I}}{d\epsilon}\right)_{\epsilon=0} \epsilon \tag{2.24}$$

Accordingly, setting $(d\tilde{I}/d\epsilon)_{\epsilon=0} = 0$, as we have done for a single-parameter family approach, is tantamount to setting the first variation of I equal to zero.† We shall use both approaches in this text for finding the extremal functions.

As a next step, we examine certain simple cases of the Euler-Lagrange equation to ascertain first integrals.

2.5 FIRST INTEGRALS OF THE EULER-LAGRANGE EQUATION

We now present four cases where we can make immediate statements concerning first integrals of the Euler-Lagrange equation as presented thus far.

Case (a). F is not a function of y–i.e., $F = F(x, y')$

In this case the Euler-Lagrange equation degenerates to the form

$$\frac{d}{dx}\left(\frac{\partial F}{\partial y'}\right) = 0$$

Accordingly, we can say for a first integral that

$$\frac{\partial F}{\partial y'} = \text{const} \tag{2.26}$$

Case (b). F is a function only of y'–i.e., $F = F(y')$

Equation (2.26) still applies because of the lack of presence of the variable y. However, now we know that the left side must be a function only of y'. Since this function must equal a constant at all times we conclude that $y' = \text{const}$ is a possible solution. This means that for this case an extremal path is simply that of a straight line.

Case (c). F is independent of the independent variable x–i.e., $F = F(y, y')$

For this case we begin by presenting an identity, which you are urged to verify. Thus noting that d/dx may be expressed here as

$$\left(\frac{\partial}{\partial x} + y'\frac{\partial}{\partial y} + y''\frac{\partial}{\partial y'}\right)$$

we can say

$$\frac{d}{dx}\left(y'\frac{\partial F}{\partial y'} - F\right) = -y'\left[\frac{\partial F}{\partial y} - \frac{d}{dx}\left(\frac{\partial F}{\partial y'}\right)\right] - \frac{\partial F}{\partial x}$$

†Note that for a two-parameter family we have from Eq. (2.22) the result:

$$\delta^{(1)}I = \left(\frac{\partial \tilde{I}}{\partial \epsilon_1}\right)_{\substack{\epsilon_1=0\\ \epsilon_2=0}} \epsilon_1 + \left(\frac{\partial \tilde{I}}{\partial \epsilon_2}\right)_{\substack{\epsilon_1=0\\ \epsilon_2=0}} \epsilon_2 \tag{2.25}$$

If F is not explicitly a function of x, we can drop the last term. Satisfaction of the Euler-Lagrange equation now means that the right side of the equation is zero, so we may conclude that

$$y' \frac{\partial F}{\partial y'} - F = C_1 \tag{2.27}$$

for the extremal function. We thus have next to solve a first-order differential equation to determine the extremal function.

Case (d). *F* is the total derivative of some function *g*(*x*, *y*)–i.e., *F* = *dg*/*dx*

It is easy to show that when $F = dg/dx$, it must satisfy identically the Euler-Lagrange equation. We first note that

$$F = \frac{\partial g}{\partial x} + \frac{\partial g}{\partial y} y' \tag{2.28}$$

Now substitute the above result into the Euler-Lagrange equation. We get

$$\frac{\partial}{\partial y}\left(\frac{\partial g}{\partial x} + \frac{\partial g}{\partial y} y'\right) - \frac{d}{dx}\left(\frac{\partial g}{\partial y}\right) = 0$$

Carry out the various differentiation processes:

$$\frac{\partial^2 g}{\partial x \, \partial y} + \frac{\partial^2 g}{\partial y^2} y' - \frac{\partial^2 g}{\partial x \, \partial y} - \frac{\partial^2 g}{\partial y^2} y' = 0$$

The left side is clearly identically zero and so we have shown that a *sufficient* condition for F to satisfy the Euler-Lagrange equation identically is that $F = dg(x, y)/dx$. It can also be shown that this is a *necessary* condition for the identical satisfaction of the Euler-Lagrange equation. It is then obvious that we can always add a term of the form dg/dx to the function F in I without changing the Euler-Lagrange equations for the functional I. That is, for any Euler-Lagrange equation there are an *infinity of functionals differing from each other in their integrands by terms of the form dg/dx.*

We shall have ample occasion to use these simple solutions. Now we consider the geodesic problem, presented earlier, for the case of the sphere.

EXAMPLE 2.2 Consider a sphere of radius R having its center at the origin of reference xyz. We wish to determine the shortest path between two points on this sphere. Using spherical coordinates R, θ, ϕ (see Fig. 2.5), a point P on the sphere has the following coordinates:

$$\begin{aligned} x &= R \sin\theta \cos\phi \\ y &= R \sin\theta \sin\phi \\ z &= R \cos\theta \end{aligned} \tag{a}$$

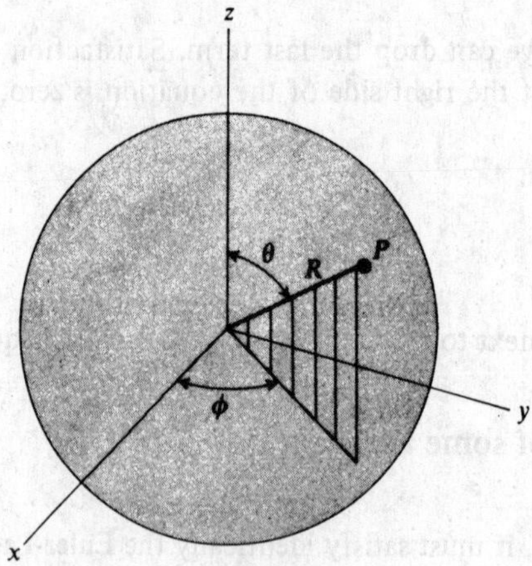

FIGURE 2.5
Spherical coordinat

The increment of distance ds on the sphere may be given as follows:

$$ds^2 = dx^2 + dy^2 + dz^2 = R^2(d\theta^2 + \sin^2\theta\, d\phi^2)$$

Hence the distance between points P and Q can be given as follows:

$$I = \int_P^Q ds = \int_P^Q R(d\theta^2 + \sin^2\theta\, d\phi^2)^{1/2} \qquad (b)$$

With ϕ and θ as independent variables, the transformation equations (a) depict a sphere. If ϕ is related to θ, i.e., is a function of θ, then clearly x, y, and z in Eqs. (a) are functions of a single parameter θ and this must represent some curve on the sphere. Accordingly, since we are seeking a curve on the sphere, we shall assume that ϕ is a function of θ in the above equation so that we can say:

$$I = R\int_P^Q \left[1 + \sin^2\theta\left(\frac{d\phi}{d\theta}\right)^2\right]^{1/2} d\theta \qquad (c)$$

We have here a functional with θ as the independent variable and ϕ as the function, with the derivative of ϕ appearing explicitly. With ϕ not appearing explicitly in the integrand, we recognize this to be case (a) discussed earlier. We may then say, using Eq. (2.26):

$$\frac{\partial}{\partial\phi'}[1 + \sin^2\theta\,(\phi')^2]^{1/2} = C_1 \qquad (d)$$

This becomes:

$$\frac{\sin^2\theta\,(\phi')}{[1 + \sin^2\theta\,(\phi')^2]^{1/2}} = C_1$$

Solving for ϕ', we get

$$\phi' = \frac{C_1}{\sin\theta\,(\sin^2\theta - C_1^2)^{1/2}}$$

Integrating, we have

$$\phi = C_1 \int \frac{d\theta}{\sin\theta\,(\sin^2\theta - C_1^2)^{1/2}} + C_2 \qquad (e)$$

We next make the following substitution for θ:

$$\theta = \tan^{-1}\frac{1}{\eta} \qquad (f)$$

This give us

$$\phi = C_1 \int \frac{[1/(1+\eta^2)]\,d\eta}{[1/(1+\eta^2)]^{1/2}\,[1/(1+\eta^2) - C_1^2]^{1/2}} + C_2$$

$$= \int \frac{d\eta}{[(1/C_1^2 - 1) - \eta^2]^{1/2}} + C_2$$

Denoting $(1/C_1^2 - 1)^{1/2}$ as $1/C_3$ and integrating we get:

$$\phi = \sin^{-1}(C_3\eta) + C_2 \qquad (g)$$

Replacing η from Eq. (*f*), we get

$$\phi = \sin^{-1}\left(C_3\frac{1}{\tan\theta}\right) + C_2$$

Hence we can say

$$\sin(\phi - C_2) = \frac{C_3}{\tan\theta}$$

This equation may next be written as follows:

$$\sin\phi\cos C_2 - \cos\phi\sin C_2 = C_3\frac{\cos\theta}{\sin\theta}$$

Hence:

$$\sin\phi\sin\theta\cos C_2 - \sin\theta\cos\phi\sin C_2 = C_3\cos\theta$$

Observing the transformation equations (*a*) we may express the above equation in terms of Cartesian coordinates in the following manner:

$$y \cos C_2 - x \sin C_2 = zC_3 \tag{h}$$

where we have canceled the term R. This is the equation of a plane surface going through the origin. The intersection of this plane surface and the sphere then gives the proper locus of points on the sphere that forms the desired extremal path. Clearly this curve is the expected great circle.

2.6 FIRST VARIATION WITH SEVERAL DEPENDENT VARIABLES

We now consider the case where we may have any number of functions with only a single *independent* variable. We shall denote the functions as $q_1, q_2, \ldots, q_n$ and the independent variable as t. (This is a notation that is often used in particle mechanics.) The functional for this case is then given as follows:

$$I = \int_{t_1}^{t_2} F(q_1, q_2, \ldots, q_n, \dot{q}_1, \dot{q}_2, \ldots, \dot{q}_n, t)\, dt \tag{2.29}$$

where $\dot{q}_1 = dq_1/dt$, etc. We wish to determine a set of functions $q_1(t), q_2(t), \ldots, q_n(t)$ which are twice differentiable and which extremize the functional I with respect to a broad class of admissible functions. We shall denote the varied functions as $\tilde{q}_i(t)$ and, as before, we shall henceforth consider the notation $q_i(t)$ to identify the extremal functions we are seeking.

We shall use the following single-parameter family of admissible functions:

$$\begin{aligned}
\tilde{q}_1(t) &= q_1(t) + \epsilon\eta_1(t)\\
\tilde{q}_2(t) &= q_2(t) + \epsilon\eta_2(t)\\
&\vdots\\
\tilde{q}_n(t) &= q_n(t) + \epsilon\eta_n(t)
\end{aligned} \tag{2.30}$$

where $\eta_1(t), \eta_2(t), \ldots, \eta_n(t)$ are arbitrary functions having proper continuity and differentiability properties for the ensuing steps. Also, these functions are equal to zero at the endpoints t_1 and t_2. Finally, when we set $\epsilon = 0$ we get back to the designated extremal functions $q_1(t), \ldots, q_n(t)$.

We now form $\tilde{I}(\epsilon)$ as follows:

$$\tilde{I}(\epsilon) = \int_{t_1}^{t_2} F(\tilde{q}_1, \tilde{q}_2, \ldots, \tilde{q}_n, \dot{\tilde{q}}_1, \dot{\tilde{q}}_2, \ldots, \dot{\tilde{q}}_n, t)\, dt$$

In order for the $q_i(t)$ to be the extremal paths, we now require that

$$\left[\frac{d\tilde{I}(\epsilon)}{d\epsilon}\right]_{\epsilon=0} = 0$$

Hence:

$$\left[\int_{t_1}^{t_2}\left(\frac{\partial F}{\partial \tilde{q}_1}\eta_1 + \cdots \frac{\partial F}{\partial \tilde{q}_n}\eta_n + \frac{\partial F}{\partial \dot{\tilde{q}}_1}\dot{\eta}_1 + \ldots \frac{\partial F}{\partial \dot{\tilde{q}}_n}\dot{\eta}_n\right) dt\right]_{\epsilon=0} = 0$$

Now setting $\epsilon = 0$ in the above expression is the same as removing the tildes from the q's and $\dot{q}$'s. Thus we have

$$\int_{t_1}^{t_2}\left(\frac{\partial F}{\partial q_1}\eta_1 + \cdots \frac{\partial F}{\partial q_n}\eta_n + \frac{\partial F}{\partial \dot{q}_1}\dot{\eta}_1 + \cdots \frac{\partial F}{\partial \dot{q}_n}\dot{\eta}_n\right) dt = 0$$

Now the functions $\eta_i(t)$ are arbitrary and so we can take all $\eta_i(t)$ except $\eta_1(t)$ equal to zero. We then have, on integrating $(\partial F/\partial \dot{q}_1)\dot{\eta}_1$ by parts:

$$\int_{t_1}^{t_2}\left(\frac{\partial F}{\partial q_1} - \frac{d}{dt}\frac{\partial F}{\partial \dot{q}_1}\right)\eta_1 \, dt = 0$$

Using the fundamental lemma of the calculus of variations, we conclude that:

$$\frac{\partial F}{\partial q_1} - \frac{d}{dt}\left(\frac{\partial F}{\partial \dot{q}_1}\right) = 0$$

We may similarly assume that η_2 is the only nonzero function and so forth to lead us to the conclusion that

$$\boxed{\frac{\partial F}{\partial q_i} - \frac{d}{dt}\left(\frac{\partial F}{\partial \dot{q}_i}\right) = 0 \qquad i = 1, 2, \ldots, n} \tag{2.31}$$

are necessary conditions for establishing the $q_i(t)$ as the extremal functions. These are again the Euler-Lagrange equations, which lead us on substitution for F to a system of second-order ordinary differential equations for establishing the $q_i(t)$. These equations may be coupled (i.e., simultaneous equations) or uncoupled, depending on the variables q_i chosen to be used in forming the functional I in Eq. (2.29).

We now illustrate the use of the Euler-Lagrange equations for a problem in particle mechanics.

EXAMPLE 2.3 We will present for use here the very important Hamilton principle in order to illustrate the multifunction problem of this section. Later we shall take the time to consider this principle in detail.

For a system of particles acted on by conservative forces, Hamilton's principle states that the *proper paths taken from a configuration at time* t_1 *to a configuration at time* t_2 *are those that extremize the following functional:*

$$I = \int_{t_1}^{t_2} (T - V)\, dt \tag{a}$$

where T is the kinetic energy of the system and therefore a function of velocity $\dot{x}_i$ of each particle, while V is the potential energy of the system and therefore a function of the coordinates x_i of the particles. We thus have here a functional of many dependent variables x_i with the presence of a single independent variable t.

Consider two identical masses connected by three springs, shown in Fig. 2.6. The masses can only move along a straight line as a result of frictionless constraints. Two independent coordinates are needed to locate the system; they are shown as x_1 and x_2. The springs are unstretched when $x_1 = x_2 = 0$.

From elementary mechanics we can say for the system:

$$\begin{aligned} T &= \tfrac{1}{2}M\dot{x}_1^2 + \tfrac{1}{2}M\dot{x}_2^2 \\ V &= \tfrac{1}{2}K_1 x_1^2 + \tfrac{1}{2}K_2(x_2 - x_1)^2 + \tfrac{1}{2}K_1 x_2^2 \end{aligned} \tag{b}$$

Hence we have for I:

$$\begin{aligned} I = \int_{t_1}^{t_2} &\{\tfrac{1}{2}M(\dot{x}_1^2 + \dot{x}_2^2) - \tfrac{1}{2}[K_1 x_1^2 + K_2(x_2 - x_1)^2 \\ &+ K_1 x_2^2]\}\, dt \end{aligned} \tag{c}$$

To extremize I we employ Eq. (2.31) as follows:

$$\begin{aligned} \frac{\partial F}{\partial x_1} - \frac{d}{dt}\left(\frac{\partial F}{\partial \dot{x}_1}\right) &= 0 \\ \frac{\partial F}{\partial x_2} - \frac{d}{dt}\left(\frac{\partial F}{\partial \dot{x}_2}\right) &= 0 \end{aligned} \tag{d}$$

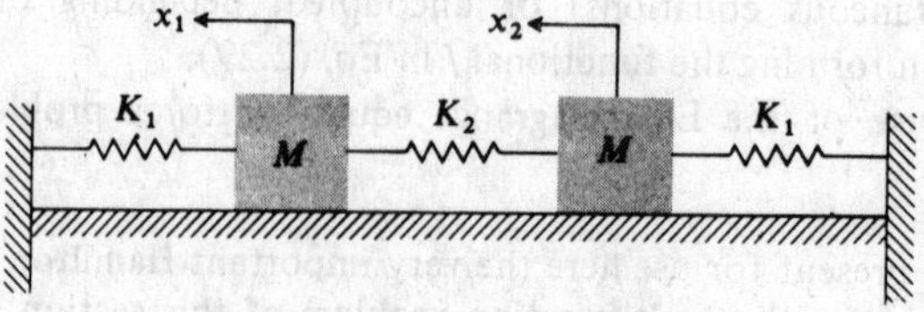

FIGURE 2.6
Spring mass system with two degrees of freedom.

Substituting, we get

$$-K_1x_1 + K_2(x_2 - x_1) - \frac{d}{dt}(M\dot{x}_1) = 0$$
$$-K_2x_2 - K_2(x_2 - x_1) - \frac{d}{dt}(M\dot{x}_2) = 0 \tag{e}$$

Rearranging, we then have

$$M\ddot{x}_1 + K_1x_1 - K_2(x_2 - x_1) = 0$$
$$M\ddot{x}_2 + K_2x_2 + K_2(x_2 - x_1) = 0 \tag{f}$$

These are recognized immediately as equations obtainable directly from Newton's laws. Thus the Euler-Lagrange equations lead to the basic equations of motion for this case. We may integrate these equations and, using initial conditions $\dot{x}_1(0)$, $\dot{x}_2(0)$, $x_1(0)$, and $x_2(0)$, we may then fully establish the subsequent motion of the system.

In this problem we could have more easily employed Newton's law directly. There are many problems, however, where it is easier to proceed by the variational approach to arrive at the equations of motion. Also, as will be seen in the text, other advantages accrue by the use of the variational method.

2.7 THE ISOPERIMETRIC PROBLEM

We now investigate the isoperimetric problem in its simplest form, where we wish to extremize the functional

$$I = \int_{x_1}^{x_2} F(x, y, y')\,dx \tag{2.32}$$

subject to the restriction that $y(x)$ have a form such that:

$$J = \int_{x_1}^{x_2} G(x, y, y')\,dx = \text{const} \tag{2.33}$$

where G is a given function.

We proceed essentially as in Section 2.3. We take $y(x)$ from here on to represent the extremal function for the functional of Eq. (2.32). We next introduce a system of varied paths $\tilde{y}(x)$ for computation of $\tilde{I}$. Our task is then to find conditions required to be imposed on $y(x)$ so as to extremize $\tilde{I}$ with respect to the admissible varied paths $\tilde{y}(x)$ while satisfying the *isoperimetric constraint* of Eq. (2.33). To facilitate the computations we shall require that the admissible varied paths also satisfy Eq. (2.33). Thus we have

$$\tilde{I} = \int_{x_1}^{x_2} F(x, \tilde{y}, \tilde{y}')\, dx \tag{2.34a}$$

$$\tilde{J} = \int_{x_1}^{x_2} G(x, \tilde{y}, \tilde{y}')\, dx = \text{const} \tag{2.34b}$$

Because of this last condition on $y(x)$ we shall no longer use the familiar single-parameter family of admissible functions, since varying ϵ alone for a single-parameter family of functions may mean that the constraint for the corresponding paths $y(x)$ is not satisfied. To allow for enough flexibility to carry out extremization while maintaining *intact* the constraining equation, we employ a two-parameter family of admissible functions of the form

$$\tilde{y}(x) = y(x) + \epsilon_1 \eta_1(x) + \epsilon_2 \eta_2(x) \tag{2.35}$$

where $\eta_1(x)$ and $\eta_2(x)$ are arbitrary functions that vanish at the endpoints x_1, x_2, and where ϵ_1 and ϵ_2 are two small parameters. Using this system of admissible functions, it is clear that $\tilde{I}$ and $\tilde{J}$ are functions of ϵ_1 and ϵ_2. Thus:

$$\tilde{I}(\epsilon_1, \epsilon_2) = \int_{x_1}^{x_2} F(x, \tilde{y}, \tilde{y}')\, dx \tag{2.36a}$$

$$\tilde{J}(\epsilon_1, \epsilon_2) = \int_{x_1}^{x_2} G(x, \tilde{y}, \tilde{y}')\, dx = \text{const.} \tag{2.36b}$$

To extremize $\tilde{I}$ when $\tilde{y} \to y$ we require [see Eq. (2.25)] that

$$\delta^{(1)}\tilde{I} = \left(\frac{\partial \tilde{I}}{\partial \epsilon_1}\right)_{\substack{\epsilon_1=0\\ \epsilon_2=0}} \epsilon_1 + \left(\frac{\partial \tilde{I}}{\partial \epsilon_2}\right)_{\substack{\epsilon_1=0\\ \epsilon_2=0}} \epsilon_2 = 0 \tag{2.37}$$

If ϵ_1 and ϵ_2 were independent of each other, we could then set each of the partial derivatives in the above equation equal to zero separately to satisfy the equation. However, ϵ_1 and ϵ_2 are related by the requirement that $\tilde{J}(\epsilon_1, \epsilon_2) = \text{const}$. The first variation of $\tilde{J}$ must be zero because of the constancy of its value, and we have

$$\left(\frac{\partial \tilde{J}}{\partial \epsilon_1}\right)_{\substack{\epsilon_1=0\\ \epsilon_2=0}} \epsilon_1 + \left(\frac{\partial \tilde{J}}{\partial \epsilon_2}\right)_{\substack{\epsilon_1=0\\ \epsilon_2=0}} \epsilon_2 = 0 \tag{2.38}$$

At this time we make use of the method of the *Lagrange multiplier*. That is, we multiply Eq. (2.38) by an undetermined constant λ (the Lagrange multiplier) and add Eqs. (2.37) and (2.38) to get

$$\left[\left(\frac{\partial \tilde{I}}{\partial \epsilon_1}\right)_{\substack{\epsilon_1=0\\ \epsilon_2=0}} + \lambda \left(\frac{\partial \tilde{J}}{\partial \epsilon_1}\right)_{\substack{\epsilon_1=0\\ \epsilon_2=0}}\right]\epsilon_1 + \left[\left(\frac{\partial \tilde{I}}{\partial \epsilon_2}\right)_{\substack{\epsilon_1=0\\ \epsilon_2=0}} + \lambda \left(\frac{\partial \tilde{J}}{\partial \epsilon_2}\right)_{\substack{\epsilon_1=0\\ \epsilon_2=0}}\right]\epsilon_2 = 0 \tag{2.39}$$

We now choose λ so that one of the two bracketed quantities is zero. Here we choose the second bracketed quantity, so that:

$$\left(\frac{\partial \tilde{I}}{\partial \epsilon_2}\right)_{\substack{\epsilon_1=0\\ \epsilon_2=0}} + \lambda \left(\frac{\partial \tilde{J}}{\partial \epsilon_2}\right)_{\substack{\epsilon_1=0\\ \epsilon_2=0}} = 0 \tag{2.40}$$

Now we may consider that ϵ_2 is the dependent variable and that ϵ_1 is the independent variable. We must then conclude from Eq. (2.39) that the coefficient of ϵ_1 is zero. Thus:

$$\left(\frac{\partial \tilde{I}}{\partial \epsilon_1}\right)_{\substack{\epsilon_1=0\\ \epsilon_2=0}} + \lambda \left(\frac{\partial \tilde{J}}{\partial \epsilon_1}\right)_{\substack{\epsilon_1=0\\ \epsilon_2=0}} = 0 \tag{2.41}$$

Thus, Eqs. (2.40) and (2.41) with the use of the Lagrange multiplier λ give us the necessary conditions for an extreme of $\tilde{I}$ while maintaining the constraint condition intact.

We can now shorten the notation by introducing $\tilde{I}^*$ as follows:

$$\tilde{I}^* = \tilde{I} + \lambda \tilde{J} \tag{2.42}$$

so that Eqs. (2.40) and (2.41) become

$$\left(\frac{\partial \tilde{I}^*}{\partial \epsilon_1}\right)_{\substack{\epsilon_1=0\\ \epsilon_2=0}} = \left(\frac{\partial \tilde{I}^*}{\partial \epsilon_2}\right)_{\substack{\epsilon_1=0\\ \epsilon_2=0}} = 0 \tag{2.43}$$

In integral form we have for $\tilde{I}^*$

$$\tilde{I}^* = \int_{x_1}^{x_2} F(x, \tilde{y}, \tilde{y}')\, dx + \lambda \int_{x_1}^{x_2} G(x, \tilde{y}, \tilde{y}')\, dx$$

$$= \int_{x_1}^{x_2} [F(x, \tilde{y}, \tilde{y}') + \lambda G(x, \tilde{y}, \tilde{y}')]\, dx = \int_{x_1}^{x_2} F^*(x, \tilde{y}, \tilde{y}')\, dx \tag{2.44}$$

where

$$F^* = F + \lambda G \tag{2.45}$$

We now apply the conditions given by Eq. (2.43), using Eq. (2.44) to replace $\tilde{I}^*$:

$$\left(\frac{\partial \tilde{I}^*}{\partial \epsilon_i}\right)_{\substack{\epsilon_1=0\\ \epsilon_2=0}} = \left[\int_{x_1}^{x_2} \left(\frac{\partial F^*}{\partial \tilde{y}} \eta_i + \frac{\partial F^*}{\partial \tilde{y}'} \eta_i'\right) dx\right]_{\substack{\epsilon_1=0\\ \epsilon_2=0}} = 0 \qquad i = 1, 2 \qquad (2.46)$$

We thus get a statement of the extremization process without the appearance of the constraint and this now becomes the starting point of the extremization process. Removing the tildes from $\tilde{y}$ and $\tilde{y}'$ is equivalent to setting $\epsilon_i = 0$, and so we may say

$$\left(\frac{\partial \tilde{I}^*}{\partial \epsilon_i}\right)_{\substack{\epsilon_1=0\\ \epsilon_2=0}} = \int_{x_1}^{x_2} \left(\frac{\partial F^*}{\partial y} \eta_i + \frac{\partial F^*}{\partial y'} \eta_i'\right) dx = 0 \qquad i = 1, 2 \qquad (2.47)$$

Integrating the second expression in the integrand by parts and noting that $\eta_i(x_1) = \eta_i(x_2) = 0$, we get

$$\int_{x_1}^{x_2} \left(\frac{\partial F^*}{\partial y} - \frac{d}{dx}\frac{\partial F^*}{\partial y'}\right) \eta_i \, dx = 0 \qquad i = 1, 2$$

Now, using the fundamental lemma of the calculus of variations, we conclude that

$$\boxed{\frac{\partial F^*}{\partial y} - \frac{d}{dx}\left(\frac{\partial F^*}{\partial y'}\right) = 0} \qquad (2.48)$$

Thus the Euler-Lagrange equation is again a necessary condition for the desired extremum, this time applied to F^* and thus including the Lagrange multiplier. This leads to a second-order differential equation for $y(x)$, the extremal function. Integrating this equation then leaves us two constants of integration plus the Lagrange multiplier. These are determined from the specified values of y at the endpoints plus the constraint condition given by Eq. (2.33).

We have thus far considered only a single constraining integral. If we have n such integrals

$$\int_{x_1}^{x_2} G_k(x, y, y')\, dx = C_k \qquad k = 1, 2, \ldots, n$$

then by using an $n + 1$ parameter family of varied paths $\tilde{y} = y + \epsilon_1 \eta_1 + \epsilon_2 \eta_2 + \cdots \epsilon_{n+1} \eta_{n+1}$ we can arrive as before at the following requirement for extremization:

$$\frac{\partial F^*}{\partial y} - \frac{d}{dx}\left(\frac{\partial F^*}{\partial y'}\right) = 0$$

where

$$F^* = F + \sum_{k=1}^{n} \lambda_k G_k$$

The λ's are again the Lagrange multipliers. Finally, for p dependent variables, i.e.,

$$I = \int_{t_1}^{t_2} F(q_1, q_2, \ldots, q_p, \dot{q}_1, \ldots, \dot{q}_p, t)\, dt$$

with n constraints

$$\int_{t_1}^{t_2} G_k(q_1, \ldots, q_p, \dot{q}_1, \ldots, \dot{q}_p, t)\, dt = C_k \qquad k = 1, 2, \ldots, n$$

the extremizing process yields

$$\frac{\partial F^*}{\partial q_i} - \frac{d}{dt}\left(\frac{\partial F^*}{\partial \dot{q}_i}\right) = 0 \qquad i = 1, \ldots, p \tag{2.49}$$

where

$$F^* = F + \sum_{1}^{n} \lambda_k G_k \tag{2.50}$$

We now illustrate the use of these equations by considering in detail the isoperimetric problem presented earlier.

EXAMPLE 2.4 Recall from Section 2.2 that the isoperimetric problem asks us to find the particular curve $y(x)$ which, for a given length L, encloses the largest area A. Expressed parametrically, we have a functional with two functions y and x; the independent variable is τ. Thus:

$$I = A = \frac{1}{2}\int_{\tau_1}^{\tau_2} \left(x \frac{dy}{d\tau} - y \frac{dx}{d\tau}\right) d\tau \tag{a}$$

The constraint relation is

$$L = \int_{\tau_1}^{\tau_2} \sqrt{\left(\frac{dx}{d\tau}\right)^2 + \left(\frac{dy}{d\tau}\right)^2}\, d\tau \tag{b}$$

We now form F^* for this case:

$$F^* = \tfrac{1}{2}(x\dot{y} - y\dot{x}) + \lambda(\dot{x}^2 + \dot{y}^2)^{1/2}$$

where we use the dot superscript to represent $(d/d\tau)$. We now set forth the Euler-Lagrange equations:

$$\frac{\dot{y}}{2} - \frac{d}{d\tau}\left[-\frac{y}{2} + \frac{\lambda\dot{x}}{(\dot{x}^2 + \dot{y}^2)^{1/2}}\right] = 0 \qquad (c)$$

$$-\frac{\dot{x}}{2} - \frac{d}{d\tau}\left[\frac{x}{2} + \frac{\lambda\dot{y}}{(\dot{x}^2 + \dot{y}^2)^{1/2}}\right] = 0 \qquad (d)$$

We next integrate Eqs. (*c*) and (*d*) with respect to τ to get

$$y - \frac{\lambda\dot{x}}{(\dot{x}^2 + \dot{y}^2)^{1/2}} = C_1 \qquad (e)$$

$$x + \frac{\lambda\dot{y}}{(\dot{x}^2 + \dot{y}^2)^{1/2}} = C_2 \qquad (f)$$

After eliminating λ between the equations we reach the following result:

$$(x - C_2)\,dx + (y - C_1)\,dy = 0 \qquad (g)$$

The last equation is easily integrated to yield

$$\frac{(x - C_2)^2}{2} + \frac{(y - C_1)^2}{2} = C_3^2 \qquad (h)$$

where C_3 is a constant of integration. Thus we get as the required curve a circle—a result that should surprise no one. The radius of the circle is $\sqrt{2}C_3$, which you may readily show [by eliminating C_1 and C_2 from Eq. (*h*), using Eqs. (*e*) and (*f*), and solving for λ] is the value of the Lagrange multiplier† λ. The constants C_1 and C_2 merely position the circle.

2.8 FUNCTION CONSTRAINTS

We now consider the functional

$$I = \int_{t_1}^{t_2} F(q_1, q_2, \ldots, q_n, \dot{q}_1, \ldots, \dot{q}_n, t)\,dt \qquad (2.51)$$

with the following m constraints on the n functions q_i:‡

$$\begin{aligned} &G_1(q_1, \ldots, q_n, \dot{q}_1, \ldots, \dot{q}_n, t) = 0 \\ &\quad\vdots \\ &G_m(q_1, \ldots, q_n, \dot{q}_1, \ldots, \dot{q}_n, t) = 0 \end{aligned} \qquad (2.52)$$

†The Lagrange multiplier is usually of physical significance.

‡In the dynamics of particles, if the constraining equations do not have derivatives the constraints are called *holomonic*.

We assume $m < n$. To extremize the functional I we may proceed by employing n one-parameter families of varied functions of the form

$$\tilde{q}_i(t) = q_i(t) + \epsilon\eta_i(t) \qquad i = 1, \ldots, n \tag{2.53}$$

where $\eta_i(t_1) = \eta_i(t_2) = 0$. Furthermore, we assume that the varied functions $\tilde{q}_i$ satisfy the constraining equation (2.52). That is,

$$G_j(\tilde{q}_1, \ldots, \tilde{q}_n, \dot{\tilde{q}}_1, \ldots, \dot{\tilde{q}}_n, t) = 0 \qquad j = 1, \ldots, m \tag{2.54}$$

We now set the first variation of I equal to zero:

$$\delta^{(1)}I = 0 = \left(\frac{\partial \tilde{I}}{\partial \epsilon}\right)_{\epsilon=0} \epsilon$$

Setting $(\partial\tilde{I}/\partial\epsilon)_{\epsilon=0}$ equal to zero, we have

$$\left[\int_{t_1}^{t_2} \left(\frac{\partial F}{\partial \tilde{q}_1}\eta_1 + \cdots + \frac{\partial F}{\partial \tilde{q}_n}\eta_n + \frac{\partial F}{\partial \dot{\tilde{q}}_1}\dot{\eta}_1 + \cdots + \frac{\partial F}{\partial \dot{\tilde{q}}_n}\dot{\eta}_n\right) dt\right]_{\epsilon=0} = 0$$

Dropping the tildes and subscript $\epsilon = 0$, we have

$$\int_{t_1}^{t_2} \left(\frac{\partial F}{\partial q_1}\eta_1 + \cdots + \frac{\partial F}{\partial q_n}\eta_n + \frac{\partial F}{\partial \dot{q}_1}\dot{\eta}_1 + \cdots + \frac{\partial F}{\partial \dot{q}_n}\dot{\eta}_n\right) dt = 0 \tag{2.55}$$

Since by assumption the varied q's satisfy the constraining equations, we can conclude since $G_j = 0$ that

$$\delta^{(1)}(G_j) = 0 = \left(\frac{\partial G_j}{\partial \epsilon}\right)_{\epsilon=0} \epsilon$$

Setting $(\partial G_j/\partial\epsilon)_{\epsilon=0} = 0$ here we have

$$\left(\frac{\partial G_j}{\partial \tilde{q}_1}\eta_1 + \cdots + \frac{\partial G_j}{\partial \tilde{q}_n}\eta_n + \frac{\partial G_j}{\partial \dot{\tilde{q}}_1}\dot{\eta}_1 + \cdots + \frac{\partial G_j}{\partial \dot{\tilde{q}}_n}\dot{\eta}_n\right)_{\epsilon=0} = 0 \qquad j = 1, 2, \ldots, m$$

Dropping the tildes and $\epsilon = 0$ we then have

$$\frac{\partial G_j}{\partial q_1}\eta_1 + \cdots + \frac{\partial G_j}{\partial q_n}\eta_n + \frac{\partial G_j}{\partial \dot{q}_1}\dot{\eta}_1 + \cdots + \frac{\partial G_j}{\partial \dot{q}_n}\dot{\eta}_n = 0 \qquad j = 1, 2, \ldots, m$$

We now multiply each of the above m equations by an *arbitrary time function* $\lambda_j(t)$, which we may call a *Lagrange multiplier function.* Adding the resulting equations we have

$$\sum_{j=1}^{m} \lambda_j(t) \left(\frac{\partial G_j}{\partial q_1} \eta_1 + \cdots + \frac{\partial G_j}{\partial q_n} \eta_n + \frac{\partial G_j}{\partial \dot{q}_1} \dot{\eta}_1 + \cdots + \frac{\partial G_j}{\partial \dot{q}_n} \dot{\eta} \right) = 0$$

Now integrate the above sum from t_1 and t_2 and add the results to Eq. (2.55). We then have

$$\int_{t_1}^{t_2} \left[\frac{\partial F}{\partial q_1} \eta_1 + \cdots + \frac{\partial F}{\partial q_n} \eta_n + \frac{\partial F}{\partial \dot{q}_1} \dot{\eta}_1 + \cdots + \frac{\partial F}{\partial \dot{q}_n} \dot{\eta}_n \right.$$

$$\left. + \sum_{j=1}^{m} \lambda_j \left(\frac{\partial G_j}{\partial q_1} \eta_1 + \cdots + \frac{\partial G_j}{\partial q_n} \eta_n + \frac{\partial G_j}{\partial \dot{q}_1} \dot{\eta}_1 + \cdots + \frac{\partial G_j}{\partial \dot{q}_n} \dot{\eta}_n \right) \right] dt = 0$$

Integrating by parts the terms with $\dot{\eta}_i$ and regrouping the results, we then have

$$\int_{t_1}^{t_2} \left(\left\{ \frac{\partial F}{\partial q_1} - \frac{d}{dt}\left(\frac{\partial F}{\partial \dot{q}_1} \right) + \sum_{j=1}^{m} \left[\lambda_j \frac{\partial G_j}{\partial q_1} - \frac{d}{dt}\left(\lambda_j \frac{\partial G_j}{\partial \dot{q}_1} \right) \right] \right\} \eta_1 \right.$$

$$+ \cdots\cdots\cdots\cdots\cdots\cdots\cdots\cdots$$

$$+ \cdots\cdots\cdots\cdots\cdots\cdots\cdots\cdots$$

$$\left. + \left\{ \frac{\partial F}{\partial q_n} - \frac{d}{dt}\left(\frac{\partial F}{\partial \dot{q}_n} \right) + \sum_{j=1}^{m} \left[\lambda_j \frac{\partial G_j}{\partial q_n} - \frac{d}{dt}\left(\lambda_j \frac{\partial G_j}{\partial \dot{q}_n} \right) \right] \right\} \eta_n \right) dt = 0 \tag{2.56}$$

Now introduce F^*, defined as

$$F^* = F + \sum_{i=1}^{m} \lambda_j(t) G_j \tag{2.57}$$

We can then rewrite Eq. (2.56) as follows:

$$\int_{t_1}^{t_2} \left[\left(\frac{\partial F^*}{\partial q_1} - \frac{d}{dt} \frac{\partial F^*}{\partial \dot{q}_1} \right) \eta_1 + \cdots + \left(\frac{\partial F^*}{\partial q_n} - \frac{d}{dt} \frac{\partial F^*}{\partial \dot{q}_n} \right) \eta_n \right] dt = 0$$

Now the η's are not independent [they are related through m equations (2.54)] and so we cannot set the coefficients of the bracketed expressions equal to zero separately. However, we can say that $(n - m)$ of the η's are independent. For the remaining m η's we now assume that the time functions $\lambda_j\,(t)$ are so chosen that the coefficients of the η's are zero. Then, since we are left with only independent η's, we can take the remaining coefficients equal to zero. In this way we conclude that

$$\frac{\partial F^*}{\partial q_i} - \frac{d}{dt}\frac{\partial F^*}{\partial \dot{q}_i} = 0 \qquad i = 1, 2, \ldots, n$$

We thus arrive at the Euler-Lagrange equations once again. However, we have now m unknown time functions λ_j to be determined with the aid of the original m constraining equations.[†]

We shall have ample opportunity to use the formulations of this section in the following chapter when we consider the Reissner functional.

2.9 A NOTE ON BOUNDARY CONDITIONS

In previous efforts to extremize I we specified the endpoints (x_1, y_1) and (x_2, y_2) through which the extremal function had to proceed. Thus, in Fig. 2.2 we asked for the function $y(x)$ going through (x_1, y_1) and (x_2, y_2) to make it extremize I relative to neighboring varied paths *also* going through these endpoints. The boundary conditions specifying y at x_1 and x_2 in this case are called the *kinematic* or *rigid boundary conditions* of the problem.

We now pose a different query. Suppose only x_1 and x_2 are given, as shown in Fig. 2.7, and we ask for the function $y(x)$ that extremizes the functional $\int_{x_1}^{x_2} F(x, y, y')\,dx$ between these limits. Thus, we *do not specify* y at x_1 and x_2 for the extremal function.

As before, we denote the extremal function in the discussion as $y(x)$ and we have shown it so labeled at some position in Fig. 2.7. A system of nearby admissible "paths" $\tilde{y}(x)$ has also been shown. Some of these paths go through the endpoints of the extremizing path while others do not. We shall extremize I with respect to such a family of admissible functions to obtain certain necessary requirements for $y(x)$ to

[†]It may be of interest to note that we used a Lagrange multiplier *constant* in conjunction with *integral constraints* because the terms $(\partial I/\partial \epsilon_1)_{\substack{\epsilon_1=0\\ \epsilon_2=0}}$, etc., in Eq. (2.39) are constant and a properly chosen constant λ is sufficient to render one of the bracketed expressions equal to zero. Here, the bracketed expressions forming coefficients of η_i in Eq. (2.56) are composed of *functions*, and the Lagrange multiplier must also be a function to do the assigned job of setting certain of the coefficients of η_i equal to zero.

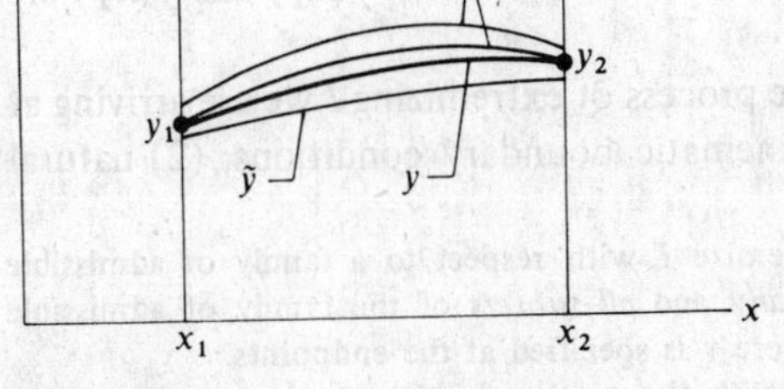

FIGURE 2.7
Varied paths need not have same end values of y.

maintain the role of the extremal function. Thus with the δ operator we arrive at the following necessary condition, using the same steps as in earlier discussions and noting that δy need no longer always be zero at the endpoints:

$$\delta^{(1)} I = 0 = \int_{x_1}^{x_2} \left[\frac{\partial F}{\partial y} - \frac{d}{dx}\left(\frac{\partial F}{\partial y'}\right) \right] \delta y \, dx + \frac{\partial F}{\partial y'} \delta y \bigg|_{x=x_2} - \frac{\partial F}{\partial y'} \delta y \bigg|_{x=x_1} \tag{2.58}$$

There is here a subset of admissible functions that goes through the endpoints (x_1, y_1) and (x_2, y_2) of the designated extremal function $y(x)$. For such a subset of admissible functions, $\delta y_1 = \delta y_2 = 0$ and we conclude that a necessary condition for the extremum of this subset of admissible functions is†

$$\int_{x_1}^{x_2} \left[\frac{\partial F}{\partial y} - \frac{d}{dx}\left(\frac{\partial F}{\partial y'}\right) \right] \delta y \, dx = 0$$

Following the familiar procedures of previous computations we then readily arrive at the Euler-Lagrange equations as a necessary condition for $y(x)$ to be an extreme:

$$\frac{\partial F}{\partial y} - \frac{d}{dx}\left(\frac{\partial F}{\partial y'}\right) = 0$$

There are now, however, additional necessary requirements for extremizing I if δy is not zero at the endpoints of the extremal function, so that a broader class of admissible functions is considered. Accordingly, from Eq. (2.58) we conclude that for such circumstances we require

$$\frac{\partial F}{\partial y'}\bigg|_{x=x_1} = 0 \tag{2.59a}$$

$$\frac{\partial F}{\partial y'}\bigg|_{x=x_2} = 0 \tag{2.59b}$$

The conditions (2.59) are termed the *natural boundary conditions*. They are the boundary conditions that must be prescribed if both the values $y(x_1)$ and $y(x_2)$ are not specified.‡

It is vital to point out now that in the process of extremizing I we are arriving at a differential equation and a set of (1) kinematic boundary conditions, (2) natural

†We remind you that if $y(x)$ is to extremize I with respect to a family of admissible functions, it must extremize I with respect to *any* and *all subsets* of the family of admissible functions. We are here considering the subset where y is specified at the endpoints.

‡In problems of solid mechanics dealing with the total potential energy we will see that the kinematic boundary conditions involve displacement conditions of the boundary, while natural boundary conditions involve force conditions at the boundary.

boundary conditions, or (3) a combination of parts of the kinematic boundary conditions and natural boundary conditions, which will facilitate extremizing the functional. We will see in Section 3.12 that for a broad class of functionals the resulting boundary-value problem [i.e., the Euler-Lagrange equation plus the boundary conditions (1), (2), or (3) above] has the properties of being *self-adjoint* and *positive definite.*† Furthermore, from the theory of partial differential equations, it is known that boundary-value problems having these properties are "properly posed." This means that a *unique solution exists*. Thus the variational process leads to a properly posed boundary-value problem. Is this an important factor? Indeed it is! We cannot take a differential equation and assign boundary conditions haphazardly or arbitrarily. Doing this will produce an improperly posed boundary-value problem, which will be meaningless.

We will see in Chapter 6, for example, that all the proper boundary conditions for plates are not physically obvious. That is, the proper boundary conditions for the Sophie Germain plate equation must be formulated through the variational process we are now developing.

This is only one of many valuable results stemming from the calculus of variations.

In more general functionals I set forth earlier and those to be considered in following sections, we find natural boundary conditions in much the same way as set forth in this section. In essence, we proceed with the variation process *without* requiring the η functions (or, by the same token, the variations δy) to be zero at the boundaries. Rather, we set equal to zero all expressions established on the boundary by the integration-by-parts procedures. The resulting conditions are the natural boundary conditions.

Note that for all the various possible boundary conditions, we work in any particular case with the same differential equation. However, the extremal function will eventually depend on the particular combination of boundary conditions we choose to employ. Usually certain kinematic boundary conditions are imposed by the constraints present in a particular problem. We must use these boundary conditions or else our extremal functions will not correspond to the problem at hand. The remaining boundary conditions are then natural ones that satisfy the requirements for the extremizing process. We shall illustrate these remarks in Example 2.5 after a discussion of higher-order derivatives.

2.10 FUNCTIONALS INVOLVING HIGHER-ORDER DERIVATIVES

We have thus far considered only first-order derivatives in the functionals. At this time we extend the work by finding extremal functions $y(x)$ for functionals having higher-order derivatives. Accordingly, we shall consider the following functional:

$$I = \int_{x_1}^{x_2} F(x, y, y', y'', y''')\, dx \tag{2.60}$$

†We shall define these properties in the next chapter.

The cases for lower- or higher-order derivatives other than y''' are easily attainable from the procedure we shall follow. Using the familiar one-parameter family of admissible functions for extremizing I, we can then say

$$\left(\frac{d\tilde{I}}{d\epsilon}\right)_{\epsilon=0} = \left[\frac{d}{d\epsilon}\int_{x_1}^{x_2} F(x, \tilde{y}, \tilde{y}', \tilde{y}'', \tilde{y}''')\,dx\right]_{\epsilon=0} = 0$$

This becomes

$$\left[\int_{x_1}^{x_2}\left(\frac{\partial F}{\partial \tilde{y}}\eta + \frac{\partial F}{\partial \tilde{y}'}\eta' + \frac{\partial F}{\partial \tilde{y}''}\eta'' + \frac{\partial F}{\partial \tilde{y}'''}\eta'''\right)dx\right]_{\epsilon=0} = 0$$

We may drop the subscript $\epsilon = 0$ along with the tildes as follows:

$$\int_{x_1}^{x_2}\left(\frac{\partial F}{\partial y}\eta + \frac{\partial F}{\partial y'}\eta' + \frac{\partial F}{\partial y''}\eta'' + \frac{\partial F}{\partial y'''}\eta'''\right)dx = 0$$

We now carry out the following series of integration by parts:

$$\int_{x_1}^{x_2}\frac{\partial F}{\partial y'}\eta'\,dx = \frac{\partial F}{\partial y'}\eta\Bigg|_{x_1}^{x_2} - \int_{x_1}^{x_2}\frac{d}{dx}\left(\frac{\partial F}{\partial y'}\right)\eta\,dx$$

$$\int_{x_1}^{x_2}\frac{\partial F}{\partial y''}\eta''\,dx = \frac{\partial F}{\partial y''}\eta'\Bigg|_{x_1}^{x_2} - \int_{x_1}^{x_2}\frac{d}{dx}\left(\frac{\partial F}{\partial y''}\right)\eta'\,dx$$

$$= \frac{\partial F}{\partial y''}\eta'\Bigg|_{x_1}^{x_2} - \frac{d}{dx}\left(\frac{\partial F}{\partial y''}\right)\eta\Bigg|_{x_1}^{x_2} + \int_{x_1}^{x_2}\frac{d^2}{dx^2}\left(\frac{\partial F}{\partial y''}\right)\eta\,dx$$

$$\int_{x_1}^{x_2}\frac{\partial F}{\partial y'''}\eta'''\,dx = \frac{\partial F}{\partial y'''}\eta''\Bigg|_{x_1}^{x_2} - \int_{x_1}^{x_2}\frac{d}{dx}\left(\frac{\partial F}{\partial y'''}\right)\eta''\,dx$$

$$= \frac{\partial F}{\partial y'''}\eta''\Bigg|_{x_1}^{x_2} - \frac{d}{dx}\left(\frac{\partial F}{\partial y'''}\right)\eta'\Bigg|_{x_1}^{x_2} + \int_{x_1}^{x_2}\frac{d^2}{dx^2}\left(\frac{\partial F}{\partial y'''}\right)\eta'\,dx$$

$$= \frac{\partial F}{\partial y'''}\eta''\Bigg|_{x_1}^{x_2} - \frac{d}{dx}\left(\frac{\partial F}{\partial y'''}\right)\eta'\Bigg|_{x_1}^{x_2} + \frac{d^2}{dx^2}\left(\frac{\partial F}{\partial y'''}\right)\eta\Bigg|_{x_1}^{x_2}$$

$$- \int\frac{d^3}{dx^3}\left(\frac{\partial F}{\partial y'''}\right)\eta\,dx$$

Now, combining the results, we get

$$-\int_{x_1}^{x_2}\left[\frac{d^3}{dx^3}\left(\frac{\partial F}{\partial y'''}\right)-\frac{d^2}{dx^2}\left(\frac{\partial F}{\partial y''}\right)+\frac{d}{dx}\left(\frac{\partial F}{\partial y'}\right)-\frac{\partial F}{\partial y}\right]\eta\,dx$$

$$+\frac{\partial F}{\partial y'''}\eta''\Bigg|_{x_1}^{x_2}-\left(\frac{d}{dx}\frac{\partial F}{\partial y'''}-\frac{\partial F}{\partial y''}\right)\eta'\Bigg|_{x_1}^{x_2}$$

$$+\left[\frac{d^2}{dx^2}\left(\frac{\partial F}{\partial y'''}\right)-\frac{d}{dx}\left(\frac{\partial F}{\partial y''}\right)+\frac{\partial F}{\partial y'}\right]\eta\Bigg|_{x_1}^{x_2}=0 \tag{2.61}$$

Since the functions $\tilde{y}$ having the properties such that $\eta(x_1)=\eta'(x_1)=\eta''(x_1)=\eta(x_2)=\eta'(x_2)=\eta''(x_2)=0$ form a subset of the admissible functions no matter what the boundary conditions may be, it is clear that a necessary condition for an extremum is

$$\int_{x_1}^{x_2}\left[\frac{d^3}{dx^3}\left(\frac{\partial F}{\partial y'''}\right)-\frac{d^2}{dx^2}\left(\frac{\partial F}{\partial y''}\right)+\frac{d}{dx}\left(\frac{\partial F}{\partial y'}\right)-\frac{\partial F}{\partial y}\right]\eta\,dx=0$$

And we can then arrive at the following Euler-Lagrange equation:

$$\frac{d^3}{dx^3}\left(\frac{\partial F}{\partial y'''}\right)-\frac{d^2}{dx^2}\left(\frac{\partial F}{\partial y''}\right)+\frac{d}{dx}\left(\frac{\partial F}{\partial y'}\right)-\frac{\partial F}{\partial y}=0 \tag{2.62}$$

If, specifically, the conditions $y(x_1)$, $y(x_2)$, $y'(x_1)$, $y'(x_2)$, $y''(x_1)$, and $y''(x_2)$ are *specified* (these are the kinematic boundary conditions of this problem), the admissible functions must have these prescribed values and these prescribed derivatives at the endpoints. Then clearly we have for this case:

$$\eta(x_1)=\eta(x_2)=\eta'(x_1)=\eta'(x_2)=\eta''(x_1)=\eta''(x_2)=0$$

Thus we see that giving the kinematic end conditions and satisfying the Euler-Lagrange equations permits the resulting $y(x)$ to be an extremal. If the kinematic conditions are not specified then we may satisfy the *natural boundary conditions* for this problem. Summarizing, we may use either of the following overall sets of requirements [see Eq. (2.61)] at the endpoints:

Kinematic		*Natural*	
y'' specified	or	$\dfrac{\partial F}{\partial y'''}=0$	(2.63*a*)
y' specified	or	$\dfrac{d}{dx}\left(\dfrac{\partial F}{\partial y'''}\right)-\dfrac{\partial F}{\partial y''}=0$	(2.63*b*)

Kinematic		*Natural*	
y specified	or	$\dfrac{d^2}{dx^2}\left(\dfrac{\partial F}{\partial y'''}\right) - \dfrac{d}{dx}\left(\dfrac{\partial F}{\partial y''}\right) + \dfrac{\partial F}{\partial y'} = 0$	(2.63*c*)

Clearly, combinations of kinematic and natural boundary conditions are also permissible.

It should be apparent that by disregarding the terms involving y''' we may use the results for a functional whose highest-order derivative is second-order. And, by disregarding both terms with y''' and those with y'', we get back to the familiar expressions used earlier where y' was the highest-order term in the functional. Furthermore, the pattern of growth from first-order derivatives on up is clearly established by the results, so that one can readily extrapolate the formulations for orders higher than three. Finally, we may directly extend the results of this problem to include functionals I with more than one function y. We simply get equations of the form (2.62) for each function. Thus, for t as the independent variable and q_1, q_2, $\ldots$, q_n as the functions in I, we have for derivatives up to order three for all functions:

$$\frac{d^3}{dt^3}\left(\frac{\partial F}{\partial q_1'''}\right) - \frac{d^2}{dt^2}\left(\frac{\partial F}{\partial q_1''}\right) + \frac{d}{dt}\left(\frac{\partial F}{\partial q_1'}\right) - \frac{\partial F}{\partial q_1} = 0$$

$$\vdots$$

$$\frac{d^3}{dt^3}\left(\frac{\partial F}{\partial q_n'''}\right) - \frac{d^2}{dt^2}\left(\frac{\partial F}{\partial q_n''}\right) + \frac{d}{dt}\left(\frac{\partial F}{\partial q_n'}\right) - \frac{\partial F}{\partial q_n} = 0 \qquad (2.64)$$

Similarly, for each q we have a set of conditions corresponding to Eqs. (2.63).

EXAMPLE 2.5 We now consider the deflection w of the centerline of a simply supported beam having a rectangular cross section and loaded in the plane of symmetry of the cross section (see Fig. 2.8). In Chapter 4 we will show through the principal of minimum total potential energy that the following functional is to be extremized to ensure equilibrium:

$$I = \int_0^L \left[\frac{EI}{2}\left(\frac{d^2 w}{dx^2}\right)^2 - qw\right] dx \qquad (a)$$

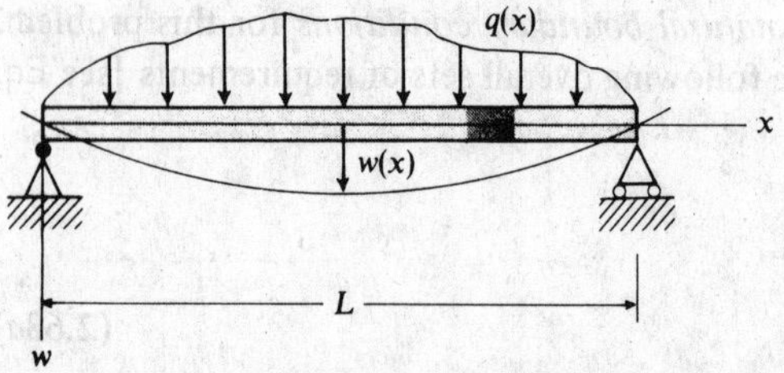

FIGURE 2.8
Simply supported beam showing deflection curve.

where the I next to E is the second moment of the cross section about the horizontal centroidal axis, and q is the transverse loading function.

The Euler-Lagrange equation for this case may be formed from Eq. (2.62) as follows:

$$-\frac{d^2}{dx^2}\left(\frac{\partial F}{\partial w''}\right)+\frac{d}{dx}\left(\frac{\partial F}{\partial w'}\right)-\frac{\partial F}{\partial w}=0 \qquad (b)$$

where

$$F=\frac{EI}{2}(w'')^2-qw \qquad (c)$$

We then get for the Euler-Lagrange equation:

$$-\frac{d^2}{dx^2}(EIw'')+\frac{d}{dx}(0)+q=0$$

Therefore

$$EIw^{\mathrm{IV}}=q \qquad (d)$$

This is the well-known beam equation from strength of materials. The possible boundary conditions needed for extremization of functional (*a*) can be readily deduced from Eqs. (2.63). Thus at the endpoints:

$$w' \text{ specified} \quad \text{or} \quad \frac{\partial F}{\partial w''}=0 \qquad (e)$$

$$w \text{ specified} \quad \text{or} \quad -\frac{d}{dx}\left(\frac{\partial F}{\partial w''}\right)+\frac{\partial F}{\partial w'}=0 \qquad (f)$$

It is clear from Fig. 2.8 that we *must* employ the kinematic boundary condition $w=0$ at the ends. In order then to carry out the extremization process properly we must require additionally that

$$\frac{\partial F}{\partial w''}=0 \quad \text{at ends} \qquad (g)$$

Substituting for F, we get

$$w''(0)=w''(L)=0 \qquad (h)$$

You may recall that this indicates that the bending moments are zero at the ends—a condition needed for the frictionless pins there. We thus see here that the natural boundary conditions have to do with forces in structural problems.

In the following section we consider the case where we have higher-order derivatives involving more than one independent variable.

2.11 A FURTHER EXTENSION

As we shall demonstrate later, the total potential energy for a plate is generally expressible as a double integral of the displacement function w and partial derivatives of w in the following form:

$$I = \iint_S F\left(x, y, w, \frac{\partial w}{\partial x}, \frac{\partial w}{\partial y}, \frac{\partial^2 w}{\partial x^2}, \frac{\partial^2 w}{\partial x\, \partial y}, \frac{\partial^2 w}{\partial y\, \partial x}, \frac{\partial^2 w}{\partial y^2}\right) dx\, dy \tag{2.65}$$

where S is the area over which the integration is carried out. Using the notation $\partial w/\partial x = w_x$, $\partial^2 w/\partial x^2 = w_{xx}$, etc., the above functional may be given as follows:

$$I = \iint_S F(x, y, w, w_x, w_y, w_{xy}, w_{yx}, w_{yy}, w_{xx})\, dx\, dy \tag{2.66}$$

We have here a functional with *more than one independent variable*. The procedure for finding the extremizing function, which we consider now to be represented as $w(x, y)$, is very similar to what we have done in the past. We use as admissible functions $\tilde{w}(x, y)$ a one-parameter family of functions defined as follows:

$$\tilde{w}(x, y) = w(x, y) + \epsilon\eta(x, y) \tag{2.67}$$

Accordingly, we have for the extremization process:

$$\left(\frac{d\tilde{I}}{d\epsilon}\right)_{\epsilon=0} = 0 = \left[\iint_S \left(\frac{\partial F}{\partial \tilde{w}}\eta + \frac{\partial F}{\partial \tilde{w}_x}\eta_x + \frac{\partial F}{\partial \tilde{w}_y}\eta_y + \frac{\partial F}{\partial \tilde{w}_{xx}}\eta_{xx} + \frac{\partial F}{\partial \tilde{w}_{xy}}\eta_{xy} + \frac{\partial F}{\partial \tilde{w}_{yx}}\eta_{yx} + \frac{\partial F}{\partial \tilde{w}_{yy}}\eta_{yy}\right) dx\, dy\right]_{\epsilon=0}$$

We may remove the tildes and the subscript notation $\epsilon = 0$ simultaneously to get the following statement:

$$\iint_S \left(\frac{\partial F}{\partial w}\eta + \frac{\partial F}{\partial w_x}\eta_x + \frac{\partial F}{\partial w_y}\eta_y + \frac{\partial F}{\partial w_{xx}}\eta_{xx} + \frac{\partial F}{\partial w_{xy}}\eta_{xy} + \frac{\partial F}{\partial w_{yx}}\eta_{yx} + \frac{\partial F}{\partial w_{yy}}\eta_{yy}\right) dx\, dy = 0 \tag{2.68}$$

The familiar integration-by-parts process can be carried out here by employing Green's theorem (see Appendix I) in two dimensions, which is given as follows:

$$\iint_S G\frac{\partial \eta}{\partial x}\, dx\, dy = -\iint_S \frac{\partial G}{\partial x}\eta\, dx\, dy + \oint \eta G \cos(n, x)\, dl \tag{2.69}$$

Recall that n is the normal to the bounding curve C of the domain S. We apply the above theorem once to the second and third terms in the integrand of Eq. (2.68) and twice to the last four terms to get

$$\iint_S \left[\frac{\partial F}{\partial w} - \frac{\partial}{\partial x}\left(\frac{\partial F}{\partial w_x}\right) - \frac{\partial}{\partial y}\left(\frac{\partial F}{\partial w_y}\right) + \frac{\partial^2}{\partial x^2}\left(\frac{\partial F}{\partial w_{xx}}\right) + \frac{\partial^2}{\partial x\,\partial y}\left(\frac{\partial F}{\partial w_{xy}}\right)\right.$$
$$\left. + \frac{\partial^2}{\partial y\,\partial x}\left(\frac{\partial F}{\partial w_{yx}}\right) + \frac{\partial^2}{\partial y^2}\left(\frac{\partial F}{\partial w_{yy}}\right)\right] \eta\, dx\, dy + \text{system of line integrals} = 0$$

We can conclude, as in earlier cases, that a necessary requirement is that

$$\frac{\partial F}{\partial w} - \frac{\partial}{\partial x}\left(\frac{\partial F}{\partial w_x}\right) - \frac{\partial}{\partial y}\left(\frac{\partial F}{\partial w_y}\right) + \frac{\partial^2}{\partial x^2}\left(\frac{\partial F}{\partial w_{xx}}\right)$$
$$+ \frac{\partial^2}{\partial y\,\partial x}\left(\frac{\partial F}{\partial w_{yx}}\right) + \frac{\partial^2}{\partial x\,\partial y}\left(\frac{\partial F}{\partial w_{xy}}\right) + \frac{\partial^2}{\partial y^2}\left(\frac{\partial F}{\partial w_{yy}}\right) = 0 \qquad (2.70)$$

This is the Euler-Lagrange equation for this case. We may deduce the natural boundary conditions from the line integrals. However, we shall exemplify this step later when we study the particular problem of classical plate theory, since the results require considerable discussion.

If we have several functions in F, we find that additional equations of the form given by Eq. (2.70) then constitute additional Euler-Lagrange equations for the problem.

2.12 CLOSURE

We have set forth a brief introduction in this chapter into elements of the calculus of variations. In short, we have considered certain classes of functionals with the view of establishing necessary conditions for finding functions that extremize the functionals. The results were ordinary or partial differential equations for the extremal functions (the Euler-Lagrange equations) as well as the establishment of the dualities of kinematic (or rigid) and natural boundary conditions. We will see later that the natural boundary conditions are not easily established without the use of the variational approach. And since these conditions are often important for properly posing particular boundary-value problems, we can then conclude that the natural boundary conditions are valuable products of the variational approach.

We also note that by the nature of the assumption leading to Eq. (2.9) we have been concerned with varied paths $\tilde{y}(x)$ which not only are close to the extremal $y(x)$ but also have derivatives $\tilde{y}'(x)$ close to $y'(x)$. Such variations are called *weak variations*. There are, however, variations that do not require closeness of the derivatives of the varied paths to that of the extremal function (see Fig. 2.9, showing such a varied path). When this is the case we say that we have *strong variations*. A more complex theory is then needed beyond the level of this text. The reader is referred to more advanced books such as those given in the footnote on page 67.

In the remainder of the text we shall employ the formulations presented in this

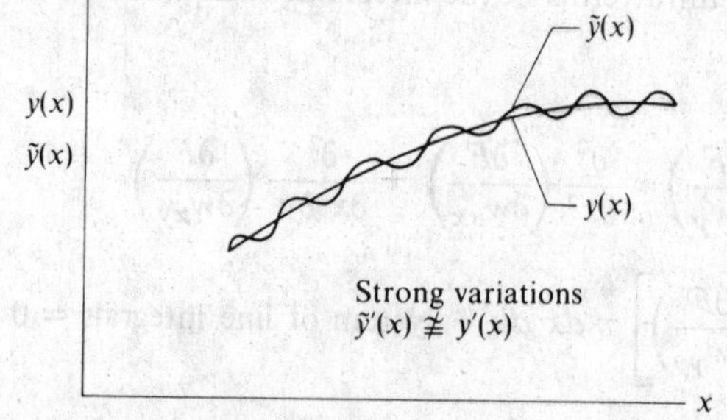

FIGURE 2.9
Strong variations.

chapter for Euler-Lagrange equations and boundary conditions when there is only a single independent variable present. The equations, you should recall, are then ordinary differential equations and the boundary conditions are prescriptions at the endpoints. For more than one independent variable we shall work out the Euler-Lagrange equations (now partial differential equations) as well as the boundary conditions (now line or surface integrals) from first principles, using the formulation $\delta^{(1)}(I) = 0$ or $(\partial\tilde{I}/\partial\epsilon)_{\epsilon=0} = 0$ as the basis of evaluations.

It is to be pointed out that this chapter does not terminate the development of the calculus of variations. In the next chapter we shall investigate the process of going from a boundary-value problem to a functional for which the differential equation corresponds to the Euler-Lagrange equations. This is the inverse of the process set forth in this chapter and leads to the useful quadratic functional. In Chapter 7 we shall set forth two basic theorems, namely the maximum theorem and the "minimax" theorem, which form the basis for key approximation procedures used for estimating eigenvalues needed in the study of vibrations and stability of elastic bodies. In addition, in Chapter 8 we shall examine the second variation. Clearly there will be a continued development of the calculus of variations as we proceed.

Our immediate task in Chapter 3 is to set forth functionals whose Euler-Lagrange equation and boundary conditions form important boundary-value problems in solid mechanics. With these functionals we will carefully set forth the so-called *energy methods* of solid mechanics. We shall illustrate the use of these methods for problems of simple trusses; we thus launch the study of structural mechanics as a by-product. In addition, we shall introduce certain valuable approximation techniques associated with the variational process that form the cornerstone of much such work throughout the text.

It may be apparent from these remarks that Chapter 3 is one of the key chapters in this text.

REFERENCES

Courant, R., "Differential and Integral Calculus," Interscience, New York, 1947.
Fox, C., "An Introduction to the Calculus of Variations," Oxford Univ. Press, New York, 1954.
Gelfand, I. M., and Fomin, S. V., "Calculus of Variations," Prentice-Hall, Englewood Cliffs, N.J., 1963.

READING

Courant, R., and Hilbert, D.: "Methods of Mathematical Physics," vol. 1, Interscience, New York, 1953.
Elsgolc, L. E.: "Calculus of Variations," Addison-Wesley, Reading, Mass., 1961.
Fox, C.: "An Introduction to the Calculus of Variations," Oxford Univ. Press, New York, 1954.
Gelfand, I. M., and Fomin, S. V.: "Calculus of Variations," Prentice-Hall, Englewood Cliffs, N.J., 1963.
Schechter, R. S.: "The Variational Method in Engineering," McGraw-Hill, New York, 1967.
Smirnov, V. I.: "Integral Equations and Partial Differential Equations," vol. 4, chap. II. Pergamon, New York, 1964.
Weinstock, R.: "Calculus of Variations," McGraw-Hill, New York, 1952.

PROBLEMS

2.1 Given the functional:

$$I = \int_{x_1}^{x_2} [3x^2 + 2x(y')^2 + 10xy]\, dx$$

What is the Euler-Lagrange equation?

2.2 What is the first variation of F in the preceding problem? What is the first variation of I in this problem?

2.3 What is the Euler-Lagrange equation for the functional

$$I = \int_{x_1}^{x_2} (3x^2 y + 2xy' + 10)\, dx$$

2.4 Justify the identity given at the outset of case (c) in Section 2.5.

2.5 Consider the functional:

$$I = \int_{x_1}^{x_2} [3(y')^2 + 4x]\, dx$$

What is the extremal function? Take $y = 0$ at $x_1 = 0$ and $y = 2$ at $x_2 = 3$.

2.6 Fermat's principle states that the path of light is such as to minimize the time passage from one point to another. What is the proper path of light in a plane medium from point (1) to point (2) (see Fig. 2.10), where the speed of light is given as (1) $v = Cy$ and (2) $v = C/y$?

2.7 Demonstrate that the shortest distance between two points in a plane is a straight line.

*2.8 Consider a body of revolution moving with speed V through rarefied gas outside the earth's atmosphere (Fig. 2.11). We wish to design the shape of the body, i.e., get $y(x)$, so as to minimize the drag. Assume that there is no friction on the body at contact with the molecules of the gas. Then one can readily show that the pressure on the body is given as

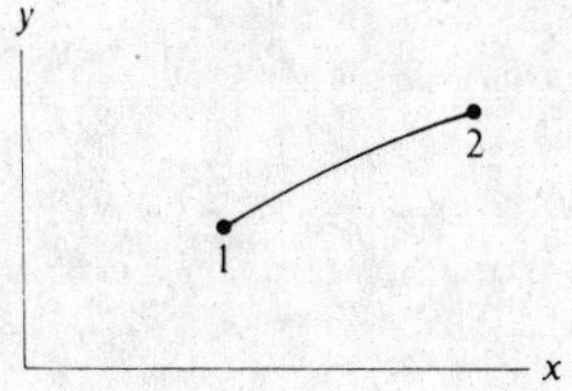

FIGURE 2.10

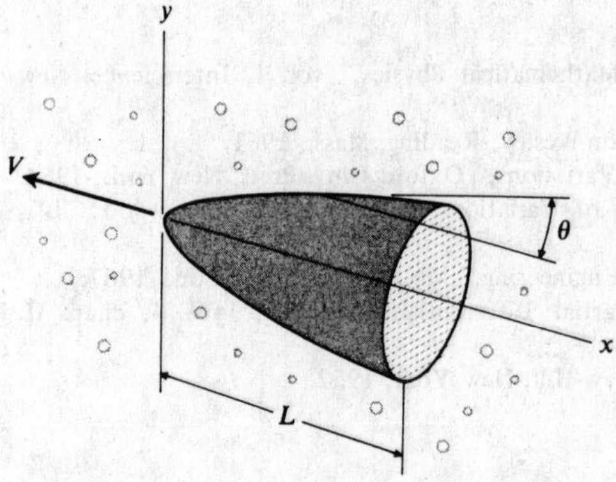

FIGURE 2.11

$$p = 2\rho V^2 \sin^2 \theta$$

(This means that the molecules are reflected *specularly*.) Show that the drag F_D for a length L of the body is given as

$$F_D = \int_0^L 4\pi\rho V^2 (\sin^3 \theta) y \left[1 + \left(\frac{dy}{dx}\right)^2\right]^{1/2} dx$$

Assume now that dy/dx is small so that

$$\sin \theta = \frac{dy/dx}{[1 + (dy/dx)^2]^{1/2}} \cong \frac{dy}{dx}$$

Show that

$$F_D = 4\pi\rho V^2 \int_0^L \left(\frac{dy}{dx}\right)^3 y \, dx$$

What is the Euler-Lagrange equation for $y(x)$ in order to minimize the drag?

2.9 Consider the problem where a curve is desired between points (x_1, y_1) and (x_2, y_2) (see Fig. 2.12) which, upon rotation about the x axis, produces a surface of revolution having a minimum area. First show that

$$y' = \frac{\sqrt{y^2 - C_1^2}}{C_1}$$

Then let $y = C_1 \cosh t$. Finally, show that

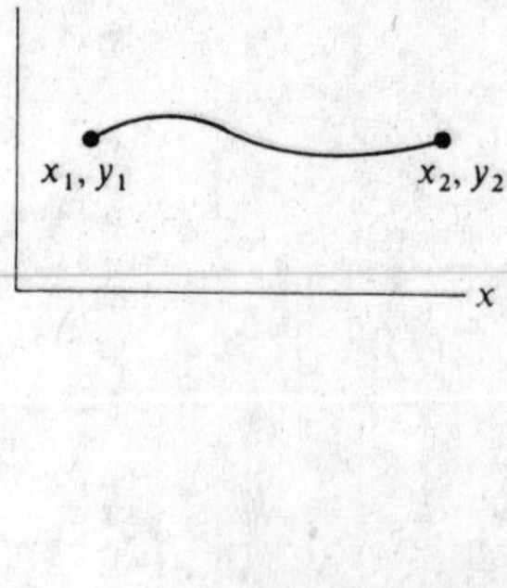

FIGURE 2.12

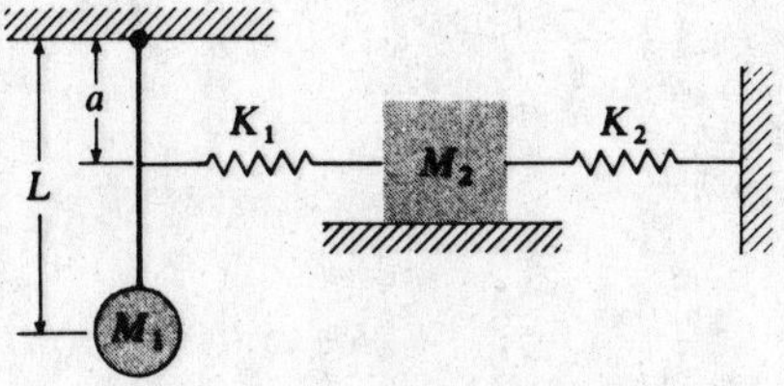

FIGURE 2.13

$$t = \frac{x + C_2}{C_1}$$

This is the parametric representation of a *catenary*.

2.10 Use the δ operator on the functional given in Example 2.1 to find the Euler-Lagrange equation for the brachistochrone problem [Eq. (*c*)].

2.11 Using Hamilton's principle, find the equations of motion for the system shown in Fig. 2.13 for small oscillations.

2.12 Using Hamilton's principle, show that

$$mR^2 \dot{\phi} \sin^2 \theta = \text{const}$$

for the *spherical pendulum* shown in Fig. 2.14. Gravity acts in the $-z$ direction. Neglect friction and the weight of the connecting link R. Use spherical coordinates θ and ϕ as shown in the diagram, and take the lowest position of m as the datum for potential energy.

2.13 Use the δ operator on the functional given in the text to find the Euler-Lagrange equations for the spring-mass system in Example 2.3.

2.14 Extend the results of Problem 2.7 to any two points in three-dimensional space.

2.15 Consider the case of a rope hanging between two points (x_1, y_1) and (x_2, y_2) as shown in Fig. 2.15. Find the Euler-Lagrange equation for the rope by minimizing the potential energy of the rope *for a given length L*. The rope is perfectly flexible. Show that

$$y = \frac{-\lambda}{wg} + \frac{C_1}{wg} \cosh \frac{x + C_2}{C_1} wg$$

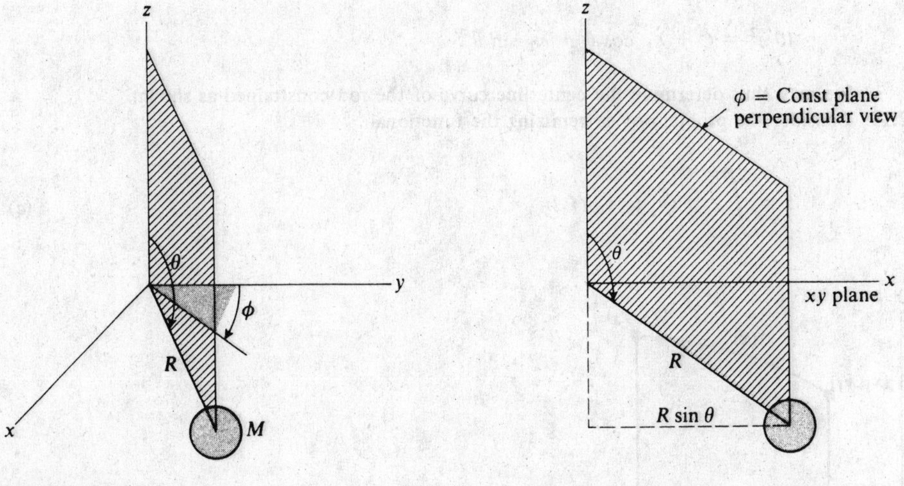

FIGURE 2.14

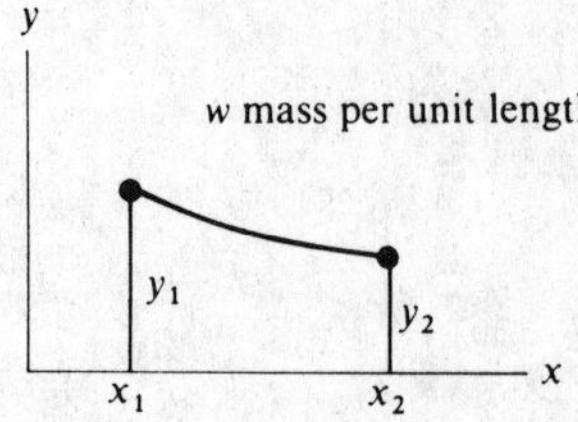

FIGURE 2.15

How are C_1, C_2, and λ obtained?

2.16 Show that the curve (Fig. 2.16) from position (x_1, y_1) to (x_2, y_2) which has *a given length L* and which maximizes the first moment of area of the crosshatched area about the x axis has a differential equation given as

$$\frac{y^2}{2} + \frac{\lambda}{\sqrt{1 + (y')^2}} = C_1$$

where λ is a Lagrange multiplier. Solve in the form of a quadrature but do not carry out the integration.

2.17 Consider two points A and B and a straight line joining these points. Of all the curves of length L connecting these points, what is the curve that maximizes the area between the curve and the line AB? What is the physical interpretation of the Lagrange multiplier λ?

2.18 Consider a uniform rod *fixed* at points (x_0, y_0) and (x_1, y_1) as shown in Fig. 2.17. The distance s measured along the rod will be considered a coordinate and the length $\int_A^B ds$ of the rod is L. Let $\theta(s)$ be the angle between the tangent to the curve and the x axis. Clearly $\theta(0) = \theta(L) = 0$ due to the constraints. We will later show that the strain energy is proportional to the integral.

$$\int_0^L \left(\frac{d\theta}{ds}\right)^2 ds$$

Express two isoperimetric constraints that link the length L and the positions (x_0, y_0) and (x_1, y_1). Extremize the above functional, using the aforementioned constraints, and show that for a first integral of Euler's equation we get

$$(\theta')^2 = C + \lambda_1 \cos\theta + \lambda_2 \sin\theta$$

We may thus determine the centerline curve of the rod constrained as shown.

2.19 Consider the problem of extremizing the functional

$$I = \int_{t_1}^{t_2} F(y, \dot{y}, \ddot{y}, t)\, dt \tag{a}$$

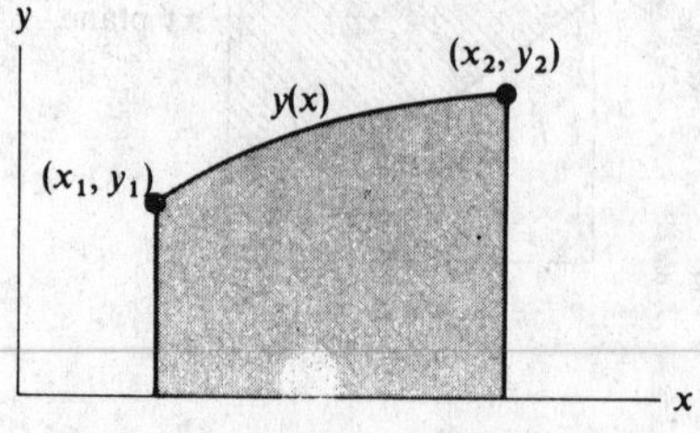

FIGURE 2.16

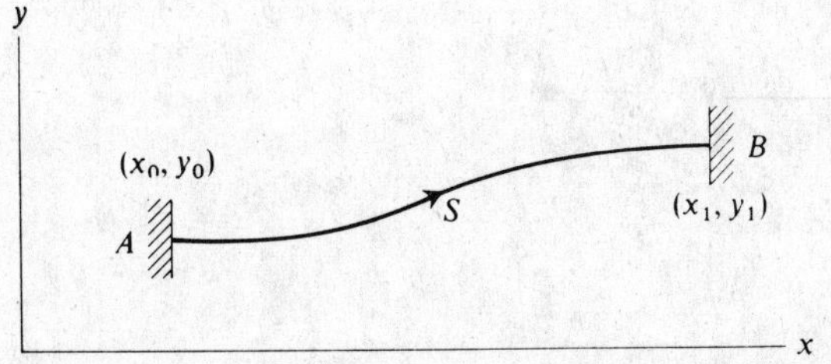

FIGURE 2.17

where y and $\dot{y}$ are specified at the endpoints. Reformulate the functional as one having only *first-order* derivatives and a *constraining equation*. Using Lagrange multiplier functions, express the Euler-Lagrange equations. Eliminate the Lagrange multiplier function and write the resulting Euler-Lagrange equation.

2.20 In the preceding problem, take

$$F = [y + 2(\ddot{y})^2 - ty]$$

What is the extremal function if $y = \dot{y} = 0$ at $t = 0$ and $y = \dot{y} = 1$ at $t = 1$. What is the Lagrange multiplier function?

2.21 Consider the brachistochrone problem in a *resistive medium* where the drag force is expressed as a given function of speed, $R(V)$, per unit mass of particle. Note there are now two functions y and V to be considered. What is the constraining equation between these functions? Show that using $\lambda(x)$, a Lagrange multiplier function, one of the Euler-Lagrange equation is

$$\frac{V\lambda'(x)}{\sqrt{1 + (y')^2}} = \frac{dH}{dV}$$

and an integral of the other is

$$\frac{Hy'}{\sqrt{1 + (y')^2}} = C + \lambda(x)g$$

where C is a constant of integration, g is the acceleration due to gravity, and H is given as

$$H = \frac{1}{V} + \lambda(x)R(V)$$

2.22 Derive the Euler-Lagrange equation and the natural boundary conditions for the following functional:

$$I = \int_{x_1}^{x_2} F\left(x, \frac{d^4 y}{dx^4}\right) dx$$

2.23 Do Problem 2.19 using the given functional F (i.e., with second-order derivative).

2.24 Derive the Euler-Lagrange equation and natural and geometric boundary conditions of beam theory using the δ operator on the functional given in the text (Example 2.5).

2.25 In our study of plates we will see (Chapter 6) that to find the equation of equilibrium for a circular plate (Fig. 2.18) loaded symmetrically by a force per unit area $q(r)$, and fixed at the edges, we must extremize the following functional:

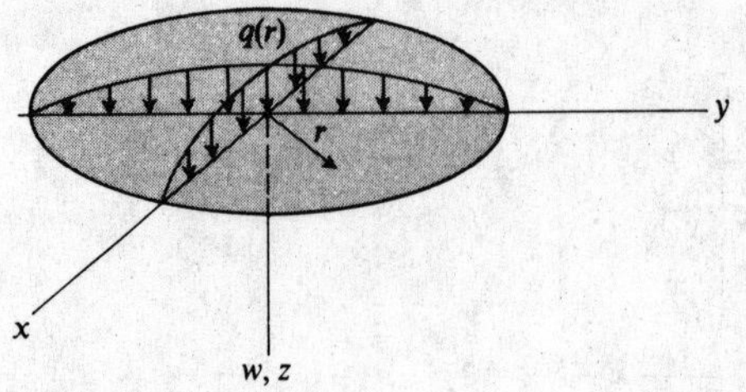

FIGURE 2.18

$$I = D\pi \int_0^a \left[r\left(\frac{d^2w}{dr^2}\right)^2 + \frac{1}{r}\left(\frac{dw}{dr}\right)^2 + 2\nu \frac{dw}{dr}\frac{d^2w}{dr^2} - \frac{2q}{D} rw \right] dr$$

where D and ν are elastic constants and w is the deflection (in the z direction) of the center plane of the plate. Show that the following is the proper governing differential equation for w:

$$r\frac{d^4w}{dr^4} + 2\frac{d^3w}{dr^3} - \frac{1}{r}\frac{d^2w}{dr^2} + \frac{1}{r^2}\frac{dw}{dr} = \frac{qr}{D}$$

3

VARIATIONAL PRINCIPLES OF ELASTICITY

3.1 INTRODUCTION

This chapter is subdivided into four parts. In Part A we shall set forth certain key principles that are related to or directly involve variational approaches. Specifically, we set forth the principles of virtual work and complementary virtual work, and from these respectively derive the principles of total potential energy and total complementary energy. Reissner's principle is also derived in this part of the chapter. Serving to illustrate certain of the aforementioned principles, we have employed a number of simple truss problems. These problems also serve the purpose of the beginning of our work in this text in structural mechanics. In Part B of the chapter we concentrate on formulations that are derivable from the principles of Part A and are particularly useful for the study of structural mechanics. These are the well-known Castigliano theorems. We continue the discussion of trusses begun in Part A, thereby presenting in this chapter a reasonably complete discussion of simple trusses. In the following chapter we extend the structural considerations begun here to beams.

Whereas in Part A of the chapter we set forth certain functionals and show that the corresponding Euler-Lagrange equations are of critical importance in solid mechanics, we take the opposite approach in Part C. Here, under certain prescribed conditions, we start with a differential equation and establish the functional for which the equation is the Euler-Lagrange equation. This reverse approach is much more difficult. All the results of this discussion apply to linear elasticity and have relevance to other fields of study. They set the stage for the very useful approximation techniques that we consider in Part D of this chapter, namely the Ritz method and the Galerkin method.

Part A
Key Variational Principles

3.2 VIRTUAL WORK

You will recall from particle mechanics that *virtual work* is defined as the work done on a particle by all the forces acting on the particle as this particle is given a small hypothetical displacement, a *virtual displacement*, which is consistent with the constraints present. The applied forces are kept constant during the virtual displacement. The concept of virtual work will now be extended to the case of a deformable body by specifying a continuous displacement field which falls within the category of small deformation and which does not violate any of the constraints of the problem under consideration. The *applied forces* are again *kept constant* during such a displacement. It should be clear that we can conveniently denote a virtual displacement by employing the variational operator δ. Thus, δu_i may represent a virtual displacement field from a given configuration. The constraints present are taken into account by imposing proper conditions on the variation. For example, consider the case of a beam clamped at both ends, as shown in Fig. 3.1. The constraints for the beam are zero displacement and zero slope at the ends, $x = 0$ and $x = L$. We can give a virtual displacement for points along the neutral axis of the beam (this we will see suffices to characterize important aspects of the deformation of the beam under usual circumstances) as follows:

$$\delta w = A\left(1 - \cos\frac{2\pi x}{L}\right)$$

Notice that both δw and $(d/dx)(\delta w)$ are zero at the ends. The virtual work δW_{virt} for the indicated applied loads then becomes

$$\delta W_{\text{virt}} = \int_0^L q(x)\left[A\left(1 - \cos\frac{2\pi x}{L}\right)\right] dx$$
$$+ F_1 A\left(1 - \cos\frac{2\pi x_1}{L}\right) + F_2 A\left(1 - \cos\frac{2\pi x_2}{L}\right)$$

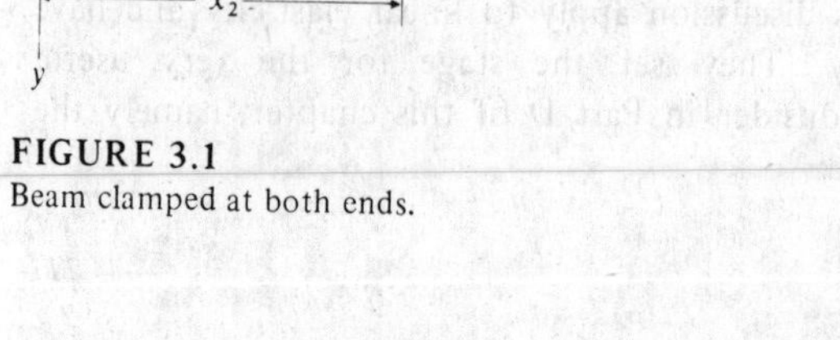

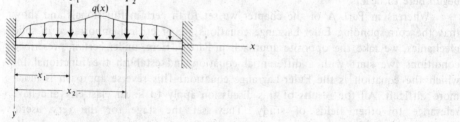

FIGURE 3.1
Beam clamped at both ends.

In a more general situation we would have as load possibilities a body force distribution B_i throughout the body as well as surface tractions $T_i^{(\nu)}$ over part of the boundary, S_1, of the body. Over the remaining part of the boundary, S_2, we have prescribed the displacement field u_i–in which case, to avoid violating the constraints we must be sure that $\delta u_i = 0$ on S_2. It should then be clear that the virtual work for such a general situation would be given as follows:

$$\delta W_{\text{virt}} = \iiint_V B_i \, \delta u_i \, dv + \oiint_S T_i^{(\nu)} \, \delta u_i \, dS \tag{3.1}$$

We note again that B_i and $T_i^{(\nu)}$ must not depend on δu_i in the computation of δW_{virt}. Also, since $\delta u_i = 0$ on S_2, we have extended the surface integral to cover the entire surface $S = S_1 + S_2$.

We now develop the principle of virtual work for a deformable body. Using successively Cauchy's formula [Eq. (1.5)] and Gauss' theorem, we rewrite the right side of the above equation as follows:

$$\begin{aligned}\iiint_V B_i \, \delta u_i \, dv + \oiint_S T_i^{(\nu)} \, \delta u_i \, dS &= \iiint_V B_i \, \delta u_i \, dv + \oiint_S \tau_{ij} \nu_j \, \delta u_i \, dS \\ &= \iiint_V B_i \, \delta u_i \, dv + \iiint_V (\tau_{ij} \, \delta u_i)_{,j} \, dv \\ &= \iiint_V (B_i + \tau_{ij,j}) \, \delta u_i \, dv + \iiint_V \tau_{ij} (\delta u_i)_{,j} \, dv\end{aligned} \tag{3.2}$$

We now introduce a *kinematically compatible strain field variation*† $\delta \epsilon_{ij}$, in the last integral as follows:

$$(\delta u_i)_{,j} = \delta(u_{i,j}) = \delta(\epsilon_{ij} + \omega_{ij}) = \delta \epsilon_{ij} + \delta \omega_{ij}$$

Because of the skew symmetry of the rotation tensor and the symmetry of the stress tensor, $\tau_{ij} \delta \omega_{ij} = 0$, and we can conclude that

$$\tau_{ij} (\delta u_i)_{,j} = \tau_{ij} \delta \epsilon_{ij}$$

Accordingly, Eq. (3.2) becomes

†The strain field variation is kinematically compatible because it is formed directly from the displacement field variation.

$$\iiint_V B_i\,\delta u_i\,dv + \oiint_S T_i^{(\nu)}\,\delta u_i\,dS = \iiint_V (\tau_{ij,j} + B_i)\,\delta u_i\,dv + \iiint_V \tau_{ij}\,\delta\epsilon_{ij}\,dv$$

We now impose the condition that we have static equilibrium. This means in the above equation that

1. The external loads B_i and $T_i^{(\nu)}$ are such that there is overall equilibrium for the body from the viewpoint of rigid-body mechanics. We say that B_i and $T_i^{(\nu)}$ are *statically compatible*. And
2. At any point in the body $\tau_{ij,j} + B_i = 0$, so that the first term on the right side of the above equation vanishes.

The resulting equation is given as follows:

$$\iiint_V B_i\,\delta u_i\,dv + \oiint_S T_i^{(\nu)}\,\delta u_i\,dS = \iiint_V \tau_{ij}\,\delta\epsilon_{ij}\,dv \tag{3.3}$$

This is the principle of virtual work for a deformable body. Another way of viewing this equation is to consider the left side as *external virtual work* and the right side as *internal virtual work*. We can then say that a necessary condition for *equilibrium* is that for *any kinematically compatible deformation* field $(\delta u_i, \delta\epsilon_{ij})$, the external virtual work, with statically compatible body forces and surface tractions, must equal the internal virtual work. (We will soon show that this is sufficient for equilibrium.) Another more useful interpretation of the principle of virtual work is as follows. The necessary requirements for equilibrium of a particular stress field τ_{ij} are that

1. B_i and $T_i^{(\nu)}$ are *statically compatible.*
2. The particular stress field τ_{ij} satisfies the virtual work equation for *any* kinematically compatible, admissible, deformation field.

This relation is in the form of an integral equation for unknowns τ_{ij}, assuming $T_i^{(\nu)}$ and B_i are known, and that δu_i and $\delta\epsilon_{ij}$ are chosen. Most significant is the fact that this mathematical relation between a deformation field and a stress field is *independent of any constitutive law* and applies to all materials within the limitations of small deformation.

We may use the principle of virtual work directly in structural problems when we can completely specify a virtual displacement field by giving the variation of a finite number of parameters. Thus for trusses we need only specify the virtual displacements of the joints; the virtual strain in the members is then readily determined by geometric considerations. Considering a *plane truss*, suppose we institute a virtual displacement δu_q on the qth pin in the x direction while keeping all other pins stationary. The method of virtual work then stipulates that

$$(P_q)_x\,\delta u_q = \iiint_V \tau_{ij}\,\delta\epsilon_{ij}\,dv \tag{3.4}$$

where P_q is the external force at pin q. The virtual strains $\delta\epsilon_{ij}$ are computed from δu_q so as to give a kinematically compatible deformation field.

Note that, aside from this equation, $(P_q)_x$ and τ_{ij} *are in no way related to* δu_q *and its associated* $\delta\epsilon_{ij}$. We go through the same procedure for δv_q, i.e., for the vertical virtual displacement of pin q (assuming no external constraint in this direction) to arrive at another equation like Eq. (3.4). Continuing, we do this at every pin i where there is no constraint in either the x or y direction, deleting the formulation for the direction where there is external constraint. We thus arrive at m equations. Now, using an appropriate *constitutive law*, we *separately* calculate the total stresses τ_{ij} stemming from possible displacements $(\bar{u}, \bar{v})_i$ for all pins i that are not externally constrained. (If a pin is constrained in the x direction but not in the y direction, then set $\bar{u} = 0$ and use only $\bar{v}$ for that pin.) There will be a total of m such displacements. And in the equations formed by virtual work of the form of Eq. (3.4), we shall replace the stresses τ_{ij} in terms of the m displacements $(\bar{u}, \bar{v})_i$ for all movable pins without regard to $\delta\epsilon_{ij}$, which is independent of these stresses. Now solve the system of m simultaneous equations for m unknown $\bar{u}$'s and $\bar{v}$'s for the movable pins (the variation terms δu_q, etc., will all cancel out in this process). We can now get the stresses τ_{ij} associated with the $\bar{u}$'s and $\bar{v}$'s. These stresses must then be the *actual* stresses and the displacements $(\bar{u}, \bar{v})_i$ must be the *actual* displacements of the pins.

Why is this so?

1. These stresses *satisfy virtual work* for an arbitrary, kinematically compatible deformation found by giving the pins i virtual displacements δu_i, δv_i consistent with the constraints. And for the actual external loads, the τ_{ij} thus satisfy the equation of equilibrium.
2. Also, the displacements $(\bar{u}, \bar{v})_i$ are related to these stresses through an appropriate constitutive law and, from the manner of their introduction, form single-valued displacement fields.

It is important to note here that with this method we may consider materials having any constitutive law as well as statically indeterminate trusses.† In succeeding

†You will recall from earlier studies that a necessary condition for static determinacy of a plane truss is that

$$m = 2j - 3 \qquad \text{where} \quad \begin{cases} m = \text{number of members} \\ j = \text{number of pins} \end{cases}$$

And for a space truss we must have

$$m = 3j - 6$$

These conditions are not sufficient, however.

discussions on trusses we will at times impose restrictions as to the linearity of the constitutive law and static determinacy of the system.

We now illustrate the approach for a relatively simple problem.

EXAMPLE 3.1 Consider a system of n pin-connected rods supporting a force P, as shown in Fig. 3.2a. Each rod q is inclined at an angle α_q with the horizontal, and load P is inclined at an angle β with the horizontal. We shall neglect body forces in this problem and determine the forces in the members for linearly elastic behavior.

Notice that the movement of joint A characterizes, through the use of trigonometry, the entire deformation of the system. Thus we now impose a virtual displacement of joint A in the x direction and we denote this virtual displacement as δ_x. The rods then change length and rotate about the fixed ends B_q. (The constraints of the problem are then in no way violated.) Each rod q has a virtual change in length given as (see Fig. 3.2b).

$$(\delta L_q) = \delta_x \cos \gamma_q \approx \delta_x \cos \alpha_q \qquad (a)$$

and so the virtual strain in each rod is simply given as

$$(\delta \epsilon_q)_x = \frac{\delta_x \cos \alpha_q}{L_q} \qquad (b)$$

Similarly, for δ_y we have

$$(\delta \epsilon_q)_y = \delta_y \frac{\sin \alpha_q}{L_q} \qquad (c)$$

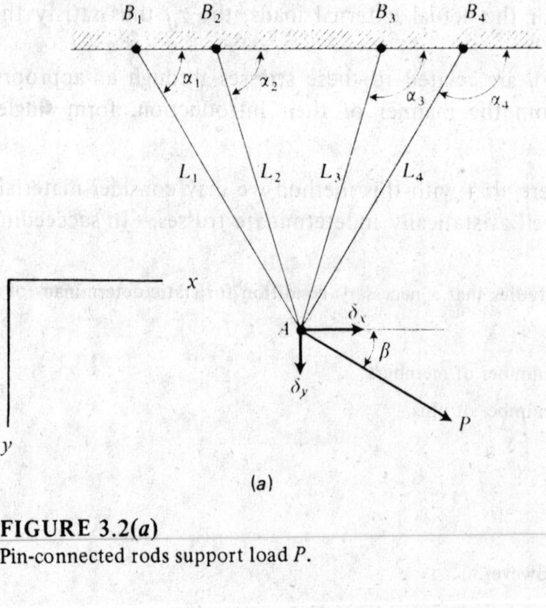

FIGURE 3.2(a)
Pin-connected rods support load P.

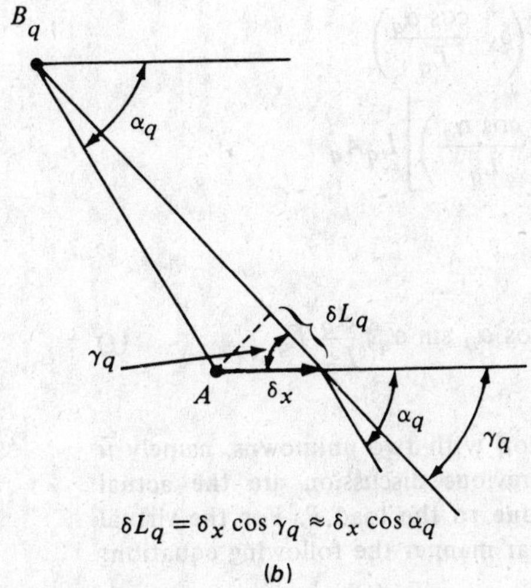

FIGURE 3.2(*b*)
Elongation of qth member.

The only pin that is not constrained is pin A, and so we assume it has displacement components $\bar{u}$ and $\bar{v}$. Upon satisfaction of the virtual work principle and Hooke's law these will be the *actual* displacement components of pin A resulting from load P. The strain $\bar{\epsilon}_q$ in the qth rod from these displacement components is then, using the same reasoning as above in Eqs. (*b*) and (*c*),

$$\bar{\epsilon}_q = \frac{\bar{u} \cos \alpha_q}{L_q} + \frac{\bar{v} \sin \alpha_q}{L_q} \tag{d}$$

Hence the stress $\bar{\tau}_q$ in the qth rod from $\bar{u}$ and $\bar{v}$ with elastic modulus E_q, is

$$\bar{\tau}_q = E_q \frac{\bar{u} \cos \alpha_q}{L_q} + E_q \frac{\bar{v} \sin \alpha_q}{L_q} \tag{e}$$

We now employ the principle of virtual work, first for δ_x, using the actual external load and the stresses $\bar{\tau}_q$ along with the aforestated virtual displacement and associated deformation field. The only surface traction force that does virtual work is P. Thus we have for Eq. (3.3) applied to this problem:

$$P \cos \beta \, \delta_x = \sum_{q=1}^{4} \iiint_V (\bar{\tau}_q)(\delta\epsilon_q)\, dv = \sum_{q=1}^{4} (\bar{\tau}_q)(\delta\epsilon_q) L_q A_q$$

Substituting from Eqs. (*b*) and (*e*) we get

$$P \cos \beta \, \delta_x = \sum_{q=1}^{4} \left[E_q \frac{\bar{u} \cos \alpha_q}{L_q} \left(\delta_x \frac{\cos \alpha_q}{L_q} \right) + E_q \frac{\bar{v} \sin \alpha_q}{L_q} \left(\delta_x \frac{\cos \alpha_q}{L_q} \right) \right] L_q A_q$$

Canceling δ_x we get†

$$P \cos \beta = \sum_{q=1}^{4} (\bar{u} \cos^2 \alpha_q + \bar{v} \cos \alpha_q \sin \alpha_q) \frac{A_q}{L_q} E_q \tag{f}$$

We have here a single algebraic equation with two unknowns, namely $\bar{u}$ and $\bar{v}$, which we know from the previous discussion are the actual displacement components of joint A due to the load P. For the virtual displacement δ_y we generate in a similar manner the following equation:

$$P \sin \beta = \sum_{q=1}^{4} (\bar{u} \sin \alpha_q \cos \alpha_q + \bar{v} \sin^2 \alpha_q) \frac{A_q}{L_q} E_q \tag{g}$$

We may solve Eqs. (*f*) and (*g*) simultaneously for the unknowns $\bar{u}$ and $\bar{v}$. The actual stresses $\bar{\tau}_q$ and actual forces in the rods are then readily computed.

The method presented, stemming from the principle of virtual work, is called the *dummy displacement* method. It is one of several related approaches called, in general, *displacement* methods, wherein deformation parameters become the key factors in the approach. We will examine other displacement approaches in this chapter. There are also other approaches that are analogous to the *displacement methods*, namely the *force methods*, wherein force parameters rather than deformation parameters become the key factors. We shall also examine the force methods in this chapter and the analogy alluded to above will then be clear.

We have shown that the satisfaction of the principle of virtual work is a *necessary* relation between the external loads and the stresses in a body in equilibrium. We can also show that the satisfaction of the princple of virtual work is *sufficient* to satisfy the equilibrium requirements of a body. We start by assuming that Eq. (3.3) is valid for a body. Rewrite the last integral of this equation in the following manner:

$$\iiint_V \tau_{ij} \, \delta\epsilon_{ij} \, dv = \iiint_V \tau_{ij} \, \delta \left(\frac{u_{i,j} + u_{j,i}}{2} \right) dv$$

†We reiterate at this time that the varied deformation $\delta\epsilon_{ij}$ in Eq. (3.3) is *not related* to stresses τ_{ij} via a constitutive law. The constitutive law used here served solely to replace τ_{ij} with the $\bar{u}$, $\bar{v}$ terms and has nothing to do with $\delta\epsilon_{ij}$ in Eq. (3.3). Indeed, the generator of $\delta\epsilon_{ij}$, namely δ_x, *cancels out* of the final equation.

$$\iiint_V \tau_{ij}\,\delta\epsilon_{ij}\,dv = \iiint_V \tau_{ij}\frac{(\delta u_i)_{,j}}{2}\,dv + \iiint_V \tau_{ij}\frac{(\delta u_j)_{,i}}{2}\,dv$$

$$= \iiint_V \tau_{ij}(\delta u_i)_{,j}\,dv$$

where we have made use of the symmetry of τ_{ij} in the last equation. We can rewrite this last expression as follows:

$$\iiint_V \tau_{ij}(\delta u_i)_{,j}\,dv = \iiint_V (\tau_{ij}\delta u_i)_{,j}\,dv - \iiint_V \tau_{ij,j}\,\delta u_i\,dv$$

Now, using the divergence theorem, we get

$$\iiint_V \tau_{ij}(\delta u_i)_{,j}\,dv = \iint_S \tau_{ij}\,\delta u_i \nu_j\,dS - \iiint_V \tau_{ij,j}\,\delta u_i\,dv$$

$$= \iint_{S_1} \tau_{ij}\,\delta u_i \nu_j\,dS - \iiint_V \tau_{ij,j}\,\delta u_i\,dv$$

where we have made use of the fact that $\delta u_i = 0$ on S_2. Now substitute these results for the last integral in Eq. (3.3). We get, on rearranging and consolidating, the terms

$$\iiint_V (\tau_{ij,j} + B_i)\,\delta u_i\,dv + \iint_{S_1} (T_i^{(\nu)} - \tau_{ij}\nu_j)\,\delta u_i\,dS = 0$$

Since δu_i is *arbitrary*, we can assume for now that $\delta u_i = 0$ on S_1 but is not zero inside the body. We must then conclude that

$$\tau_{ij,j} + B_i = 0 \qquad \text{in } V \tag{3.5a}$$

By the same reasoning we can also conclude that

$$T_i^{(\nu)} - \tau_{ij}\nu_j = 0 \qquad \text{on } S_1 \tag{3.5b}$$

Therefore

$$T_i^{(\nu)} = \tau_{ij}\nu_j \qquad \text{on } S_1 \tag{3.5c}$$

We have thus generated Newton's law for equilibrium at any point inside the boundary and Cauchy's formula, which ensures equilibrium at the boundary. We accordingly conclude that satisfaction of the principle of virtual work is both *necessary* and *sufficient* for equilibrium.

3.3 THE METHOD OF TOTAL POTENTIAL ENERGY

The method of virtual work is very valuable since it is valid for any constitutive law. We shall now develop from it the very useful concept of *total potential energy*, which applies only to elastic bodies. Accordingly, we shall now restrict the discussion to elastic continua (not necessarily *linear* elastic continua). As discussed in Chapter 2, for such cases there exists a strain energy density (i.e., strain energy per unit volume) function $\mathcal{U}$, such that

$$\tau_{ij} = \frac{\partial \mathcal{U}}{\partial \epsilon_{ij}} \tag{3.6}$$

where $\mathcal{U}$ is a function of the strains at a point and is assumed to be a positive definite function.† Whereas in the previous section the virtual displacement field δu_i was a priori not related to the stress field τ_{ij} when applying the principle of virtual work, we now *link these fields* through a constitutive law which is in accord with Eq. (3.6). Thus, replacing τ_{ij} in Eq. (3.3) via Eq. (3.6), we get

$$\begin{aligned}\iiint_V B_i\,\delta u_i\,dv + \iint_{S_1} T_i^{(\nu)}\,\delta u_i\,dS &= \iiint_V \frac{\partial \mathcal{U}}{\partial \epsilon_{ij}}\,\delta\epsilon_{ij}\,dv \\ &= \iiint_V \delta^{(1)}\mathcal{U}\,dv = \delta^{(1)}\left(\iiint_V \mathcal{U}\,dv\right) \\ &= \delta^{(1)}U \end{aligned}\tag{3.7}$$

where we have interchanged the variation and the integration operations. Equation (3.7) equates the virtual work on the body with the variation of the strain energy of a body. We now *define* the potential energy V of the applied loads as a function of displacement fields u_i and the applied loads as follows:

$$V = -\iiint_V B_i u_i\,dv - \iint_{S_1} T_i^{(\nu)} u_i\,dS \tag{3.8}$$

With B_i and $T_i^{(\nu)}$ prescribed, we have for the first variation of V

$$\delta^{(1)}V = -\iiint_V B_i \frac{\partial u_i}{\partial u_j}\,\delta u_j\,dv - \iint_{S_1} T_i^{(\nu)} \frac{\partial u_i}{\partial u_j}\,\delta u_j\,dS$$

†Recall that $\mathcal{U}$ represents the specific internal energy for an isentropic process and the specific free energy for an isothermal process. It can be shown through thermodynamic considerations to be a positive definite function.

Noting that $\partial u_i/\partial u_j = \delta_{ij}$ and letting $j = i$ to take into account the Kronecker delta terms, we have

$$\delta^{(1)}V = -\iiint_V B_i\,\delta u_i\,dv - \iint_{S_1} T_i^{(\nu)}\,\delta u_i\,dS \tag{3.9}$$

where δu_i are virtual displacement fields. Combining Eqs. (3.7) and (3.9), we then get

$$\boxed{\delta^{(1)}(U + V) = 0} \tag{3.10}$$

The quantity $(U + V)$, which we shall denote as π, is called the *total potential energy of the body* and is given as

$$\boxed{\pi = U - \iiint_V B_i u_i\,dv - \iint_{S_1} T_i^{(\nu)} u_i\,dS} \tag{3.11}$$

Equation (3.10) is known as the *principle of total potential energy.* An interpretation of this law may be given as follows.

The *necessary* requirements for *equilibrium* of a particular stress field τ_{ij} are that

1. B_i and $T_i^{(\nu)}$ are *statically compatible.*
2. The *deformation field,* to which the field τ_{ij} is related through a constitutive law for *elastic* behavior, *extremizes* π with respect to all other kinematically compatible, admissible deformation fields.

We have shown that the extremization of the total potential energy with respect to admissible deformation fields is necessary for equilibrium to exist between the forces and the stresses in a body. Just as in the method of virtual work, we now show it to be a sufficient condition for equilibrium.

Note that the total potential energy principle has brought us back to the variational calculus of Chapter 2. The functional is π with independent variables x, y, z and with function u_i to be varied. We will now show that the Euler-Lagrange equations for this functional are the familiar equations of equilibrium, and the boundary conditions comprise Cauchy's formula. On accomplishing this, we will have proved that satisfying the total potential energy principle is not only *necessary* for equilibrium, it is *sufficient* for equilibrium.

We start by assuming that

$$\delta^{(1)}(\pi) = \delta^{(1)}(U + V) = 0$$

Noting that $U = \iiint \mathcal{U}\,dv$, where $\mathcal{U}$ is a function of ϵ_{ij}, and observing Eq. (3.9) for the first variation of V, we have:†

$$\delta^{(1)}\pi = 0 = \iiint_V \frac{\partial \mathcal{U}}{\partial \epsilon_{ij}}\,\delta\epsilon_{ij}\,dv - \iiint_V B_i\,\delta u_i\,dv - \oiint_S T_i^{(\nu)}\,\delta u_i\,dS \tag{3.12}$$

Consider the first integral. Using a constitutive law for elastic bodies and the formulations of ϵ_{ij} in terms of u_i, we have

$$\iiint_V \frac{\partial \mathcal{U}}{\partial \epsilon_{ij}}\,\delta\epsilon_{ij}\,dv = \iiint_V \tau_{ij}\,\delta\left[\frac{1}{2}(u_{i,j}+u_{j,i})\right]dv$$

$$= \iiint_V \left[\frac{\tau_{ij}}{2}\,\delta(u_{i,j}) + \frac{\tau_{ij}}{2}\,\delta(u_{j,i})\right]dv \tag{3.13}$$

Replacing τ_{ij} in the last expression by τ_{ji} (permissible by symmetry of the stress tensor), we have the expression

$$\frac{\tau_{ji}}{2}\,\delta(u_{j,i})$$

Noting that j and i are dummy indices, we can interchange j and i so that instead of having $\frac{1}{2}\tau_{ji}\;\delta(u_{j,i})$ we can replace it by $\frac{1}{2}\tau_{ij}\;\delta(u_{i,j})$. Going back to Eq. (3.13), we can now combine terms on the right side of the equation.

$$\iiint_V \frac{\partial \mathcal{U}}{\partial \epsilon_{ij}}\,\delta\epsilon_{ij}\,dv = \iiint_V \tau_{ij}\,\delta(u_{i,j})\,dv = \iiint_V \tau_{ij}(\delta u_i)_{,j}\,dv$$

$$= \iiint_V (\tau_{ij}\,\delta u_i)_{,j}\,dv - \iiint_V \tau_{ij,j}\,\delta u_i\,dv$$

$$= \oiint_S \tau_{ij}\,\delta u_i \nu_j\,dS - \iiint_V \tau_{ij,j}\,\delta u_i\,dv$$

Note we have used the fact that a delta operator and derivative operator are commutative, and we have used Gauss's theorem. We then can say for Eq. (3.12):

$$\oiint_S \tau_{ij}\nu_j\,\delta u_i\,dA - \iiint_V \tau_{ij,j}\,\delta u_i\,dv - \iiint_V B_i\,\delta u_i\,dv - \oiint_S T_i^{(\nu)}\,\delta u_i\,dS = 0$$

Now collect terms:

†We may use $S = S_1 + S_2$ instead of S_1 in the surface integral since u_i being prescribed for S_2 requires $\delta u_i = 0$ on S_2.

$$\oiint_S (\tau_{ij}\nu_j - T_i^{(\nu)})\,\delta u_i\,dS - \iiint_V (\tau_{ij,j} + B_i)\,\delta u_i\,dv = 0$$

This statement must be true for all δu_i. Suppose $\delta u_i = 0$ on the boundary but not inside the boundary. Then it follows that

$$\tau_{ij,j} + B_i = 0 \tag{3.14a}$$

Further, it follows that

$$T_i^{(\nu)} = \tau_{ij}\nu_j \tag{3.14b}$$

Thus we see that the Euler-Lagrange equations are the equations of equilibrium and the boundary conditions are included in the Cauchy formula. We thus have proved that the principle $\delta^{(1)}\pi = 0$ is *sufficient* for equilibrium.

We shall now show that the total potential energy (π) is actually a local *minimum* for the equilibrium configuration under loads B_i and $T_i^{(\nu)}$ compared with the total potential energy corresponding to neighboring admissible configurations with the same B_i and $T_i^{(\nu)}$. We accordingly examine the difference between the total potential energies of the equilibrium state and an admissible neighboring state having displacement field $(u_i + \delta u_i)$ and a corresponding strain field $(\epsilon_{ij} + \delta\epsilon_{ij})$. Thus,

$$(\pi)_{\epsilon_{ij}+\delta\epsilon_{ij}} - (\pi)_{\epsilon_{ij}} = \iiint_V [\mathcal{U}(\epsilon_{ij} + \delta\epsilon_{ij}) - \mathcal{U}(\epsilon_{ij})]\,dv - \iiint_V B_i\,\delta u_i\,dv - \iint_{S_1} T_i^{(\nu)}\,\delta u_i\,dS \tag{3.15}$$

Now expand $\mathcal{U}(\epsilon_{ij} + \delta\epsilon_{ij})$ as a Taylor series:

$$\mathcal{U}(\epsilon_{ij} + \delta\epsilon_{ij}) = \mathcal{U}(\epsilon_{ij}) + \left(\frac{\partial\mathcal{U}}{\partial\epsilon_{ij}}\right)\delta\epsilon_{ij} + \frac{1}{2!}\left(\frac{\partial^2\mathcal{U}}{\partial\epsilon_{ij}\,\partial\epsilon_{kl}}\right)\delta\epsilon_{ij}\,\delta\epsilon_{kl} + \cdots \tag{3.16}$$

Substituting into Eq. (3.15) and rearranging terms, we get

$$(\pi)_{\epsilon_{ij}+\delta\epsilon_{ij}} - (\pi)_{\epsilon_{ij}} = \iiint_V \left(\frac{\partial\mathcal{U}}{\partial\epsilon_{ij}}\right)\delta\epsilon_{ij}\,dv - \iiint_V B_i\,\delta u_i\,dv - \iint_{S_1} T_i^{(\nu)}\,\delta u_i\,dS + \iiint_V \frac{1}{2!}\left(\frac{\partial^2\mathcal{U}}{\partial\epsilon_{ij}\,\partial\epsilon_{kl}}\right)\delta\epsilon_{ij}\,\delta\epsilon_{kl}\,dv + \cdots$$

The first three terms on the right side clearly give a net value of zero for all admissible configurations as a direct result of the principle of total potential energy. Now considering only the expression in the series having products of terms $\delta\epsilon_{ij}$ and calling this the *second variation* of π, denoted at $\delta^{(2)}\pi$, we get:†

$$\delta^{(2)}\pi = \frac{1}{2}\iiint_V \frac{\partial^2 \mathcal{U}}{\partial\epsilon_{ij}\,\partial\epsilon_{kl}}\,\delta\epsilon_{ij}\,\delta\epsilon_{kl}\,dv \tag{3.17}$$

We will demonstrate that the integrand of the expression on the right side of this equation is $\mathcal{U}(\delta\epsilon_{ij})$ for the case where $\epsilon_{ij} = 0$. Consider for this purpose that $\epsilon_{ij} = 0$ in Eq. (3.16). Then the first term of the series is a constant throughout the body and is taken to be zero, so as to have the strain energy vanish in the unstrained state. By definition, $\partial\mathcal{U}/\partial\epsilon_{ij}$ in the next expression of the series is τ_{ij}. Since τ_{ij} corresponds here to the unstrained state, we may take it to be zero everywhere in the body. We see then that up to second-order terms:

$$\mathcal{U}(\delta\epsilon_{ij}) = \frac{1}{2!}\left(\frac{\partial^2\mathcal{U}}{\partial\epsilon_{ij}\,\partial\epsilon_{kl}}\right)_{\epsilon_{ij}=0}\delta\epsilon_{ij}\,\delta\epsilon_{kl}$$

We may thus express Eq. (3.17) as follows for the case $\epsilon_{ij} = 0$:

$$\delta^{(2)}(\pi) = \iiint_V \mathcal{U}(\delta\epsilon_{ij})\,dv$$

But we indicated earlier that $\mathcal{U}$ is a positive definite function of strain, and so the second variation of the total potential energy is positive. We may thus conclude that the total potential energy is a minimum for the equilibrium state $\epsilon_{ij} = 0$ when compared to all other neighboring admissible deformation fields. We now extend this conclusion by stating, without proof,‡ that the total potential energy is minimum for *any* equilibrium deformation field ϵ_{ij} compared to neighboring admissible deformation fields.

In the previous section we employed the dummy displacement method for a simple truss problem stemming from the principle of virtual work. We now examine the same problem to illustrate the use of a second displacement method, namely one stemming from the total potential energy principle.

EXAMPLE 3.2 Find the deflection of pin A in Fig. 3.2a from the force P, assuming the same condition as given in Example 3.1.

We may set forth kinematically compatible deformation fields in the problem by giving pin A virtual displacements δ_x and δ_y. The potential energy of the external forces for such displacements then becomes

$$V = -(P\cos\beta)\,\delta_x - (P\sin\beta)\,\delta_y \tag{a}$$

†We shall consider the second variation in Chapter 8, where it will have more immediate use.

‡In Problem 3.6 you will be asked to show the proof of this minimum for linear elastic materials without assuming $\epsilon_{ij} = 0$.

The strain energy U is a function of δ_x and δ_y and may be given as follows:

$$U = \iiint_V \left(\int \tau_{ij} \, d\epsilon_{ij} \right) dv$$

$$= \sum_{p=1}^{4} \left[\int (E_p \epsilon_p \, d\epsilon_p) \right] A_p L_p = \sum_{p=1}^{4} E_p \frac{\epsilon_p^2}{2} A_p L_p \qquad (b)$$

From simple trigonometric relations we may express ϵ_p for small deformation as follows:

$$\epsilon_p = \frac{\delta_x \cos \alpha_p}{L_p} + \frac{\delta_y \sin \alpha_p}{L_p} \qquad (c)$$

Hence

$$U = \sum_{p=1}^{4} \frac{E_p A_p}{2L_p} (\delta_x \cos \alpha_p + \delta_y \sin \alpha_p)^2 \qquad (d)$$

We may then give π as follows:

$$\pi = \sum_{p=1}^{4} \frac{E_p A_p}{2L_p} (\delta_x \cos \alpha_p + \delta_y \sin \alpha_p)^2 - P(\delta_x \cos \beta + \delta_y \sin \beta) \qquad (e)$$

The principle of total potential energy then requires that π be an extremum with respect to the kinematically admissible deformation fields characterized by δ_x and δ_y. Hence:

$$\delta^{(1)}\pi = \frac{\partial \pi}{\partial \delta_x} \delta(\delta_x) + \frac{\partial \pi}{\partial \delta_y} \delta(\delta_y) = 0$$

Since δ_x and δ_y are independent of each other, we conclude that

$$\frac{\partial \pi}{\partial \delta_x} = 0 \qquad \frac{\partial \pi}{\partial \delta_y} = 0 \qquad (f)$$

Substitute Eq. (*e*) for π in Eqs. (*f*). We then get two simultaneous equations for δ_x and δ_y. By having extremized π, the resulting values of δ_x and δ_y must then be related through the constitutive law used (Hooke's law) to stresses that *must satisfy equilibrium*. Accordingly, the resulting δ_x and δ_y must be the actual displacements of the movable pin, and the related stresses are the actual stresses. Hence, δ_x and δ_y must correspond, respectively, to $\bar{u}$ and $\bar{v}$ of the dummy displacement method. With this in mind, you may readily show that the aforementioned equations are identical to Eqs. (*f*) and (*g*) of Example 3.1.

The procedure presented here can readily be extended to apply to more complex, trusses. We simply set forth kinematically compatible deformations of the truss by imagining that every movable pin (i.e., pins not constrained) has deflection components in the x and y directions like δ_x and δ_y for pin A of this problem. We shall not illustrate such a case at this time. In Part B of the chapter we shall consider a third displacement method involving Castigliano's first theorem and at that time we shall consider such a truss. The formulations are very similar to the procedure we would follow here, as you will then see.

We will now interrupt our series of examples of truss problems to consider the problem of the equilibrium configuration of a flexible string. This example will illustrate an important use of the total potential energy functional that will pervade much of the text. Later in Section 3.4, when we resume our discussion of trusses, we shall begin a discussion of the so-called force methods for trusses.

EXAMPLE 3.3 We now demonstrate the use of the method of total potential energy to derive the equation of equilibrium for a perfectly flexible string having a high value of initial tension T and loaded transversely by a loading $q(x)$ small enough so as not to cause large deflections (see Fig. 3.3). Under these conditions we can assume that there is a uniform tension T all along the string. The string is of length L and horizontal before the load $q(x)$ is applied (i.e., we neglect gravity). It is simply supported at the ends.

We first compute the strain energy for the string. For convenience, we shall consider that U is composed of two parts: a strain energy U_1 due to the original tension in the string before application of loading $q(x)$, and a strain energy U_2 due to the stretching of the string as a result of the transverse loading. The term U_2 can be computed easily by multiplying the tensile force in the string, which we assume is constant, by the total elongation of the string as a result of the transverse loading. Thus, using Y to denote the deflection of the string,

$$U_2 = T \int_0^L (ds - dx) = T \int_0^L (\sqrt{dx^2 + dY^2} - dx)$$

$$= T \int_0^L \left[\sqrt{1 + \left(\frac{dY}{dx}\right)^2} - 1 \right] dx$$

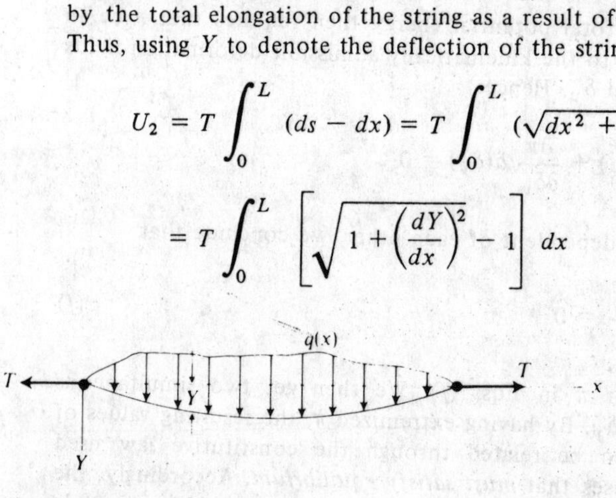

FIGURE 3.3
Simply supported flexible string.

Since $(dY/dx)^2$ is less than unity here we can expand the root as a power series in the last expression. Because dY/dx is small we need only retain the first two terms for the accuracy desired. We then have

$$U_2 = T\int_0^L \left\{\left[1 + \frac{1}{2}\left(\frac{dY}{dx}\right)^2 + \cdots\right] - 1\right\} dx \approx \frac{T}{2}\int_0^L \left(\frac{dY}{dx}\right)^2 dx$$

The strain energy for the string can then be given as follows:

$$U = U_1 + \frac{T}{2}\int_0^L \left(\frac{dY}{dx}\right)^2 dx$$

Now we can employ the principle of total potential energy as follows:

$$\delta^{(1)}\left[U_1 + \frac{T}{2}\int_0^L \left(\frac{dY}{dx}\right)^2 dx - \int_0^L qY\,dx\right] = 0$$

Since $\delta^{(1)}U_1 = 0$ we get

$$\delta^{(1)}\left\{\int_0^L \left[\frac{T}{2}\left(\frac{dY}{dx}\right)^2 - qY\right] dx\right\} = 0$$

We thus arrive at the classic variational problem discussed at length in the previous chapter. Employing the Euler-Lagrange equation $\partial F/\partial Y - (d/dx)(\partial F/\partial Y') = 0$, we get

$$\frac{d^2Y}{dx^2} = -\frac{q(x)}{T}$$

Since Y, being single-valued and continuous (as required by the operations we have carried out on it), represents a kinematically compatible deformation, it is clear in accordance with the principle of total potential energy that the above equation represents a requirement for equilibrium for the simple constitutive law we have used. Indeed, it is the well-known equation of equilibrium for the loaded string.

The preceding problem is a simple illustration of a use of the method of total potential energy–simple primarily because a tacitly assumed vanishingly small cross section of the string rendered the specification of Y sufficient for fully specifying the deformed state of the string. Then using in effect a constitutive law that maintains the tension in the string independent of the elongation, for reasons set forth in the example, we were able to arrive at a functional in terms of Y valid for small deformation of the string. The extremization of this functional then gave the differential equation of equilibrium for small deformation of the simply supported string.†

In the ensuing chapters on beams and plates we shall make use of the method of total potential energy in a manner paralleling this simple example. We shall propose by physical arguments a simplified mode of deformation that permits expression of the three displacement field components u_1, u_2, and u_3 in terms of a single function. For beams this function will be the deflection $w(x)$ of the neutral axis of the beam (as per the discussion in Section 3.2) with the assumption that plane sections remain both plane and perpendicular to the neutral axis on deformation; for plates it will be the deflection $w(x, y)$ of the midplane of the plate with the assumption that initially vertical line segments remain both straight and perpendicular to the midplane surface on deformation. This corresponds to using $Y(x)$ for the string with the tacit assumption that sections of the string remain normal to $Y(x)$. Then using a constitutive law, which may be some approximation of the actual one as in the case of the string, we may express the total potential energy functional in terms of w. Extremization of the total potential energy then yields a differential equation for w ensuring equilibrium, plus the kinematic and natural boundary conditions for the problem. We thus will use the method of total potential energy to generate a boundary-value problem for the particular structural body and loads in terms of a single function w.

But, after the insertions of physically inspired arguments to get u_i in terms of w (assuming for the moment we have used the correct constitutive law), what is the

†Actually the variation process in this case also yields another set of permissible boundary conditions for the energies used in arriving at the total potential energy, namely the natural boundary conditions. In this case, it is thus also permitted to have

$$\frac{dY}{dx} = 0 \qquad \text{at the ends}$$

This case corresponds to a string connected at the ends by frictionless rings to vertical supports (see Fig. 3.4). Now because no external vertical force can be considered at the ends (otherwise we would have had to include a virtual work contribution at the ends) it is necessary that $q(x)$ have a zero resultant in order to have a statically compatible external load system required by the method of total potential energy. Physically, it is obvious that this must be so to maintain equilibrium. Finally, we can also have $Y = 0$ at one end and $Y' = 0$ at the other.

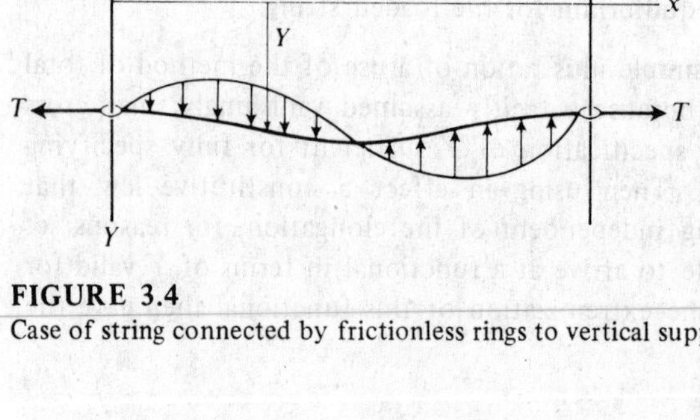

FIGURE 3.4
Case of string connected by frictionless rings to vertical supports at ends.

meaning and significance of w satisfying the resulting Euler-Lagrange equations plus boundary conditions? To best answer this, note that in expressing u_i in terms of w in the total potential energy we are *restricting* the admissible displacements u_i for the problem to a *subclass* of all those possible and therefore the extremization is carried out with respect to a smaller class of functions than if the arguments leading to w were not employed. This, in turn, means we are tacitly including in the body certain constraints that, because they do not appear in the calculations, do no work during deformation of the body. The resulting solutions w of the equations of equilibrium and boundary conditions for the variational process then represent equilibrium states for the body with the aforementioned constraints present. Thus, by going over from u_i to w to facilitate handling of the problem, we are effectively considering a *different* system–actually a *stiffer* and more highly constrained system. Now if the simplification of the deformation process leading to w is wisely made, the results for the stiffer system are very close under certain conditions to those for the actual system.

As for the use of an incorrect constitutive law in reaching the total potential energy functional, it simply means that the solution for the stiffer system corresponds to a state of equilibrium for that constitutive law. One must then demonstrate that some or all of the results of the generated boundary-value problems are reasonably close to the actual case.

With these remarks, the reader has a preview of much that will be presented in later chapters. We will show later in this chapter that the total potential energy not only permits means of arriving at equations of equilibrium and of establishing proper boundary conditions, but also affords means of developing approximate solutions.

3.4 COMPLEMENTARY VIRTUAL WORK

In the previous sections we focused our attention on varying the displacement field while keeping the external forces fixed to arrive at useful variational principles. Now we investigate the process of varying in some way the stress field and external forces while holding displacement fields fixed. Specifically, we allow here as admissible variations of stress and external loads only those which satisfy the equations of equilibrium inside the body and on the boundary. That is, we require

$$(\delta\tau_{ij})_{,j} + \delta B_i = 0 \qquad \text{in } V \tag{3.18a}$$

$$(\delta\tau_{ij})\nu_j = \delta T_i^{(\nu)} \qquad \text{on } S_2 \tag{3.18b}$$

We now define the *complementary virtual work* δW^* as follows:

$$\delta W^* = \iiint_V u_i\, \delta B_i\, dv + \iint_{S_2} u_i\, \delta T_i^{(\nu)}\, dS \tag{3.19}$$

where u_i is any displacement field and δB_i and $\delta T_i^{(\nu)}$ satisfy Eqs. (3.18). The principle of complementary virtual work can be developed from the above equations in a manner analogous to that used for the principle of virtual work. Thus:

$$\iiint_V u_i\,\delta B_i\,dv + \oint\!\!\oint_S u_i\,\delta T_i^{(\nu)}\,dS = \iiint_V u_i\,\delta B_i\,dv + \oint\!\!\oint_S u_i(\delta\tau_{ij})\nu_j\,dS$$

$$= \iiint_V u_i\,\delta B_i\,dv + \iiint_V (u_i\,\delta\tau_{ij})_{,j}\,dv$$

$$= \iiint_V u_i\,[\delta B_i + (\delta\tau_{ij})_{,j}]\,dv$$

$$+ \iiint_V u_{i,j}\,\delta\tau_{ij}\,dv$$

Now employing Eq. (3.18*a*) we may drop the first integral on the right side of the equation. Next if we replace $u_{i,j}$ by $(\epsilon_{ij} + \omega_{ij})$ we *ensure that the strain field* ϵ_{ij} *is a kinematically compatible one.* Noting the symmetry and skew-symmetry properties of τ_{ij} and ω_{ij}, respectively, we then arrive at the following desired statement, which is the *principle of complementary work*:

$$\boxed{\iiint_V u_i\,\delta B_i\,dv + \oint\!\!\oint_S u_i\,\delta T_i^{(\nu)}\,dS = \iiint_V \epsilon_{ij}\,\delta\tau_{ij}\,dv} \tag{3.20}$$

Note that as in the principle of virtual work, the above relation in no way involves a constitutive law of any kind. We have here a *relation in the form of an integral equation between a kinematically compatible deformation field and any statically compatible stress and force field* ($\delta\tau_{ij}$, $\delta T_i^{(\nu)}$ and δB_i). We may now say that:

A *necessary* requirement for a *kinematically compatible* deformation is that u_i and ϵ_{ij} satisfy the complementary virtual work equation for *any* statically compatible body force field δB_i and traction field $\delta T_i^{(\nu)}$ for which an associated stress field $\delta\tau_{ij}$ satisfies equilibrium. We will next show that the above condition is *sufficient* for kinematic compatibility.

We have shown that the principle of complementary virtual work is a *necessary* condition for a kinematically compatible deformation. That is, a kinematically compatible deformation field (u_i, ϵ_{ij}) must satisfy Eq. (3.20) for any admissible load and stress field. We can also show that the principle of complementary energy is *sufficient* for kinematic compatibility. We shall demonstrate this for the special case of plane stress, where you will recall from Chapter 1 the use of an Airy stress function Φ defined as:

$$\tau_{xx} = \frac{\partial^2 \Phi}{\partial y^2} \qquad \tau_{yy} = \frac{\partial^2 \Phi}{\partial x^2} \qquad \tau_{xy} = -\frac{\partial^2 \Phi}{\partial x\,\partial y} \tag{3.21}$$

was sufficient for satisfying the equations of equilibrium. We shall assume that the principle of complementary virtual work applies and we shall for simplicity use as admissible force and stress distributions those for which B_i and $T_i^{(v)}$ are zero and only the internal stresses are varied by varying the stress function ϕ. That is, we shall use a subset of the admissible load and stress functions. Then expressing Eq. (3.20) for this case we get

$$0 = \iiint_V \epsilon_{ij}\,\delta\tau_{ij}\,dv = \iiint_V \left(\epsilon_{xx}\frac{\partial^2\,\delta\Phi}{\partial y^2} - 2\epsilon_{xy}\frac{\partial^2\,\delta\Phi}{\partial x\,\partial y} + \epsilon_{yy}\frac{\partial^2\,\delta\Phi}{\partial x^2}\right) dv$$

We now employ Green's theorem [see Appendix Eq. (I.31)] to accomplish integrations by parts to reach the following result:

$$\iiint_V \left(\frac{\partial^2\epsilon_{xx}}{\partial y^2} - 2\frac{\partial^2\epsilon_{xy}}{\partial x\,\partial y} + \frac{\partial^2\epsilon_{yy}}{\partial x^2}\right)\delta\Phi\,dv + \text{surface integrals} = 0$$

Since δ_Φ is arbitrary on the boundaries and inside, we can conclude from the above that

$$\frac{\partial^2\epsilon_{xx}}{\partial y^2} - 2\frac{\partial^2\epsilon_{xy}}{\partial x\,\partial y} + \frac{\partial^2\epsilon_{yy}}{\partial x^2} = 0 \tag{3.22}$$

But this equation is recognized as the kinematic compatibility equation for plane stress. We thus conclude that Eq. (3.20) is sufficient for the satisfaction of kinematic compatibility at least for simply connected domains, and may be used in place of the familiar compatibility equations.

As in the case of the principle of virtual work we can use the principle of complementary virtual work for solving certain structural problems directly. In this case we can very easily find deflections of any particular joint in any particular direction for statically determinate structures with no restrictions as to the nature of the constitutive law. The following example illustrates the straightforward procedure that one follows for such problems and is the first of the *force methods* that we will examine.

EXAMPLE 3.4 Consider a simple pin-connected structure as shown in Fig. 3.5. We wish to determine the deflection of joint "a" in the direction s as a result of the loads P_1 and P_2. The materials in the structure follow linear elastic constitutive laws.

In approaching this problem keep in mind that δB_i, $\delta T_i^{(v)}$, and $\delta\tau_{ij}$ represent *any* system of loads and stresses that satisfy Eqs. (3.18). Accordingly, we shall choose for external loading a unit force on joint "a" in the direction of s with supporting forces A_x, A_y, and B_y computed from equilibrium considerations of rigid-body mechanics for this unit load (see Fig. 3.6). The left side of Eq. (3.20) then becomes simply u_s since $\delta T_i^{(v)}$ is just the unit force at joint "a" (we have not included body forces in our loading). Note u_i are zero in the directions of the supporting force components.

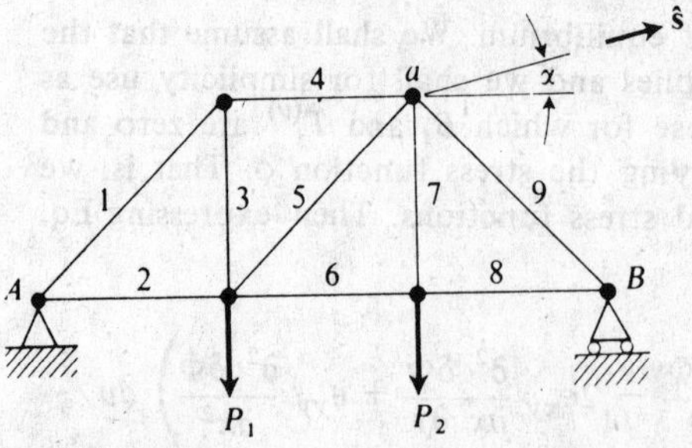

FIGURE 3.5
Simple truss.

Now going to the right side of the equation, we need $\delta\tau_{ij}$ for each member. Since we have a statically determinate truss here, this is readily established by first computing the virtual forces $(F_V)_i$ in each member from the unit force at "a" using equilibrium considerations of rigid-body mechanics. We have then for each member

$$\delta\tau_q = \frac{(F_V)_q}{A_q}$$

As for ϵ_{ij} we shall here employ *actual* strains computed from the *actual* loads P_1 and P_2 and the appropriate reactions. Since the members are tension and compression members each with a uniform tensile or compressive strain along the axis of the member, we can say from elementary strength of materials for such strains ϵ:†

$$\epsilon_q = \frac{F_q}{A_q E_q}$$

†We reiterate that the varied stress field $\delta\tau_{ij}$ is in no way related through a constitutive law in Eq. (3.20) to the strain field ϵ_{ij} in that equation. The constitutive law used here, namely Hooke's law, was for the sole purpose of computing the strains ϵ_{ij} in Eq. (3.20) from the *actual* stresses τ_{ij} and has nothing to do with the varied stresses $\delta\tau_{ij}$.

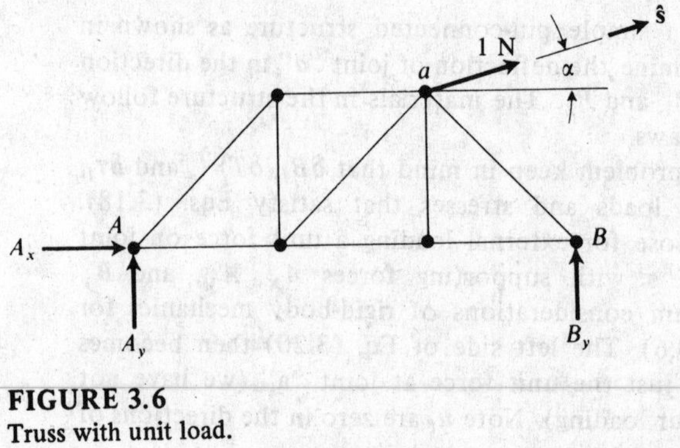

FIGURE 3.6
Truss with unit load.

where F_q is the actual force in the qth member computed by rigid-body mechanics.

We may then employ Eq. (3.20) to compute u_s as follows:

$$u_s = \sum_{q=1}^{9} \int_0^{L_q} \frac{F_q}{A_q E_q} \left[\frac{(F_V)_q}{A_q} \right] (dx\, A_q) = \sum_{q=1}^{9} \frac{F_q (F_V)_q}{A_q E_q} L_q$$

Since we have used *actual strains* on the right side of the equation, we conclude that u_s must be the *actual displacement* associated with these strains as a result of satisfying compatibility via the principle of complementary virtual work. Thus, by using a convenient varied force system designed to "isolate" the desired displacement and by using easily computed actual strains within the structure, we are able to compute directly the desired actual displacement. Understandably, this method is called the *unit (dummy) load method.*

The dummy load method is analogous to the dummy displacement method in this respect. In the former we imposed certain convenient loadings on the structure and used the method of complementary virtual work, while in the latter, you will recall, we imposed certain deflections on the structure and used the method of virtual work.

We could have done this problem without the use of energy methods by first computing the elongation or compression of each member and then employing geometric and trigonometric considerations, proceeding to ascertain the resulting deflection of the joint. An effort to do this yourself will show most effectively the vast superiority of the energy approach.

Just as we were able to reformulate the principle of virtual work as a variational principle involving the total potential energy, so we will next reformulate the principle of complementary virtual work as a variational principle in terms of the so-called total complementary energy.

3.5 PRINCIPLE OF TOTAL COMPLEMENTARY ENERGY

You will recall from Chapter 1 the concept of the complementary energy density function $\mathfrak{u}^*$ defined for elastic bodies as a function of stress such that

$$\frac{\partial \mathfrak{u}^*}{\partial \tau_{ij}} = \epsilon_{ij} \tag{3.23}$$

Up to this time we have not related a priori the admissible stress and force fields with the deformation field in the principle of complementary energy. We shall now *link these fields* in Eq. (3.20) via a constitutive law embodied in Eq. (3.23) and thus limit the forthcoming results to elastic (not necessarily linearly elastic) bodies. Thus we express Eq. (3.20) as follows:

$$\iiint_V u_i\, \delta B_i\, dv + \oiint_S u_i\, \delta T_i^{(\nu)}\, dS = \iiint_V \left(\frac{\partial \mathrm{u}^*}{\partial \tau_{ij}}\right) \delta\tau_{ij}\, dv \tag{3.24}$$

The right side of this equation is simply $\delta^{(1)}U^*$, the first variation of the complementary energy for the body. Next we introduce a complementary potential function V^* defined such that

$$V^* = -\iiint_V u_i B_i\, dv - \oiint_S u_i T_i^{(\nu)}\, dS \tag{3.25}$$

for which the first variation $\delta^{(1)}V^*$ is given as follows:

$$\delta^{(1)}V^* = -\iiint_V u_i\, \delta B_i\, dv - \oiint_S u_i\, \delta T_i^{(\nu)}\, dS \tag{3.26}$$

Note that the variation of the traction force on S_1, where the traction is prescribed, is taken as zero. Also, in order to use $\delta^{(1)}V^*$ in Eq. (3.26) the varied body forces and surface tractions must satisfy overall requirements of equilibrium for the body. We can then give Eq. (3.24) in the following form:

$$\boxed{\delta^{(1)}(U^* + V^*) = 0} \tag{3.27}$$

We shall call $(U^* + V^*)$ the *total complementary energy* and denote it as π^*. Thus we can say

$$\boxed{\delta^{(1)}\pi^* = 0} \tag{3.28}$$

where

$$\boxed{\pi^* = U^* - \iiint_V u_i B_i\, dv - \oiint_S u_i T_i^{(\nu)}\, dS} \tag{3.29}$$

The principle of total complementary energy has the following meaning. The necessary requirements for ϵ_{ij} to be *kinematically compatible* are that

1. δB_i and $\delta T_i^{(\nu)}$ are statically compatible.
2. The stress field $\delta\tau_{ij}$, related to ϵ_{ij} by an elastic-type constitutive law, must extremize π^* with respect to all other stress fields satisfying equilibrium.

It may furthermore be shown that the total complementary energy is a minimum for the proper stress field.

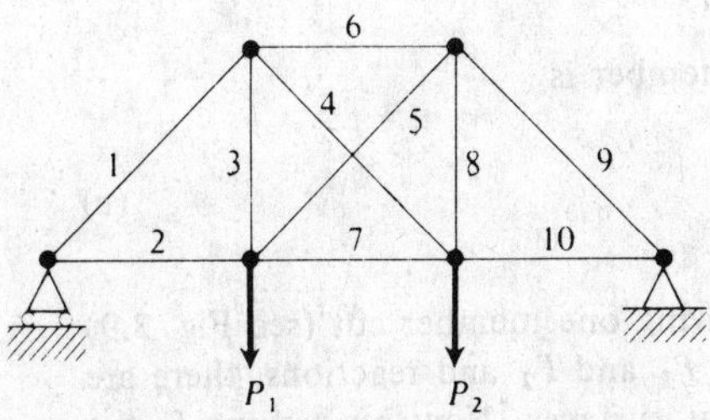

FIGURE 3.7
Statically indeterminate truss.

We now illustrate the use of this principle for finding the forces in a statically indeterminate structure. It is yet another force method.

EXAMPLE 3.5 Shown in Fig. 3.7 is a simple *statically indeterminate* plane truss. We wish to determine the forces in the members, assuming linear elastic behavior.

To solve this problem we will cut member 5 and replace the cut by forces f_5, as shown in Fig. 3.8. Furthermore, we will delete all external loads. What is left then is a statically determinate truss with equal and opposite forces f_5 applied as shown in the diagram. We can readily solve for the forces in each of the uncut members (note the reaction forces are zero) and we shall denote the forces as f_i. Clearly (for a linear system) each such force will be directly proportional to f_5 such that:

$$f_i = C_i f_5 \qquad i = 1, 2, \ldots, 10 \tag{a}$$

where $C_5 = 1$. Now delete the member 5, thus rendering the truss statically determinate, and solve for the forces in the remaining nine members stemming from the applied loads P_1 and P_2. We shall call these forces F_i. We take $F_5 = 0$. Now add to these forces those given by Eq. (*a*), thus including the contribution from the omitted member 5 (Fig. 3.8). The total force in the pth member can then be given as

$$F_p + C_p f_5$$

The stress in the pth member is then†

$$\tau_p = \frac{F_p + C_p f_5}{A_p} \tag{b}$$

†The fact that we are superposing stresses means that the procedure is valid only for linear elastic materials.

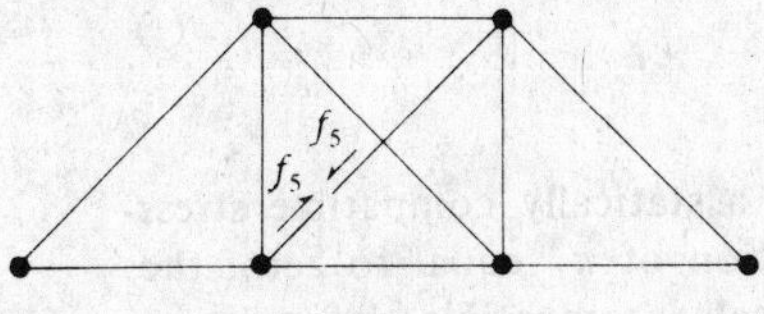

FIGURE 3.8
Truss with cut member and no external loads.

The corresponding strain in the pth member is

$$\epsilon_p = \frac{F_p + C_p f_5}{A_p E_p} \tag{c}$$

We may thus think of the truss as having one member cut (see Fig. 3.9) so that in addition to traction forces P_1 and P_2 and reactions, there are also two traction forces f_5 at the cut surfaces. Now *by varying* f_5 *we have available a family of admissible stress fields as given by Eq. (b).* The complementary strain energy function for such a family of admissible stresses then is

$$U^* = \iiint_V \int \epsilon_{ij}\, d\tau_{ij}\, dv = \sum_{p=1}^{10} \int \left(\frac{F_p + C_p f_5}{E_p A_p}\right) d\left(\frac{F_p + C_p f_5}{A_p}\right) A_p L_p$$

$$= \sum_{p=1}^{10} \frac{1}{2}\left(\frac{F_p + C_p f_5}{A_p}\right)^2 \frac{1}{E_p} A_p L_p = \frac{1}{2} \sum_{p=1}^{10} \frac{(F_p + C_p f_5)^2}{A_p E_p} L_p \tag{d}$$

The total complementary energy principle can then be given as follows, for no body forces:

$$\delta^{(1)} U^* - \iint_{S_2} u_i\, \delta T_i^{(\nu)}\, dS = 0 \tag{e}$$

The only traction force that is varied for the family of admissible stresses we have decided to use is f_5 at the two surfaces exposed by the cut (Fig. 3.8). However, since these surfaces undergo the *same* displacement and since the force f_5 on one surface is *opposite* in direction to that of the other surface, clearly the surface integral in the above equation is zero. We can accordingly say for the principle of total complementary energy:

$$\delta^{(1)} \left[\frac{1}{2} \sum_{p=1}^{10} \frac{(F_p + C_p f_5)^2}{A_p E_p} L_p\right] = 0$$

Therefore

$$\frac{d}{df_5}\left[\frac{1}{2} \sum_{p=1}^{10} \frac{(F_p + C_p f_5)^2}{A_p E_p} L_p\right] = 0 \tag{f}$$

Any value of f_5 would have led to a statically compatible stress field. Now, having rendered the first variation of π^* equal to zero, the resulting f_5 must be tied to a kinematically compatible deformation

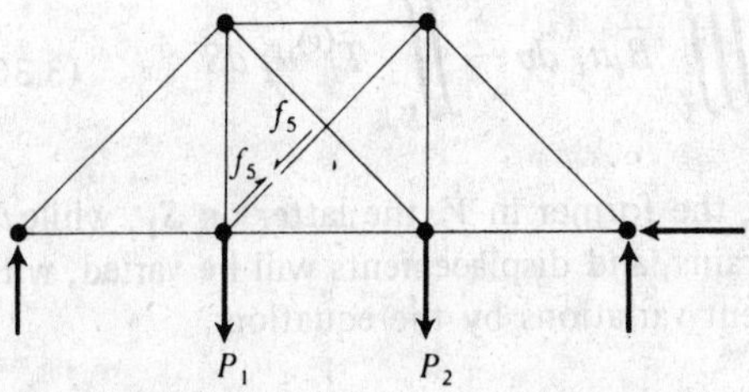

FIGURE 3.9
Truss with cut member and external loads.

field through the appropriate constitutive law. Accordingly, the computed value of f_5 from Eq. (*f*) must then be the actual force in member 5 of the truss; the actual forces in the other members can then be determined from Eq. (*b*).

Although we carried out the discussion in terms of a special simple truss without involving ourselves with numbers, the reader should have no trouble using the approach for other configurations. If there are two or more redundant members, these are cut in the manner in which we cut member 5 here. We may then reach a total complementary potential energy π^* having two or more force parameters. Extremizing π^* with respect to these parameters permits us to form enough equations to solve for these force parameters and thus solve for the actual forces in the truss.

Note that this method is analogous to the one presented in Example 3.2 illustrating the use of the total potential energy principle. That is, in the present undertaking we extremized the total complementary energy with respect to statically compatible stress fields (found by adjusting f_5 in cut members, etc.) while in the earlier undertaking we extremized the total potential energy with respect to kinematically compatible deformations found by giving displacements to movable pins.

*3.6 STATIONARY PRINCIPLES; REISSNER'S PRINCIPLE

We have presented two minimum principles, namely the principles of minimum total potential energy and the principle of minimum total complementary energy. We shall now present a principle which is set forth only as extrema–the exact nature of the extremum is not known. This and other such principles are becoming increasingly more important in structural mechanics and are called *stationary principles.*

We consider a functional given by E. Reissner in 1950, the extremization of which yields the equations of equilibrium, Cauchy's formula, a constitutive relation and the appropriate boundary condition–provided only that (1) we employ the classical strain-displacement relations of elasticity (this assures kinematic compatibility) in the variational process and (2) we take $\tau_{ij} = \tau_{ji}$ (symmetry). Thus we consider the functional

$$I_R = \iiint_V [\tau_{ij}\epsilon_{ij} - \mathfrak{U}^*(\tau_{ij})]\, dv - \iiint_V \bar{B}_i u_i \, dv - \iint_{S_1} \bar{T}_i^{(\nu)} u_i \, dS \tag{3.30}$$

The quantities $\bar{B}_i$ and $\bar{T}_i^{(\nu)}$ are prescribed, the former in V, the latter on S_1, while $\bar{u}_i$ is prescribed on S_2. Now the stresses, strains, and displacements will be varied, with strain variations related to the displacement variations by the equation:

$$\delta\epsilon_{ij} = \tfrac{1}{2}(\delta u_{i,j} + \delta u_{j,i}) \tag{3.31}$$

Carrying out the variations, we get

$$\delta^{(1)} I_R = \iiint_V \left(\delta\tau_{ij}\epsilon_{ij} + \tau_{ij}\,\delta\epsilon_{ij} - \frac{\partial \mathfrak{U}^*}{\partial \tau_{ij}}\,\delta\tau_{ij}\right) dv$$

$$- \iiint_V \bar{B}_i\,\delta u_i\, dv - \iint_{S_1} \bar{T}_i^{(\nu)}\,\delta u_i\, dS \tag{3.32}$$

Note that in view of Eq. (3.31), the second term on the right of the above equation can be recast as follows:

$$\iiint_V \tau_{ij}\,\delta\epsilon_{ij}\, dv = \iiint_V \tau_{ij}\,\delta\left(\frac{u_{i,j} + u_{j,i}}{2}\right) dv = \iiint_V \tau_{ij}\,\delta u_{i,j}\, dv$$

where we have used the symmetry of the stress tensor in the last step. Continuing, we have

$$\iiint_V \tau_{ij}\,\delta\epsilon_{ij}\, dv = \iiint_V (\tau_{ij}\,\delta u_i)_{,j}\, dv - \iiint_V (\tau_{ij,j}\,\delta u_i)\, dv$$

Using Gauss's theorem, we then get

$$\iiint_V \tau_{ij}\,\delta\epsilon_{ij}\, dv = \oint\!\!\oint_{S=S_1+S_2} \tau_{ij}\nu_j\,\delta u_i\, dS - \iiint_V \tau_{ij,j}\,\delta u_i\, dv$$

Now use the above result in Eq. (3.32). We get, on regrouping the terms, the first variation of the Reissner functional:

$$\delta^{(1)} I_R = \iiint_V \left[\left(\epsilon_{ij} - \frac{\partial \mathfrak{U}^*}{\partial \tau_{ij}}\right)\delta\tau_{ij} - (\tau_{ij,j} + \bar{B}_i)\,\delta u_i\right] dv$$

$$- \iint_{S_1} (\bar{T}_i^{(\nu)} - \tau_{ij}\nu_j)\,\delta u_i\, dS - \iint_{S_2} \tau_{ij}\nu_j\,\delta u_i\, dS \tag{3.33}$$

If we require $\delta^{(1)} I_R$ to vanish for independent variations $\delta\tau_{ij}$ and δu_i both in V and on $S = S_1 + S_2$, we obtain the following Euler-Lagrange equations and boundary conditions:

$$\epsilon_{ij} = \frac{\partial \mathcal{U}^*}{\partial \tau_{ij}} \quad \text{in } V; \qquad \tau_{ij,j} + \bar{B}_i = 0 \quad \text{in } V \tag{3.34}$$

$$\bar{T}_i^{(\nu)} = \tau_{ij}\nu_j \quad \text{on } S_1; \qquad u_i = \bar{u}_i \quad \text{on } S_2 \tag{3.35}$$

Thus both the stress-strain law and the equations of equilibrium are derivable from the Reissner functional, as are stress and displacement boundary conditions. Recall that the strain-displacement relations are built into the variational process via Eq. (3.31).

Although the nature of the extremum is not known for the Reissner principle, we wish to point out a distinct advantage of this principle. You will recall in the potential energy principle we vary displacements in order to arrive at an equilibrium configuration. In approximation techniques (as pointed out in Section 3.3) we shall employ approximate displacement fields to obtain approximate equilibrium configurations. As will be shown, this can lead to results very closely resembling real displacement fields. However, the corresponding stress field may be considerably in error. This occurs because of the rapid deterioration of accuracy of an approximate solution, such as the displacement field, when differentiations are required to get other results, such as the stress field. Furthermore, approximate techniques based on the complementary energy principles lead to the converse problem; while we may get good approximations as to stress fields, we often obtain poor results for displacement fields owing to the fact that we must solve further differential equations in the process. By contrast, the Reissner principle allows for the arbitrary variation of *both* stress and displacements. We can, for this reason, handle both quantities independently in approximation procedures to achieve good results simultaneously in both categories.† We shall illustrate the use of the Reissner principle in the following chapter.

Part B
The Castigliano Theorems and Structural Mechanics

3.7 PRELIMINARY REMARKS

In Part A we presented a number of key principles that will form the basis of study for much of this text. And to illustrate these principles in a most simple way we considered certain problems concerning both statically determinate and indeterminate simple trusses. We have thus already launched our effort in structural mechanics.

†Nevertheless, we note the apparent preference for the potential and complementary energy principles in the technical literature. This preference is probably due to the greater physical appeal of dealing with a *minimum* of an *energy* quantity. There is, however, a growing tendency to use the Reissner principle in finite element approaches to structural mechanics.

In Part B we continue in this effort in a more formal way by deriving the Castigliano theorems directly from the principles of the previous sections and applying these theorems to various kinds of truss problems. In Chapter 4 we will find that these theorems are useful for the study of beams.

3.8 THE FIRST CASTIGLIANO THEOREM

We may now employ the principle of total potential energy to formulate the first Castigliano theorem. Let us consider an elastic body maintained at all times in a state of equilibrium by a system of rigid supports (see Fig. 3.10*a*) against external loads. Consider a system of N external, independent point force vectors and point couple vectors. We shall denote both force and couple vectors as Q_i. At each point force Q_i we identify Δ_i as the displacement component in the direction of Q_i of the movement of this point resulting from deformation of the body. And at each point of application of couple Q_i we identify Δ_i as the rotation component of this point in the direction of Q_i resulting from deformation of the body. We shall call Δ_i the *generalized displacements*. We thus have N generalized displacements Δ_i and we

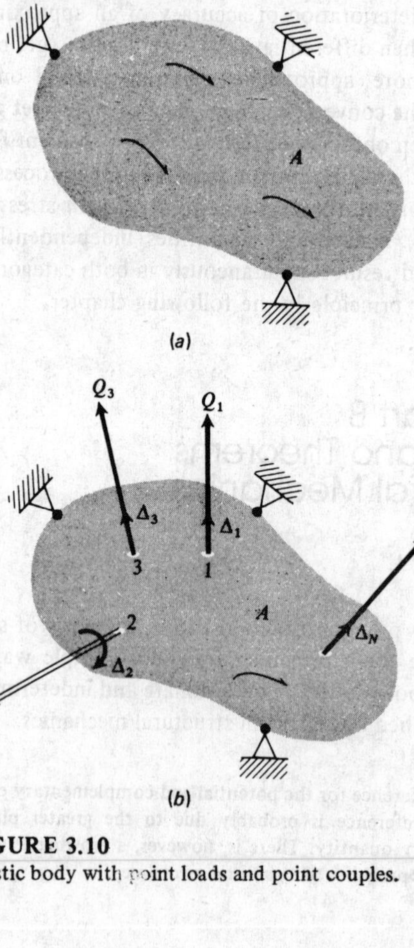

FIGURE 3.10
Elastic body with point loads and point couples.

assume next that we can give the strain energy of the body U in terms of these quantities. The total potential energy can then be given as follows:

$$\pi = U(\Delta_i) - \sum_{k=1}^{N} Q_k \Delta_k \tag{3.36}$$

We see that the total potential energy under the aforestated circumstances depends on the parameters Δ_i, which characterize the deformation as far as π is concerned, and the associated external loads and couples. Accordingly, for *equilibrium* we require that the first variation of π, found by varying Δ_i, be zero. Thus:

$$\delta^{(1)}\pi = \sum_{i=1}^{N} \left(\frac{\partial \pi}{\partial \Delta_i} \delta \Delta_i \right) = 0$$

Therefore

$$\sum_{i=1}^{N} \left(\frac{\partial U}{\partial \Delta_i} \delta \Delta_i - Q_i \delta \Delta_i \right) = 0 \tag{3.37}$$

We may next express the above equation as follows:

$$\sum_{i=1}^{N} \left(\frac{\partial U}{\partial \Delta_i} - Q_i \right) \delta \Delta_i = 0 \tag{3.38}$$

We may consider the $\delta\Delta_i$ to be independent of each other and so we conclude that

$$\boxed{Q_i = \frac{\partial U}{\partial \Delta_i} \qquad i = 1, \ldots, N} \tag{3.39}$$

This is the well-known *first Castigliano theorem*. By computing U as a function of Δ_i from elasticity considerations we can then readily determine the required force or torque Q_n for a particular generalized displacement Δ_i. The following example illustrates how we may use this theorem.

EXAMPLE 3.6 We consider now a very simple problem, shown in Fig. 3.11. We wish to determine what force P is needed to cause joint A to descend a given distance δ. We assume the members are composed of linear elastic material and are identical in every way.

We could do this problem by computing the necessary elongation in each bar to achieve the deflection δ and then, using Hooke's law, we could determine the force in each member. Finally, considering the pin A as a free body, we could then evaluate P. We shall instead illustrate

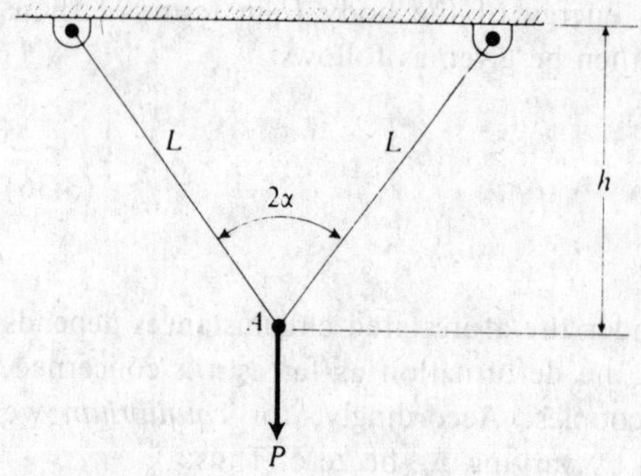

FIGURE 3.11
Force P causes A to descend distance δ.

the use of the Castigliano first theorem here to accomplish the same task. Clearly δ is the sole generalized coordinate present and so we must determine U as a function of δ. This is readily done once we see that the new length L' of the members is given (for reasonably small *changes* in the angle α) as (see Fig. 3.12):

$$L' = \frac{h}{\cos\alpha} + \delta\cos\alpha \tag{a}$$

The strain in each bar is then

$$\epsilon = \frac{\delta\cos\alpha}{h/\cos\alpha} = \frac{h\delta}{L^2} \tag{b}$$

The strain energy of the system U can now easily be determined as

$$U = 2\left(\iiint_V \int_\epsilon \tau\, d\epsilon\, dv\right) = 2\left(\int_\epsilon E\epsilon\, d\epsilon\right) AL = E\epsilon^2 AL$$

$$= E\left(\frac{h\delta}{L^2}\right)^2 A\left(\frac{h}{\cos\alpha}\right) = \frac{EA\,\delta^2}{h}\cos^3\alpha \tag{c}$$

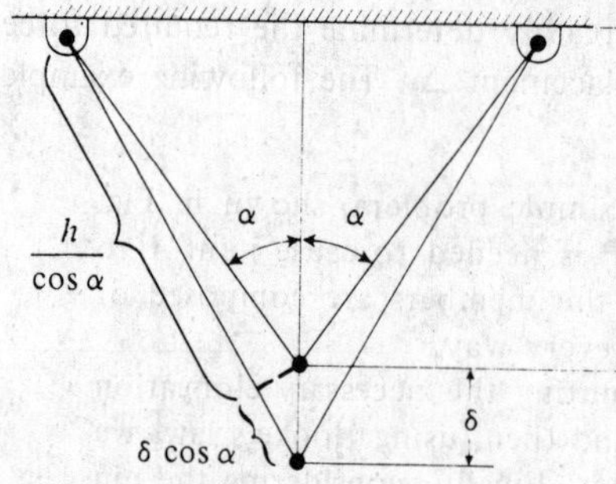

FIGURE 3.12
System showing deformed geometry.

where A is the cross-sectional area of the members. Hence:

$$P = \frac{\partial U}{\partial \delta} = \frac{2EA\,\delta \cos^3 \alpha}{h} \tag{d}$$

We will now use the Castigliano theorem to present the *stiffness method*† of solving *linearly elastic*, statically indeterminate trusses. Recall from Section 3.2 that we could readily establish kinematically compatible deformation fields by giving those pins that are not fully constrained (we shall call them movable pins) virtual displacement components in the x and y directions. Suppose we consider n such displacement components, denoting them as $\Delta_1, \Delta_2, \ldots, \Delta_n$. It will be understood that these deflections take the truss from an undeformed geometry to a kinematically compatible deformed geometry. The strain for any member, say the pth member, can be found by superposing strains developed from the separate displacements of the movable pins.‡ We thus may say

$$\epsilon_p = \sum_{s=1}^{n} \kappa_{ps}\,\Delta_s \tag{3.40}$$

where the κ_{ps} are constants, the first subscript of which refers to the member on which the strain is being considered and the second subscript refers to the virtual displacement. (Thus κ_{12} is the strain in member 1 per unit deflection corresponding to Δ_2.) A strain energy expression can then be given in terms of the displacements Δ_p as follows for a truss with M members:

$$\begin{aligned} U &= \iiint_V \int_\epsilon \tau\, d\epsilon\, dv = \sum_{p=1}^{M} \int_\epsilon (E_p \epsilon_p)\, d\epsilon_p (A_p L_p) \\ &= \sum_{p=1}^{M} \frac{E_p \epsilon_p^2}{2} A_p L_p = \sum_{p=1}^{M} \frac{E_p}{2} \left(\sum_{s=1}^{n} \kappa_{ps}\,\Delta_s \right)^2 A_p L_p \end{aligned} \tag{3.41}$$

Now employ Castigliano's first theorem. We can then say

$$\frac{\partial U}{\partial \Delta_r} = P_r = \sum_{p=1}^{M} E_p \left(\sum_{s=1}^{n} \kappa_{ps}\,\Delta_s \right) \kappa_{pr} A_p L_p = \sum_{p=1}^{M} \sum_{s=1}^{n} E_p A_p L_p \kappa_{ps} \kappa_{pr}\,\Delta_s$$

$$r = 1, 2, \ldots, n$$

†The method to be developed corresponds to the dummy displacement method employed in conjunction with the principle of virtual work.

‡It is because we are using here the *superposition principle* that we are restricted to *linear elastic behavior*.

We have thus a set of n simultaneous equations involving the forces P_r, in the direction of the displacements Δ_r, and these displacements. These equations may be written as follows:

$$P_r = \sum_{s=1}^{n} K_{rs}\,\Delta_s \tag{3.42a}$$

where

$$K_{rs} = \left(\sum_{p=1}^{M} E_p A_p L_p \kappa_{pr} \kappa_{ps} \right) \tag{3.42b}$$

The constants K_{rs} are called *stiffness constants*. We may interpret K_{rs} as the force in the direction of P_r needed per unit deflection Δ_s. Now the forces P_r in Eq. (3.42*a*) are those needed to maintain the deflections Δ_s. If we know K_{rs} and if we insert the *known values* of the external loads, we arrive at a system of simultaneous equations for directly determining the Δ_s, which clearly now must represent the *actual displacements* of the movable pins. The forces in the truss members may then be readily computed and the truss has been solved completely. We will pursue this procedure in detail in Chapter 9 when we introduce the method of finite elements.

In order to cast Eq. (3.39) in another useful form, we now introduce the so-called *generalized coordinates* q_i, as any set of variables that *uniquely establish the configuration of a body or system of bodies in space*. These variables need not have the units of length or rotation. Suppose as a result of assumptions as to deformation that there are N variables that serve as generalized coordinates for a particular body loaded by forces $P_1, P_2, \ldots$. Then we can give the differentials of the potential energy V of these forces, for use in the total potential energy functional, in the following manner:

$$dV = -\sum_{i=1}^{N} f_i(P_1, P_2, \ldots)\, dq_i$$

where f_i are functions of the external loads. We now define the *generalized force* $(Q_i)_g$ as that function f_i that accompanies dq_i in the above formulation. Clearly the generalized force may have units other than that of ordinary forces or torques. Expressing U as a function of the generalized coordinates, the total potential energy principle then becomes

$$\delta^{(1)}\pi = \sum_{i=1}^{N} \frac{\partial U}{\partial q_i}\,\delta q_i - \sum_{i=1}^{N} (Q_i)_g\,\delta q_i = 0$$

As before, we reach the result:

$$\frac{\partial U}{\partial q_i} = (Q_i)_g \tag{3.43}$$

now in terms of generalized forces and generalized coordinates.

As an example of generalized forces, consider the frame shown in Fig. 3.13*a* loaded by a single force F_1 and a couple moment M_2. Then Δ_1 and Δ_2 shown in Fig. 3.13*a* would be generalized displacements for these loads. But it is difficult to compute U for these displacements. Instead, we present now a means of describing the deformed geometry of the frame approximately in terms of three parameters called frame parameters. Accordingly, we consider that corners C and D of the frame remain right angles and merely rotate as rigid corners by the amounts α and β, respectively, as shown in Fig. 3.13*b*. Furthermore, we will assume that these corners have the same horizontal movement γ. At all other positions of the frame we can consider deformation consistent with simple beam theory. We are thus describing the deformation with three parameters—the so-called frame parameters α, β, and γ. We may consider these to be generalized displacements describing the deformation of the frame. What then are the generalized forces $(Q_i)_g$ accompanying these generalized displacements? For this, we go to Castigliano's first theorem and, after expressing the strain energy U in terms of α, β, and γ, we can say

$$\begin{aligned}
\frac{\partial U(\alpha, \beta, \gamma)}{\partial \alpha} &= (Q_\alpha)_g \\
\frac{\partial U(\alpha, \beta, \gamma)}{\partial \beta} &= (Q_\beta)_g \\
\frac{\partial U(\alpha, \beta, \gamma)}{\partial \gamma} &= (Q_\gamma)_g
\end{aligned} \tag{3.44}$$

How do we determine these generalized Q's beyond using the above formulations? We can equate the work done W_k by F_1 and M_2 for the deformed frame to that formulated with $(Q_\alpha)_g$, $(Q_\beta)_g$, and $(Q_\gamma)_g$. Thus we have for linear elastic behavior:

$$W_k = \tfrac{1}{2} F_1 w_1 + \tfrac{1}{2} M_2 w_2' = \tfrac{1}{2}(Q_\alpha)_g \alpha + \tfrac{1}{2}(Q_\beta)_g \beta + \tfrac{1}{2}(Q_\gamma)_g \gamma$$

where w_1 is found from the deflection curve for the left upright member at F_1, and w_2' is the slope of the right upright member at M_2. Now w_1 and w_2' can be expressed in terms of the frame parameters. This means that W_k in the preceding equation is available as a function of the frame parameters. Hence, we can now get the generalized forces Q_q in terms of the frame parameters by differentiating the above equations as follows:

$$\begin{aligned}
\frac{\partial W_k}{\partial \alpha} &= \frac{1}{2}(Q_\alpha)_g \\
\frac{\partial W_k}{\partial \beta} &= \frac{1}{2}(Q_\beta)_g
\end{aligned} \tag{3.45}$$

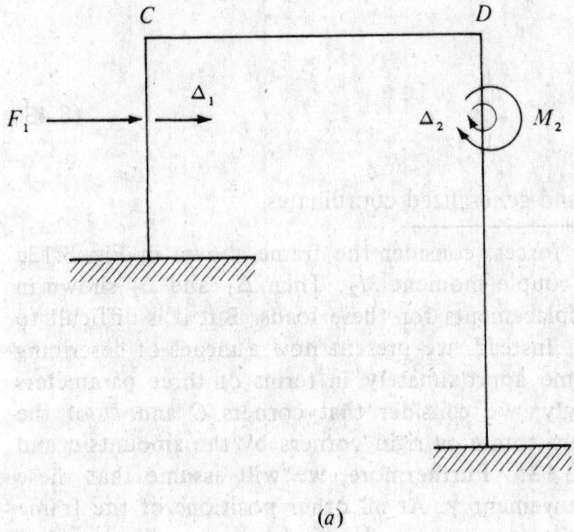

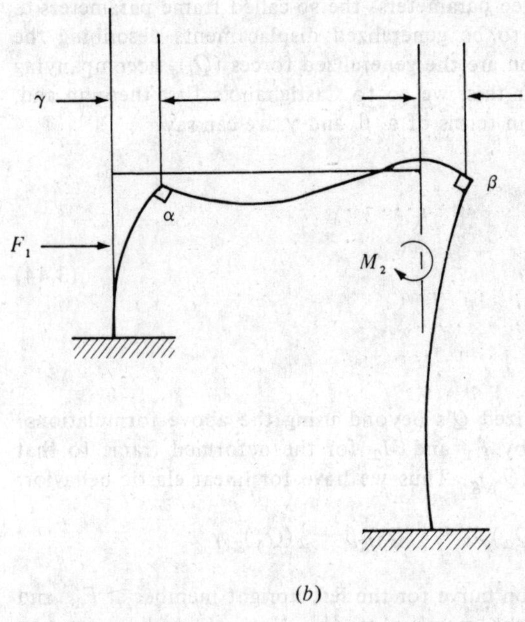

FIGURE 3.13
Frame showing deformation and frame parameters.

$$\frac{\partial W_k}{\partial \gamma} = \frac{1}{2}(Q_\gamma)_g \tag{3.45}$$
(Cont.)

If we use these Q's in Eq. (3.44) (from the first Castigliano theorem) with U expressed in terms of α, β, and γ, we can then solve for the proper α, β, and γ for specific values of P_1 and M_2. We thus may determine an approximation of the deformation of the frame.

3.9 THE SECOND CASTIGLIANO THEOREM

We may derive the second Castigliano theorem from the principle of total complementary energy. This theorem will accordingly afford us another force method for solving structural problems.

Consider again an elastic body acted on by a system of N point forces and point couples, which we represent as Q_i. This body is maintained in equilibrium by a system of rigid supports (Fig. 3.10). We suppose that we can vary Q_i *independently* to form δQ_i. We again employ the generalized displacement components Δ_i, described in the previous section, stemming from the actual deformation of the body resulting from Q_i. The complementary potential of Q_i may then be given as

$$V^* = -\sum_{k=1}^{N} \Delta_k Q_k \tag{a}$$

so that

$$\delta^{(1)} V^* = -\sum_{k=1}^{N} \Delta_k \, \delta Q_k \tag{b}$$

Now the total complementary energy π^* becomes

$$\pi^* = U^*(Q_i) + V^* \tag{3.46}$$

where, you will note, we have expressed the complementary energy as a function of Q_i. The principle of total complementary energy may now be used to ensure kinematic compatibility as follows:

$$\delta^{(1)}\pi^* = 0$$

therefore

$$\sum_{i=1}^{N} \left(\frac{\partial U^*}{\partial Q_i} \, \delta Q_i - \Delta_i \, \delta Q_i \right) = 0 \tag{3.47}$$

Collecting terms, we get

$$\sum_{i=1}^{N} \left[\left(\frac{\partial U^*}{\partial Q_i} - \Delta_i \right) \delta Q_i \right] = 0 \tag{3.48}$$

Since we have assumed that the δQ_i are independent of each other, we can conclude that

$$\boxed{\Delta_i = \frac{\partial U^*}{\partial Q_i}} \tag{3.49}$$

This is the *second Castigliano theorem*. Thus, if we can express U^* in terms of Q_i we can compute the values of generalized displacements. For trusses this is sufficient for determining the deformation of the entire truss.

If the body is linearly elastic you will recall from Chapter 1 that

$$U^* = U$$

Accordingly, the second Castigliano theorem for such cases may be given as follows:

$$\boxed{\Delta_i = \frac{\partial U}{\partial Q_i}} \tag{3.50}$$

We shall now illustrate the use of the second Castigliano theorem for truss problems. These are further examples of force methods.

EXAMPLE 3.7 Consider a simple structural system composed of two bars (see Fig. 3.14) and supporting a load P equal to 4×10^4 N. The bars each have a cross-sectional area of 500 mm^2 and are made of a material having a nonlinear elastic stress-strain behavior (see Fig. 3.15) that may be approximately represented by the following relation:

$$\tau = 6 \times 10^9 \, \epsilon^{1/2} \tag{a}$$

We are to determine the vertical deflection Δ of joint B. Note that Δ is a generalized displacement for this problem.

The system is statically determinate; we can compute the forces in each member by considering joint B as a free body. Thus we get:

$$\begin{aligned} P_{(BC)} &= 2P \text{ N} \qquad &\text{therefore } \tau_{BC} &= 4000P \text{ Pa} \\ P_{(AB)} &= -1.732P \text{ N} \qquad &\text{therefore } \tau_{AB} &= -3464P \text{ Pa} \end{aligned} \tag{b}$$

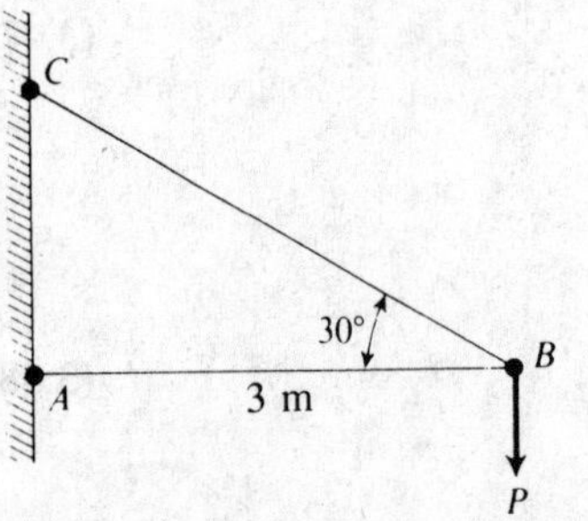

FIGURE 3.14
Two-bar structure.

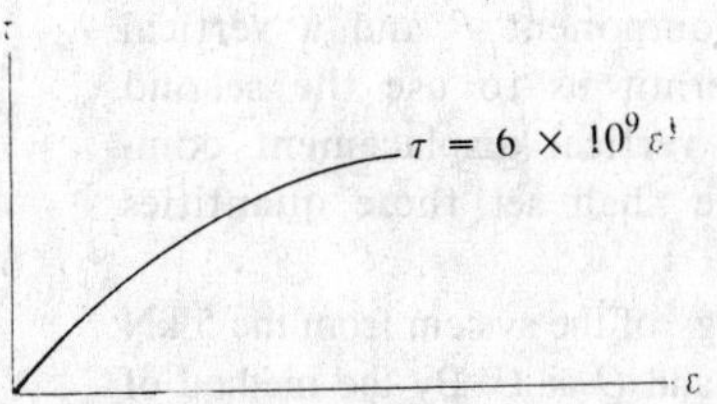

FIGURE 3.15
Nonlinear elastic stress-strain law.

The complementary strain energy can now be determined as a function of P as follows:

$$U^* = \iiint_V \int_{\tau_{ij}} \epsilon_{ij}\, d\tau_{ij}\, dv = \iiint_V \int_0^{4000P} \left(\frac{\tau_{BC}}{6 \times 10^9}\right)^2 d\tau_{BC}\, dv$$

$$+ \iiint_V \int_0^{-3464P} \left(\frac{\tau_{AB}}{6 \times 10^9}\right)^2 d\tau_{AB}\, dv = \left(\frac{1}{6 \times 10^9}\right)^2 (5 \times 10^{-4})$$

$$\cdot \left(\int_0^{3.464} \frac{\tau_{BC}^3}{3}\Bigg|_0^{4000P} dL + \int_0^3 \frac{\tau_{AB}^3}{3}\Bigg|_0^{-3464P} dL \right) = 4.491 \times 10^{-13} P^3$$

Now employing the Castigliano theorem we get

$$\Delta = \frac{\partial U^*}{\partial P} = 1.347 \times 10^{-12} P^2$$

For a load of 40,000 N we have for Δ

$$\Delta = 0.0022 \text{ m}$$

EXAMPLE 3.8 Shown in Fig. 3.16 is a statically determinate simple truss, loaded by concentrated loads at pins D and B. What is the total deflection of pin C as a result of these loads? The members are made of linear elastic material.

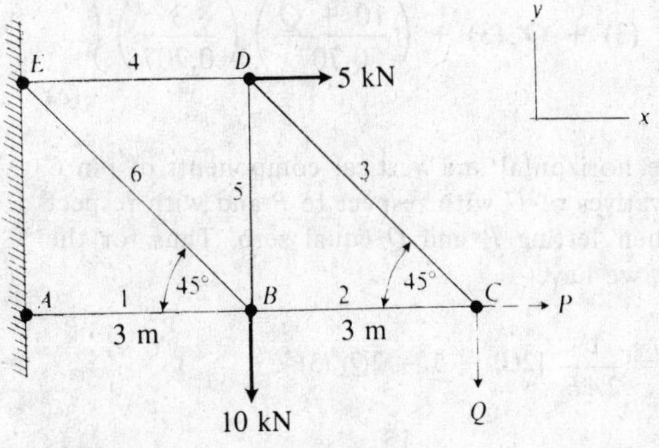

FIGURE 3.16
Setup to find deflection of joint C.

We shall assume a horizontal force component P and a vertical force component Q at pin C so as to permit us to use the second Castigliano theorem for the horizontal and vertical displacement components there. At the appropriate time we shall set these quantities equal to zero.

Our first step is to determine the strain energy of the system from the 5 kN load at D, the 10 kN load at B, and the loads P and Q at C. By the method of joints, we have for the forces in the members of the truss

$$\begin{aligned} AB &= P - 2(5 + Q) && \text{tension} \\ BC &= P - Q && \text{tension} \\ CD &= \frac{Q}{0.707} && \text{tension} \\ DE &= 5 + Q && \text{tension} \\ DB &= Q && \text{compression} \\ EB &= \frac{10 + Q}{0.707} && \text{tension} \end{aligned} \qquad (a)$$

We now determine U in the following way:

$$U = \iiint_V \int_\epsilon \tau \, d\epsilon \, dv = \sum_p \int \tau_p \left(\frac{d\tau_p}{E_p}\right) A_p L_p$$

$$= \sum_p \frac{\tau_p^2}{2E_p} A_p L_p = \sum_p \frac{F_p^2 L_p}{2A_p E_p} \qquad (b)$$

Taking A_p and E_p as having the same value for each member, we get

$$\mathrm{U} = \frac{1}{2AE}\left\{[P - 2(5 + Q)]^2 (3) + (P - Q)^2 (3) + \left(\frac{Q}{0.707}\right)^2 \left(\frac{3}{0.707}\right) + (5 + Q)^2 (3) + Q^2 (3) + \left(\frac{10 + Q}{0.707}\right)^2 \left(\frac{3}{0.707}\right)\right\} \qquad (c)$$

We may now compute the horizontal and vertical components of pin C by first taking partial derivatives of U with respect to P and with respect to Q, respectively, and then letting P and Q equal zero. Thus for the horizontal component Δ_H, we have

$$\Delta_H = \left(\frac{\partial U}{\partial P}\right)_{P=Q=0} = \frac{1}{2AE} [2(P - 5 - 2Q) (3) + [2(P - Q) (3)]_{P=Q=0} = -\frac{15}{AE} \text{ m} \qquad (d)$$

The minus sign above indicates that the horizontal deflection component of the pin is opposite in sense to the direction of the force P shown in Fig. 3.16. Now we get Δ_V;

$$\Delta_V = \left(\frac{\partial U}{\partial Q}\right)_{P=Q=0} = \frac{1}{2AE}\left[2(P-5-2Q)(3)(-2) + 2(P-Q)(3)(-1) + \frac{2Q}{(0.707)^2}\left(\frac{3}{0.707}\right) + 2(5+Q)(3) + 6Q + \frac{2(10+Q)}{(0.707)^2}\frac{3}{0.707}\right]_{P=Q=0}$$

therefore

$$\Delta_V = \frac{129.9}{AE}\ \text{m} \tag{e}$$

The deflection at pin C can then be given as

$$\Delta_C = \frac{1}{AE}\,(-15\mathbf{i} + 129.9\mathbf{j})\ \text{m} \tag{f}$$

3.10 SUMMARY CONTENTS FOR PARTS A AND B

We have thus far set forth two analogous sets of principles and theorems, namely those generating displacement methods and those generating force methods. We now list certain aspects of these results side by side in Table 3.1.

We shall proceed further in this text with trusses when we consider them in conjunction with the method of finite elements. It should be obvious that matrix methods can be effectively used in truss problems. Indeed, any further work in this area would most certainly center around matrix methods (see Martin, 1966).

In Part A we defined certain functionals and then went on to show that the Euler-Lagrange equations were important equations in solid mechanics. Thus the total potential energy functional generated the equations of equilibrium while the total complementary energy functional generated the compatibility equations. In Part C of this chapter we shall consider the reverse process in that we shall start with a boundary-value problem and attempt to generate a functional whose Euler-Lagrange equations and boundary conditions comprise the given boundary-value problem. This material will form additional background for later undertakings in Parts II and III of the text. It will also help us to understand the approximation techniques that we shall introduce in Part D of this chapter.

Part C
Quadratic Functionals

3.11 SELF-ADJOINT AND POSITIVE DEFINITE OPERATORS

In previous sections we have presented functionals whose Euler-Lagrange equations were important equations of solid mechanics. In this section we shall begin with a

TABLE 3.1
Summary of Key Energy Methods

Displacement methods	Force methods	Comments
Principle of virtual work $\iiint_V B_i\, \delta u_i\, dv + \iint_S T_i^{(\nu)}\, \delta u_i\, ds = \iiint_V \tau_{ij}\, \delta \epsilon_{ij}\, dv$ Dummy displacement method for trusses	*Principle of complementary virtual work* $\iiint_V u_i\, \delta B_i\, dr + \iint_S u_i\, \delta T_i^{(\nu)}\, ds = \iiint_V \epsilon_{ij}\, \delta \tau_{ij}\, dv$ Dummy load method for trusses	No restriction on the constitutive law Infinitesimal deformation
Principle of total potential energy $\delta^{(1)}(\pi) = 0$ Form kinematically compatible deformation fields for variation process. Move pins for trusses.	*Principle of total complementary energy* $\delta^{(1)}(\pi^*) = 0$ Form statically compatible stress fields for variation process by cutting members and replacing cuts by forces which may be varied.	For elastic bodies only
First Castigliano theorem $Q_i = \frac{\partial U}{\partial \Delta_i}$ Stiffness methods	*Second Castigliano theorem* $\Delta_i = \frac{\partial U^*}{\partial Q_i}$ Flexibility methods	Generalized point forces and displacement components in the direction of generalized loads

differential equation which is one of a certain class of differential equations and we shall show how under certain conditions we can reach the appropriate functional for which this equation is the Euler-Lagrange equation. That is, we shall be moving in a direction *opposite* to that presented earlier in this chapter and in Chapter 2. And our results will be valid for many aspects of solid mechanics and will have validity transcending this field. We thus have the opportunity here of opening the discussion toward other areas of mathematical physics [for an excellent exposition of this approach in detail see Mikhlin (1964)]. In order to achieve this greater generality we shall need now to introduce several definitions. (These definitions will also be needed for other purposes in Chapters 7 and 17.)

We will represent a differential equation in the discussion as follows:

$$Lu = f \tag{3.51}$$

Here L is a linear operator acting on the dependent variable u (or variables $u, v \ldots$) and f, the so-called driving (or forcing) function, represents a function of the independent variables. Examples of *linear* operators are the harmonic operator ∇^2 and the biharmonic operator ∇^4. A *nonlinear* operator is

$$L = \frac{d^2}{dx^2} + \sin$$

Next we define the *inner product* of two functions g and h over the domain of the problem V as follows:

$$\langle g, h \rangle \equiv \text{INNER PRODUCT OF } g \text{ AND } h \equiv \iiint_V gh\, dv \tag{3.52a}$$

The inner product of two vector fields **A** and **B** over the domain of the problem V is defined as:

$$\langle \mathbf{A}, \mathbf{B} \rangle \equiv \iiint_V \mathbf{A} \cdot \mathbf{B}\, dv \tag{3.52b}$$

We will limit the discussion now to operators L that are *self-adjoint* (or *symmetric*). Such operators have the following property:

$$\langle Lu, v \rangle = \langle u, Lv \rangle \tag{3.53}$$

where u and v are any two functions that satisfy the same appropriate boundary conditions. To illustrate the establishment of the self-adjoint property for an operator L along with the appropriate boundary conditions, we shall now consider two cases for operators involved in the theory of beams (Chapter 4) and the theory of plates (Chapter 6) respectively.

Case 1.

$$L = \frac{d^2}{dx^2}\left[g(x) \frac{d^2}{dx^2} \right]$$

We proceed to investigate the conditions under which Eq. (3.53) is valid for the above operator. Thus we first examine the left side of Eq. (3.53). Considering the domain of the problem to extend from $x = 0$ to $x = L$ we have

$$\langle Lu, v \rangle = \int_0^L \left[\frac{d^2}{dx^2} g(x) \frac{d^2 u}{dx^2} \right] v\, dx \tag{3.54}$$

Integrating by parts twice on the right side we get

$$\int_0^L \left(\frac{d^2}{dx^2} g(x) \frac{d^2 u}{dx^2} \right) v\, dx = \frac{d}{dx}\left[g(x) \frac{d^2 u}{dx^2} \right] v \Bigg|_0^L \tag{3.55}$$

$$-g(x)\frac{d^2u}{dx^2}\frac{dv}{dx}\bigg|_0^L + \int_0^L g(x)\frac{d^2u}{dx^2}\frac{d^2v}{dx^2}\,dx \qquad \text{(3.55) (Cont.)}$$

Similarly, considering the right side of Eq. (3.53) we get

$$\int_0^L u\left[\frac{d^2}{dx^2}g(x)\frac{d^2v}{dx^2}\right]dx = \frac{d}{dx}\left[g(x)\frac{d^2v}{dx^2}\right]u\bigg|_0^L$$

$$-g(x)\frac{d^2v}{dx^2}\frac{du}{dx}\bigg|_0^L + \int_0^L g(x)\frac{d^2u}{dx^2}\frac{d^2v}{dx^2}\,dx \qquad (3.56)$$

Equating the right sides of Eqs. (3.55) and (3.56) and canceling the integrals we get, as a requirement of the self-adjoint property on moving all terms to the left side of the equation,

$$\left\{\frac{d}{dx}\left[g(x)\frac{d^2u}{dx^2}\right]v - g(x)\left(\frac{d^2u}{dx^2}\right)\left(\frac{dv}{dx}\right) - \frac{d}{dx}\left[g(x)\frac{d^2v}{dx^2}\right]u + g(x)\left(\frac{d^2v}{dx^2}\right)\left(\frac{du}{dx}\right)\right\}_0^L = 0$$

Using primes for derivatives and rearranging, we get the following requirement:

$$[v(gu'')' - u(gv'')' + g(v''u' - u''v')]\bigg|_0^L = 0 \qquad (3.57)$$

Thus the above equation contains the appropriate boundary conditions for the operator

$$\frac{d^2}{dx^2}\left(g\,\frac{d^2}{dx^2}\right)$$

to be self-adjoint. Clearly beams that are pin-ended, fixed at the ends,† or free at the ends satisfy the above requirements as do beams with combinations of these end conditions.

Case 2.

$$L = \nabla^2[g(x, y)\nabla^2] \qquad \text{where } \nabla^2 = \frac{\partial^2}{\partial x^2} + \frac{\partial^2}{\partial y^2}$$

We proceed as before by considering the left side of Eq. (3.53) as follows

†We will see in the next chapter that the following are the appropriate boundary conditions, using the function u here:

Simple supports	$u = u'' = 0$
Fixed end	$u = u' = 0$
Free end	$u'' = (gu'')' = 0$

$$\iint_S [\nabla^2(g\nabla^2 u)]v\, dA = \iint_S \{\nabla\cdot[v\nabla(g\nabla^2 u)] - \nabla v\cdot\nabla(g\nabla^2 u)\}\, dA \qquad (3.58)$$

where the right side may be verified by carrying out the divergence operator for the first term using the chain rule. Now, using a well-known vector identity involving the dot product of two gradients, we have for the last term in the above equation:

$$\nabla v\cdot\nabla(g\nabla^2 u) = \nabla\cdot[(g\nabla^2 u)(\nabla v)] - g\nabla^2 u\nabla^2 v$$

Substituting the above result into Eq. (3.58) we get

$$\iint_S [\nabla^2(g\nabla^2 u)]v\, dA = \iint_S \{\nabla\cdot[v\nabla(g\nabla^2 u) - (g\nabla^2 u)(\nabla v)] + g\nabla^2 u\nabla^2 v\}\, dA \qquad (3.59)$$

Similarly, for the right side of Eq. (3.53) we get

$$\iint_S u[\nabla^2(g\nabla^2 v)]\, dA = \iint_S \{\nabla\cdot[u\nabla(g\nabla^2 v) - (g\nabla^2 v)(\nabla u)] + g\nabla^2 u\nabla^2 v\}\, dA$$

Equating the right sides of the two equations above we get the following result:

$$\iint_S \nabla\cdot[v\nabla(g\nabla^2 u) - u\nabla(g\nabla^2 v) + (g\nabla^2 v)\nabla u - (g\nabla^2 u)\nabla v]\, dA = 0$$

Finally, employing the divergence theorem we have

$$\oint_\Gamma [v\nabla(g\nabla^2 u) - u\nabla(g\nabla^2 v) + (g\nabla^2 v)\nabla u - (g\nabla^2 u)\nabla v]\cdot\mathbf{n}\, ds = 0$$

where $\mathbf{n}$ is the outward normal from the boundary. Since $\nabla(\ )\cdot\mathbf{n} = \partial(\ )/\partial n$ we can say for the above:

$$\oint_\Gamma \left[v\frac{\partial}{\partial n}(g\nabla^2 u) - u\frac{\partial}{\partial n}(g\nabla^2 v) + g\nabla^2 v\frac{\partial u}{\partial n} - g\nabla^2 u\frac{\partial v}{\partial n}\right] ds = 0 \qquad (3.60)$$

This equation is the appropriate condition for the self-adjoint property of the operator $\nabla^2(g\nabla^2)$. For a clamped plate $w = \partial w/\partial n = 0$ (w is the deflection) on the boundary and for a simply supported rectangular plate we will later see that

$$w = \frac{\partial^2 w}{\partial x^2} = \frac{\partial^2 w}{\partial y^2} = 0$$

on the boundaries. Accordingly, the above condition is satisfied for such supports.

We now impose one more requirement on the operator L. We now require that L be *positive definite*. This means that

$$\langle Lu, u\rangle \geqslant 0 \tag{3.61}$$

for functions u satisfying appropriate boundary conditions dependent for their form on L, and that the equality holds *only* when $u = 0$ everywhere in the domain.

We now go back to *case 1* to consider the operator

$$L \equiv \frac{d^2}{dx^2}\left[g(x)\frac{d^2}{dx^2}\right]$$

for this property. It will now be assumed here that $g(x) > 0$ everywhere in the domain—a fact consistent with beam theory, to which the operator can be applied. Then using Eq. (3.55) with v replaced by u we may say

$$\langle Lu, u\rangle = \int_0^L g\left(\frac{d^2u}{dx^2}\right)^2 dx + [(gu'')'u - gu''u']_0^L$$

The bracketed quantity is clearly zero for beams with simple, fixed, or free supports, or any mix of these. Since g is never negative we conclude that

$$\langle Lu, u\rangle \geqslant 0$$

Now assume the equality is imposed in the above equation. To show that L is positive definite we must show that $u = 0$ everywhere for such a situation. Accordingly, we consider

$$\int_0^L g\left(\frac{d^2u}{dx^2}\right)^2 dx = 0$$

This means that

$$\frac{d^2u}{dx^2} = 0$$

therefore

$$u = C_1 x + C_2$$

in the domain. For simply supported beams, and beams cantilevered at one or both ends, $C_1 = C_2 = 0$ and indeed $u = 0$ everywhere. Clearly for such boundary conditions:

$$\frac{d^2}{dx^2}\, g(x)\, \frac{d^2}{dx^2}$$

is positive definite. Those functions u satisfying the boundary conditions for which L is both self-adjoint and positive definite will be called the *field of definition* of L.

As for *case 2*, where the operator $L \equiv \nabla^2[g\,\nabla^2]$ is vital for the analysis of plates, we can consider Eq. (3.59) with v replaced by u. We get

$$\langle u, \nabla^2(g\nabla^2 u)\rangle = \iint_S \{\nabla\cdot[u\nabla(g\nabla^2 u) - (g\nabla^2 u)\nabla u] + g(\nabla^2 u)^2\}\, dA$$

Now, using the divergence theorem in the above equation, we get

$$\langle u, \nabla^2(g\nabla^2 u)\rangle = \iint_S g(\nabla^2 u)^2\, dA + \oint_\Gamma \left[u\frac{\partial}{\partial n}(g\nabla^2 u) - (g\nabla^2 u)\frac{\partial u}{\partial n}\right] ds$$

It is clear that for a clamped plate $u = \partial u/\partial n = 0$ on the boundary and the line integral vanishes. We will see that this integral vanishes also for rectangular, simply supported plates. Accordingly, for a positive function $g(x, y)$ as appears in plate theory we can conclude that

$$\langle u, \nabla^2(g\nabla^2 u)\rangle = \iint_S g(\nabla^2 u)^2\, dA \geqslant 0$$

If we examine the case where the equality holds then we conclude that

$$\nabla^2 u = 0$$

in the domain–i.e., u is harmonic† for this case. We will now use a well-known theorem from potential theory (Kellogg, 1928) which says that at any point x, y in the domain having Γ as the boundary, the value of a harmonic function u is zero inside the boundary if u is zero on the boundary. We can then conclude for the clamped or simply supported plate that since $u(x, y) = 0$ on Γ, $u = 0$ inside Γ. We have thus proved that the operator $\nabla^2(g\,\nabla^2)$ is positive definite for such cases.

We have shown that two operators that are vital in the study of structural mechanics are self-adjoint and positive definite for homogeneous boundary conditions. Actually, one can show‡ that the operator L for the Navier equation is also self-adjoint and positive definite for homogeneous boundary conditions employed in the theory of elasticity. We can then conclude that the key equations of structural mechanics and elasticity have self-adjoint and positive definite operators for all physically meaningful boundary conditions.

Other operators such as the Laplacian operator and the operator

†Harmonic functions are those satisfying Laplace's equation $\nabla^2\phi = 0$. The theory of these and related functions is often called potential theory.

‡See Problem 3.18.

$$\sum_{k=0}^{m} (-1)^k \frac{d^k}{dx^k}\left[p_k(x)\frac{d^k}{dx^k}\right]$$

are similarly self-adjoint and positive definite for certain boundary conditions and we can conclude that this class of operators is rather large, encompassing many areas of study outside solid mechanics.

3.12 QUADRATIC FUNCTIONALS WITH HOMOGENEOUS BOUNDARY CONDITIONS

At this time, we introduce a functional $I(u)$, which we denote as a *quadratic functional,* related to the equation $Lu = f$ for *homogeneous* boundary conditions. We define this functional as follows:

$$I(u) = \langle Lu, u\rangle - 2\langle u, f\rangle \tag{3.62}$$

The significance of $I(u)$ stems from the following theorems.

THEOREM. If the equation $Lu = f$ has a solution and if L is a *self-adjoint, positive definite* operator, then the function u from the field of definition of operator L that *minimizes* $I(u)$ is the function that is the solution of the differential equation $Lu = f$.

PROOF. Let us say that u_0 minimizes $I(u)$. Hence consider a one-parameter family of functions $u_0 + \epsilon\eta$, where η belongs to the field of definition of L and thus satisfies homogeneous boundary conditions. We then have

$$I(u_0 + \epsilon\eta) - I(u_0) \geqslant 0 \tag{3.63}$$

Now using the definition of the functional I in the above expression:

$$\langle L(u_0 + \epsilon\eta), (u_0 + \epsilon\eta)\rangle - 2\langle (u_0 + \epsilon\eta), f\rangle - \langle Lu_0, u_0\rangle + 2\langle u_0, f\rangle \geqslant 0$$

Using the linear property of the operator L, this becomes

$$\langle Lu_0, u_0\rangle + \langle Lu_0, \epsilon\eta\rangle + \langle \epsilon L\eta, u_0\rangle + \epsilon^2\langle L\eta, \eta\rangle - 2\langle u_0, f\rangle$$
$$- 2\langle \epsilon\eta, f\rangle - \langle Lu_0, u_0\rangle + 2\langle u_0, f\rangle \geqslant 0$$

On canceling terms and regrouping, this may be written as

$$[\langle Lu_0, \eta\rangle + \langle L\eta, u_0\rangle - 2\langle \eta, f\rangle]\epsilon + \langle L\eta, \eta\rangle\epsilon^2 \geqslant 0$$

Now using the self-adjoint property of the operator L, we note that $\langle Lu_0, \eta\rangle = \langle L\eta, u_0\rangle$, and so we have

$$2[\langle Lu_0, \eta\rangle - \langle \eta, f\rangle]\epsilon + \langle L\eta, \eta\rangle\epsilon^2 \geqslant 0$$

The left side of the inequality, which we denote as $g(\epsilon)$, is *quadratic* in ϵ and must have a plot such as "a" in Fig. 3.17, where there are no roots, or "b", where there is only a single root. Two real roots are not possible since $g(\epsilon) \geqslant 0$. Hence the discriminant of this quadratic must be negative (for no real roots) or zero (for a single real root). Thus we can say for the discriminant, since $\langle Lu_0, \eta\rangle - \langle \eta, f\rangle = \langle (Lu_0 - f), \eta\rangle$, that

$$4[\langle (Lu_0 - f), \eta\rangle^2 - \langle L\eta, \eta\rangle(0)] \leqslant 0$$

Since the expressions Lu_0, f, and η are real, this means on taking the square root that

$$\langle (Lu_0 - f), \eta\rangle \equiv \iiint_V (Lu_0 - f)\eta \, dv = 0$$

Using the fundamental lemma of the calculus of variations† we conclude

$$Lu_0 - f = 0$$

and so we see that u_0 satisfies the differential equation.

We now examine the *converse* of the preceding theorem.

THEOREM. If the solution to $Lu = f$ is u_0, where L is positive definite and self-adjoint, then u_0 must minimize $I(u)$ with respect to all u's that are in the field of definition of L.

PROOF. We express $I(u)$ with Lu_0 replacing f as follows:

$$I(u) = \langle Lu, u\rangle - 2\langle Lu_0, u\rangle$$

We add and subtract the quantity $\langle Lu_0, u_0\rangle$ in order to rearrange the above equation. Thus

†It is because the condition $\eta = 0$ at the boundaries is required to apply the fundamental lemma that we have to limit ourselves to homogeneous boundary conditions.

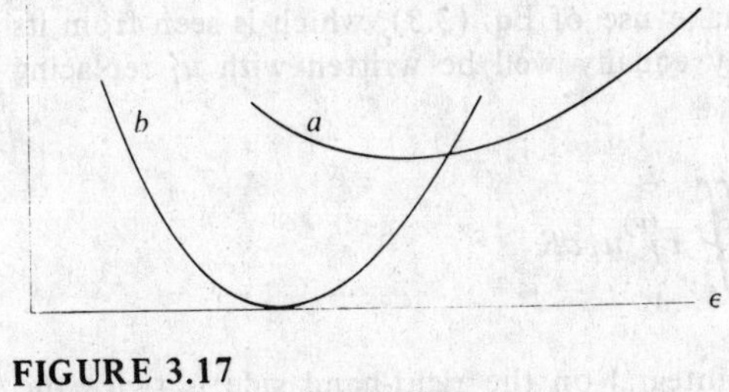

FIGURE 3.17
Only possible curves for $g(\epsilon)$ versus ϵ.

$$I(u) = \langle Lu, u\rangle - 2\langle Lu_0, u\rangle + \langle Lu_0, u_0\rangle - \langle Lu_0, u_0\rangle$$
$$= \langle L(u-u_0), (u-u_0)\rangle - \langle Lu_0, u_0\rangle$$

You may readily verify the last step by making use of the self-adjoint property of the operator L. Since L is positive definite, *it is clear from the above that when* $u = u_0$ *the first expression on the right reaches its smallest value, zero, and the right side becomes a minimum*. We thus have proved that the solution to the equation $Lu = f$ minimizes the corresponding quadratic functional $I(u)$.

We now can conclude from the first theorem that, for the conditions specified on L, the extremal function for the functional $I(u)$ satisfies the differential equation $Lu = f$ and so the latter is the Euler-Lagrange equation for the particular functional. It is to be noted that in both Chapter 2 and this chapter we started with the functional and showed how to get the appropriate differential equation and boundary conditions. Now under certain conditions we can produce the appropriate functional for a given differential equation and boundary conditions. We have thus presented the reverse process. Why would we want the functional for a given differential equation? We will soon see as a result of the second theorem that for producing *approximate* solutions to the differential equation it is sometimes more feasible to work with the corresponding quadratic functional with a view toward minimizing its value with respect to an admissible class of functions.

We now show that the potential energy is actually half of the quadratic functional for Navier's equation for homogeneous boundary conditions. We rewrite this equation [Eq. (1.106)] in vector notation. One may show that this equation may be expressed as follows with the aid of vector identities:

$$-[(\lambda + 2G)\nabla(\nabla \cdot \mathbf{u}) - G\nabla \times (\nabla \times \mathbf{u})] = \mathbf{B} \tag{3.64}$$

We have here a vector operator acting on a vector, and we denote this as $\mathbf{Lu}$ for the left side of the equation. Thus we have for the above

$$\mathbf{Lu} = \mathbf{B} \tag{3.65}$$

where

$$\mathbf{L} = -[(\lambda + 2G)\nabla(\nabla \cdot\ ) - G\nabla \times (\nabla \times\ )]$$

In the ensuing development, we shall make use of Eq. (3.3), which is seen from its development to be an identity and may equally well be written with u_i replacing δu_i and with ϵ_{ij} replacing $\delta\epsilon_{ij}$ as follows:

$$\iiint_V B_i u_i \, dv = \iiint_V \tau_{ij}\epsilon_{ij} \, dv - \oiint_S T_i^{(\nu)} u_i \, dS$$

Noting from earlier work that the first integral on the right-hand side is twice the strain energy for linear elastic bodies and replacing $\mathbf{B}$ by $\mathbf{Lu}$ from Eq. (3.65), we get

$$\iiint_V \mathbf{Lu}\cdot\mathbf{u}\, dv = 2\iiint_V \mathfrak{U}\, dv - \oint\!\!\oint_S \mathbf{u}\cdot\mathbf{T}^{(\nu)}\, dS \tag{3.66}$$

We now give the quadratic functional for the Navier equation, making use of the identity (3.66) and replacing **f** by **B**:

$$I(\mathbf{u}) = \langle \mathbf{Lu}, \mathbf{u}\rangle - 2\langle \mathbf{f}, \mathbf{u}\rangle = \left(2\iiint_V \mathfrak{U}\, dv - \oint\!\!\oint_S \mathbf{u}\cdot\mathbf{T}^{(\nu)}\, dS\right) - 2\iiint_V \mathbf{B}\cdot\mathbf{u}\, dv$$

With $\mathbf{T} = \mathbf{0}$ on S_1 and $\mathbf{u} = \mathbf{0}$ on S_2 (homogeneous boundary conditions) the surface integral in the above equation vanishes and we have

$$\langle \mathbf{Lu}, \mathbf{u}\rangle - 2\langle \mathbf{f}, \mathbf{u}\rangle = 2\left(\iiint_V \mathfrak{U}\, dv - \iiint_V \mathbf{B}\cdot\mathbf{u}\, dv\right)$$

We thus get for this case an expression proportional to the total potential energy. For nonhomogeneous boundary conditions, one has to extremize a more complex quadratic functional in that additional terms are needed. We merely point out at this time that for more general conditions for the case at hand we get as the functional $I(u)$ the expression

$$I(u) = 2\left(\iiint_V \mathfrak{U}\, dv - \iiint_V B_i u_i\, dv - \oint\!\!\oint_S T_i^{(\nu)} u_i\, dS\right)$$

which again is just twice the total potential energy.

We wish next to point out that there can be an infinity of functionals mutually related in a certain way for a given Euler-Lagrange equation. Thus the functional

$$I = \int_{x_1}^{x_2} F(x, y, y', \ldots, y^n)\, dx$$

can be altered, without changing the Euler-Lagrange equations, by adding to the integrand an expression of the form

$$(\text{Const.})\, \frac{dG(x, y, y', \ldots, y^{(n-1)})}{dx}$$

Also the functional

$$I = \iiint_V F(x, y, z, w, w_x, w_y, w_z)\, dx\, dy\, dz$$

has the same Euler-Lagrange equation if the following "divergence-like" expression is added to the integrand:

$$\alpha \frac{\partial A}{\partial x} + \beta \frac{\partial B}{\partial y} + \gamma \frac{\partial C}{\partial z}$$

where α, β, and γ are constants and A, B, and C are functions of the variables x, y, z, and w. Note that these functions have only w and not w_x, w_y, and w_z as in the functional I. For functionals I with partial derivatives of higher-order than unity, it is not in general possible to choose A, B, and C with the derivatives appearing in these functions having a lower order than in the original functional (Courant and Hilbert, 1953, pp. 194, 195).

3.13 VARIATIONAL PROCESS WITH NONHOMOGENEOUS BOUNDARY CONDITIONS

We consider once again a linear operator L forming the familiar equation:

$$Lu = f \tag{3.67}$$

This time, in contrast to what we had in Section 3.12, we will now have *nonhomogeneous* boundary conditions. That is:

$$\begin{aligned} H_1(u)|_S &= h_1 \\ &\vdots \\ H_r(u)|_S &= h_r \end{aligned} \tag{3.68}$$

where the H_i are linear operators and the h_i are functions of the independent variables. The subscript S indicates specification on the boundary S. The number r of boundary conditions depends on the order of the differential operator. We wish to formulate the quadratic functional for this kind of boundary-value problem.

As a first step, we shall assume the *existence* of a function ψ of the *independent variables* satisfying identically the same boundary conditions specified above for the function u. That is:

$$\begin{aligned} H_1(\psi)|_S &= h_1 \\ &\vdots \\ H_r(\psi)|_S &= h_r \end{aligned} \tag{3.69}$$

We will not have to ascertain ψ—only its existence is of significance.

We next form a function v in the form

$$v = u - \psi \tag{3.70}$$

Hence, we can say

$$Lv = Lu - L\psi = f - L\psi \tag{3.71}$$

The differential equation for v given above has as a driving function $(f - L\psi)$. Note that since u and ψ satisfy identically the same boundary conditions on S, it follows from Eq. (3.70) that the boundary conditions for v on S are *homogeneous*.

We will next assume that the operator L is positive definite and self-adjoint for the dependent variable v and its homogeneous boundary conditions. Hence we can form the familiar quadratic functional of Section 3.12 as follows:

$$I(v) = \langle Lv, v\rangle - 2\langle v, (f - L\psi)\rangle \tag{3.72}$$

Now replace v by using Eq. (3.70) and carry out the indicated operations:

$$\begin{aligned} I(v) &= \langle L(u-\psi), (u-\psi)\rangle - 2\langle (u-\psi), (f-L\psi)\rangle \\ &= \langle Lu, u\rangle - \langle Lu, \psi\rangle - \langle L\psi, u\rangle + \langle L\psi, \psi\rangle \\ &\quad - 2\langle u, f\rangle + 2\langle u, L\psi\rangle + 2\langle \psi, f\rangle - 2\langle \psi, L\psi\rangle \end{aligned}$$

Canceling terms and rearranging the equation, we get

$$I(v) = \langle Lu, u\rangle - 2\langle u, f\rangle - \langle L\psi, \psi\rangle + 2\langle \psi, f\rangle - \langle Lu, \psi\rangle + \langle u, L\psi\rangle \tag{3.73}$$

Now you may be tempted to drop the last two expressions by cancellation on the argument that L is self-adjoint. But you cannot do this because L is only self-adjoint for homogeneous boundary conditions, while ψ and u have non-homogeneous boundary conditions. We will extract these last two expressions of Eq. (3.73) and put them in integral form. That is:

$$\langle u, L\psi\rangle - \langle \psi, Lu\rangle = \iiint_\Omega (uL\psi - \psi Lu)\, dv$$

We shall replace the right side of the above equation by extrapolating Green's formula, presented in Appendix I as Eq. (I.38), to apply to operators other than the Laplacian operator. That is, we say that

$$\langle u, L\psi\rangle - \langle \psi, Lu\rangle = \oint\!\!\oint_S [uQ(\psi) - \psi Q(u)]\, dS = \oint\!\!\oint_S R(u, \psi)\, dS \tag{3.74}$$

where Q is an operator that usually can be determined and R is a function of u and

ψ. Further, we can often express $R(u, \psi)$ as the sum of a function $P(u, h_1, \ldots, h_r)$ plus a separate function $E(\psi, h_1, \ldots, h_r)$. That is,

$$R(u, \psi) = P(u, h_1, \ldots, h_r) + E(\psi, h_1, \ldots, h_r) \tag{3.75}$$

This separation may be accomplished with the aid of Eqs. (3.67) and (3.68). Equation (3.73) can now be written as

$$I(v) = \langle Lu, u\rangle - 2\langle u, f\rangle + \oiint_S P(u)\, dS + \left[-\langle L\psi, \psi\rangle + \oiint_S E(\psi)\, dS + 2\langle \psi, f\rangle\right] \tag{3.76}$$

If the function ψ exists as assumed earlier, then the bracketed expression is a constant. However, since the function may not be known, it follows that this constant may not be known. Equation (3.76) can accordingly be expressed as follows:

$$I(v) = \langle Lu, u\rangle - 2\langle u, f\rangle + \oiint_S P(u)\, dS + \text{Const.} \tag{3.77}$$

Extremizing $I(v)$ with respect to v for homogeneous boundary conditions is then equivalent to extremizing $J(u)$, which is the right side of Eq. (3.77) without the constant, with respect to u for the nonhomogeneous boundary condition (3.68). Thus we have shown for the boundary-value problem:

$$\begin{gathered} Lu = f \\ \left\{\begin{aligned} &H_1(u)|_S = h_1 \\ &\quad\vdots \\ &H_r(u)|_S = h_r \end{aligned}\right. \end{gathered} \tag{3.78}$$

the appropriate quadratic functional to extremize with respect to admissible functions u is

$$\boxed{J(u) = \langle Lu, u\rangle - 2\langle u, f\rangle + \oiint_S P(u)\, dS} \tag{3.79}$$

3.14 THE DIRICHLET AND VON NEUMANN BOUNDARY-VALUE PROBLEMS

We now find the quadratic functional for Laplace's equation, which is given as:

$$-\nabla^2 \phi = 0$$

where ϕ is specified on the boundary in the form

$$\phi|_S = h \tag{3.81}$$

This is the *Dirichlet* problem.

We go back to Eq. (3.74) in order to find the function $R(\phi, \psi)$, where we assume the existence of a function ψ having the same boundary conditions prescribed by Eq. (3.81) for ϕ. Since $L = -\nabla^2$, we can use Appendix I Eq. (I.38), Green's theorem, directly. Thus, from Eq. (3.74) we have

$$\langle\phi, L\psi\rangle - \langle\psi, L\phi\rangle = -\iiint_V (\phi\nabla^2\psi - \psi\nabla^2\phi)\, dv = -\oiint_S \left(\phi\frac{\partial\psi}{\partial\nu} - \psi\frac{\partial\phi}{\partial\nu}\right) dS \tag{3.82}$$

Noting that both ϕ and ψ on the boundary S are given by the function h from Eq. (3.81), we can say for the above equation

$$\langle\phi, L\psi\rangle - \langle\psi, L\phi\rangle = -\oiint_S \left(h\frac{\partial\psi}{\partial\nu} - h\frac{\partial\phi}{\partial\nu}\right) dS$$

Thus

$$R(\phi, \psi) = h\frac{\partial\phi}{\partial\nu} - h\frac{\partial\psi}{\partial\nu} \tag{3.83}$$

Also, on noting Eq. (3.75), we see that

$$P = h\frac{\partial\phi}{\partial\nu} \tag{3.84a}$$

$$E = -h\frac{\partial\psi}{\partial\nu} \tag{3.84b}$$

The desired quadratic functional $J(u) \equiv I(\phi)$ becomes [see Eq. (3.79)]

$$I(\phi) = \langle L\phi, \phi\rangle - 2\langle\phi, 0\rangle + \oiint_S h\frac{\partial\phi}{\partial\nu}\, dS + \text{Const.} \tag{3.85}$$

Hence, deleting the constant,

$$I(\phi) = -\iiint_V \phi\nabla^2\phi\, dv + \oiint_S h\frac{\partial\phi}{\partial\nu}\, dS \tag{3.86}$$

Finally we go to Eq. (I.36) in Appendix I and let $u = \phi$ in the formulation. We get as a result

$$\iiint_V \nabla\phi \cdot \nabla\phi \, dv = -\iiint_V \phi \nabla^2 \phi \, dv + \oiint_S \phi \frac{\partial \phi}{\partial \nu} \, dS$$

Solving for $-\iiint_V \phi \nabla^2 \phi \, dv$, we get from the above equation

$$-\iiint_V \phi \nabla^2 \phi \, dv = \iiint_V (\nabla\phi)^2 \, dv - \oiint_S \phi \frac{\partial \phi}{\partial \nu} \, dS \tag{3.87}$$

Now substitute the above result in Eq. (3.86). Noting that $\phi = h$ on S, we arrive at the following result after canceling terms:

$$\boxed{I(\phi) = \iiint_V (\nabla\phi)^2 \, dv} \tag{3.88}$$

We thus have arrived at the desired functional for the Dirichlet problem.

We next consider the *von Neumann* problem, where $\partial\phi/\partial\nu$ is specified on the boundary S as $p(x, y)$ and where $-\nabla^2\phi = 0$ again. Now go back to Eq. (3.83). On denoting $\partial\psi/\partial\nu = p(x, y)$ in this equation, we have for $R(\phi, \psi)$

$$R(\phi, \psi) = \psi p - \phi p \tag{3.89}$$

We thus conclude immediately that

$$P = -\phi p \tag{3.90a}$$

$$E = \psi p \tag{3.90b}$$

Again go back to Eq. (3.79) for $I(\phi)$. Using Eq. (3.87) to replace $\langle \phi, L\phi \rangle$ and Eq. (3.90*a*) to replace P, we get, on noting that $\partial\psi/\partial\nu = p(x, y)$ on the boundary:

$$I(\phi) = \iiint_V (\nabla\phi)^2 \, dv - \oiint_S \phi p \, dS - \oiint_S \phi p \, dS$$

Therefore,

$$\boxed{I(\phi) = \iiint_V (\nabla\phi)^2 \, dv - 2\oiint_S \phi p \, dS} \tag{3.91}$$

Part D
Approximate Methods

3.15 INTRODUCTORY COMMENT

We have presented by various arguments a number of functionals whose Euler-Lagrange equations were of particular interest in solid mechanics in Part A, and in Part C we showed how, for a certain rather broad class of ordinary and partial differential equations, we could formulate the appropriate functional for which the equation was the corresponding Euler-Lagrange equation. Thus in either case we have available the functional and the Euler-Lagrange equations. The principal approach in Part D of this chapter will be to work with the functional for the purpose of finding approximate solutions to the corresponding differential equation.

We shall first consider the Ritz method as it is applied to the total potential energy functional. This method may also be used, however, for other functionals such as the total complementary energy and the quadratic functionals discussed in the previous section. We will then briefly consider the Galerkin method which, while not being a variational process, is nevertheless so closely related to the Ritz method and so useful as to warrant discussion at this time. We shall cover other approximation methods, such as the method of Trefftz and the method of Kantorovich, later in the text. They will be presented at a time when there will be suitable motivation and background to appreciate their particular characteristics.

3.16 THE RITZ METHOD

One of the most useful approximate methods stemming from variational considerations is the Ritz method, wherein for the present we employ the following approximate displacement field components for expressing the total potential energy:

$$u_n = \phi_0(x, y, z) + \sum_{i=1}^{n} a_i\phi_i(x, y, z)$$

$$v_n = \psi_0(x, y, z) + \sum_{i=1}^{n} b_i\psi_i(x, y, z) \tag{3.92}$$

$$w_n = \gamma_0(x, y, z) + \sum_{i=1}^{n} c_i\gamma_i(x, y, z)$$

The functions with the subscript zero satisfy the boundary conditions on S_2 and thus are in the field of definition of L while the remaining $3n$ functions are zero there. The coefficients a_i, b_i, c_i are undetermined. The scheme for the Ritz method is to choose the values of these coefficients so as to minimize the total potential energy. When the values of the coefficients are thus determined they are called *Ritz coefficients*. It can be shown (Trefftz, 1928) that if $n \to \infty$ in this process, then u_n, v_n, and w_n converge in energy† to the exact solution for the problem provided the functions ϕ_i, ψ_i, and γ_i are complete.‡ When n is small, we can still reach very good approximations to the exact solution if a judicious selection of the functions $\phi_0, \phi_1, \ldots, \phi_n$, $\psi_0, \psi_1, \ldots, \psi_n$, and $\gamma_0, \gamma_1, \ldots, \gamma_n$ is made.

The procedure is to use u_n, v_n, and w_n to formulate an approximate total potential energy $\tilde{\pi}$, which then becomes a function of the $3n$ undetermined constants. We minimize $\tilde{\pi}$ by imposing the $3n$ requirements:

$$\frac{\partial \tilde{\pi}}{\partial a_i} = 0$$

$$\frac{\partial \tilde{\pi}}{\partial b_i} = 0 \qquad i = 1, 2, \ldots, n$$

$$\frac{\partial \tilde{\pi}}{\partial c_i} = 0$$

This yields $3n$ equations for the unknown coefficients. The solution of these equations for these coefficients then yields the aforementioned Ritz coefficients.

Suppose next that the boundary conditions are homogeneous, as for example the condition

$$\alpha u_i + \beta \frac{\partial u_i}{\partial n} = 0$$

with α and β as constants and n as a normal direction to the boundary. Then if a function $\phi_0(x, y, z)$ satisfies the boundary conditions of the problem, so will $K\phi_0(x, y, z)$, where K is an arbitrary constant. If we go back to Eq. (3.92) we can set forth the following possible arrangement for approximate displacement field components u_n, v_n, and w_n to be used in the Ritz method:

$$u_n = \left(\sum_i^n a_i \right) \phi_0(x, y, z) + \sum_i^n a_i \phi_i(x, y, z) \tag{3.93}$$

†A sequence of functions u_n may be said to converge to a function u in many different senses. Thus there may be *uniform convergence* where in the interval the quantity $[(u_n)_{\max} - (u)_{\max}]$ can be made arbitrarily small by increasing n. There is *convergence in the mean* whereby $\iiint [u_n - u]^2 \, dv$ can be made arbitrarily small by increasing n. The convergence referred to here, namely that of *energy*, requires that $\langle L(u - u_n), (u - u_n) \rangle$ which is in unabridged notation $\iiint [L(u - u_n)] [u - u_n] \, dv$, can be made arbitrarily small by increasing n. We will see in Chapter 7 that $\langle Lu, u \rangle$ actually does relate to the energy of the system and so this convergence does have physical significance.

‡A system of functions can be *complete* in several ways. A series of functions ϕ_i may be complete if, for any function f, a set of constants c_i can be found so that $\Sigma \, c_i \phi_i$ converges (1) uniformly, (2) in the mean, or (3) in energy to this function. The nature of the completeness must, of course, be stated in a discussion.

$$v_n = \left(\sum_i^n b_i\right)\psi_0(x, y, z) + \sum_i^n b_i\psi_i(x, y, z)$$

$$w_n = \left(\sum_i^n c_i\right)\gamma_0(x, y, z) + \sum_i^n c_i\gamma_i(x, y, z) \tag{3.93 (Cont.)}$$

This arrangement has the same properties as that given by Eq. (3.92). Now rearranging we get

$$u_n = \sum_i^n a_i(\phi_0 + \phi_i) = \sum_i^n a_i\Phi_i$$

$$v_n = \sum_i^n b_i(\psi_0 + \psi_i) = \sum_i^n b_i\Psi_i \tag{3.94}$$

$$w_n = \sum_i^n c_i(\gamma_0 + \gamma_i) = \sum_i^n c_i\Gamma_i$$

The functions Φ_i, Ψ_i, and Γ_i clearly satisfy the boundary conditions of the problem. For homogeneous boundary conditions we know, making use of the convergence of the form given by Eq. (3.92), that this sequence, with the undetermined coefficients chosen to minimize $\tilde{\pi}$, will converge in energy to the exact solution u_i in the limit as $n \to \infty$, provided the functions Φ_i, Ψ_i, and Γ_i are complete sets with respect to energy.

We shall have ample opportunity to employ the Ritz method in succeeding chapters. At that time we will, by example, discuss how best to choose the functions (usually called coordinate functions) for expeditious use in the Ritz method for small values of n. For such problems we will have restricted the admissible virtual displacements so that only one function, w, appears in π rather than the three functions u_1, u_2, and u_3. (This was discussed, you will recall, in Section 3.3.) The Ritz method in such cases will be given as

$$w = \sum_i^n a_i\Phi_i$$

where Φ_i satisfies the homogeneous boundary conditions for w.

We now turn to the remarks of Section 3.11 concerning the quadratic functionals. We may employ the Ritz method for minimizing the quadratic functional and thereby approximate the solution of the corresponding Euler-Lagrange equation. The proof of convergence of the process in this general case again requires that the coordinate functions Φ_i be complete in energy. The proof of such completeness (this places further limitations on the operators L) is a difficult one (Mikhlin, 1964).

EXAMPLE 3.9 Consider the following ordinary differential equation:

$$x^2 \frac{d^2 y}{dx^2} + 2x \frac{dy}{dx} = 6x \tag{a}$$

We wish to solve this equation approximately, using the Ritz method for the following end conditions:

$$y(1) = y(2) = 0 \tag{b}$$

Since an exact solution here is readily available we can then compare results.

We must first find the functional associated with this boundary-value problem. Note in this regard that the equation can be written as follows:

$$-\frac{d}{dx}\left(x^2 \frac{dy}{dx}\right) = -6x \tag{c}$$

We leave it for you to show that the operator

$$L = -\left(\frac{d}{dx}\, x^2 \frac{d}{dx}\right)$$

is a positive definite, symmetric operator for the given boundary conditions. And so we have for the functional I:

$$I = -\int_1^2 \left\{ \frac{d}{dx}\left[x^2 \frac{dy}{dx}\right] y - 12yx \right\} dx$$

$$= -\int_1^2 \left(x^2 y \frac{d^2 y}{dx^2} + 2xy \frac{dy}{dx} - 12xy \right) dx \tag{d}$$

We now choose a single coordinate function satisfying the boundary conditions:

$$y_1 = C_1 \Phi_1 = C_1(x-1)(x-2) = C_1(x^2 - 3x + 2) \tag{e}$$

Substituting y_1 into the functional, we get

$$\tilde{I} = -\int_1^2 \left[x^2 \Phi_1 \left(\frac{d^2 \Phi_1}{dx^2}\right) C_1^2 + 2x\Phi_1 \frac{d\Phi_1}{dx} C_1^2 - 12x\Phi_1 C_1 \right] dx$$

Extremizing $\tilde{I}$ with respect to C_1 we get, on canceling -2 from the result,

$$\int_1^2 \left(x^2\Phi_1 \frac{d^2\Phi_1}{dx^2} C_1 + 2x\Phi_1 \frac{d\Phi_1}{dx} C_1 - 6x\Phi_1\right) dx = 0 \qquad (f)$$

Substituting for Φ_1 we have

$$\int_1^2 [(3x^4 - 12x^3 + 15x^2 - 6x)C_1 - 3(x^3 - 3x^2 + 2x)]\, dx = 0$$

Integrating we get

$$[(\tfrac{3}{5}x^5 - 3x^4 + 5x^3 - 3x^2)C_1 - (\tfrac{3}{4}x^4 - 3x^3 + 3x^2)]_1^2 = 0$$

and therefore

$$C_1 = 1.875$$

The approximate solution from the Ritz method is then

$$y_1 = 1.875(x^2 - 3x + 2)$$

We may get an exact solution readily for this equation by noting that it is an Euler-Cauchy type (Golomb and Shanks, 1965). By making the following change of variable:

$$x = e^t \qquad t = \ln x$$

we can get an equation having constant coefficients. Thus with t as the independent variable we have

$$\frac{d^2y}{dt^2} + \frac{dy}{dt} = 6e^t$$

The complementary solution is readily found to be

$$y_c = Ae^{-t} + B$$

A particular solution is seen by inspection to be

$$y_p = 3e^t$$

Hence the exact solution is

$$y_{\text{ex}} = Ae^{-t} + B + 3e^t = \frac{A}{x} + B + 3x$$

To satisfy the boundary condition we require that

$$0 = A + B + 3$$

$$0 = \frac{A}{2} + B + 6$$

Hence $A = 6$ and $B = -9$. The exact solution for the problem is then

$$y_{ex} = \frac{6}{x} + 3x - 9 \qquad (g)$$

To compare methods we examine results at position $x = 1.5$. We get for the approximate and exact solutions

$$y_1(1.5) = -0.469 \qquad y_{ex}(1.5) = -0.50$$

To achieve greater accuracy we may form a more general coordinate function according to the following pattern:

$$y_2 = (x - 1)(x - 2)(C_1 + C_2 x)$$

The following results give the values of $y_2(1.5)$, showing an improvement in accuracy:

$$y_2(1.5) = -0.50895 \qquad y_{ex}(1.5) = -0.50$$

We will have much opportunity to employ the Ritz method in succeeding chapters.

3.17 GALERKIN'S METHOD

We now examine another method for finding an approximate solution to a differential equation–namely the Galerkin method. This method involves direct use of the differential equation; it does not require the existence of a functional. For this reason the method has a broader range of application than does the Ritz method. Yet we will soon see that in the area of solid mechanics the methods are closely related.

Consider the linear equation

$$Lu = f \qquad (3.95)$$

where the boundary conditions are *homogeneous*. The right side may be thought of as a forcing function of some sort and we may formulate a "virtual work" expression for this function as follows:

$$(\delta W)_1 = \iiint_V f\, \delta u\, dv \qquad (3.96)$$

where δu is a "virtual displacement" consistent with the constraints. It must also be true from Eq. (3.95) that

$$\iiint_V (Lu)\,\delta u\,dv = \iiint_V f\,\delta u\,dv \tag{3.97}$$

for any "virtual displacement field" δu consistent with the constraints. However, if we use an approximate $\tilde{u}$ for u in expressing Lu in Eq. (3.97), then the two expressions for "virtual work" in Eq. (3.97) will no longer be equal. And this lack of equality may be in some way a measure of the departure of $\tilde{u}$ from the exact solution u. We now express $\tilde{u}$ as follows:

$$\tilde{u} = \sum_{1}^{n} a_i \Phi_i$$

where the functions Φ_i, again called the coordinate functions, satisfy all the boundary conditions of the problem. Then we can generally say

$$\iiint_V (L\tilde{u})\,\delta u\,dv \neq \iiint_V f\,\delta u\,dv$$

However, we can, for any given virtual displacement, force an equality in the above statement by properly adjusting the constants a_i. With n constants a_i we can actually force an equality for n different virtual displacements. Forcing such an equality for these virtual displacement fields brings $\tilde{u}$ closer to the correct function u. Indeed, if we had an infinite set of properly selected coordinate functions we might conceivably force an equality for *any* admissible displacement field, and thus with the left sides of Eqs. (3.96) and (3.97) becoming equal, presumably we would expect $\tilde{u}$ to converge in some way in the limit to u, the exact solution. We shall not involve ourselves here in considerations of convergence criteria for this method; they are quite complicated and beyond the level of this text. In practice, a judiciously chosen finite number of functions Φ_i, to represent $\tilde{u}$, can lead to good approximations to the exact solution of the problem. In this format the method is called the *method of weighted residuals*, where the n functions δu are the weighting functions. If, however, for the n functions δu we use instead the coordinate functions Φ_i, we call the method *Galerkin's method*. It is this method that we shall follow. Thus, for the above equation we have

$$\iiint_V (L\tilde{u})\Phi_i\,dv = \iiint_V f\Phi_i\,dv \qquad i = 1, 2, \ldots, n$$

therefore

$$\iiint_V (L\tilde{u} - f)\Phi_i\,dv = 0 \tag{3.98}$$

[Another viewpoint is to say that we have required that $(L\tilde{u} - f)$ be orthogonal to the ith coordinate function Φ_i.] This gives us n equations to solve for the undetermined constants a_i. In Problem 3.22 we have asked the reader to show that *for the theory of elasticity the Ritz coefficients are identical to the coefficients found by the Galerkin method for the same system of coordinate functions* Φ_i. *Thus these methods are equivalent to each other in this area of physics.* There is some advantage in using the Galerkin method for problems of linear elasticity over the Ritz method in that the equations for the coefficients a_i are reached more directly via Eq. (3.98) than in the Ritz method, where one deals first with the functional.

We now demonstrate the use of Galerkin's method.

EXAMPLE 3.10 We shall reconsider the previous example with a view toward employing the Galerkin method.

For this purpose we again choose a single coordinate function Φ_1 to be the same function of Example 3.9. That is:

$$\Phi_1 = (x-1)(x-2) = (x^2 - 3x + 2) \tag{a}$$

Now we can give $\tilde{y}$ as follows:

$$\tilde{y} = C_1\Phi_1 = C_1(x^2 - 3x + 2) \tag{b}$$

Employing next Eq. (3.98) we get

$$\int_1^2 \left(x^2 \frac{d^2\tilde{y}}{dx^2} + 2x\frac{d\tilde{y}}{dx} - 6x \right) \Phi_1 \, dx = 0 \tag{c}$$

Hence

$$\int_1^2 \left(x^2 \Phi_1 \frac{d^2\Phi_1}{dx^2} C_1 + 2x\Phi_1 \frac{d\Phi_1}{dx} C_1 - 6x\Phi_1 \right) dx = 0 \tag{d}$$

But the above formulation is identical to Eq. (*f*) of Example 3.9. Hence the result for C_1 is identical for both the Ritz and Galerkin methods.

One can show (Kantorovich and Krylov, 1964) that the Ritz and Galerkin methods are identical for self-adjoint, second-order ordinary differential equations with homogeneous and even nonhomogeneous boundary conditions on the dependent variable. (It is clear from Examples 3.9 and 3.10 that the Galerkin method provides a more direct way of setting up the equations for determining the coefficients.) Also, whereas the Ritz method requires the differential equation to have a functional, the Galerkin method poses no such restriction.

3.18 CLOSURE

In this key chapter we have presented variational and related principles that form the basis of this text. And in doing so, we have inserted a self-contained discussion of trusses forming the first of a continuing study of structures. Finally, we have set forth certain mathematical notions and a theorem concerning functionals that leads us smoothly into the area of approximation techniques, forming one of the chief benefits of variational considerations.

In the following chapter we will be able to apply many of the principles and techniques presented in this chapter. In so doing, we will be able to continue the development of structural mechanics as we consider beams. Finally, in Part III we will find this chapter of great value in studying finite elements.

REFERENCES

Courant, R., and Hilbert, D., "Methods of Mathematical Physics," vol. 1, Interscience, New York, 1953.

Golomb, M., and Shanks, M., "Elements of Ordinary Differential Equations," McGraw-Hill, New York, 1965, chap. 7.

Kantorovich, L. V., and Krylov, V. I., "Approximate Methods of Higher Analysis," Interscience, New York, 1964.

Kellogg, O. D., "Potential Theory," Dover, Mineola, N.Y., 1928.

Martin, H. C., "Introduction to Matrix Methods of Structural Analysis," McGraw-Hill, New York, 1966.

Mikhlin, S. G., "Variational Methods in Mathematical Physics," Macmillan, New York, 1964.

Trefftz, E., "Handbüch der Physik," Springer-Verlag, Berlin, 1928.

READING

Arthurs, A. M.: "Complementary Variational Methods," Oxford Univ., 1970.

Biezeno, and Grammel: "Engineering Dynamics," vol. 1. Blackie and Son, London, 1956.

Forray, M.: "Variational Calculus in Science and Engineering," McGraw-Hill, New York, 1968.

Hoff, N. J.: "The Analysis of Structures," Wiley, New York, 1956.

Kantorovich, and Krylov: "Approximate Methods in Higher Analysis," Interscience, New York, 1964.

Lanczos, C.: "The Variational Principles of Mechanics," Univ. of Toronto Press, 1957.

Langhaar, H. L.: "Energy Methods in Applied Mechanics," Wiley, New York, 1962.

Mikhlin, S. G.: "Variational Methods in Mathematical Physics," Macmillan, New York, 1964.

Moiseiwitch, B. L.: "Variational Principles," Interscience, New York, 1966.

Oden, J. T.: "Mechanics of Elastic Structures," McGraw-Hill, New York, 1967.

Shames, I. H.: "Mechanics of Deformable Solids," Krieger Publishers, Melbourne, Fla., 1964.

Sokolnikoff, I. S.: "The Mathematical Theory of Elasticity," McGraw-Hill, New York, 1956.

Temple, G., and Bickley, W. G.: "Rayleigh's Principle," Dover, Mineola, N.Y., 1956.

Washizu, K.: "Variational Methods in Elasticity and Plasticity," Pergamon, New York, 1968.

PROBLEMS

3.1 Use the method of virtual work to find the forces in the members shown in Fig. 3.18. What is the horizontal movement of pin A? Assume linear elastic behavior with each member having the same modulus of elasticity and the same cross-sectional area.

3.2 Do the previous problem for the case of nonlinear elastic behavior where for all members

$$|\tau| = |\epsilon^{0.8}| \times 15 \times 10^9$$

All members are identical in cross-sectional area. Set up the equations only.

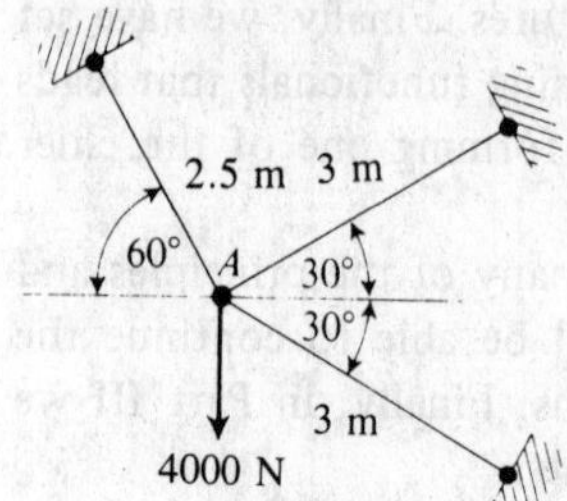

FIGURE 3.18

3.3 Show by extremizing π as given by Eq. (3.11) that we can arrive at the equations of equilibrium and Cauchy's formula.

3.4 Do Problem 3.1 by the method of total potential energy.

3.5 Consider a perfectly flexible membrane simply supported at the boundary (see Fig. 3.19) on the xy plane and with a tensile force T per unit length which is everywhere constant in the membrane. If a normal pressure distribution $q(x, y)$ is applied to cause small deflection of the membrane, show that the static deflection $w(x, y)$ of this membrane is

$$\frac{\partial^2 w}{\partial x^2} + \frac{\partial^2 w}{\partial y^2} = -\frac{q}{T} \qquad (a)$$

Hint: Show that the area of the surface of the membrane in its deformed configuration is given as (from differential geometry)

$$A = \iint_S \left[1 + \left(\frac{\partial w}{\partial x}\right)^2 + \left(\frac{\partial w}{\partial y}\right)^2\right]^{1/2} dx\, dy \qquad (b)$$

Hence, the strain energy is $T\,\Delta A$, where ΔA is the *change* in area of the membrane from deformation. Now show, using a binomial expansion, that

$$U = \frac{1}{2} T \iint_S \left[\left(\frac{\partial w}{\partial x}\right)^2 + \left(\frac{\partial w}{\partial y}\right)^2\right] dx\, dy \qquad (c)$$

Then extremize π.

*3.6 We have shown in this text that with $\delta^{(1)}\pi = 0$ we have

$$\pi(\epsilon_{ij} + \delta\epsilon_{ij}) - \pi(\epsilon_{ij}) = \frac{1}{2!}\iiint_V \frac{\partial^2 U_0}{\partial \epsilon_{ij}\,\partial \epsilon_{kl}}\,\delta\epsilon_{kl}\,\delta\epsilon_{ij}\,dv + \cdots$$

Consider the stress-strain law in conjunction with the variation of τ_{ij}. Show that for any ϵ_{ij} the first term on the right side of the above equation is equal to

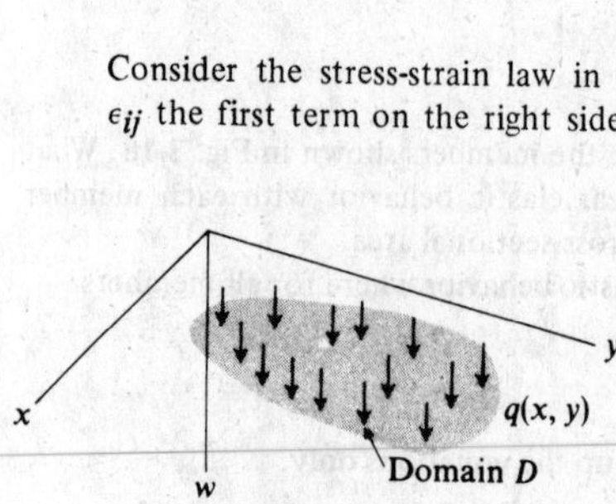

FIGURE 3.19

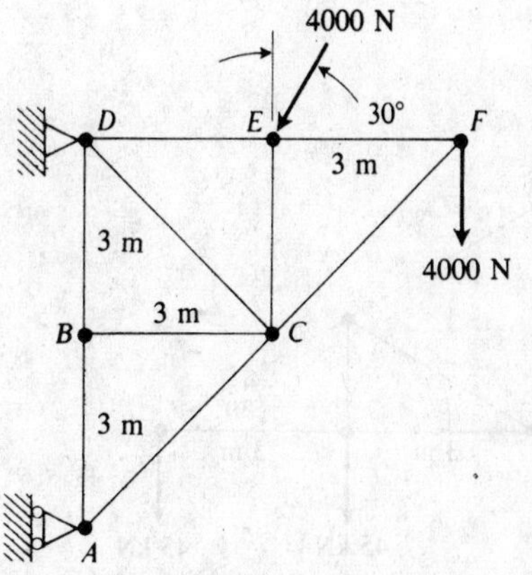

FIGURE 3.20

$$\frac{1}{2!}\iiint_V \delta\tau_{ij}\,\delta\epsilon_{ij}\,dv$$

Explain why the above expression is positive definite for linear elastic materials. Thus the minimum nature of the total potential energy functional is proved for all values of ϵ_{ij} for linear elastic materials.

*3.7 For constant tension T we have derived the governing equation of equilibrium of a transversely loaded string (Example 3.3). Derive the analogous equation for the case where the tension is not fixed, i.e., $T = T(x)$. With $H(x) = \int q(x)\,dx$ and $T(x) = -1/y'\,[H(x) + C_1]$, show that

$$q + \frac{q}{(y')^2\sqrt{1+(y')^2}} - 2\,\frac{\int q\,dx + C_1}{(y')^3\sqrt{1+(y')^2}} - \frac{\int q\,dx + C_1}{(y')[1+(y')^2]^{3/2}} = 0$$

3.8 Use the dummy load method to compute the vertical movement of joint C of the truss in Fig. 3.20 due to the external loads. Assume linear elastic behavior with all members identical in cross section and having the same modulus of elasticity.

3.9 What is the movement of pin D in the horizontal direction in the truss shown in Fig. 3.21. The members are identical in cross section and have the same modulus of elasticity.

3.10 Find the forces in the truss shown in Fig. 3.22. Assume linear elastic behavior with all members identical in cross section and with the same modulus of elasticity. Use the method of total complementary energy.

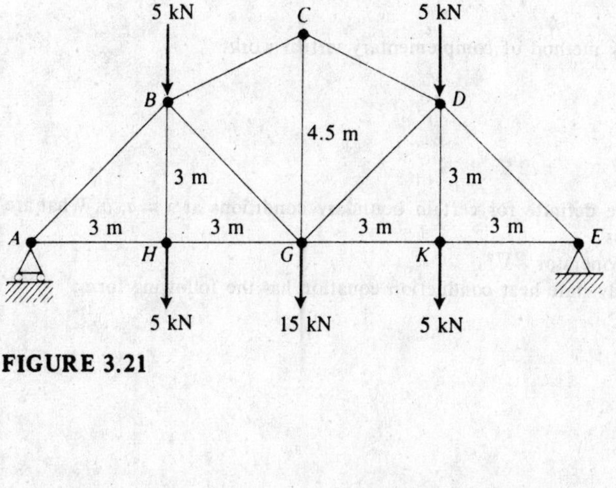

FIGURE 3.21

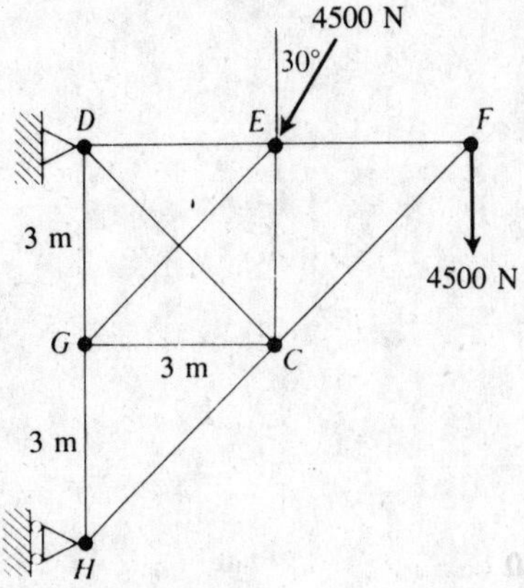

FIGURE 3.22

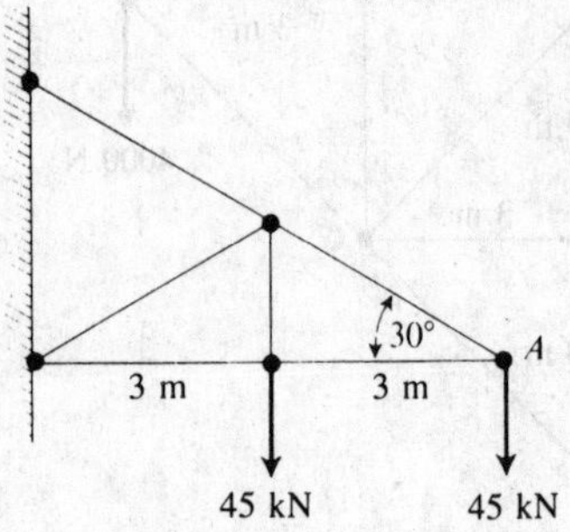

FIGURE 3.23

*3.11 Show that $\delta^{(T)}I_R$ can be given as follows [see Eqs. (3.30) and (3.32)].

$$\delta^{(T)}(I_R) = \delta^{(1)}I_R + \iiint \left[\delta\tau_{ij} \frac{1}{2!} \frac{\partial^2 \mathfrak{U}^*}{\partial\tau_{kl}\,\partial\tau_{ij}} \delta\tau_{kl} + \delta\epsilon_{ij} \frac{1}{2!} \frac{\partial^2 \mathfrak{U}}{\partial\epsilon_{kl}\,\partial\epsilon_{ij}} \delta\epsilon_{kl} \right.$$

$$\left. - \frac{1}{2!} \frac{\partial^2 \mathfrak{U}^*}{\partial\tau_{ij}\,\partial\tau_{kl}} \delta\tau_{ij}\,\delta\tau_{kl} + \cdots \right]$$

Take $\delta^{(1)}R = 0$. Using a constitutive law, show that

$$\delta^{(T)}(I_R) = \tfrac{1}{2} \iiint_V \delta\tau_{ij}\,\delta\epsilon_{ij}\,dv + \cdots$$

Even for linear elastic materials, why can't we argue that the expression on the right side of the equation is positive definite? Accordingly, we *cannot show* that the Reissner principle is a *minimum* principle.

3.12 Using the second Castigliano theorem, find the deflection in the vertical direction for pin A of the truss, in Fig. 3.23. Each member has a cross-sectional area A and is nonlinear-elastic with a stress-strain law given as

$$|\tau| = 7 \times 10^9\,|\epsilon^{0.8}|$$

Also solve the problem by method of complementary virtual work.

3.13 Show that the operator

$$-\frac{d}{dx}\left(x^2 \frac{d}{dx}\right)$$

is self-adjoint and positive definite for certain boundary conditions at $x = a, b$. What are those boundary conditions?

3.14 Do Problem 3.13 for the operator $-\nabla^2$.

3.15 The one-dimensional steady-state heat conduction equation has the following form:

$$\frac{d^2 T}{dx^2} + \psi(x) = 0$$

where T is the temperature and ψ is a specified function representing a heat source distribution. If $T = 0$ at the endpoints show that the quadratic functional can be found directly to be

$$I = \int_a^b (\tfrac{1}{2}T_x^2 - T\psi)\, dx$$

where $T_x = dT/dx$.

3.16 Consider the following functionals:

$$\iiint_V H(x, y, z, w, w_x, w_y)\, dx\, dy\, dz \tag{a}$$

$$\iiint_V \left[H(x, y, z, w, w_x, w_y) + \alpha \frac{\partial A(x, y, z, w)}{\partial x} + \beta \frac{\partial B(x, y, z, w)}{\partial y} + \gamma \frac{\partial C(x, y, z, w)}{\partial z} \right] dx\, dy\, dz \tag{b}$$

where α, β, and γ are constants. Show that you get the same Euler-Lagrange equation for both functionals but that the boundary conditions are different.

3.17 Starting with the partial differential equation for the deflection of the flexible membrane of Problem 3.5 in the form

$$-\nabla^2 w = \frac{q}{T}$$

formulate the corresponding quadratic functional, using Eq. I.36 (Appendix I) in two dimensions to integrate by parts. Compare the quadratic functional with the total potential energy functional for this problem. Need they be the same? See Eq. (*c*) of Problem 3.5.

*3.18 (*a*) The Navier equation for linear elasticity can be put in the form [see Eqs. (1.106) and (1.98)]:

$$-G\left(\frac{1}{1-2\nu}\, u_{i,ij} + u_{j,ii}\right) = B_j \tag{a}$$

Consider the traction field $T_i^{(\nu)}$ associated with displacement field u_i to be denoted as $\overset{u}{T}{}_i^{(\nu)}$. From Cauchy's formula and Hooke's law [see Eqs. (1.96) and (1.98)] we can say that

$$\overset{u}{T}{}_i^{(\nu)} = \tau_{ij} n_j = G\left[\frac{2\nu}{1-2\nu}\, u_{m,m}\, \delta_{ij} + (u_{i,j} + u_{j,i})\right] n_j \tag{b}$$

Let $\overset{u}{e} = u_{i,i}$ and $2\overset{u}{\epsilon}_{ij} = (u_{i,j} + u_{j,i})$. Show for the operator of Eq. (*a*) that by adding and subtracting $\overset{u}{e}_{,j} v_j$ we can arrive at the following result on employing the divergence theorem:

$$(\mathbf{Lu}, \mathbf{v}) = G \iiint_V \left(\frac{2\nu}{1-2\nu} \overset{u}{e}\overset{v}{e} + 2\overset{u}{\epsilon}_{ij}\overset{v}{\epsilon}_{ij} \right) dv$$

$$- G \iint_S \left(\frac{2\nu}{1-2\nu} \overset{u}{e} v_j \delta_{ij} + 2\overset{u}{\epsilon}_{ij} v_j \right) n_j \, dS \tag{c}$$

where n_j are components of the unit normal to the bounding surface s. Show next on using Eq. (*b*) that

$$(\mathbf{Lu}, \mathbf{v}) - (\mathbf{Lv}, \mathbf{u}) = \iint_S \overset{v}{T}{}_i^{(\nu)} u_i \, dS - \iint_S \overset{u}{T}{}_i^{(\nu)} v_i \, dS \tag{d}$$

The well-known *Maxwell-Betti reciprocal theorem* states that for *linear elastic solids* the work done by a system of forces A, through displacements caused by a second system of forces B, equals the work done by the second system of forces B through displacements caused by the system of forces A. Using $\mathbf{B}$ to replace $\mathbf{Lu}$, show that the Maxwell-Betti reciprocal theorem is obtained from result (*d*). What are the implications for the adjointness of the operator L from Navier's equation in light of the Maxwell-Betti reciprocity theorem? Show for self-adjoint operators $\mathbf{L}$ that

$$\iiint (\overset{u}{B}_i v_i - \overset{v}{B}_i u_i) \, dv = 0$$

(*b*) From the above result show that

$$(\mathbf{Lu}, \mathbf{u}) = -\iint_S \overset{u}{T}{}_i^{(\nu)} u_i \, dS + 2G \iiint_V \left[\frac{\nu}{1-2\nu} (u_{m,m})^2 + \epsilon_{ij}\epsilon_{ij} \right] dv$$

Discuss the implications of this result for the positive-definiteness of the operator L. Are there situations where $(\mathbf{Lu}, \mathbf{u}) = 0$ while $\mathbf{u} \neq \mathbf{0}$? What do they mean? What occurs if rigid-body motions $\overset{R}{u_i} = u_i^0 + \omega_{ij} x_j$ are substituted above?

3.19 Show for the differential equation $Lu = f$, that if L is self-adjoint then the Galerkin solution equation is obtainable as the *first variation* of the quadratic functional $I(u)$.

3.20 Consider the differential equation:

$$x^3 \frac{d^4 y}{dx^4} + 6x^2 \frac{d^3 y}{dx^3} + 6x \frac{d^2 y}{dx^2} = 10x$$

If $y' = 0$ and $y = 0$ at $x = 1$ and $x = 3$ solve this boundary-value problem by a one-parameter approximation using the Ritz method.

3.21 Do Problem 3.20 using the Galerkin method.

*3.22 Starting from the Navier equations, show that for the Galerkin method the required equations to determine the undetermined coefficients a_n are

$$\iiint_V \left(\nabla^2 \tilde{u}_1 + \frac{1}{1-2\nu} \frac{\partial \epsilon}{\partial x} + \frac{B_x}{G} \right) \Phi_i \, dv = 0 \tag{a}$$

$$\iiint_V \left(\nabla^2 \tilde{u}_2 + \frac{1}{1-2\nu} \frac{\partial \epsilon}{\partial y} + \frac{B_y}{G} \right) \Psi_i \, dv = 0 \tag{b}$$

$$\iiint_V \left(\nabla^2 \tilde{u}_3 + \frac{1}{1-2\nu}\frac{\partial \epsilon}{\partial z} + \frac{B_z}{G}\right)\Gamma_i\,dv = 0 \tag{c}$$

where $\epsilon = \epsilon_{xx} + \epsilon_{yy} + \epsilon_{zz} = \epsilon_{ii}$. Now express the strain energy of a body as follows:

$$\tilde{\mathfrak{u}} = G\left[\tilde{\epsilon}_{xx}^2 + \tilde{\epsilon}_{yy}^2 + \tilde{\epsilon}_{zz}^2 + \frac{\tilde{\epsilon}^2}{1-2\nu} + 2(\tilde{\epsilon}_{yz}^2 + \tilde{\epsilon}_{zx}^2 + \tilde{\epsilon}_{xy}^2)\right]$$

Formulate π, the total potential energy of the system. Noting that

$$\tilde{u}_1 = \Sigma\, a_i \Phi_i$$

$$\tilde{u}_2 = \Sigma\, b_i \Psi_i$$

$$\tilde{u}_3 = \Sigma\, c_i \Gamma_i$$

set $\partial\tilde{\pi}/\partial a_i = 0$ and reach the following result:

$$-2G\iiint_V \left[\left(\tilde{\epsilon}_{xx} + \frac{\tilde{\epsilon}}{\nu - 2}\right)\frac{\partial \Phi_i}{\partial x} + \tilde{\epsilon}_{yx}\frac{\partial \Phi_i}{\partial y} + \tilde{\epsilon}_{zx}\frac{\partial \Phi_i}{\partial z}\right] dv$$

$$-\iiint_V B_x \Phi_i\,dv - \iint_S T_x^{(\nu)}\Phi_i\,dS = 0$$

Integrate by parts for the first integral [i.e., Eq. (I.31), Appendix I] to get the result:

$$2G\iint_S \left[\left(\tilde{\epsilon}_{xx} + \frac{\tilde{\epsilon}}{1-2\nu}\right)a_{nx} + \tilde{\epsilon}_{yx}a_{ny} + \tilde{\epsilon}_{zx}a_{nz} - \frac{T_x^{(\nu)}}{2G}\right]\Phi_i\,dS$$

$$-\,2G\iiint_V \left[\frac{\partial[\tilde{\epsilon}_{xx} + \tilde{\epsilon}/(1-2\nu)]}{\partial x} + \frac{\partial \tilde{\epsilon}_{yx}}{\partial y} + \frac{\partial \tilde{\epsilon}_{zx}}{\partial z} + \frac{B_x}{2G}\right]\Phi_i\,dv = 0$$

Get the integrand of the first expression in terms of stresses and show that it must be zero if the boundary conditions are maintained everywhere. Finally, show that the second integral may be written in terms of $\tilde{u}$ and $\tilde{\epsilon}$ so that we have

$$\iiint_V \left(\nabla^2 \tilde{u} + \frac{1}{1-2\nu}\frac{\partial \tilde{\epsilon}}{\partial x} + \frac{B_x}{G}\right)\Phi_i\,dv = 0$$

This equation is identical to Eq. (*a*). By performing the preceding steps for $\partial\tilde{\pi}/\partial b_i$ and $\partial\tilde{\pi}/\partial c_i$ we can reach Eqs. (*b*) and (*c*). We thus show that the Galerkin method and the Ritz method give the *same set of simultaneous equations for the undetermined constants* when used with the Navier equation.

II

STRUCTURAL MECHANICS

Using the energy methods formulated in Part I, we develop in Part II the theory of beams, including the Timoshenko beam; the theory of plates, including the Mindlin plate theory; the theory of torsion; the dynamics of beams and plates; and finally, the theory of elastic stability. In each area of study, we do the following:

1. Derive the boundary-value problem variationally, using carefully formulated quadratic functionals, thus ensuring proper kinematic and natural boundary conditions for properly posed boundary-value problems.
2. Discuss the validity of the resulting formulations in view of the simplifications of geometry, of the kinematics of deformation, and of the choice of constitutive law instituted in choosing the quadratic functional.
3. Present some exact solutions of the boundary-value problem.
4. Present approximate solutions, using the methods of Ritz, Rayleigh, and Kantorovich, and compare results with the exact solutions. These approximate solutions yield functions covering the entire domain of the problem.

It is to be noted that additional considerations of the calculus of variations will be presented in Chapter 7, pertaining to eigenfunctions and eigenvalues in relation to the Rayleigh-Ritz method of approximation.

Part II is thus a self-contained treatment of important areas of structural mechanics. At the same time, Part II makes readily available the background material needed for the finite element study in Part III.

4

BEAMS

4.1 INTRODUCTION

In the preceding chapter, while developing as the primary effort certain variational principles of mechanics, we entered into a discussion of trusses in order both to illustrate certain aspects of the theory and to present a discussion of the most simple class of structures. We could take on this dual task at this stage because the stress and deformation of any one single member of a truss is a very simple affair. That is, the only stress on any section (away from the ends)† is a uniform normal stress given as F/A, while the strain at any section is that of normal strain directly available from F/A through an appropriate constitutive law.

We shall now consider a more complex structural member, the beam, wherein we will first develop approximate equations for determining stress and deformation of the beam via the method of total potential energy. The theory that we shall propose is called the *technical* or *engineering theory of beams.* We shall then consider approximate solutions to the deformation of beams by using the Ritz method and a method stemming from the Reissner principle. Next we shall use Castigliano's second theorem to consider statically indeterminate supporting force systems for beams.

4.2 TECHNICAL THEORY OF BEAMS

In Chapter 1 we first presented the full theory of linear elasticity, and we then made simplifying assumptions as to stress distributions (plane stress) by stating that certain stresses were to be taken as zero or constant in the domain of interest. This decreased the number of dependent variables to be dealt with and permitted the solution of several problems of interest. We pointed out in Section 3.3 that in the structural

†At the ends, in reality, due to friction of the supports and complicated boundaries of the member, it is unlikely that simple uniaxial tension or compression will exist.

studies that follow we make the assumptions primarily about displacement fields. Also, we usually make assumptions about certain aspects of the constitutive law to be employed. The variational process, as it relates to the total potential energy, will then give us the proper equations of equilibrium and the proper boundary conditions for a problem like the one we start with, except that it now has certain internal constraints (as a result of the aforementioned displacement assumptions) and it now behaves according to our assumed constitutive law. Most important, by making appropriate assumptions for the displacement field, we will be able to reduce the number of variables significantly and thus facilitate actual computations. For the technical theory of beams, we shall reduce the problem to a single dependent variable $w(x)$, which represents the deflection of the centerline of the beam; thus we are considering a curve in space to represent the beam. Similarly, in a later chapter, we shall be able to employ the considerations of a surface in space to represent a plate. How do we know when the equations we thus derive from variational considerations reasonably depict the actual problem? We can answer this by checking results for simple cases with those that stem from the full theory of elasticity, and also we can check results from our theory with those from experiments.

Consider then a prismatic beam of rectangular cross section, as shown in Fig. 4.1. We assume that the dimensions h and b are both much smaller than the length L. We take x as the centerline of the beam. The loading $q(x)$ is assumed in the plane of symmetry of the cross section, i.e., in the xz plane, and is directed transversely to the beam. (Such loadings might entail point forces or couples.) Axial tensile loading is also possible along the centroidal axis (taken as the x axis) of the beam.

We now present a simplified view of the deformation to be expected from these loads. In the x direction, we imagine a *stretching* action from the axial load wherein each section undergoes a displacement in the x direction, which is a function of the position x of the section. Thus, using the notation $\mathbf{u} = u_1\mathbf{i} + u_2\mathbf{j} + u_3\mathbf{k}$, the strain at any section due to the axial tensile loads is $d[u_1(x)]_s/dx$, where the subscript s refers to the axial stretching action. In addition, we superpose a *bending* contribution from the transverse loads. To assess this contribution consider the line in the beam in the undeformed geometry coinciding with the x axis. As a result of deformation from the transverse loads, this line deforms into a curve whose position relative to the reference is denoted as $w(x)$, as shown in Fig. 4.2. For small deformations, a line element dx in

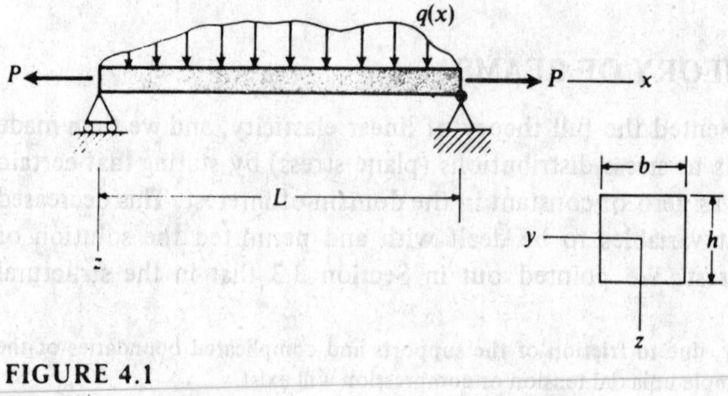

FIGURE 4.1
Simply supported beam with axial and transverse loads.

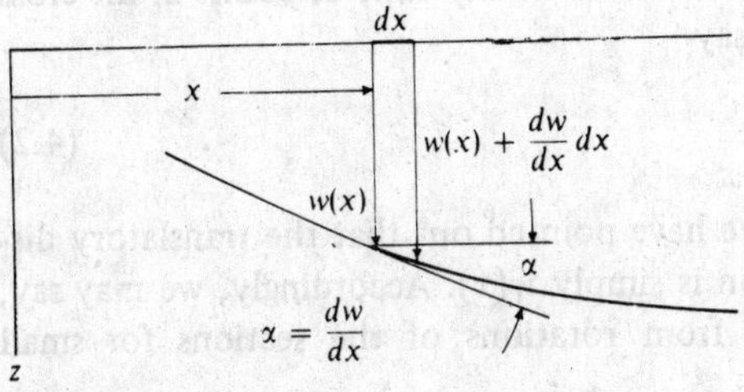

FIGURE 4.2
Deflection curve of beam.

the undeformed geometry is assumed to remain at coordinate x but translates in the z direction an amount $w(x)$ and rotates in the xz plane an amount given by the slope dw/dx of the deflection curve at coordinate x. We now assume that plane sections such as ab in Fig. 4.3, originally normal to the centerline of the beam in the undeformed geometry, remain plane and normal to the centerline (see line $a'b'$) in the deformed geometry as the beam bends from the loads. Furthermore, in the planes of these sections we assume there is no stretching or shortening; they are assumed to act like rigid surfaces. Thus, as a result of bending, sections of the beam translate vertically an amount $w(x)$ to new positions and, in addition, rotate by the amount dw/dx. Accordingly, in considering the bending displacement in the x direction of a point in a cross section of a beam at position z below the centroidal axis in the undeformed geometry, we get for small deflections

$$(u_1)_{\text{bend}} = -z\frac{dw(x)}{dx}$$

The total displacement u_1 (i.e., the total displacement in the x direction) for the aforementioned point is then

$$u_1 = (u_1)_s - z\frac{dw(x)}{dx}$$

while the corresponding strain is then

$$\epsilon_{xx} = \frac{d[u_1(x)]_s}{dx} - z\frac{d^2w(x)}{dx^2} \tag{4.1}$$

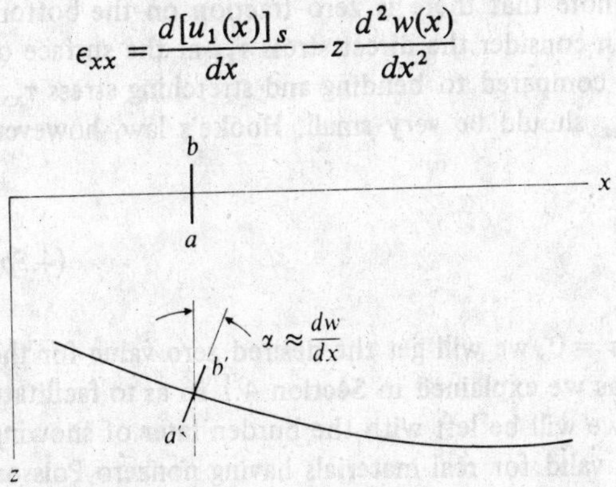

FIGURE 4.3
Rotation of vertical line segment.

There is no relative motion in the y direction at any time of points in the cross section of the beam. Accordingly, we can say

$$\epsilon_{yy} = 0 \tag{4.2}$$

We may conclude that $u_2 = 0$. Finally, we have pointed out that the translatory displacement of the section in the z direction is simply $w(x)$. Accordingly, we may say, neglecting vertical movement stemming from rotations of the sections for small deformation,

$$u_3 = w(x) \tag{4.3}$$

We now summarize the results presented for the proposed displacement field:

$$\boxed{\begin{aligned} u_1 &= [u_1(x)]_s - z\frac{dw(x)}{dx} \\ u_2 &= 0 \\ u_3 &= w(x) \end{aligned}} \tag{4.4}$$

It is clear that for the above displacement field all strains except ϵ_{xx} are zero. Also note that the bending deformation is now given in terms of the deformation of the centerline of the beam.

The assumed deformation field has obvious errors even for small deformation. For instance, in the case of a thin beam ($b \ll L$) and with no loading on the sides of the beam, we would expect intuitively that the stresses τ_{yy} would be very close to zero. However, it is clear for a Hookean material that the stress τ_{yy} as a result of our assumptions must be, in accordance with Eqs. (1.96) and (1.98*b*),

$$\tau_{yy} = \frac{\nu E}{(1+\nu)(1-2\nu)}\epsilon_{xx} \tag{4.5a}$$

Furthermore, in considering τ_{zz} we note that there is zero traction on the bottom surface of the beam and that we can consider the direct stress τ_{zz} at the surface of contact for q to be negligibly small compared to bending and stretching stress τ_{xx}. Accordingly, for $h \ll L$, the stress τ_{zz} should be very small. Hooke's law, however, gives τ_{zz} as follows:

$$\tau_{zz} = \frac{\nu E}{(1+\nu)(1-2\nu)}\epsilon_{xx} \tag{4.5b}$$

If we use a constitutive law wherein $\nu = 0$, we will get the desired zero value for the stresses τ_{yy} and τ_{zz}. We will do this as we explained in Section 4.1 so as to facilitate the development of the theory, but we will be left with the burden later of showing that the results are at least partially valid for real materials having nonzero Poisson ratios.

It is also to be noted that the theory presented thus far yields a zero value for τ_{xz}. For beams with nonzero lateral loading, we know from earlier work in strength of materials that there will be a nonzero stress τ_{xz} over much of the beam. However, this shear stress and the accompanying shear deformation must be small for the class of problems we are studying, and despite the incorrect result from our theory for τ_{xz}, we still get a good evaluation of the far more significant stress τ_{xx} and the quantity $w(x)$.†

The theory presented for evaluating τ_{xx} and $w(x)$ is called the *technical theory of beams.* We shall in succeeding sections make efforts to assess the accuracy of the technical theory of beams. It gives us effectively a one-dimensional stress problem to consider, and we shall study the variational aspects of this problem in the next section.

4.3 DEFLECTION EQUATIONS FOR THE TECHNICAL THEORY OF BEAMS

As pointed out in the preceding section, we can formulate a technical theory of beams as a one-dimensional problem. In computing the strain energy U for the beam we will have for linear elastic behavior:

$$U = \frac{1}{2} \iiint_V \tau_{ij}\, \epsilon_{ij}\, dv = \frac{1}{2}\, b \int_0^L \int_{-h/2}^{h/2} \tau_{xx}\, \epsilon_{xx}\, dz\, dx \tag{4.6}$$

We shall wish to make use of the method of minimum total potential energy and so we next examine the potential of the applied loads to do work. At position $x = L$ the applied axial load P (see Fig. 4.1) is in the same direction as the positive displacement $[u_1(l)]_s$, while at $x = 0$ the applied load P is in the opposite direction to the positive displacement $[u_1(0)]_s$. Accordingly, we have for loads P the following result:

$$V_p = -\{P[u_1(L)]_s - P[u_1(0)]_s\} = -P \int_0^L \frac{d}{dx}(u_1)_s\, dx$$

In consider the work of loading $q(x)$, we assume that this loading acts at the centroidal axis of the beam and not on the top surface. This means that $w(x)$ characterizes the movement of the points of loading for $q(x)$, a conclusion consistent with the deformation assumption. Accordingly, since q and w are both positive when directed downward, we have

$$V_q = -\int_0^L q(x)w(x)\, dx \tag{4.7}$$

†Note because we are neglecting the shear deformation associated with τ_{xz}, $w(x)$ represents deflection due only to bending.

The total potential energy for the problem then becomes

$$\pi = \frac{1}{2} b \int_0^L \int_{-h/2}^{h/2} \tau_{xx} \epsilon_{xx} \, dx \, dz - P \int_0^L \frac{d}{dx} (u_1)_s \, dx - \int_0^L q(x) w(x) \, dx$$

In computing the total potential energy we shall use Hooke's law to replace τ_{xx}. Maintaining the earlier assumption that $\nu = 0$, we can then give τ_{xx} simply as $E\epsilon_{xx}$. Now replace ϵ_{xx} in the above equation in terms of the displacement field given by Eq. (4.1). Thus:

$$\pi = \frac{1}{2} \int_0^L \int_{-h/2}^{h/2} bE \left[\frac{d(u_1)_s}{dx} - z \frac{d^2 w}{dx^2} \right]^2 dz \, dx - P \int_0^L \frac{d(u_1)_s}{dx} \, dx$$

$$- \int_0^L qw \, dx$$

If we carry out the indicated squaring operation and note the following integrals:

$$\int_{-h/2}^{h/2} (1, z, z^2) \, dz = \left(h, 0, \frac{h^3}{12} \right)$$

we easily obtain the following result:

$$\pi = \int_0^L \left\{ \frac{EA}{2} \left[\frac{d(u_1)_s}{dx} \right]^2 + \frac{EI}{2} \left(\frac{d^2 w}{dx^2} \right)^2 - P \frac{d(u_1)_s}{dx} - qw \right\} dx \tag{4.8}$$

We have then a functional with one independent variable x and involving the function $(u_1)_s$, with a derivative of order unity, and the function w, with a derivative of order two. The Euler-Lagrange equations for finding the extremal functions are readily established with the aid of Eq. (2.64).† Thus for the stretching displacement $(u_1)_s$ we get

$$\frac{d}{dx} \left[EA \frac{d(u_1)_s}{dx} - P \right] = 0 \tag{4.9}$$

while for the bending displacement w we get

$$\frac{d^2}{dx^2} \left(EI \frac{d^2 w}{dx^2} \right) = q \tag{4.10}$$

†You will be asked as an exercise (Problem 4.1) to find these results using the δ operator.

As for the boundary conditions [see Eq. (2.63)], we may specify the kinematic conditions at the ends of the beam. This means that $(u_1)_s$ is specified at an end, as are w and w'. If we wish to employ the natural boundary conditions to establish an extremum we see from Eq. (2.63) that for the function $(u_1)_s$ we get

$$\frac{\partial F}{\partial (u_1)_s'} = EA\frac{d(u_1)_s}{dx} - P = 0 \quad \text{at } x = 0, x = L \tag{4.11}$$

and in place of specifying w itself we have the natural boundary conditions which, in accordance with Eqs. (2.63) and (4.8), become

$$-\frac{d}{dx}\left(\frac{\partial F}{\partial w''}\right) + \frac{\partial F}{\partial w'} = -\frac{d}{dx}\left(EI\frac{d^2 w}{dx^2}\right) = 0 \quad \text{at } x = 0, x = L \tag{4.12}$$

Finally, in place of specifying w' at the ends we see that

$$-\frac{\partial F}{\partial w''} = EI\frac{d^2 w}{dx^2} = 0 \quad \text{at } x = 0, x = L \tag{4.13}$$

We now summarize the boundary conditions as follows:

Kinematic boundary conditions		*Natural boundary conditions*
$(u_1)_s$ prescribed	or	$EA\dfrac{d(u_1)_s}{dx} = P$
$\dfrac{dw}{dx}$ prescribed	or	$EI\dfrac{d^2 w}{dx^2} = 0$
w prescribed	or	$\dfrac{d}{dx}\left(EI\dfrac{d^2 w}{dx^2}\right) = 0$ (4.14)

We shall now give physical interpretations for the natural boundary conditions. Let us first define an axial stress resultant N and a moment M as†

$$N = \int_{-h/2}^{h/2} \tau_{xx} b\, dz \tag{4.15a}$$

$$M = \int_{-h/2}^{h/2} \tau_{xx} zb\, dz \tag{4.15b}$$

Now expressing τ_{xx} as $E\epsilon_{xx}$ and using Eq. (4.1) for the strain, we find after simple calculations that

†We are using the same convention for signs of V and M as we used for stress. Thus, if the area normal of the beam section points in a positive coordinate direction, then positive V and M must have vectorial representations in positive coordinate directions.

$$N = EA\frac{d(u_1)_s}{dx} \tag{4.16a}$$

$$M = -EI\frac{d^2w}{dx^2} \tag{4.16b}$$

We thus have an interpretation of the first two of the natural boundary conditions of Eqs. (4.14). Our next task is to examine the third natural boundary condition. Thus, consider an element of the beam as shown in Fig. 4.4. The shear force V is the stress resultant due to τ_{xz}. That is:

$$V = \int_{-h/2}^{h/2} \tau_{xz}\, b\, dz$$

We wish next to relate V and M. One way of computing the relationship between V and M as well as the distribution of shear stress in the cross section is to take moments about the point "a" in Fig. 4.4, and then resort to the classical $\tau = VQ/Ib$ formula of strength of materials. We shall instead determine the shear stress distribution from the first equation of equilibrium of elasticity theory, and then evaluate the shear force resultant.

The equation of equilibrium in the x direction in the absence of body forces and with $\tau_{xy} = 0$ is

$$\frac{\partial \tau_{xx}}{\partial x} + \frac{\partial \tau_{xz}}{\partial z} = 0$$

Let us now evaluate the bending stress τ_{xx} due to the bending deflection $w(x)$. From Eq. (4.1) and the one-dimensional constitutive law,

$$\tau_{xx} = E\epsilon_{xx} = -Ez\frac{d^2w}{dx^2}$$

The rate of change of curvature d^2w/dx^2 can be eliminated from the above equation by using Eq. (4.16b) to yield the classical flexure formula

$$\tau_{xx} = \frac{Mz}{I} \tag{4.17}$$

Substituting this bending stress distribution into the elasticity equilibrium equation and integrating with respect to z yields for τ_{xz}:

$$\tau_{xz} = -\int \frac{dM(x)}{dx}\frac{z}{I}\,dz + g(x) = -\frac{dM(x)}{dx}\left(\frac{z^2}{2I}\right) + g(x)$$

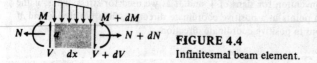

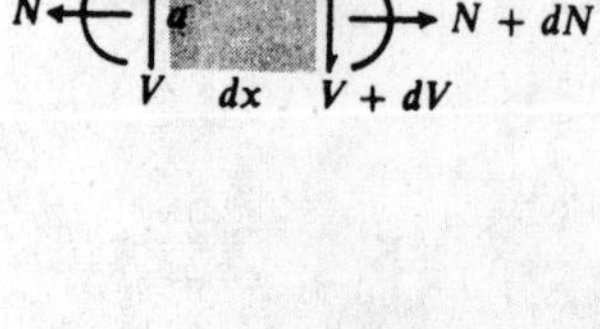

FIGURE 4.4
Infinitesmal beam element.

where $g(x)$ is an arbitrary function of x. Since the shear stress vanishes at $z = \pm h/2$, as we are applying only normal loading to the beam, we can determine $g(x)$ and find that

$$\tau_{xz} = -\frac{dM}{dx}\left(\frac{1}{2I}\right)\left(z^2 - \frac{h^2}{4}\right) \tag{4.18}$$

The shear resultant V can then be found as follows:

$$V = \int_{-h/2}^{h/2} \tau_{xz}\, b\, dz = \int_{-h/2}^{h/2} -\frac{dM}{dx}\frac{b}{2I}\left(z^2 - \frac{h^2}{4}\right) dz = \frac{dM}{dx} \tag{4.19}$$

Hence we have derived another classical formula between the shear and moment resultants. We can combine Eqs. (4.16*b*) and (4.19) to relate the shear to the bending deflection, i.e.,

$$V = -\frac{d}{dx}\left(EI\frac{d^2w}{dx^2}\right) \tag{4.20}$$

Using Eqs. (4.16) and (4.20), we can now go back to the natural boundary conditions as given by Eq. (4.14) and restate all of them in terms of forces and moments. We now give the boundary conditions as follows (at $x = 0, x = L$):

Kinematic boundary conditions		*Natural boundary conditions*	
$(u_1)_s$ prescribed	or	$N = P$	
$\dfrac{dw}{dx}$ prescribed	or	$M = 0$	
w prescribed	or	$V = 0$	(4.21)

We can now state the boundary conditions as follows:

1. Either the longitudinal displacement or the total axial force is prescribed at the ends;
2. Either the slope of the deflected centerline is prescribed at the ends or the total bending moment is zero there; and finally
3. Either the vertical displacement is prescribed at the ends or the shear force is zero there.

Also, a combination of (2) and (3) will do.

Now we will note an important characteristic of the equations we have just derived. If we examine the differential equations [Eqs. (4.9) and (4.10)] and the boundary conditions [Eqs. (4.14) or (4.21)], we see that the *stretching problem* [involving the displacement $(u_1)_s$ and the axial stress resultant N] is completely solvable independent of the bending problem. That is, the stretching and bending problems are *uncoupled*. This is a direct consequence of the kinematic assumptions we have made [Eq. (4.4)]. Because the stretching problem is very simple, we shall concern ourselves in the sequel only with the bending problem.

4.4 SOME JUSTIFICATIONS FOR THE TECHNICAL THEORY OF BEAMS

For the purpose of demonstrating the validity of the technical theory of beams, we now consider a simple problem that has already been investigated (Problem 1.26) in Chapter 1 as an example of plane stress. Thus we show in Fig. 4.5 a simply supported beam loaded uniformly in the transverse direction. We shall place the origin of the reference at the center of the beam as shown in the diagram. The obvious boundary conditions that we shall use for this problem at $x = \pm(L/2)$ are that the deflection w is zero and that the bending moment M is zero. The latter is accomplished by requiring that $d^2w/dx^2 = 0$ at the ends [see Eq. (4.16*b*)].

The equation for the deflection of the centerline [see Eq. (4.10)] may then be written as follows for uniform properties of the beam and uniform shape of the cross section:

$$\frac{d^4 w}{dx^4} = \frac{q_0}{EI}$$

Integrating four times we get

$$w = \frac{q_0 x^4}{24EI} + \frac{C_1 x^3}{6} + \frac{C_2 x^2}{2} + C_3 x + C_4$$

We may determine the constants of integration from the boundary conditions. Note first that

$$\frac{d^2 w}{dx^2} = \frac{q_0 x^2}{2EI} + C_1 x + C_2$$

Hence at $x = \pm L/2$ we have

$$0 = \frac{q_0 L^2}{8EI} \pm C_1 \frac{L}{2} + C_2$$

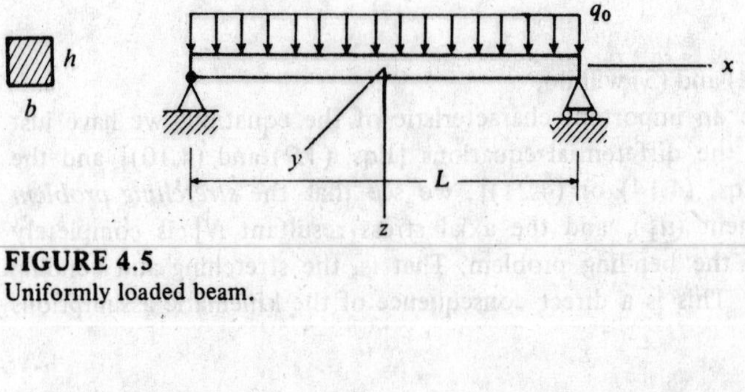

FIGURE 4.5
Uniformly loaded beam.

Clearly, C_1 must be zero for the above to be correct and $C_2 = -q_0 L^2/8EI$. Also, setting $w = 0$ at $x = \pm L/2$, we get

$$0 = \frac{q_0 L^4}{384EI} \pm \frac{C_3 L}{2} - \frac{q_0 L^2}{8EI}\left(\frac{L^2}{4}\right) + C_4$$

Again it is immediately apparent from above that $C_3 = 0$ and that $C_4 = 5q_0L^4/(384EI)$. We then have for the deflection curve

$$w = \frac{q_0 x^4}{24EI} - \frac{q_0 L^2}{8EI}\frac{x^2}{2} + \frac{5q_0L^4}{384EI} = \frac{q_0 L^4}{24EI}\left[\left(\frac{x}{L}\right)^4 - \frac{3}{2}\left(\frac{x}{L}\right)^2 + \frac{5}{16}\right] \tag{4.22}$$

As a first step in evaluating the validity of the above result, we show that there is inner consistency in the theory. Specifically, we have neglected the shear strain energy in the development of the theory, so we shall now examine the shear strain energy that stems from the resulting formulation of the theory. If there is consistency, the shear strain energy from this formulation must also be small. Denoting the shear strain energy as U_{shear}, we have

$$U_{\text{shear}} = \frac{1}{2}\iiint_V \tau_{xz}\gamma_{xz}\, dv = \frac{1}{2G}\iiint_V \tau_{xz}^2\, dv$$

Using Eqs. (4.18) and noting for the problem that $dM/dx = -q_0 x$

$$U_{\text{shear}} = \frac{1}{2G}\int_{-L/2}^{L/2}\int_{-h/2}^{h/2}\left[-\frac{q_0 x}{2I}\left(\frac{h^2}{4} - z^2\right)\right]^2 b\, dz\, dx$$

$$= \frac{q_0^2 L^5}{240EI}\left[2(1+\nu)\left(\frac{h}{L}\right)^2\right] \tag{4.23}$$

where we have replaced G by the expression $E/[2(1+\nu)]$.

Now compute the strain energy due to bending. We have, using the portion of Eq. (4.8) that is applicable and substituting from Eq. (4.22),

$$U_{\text{bend}} = \frac{EI}{2}\int_{-L/2}^{L/2}\left(\frac{d^2 w}{dx^2}\right)^2 dx = \frac{q_0^2 L^5}{240EI} \tag{4.24}$$

Taking the ratio of Eqs. (4.23) and (4.24) for comparison purposes, we have

$$\frac{U_{\text{shear}}}{U_{\text{bend}}} = 2(1+\nu)\left(\frac{h}{L}\right)^2$$

Thus we see that the ratio of the shear strain energy to the bending strain energy is proportional to (h/L) squared. Hence for long slender beams the shear strain energy is very small compared to the bending strain energy. This result is entirely consistent with the assumption made in developing the technical theory.

We may also readily show here that the ratio of the maximum magnitude of the shear stress τ_{xz} to that of the normal stress τ_{xx} is proportional to (h/L). To do this, we return to Eq. (4.18) and, introducing the dimensionless variables

$$\xi = \frac{x}{L} \qquad \eta = \frac{z}{h}$$

we get, on noting that $dM/dx = -q_0 x$:

$$\tau_{xz} = -\frac{q_0 L^2 h}{2I}\left(\frac{h}{L}\right)\left(\frac{1}{4} - \eta^2\right)\xi \tag{4.25a}$$

Now expressing $\tau_{xx} = E\epsilon_{xx} = -Ez(d^2w/dx^2)$ as per Eq. (4.1), and using Eq. (4.22) for w, we get

$$\tau_{xx} = -\frac{q_0 L^2 h}{2I}\left(\xi^2 - \frac{1}{4}\right)\eta \tag{4.25b}$$

Note that $|\tau_{xz}|_{\max}$ occurs at $\xi = \frac{1}{2}$ and $\eta = 0$ while $|\tau_{xx}|_{\max}$ occurs at $\xi = 0$ and $\eta = \frac{1}{2}$. We thus have for these quantities

$$|\tau_{xz}|_{\max} = \frac{qL^2 h}{2I}\left(\frac{h}{L}\right)\left(\frac{1}{4} - 0\right)\left(\frac{1}{2}\right)$$

$$|\tau_{xx}|_{\max} = -\frac{qL^2 h}{2I}\left(0 - \frac{1}{4}\right)\left(\frac{1}{2}\right)$$

Clearly the ratio $|\tau_{xx}|_{\max}/|\tau_{xz}|_{\max}$ is (L/h) and we see that τ_{xx} is the important stress for long slender beams. Hence considering τ_{xz} small in the development of the theory is consistent with the results stemming from the theory.

The test presented in the example is an indirect one for our purposes. It shows no internal consistency. However, we need further evidence that it is correct, since the absence of internal inconsistencies is not sufficient for justification that the result is correct.

For this purpose we shall now compare the results given here for the simply supported, uniformly loaded beam with the results presented in Problem 1.26 for the same beam problem as developed from the theory of elasticity. Thus we now rewrite the plane stress solution of the aforementioned problem as given in Problem 1.26:

$$\tau_{xx} = -\frac{q_0}{2I}\left[x^2 - \left(\frac{L}{2}\right)^2\right]z + \frac{q_0}{2I}\left(\frac{2}{3}z^3 - \frac{h^2}{10}z\right)$$

$$\tau_{zz} = -\frac{q_0}{2I}\left(\frac{1}{3}z^3 - \frac{h^2}{4}z + \frac{h^3}{12}\right)$$

$$\tau_{xz} = -\frac{q_0}{2I}\left(\frac{h^2}{4} - z^2\right)x \tag{4.26}$$

Introducing the dimensionless variables ξ and η, we get for the above equations

$$\tau_{xx} = -\frac{q_0 L^2 h}{2I}\left[\left(\xi^2 - \frac{1}{4}\right)\eta + \left(\frac{h}{L}\right)^2\left(\frac{2}{3}\eta^3 - \frac{1}{10}\eta\right)\right] \tag{4.27a}$$

$$\tau_{zz} = -\frac{q_0 L^2 h}{2I}\left(\frac{h}{L}\right)^2\left(\frac{1}{3}\eta^3 - \frac{1}{4}\eta + \frac{1}{12}\right) \tag{4.27b}$$

$$\tau_{xz} = -\frac{q_0 L^2 h}{2I}\left(\frac{h}{L}\right)\left(\frac{1}{4} - \eta^2\right)\xi \tag{4.27c}$$

Now considering τ_{xx} we see that it consists of the stress given by the technical theory [see Eq. (4.25)] plus a corrective term containing the multiplicative parameter $(h/L)^2$. Thus the technical theory gives good results for τ_{xx} when compared to the exact theory for long slender beams, where $(h/L)^2$ is very small. Also, τ_{zz} and τ_{xz} are smaller than τ_{xx} for long slender beams.

From the results of the plane stress solution (Problem 1.27), we have for the deflection of the centerline of the simply supported, uniformly loaded beam:

$$w(x,0) = \frac{q_0 L^4}{24EI}\left[\frac{5}{16} - \frac{3}{2}\left(\frac{x}{L}\right)^2 + \left(\frac{x}{L}\right)^4\right] + \frac{q_0 L^4}{24EI}\left(\frac{h}{L}\right)^2\left(\frac{12}{5} + \frac{3\nu}{2}\right)\left[\frac{1}{4} - \left(\frac{x}{L}\right)^2\right] \tag{4.28}$$

We see that Eq. (4.28) contains the elementary solution [see Eq. (4.22)] plus a "corrective" term multiplied by $(h/L)^2$. Again we see that for long slender beams the results of elasticity theory reduce to those of elementary beam theory.

Note from Eqs. (4.27) that for long slender beams the stress τ_{zz}, because of the presence of term $(h/L)^2$, can be expected generally to be considerably smaller than τ_{xz}, which has in its expression the term (h/L). In the following section, in which we present a more accurate theory, we accordingly take into account shear effects while still setting $\tau_{zz} = 0$.

We have thus demonstrated via this problem both the inner consistency of beam theory as well as the correlation of certain of its results with appropriate limiting cases of the elasticity solution.

4.5 TIMOSHENKO BEAM THEORY

The technical theory of beams presented in the previous section does not include the effects of shear deformation. For short stubby beams this contribution clearly cannot be neglected, and for this reason we now present the Timoshenko theory of beams as a means of accounting for the effects of shear in a simple manner.

Let us accordingly consider shear deformation *alone*. We will assume that line elements such as *ab* in Fig. 4.6 *normal* to the centerline of the beam in the undeformed state move only in the vertical direction and also *remain vertical* during deformation. Line elements *tangent* to the centerline meanwhile undergo a rotation $\beta(x)$, as has

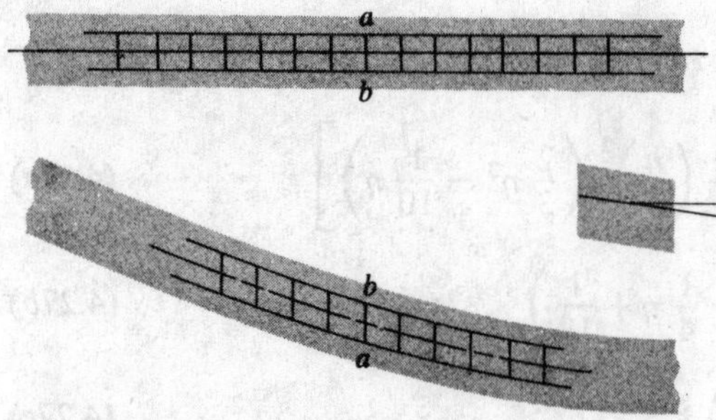

FIGURE 4.6
Shear deformation in a beam.

been shown in an exaggerated manner in the diagram. [$\beta(x)$ accordingly gives the shear angle γ_{xz} at points along the centerline.] The total slope dw/dx of the centerline stemming respectively from shear deformation and bending deformation can then be given as the sum of two parts in the following way:

$$\frac{dw}{dx} = \psi(x) + \beta(x) \tag{4.29}$$

where $\psi(x)$ is the rotation of line elements along the centerline due to bending only. We now make a further assumption which will easily be recognized as being incorrect but which we shall later adjust. We assume that the shear strain is the same at *all* points over a given cross section of the beam. That is, the angle $\beta(x)$, used heretofore for rotation of elements along the centerline, is considered to measure the shear angle at all points in the cross section of the beam at position x. This has been shown in Fig. 4.7. Such an assumed action means that the shear stress is uniform through the thickness–a result that is not possible when only normal loads act on the upper surface and no traction whatever exists on the lower surface.† Such an assumed behavior nevertheless will greatly facilitate computations, and the error incurred can and will be reasonably taken into account later. The displacement field can now be

†The shear stress must then be zero at the upper and lower boundary surfaces.

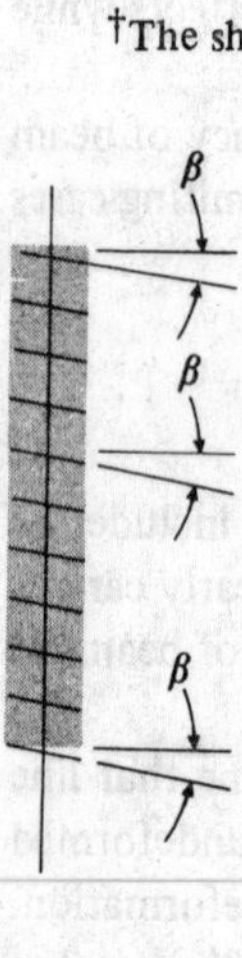

FIGURE 4.7
Uniform shear angle along a cross section.

considered the superposition of the bending action set forth in Section 4.2 and the aforestated shearing action. Clearly the axial displacement u_1 is then due only to bending action. Employing Eq. (4.29) we can thus state:

$$u_1(x,y,z) = -z\psi(x) = -z\left[\frac{dw}{dx} - \beta(x)\right]$$

$$u_2(x,y,z) = 0$$

$$u_3(x,y,z) = w(x) \tag{4.30}$$

It then follows from strain-displacement relations that

$$\epsilon_{xx} = -z\,\frac{d\psi}{dx} \tag{4.31a}$$

$$\epsilon_{xz} = \tfrac{1}{2}\beta(x) \tag{4.31b}$$

Now making use of the approximation $\tau_{xx} = E\epsilon_{xx}$ (i.e., using $\nu = 0$) and using Eqs. (4.31) we find, employing Fig. 4.1,

$$M = \int_{-h/2}^{h/2} \tau_{xx} zb\, dz = -EI\,\frac{d\psi}{dx} \tag{4.32a}$$

$$V = \int_{-h/2}^{h/2} \tau_{xz} b\, dz = \tau_{xz} \int_{-h/2}^{h/2} b\, dz = \tau_{xz} A = GA\beta \tag{4.32b}$$

The assumption of a uniform shear stress through the thickness leads, from the above formulation, to the following result for such a stress:

$$\tau_{xz} = \frac{V}{A} = G\beta(x) \tag{4.33}$$

Actually, a more proper statement for τ_{xz} would appear as follows:

$$\tau_{xz} = G\beta(x,z) \tag{4.34}$$

where a variation with z is taken into account. However, a procedure such as one taken in Eq. (4.34) would complicate the problem and remove the simplicity inherent in a one-dimensional beam approach. In order to recognize the nonuniform shear stress distribution at a section while still retaining the one-dimensional approach, we modify Eq. (4.34) by introducing a factor k as follows:

$$\tau_{xz} = kG\beta(x) \tag{4.35}$$

Hence, on using the above result in Eq. (4.33), we have

$$V(x) = kGA\beta(x) \tag{4.36}$$

TABLE 4.1
***k* Factors for Various Cross Sections**

CIRCLE

$$k = \frac{6(1 + \nu)}{7 + 6\nu}$$

THIN WALLED TUBE

$$k = \frac{2(1 + \nu)}{4 + \nu}$$

HOLLOW CYLINDER

$$k = \frac{6(1 + \nu)(1 + m)^2}{(7 + 6\nu)(1 + m)^2 + (20 + 12\nu)m^2}$$

$$m = b/a$$

THIN-WALLED SQUARE TUBE

$$k = \frac{20(1 + \nu)}{48 + 39\nu}$$

RECTANGLE

$$k = \frac{10(1 + \nu)}{12 + 11\nu}$$

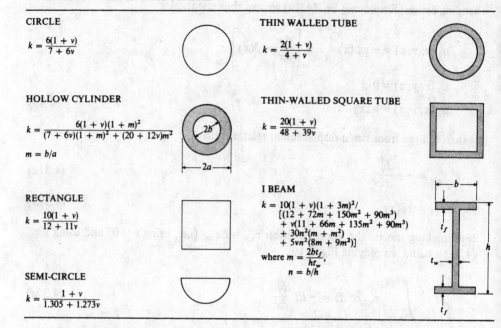

I BEAM

$$k = 10(1 + \nu)(1 + 3m)^2/ [(12 + 72m + 150m^2 + 90m^3) + \nu(11 + 66m + 135m^2 + 90m^3) + 30n^2(m + m^2) + 5\nu n^2(8m + 9m^2)]$$

$$\text{where } m = \frac{2bt_f}{ht_w}, \quad n = b/h$$

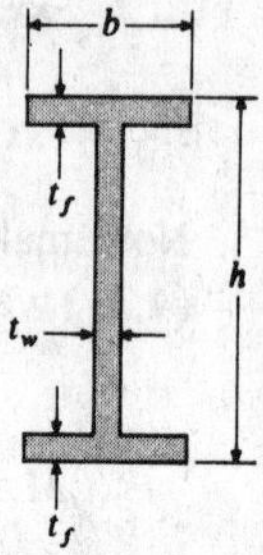

SEMI-CIRCLE

$$k = \frac{1 + \nu}{1.305 + 1.273\nu}$$

There are virtually as many definitions† of k as there are published papers on the Timoshenko beam. A recent one by Cowper (1966) gives an interesting review (as well as a new definition). Suffice it here to mention that it is a function of the cross section and, depending on the interpretation, may also be a function of the Poisson ratio ν. In Table 4.1 we have shown a listing of k taken from Cowper's paper.

We now find the total potential energy for the assumed deformation field of the Timoshenko beam. From the work of the previous section, it is clear that

$$\pi = \frac{1}{2}\int_0^L\int_{-h/2}^{h/2} \tau_{xx}\,\epsilon_{xx}\,dx\,b\,dz + \int_0^L\int_{-h/2}^{h/2} \tau_{xz}\,\epsilon_{xz}\,dx\,b\,dz - \int_0^L qw\,dx$$

Using $\tau_{xx} = E\epsilon_{xx}$ and using Eq. (4.31a) in the first integral while employing $\tau_{xz} = kG\beta$ and Eq. (4.31b) in the second integral, we get

$$\pi = \frac{EI}{2}\int_0^L \left(\frac{d\psi}{dx}\right)^2 dx + \frac{GA}{2}\int_0^L k\beta^2\,dx - \int_0^L qw\,dx$$

$$= \frac{EI}{2}\int_0^L \left(\frac{d\psi}{dx}\right)^2 dx + \frac{kGA}{2}\int_0^L \left(\frac{dw}{dx} - \psi\right)^2 dx - \int_0^L qw\,dx$$

†Among the many definitions are those stemming from relating the maximum shear stress through the thickness, as developed from a more exact solution, to the approximation (4.35), or those from matching of certain wave speeds from the dynamics of a Timoshenko beam to more accurate results of elasticity theory.

wherein we have employed Eq. (4.29) to eliminate β in the second integral. Collecting terms, we have for the total potential energy

$$\pi = \int_0^L \left[\frac{EI}{2}\left(\frac{d\psi}{dx}\right)^2 + \frac{kGA}{2}\left(\frac{dw}{dx} - \psi\right)^2 - qw \right] dx \tag{4.37}$$

Thus the functional has two functions, ψ and w. The Euler-Lagrange equations for this case are then easily deduced from Eq. (2.64).† Thus we first have

$$\frac{d}{dx}\left(EI\frac{d\psi}{dx}\right) + kGA\left(\frac{dw}{dx} - \psi\right) = 0 \tag{4.38a}$$

where for constant E and I we get

$$EI\left(\frac{d^2\psi}{dx^2}\right) + kGA\left(\frac{dw}{dx} - \psi\right) = 0 \tag{4.38b}$$

Also,

$$\frac{d}{dx}\left[kGA\left(\frac{dw}{dx} - \psi\right)\right] + q = 0 \tag{4.39a}$$

where for constant G and A we get

$$kGA\left(\frac{d\psi}{dx} - \frac{d^2w}{dx^2}\right) = q \tag{4.39b}$$

Also, from Eq. (2.63) adjusted for several dependent variables it becomes clear that for the end conditions at $x = 0$ and $x = L$ we have:

$$\begin{aligned} &1.\ \text{Either } \psi \text{ is specified or } EI\frac{d\psi}{dx} = 0 \\ &2.\ \text{Either } w \text{ is specified or } kGA\left(\frac{dw}{dx} - \psi\right) = 0 \end{aligned} \tag{4.40a}$$

Considering Eqs. (4.32*a*) and (4.32*b*), with β replaced in the latter by $(dw/dx - \psi)$ according to Eq. (4.29) and using the factor k, we may give the above results as follows:

$$\begin{aligned} &1.\ \text{Either } \psi \text{ is specified or } M = 0 \text{ at } x = 0, L \\ &2.\ \text{Either } w \text{ is specified or } V = 0 \text{ at } x = 0, L \end{aligned} \tag{4.40b}$$

†As an exercise you will be asked to verify the following results by carrying out the extremization process.

Similarly, making the same substitutions, it is seen that the Euler-Lagrange equations [Eqs. (4.38*a*) and (4.39*a*)] reduce to relations that are no doubt familiar to you from earlier course work in strength of materials. Thus:

$$\frac{dM}{dx} = V \tag{4.41a}$$

$$\frac{dV}{dx} = -q \tag{4.41b}$$

We may ***uncouple*** the Euler-Lagrange equations by differentiating Eq. (4.38*b*) with respect to x and then substituting from Eq. (4.39*b*). We get for this calculation

$$\boxed{EI\frac{d^3\psi}{dx^3} = q} \tag{4.42}$$

To get the uncoupled differential equation for w first differentiate Eq. (4.39*b*) twice with respect to x and solve for ψ''' to get

$$\psi''' = \frac{q''}{kGA} + w^{\text{IV}} \tag{4.43}$$

Now differentiate Eq. (4.38*b*) with respect to x and replace ψ''' by using the above equation. Also replace the expression $kGA(\psi' - w'')$ by using Eq. (4.39*b*). We then get, on rearranging terms, the desired equation for w:

$$\boxed{EIw^{\text{IV}} = q - \frac{EI}{kGA}q''} \tag{4.44}$$

Notice the resemblance of this equation to the corresponding deflection equation from the technical theory of beams.

To compare the Timoshenko beam theory with the technical theory and the plane stress theory, we examine again the simply supported beam carrying a uniform load q_0. Equations (4.42) and (4.44) become for this case:†

$$EI\frac{d^3\psi}{dx^3} = q_0 \tag{4.45a}$$

$$EI\frac{d^4w}{dx^4} = q_0 \tag{4.45b}$$

†Note before proceeding further that solutions of the uncoupled equations must be checked in the coupled form since the uncoupled equations form a higher-order set of equations, thus having as an outcome extraneous solutions.

We require that $w = 0$ and that $M = 0$ at the ends. The latter condition in turn means [see Eq. (4.32*a*)] that $d\psi/dx = 0$ at the ends. We shall proceed by carrying out the quadratures for ψ as follows:

$$\psi = \frac{1}{EI}\left(\frac{q_0 x^3}{6} + \frac{C_1 x^2}{2} + C_2 x + C_3\right) \tag{4.46}$$

If we place the origin of the reference at the center of the beam, it should be clear from the symmetry of the problem that ψ is an *odd* function and so we set $C_1 = C_3 = 0$. Furthermore, the end condition $d\psi/dx = 0$ at $x = \pm L/2$ gives us C_2 as

$$0 = \frac{1}{EI}\left(\frac{q_0 x^2}{2} + C_2\right)_{x=\pm L/2}$$

therefore

$$C_2 = -\frac{q_0 L^2}{8}$$

We then have for ψ:

$$\psi = \frac{1}{EI}\left(\frac{q_0 x^3}{6} - \frac{q_0 L^2}{8} x\right)$$

To get w most simply we will substitute the above result in Eq. (4.38*b*) as follows:

$$-q_0 x + kGA\left[\frac{1}{EI}\left(\frac{q_0 x^3}{6} - \frac{q_0 L^2}{8} x\right) - \frac{dw}{dx}\right] = 0$$

Rearranging, we get

$$\frac{dw}{dx} = -\frac{q_0 x}{kGA} + \frac{1}{EI}\left(\frac{q_0 x^3}{6} - \frac{q_0 L^2}{8} x\right)$$

Integrating, we have

$$w = -\frac{q_0 x^2}{2kGA} + \frac{1}{EI}\left(\frac{q_0 x^4}{24} - \frac{q_0 L^2 x^2}{16}\right) + C$$

Setting $w = 0$ at $x = \pm L/2$ we get

$$-\frac{q_0 L^2}{8kGA} + \frac{1}{EI}\left(\frac{q_0 L^4}{384} - \frac{q_0 L^4}{64}\right) + C = 0$$

therefore

$$C = \frac{q_0 L^2}{8kGA} + \frac{5q_0 L^4}{384EI}$$

We thus have for w:

$$w = \frac{q_0}{2kGA}\left(\frac{L^2}{4} - x^2\right) + \frac{q_0}{EI}\left(\frac{x^4}{24} - \frac{L^2x^2}{16} + \frac{5L^4}{384}\right)$$

Rearranging, we have

$$w = \frac{q_0L^4}{24EI}\left[\left(\frac{x}{L}\right)^4 - \frac{3}{2}\left(\frac{x}{L}\right)^2 + \frac{5}{16}\right] - \frac{q_0L^2}{2k[E/2(1+\nu)]bh}\left[\left(\frac{x}{L}\right)^2 - \frac{1}{4}\right]$$

where we have replaced G by $E/(2(1+\nu))$ and A by bh in the last expression. Finally, we give w as follows, on introducing $I = \frac{1}{12}bh^3$ into the last expression:

$$w = \frac{q_0L^4}{24EI}\left\{\left[\left(\frac{x}{L}\right)^4 - \frac{3}{2}\left(\frac{x}{L}\right)^2 + \frac{5}{16}\right] - \left(\frac{h}{L}\right)^2 \frac{2(1+\nu)}{k}\left[\left(\frac{x}{L}\right)^2 - \frac{1}{4}\right]\right\} \tag{4.47}$$

Comparing this result with that of the technical theory of beams [Eq. (4.22)], we see that the first part above is identical to the latter. Clearly the second part of the above result corresponds to the contribution of shear deformation. Now compare the above result with that of plane stress [Eq. (4.28)]. We see that the first expression in the bracket is identical to the dominant expression in the plane stress solution, while the second expression above (shear deformation) is of the same order of magnitude as the remaining part of the plane stress solution. In fact, if we choose Cowper's value of k for the rectangular cross section from Table 4.1, namely

$$k = \frac{10(1+\nu)}{12+11\nu}$$

we find the ratio of these second-order contributions to be

$$\frac{w_{\text{shear-Timo}}}{w_{\text{shear-plane stress}}} = \frac{24+22\nu}{24+15\nu}$$

This number ranges from 1.00 to 1.11 as the Poisson ratio ranges from 0 to $\frac{1}{2}$.

Thus we have a rather simple theory that includes shear deformation contribution with what would appear to be reasonably good accuracy. And so we may now be able to ascertain the stress τ_{xx} and the deflection curve $w(x)$ for short as well as long beams.

You will note that we have used the variational approach up to this time to formulate from the total potential energy functional the differential equations of interest and the natural boundary conditions. Now we will work directly with the functional in order to present approximate solutions to the technical beam equations.

4.6 COMMENTS ON THE RITZ METHOD

In Chapter 3 we pointed out that a limiting sequence of linearly independent functions u_n that is complete in energy and that minimizes the quadratic functional I

converges in energy to a function u_0, the solution to the differential equation associated with I. However, we indicated that by a judicious choice of functions, only a very few of them could yield, via the Ritz method, a very good approximation to the solution to the differential equation. In this section we shall demonstrate by example how this may be accomplished.

Accordingly, we consider a simply supported beam with a concentrated load P, as shown in Fig. 4.8. The solution of the differential equation derived from the technical theory of beams [Eq. (4.10)] is easily found in this to be, for $0 \leqslant x \leqslant L/2$,

$$w_{ex} = \frac{PL^3}{48EI}\left[3\left(\frac{x}{L}\right) - 4\left(\frac{x}{L}\right)^3\right] \tag{4.48}$$

where the subscript "ex" indicates that the solution is exact for the given differential equation. And the total potential energy for the problem is

$$\pi = \frac{EI}{2}\int_0^L \left(\frac{d^2w}{dx^2}\right)^2 dx - Pw(x)|_{x=L/2} \tag{4.49}$$

Now we will employ one function for the Ritz approach. Consider the following:

$$w_1 = A \sin\frac{\pi x}{L} \tag{4.50}$$

where A is an undetermined constant. Notice from Eq. (4.14) that the function $\sin \pi x/L$ satisfies a kinematic boundary condition ($w = 0$ at the ends) and a natural boundary condition ($w'' = 0$ at the ends) and thus fully satisfies the requirements imposed on the extremizing functions by the variational process. Furthermore, the assumed deflection curve has a shape that is "reasonable" in that it is symmetric about the center of the beam. Now substitute w_1 into Eq. (4.49). We get

$$\pi = \frac{EI}{2}A^2\left(\frac{\pi}{L}\right)^4\frac{L}{2} - PA \tag{4.51}$$

To establish the proper value of A for the Ritz method we differentiate with respect to A and set the result equal to zero. Thus:

$$\frac{EIL}{2}A\left(\frac{\pi}{L}\right)^4 - P = 0$$

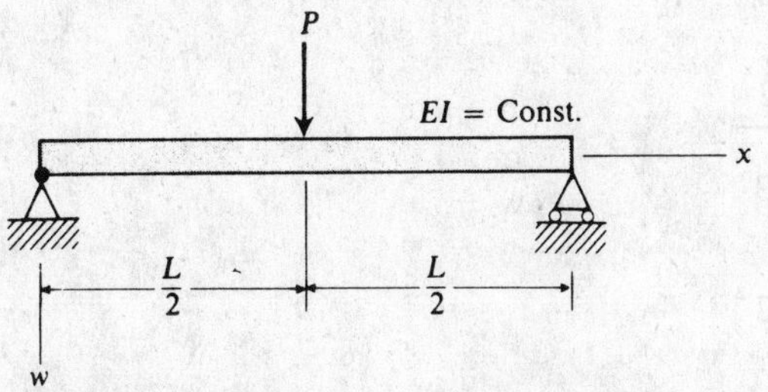

FIGURE 4.8
Simply supported beam with concentrated load.

therefore

$$A = \frac{2PL^3}{\pi^4 EI}$$

We then have for the corresponding approximate function w_{app}:

$$w_{app} = \frac{2PL^3}{\pi^4 EI} \sin \frac{\pi x}{L} \tag{4.52}$$

To compare the results from Eqs. (4.52) and (4.48) we examine the deflection at points $x = L/4$ and $x = L/2$, as given below:

	$x = L/4$	$x = L/2$
$\frac{EIw_{ex}}{PL^3}$	0.01432	0.02083
$\frac{EIw_{app}}{PL^3}$	0.01452	0.02053

We see that the results are very close indeed.

Now consider as the function w_1 the following form:

$$w_1 = B(x^3 - L^2 x) \tag{4.53}$$

Note that the kinematic boundary condition $w = 0$ is satisfied at both ends, but only at the left end is the natural boundary condition $w'' = 0$ satisfied. A full set of boundary conditions as required by the variational process has not accordingly been satisfied [see Eq. (4.14)]. Then by substituting the above result into the total potential energy and extremizing the result with respect to B we get

$$w_{app} = \frac{PL^3}{32EI}\left[\frac{x}{L} - \left(\frac{x}{L}\right)^3\right] \tag{4.54}$$

It is easy to see from the data below that the above result does not compare well with the exact solution.

	$x = L/4$	$x = L/2$
$\frac{EIw_{ex}}{PL^3}$	0.01432	0.02083
$\frac{EIw_{app}}{PL^3}$	0.00732	0.01172

We shall ascribe the reason for this poor result to the fact that one of the needed boundary conditions is not satisfied and the fact that the function $x^3 - L^2 x$ is not symmetric about $x = L/2$, a condition one would expect as a result of the loading and geometry of the problem. We will remedy these two conditions now and we will see a remarkable improvement in the accuracy. To do this, we express w_1 as follows:†

$$w_1 = B(3L^2x - 4x^3) \qquad 0 \leqslant x \leqslant \frac{L}{2} \tag{4.55}$$

with the understanding that the function for $L/2 \leqslant x \leqslant L$ is given so as to form an even function w_1 over the interval. Now, substituting into the expression for the total potential energy and extremizing with respect to B, we get the following result:

$$w_1 = \frac{PL^3}{48EI}\left[3\left(\frac{x}{L}\right) - 4\left(\frac{x}{L}\right)^3\right] \tag{4.56}$$

Indeed, by comparing the above result with Eq. (4.48) we see that we have arrived at the exact solution as a result of the changes made above.

We next conclude from these calculations that for good results with few functions, it is important to satisfy a *full set* of boundary conditions set forth by the extremization process, and in addition take into consideration symmetry in choosing the functions.

To further bring out the importance of satisfying a full set of boundary conditions, we next use a third function $[1 - \cos(2\pi x/L)]$ so that:

$$w_1 = A\left(1 - \cos\frac{2\pi x}{L}\right) \tag{4.57}$$

Inspection of this function shows clearly that at the ends the kinematic boundary condition $w = 0$ is the only one satisfied. We have accordingly not satisfied a complete set of conditions. Now substituting into the total potential energy and extremizing with respect to A we get

$$w_{\text{app}} = \frac{PL^3}{4\pi^4 EI}\left(1 - \cos\frac{2\pi x}{L}\right) \tag{4.58}$$

The deflection at $x = L/4$ and $x = L/2$ for the above approximation and for the exact solution of the differential equation are given below:

	$x = L/4$	$x = L/2$	
$EI\dfrac{w_{\text{ex}}}{PL^3}$	0.01432	0.02083	(4.59*a*)
$EI\dfrac{w_{\text{app}}}{PL^3}$	0.00257	0.00513	(4.59*b*)

†The coefficients 3 and 4 have been chosen to give w_1 a zero slope at $x = L/2$, as required of a beam with a continuous slope and symmetry about $x = L/2$.

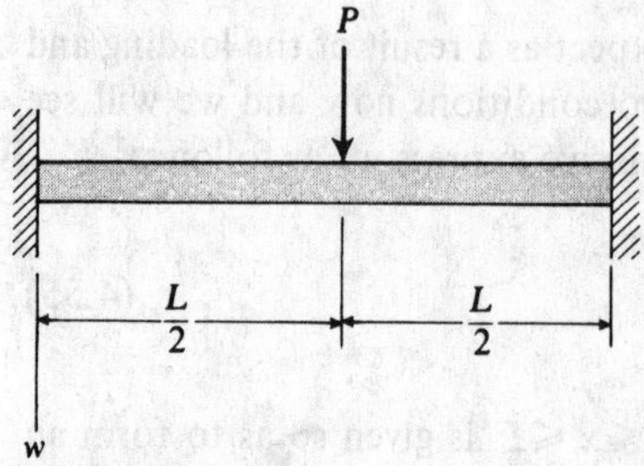

FIGURE 4.9
Fixed-end beam.

There is obviously very poor correlation here. If we had used Eq. (4.57) for the *fixed-end beam* in Fig. 4.9, we see that the function would have satisfied a ***complete set*** of boundary conditions for this problem (two kinematic boundary conditions in this case) since $w_1 = w_1' = 0$ at the ends. The results for an exact solution of the fixed-end beam problem are given below for comparison:

	$x = L/4$	$x = L/2$
$EI\dfrac{w_{ex}}{PL^3}$	0.00261	0.00522

(4.60)

We see by comparing Eq. (4.59*b*) and the above result that, with a complete set of boundary conditions satisfied, we again have excellent correlation between the results from the exact solution of the differential equation and the approximate solution.

4.7 THE RITZ METHOD FOR A SERIES SOLUTION

In previous sections we investigated how one might best proceed via the Ritz method to produce good approximate solutions by employing one or several coordinate functions. Now we consider the Ritz method once again for the purpose of generating infinite series solutions to certain beam problems. We shall accordingly now consider beams supported by a continuous elastic foundation.

We examine the case of a beam simply supported at the ends and having a uniform elastic foundation between the ends. This has been shown in Fig. 4.10, where you will note a uniform load q_0 has been indicated as acting on the beam. The elastic

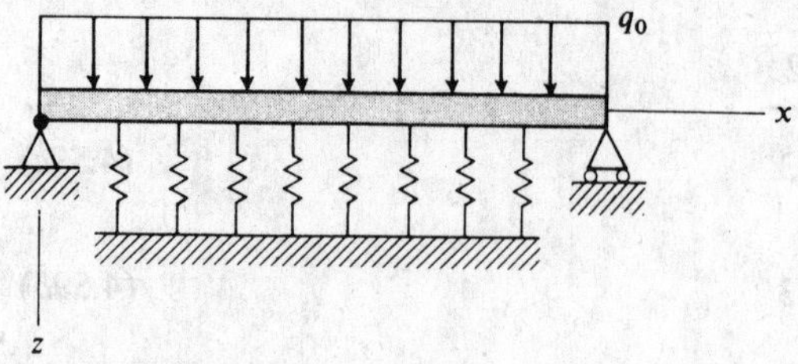

FIGURE 4.10
Beam with uniform elastic foundation.

foundation is indicated graphically as a series of springs. The force developed by the elastic foundation per unit length of the foundation and per unit deflection at a point is called the *foundation modulus* and is denoted as k. The force distribution on the beam from the elastic foundation is then given as kw. The potential energy of the foundation for a given deflection w is computed as follows:

$$V_{\text{found}} = \int_0^L \left(\int_0^w kw\, dw \right) dx = \int_0^L \frac{1}{2} kw^2\, dx \tag{4.61}$$

We can then give the total potential energy for bending of the beam, using results from previous sections:

$$\pi = \frac{EI}{2} \int_0^L [w''(x)]^2\, dx + \frac{1}{2} k \int_0^L w^2\, dx - q_0 \int_0^L w\, dx$$

$$= \int_0^L \left[\frac{EI}{2} (w'')^2 + \frac{1}{2} kw^2 - q_0 w \right] dx \tag{4.62}$$

It will be left to the reader to show [see Eq. (2.62)] that the following equation is the proper Euler-Lagrange equation for this functional:

$$EI \frac{d^4 w}{dx^4} + kw = q_0 \tag{4.63}$$

We can get an exact solution to this differential equation satisfying a kinematic boundary condition at each end ($w = 0$) and a natural boundary condition at each end ($w'' = 0$). Using as trial solutions e^{px}, you may readily deduce that the complementary solution is

$$w_c = A \sinh \lambda x \sin \lambda x + B \cosh \lambda x \sin \lambda x + C \sinh \lambda x \cos \lambda x + D \cosh \lambda x \cos \lambda x$$

where

$$\lambda = \sqrt[4]{\frac{k}{4EI}}$$

and is the reciprocal of the so-called characteristic length. The particular solution is, by inspection,

$$w_p = \frac{q_0}{k}$$

Submitting the combined solution to the aforementioned boundary conditions leads us to the following result:

$$w = \frac{q_0}{k}\left[1 - \frac{\cosh \lambda x \cos \lambda(x-L) + \cosh \lambda(x-L) \cos \lambda x}{\cosh \lambda L + \cos \lambda L}\right] \tag{4.64}$$

The exact bending moment is then easily computed as [see Eq. (4.16)]:

$$M = -EIw'' = -\frac{q_0}{2\lambda^2} \frac{\sinh \lambda x \sin \lambda(x-L) + \sinh \lambda(x-L) \sin \lambda x}{\cosh \lambda L + \cos \lambda L} \tag{4.65}$$

By way of comparison we now use the Ritz method for the functional given by Eq. (4.62). We employ a single function which satisfies a complete set of boundary conditions:

$$w_1 = a_1 \sin \frac{\pi x}{L}$$

A straightforward extremization of the functional with respect to a_1 then leads to the following approximation:

$$\begin{aligned} w_{\text{app}} &= \frac{4q_0L^4/(\pi^5 EI)}{1 + kL^4/(\pi^4 EI)} \sin \frac{\pi x}{L} \\ &= \frac{(16/\pi^5)(q_0/k)(L\lambda)^4}{1 + (4/\pi^4)(\lambda L)^4} \sin \frac{\pi x}{L} \end{aligned} \tag{4.66}$$

with the corresponding moment distribution:

$$\begin{aligned} M_{\text{app}} &= \frac{4q_0L^2/\pi^3}{1 + kL^4/(\pi^4 EI)} \sin \frac{\pi x}{L} \\ &= \frac{4q_0L^2/\pi^3}{1 + (4/\pi^4)(\lambda L)^4} \sin \frac{\pi x}{L} \end{aligned} \tag{4.67}$$

Now let us increase the number of terms in the series as follows:

$$w_p = \sum_{n=1}^{p} a_n \sin \frac{n\pi x}{L} \tag{4.68}$$

The functions $\phi_n = \sin(n\pi x/L)$ are eigenfunctions† of the operator $[d^4/dx^4 + k/EI]$ for the boundary conditions of this problem and are linearly independent. As such, one can show that they are complete in energy (see footnote on page 168). We know from our discussion in Chapter 3 that the partial sums w_p form a minimizing sequence for this functional when the coefficients are chosen to extremize the value of the functional with respect to the a's (that is, the a's are the Ritz coefficients). Hence, with the a's so chosen we can say that

†We shall consider eigenfunctions in Chapter 7.

$$\lim_{n\to\infty} w_n = \sum_{n=1}^{\infty} (a_n)_{\text{Ritz}} \sin\frac{n\pi x}{L} \equiv \text{solution to differential equation} \tag{4.69}$$

To get the Ritz coefficients, first substitute Eq. (4.69) into Eq. (4.62). We then get the following result:

$$\pi = \sum_{n=1,3,5,\ldots}^{\infty} \left\{ a_n^2 \left[\left(\frac{EI}{2}\right)\left(\frac{L}{2}\right)\left(\frac{n\pi}{L}\right)^4 + \frac{kL}{4} \right] - a_n \frac{2q_0L}{n\pi} \right\}$$

$$+ \sum_{n=2,4,\ldots}^{\infty} \left\{ a_n^2 \left[\left(\frac{EI}{2}\right)\left(\frac{L}{2}\right)\left(\frac{n\pi}{L}\right)^4 + \frac{kL}{4} \right] \right\}$$

wherein we have used the orthogonal properties of the eigenfunctions in carrying out the integrations. Noting that

$$\frac{\partial}{\partial a_n}(\pi) = 0$$

we get the Ritz coefficients to be, for *odd* values of n,

$$a_n = \frac{4q_0L^4/(n^5\pi^5 EI)}{1 + kL^4/(n^4\pi^4 EI)}$$

The even a_n's are zero. Accordingly, the correct solution to the differential equation is given as

$$w = \sum_{n=1,3,5,\ldots}^{\infty} \frac{4q_0L^4/(n^5\pi^5 EI)}{1 + kL^4/(n^4\pi^4 EI)} \sin\frac{n\pi x}{L} \tag{4.70}$$

Let us suppose that we can differentiate the above series term by term. We get

$$M = \sum_{n=1,3,5,\ldots}^{\infty} \frac{4q_0L^2/(n\pi)^3}{1 + kL^4/(n^4\pi^4 EI)} \sin\frac{n\pi x}{L} \tag{4.71}$$

It is immediately clear from inspection of the coefficients that both series converge very rapidly. We present a table showing the exact results for w and M at the center of the beam along with approximate results, for $n = 1$, $n = 1, 3$, etc. and for a given value of $kL^4/EI = 4$.

	Exact	$n = 1$	$n = 1, 3$	$n = 1, 3, 5$	$n = 1, 3, 5, 7$
$\frac{EIw(L/2)}{q_0L^4}$	0.012505	0.012556	0.012502	0.012506	0.012505
$\frac{M(L/2)}{q_0L^2}$	0.119914	0.123918	0.119142	0.120174	0.119798

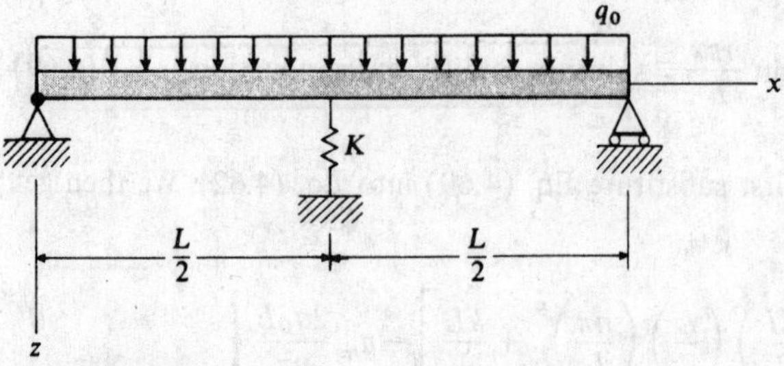

FIGURE 4.11
Beam with an elastic midsupport.

You will note in the previous calculations that we were able to compute the Ritz coefficients *individually* in a direct manner. That is, the algebraic equations for a_n were independent of each other. We next consider a problem where this is not the case. We have shown in Fig. 4.11 a simply supported beam having an elastic support at the center. This support is represented by a linear spring having a spring constant K. The total potential energy for this system, assuming that the spring support is undeformed when there is no load on the beam, is given as follows:

$$\pi = \frac{EI}{2}\int_0^L [w''(x)]^2\,dx + \frac{1}{2}K[(w)|_{x=L/2}]^2 - q_0\int_0^L w(x)\,dx$$

We will now demonstrate that the spring introduces into calculations for the Ritz method a *coupling* between the coefficients. Thus suppose we assume for w the following:

$$w = a_1 \sin\frac{\pi x}{L} + a_3 \sin\frac{3\pi x}{L}$$

Substitution of the above into the expression for the total potential energy then gives us:

$$\pi = \frac{EIL}{4}\left[a_1^2\left(\frac{\pi}{L}\right)^4 + a_3^2\left(\frac{3\pi}{L}\right)^4\right] + \frac{1}{2}K(a_1 - a_3)^2 + \frac{2q_0L}{\pi}\left(a_1 + \frac{a_3}{3}\right)$$

Extremizing the total potential energy with respect to a_1 and a_3 then gives us the following pair of equations

$$a_1\left(1 + \frac{2KL^3}{EI\pi^4}\right) - \frac{2KL^3}{EI\pi^4}a_3 = \frac{4q_0L^4}{EI\pi^5}$$

$$-\frac{2KL^3}{EI\pi^4}a_1 + \left(3^4 + \frac{2KL^3}{EI\pi^4}\right)a_3 = -\frac{4q_0L^4}{3EI\pi^5}$$

We see that for the determination of a_1 and a_3 the equations are now *simultaneous* equations. An interesting question that we shall now pursue centers around the calculation of the a's in the above problem when an infinite series is used to describe the deflection. That is, let

$$w = \sum_{n=1,3,5,\ldots}^{\infty} a_n \sin \frac{n\pi x}{L} \tag{4.72}$$

where we use odd values of "n" to take into account the symmetry of the problem about the center of the beam.† The total potential energy then becomes

$$\pi = \sum_{n=1,3,5,\ldots}^{\infty} \left[a_n^2 \frac{EIL}{4} \left(\frac{n\pi}{L} \right)^4 - a_n \frac{2q_0 L}{n\pi} \right] + \frac{K}{2} \left(\sum_{n=1,3,5,\ldots}^{\infty} a_n \sin \frac{n\pi}{2} \right)^2 \tag{4.73}$$

Now extremizing with respect to the coefficients we get for a_n the following result:

$$a_n = \frac{4q_0 L^4}{\pi^5 EI} \frac{1}{n^5} - \frac{2KL^3}{\pi^4 EI} \frac{1}{n^4} \sin \frac{n\pi}{2} \left(\sum_{s=1,3,5,\ldots}^{\infty} a_s \sin \frac{s\pi}{2} \right) \tag{4.74}$$

We see here that the value of a_n depends on the values of all the other a's. That is, the above represents one of an infinite set of linear simultaneous algebraic equations. To solve this set we introduce the following notation:

$$S = \sum_{s=1,3,5,\ldots}^{\infty} a_s \sin \frac{s\pi}{2} \tag{4.75}$$

Now multiply Eq. (4.74) by $\sin (n\pi/2)$ and sum to infinity over odd n. We then have, using the above notation,

$$S = \frac{4q_0 L^4}{\pi^5 EI} \sum_{n=1,3,5,\ldots}^{\infty} \frac{\sin (n\pi/2)}{n^5} - \frac{2KL^3}{\pi^4 EI} \left(\sum_{n=1,3,5,\ldots}^{\infty} \frac{1}{n^4} \right) S \tag{4.76}$$

where we have used the fact that, for odd "n," $\sin^2 (n\pi/2) = 1$. The infinite sums in the above equation are readily obtained from tables. Thus:

$$\sum_{n=1,3,5,\ldots}^{\infty} \frac{\sin (n\pi/2)}{n^5} = \frac{5\pi^5}{(16)(19)}$$

$$\sum_{n=1,3,5,\ldots}^{\infty} \frac{1}{n^4} = \frac{\pi^4}{96}$$

Solving for S in Eq. (4.76), using the preceding sums, we get

$$S = \frac{5q_0 L^4 /384EI}{1 + KL^3 /48EI}$$

†Note that the origin of coordinates is at the left end of the beam (see Fig. 4.11).

We can now go back to Eq. (4.74) and, using the above result to replace the infinite sum on the right side of the equation, we can solve for a_n. The solution for w can then be given as

$$w = \frac{4q_0L^4}{\pi^5 EI} \sum_{n=1,3,5,\ldots}^{\infty} \frac{\sin(n\pi x/L)}{n^5} - \frac{2KSL^3}{\pi^4 EI} \sum_{n=1,3,5,\ldots}^{\infty} \left(\frac{\sin(n\pi/2)\sin(n\pi x/L)}{n^4} \right)$$

4.8 USE OF THE REISSNER PRINCIPLE

We have been considering the Ritz method for estimating the dependent variable w of the differential equation. Let us go back to the simply supported beam with the single point load P with the use of very few functions. We shall this time be concerned with finding the bending moment M at any section. Since from the flexure formula the stress at any point is proportional to the bending moment, we will in effect also be making inquiries as to the bending stress. The bending moment M was shown [Eq. (4.16*b*)] to be equal to $-EIw''$ and so we have, using the approximate solution found in Section 4.6 for a single coordinate function, $\sin(\pi x/L)$ [see Eq. (4.52)]

$$M_{\text{app}} = \frac{2PL}{\pi^2} \sin\frac{\pi x}{L} \tag{4.77}$$

The exact result from rigid body considerations is:

$$M_{\text{ex}} = \frac{Px}{2} \tag{4.78}$$

In the following table we compare these results at positions $x = L/4$ and $x = L/2$:

	$x = L/4$	$x = L/2$
$\frac{M_{\text{ex}}}{PL}$	0.125	0.250
$\frac{M_{\text{app}}}{PL}$	0.143	0.203

Note that, unlike the earlier comparison of the deflection that showed excellent agreement, we now get poor agreement for moments, and hence stresses. The reason for this poor correlation is that in order to reach the approximate stresses we carry out a twofold differentiation on an approximate deflection function. The differentiation process is not a smoothing operation as is integration, and small differences in the shape of the curve w_{app} as compared to w_{ex} result in large differences in the values of w''_{app} as compared to w''_{ex}. It is clear that the Ritz method with few functions generally cannot be expected to give good results for stress.

We may ameliorate this difficulty by employing the Reissner functional, wherein we have several dependent variables to adjust independently of each other. We shall show for this problem that we can then pick moments and displacements independently. Accordingly, we now restate the Reissner functional as given earlier [Eq. (3.30)]

$$I_R = \iiint_V [\tau_{ij}\epsilon_{ij} - \mathfrak{U}^*(\tau_{ij})]\, dv - \iiint_V \bar{B}_i u_i\, dv - \iint_{S_1} \bar{T}_i^{(\nu)} u_i\, ds \tag{4.79}$$

You will recall that the barred quantities are prescribed. For the one-dimensional technical beam theory that we have presented, with a zero body force distribution, the first integral may be reduced to

$$\iiint_V [\tau_{ij}\,\epsilon_{ij} - \mathcal{U}^*(\tau_{ij})]\, dv = \iiint_V \left(\tau_{xx}\,\epsilon_{xx} - \frac{1}{2E}\,\tau_{xx}^2\right) dv \tag{4.80}$$

where we take all other $\tau_{ij} = 0$ in the calculation, as explained earlier. The only traction forces are P at the center of the beam and the supporting forces at the ends of the beam. The traction everywhere else is zero. Since u_i in the direction of the supporting forces is zero at the ends, we accordingly get for the third integral in Eq. (4.79) the expression $Pw(x)|_{x=L/2}$.

Consequently, the Reissner functional becomes

$$I_R = \iiint_V \left(\tau_{xx}\,\epsilon_{xx} - \frac{1}{2}\,\frac{\tau_{xx}^2}{E}\right) dv - Pw(x)|_{x=L/2} \tag{4.81}$$

Now we employ for ϵ_{xx} the following relation [see Eq. (4.1) applied to bending]:

$$\epsilon_{xx} = -z\,\frac{d^2 w}{dx^2} \tag{4.82}$$

and for τ_{xx} the well-known flexure formula (as developed in Section 4.3):

$$\tau_{xx} = \frac{Mz}{I} \tag{4.83}$$

We then get

$$\begin{aligned} I_R &= -\iiint_V \left[\frac{1}{2E}\left(\frac{Mz}{I}\right)^2 + \frac{Mz}{I}\left(z\,\frac{d^2 w}{dx^2}\right)\right] dv - Pw(x)|_{x=L/2} \\ &= -\iiint_V \left(\frac{M^2}{2EI^2} + \frac{M}{I}\,\frac{d^2 w}{dx^2}\right) z^2\, dv - Pw(x)|_{x=L/2} \\ &= -2\int_0^{L/2}\int_{-h/2}^{h/2} \left(\frac{M^2}{2EI^2} + \frac{M}{I}\,w''\right) z^2 b\, dz\, dx - Pw(x)|_{x=L/2} \end{aligned}$$

where b is the width of the beam. Note we have used $2\int_0^{L/2}$ to replace $\int_0^L$ in the last equation because of the symmetry of deformation about position $x = L/2$. Now integrating with respect to z we get

$$\begin{aligned} I_R &= -2\int_0^{L/2} \left(\frac{M^2}{2EI^2} + \frac{M}{I}\,w''\right) \frac{bh^3}{12}\, dx - Pw(x)|_{x=L/2} \\ &= -2\int_0^{L/2} \left(\frac{M^2}{2EI} + Mw''\right) dx - Pw(x)|_{x=L/2} \end{aligned} \tag{4.84}$$

The functional has M and w as the variables. We shall use the Ritz method for extremizing this functional by employing the following functions:

$$w = A \sin \frac{\pi x}{L} \tag{4.85a}$$

$$M = Bx, \quad 0 \leqslant x \leqslant \frac{L}{2} \tag{4.85b}$$

Note the proposed equation for M satisfies the end conditions for M [see Eq. (4.21)] and is to be used as a symmetric function. Substituting these functions into the functional then gives us

$$I_R = 2AB - \frac{B^2 L^3}{24EI} - PA$$

Extremizing I_R with respect to both A and B, we get immediately the results:

$$A = \frac{PL^3}{48EI}$$

$$B = \frac{P}{2}$$

The equations for moment and deflection become, for $0 \leqslant x \leqslant L/2$,

$$M = \frac{Px}{2}$$

$$w = \frac{PL^3}{48EI} \sin \frac{\pi x}{L} \tag{4.86}$$

The moment distribution is now exact, while the deflection approximation is now even better than the calculation made at the outset of the previous section, where the total potential energy was used for the same deflection function.

4.9 ADDITIONAL PROBLEMS IN BENDING OF BEAMS

We have used the total potential energy principle in conjunction with certain assumptions as to the displacement field to formulate the deflection equations of the technical theory of beams. After some justifications for the theory, we used the total potential energy in conjunction with the Ritz method to generate approximate solutions to beam problems. In this regard the use of the Reissner functional was also illustrated. We shall now utilize certain of the results from the technical theory of beams to illustrate the determination of deflections of loading points on beams via the second Castigliano theorem. Also, we shall determine shear force and bending moment distributions in beams supported in a statically indeterminate manner by using the total complementary energy principle.

Consider first beams supported in a statically determinate manner with zero axial force and with discrete vertical loads P_i. The strain energy of the beam in the absence of axial forces is easily found from Eq. (4.8) by setting $[(u)_1]s = 0$ and dropping the loading term q. We then get

$$U = \int_0^L \frac{EI}{2}\left(\frac{d^2w}{dx^2}\right)^2 dx = \frac{1}{2}\int_0^L \frac{M^2}{EI}\,dx \tag{4.87}$$

where we have again used Eq. (4.16*b*) to reach the last expression. The complementary strain energy for the beam is also given as above because we have assumed linear elastic behavior. The deflection of the loading point of force P_i in the direction of force P_i is thus given as follows, according to the second Castigliano theorem:

$$\Delta_i = \frac{\partial U}{\partial P_i} = \int_0^L \frac{M(\partial M/\partial P_i)}{EI}\,dx \tag{4.88}$$

We illustrate the use of this formulation in the following example.

EXAMPLE 4.1 A cantilever beam is shown in Fig. 4.12 loaded with a point couple M_0 at the end. What are the deflection at position $x = L/2$ and the slope of the deflection curve at this position?

To get the deflection, we first place a force P at $x = L/2$, later to be set equal to zero. This has been shown in Fig. 4.13. The bending moment along the beam is, from simple equilibrium considerations,

$$M = M_0 - P\left(x - \frac{L}{2}\right)\left[u\left(x - \frac{L}{2}\right)\right] \tag{a}$$

where $[u(x - L/2)]$ is a unit step function starting at $x = L/2$. We then get for $\partial M/\partial P$ from above:

$$\frac{\partial M}{\partial P} = -\left(x - \frac{L}{2}\right)\left[u\left(x - \frac{L}{2}\right)\right] \tag{b}$$

Substituting into Eq. (4.88) we get for the deflection Δ_P in the z direction of the point $x = L/2$:

$$\Delta_P = \frac{1}{EI}\int_0^L \left\{-M_0 + P\left(x - \frac{L}{2}\right)\left[u\left(x - \frac{L}{2}\right)\right]\right\}$$
$$\times\left\{\left(x - \frac{L}{2}\right)\left[u\left(x - \frac{L}{2}\right)\right]\right\} dx$$

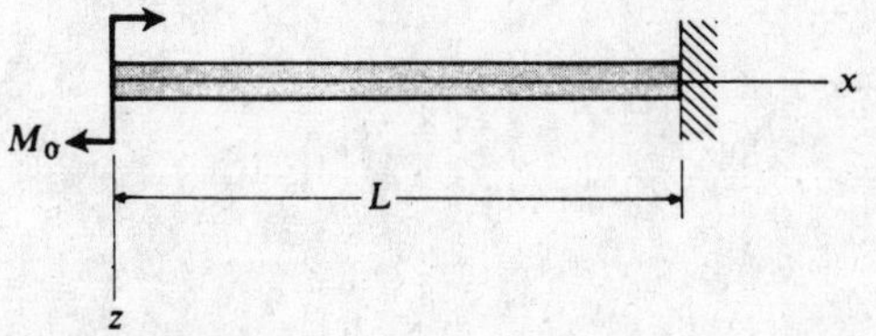

FIGURE 4.12
Cantilever beam with point couple load.

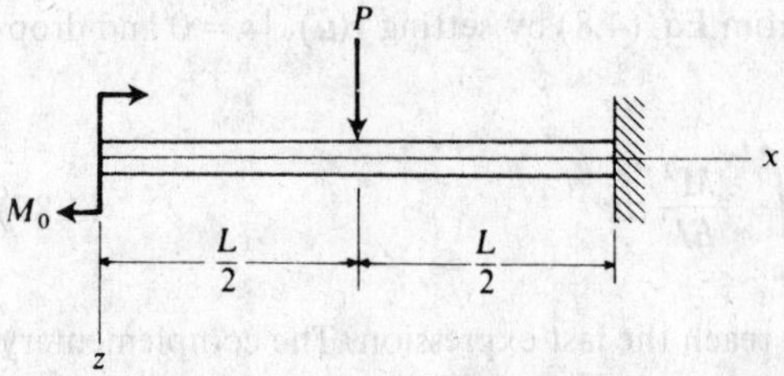

FIGURE 4.13
Cantilever beam with point couple and point force.

Setting $P = 0$ now we get for the above

$$\Delta_P = \frac{1}{EI}\int_0^L (-M_0)\left(x - \frac{L}{2}\right)\left[u\left(x - \frac{L}{2}\right)\right] dx$$

$$= \frac{1}{EI}\int_{L/2}^L (-M_0)\left(x - \frac{L}{2}\right) dx$$

Integrating and inserting limits

$$\Delta_P = -\frac{M_0 L^2}{8EI}$$

The minus sign indicates, for the z axis pointing downward, that the point moves upward.

To get the slope at $x = L/2$ we next add to the loading of Fig. 4.12 a point couple C, as shown in Fig. 4.14. We then have for M

$$M = M_0 + C\left[u\left(x - \frac{L}{2}\right)\right]$$

therefore

$$\frac{\partial M}{\partial C} = \left[u\left(x - \frac{L}{2}\right)\right]$$

Going to Eq. (4.88) we get the result

$$\Delta_C = \frac{1}{EI}\int_0^L \left\{M_0 + C\left[u\left(x - \frac{L}{2}\right)\right]\right\}\left[u\left(x - \frac{L}{2}\right)\right] dx$$

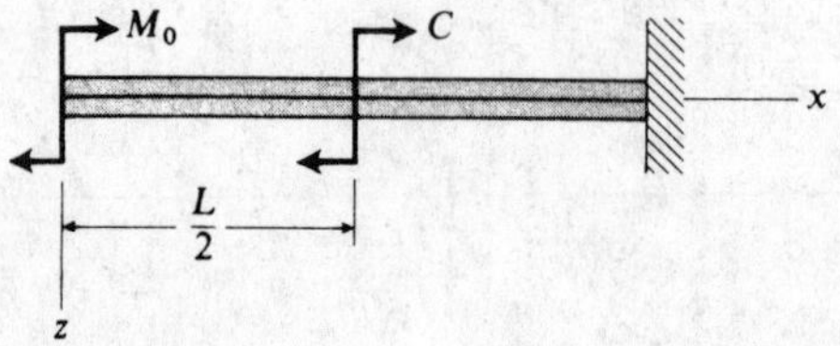

FIGURE 4.14
Cantilever beam with two point couple loads.

Setting $C = 0$ now we get

$$\Delta_C = \frac{1}{EI}\int_{L/2}^{L} M_0\, dx = \frac{M_0 L}{2EI}$$

The above result must, of course, be in radians.

As the next step we consider a beam that is supported in a statically indeterminate manner, such as the beam shown in Fig. 4.15. To determine the supporting force system for such a beam and thereby to pave the way for determination of deflections of the beam, we may proceed as follows. Consider as unknown external loads as many of the supporting forces and torques as are necessary to render the beam statically determinate in the remaining supporting forces and torques. (That is, we choose as unknown forces and torques a set of *redundant constraints.*†) Determine the remaining constraining forces in terms of the redundant constraints via rigid body equilibrium equations. We next compute the total complementary energy in terms of these redundant constraints. Since the redundant constraints have been chosen so as to result in a statically determinate system, it is clear that for any set of values we wish to assign to these constraints, there exists for the given external loads a stress field in the body that is statically compatible. Hence by varying the redundant constraints we may generate an infinity of statically compatible stress fields in the body. Thus we may appropriately vary the total complementary energy by varying the redundant constraints. By then extremizing the total complementary energy with respect to the redundant constraints we single out the *particular* stress field that is linked through the constitutive law employed to a *kinematically compatible* strain field. We thus reach the appropriate values of the redundant constraints.

Note that the process is valid for nonlinear elastic behavior and, unlike computations employing the Castigliano theorems, is not restricted to point loads. We next illustrate the procedure in the following example.

EXAMPLE 4.2 We have shown in Fig. 4.15 a cantilever beam uniformly loaded and having two additional supports beyond the base. This is clearly a statically indeterminate support system with two redundant constraints. Assume linear elastic behavior.

†Constraints not needed for equilibrium.

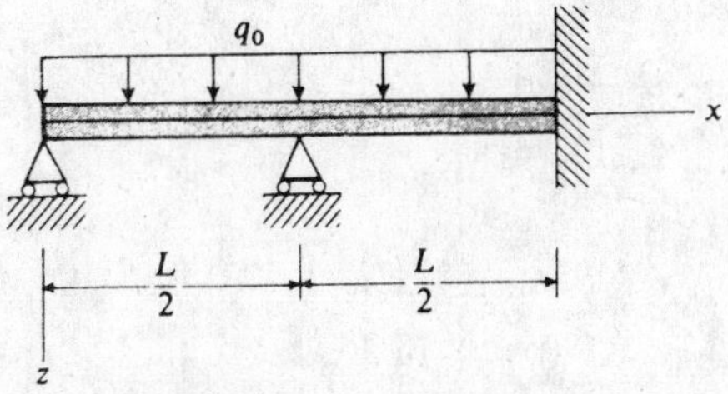

FIGURE 4.15
Statically indeterminate cantilever.

To solve the supporting force system we choose the two roller supports as the redundant constraints and show these as unknown external forces R_1 and R_2 in Fig. 4.16. We next compute the total complementary strain energy π^*. We note that

$$\pi^* = U^* + V^*$$

where

$$V^* = -\iiint_V u_i B_i \, dv - \iint_{S_2} u_i T_i^{(\nu)} \, dS$$

Note that for the problem at hand $B_i = 0$ and the surface integral above is zero since for the region S_2 (i.e., at the supports) $u_i = 0$. Accordingly, $V^* = 0$. And since we have linear elastic behavior, we can use U in place of U^* to compute π^*. Thus we have for this case, using Eq. (4.87),

$$\pi^* = \frac{1}{2}\int_0^L \frac{M^2}{EI}\, dx$$

The moment M at any section of the beam in terms of R_1 and R_2 is

$$M(x) = R_1 x - \frac{q_0 x^2}{2} + R_2 \left(x - \frac{L}{2}\right)\left[u\left(x - \frac{L}{2}\right)\right]$$

The total complementary energy then is

$$\pi^* = \frac{1}{2EI}\int_0^L \left\{R_1 x - \frac{q_0 x^2}{2} + R_2 \left(x - \frac{L}{2}\right)\left[u\left(x - \frac{L}{2}\right)\right]\right\}^2 dx \qquad (a)$$

To extremize π^* we require that

$$\frac{\partial \pi^*}{\partial R_1} = 0 \qquad (b)$$

$$\frac{\partial \pi^*}{\partial R_2} = 0 \qquad (c)$$

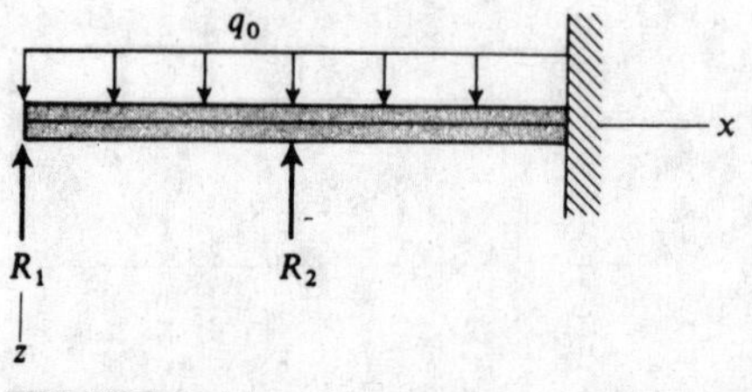

FIGURE 4.16
Cantilever beam showing redundant constraints.

Employing Eq. (*a*) for π^* in Eq. (*b*) we get

$$\frac{1}{2EI}\int_0^L 2\left\{R_1 x - \frac{q_0 x^2}{2} + R_2\left(x-\frac{L}{2}\right)\left[u\left(x-\frac{L}{2}\right)\right]\right\} x\,dx = 0$$

This becomes

$$\int_0^L \left(R_1 x^2 - \frac{q_0 x^3}{2}\right) dx + \int_{L/2}^L R_2\left(x-\frac{L}{2}\right) x\,dx = 0$$

Integrating, we get

$$\frac{R_1 L^3}{3} - \frac{q_0 L^4}{8} + \frac{5}{48}L^3 R_2 = 0$$

therefore

$$16R_1 + 5R_2 = 6q_0 L \tag{d}$$

Now doing the same for Eq. (*c*) we get

$$5R_1 + 2R_2 = \tfrac{17}{8} q_0 L \tag{e}$$

Solving Eqs. (*d*) and (*e*) simultaneously we have

$$R_1 = \tfrac{11}{56} q_0 L \qquad R_2 = \tfrac{4}{7} q_0 L \tag{f}$$

We thus have determined the supporting force system and can then proceed to evaluate moments, stresses, and deflections, using the results of the technical theory of beams.

The approach presented may be used for *nonlinear elastic behavior* and for *prescribed movement of the supports* corresponding to the redundant constraints. The total complementary energy would then be expressed as follows:

$$\pi^* = \iiint_V \int_{\tau_{ij}} \epsilon_{ij}\,d\tau_{ij}\,dv - \sum_i R^{(i)}\bar{u}^{(i)} \tag{4.89}$$

where $R^{(i)}$ is the ith redundant constraint and $\bar{u}^{(i)}$ is the prescribed displacement in the direction of the ith redundant constraint. In the case of a beam of rectangular cross section the above equation becomes

$$\pi^* = \int_{-h/2}^{h/2}\int_0^L \int_{\tau_{xx}} \epsilon_{xx}\,d\tau_{xx}\,b\,dz\,dx - \sum_i R^{(i)}\bar{u}^{(i)}$$

Now using the appropriate constitutive law for the problem we may get τ_{xx} as a function of ϵ_{xx} so that the above equation may be written as

$$\pi^* = \int_{-h/2}^{h/2} \int_0^L G(\epsilon_{xx}) b \, dz \, dx - \sum_i R^{(i)} \bar{u}^{(i)} \tag{4.90}$$

where $G(\epsilon_{xx})$ is the function resulting from integration with respect to strain. Assuming the deformation model set forth in Section 4.2 (plane sections remain plane) we may give ϵ_{xx} as

$$\epsilon_{xx} = -z \frac{d^2 w}{dx^2}$$

Next, integrating with respect to z, we express π^* as follows:

$$\pi^* = \int_0^L H\left(h, b, \frac{d^2 w}{dx^2}\right) dx - \sum_i R^{(i)} \bar{u}^{(i)} \tag{4.91}$$

where H is some function of h, b, and dw^2/dx^2.

We turn at this point to the computation of M at a section. Thus we have

$$M = \int_{-h/2}^{h/2} \tau_{xx} z b \, dz$$

Employing the constitutive law to replace τ_{xx} by a function of ϵ_{xx} and then using the plane sections assumption to replace ϵ_{xx} in terms of $-z \, dw^2/dx^2$, we then have on carrying out integration with respect to z:

$$M = K\left(h, b, \frac{d^2 w}{dx^2}\right) \tag{4.92}$$

The idea is first to solve for $d^2 w/dx^2$ in terms of M, h, and b in the above equation and to substitute this result into the function H in the complementary energy [Eq. (4.91)]. Next we express M in terms of the external loads and redundant constraints $R^{(i)}$ from simple equilibrium considerations. Finally, by replacing M in Eq. (4.91), using this result, we have π^* as a function of the redundant constraints and we are ready to carry out the extremization process.

We have presented several problems as assignments wherein you may work out the details of the procedure outlined above for specific situations.

4.10 CLOSURE

In this chapter we have dealt with only one independent variable. The problems were *one-dimensional* from this point of view. The results of Chapters 2 and 3 sufficed our needs in this chapter. We now turn our attention in Chapter 5 to the torsion of shafts. Here we must employ two independent variables. As a consequence we will introduce several new approximation procedures stemming from variational considerations. We thus once again combine developments of structural mechanics with the continued development of variational methods.

REFERENCE

Cowper, G. R., *J. Appl. Mech.*, June 1966, p. 335.

READING

Hoff, N. J.: "The Analysis of Structures," Wiley, New York, 1956.
Kinney, J. S.: "Indeterminate Structural Analysis," Addison-Wesley, Reading, Mass., 1957.
Langhaar, H. L.: "Energy Methods in Applied Mechanics," Wiley, New York, 1962.
Oden, J. T.: "Mechanics of Elastic Structures," McGraw-Hill, New York, 1967.
Rivello, R. M.: "Theory and Analysis of Flight Structures," McGraw-Hill, New York, 1969.
Vankatramen, B., and Patel, S. A.: "Structural Mechanics with Introductions to Elasticity and Plasticity," McGraw-Hill, New York, 1970.

PROBLEMS

4.1 Carry out the extremization of π for the beam [see Eq. (4.8)] by using the delta operator and verify the results [Eqs. (4.9), (4.10), and (4.14)] that were simply taken from the formulations of Chapter 2.

4.2 What is the deflection curve for a simply supported beam (Fig. 4.17) carrying a triangular loading distribution according to the Timoshenko theory of beams? The cross section of the beam is rectangular. What is the ratio of the deflection at the midpoint according to the Timoshenko theory of beams to that of the technical theory of beams? Take $\nu = 0.3$.

4.3 Using the flexure formula $\tau_{xx} = Mz/I$ and the simple result $\tau_{xz} = V/A$, show that the Reissner functional for the Timoshenko beam is given as

$$I_R = \int_0^L \left[-M(w'' - \beta') + \beta V - \frac{M^2}{2EI} - \frac{V^2}{2kAG} - qw \right] dx$$

where we use $\gamma_{xz} = V/kGA$ from Eqs. (4.36) and (4.31*b*) in the second complementary energy term. Extremizing this functional, show that we arrive at Eqs. (4.41*a*) and (4.41*b*), the differential equations of equilibrium for the Timoshenko beam, as well as Eqs. (4.32*a*) and (4.33), which are effectively stress-strain relations. Also show that the boundary conditions (4.40*b*) are satisfied. We have thus derived Timoshenko beam theory from the Reissner functional.

4.4 For the beam shown in Fig. 4.18, form one-term and two-term approximate solutions. Compute the deflection under the load for each solution and compare these results with those from the technical theory of beams.

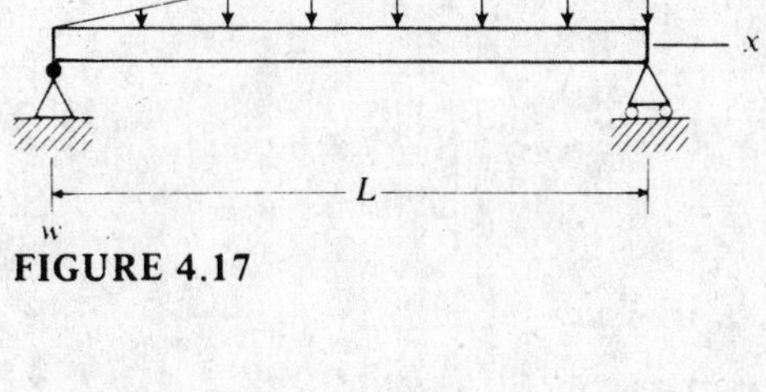

FIGURE 4.17

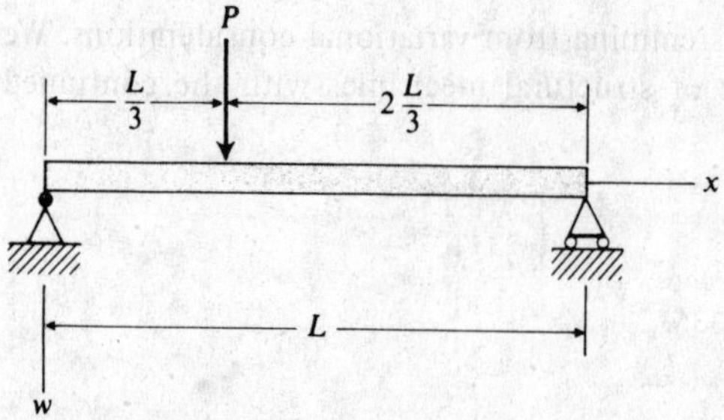

FIGURE 4.18

4.5 Devise a function that satisfies a complete set of boundary conditions for the cantilever beam shown in Fig. 4.19. Calculate the maximum deflection via the Ritz method and compare this with the result from the technical theory of beams. Take $P = 5000$ N, $q_0 = 150$ N/m and $L = 6$ m.

4.6 Obtain an approximate solution for a *clamped-clamped* beam, under a concentrated load at $x = \xi$. Show how this may be used to obtain the deflection of a uniformly loaded beam, and of a beam loaded by: $q(x) = 0,\ 0 \leqslant x \leqslant \epsilon$; $q(x) = q_0,\ \epsilon \leqslant x \leqslant (L - \epsilon)$; $q(x) = 0,\ (L - \epsilon) \leqslant x \leqslant L$. Show how the last solution can be made to yield the deflection of a centrally loaded beam.

*4.7 We wish to show that the flexure formula from the technical theory of beams is exact provided the bending moment M is a constant or a linear function of x along the beam.

Using equilibrium in the x direction, the flexure formula for τ_{xx}, and Hooke's law for the shear stresses, show that

$$\frac{y}{I}\frac{\partial M}{\partial x} + G\left[\left(\frac{\partial^2 u_x}{\partial y^2} + \frac{\partial^2 u_x}{\partial z^2} - \frac{\partial^2 u_x}{\partial x^2}\right) + \frac{\partial}{\partial x}(\text{div}\,\mathbf{u})\right] = 0 \qquad (a)$$

Show from Hooke's law, with the aid of the flexure formula, that for beams:

$$\text{div}\,\mathbf{u} = \frac{My}{I\lambda} - \frac{2G}{\lambda}\frac{\partial u_x}{\partial x} \qquad (b)$$

Accordingly, combine Eqs. (*a*) and (*b*) to reach:

$$\left(\frac{y}{GI} + \frac{y}{I\lambda}\right)\frac{\partial M}{\partial x} - \left(1 + \frac{2G}{\lambda}\right)\frac{\partial^2 u_x}{\partial x^2} = -\left(\frac{\partial^2 u_x}{\partial y^2} + \frac{\partial^2 u_x}{\partial z^2}\right) \qquad (c)$$

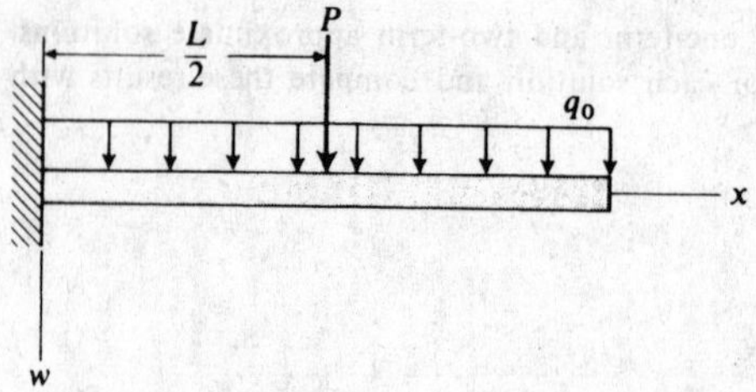

FIGURE 4.19

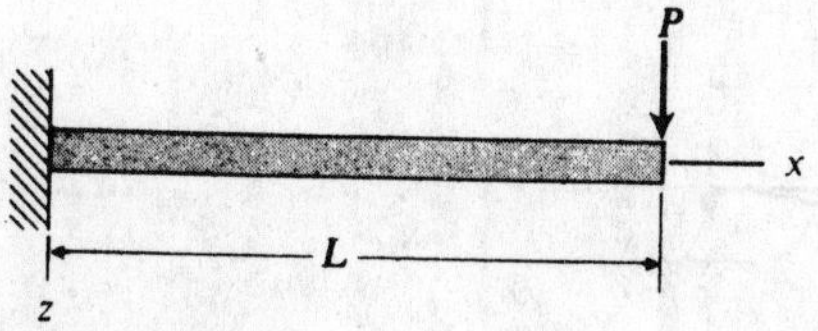

FIGURE 4.20

Using the flexure formula again show that

$$u_x = \frac{y}{EI}\int M\,dx + \chi(y, z) \tag{d}$$

where χ is an arbitrary function of y and z. Using Eq. (*d*) in Eq. (*c*) show that the following equation holds:

$$D\frac{\partial M}{\partial x} = -\frac{1}{y}\left(\frac{\partial^2 \chi}{\partial y^2} + \frac{\partial^2 \chi}{\partial z^2}\right)$$

where D is a constant. Why can you conclude from this equation that

$$M = C_1 x + C_2$$

where C_1 and C_2 are constants? We have thus shown that for the flexure formula to give exact results it is necessary that the bending moment be a linear function of x.

4.8 Give an approximation for the bending moment distribution and the deflection curve for the cantilever beam in Fig. 4.20 using the Reissner functional.

4.9 Express the Reissner functional for the cantilever beam shown in Fig. 4.21 having a triangular axial force distribution and a uniform normal loading distribution. Get an approximate solution to the bending moment, the vertical deflection curve w, and the horizontal stretching curve u_s using the Reissner functional. Compare with results from the technical theory of beams at $x = L$. Assume linear elastic behavior.

4.10 For the continuous beam shown in Fig. 4.22, find the supporting forces and the magnitude and direction of the centerline deflection Δ. Assume a uniform EI.

4.11 Determine the force of the spring for the beam shown in Fig. 4.23 using energy methods. First consider the spring force as the redundant quantity and then consider the support

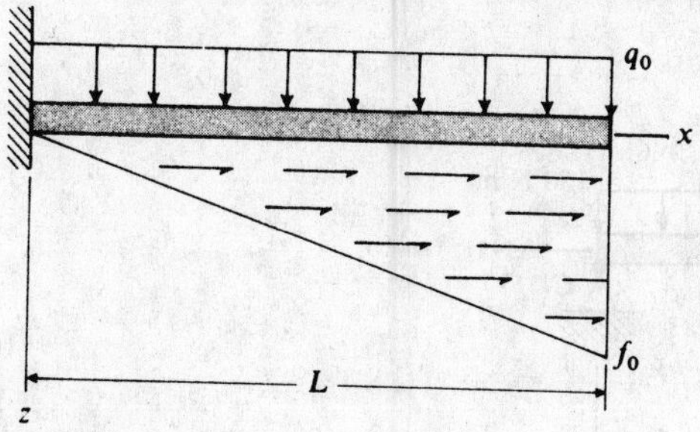

FIGURE 4.21

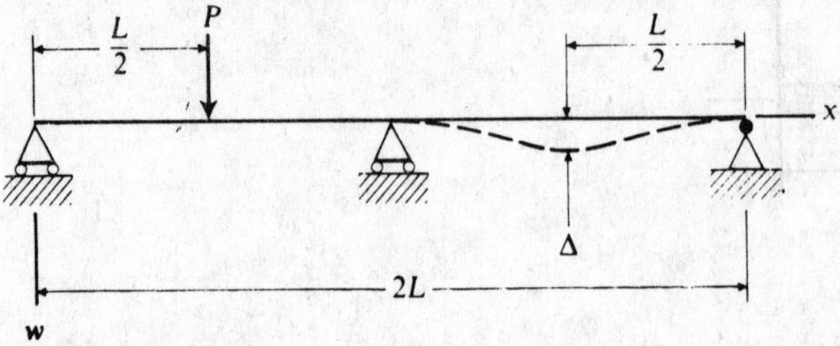

FIGURE 4.22

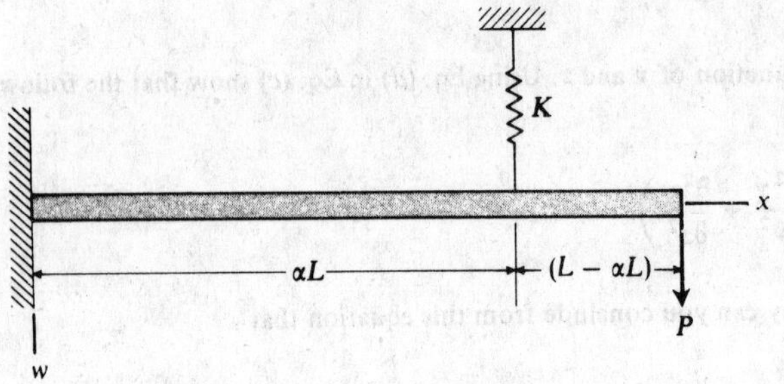

FIGURE 4.23

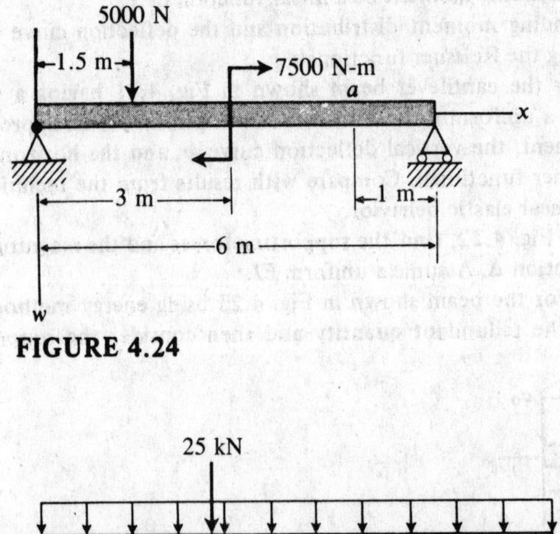

FIGURE 4.24

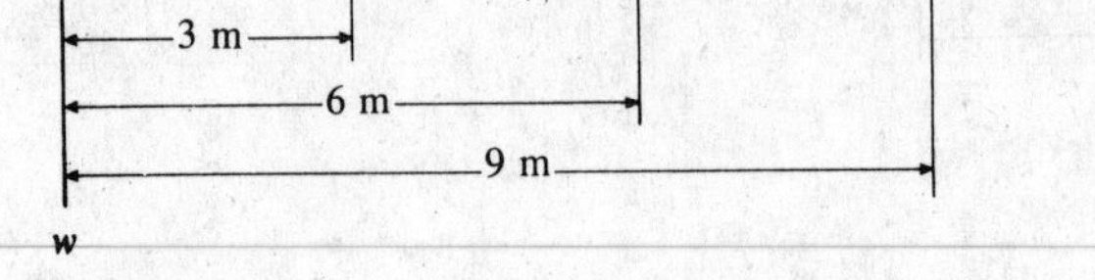

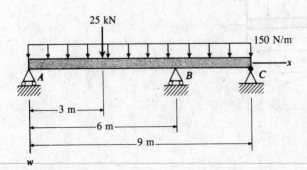

FIGURE 4.25

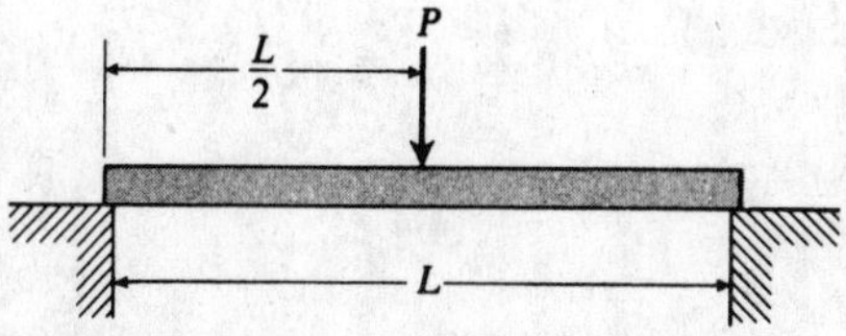

FIGURE 4.26

force as the redundant quantity. What are the limiting cases that occur when $K \to 0$ and $K \to \infty$? Also, what are the limiting cases for $K = \infty$ when $\alpha \to 0$ and when $\alpha \to 1$? Assume uniform EI.

4.12 Find the deflection and slope at point a of the simply supported beam shown in Fig. 4.24 using the second Castigliano theorem. Assume linear elastic behavior.

4.13 Find the supporting forces in the beam of Fig. 4.25 if support B sags 3 mm as a result of the loading. Assume linear elastic behavior.

4.14 It is desired to limit the centerline deflection of the spandrel beam (Fig. 4.26) to $(1/288)L$, where L is the length. Expressed in terms of the usual beam parameters (E, I, L, P), how small must the load P be so as not to violate the constraint?

4.15 How large a moment is required at the free end of a cantilever beam to provide a slope there of $\theta(x = L) = 0.036$ mm/mm? Let the beam have the properties $E = 2.1 \times 10^{11}$ Pa, $L = 2.5$ m, $I = 4 \times 10^{-7}$ m^4. How does the slope vary along the beam?

*4.16 Starting with the reciprocal theorem (Problem 1.22) and using the basic equations of beam theory (strain-displacement, equilibrium) develop the reciprocal theorem for beams in the form

$$\int_0^L q^{(1)}(x)w^{(2)}(x)\,dx = \int_0^L q^{(2)}(x)w^{(1)}(x)\,dx$$

*4.17 Using the result of Problem 4.16 and the Dirac delta function, show that the deflection at a point at $x = \xi_2$ produced by a load P at $x = \xi_1$ is identically equal to the deflection at $x = \xi_1$ produced by the load P at ξ_2.

*4.18 The beam shown in Fig. 4.27 is made of material that behaves nonlinearly. In tension the stress-strain relation is given as

$$\tau_{xx} = 2.1 \times 10^{11} (\epsilon_{xx})^{3/4}$$

while in compression the stress-strain law is given as

$$\tau_{xx} = -2.1 \times 10^{11} (-\epsilon_{xx})^{3/4}$$

Formulate an equation for finding one of the supporting forces. Solve for this supporting force numerically with the aid of a computer for $b = 50$ mm and $d = 100$ mm.

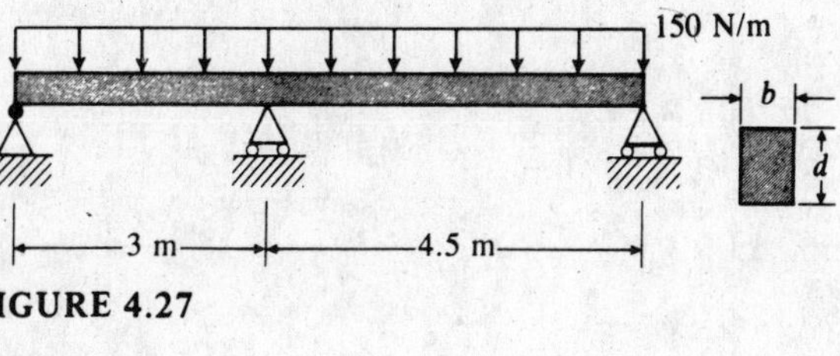

FIGURE 4.27

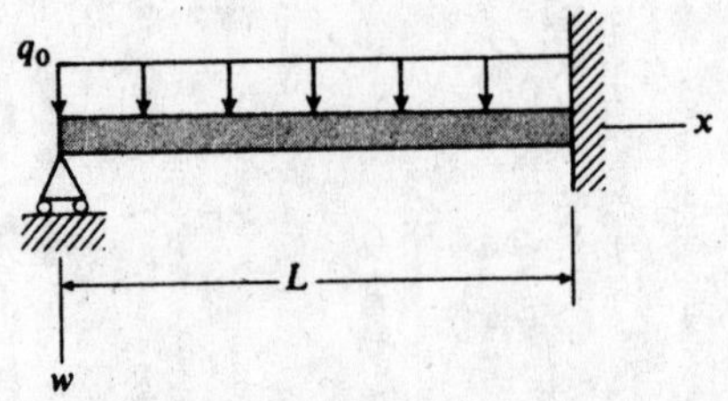

FIGURE 4.28

4.19 The material in the beam shown in Fig. 4.28 behaves according to the following stress-strain law:

$$\tau_{xx} = C_1 \sinh (C_2 \epsilon_{xx})$$

Find π^* as a function of $d^2 w/dx^2$. Now using the left support R as the redundant constraint find an equation involving $d^2 w/dx^2$, x, and R stemming from the bending moment at a section. Explain how with the aid of a computer you could then find an approximation of the proper value of R.

5

TORSION

5.1 INTRODUCTION

In this chapter we shall consider the St. Venant theory of torsion for uniform prismatic elastic rods loaded by twisting couples at the ends of the rod (see Fig. 5.1). You will recall that for the special case of a circular cross section, it is assumed in strength of materials (and shown valid in the theory of elasticity) that cross sections of the rod merely rotate as rigid surfaces under the action of the twisting couples. Thus if θ is the angle of rotation for a cross section at a position z, then we have the following result for small deformations (see Fig. 5.2):

$$u = -r\theta \sin\beta = -y\theta$$
$$v = r\theta \cos\beta = x\theta$$
$$w = 0$$

We now introduce the *rate of twist*, $\alpha = d\theta/dz$, which is a constant for any given problem with end couples. Taking $\theta = 0$ when $z = 0$, we conclude that $\theta = \alpha z$ and we can give the above equations as follows:

$$\begin{aligned} u &= -\alpha yz \\ v &= \alpha xz \\ w &= 0 \end{aligned} \tag{5.1}$$

For an *arbitrary cross section* we must abandon the assumption that plane sections remain plane (i.e., $w = 0$) and introduce a function $\kappa(x, y)$ to account for "warping" in the z direction. Thus:

$$\begin{aligned} u &= -\alpha yz \\ v &= \alpha xz \\ w &= \kappa(x, y) \end{aligned} \tag{5.2}$$

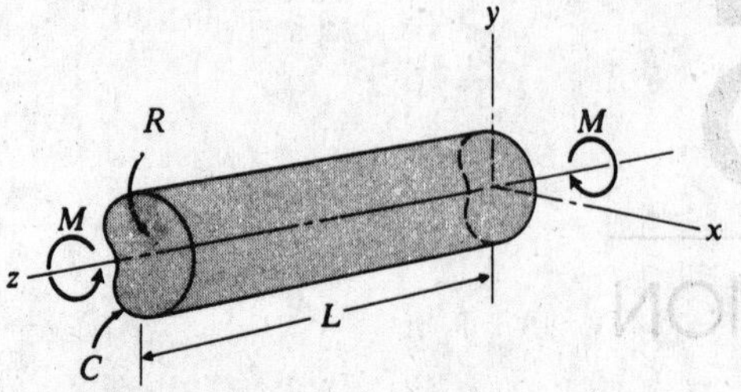

FIGURE 5.1
Prismatic rod under torsion.

Note that the warping shape of a cross section is not a function of z; all sections have the same deformed shape. We shall use the above model for torsional deformation. The corresponding strain field is

$$\begin{aligned} \epsilon_{xx} &= 0 \qquad \epsilon_{xy} = 0 \\ \epsilon_{yy} &= 0 \qquad \epsilon_{xz} = \frac{1}{2}\left(-\alpha y + \frac{\partial \kappa}{\partial x}\right) \\ \epsilon_{zz} &= 0 \qquad \epsilon_{yz} = \frac{1}{2}\left(\alpha x + \frac{\partial \kappa}{\partial y}\right) \end{aligned} \tag{5.3}$$

We now examine the principle of total potential energy for this assumed deformation field.

5.2 TOTAL POTENTIAL ENERGY; EQUATION FOR TORSION

The total potential energy for torsion can readily be given for linear elastic behavior:

$$\pi = \iiint_V (\tau_{xz}\,\epsilon_{xz} + \tau_{yz}\,\epsilon_{yz})\,dv - ML\alpha \tag{5.4}$$

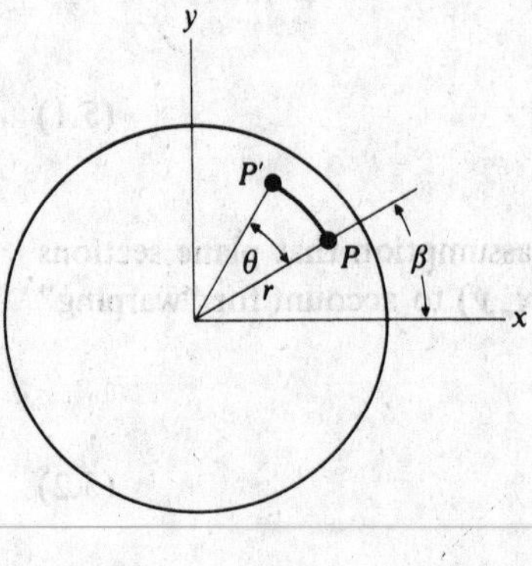

FIGURE 5.2
Torsion rod with circular cross section.

Hence, from the principle of total potential energy,

$$\delta^{(1)}\pi = \delta^{(1)}\left\{\iiint_V (\tau_{xz}\epsilon_{xz} + \tau_{yz}\epsilon_{yz})\,dv\right\} - ML\,\delta\alpha = 0$$

Using Hooke's law directly, $\tau_{xz} = 2G\epsilon_{xz}$ and $\tau_{yz} = 2G\epsilon_{yz}$, the above equation becomes

$$\delta^{(1)}\pi = 2\delta^{(1)}\left\{\iiint_V G(\epsilon_{xz}^2 + \epsilon_{yz}^2)\,dv\right\} - ML\,\delta\alpha = 0$$

Substituting from Eq. (5.3) we get

$$\delta^{(1)}\pi = \delta^{(1)}\left\{\frac{1}{2}G\iiint\left[\left(\frac{\partial\kappa}{\partial x} - y\alpha\right)^2 + \left(\frac{\partial\kappa}{\partial y} + x\alpha\right)^2\right]dv\right\} - ML\,\delta\alpha = 0$$

$$= \iiint_V G\left[\left(\frac{\partial\kappa}{\partial x} - y\alpha\right)\left(\frac{\partial(\delta\kappa)}{\partial x} - y\,\delta\alpha\right)\right.$$

$$\left. + \left(\frac{\partial\kappa}{\partial y} + x\alpha\right)\left(\frac{\partial(\delta\kappa)}{\partial y} + x\,\delta\alpha\right)\right]dv - ML\,\delta\alpha = 0$$

Collecting terms we obtain

$$\left\{\iiint_V G\left[\left(\frac{\partial\kappa}{\partial x} - y\alpha\right)(-y) + \left(\frac{\partial\kappa}{\partial y} + x\alpha\right)(x)\right]dv - ML\right\}\delta\alpha$$

$$+ \iiint_V G\left[\left(\frac{\partial\kappa}{\partial x} - y\alpha\right)\frac{\partial}{\partial x}(\delta\kappa) + \left(\frac{\partial\kappa}{\partial y} + x\alpha\right)\frac{\partial}{\partial y}(\delta\kappa)\right]dv = 0$$

Integrating over z from 0 to L for both integrals and then using Green's theorem in the second integral (in order to integrate by parts), we find that

$$\left\{\iint_R LG\left[\left(\frac{\partial\kappa}{\partial x} - y\alpha\right)(-y) + \left(\frac{\partial\kappa}{\partial y} + \alpha x\right)(x)\right]dA - ML\right\}\delta\alpha$$

$$- GL\iint_R\left[\frac{\partial}{\partial x}\left(\frac{\partial\kappa}{\partial x} - y\alpha\right) + \frac{\partial}{\partial y}\left(\frac{\partial\kappa}{\partial y} + x\alpha\right)\right]\delta\kappa\,dA$$

$$+ GL\oint_C\left[\left(\frac{\partial\kappa}{\partial x} - y\alpha\right)a_{nx} + \left(\frac{\partial\kappa}{\partial y} + \alpha x\right)a_{ny}\right]\delta\kappa\,ds = 0$$

Since $\delta\alpha$ and $\delta\kappa$ are independent, it is clear that

$$\iint_R LG\left[\left(\frac{\partial \kappa}{\partial x}-y\alpha\right)(-y)+\left(\frac{\partial \kappa}{\partial y}+\alpha x\right)(x)\right] dA - ML = 0$$

therefore

$$M = G\iint_R \left[x\,\frac{\partial \kappa}{\partial y}-y\,\frac{\partial \kappa}{\partial x}+\alpha(x^2+y^2)\right] dA \tag{5.5a}$$

Also:

$$GL\left[\frac{\partial}{\partial x}\left(\frac{\partial \kappa}{\partial x}-y\alpha\right)+\frac{\partial}{\partial y}\left(\frac{\partial \kappa}{\partial y}+x\alpha\right)\right]=0$$

therefore

$$\frac{\partial^2 \kappa}{\partial x^2}+\frac{\partial^2 \kappa}{\partial y^2}=0 \tag{5.5b}$$

Finally, we have the following natural boundary condition:

$$\left(\frac{\partial \kappa}{\partial x}-y\alpha\right)a_{nx}+\left(\frac{\partial \kappa}{\partial y}+x\alpha\right)a_{ny}=0 \quad \text{on boundary} \tag{5.5c}$$

Now let κ be given as $\alpha\phi$, where ϕ is called the *warping function.* Thus

$$\kappa = \alpha\phi \tag{5.6}$$

Then on substitution into Eqs. (5.5*b*), (5.5*a*), and (5.5*c*), we get the following equations:

$$\nabla^2\phi = 0 \tag{5.7a}$$

$$M = G\alpha\iint_R \left[x\,\frac{\partial \phi}{\partial y}-y\,\frac{\partial \phi}{\partial x}+(x^2+y^2)\right] dA \tag{5.7b}$$

$$\left(\frac{\partial \phi}{\partial x}-y\right)a_{nx}+\left(\frac{\partial \phi}{\partial y}+x\right)a_{ny}=0 \quad \text{(on boundary)} \tag{5.7c}$$

At this stage of the discussion we have used the total potential energy associated with a proposed deformation that involves the unknown warping function ϕ as well as the rate of twist α. The extremization of π has led to equations of equilibrium for the proposed system in terms of ϕ and has given us as well the appropriate natural boundary conditions. It is well known from experimental and exact analyses that the equations, when solved for the aforementioned system, give results that are very close to the actual case. The next step in our study would be to consider methods arising from the variational approach to reach approximate solutions to the torsion problem. We shall find that it is often desirable in this regard to use the total complementary energy functional for torsion. Accordingly, we now consider this functional and the associated boundary-value problem.

5.3 THE TOTAL COMPLEMENTARY ENERGY FUNCTIONAL

For linear elastic behavior there are only two nonzero stresses associated with the deformation proposed in Section 5.1. They are the shear stresses τ_{xz} and τ_{yz}. Accordingly, the equation of equilibrium for such a situation is (in the absence of body forces):

$$\frac{\partial \tau_{xz}}{\partial x} + \frac{\partial \tau_{yz}}{\partial y} = 0 \tag{5.8}$$

As in the case of plane stress, we may satisfy this equation exactly if we express the stresses in terms of a *stress function* Ψ as follows:

$$\tau_{xz} = G\alpha \frac{\partial \Psi}{\partial y} \qquad \tau_{yz} = -G\alpha \frac{\partial \Psi}{\partial x} \tag{5.9}$$

Thus working with Ψ permits the employment at all times of a *statically compatible* stress field. On the lateral boundary we have $T^{(\nu)} = 0$ and so the following condition of equilibrium must be satisfied there:

$$\tau_{zx} a_{nx} + \tau_{zy} a_{ny} = 0$$

therefore

$$G\alpha \left(\frac{\partial \Psi}{\partial y} a_{nx} - \frac{\partial \Psi}{\partial x} a_{ny} \right) = 0 \tag{5.10}$$

From Fig. 5.3 we see that

$$a_{nx} = \frac{dy}{ds} \qquad a_{ny} = -\frac{dx}{ds} \tag{5.11}$$

Hence on substitution into Eq. (5.10) we get

$$\frac{\partial \Psi}{\partial x} \frac{dx}{ds} + \frac{\partial \Psi}{\partial y} \frac{dy}{ds} = 0$$

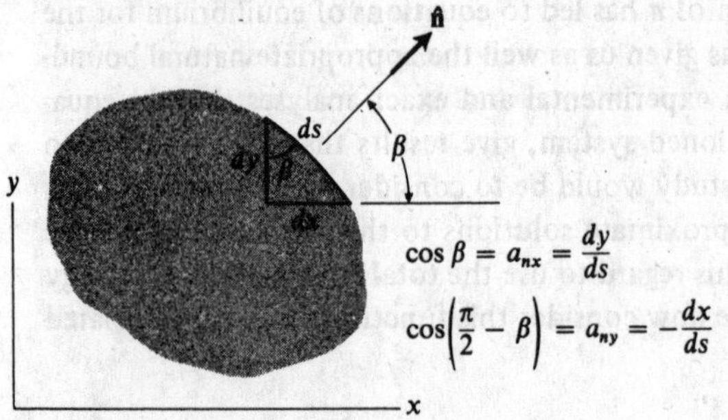

FIGURE 5.3
Infinitesimal element at boundary of section.

therefore

$$\frac{d\Psi}{ds} = 0 \quad \text{on boundary} \tag{5.12}$$

We conclude that Ψ must be a constant on the lateral boundary. For a simply connected domain there is no loss in generality in taking $\Psi = 0$ on the boundary.

We are now ready to express the complementary energy functional in terms of Ψ. Note that the tractions have been specified as zero on the lateral periphery of the body, but only resultants M have been specified at the ends. We shall for convenience consider that the end $z = 0$ is fixed so that $u = v = 0$ there. Furthermore, we shall later show that specifying M at the ends is equivalent to specifying the rate of twist α and is thus equivalent to specifying the displacement field at the end $z = L$. Hence region S_2 consists of the end faces of the shaft, and with $u_i = 0$ at $z = 0$, actually only the end at $z = L$ need be considered. Thus we have for the total complementary energy

$$\pi^* = \iiint_V (\tau_{xz}\,\epsilon_{xz} + \tau_{yz}\,\epsilon_{yz})\,dv - \left(\iint_R T_i^{(\nu)} u_i\,dA\right)_{\text{at } z=L} \tag{5.13}$$

We now employ Hooke's law to replace the strain terms in the first integral and, using $L = 1$, integrate with respect to z for this integral. In the second integral carry out the inner product using stress components for $T_i^{(\nu)}$ and displacement components given by Eqs. (5.2) for u_i. We then get

$$\pi^* = \frac{1}{2G}\iint_R (\tau_{xz}^2 + \tau_{yz}^2)\,dA - \alpha\left[\iint_R (-\tau_{xz}\,y + \tau_{yz}\,x)\,dA\right]_{z=1} \tag{5.14}$$

Since the integrand of the second integral, like that of the first, is not a function of z, it applies for all values of z and so we may delete the subscript notation $z = 1$. Now introducing the stress function Ψ, we obtain the following result for π^*:

$$\pi^* = \frac{G\alpha^2}{2} \iint_R \left[\left(\frac{\partial \Psi}{\partial x} \right)^2 + \left(\frac{\partial \Psi}{\partial y} \right)^2 \right] dA$$

$$- \iint_R G\alpha^2 \left(-\frac{\partial \Psi}{\partial y} y - \frac{\partial \Psi}{\partial x} x \right) dA$$

Also, integrate the last surface integral by parts. We find

$$\pi^* = \frac{G\alpha^2}{2} \iint_R \left[\left(\frac{\partial \Psi}{\partial x} \right)^2 + \left(\frac{\partial \Psi}{\partial y} \right)^2 \right] dA - \iint_R 2G\alpha^2 \Psi \, dA$$

$$+ \oint_C G\alpha^2 \Psi (y a_{ny} + x a_{nx}) \, ds$$

Noting that Ψ is zero on the lateral boundary, we now collect terms:

$$\pi^* = \frac{G\alpha^2}{2} \iint_R \left[\left(\frac{\partial \Psi}{\partial x} \right)^2 + \left(\frac{\partial \Psi}{\partial y} \right)^2 - 4\Psi \right] dA \tag{5.15a}$$

$$= \frac{G\alpha^2}{2} \iint_R [(\nabla \Psi)^2 - 4\Psi] \, dA \tag{5.15b}$$

Now we are in a position to carry out the variational process for π^*, according to the principle of total complementary energy:

$$\delta^{(1)} \pi^* = 0 = \frac{G\alpha^2}{2} \iint_R \left(2 \frac{\partial \Psi}{\partial x} \frac{\partial \delta \Psi}{\partial x} + 2 \frac{\partial \Psi}{\partial y} \frac{\partial \delta \Psi}{\partial y} - 4 \delta \Psi \right) dA$$

After integrating by parts and dropping $G\alpha^2$ we see that

$$\iint_R \left(-\frac{\partial^2 \Psi}{\partial x^2} - \frac{\partial^2 \Psi}{\partial y^2} - 2 \right) \delta \Psi \, dA + \oint_C \left(\frac{\partial \Psi}{\partial x} a_{nx} + \frac{\partial \Psi}{\partial y} a_{ny} \right) \delta \Psi \, ds = 0$$

Since Ψ is specified on the boundary then $\delta\Psi$ must be zero there, and so we have for the above equation

$$\iint_R (\nabla^2 \Psi + 2) \delta \Psi \, dA = 0$$

Now, using the fundamental lemma, we have the result

$$\nabla^2 \Psi = -2 \tag{5.16}$$

We would expect from considerations of the total complementary energy that Eq. (5.16) is a kinematic compatibility equation of some sort. To investigate this† examine the differential of the displacement component w as follows:

$$dw = \frac{\partial w}{\partial x}\,dx + \frac{\partial w}{\partial y}\,dy \tag{5.17}$$

Note that

$$\epsilon_{xz} = \frac{1}{2}\left(\frac{\partial w}{\partial x} + \frac{\partial u}{\partial z}\right) = \frac{1}{2}\left(\frac{\partial w}{\partial x} - \alpha y\right)$$

$$\epsilon_{yz} = \frac{1}{2}\left(\frac{\partial w}{\partial y} + \frac{\partial v}{\partial z}\right) = \frac{1}{2}\left(\frac{\partial w}{\partial y} + \alpha x\right)$$

where we have used Eq. (5.2) for u and v. Hence we have from the above for $\partial w/\partial x$ and $\partial w/\partial y$

$$\frac{\partial w}{\partial x} = 2\epsilon_{xz} + \alpha y = \frac{\tau_{xz}}{G} + \alpha y = \alpha\left(\frac{\partial \Psi}{\partial y} + y\right)$$

$$\frac{\partial w}{\partial y} = 2\epsilon_{yz} - \alpha x = \frac{\tau_{yz}}{G} - \alpha x = \alpha\left(-\frac{\partial \Psi}{\partial x} - x\right)$$

Substituting the above results into Eq. (5.17) we get

$$dw = \alpha\left(\frac{\partial \Psi}{\partial y} + y\right)dx - \alpha\left(\frac{\partial \Psi}{\partial x} + x\right)dy$$

Now integrate over any closed path Γ in the domain including the boundary. We get

$$\oint_\Gamma dw = \oint_\Gamma \left[\alpha\left(\frac{\partial \Psi}{\partial y} + y\right)dx - \alpha\left(\frac{\partial \Psi}{\partial x} + x\right)dy\right]$$

$$= \oint_\Gamma \left[\alpha\left(\frac{\partial \Psi}{\partial y} + y\right)\frac{dx}{ds} - \alpha\left(\frac{\partial \Psi}{\partial x} + x\right)\frac{dy}{ds}\right]ds$$

Employing Eqs. (5.11), we may restate the above result as follows:

$$\oint_\Gamma dw = -\oint_\Gamma \alpha\left[\left(\frac{\partial \Psi}{\partial y} + y\right)a_{ny} + \left(\frac{\partial \Psi}{\partial x} + x\right)a_{nx}\right]ds$$

Now employ the integration by parts formula [Eq. (I.32), Appendix I] on the right side of the equation to get

$$\oint_\Gamma dw = -\alpha\iint_R \left[\left(\frac{\partial^2 \Psi}{\partial y^2} + 1\right) + \left(\frac{\partial^2 \Psi}{\partial x^2} + 1\right)\right]dA$$

$$= -\alpha\iint_R (\nabla^2 \Psi + 2)\,dA$$

†Recall we have done this for the case of plane stress in Section 3.4.

If we are to have kinematic compatibility, then $\oint dw = 0$ to prevent a dislocation in w. This then means that the integrand in the above surface integral must be zero. But this requirement is simply Eq. (5.16), showing that the latter equation is, in fact, a kinematic compatibility requirement.

There is a constraint that we must impose on Ψ to give the resultant torque on the end surface its proper value M. That is:

$$M = \iint_R (-y\tau_{zx} + x\tau_{zy})\, dA = -\iint_R G\alpha\left(y\frac{\partial \Psi}{\partial y} + x\frac{\partial \Psi}{\partial x}\right) dA$$

Integrating by parts, we get

$$M = 2G\alpha \iint_R \Psi\, dA - \oint_\Gamma G\alpha(ya_{ny} + xa_{nx})\Psi\, ds$$

Since $\Psi = 0$ on the lateral boundary, we obtain the following constraining equation to go with Eq. (5.16)†:

$$M = 2G\alpha \iint_R \Psi\, dA$$

To summarize, we have from the principle of total complementary energy the following boundary-value problem:

$$\nabla^2 \Psi = -2 \tag{5.18a}$$

$$\Psi = 0 \quad \text{on lateral boundary} \tag{5.18b}$$

$$M = 2G\alpha \iint_R \Psi\, dA \tag{5.18c}$$

The idea is to find a solution Ψ from Eq. (5.18*a*) satisfying Eq. (5.18*b*) in terms of the parameter α. Then going to Eq. (5.18*c*) we can determine α in terms of M.

We have thus presented two systems of equations for torsion–one that deals with the *warping function* ϕ [see Eq. (5.7)] and the present case where we employ the *stress function* Ψ. We will now set forth yet a third statement of the boundary-value problem, using the *conjugate function* ψ. Since ϕ is a harmonic function, we know that there is a conjugate function ψ, also harmonic and related to ϕ through the Cauchy-Riemann equations. That is:

†We can now see from the following equation that specifying the torques M at the ends of the shaft is equivalent to specifying the rate of twist α. That is, M and α play equivalent roles in the mathematical aspects of the problem.

$$\frac{\partial \phi}{\partial x} = \frac{\partial \psi}{\partial y}$$
$$\frac{\partial \phi}{\partial y} = -\frac{\partial \psi}{\partial x} \tag{5.19}$$

We leave it for you as an exercise to show that in terms of ψ we may represent the torsion problem as follows:

$$\nabla^2 \psi = 0 \tag{5.20a}$$
$$\psi = \tfrac{1}{2}(x^2 + y^2) \qquad \text{on boundary} \tag{5.20b}$$
$$M = G\alpha \iint_R \left(-x \frac{\partial \psi}{\partial x} - y \frac{\partial \psi}{\partial y} + x^2 + y^2 \right) dA \tag{5.20c}$$

Also, we can show readily that the equation

$$\psi = \Psi + \tfrac{1}{2}(x^2 + y^2) \tag{5.21}$$

relates the conjugate function ψ to the stress function Ψ.

5.4 APPROXIMATE SOLUTIONS FOR LINEAR ELASTIC BEHAVIOR VIA THE RITZ METHOD

We shall now use the total complementary energy coupled with the Ritz method to formulate approximate solutions to the torsion problem for a linear elastic body.

We consider first a rectangular cross section having dimensions $2a$ and $2b$, as shown in Fig. 5.4. As a first choice for a coordinate function we consider

$$\Psi_1 = C_1(a^2 - x^2)(b^2 - y^2) \tag{5.22}$$

Note that the boundary condition [Eq. (5.18*b*)] is satisfied and the function is even with respect to x and y, as would be expected of the solution. Now substitute into the

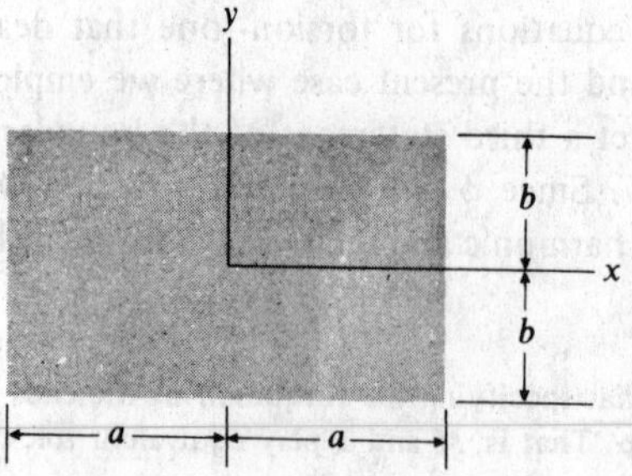

FIGURE 5.4
Rectangular cross section.

TABLE 5.1
Exact and Approximate Torsion Data

b/a	$(D/G)_{\text{app}}$	$(D/G)_{\text{exact}}$
1	2.22	2.25
2	7.12	7.32
3	12.00	12.60
4	16.73	18.60
5	21.37	23.30

total complementary energy functional [Eqs. (5.15)]. Extremizing with respect to C_1, we get for this constant

$$C_1 = \frac{5}{4} \frac{1}{b^2 + a^2}$$

The *torsional rigidity D*, defined as M/α, is then given by the following formula [see Eq. (5.18*c*)]:

$$D = 2G \iint_R \Psi \, dA = \frac{40}{9} \frac{a^3 b^3}{b^2 + a^2} G \tag{5.23}$$

For the maximum stress (occurring at the middle of the widest edge of the section) we get for $b > a$:

$$\tau_{\max} = \frac{5}{2} \frac{b^2 a}{b^2 + a^2} \alpha G \tag{5.24}$$

We may compare results of the above procedure with those of an exact solution from the theory of elasticity in the form of an infinite series (see Timoshenko and Goodier, 1951). Taking $a = 1$ and computing D/G for various values b/a, we have the results in Table 5.1. Notice that the approximation obtained is *from below.* (We shall soon present a way of formulating an approximation for *D from above* by the Trefftz method so as to provide a means of *bracketing* the desired result.) To improve the result we may use a more complex function as follows:

$$\Psi = (x^2 - a^2)(y^2 - b^2)(C_1 + C_2 x^2 y^2 + C_3 x^4 y^4 + C_4 x^6 y^6 + \cdots) \tag{5.25}$$

Employing this expression with only three unknown coefficients and extremizing the complementary potential energy expression [Eq. (5.15)] with respect to the C_i, we obtain the following equations for the C's for the rectangular section:

$$\begin{bmatrix} 5.69 & 0.162 & 0.0232 \\ 0.162 & 0.0851 & 0.0246 \\ 0.0232 & 0.0246 & 0.0112 \end{bmatrix} \begin{Bmatrix} C_1 \\ a^2 b^2 C_2 \\ a^4 b^4 C_3 \end{Bmatrix} = \frac{1}{a^2 + b^2} \begin{Bmatrix} 7.111 \\ 0.284 \\ 0.0522 \end{Bmatrix}$$

The improvement of the solution is illustrated in Table 5.2.

TABLE 5.2
Torsion Data for Varying Numbers of Parameters

Ψ	$(a^2 + b^2)D/a^3 b^3 G$
1. Parameter C_1	4.4444
2. Parameters C_1, C_2	4.4854
3. Parameters C_1, C_2, C_3	4.4856
Exact	4.5000

5.5 THE METHOD OF TREFFTZ; UPPER BOUND FOR TORSIONAL RIGIDITY

In the linear elastic problems considered in Section 5.4 we used the Ritz method and illustrated via the problem of the rectangular cross section that we could then approximate the torsional rigidity *from below.* We now set forth a method whereby the torsional rigidity is approximated *from above.* This is the method of Trefftz. By this means we can then *bracket* the correct value of the torsional ridigity.

For this purpose, we now consider again the conjugate function ψ for which the boundary-value problem in torsion is given by Eqs. (5.20). (You may recognize this to be the ***Dirichlet*** problem, wherein the function is harmonic and is specified on the boundary.) In Problem 3.14 we showed for homogeneous boundary conditions that $-\nabla^2$ is both self-adjoint and positive definite so that the quadratic functional I for this boundary-value problem is

$$I(\psi) = -\langle \psi, \nabla^2 \psi \rangle = -\iint_R \left(\psi \frac{\partial^2 \psi}{\partial x^2} + \psi \frac{\partial^2 \psi}{\partial y^2} \right) dA$$

Integrating by parts and using the homogeneous boundary conditions, we get

$$I(\psi) = \iint_R \left[\left(\frac{\partial \psi}{\partial x} \right)^2 + \left(\frac{\partial \psi}{\partial y} \right)^2 \right] dA \tag{5.26}$$

We have shown in Section 3.14 that for nonhomogeneous boundary conditions the quadratic functional again has the above form. Thus we may state the boundary-value problem for the conjugate function ψ as follows:

$$-\nabla^2 \psi = 0 \quad \text{in } R \tag{5.27a}$$

$$\psi = \tfrac{1}{2}(x^2 + y^2) \quad \text{on } C \tag{5.27b}$$

$$I(\psi) = \iint_R (\psi_x^2 + \psi_y^2)\, dA = \iint_R (\nabla \psi)^2\, dA \tag{5.27c}$$

We now prove the following theorem by Weinstein, which is vital for the Trefftz method.

THEOREM If Θ is a harmonic function in R and ψ is the solution to a Dirichlet problem for this domain, then

$$I(\Theta) \leqslant I(\psi) \tag{5.28}$$

provided that on the boundary

$$\oint_C (\psi - \Theta) \frac{d\Theta}{d\nu} \, ds = 0 \tag{5.29}$$

The *proof* is quite straightforward. Define η as follows:

$$\eta = \psi - \Theta \tag{5.30a}$$

therefore

$$\psi = \Theta + \eta \tag{5.30b}$$

Clearly η is a harmonic function. Now examine $I(\psi)$ as follows, using Eqs. (5.27*c*) and (5.30*b*):

$$\begin{aligned} I(\psi) &= \iint_R [(\Theta_x + \eta_x)^2 + (\Theta_y + \eta_y)^2] \, dA \\ &= \iint_R (\Theta_x^2 + \Theta_y^2) \, dA + \iint_R (\eta_x^2 + \eta_y^2) \, dA \\ &\quad + 2 \iint_R (\eta_x \Theta_x + \eta_y \Theta_y) \, dA \\ &= I(\Theta) + I(\eta) + 2 \iint_R (\eta_x \Theta_x + \eta_y \Theta_y) \, dA \end{aligned}$$

Now using Green's theorem in two dimensions [see Eq. (I.35), Appendix I] for the last integration, we get

$$\iint_R (\Theta_x \eta_x + \Theta_y \eta_y) \, dA = -\iint_R \eta \, \nabla^2 \Theta \, dA + \oint_C \eta \frac{d\Theta}{d\nu} \, ds$$

But $\nabla^2 \Theta = 0$, and by virtue of Eqs. (5.29) and (5.30*a*) the line integral is zero. Hence we get

$$I(\psi) = I(\Theta) + I(\eta)$$

Since $I(\eta)$, $I(\Theta)$, and $I(\psi)$ are positive definite functionals, we can conclude that

$$I(\Theta) \leqslant I(\psi)$$

thus proving the theorem.

Suppose next we form a parameter-laden sum of linearly independent functions to represent Θ in the following way:

$$\Theta = \sum_{i}^{n} a_i v_i \tag{5.31}$$

where functions v_i must be harmonic. To ensure that $I(\Theta) \leqslant I(\psi)$, we select the constants a_i to satisfy the condition given by Eq. (5.29) for the problem at hand. That is,

$$\oint_C \left(\psi - \sum_{i=1}^{n} a_i v_i\right) \frac{d}{dv}\left(\sum_{j=1}^{n} a_j v_j\right) ds = 0$$

therefore, noting Eq. (5.27*b*),

$$\oint_C \left[\frac{1}{2}(x^2 + y^2) - \sum_{i=1}^{n} a_i v_i\right]\left(\sum_{j=1}^{n} a_j \frac{dv_j}{dv}\right) ds = 0$$

To satisfy the above equation it is sufficient to require that the coefficients of each a_j be zero. That is:

$$\boxed{\oint_C \left[\frac{1}{2}(x^2 + y^2) - \sum_{i=1}^{n} a_i v_i\right] \frac{dv_j}{dv} ds = 0 \qquad j = 1, 2, \ldots, n} \tag{5.32}$$

We have here a system of n equations with n unknowns a_i. One can prove that if the boundary is such that $\frac{1}{2}(x^2 + y^2) \neq 0$ on the boundary and if the functions v_i are linearly independent then there is a *unique* set of constants a_i that can be found from the above set of equations (see Sokolnikoff, 1956). Now letting $n \to \infty$ one would expect the set of a_i determined from Eq. (5.32) to be such that $\Sigma_{i=1}^{n} a_i v_i$ approaches $\frac{1}{2}(x^2 + y^2)$ on the *boundary* in some manner. Indeed it can be shown (as usual we shall not get into the convergence question) that if functions v_i are complete in the domain except at isolated points† then $\Sigma_{i=1}^{n} a_i v_i$ converges to ψ uniformly in the

†That is, if we can approximate any harmonic function arbitrarily closely at all but isolated points by the sum

$$\sum_{i=1}^{n} a_i v_i$$

by properly choosing the a's and making n large enough.

region R. This means that $I(\Theta)$ can be made arbitrarily close to $I(\psi)$ everywhere by choosing n large enough, and Eq. (5.28) adds the fact that this approach must be *from below*.

There is a relation between an approximate torsional rigidity using $\Sigma a_i v_i$ as formulated above and the actual torsional rigidity. For this consideration examine $I(\psi)$ while replacing ψ by $[\Psi + \frac{1}{2}(x^2 + y^2)]$ in accordance with Eq. (5.21). Thus we get

$$I(\psi) = \iint_R \{\nabla[\Psi + \tfrac{1}{2}(x^2 + y^2)]\}^2 \, dA$$

$$= \iint_R [(\nabla\Psi)^2 + \nabla\Psi\cdot\nabla(x^2 + y^2) + (x^2 + y^2)] \, dA \tag{5.33}$$

Now from Green's theorem [Eq. (I.36), Appendix I] we have for the second expression of the integrand on the right side of the above equation:

$$\iint_R \nabla\Psi\cdot\nabla(x^2 + y^2)\, dA = -\iint_R \Psi\nabla^2(x^2 + y^2)\, dA + \oint_C \Psi \frac{\partial}{\partial\nu}(x^2 + y^2)\, ds$$

Since $\Psi = 0$ on the boundary, the line integral vanishes. The surface integral on the right side of the equation is simply $\iint 4\Psi\, dA$. We then get on substitution into Eq. (5.33)

$$I(\psi) = \iint_R [(\nabla\Psi)^2 - 4\Psi] \, dA + \iint_R (x^2 + y^2)\, dA \tag{5.34}$$

But the last integral is simply the *polar moment of inertia* of the cross section and we denote it simply as J. Thus we can say from Eq. (5.15*b*)

$$I(\psi) = \frac{2}{G\alpha^2}\pi^* + J$$

therefore

$$\pi^* = \frac{G\alpha^2}{2}[I(\psi) - J] \tag{5.35}$$

We shall now find another formulation of π^* in terms of the torsional stiffness D. The torsional stiffness has been shown to be Eq. (5.18*c*)]

$$\frac{M}{\alpha} = D = 2G\iint_R \Psi\, dA \tag{5.36}$$

But using Green's theorem [Eq. (I.36), Appendix I] in two dimensions with ψ and ϕ replaced by Ψ, while noting that $\Psi = 0$ on the boundary and that $\nabla^2 \Psi = -2$ everywhere, the above result can be given as

$$D = G \iint_R (\nabla\Psi)^2 \, dA \tag{5.37}$$

We may now give π^* in terms of D. Thus, employing Eqs. (5.37) and (5.36) in Eq. (5.15b), we get another formulation of π^*:

$$\pi^* = \frac{G\alpha^2}{2}\left(\frac{D}{G} - 4\,\frac{D}{2G}\right) = -\frac{D\alpha^2}{2} \tag{5.38}$$

Equating the right sides of Eq. (5.38) and Eq. (5.35) and solving for D we get

$$D = G[J - I(\psi)] \tag{5.39}$$

Now consider a quantity that we denote as $\tilde{D}_{\text{Trefftz}}$ formed by using the right side of the above equation with ψ replaced by Θ in the functional I. That is:

$$\boxed{\tilde{D}_{\text{Trefftz}} = G[J - I(\Theta)]} \tag{5.40}$$

We have already shown by the Weinstein theorem that, for properly chosen a's, $I(\Theta) \leqslant I(\psi)$ and so it is clear on comparing the right sides of the above equations for D and $\tilde{D}_{\text{Trefftz}}$ that

$$\tilde{D}_{\text{Trefftz}} \geqslant D \tag{5.41}$$

We see that $\tilde{D}_{\text{Trefftz}}$ *forms an approximation from above of the torsional rigidity.* Furthermore, because $I(\Theta)$ converges to $I(\psi)$, we can make $\tilde{D}_{\text{Trefftz}}$ arbitrarily close to D. Hence by a judicious selection of functions v_i and with the proper values of coefficients a_i, we can compute $\tilde{D}_{\text{Trefftz}}$ by using Eq. (5.40).

We have shown earlier (Section 5.4) that, for the example considered, the torsional rigidity approximation reached by the Ritz method was an approximation from below. We can now readily see that the torsional rigidity computed in this way must always be an approximation from below. Note from Eq. (5.38) that π^* must be a negative number for torsion since D must be positive.† Now an approximate total complementary energy $\tilde{\pi}^*_{\text{Ritz}}$ (via the Ritz method) must be greater than π^*. This means that $\tilde{\pi}^*_{\text{Ritz}}$ has a smaller negative value than π^*. Hence the value of torsional stiffness D that would be obtained with the Ritz approximation for the total complementary energy would, according to Eq. (5.38), perforce be a ***smaller*** positive

†Remember $D = M/\alpha$. It is clear that M and α must have the same sense and so M/α must be positive.

number than the exact value of D. Thus we conclude that the Ritz approximation for the torsional stiffness must be *from below.*

Thus by using the Ritz approach and the Trefftz approach here we can *bracket* the exact result for the torsional rigidity. We illustrate this procedure now by going back to the rectangular shaft discussed in Section 5.4. To generate suitable harmonic functions for the calculation, we may consider the real and imaginary parts of the analytic function† $(x+iy)^n$. For various values of n we then select harmonic functions that are even functions with respect to coordinates x and y (see Fig. 5.4). Thus for $n=2$ we have (x^2-y^2). Hence we may consider for Θ the function

$$\Theta = a_1(x^2 - y^2) \tag{5.42}$$

To get a_1 we employ Eq. (5.29) as follows to satisfy requirements of Weinstein's theorem:

$$\oint_C \left[\frac{1}{2}(x^2+y^2) - a_1(x^2-y^2)\right] a_1 \frac{d}{dv}(x^2-y^2)\,ds = 0 \tag{5.43}$$

This becomes, on canceling out a_1 and considering the boundary:

$$\int_{-a}^{+a} [\tfrac{1}{2}(x^2+b^2) - a_1(x^2-b^2)](2y)|_{y=-b}\,dx$$

$$+ \int_{-b}^{+b} [\tfrac{1}{2}(a^2+y^2) - a_1(a^2-y^2)](2x)|_{x=a}\,dy$$

$$+ \int_{+a}^{-a} [\tfrac{1}{2}(x^2+b^2) - a_1(x^2-b^2)](-2y)|_{y=b}(-dx)$$

$$+ \int_{+b}^{-b} [\tfrac{1}{2}(a^2+y^2) - a_1(a^2-y^2)](-2x)|_{x=-a}(-dy) = 0$$

We thus have for a_1

$$a_1 = \frac{a^2-b^2}{2(a^2+b^2)}$$

To get the torsional rigidity we have

$$\tilde{D}_{\text{Trefftz}} = G[J - I(\Theta)] = G\left(\int_{-b}^{+b}\int_{-a}^{+a}(x^2+y^2)\,dx\,dy\right.$$

$$\left. - \int_{-b}^{+b}\int_{-a}^{+a}\{\nabla[a_1(x^2-y^2)]\}^2\,dx\,dy\right) = \frac{16Ga^3b^3}{3(a^2+b^2)}$$

†Such a set of polynomials can be shown to be complete.

From the Ritz method, meanwhile, we have [see Eq. (5.23)]

$$\tilde{D}_{\text{Ritz}} = \frac{40a^3b^3}{9(a^2+b^2)}G = \frac{13.3Ga^3b^3}{3(a^2+b^2)}$$

We may conclude that the torsional stiffness is bracketed as follows:

$$\left[\tilde{D}_{\text{Ritz}} = \frac{13.3a^3b^3}{3(a^2+b^2)}G\right] \leqslant \left[D_{\text{exact}} = \frac{13.5a^3b^3}{3(a^2+b^2)}G\right]$$

$$\leqslant \left[\tilde{D}_{\text{Trefftz}} = \frac{16a^3b^3}{3(a^2+b^2)}G\right]$$

We show in Table 5.3 torsional rigidities computed by the Ritz method for a function Ψ_1 given as

$$\Psi_1 = C_1(a^2 - x^2)(b^2 - y^2)$$

as well as torsional rigidities from the Trefftz method with the following functions:

$$\Theta_1 = a_1(x^2 - y^2)$$

$$\Theta_2 = a_1(x^2 - y^2) + a_2(x^4 - 6x^2y^2 + y^4)$$

$$\Theta_3 = a_1(x^4 - 6x^2y^2 + y^4)$$

Also, a set of exact results is included for comparison. The torsional rigidity terms are given for $a = 1$ and as a function of the ratio a/b.

5.6 THE METHOD OF KANTOROVICH

A serious shortcoming of the Ritz method as well as the Galerkin method is that the results obtained have a strong dependence on the coordinate functions chosen. This was made most apparent in Chapter 4 on beams. The method of Kantorovich, which we now consider, will decrease this dependence of the results on the choice of the coordinate function, thereby making the process more effective. However, this gain will not be reached without additional computational efforts.

We shall consider in this regard functionals of the form:

$$I(w) = \iint_R F(x, y, w, w_x, w_y)\,dx\,dy \tag{5.44}$$

Now in the Kantorovich method we shall again use coordinate functions, which we now denote as H_p, as we did in the Ritz method, but instead of using undetermined

TABLE 5.3
Torsional Rigidities Using the Ritz and the Trefftz Methods

a/b	Ritz $\Phi_1 = C_1(a^2 - x^2)(b^2 - y^2)$	Trefftz $\Theta_1 = a_1(x^2 - y^2)$	$\Theta_2 = a_1(x^2 - y^2) + a_2(x^4 - 6x^2y^2 - y^4)$	$\Theta_3 = a_1(x^4 - 6x^2y^2 + y^4)$	Exact Timoshenko
1.00	2.22	2.67	2.25	2.25	2.25
1.20	3.14	3.78	3.32	3.45	3.19
1.50	4.62	5.54	5.50	6.35	4.70
2.00	7.11	8.53	10.00	13.01	7.83
2.50	9.58	11.49	14.80	20.81	9.96
3.00	12.00	14.40	19.59	29.39	12.62
4.00	16.73	20.08	28.89	49.17	17.98
5.00	21.37	25.64	37.89	73.60	23.30
10.00	44.00	52.80	80.84	314.5	49.92

constants a_i, we shall now use undetermined coefficient *functions* $C_p(x)$. Thus for the dependent variable w we form the nth partial sum:

$$w_n = H_0(x,y) + \sum_{p=1}^{n} C_p(x)H_p(x,y) \tag{5.45}$$

H_0 has the specified value of w on the boundary and functions $H_p(x,y)$ are zero on part or all of the boundary; where H_p is not zero the coefficient $C_p(x)$ must be zero on the boundary. We may susbstitute the above partial sum into Eq. (5.44) for the dependent variable w. Since F is specified and the H's are known, the functional $I(w_n)$, on carrying out the integration with respect to y, becomes one involving x as the integration variable with the C's as functions. Thus we get

$$I(w_n) = \int_{x_1}^{x_2} F[x, C_1(x), \ldots, C_p(x), \ldots, C_n(x), C_1'(x), \ldots, C_p'(x), \ldots, C_n'(x)]\, dx \tag{5.46}$$

The procedure is now to invoke the variational process to establish the Euler-Lagrange equations for the functions $C_p(x)$, rendering them extremals for the functional $I(w_n)$. The resulting function w_n is then the desired approximation to the problem according to this method.

Let us in particular consider the *torsion problem*, using the stress function Ψ as the dependent variable. The boundary-value problem for this case was shown to be

$$\begin{aligned} &\nabla^2 \Psi = -2 \\ &\Psi = 0 \quad \text{on boundary} \end{aligned} \tag{5.47}$$

We have shown in Section 5.3 that this boundary-value problem could be reached by extremizing the total complementary energy π^* as given by Eq. (5.15*b*). Dropping the constant $G\alpha^2/2$, the functional I that we may consider for the Kantorovich method as applied to this boundary-value problem is

$$I = \iint_R [(\nabla\Psi)^2 - 4\Psi]\, dx\, dy \tag{5.48}$$

In considering w_n [see Eq. (5.45)] to replace the dependent variable Ψ above, we may delete the function H_0 in this case and employ the following partial sum, where we use the notation Ψ_n in place of w_n:

$$\Psi_n = \sum_{p=1}^{n} C_p(x)H_p(x,y) \tag{5.49}$$

Now form varied functions $\tilde{C}_p(x)$ as follows:

$$\tilde{C}_p(x) = C_p(x) + \epsilon_p \eta_p(x) \tag{5.50}$$

with the condition that $\eta_p = 0$ on the boundary. Now employ $\tilde{\Psi}_n$, with $\tilde{C}_p$ as given above, in the functional (5.48). We then find $C_p(x)$ by carrying out the extremization

$$\left(\frac{\partial \tilde{I}}{\partial \epsilon_p}\right)_{\substack{\epsilon_1=0\\ \vdots \\ \epsilon_n=0}} = 0 \qquad p = 1, 2, \ldots, n$$

We have for $\tilde{I}$ from Eq. (5.48) the following result:

$$\tilde{I} = \iint_R \tilde{F}\, dx\, dy = \iint_R \left[\left(\frac{\partial \tilde{\Psi}_n}{\partial x}\right)^2 + \left(\frac{\partial \tilde{\Psi}_n}{\partial y}\right)^2 - 4\tilde{\Psi}_n\right] dx\, dy$$

where

$$\tilde{\Psi}_n = \sum_{p=1}^{n} \tilde{C}_p(x) H_p(x, y) = \sum_{p=1}^{n} [C_p(x) + \epsilon_p \eta_p(x)] H_p(x, y) \tag{5.51}$$

Hence

$$\tilde{F} = \left\{\sum_{p=1}^{n} [(\tilde{C}_p)(H_p)_x + (\tilde{C}_p)_x(H_p)]\right\}^2 + \left[\sum_{p=1}^{n} (\tilde{C}_p)(H_p)_y\right]^2$$
$$-4 \sum_{p=1}^{n} (\tilde{C}_p)(H_p) \tag{5.52}$$

where $(H_p)_x = \partial(H_p)/\partial x$ and so on. Now:

$$\left(\frac{\partial \tilde{I}}{\partial \epsilon_\alpha}\right)_{\substack{\epsilon_1=0\\ \vdots \\ \epsilon_n=0}} = 0 = \iint_R \left[\frac{\partial \tilde{F}}{\partial \tilde{C}_\alpha} \eta_\alpha + \frac{\partial \tilde{F}}{\partial (\tilde{C}_\alpha)_x} (\eta_\alpha)_x\right]_{\substack{\epsilon_1=0\\ \vdots \\ \epsilon_n=0}} dx\, dy$$

$$= \iint_R \left[\frac{\partial \tilde{F}}{\partial \tilde{C}_\alpha} - \frac{\partial}{\partial x} \frac{\partial \tilde{F}}{\partial (\tilde{C}_\alpha)_x}\right]_{\substack{\epsilon_1=0\\ \vdots \\ \epsilon_n=0}} \eta_\alpha\, dx\, dy$$

$$+ \text{line integral} \qquad \alpha = 1, 2, \ldots, n$$

Substituting from Eq. (5.52) we get

$$\iint_R \left(2 \sum_p [(C_p)(H_p)_x + (C_p)_x(H_p)] (H_\alpha)_x + 2 \left[\sum_p (C_p)(H_p)_y \right] (H_\alpha)_y \right.$$
$$\left. - 4H_\alpha - 2 \frac{\partial}{\partial x} \left\{ \sum_p [(C_p)(H_p)_x + (C_p)_x(H_p)] H_\alpha \right\} \right) \eta_\alpha \, dx \, dy$$
$$+ \text{line integral} = 0 \qquad \alpha = 1, 2, \ldots, n$$

where we have dropped the tildes and the conditions $\epsilon_1 = \epsilon_2 = \cdots \epsilon_n = 0$ from the notation. Next, carry out the differentiation of the indicated $\partial/\partial x$ above over the bracketed quantities, using the product rule. We see by inspection that one of the resulting expressions is minus the first expression in the surface integral and accordingly we can cancel these expressions. We then have, on dividing through by 2 and rearranging the terms,

$$\iint_R \left(\left\{ -\frac{\partial}{\partial x} \sum_p [C_p(H_p)_x + (C_p)_x(H_p)] \right\} H_\alpha + \sum_p C_p(H_p)_y(H_\alpha)_y - 2H_\alpha \right) \eta_\alpha \, dx \, dy$$
$$+ \text{line integral} = 0 \qquad \alpha = 1, 2, \ldots, n$$

Note that the expression

$$\sum_p [C_p(H_p)_x + (C_p)_x H_p]$$

is simply

$$\frac{\partial}{\partial x}\left(\sum_p C_p H_p \right)$$

which is $(\Psi_n)_x$. We then have

$$\iint_R \left\{ -(\Psi_n)_{xx} H_\alpha + \sum_p C_p(H_p)_y(H_\alpha)_y - 2H_\alpha \right\} \eta_\alpha \, dx \, dy$$
$$+ \text{line integral} = 0 \qquad \alpha = 1, 2, \ldots, n \tag{5.53}$$

Now integrate the second expression by parts using Green's theorem as follows:

$$\iint_R \sum_p C_p(H_p)_y(H_\alpha)_y \eta_\alpha \, dx \, dy = - \iint_R \frac{\partial}{\partial y} \left\{ \eta_\alpha \sum_p C_p(H_p)_y \right\} H_\alpha \, dx \, dy$$
$$+ \text{line integral} \qquad \alpha = 1, 2, \ldots, n$$

Recalling that C_p and η_α are functions only of x, we can express the above equation as

$$\iint_R \sum_p C_p(H_p)_y(H_\alpha)_y \eta_\alpha \, dx\, dy = -\iint_R \left(\frac{\partial^2}{\partial y^2}\sum_p C_p H_p\right)\eta_\alpha H_\alpha \, dx\, dy$$

$$+ \text{ line integral}$$

$$= -\iint_R (\Psi_n)_{yy} H_\alpha \eta_\alpha \, dx\, dy + \text{line integral} \qquad \alpha = 1, 2, \ldots, n$$

Substituting the above result into Eq. (5.53), we reach the following result on setting the line integral equal to zero, as required by the extremization process:

$$\iint_R (\nabla^2 \Psi_n + 2) H_\alpha(x, y)\eta_\alpha(x)\, dx\, dy = 0 \qquad \alpha = 1, \ldots, n \tag{5.54}$$

Now restricting ourselves to boundaries of the type shown in Fig. 5.5, we can rewrite the above equation as follows:

$$\int_{-x_1}^{x_2} \eta_\alpha(x)\, dx \int_{y=g(x)}^{y=h(x)} (\nabla^2 \Psi_n + 2) H_\alpha(x, y)\, dy = 0 \qquad \alpha = 1, 2, \ldots, n \tag{5.55}$$

Since the η's are arbitrary we can conclude that†

$$\boxed{\int_{y=g(x)}^{y=h(x)} (\nabla^2 \Psi_n + 2) H_\alpha(x, y)\, dy = 0 \qquad \alpha = 1, 2, \ldots, n} \tag{5.56}$$

†Note that by this process the problem has been altered so as to require the solution of a *Galerkin integral* (see Section 3.15). We shall make use of this transition in the Kantorovich method at later times in the text.

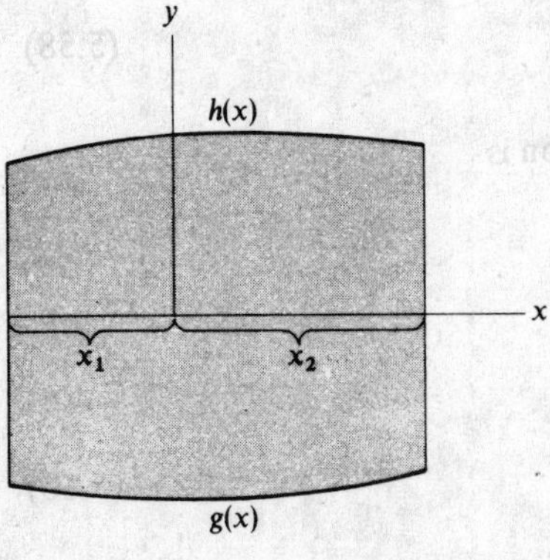

FIGURE 5.5
Section showing desired upper and lower boundaries.

When we carry out the integration with respect to y in the above n equations, we obtain n ordinary differential equations for the desired functions $C_p(x)$. The constants of integration for the solution $C_p(x)$ are determined so as to maintain Ψ_n, the approximate stress function, equal to zero over the entire boundary.

We illustrate the procedure for the Kantorovich method by considering the torsion of the shaft with a rectangular cross section (see Fig. 5.4) as presented by Kantorovich and Krylov (1964). We consider first a one-term approximating function of the form

$$\Psi_1 = [C_1(x)](b^2 - y^2) \tag{5.57a}$$

therefore

$$H_1 = b^2 - y^2 \tag{5.57b}$$

Notice that H_1 satisfies the boundary conditions on two horizontal edges of the cross section; accordingly, we will have to adjust $C_1(x)$ to satisfy the boundary condition on the vertical edges. Substituting into Eq. (5.56) we get

$$\int_{-b}^{+b} \left[C_1(-2) + (b^2 - y^2)\frac{d^2 C_1}{dx^2} + 2 \right] (b^2 - y^2)\, dy = 0$$

therefore

$$\int_{-b}^{+b} \left[2b^2(1 - C_1) - 2y^2(1 - C_1) + \frac{d^2 C_1}{dx^2}(b^4 - 2b^2y^2 + y^4) \right] dy = 0$$

Integrating, we get

$$2b^2(1 - C_1)(2b) - 2(1 - C_1)\frac{2b^3}{3} + \frac{d^2 C_1}{dx^2}\left(2b^5 - \frac{4b^5}{3} + \frac{2b^5}{5}\right) = 0$$

therefore

$$\frac{d^2 C_1(x)}{dx^2} - \frac{5}{2b^2} C_1(x) = -\frac{5}{2b^2} \tag{5.58}$$

The general solution to the above differential equation is

$$C_1(x) = A_1 \sinh \frac{\lambda_1 x}{a} + B_1 \cosh \frac{\lambda_1 x}{a} + 1 \tag{5.59}$$

where

$$\frac{\lambda_1}{a} = \sqrt{\frac{5}{2b^2}} \tag{5.60}$$

Now we must determine A_1 and B_1 to satisfy the boundary conditions at $x = \pm a$. We thus find from Eq. (5.59) that

$$C_1(\pm a) = 0 = A_1 \sinh(\pm\lambda_1) + B_1 \cosh(\pm\lambda_1) + 1$$

Since the hyperbolic sine function is an odd function and the hyperbolic cosine is an even function we set $A_1 = 0$ and solve for B_1 from the above to get

$$B_1 = -\frac{1}{\cosh\lambda_1}$$

We can then give Ψ_1 from this calculation as follows, indicating it now as $(\Psi_1)_{\text{Kant}}$:

$$(\Psi_1)_{\text{Kant}} = \left[1 - \frac{\cosh(\lambda_1 x/a)}{\cosh\lambda_1}\right](b^2 - y^2) \tag{5.61}$$

It will be of interest to compare the results of the Ritz method with the above results for the shaft with a rectangular cross section. For a one-parameter approach we obtained for the Ritz method (see beginning of Section 5.4) the following result for Ψ_1:

$$(\Psi_1)_{\text{Ritz}} = \left[\frac{5}{4}\left(\frac{1}{b^2 + a^2}\right)(a^2 - x^2)\right](b^2 - y^2) \tag{5.62}$$

Actually, the coefficients of $(b^2 - y^2)$ in Eqs. (5.61) and (5.62) are closely related in that the latter is essentially the first term in a power series expansion of the former.

We now compute the approximation for the torsional stiffness from the one-term Kantorovich calculation. Thus:

$$D_1 = 2G\iint_R \Psi_1\, dA = 2G\int_{-b}^{+b}\int_{-a}^{+a}\left[1 - \frac{\cosh(\lambda_1 x/a)}{\cosh\lambda_1}\right](b^2 - y^2)\, dx\, dy$$

Integrating, we get

$$D_1 = \frac{16}{3}Gb^3 a\left(1 - \frac{1}{\lambda_1}\tanh\lambda_1\right) \tag{5.63}$$

In Table 5.4 we compare the results from above with the corresponding results from the Ritz method as well as the exact results from the theory of elasticity. This is done for various values of a and b.

Increased accuracy can be achieved with the Kantorovich method by using for Ψ_n the following two-parameter approximation:

$$\Psi_2 = [C_1(x)](b^2 - y^2) + [C_2(x)](b^2 - y^2)y^2 \tag{5.64}$$

TABLE 5.4
Torsional Stiffness Data for the Methods of Ritz and Kantorovich

a	b	D/G (Ritz)	D/G (Kantorovich)	D/G (exact)
1	1	2.222	2.234	2.249
2	1	7.111	7.305	7.317
3	1	12.000	12.627	12.639
4	1	16.732	17.960	17.972
5	1	21.367	23.293	23.305
6	1	25.945	28.626	28.633
7	1	30.488	33.960	33.971
8	1	35.008	39.293	39.305

5.7 CLOSURE

At the outset of this chapter we set forth various forms of the boundary-value problem for torsion, using the total potential energy and the total complementary energy functionals. We then employed the familiar Ritz method to find approximate solutions to particular linear elastic and nonlinear elastic problems. Now the chief weakness of the Ritz method is that the results obtained depend heavily on the chosen coordinate function or functions. We discussed this in Chapter 4 on beams. Because we are now dealing with functionals with two dependent variables, we can compensate for this difficulty at least along one direction by using the Kantorovich method. Here the form of the function, at least as far as one coordinate is concerned, is altered by the variational process itself, and hence the results obtained are less dependent, at least for the aforementioned coordinate, on the initial choice of the coordinate function. There is, of course, additional labor involved for these gains. We shall have occasion to further illustrate the use of the Kantorovich method in later chapters when we consider plates and elastic stability.

A desirable goal in approximation calculations is to bracket the result sought—that is, find upper and lower bounds for the desired result. In this chapter we have been able to use the Ritz method in conjunction with the total potential energy and the Trefftz method to formulate lower and upper bounds, respectively, for the torsional rigidity.

We now turn to the subject of plates, where we shall again formulate appropriate equations from the variational approach and present approximate solutions to these equations, again employing variational and closely related methods.

REFERENCES

Kantorovich, L. V., and Krylov, V. I., "Approximate Methods of Higher Analysis," Interscience, New York, 1964.

Sokolnikoff, I. S., "Mathematical Theory of Elasticity," McGraw-Hill, New York, 1956, p. 427.

Timoshenko, S., and Goodier, J. N., "Theory of Elasticity," McGraw-Hill, New York, 1951, chap. 11.

READING

Sokolnikoff, I. S., "Mathematical Theory of Elasticity," McGraw-Hill, New York, 1956.
Kantorovich, L. V., and Krylov, V. I., "Approximate Methods of Higher Analysis," Interscience, New York, 1964.
Washizu, K., "Variational Methods in Elasticity and Plasticity," Pergamon, New York, 1968.
Timoshenko, S., and Goodier, J. N., "Theory of Elasticity," McGraw-Hill, New York, 1951.

PROBLEMS

5.1 Verify Eqs. (5.20) and (5.21)

5.2 Find the total potential energy in terms of ϕ, starting with Eq. (5.4) and using Eqs. (5.3), (5.6), and (5.7*b*). Now express the quadratic functional for Eq. (5.7*a*). Show that $\pi(\phi)$ and $I(\phi)$ differ by the inclusion of a "divergence-like" expression in the volume integrand. Do $\pi(\phi)$ and $I(\phi)$ have the same Euler-Lagrange equations? Compare the boundary conditions for each case.

5.3 Estimate the torsional rigidity of a rectangular shaft $2a \times 2b$, using as a coordinate function $\cos(\pi x/2a)\cos(\pi y/2b)$. What is the maximum stress? Get D/G for $a/b = 1$ and $a = 1$. Use the stress function Ψ.

*5.4 Consider linear elastic torsion of a shaft with a triangular cross section, as shown in Fig. 5.6. Show that the function

$$\Psi_1 = C_1\left(x + \frac{a}{3}\right)\left[y - \frac{1}{\sqrt{3}}\left(\frac{2a}{3} - x\right)\right]\left[y + \frac{1}{\sqrt{3}}\left(\frac{2a}{3} - x\right)\right]$$

satisfies the boundary conditions for the stress function. Using the Ritz method, show that the torsional rigidity (for $\nu = 0.3$ and $E = 2.1 \times 10^{11}$ Pa) is $3.11 \times 10^9\, a^4$ m-N/rad (for "a" in meters).

5.5 Form an approximation for the torsional rigidity of a shaft with a cross section as shown in Fig. 5.7. The upper surface is described by a fourth-degree polynomial.

5.6 The cross section (Fig. 5.8) is parabolic on top and bottom. What are the equations of the upper and lower curves? Find an approximation of the torsional rigidity by the Ritz method, as explained in Section 5.4.

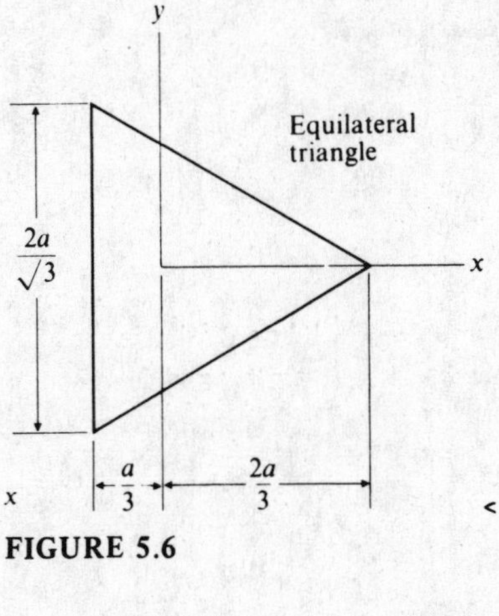

FIGURE 5.6

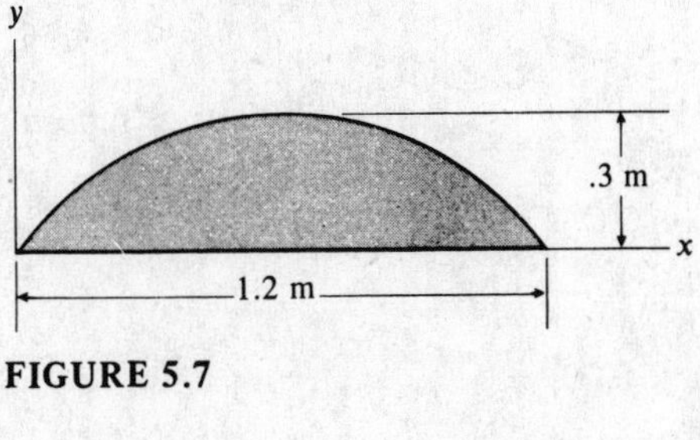

FIGURE 5.7

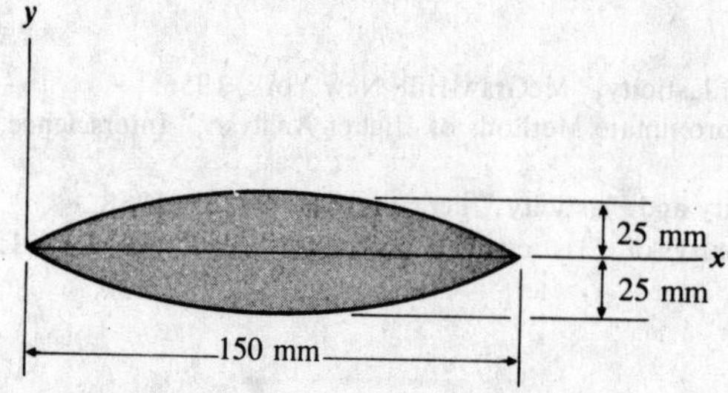

FIGURE 5.8

5.7 Show that by extremizing the functional

$$I = \iint \left[\nabla \left(\psi - \sum_{i=1}^{n} a_i v_i \right) \right]^2 dA$$

with respect to a_i we can arrive at Eq. (5.32) (which is very important in the Trefftz method). Thus the important criterion for establishing the a's for the Trefftz method can be reached by a variational process.

6

CLASSICAL THEORY OF PLATES

6.1 INTRODUCTION

In Chapter 4 on beams, we were concerned with structures having the distinguishing feature that one of the geometric dimensions dominated the configuration. This feature permitted us to make vast simplifications in that we could model the three-dimensional body by a curve and thus sharply reduce the number of variables of the problem while still obtaining important information with considerable accuracy.

We shall now consider bodies where more than one geometric dimension dominates the configuration. Specifically, we are interested in bodies bounded by surfaces whose lateral dimensions are large compared to the separation between these surfaces. When the bounding surfaces are flat the body is called a *plate*; when the surfaces are curved the body is called a *shell.* As examples of such bodies we have:

1. The hull of a ship,
2. The cover of an airplane wing,
3. A sheet pile retaining wall, and
4. Reinforced concrete roofs used in modern architecture.

It is clear from this list alone that the subject of this chapter has considerable significance for structural technology. We will develop a theory of plates, consider its validity, and then solve a number of problems. You will note that whereas we ended up working with a curve in the earlier structural considerations of Chapter 4, in the present undertaking we shall end up working with a surface.

6.2 KINEMATICS OF THE DEFORMATION OF PLATES

We shall now propose a simple mode of deformation for the plate akin to that for deformation of beams which led to the technical theory of beams. As a result of such simplifications we will need only to consider the deformation of the midplane of the plate in order to find certain significant information for the structure as a whole. Such simplifications entail the inclusion of hidden constraints in the body, as discussed in previous chapters, and the equations of equilibrium resulting from the variational process applied to the total potential energy are accordingly those for a "stiffer" system than the actual case.

We have shown a portion of a plate of thickness h in Fig. 6.1, where the xy coordinate plane corresponds to the *midplane* or *middle surface* in the undeformed geometry. On the top face of the plate we have indicated a normal load distribution $q(x, y)$, while at the edge we have shown a shear force distribution Q and a bending moment distribution M. These quantities will be discussed in the next section. Note that the bounding curve of the plate as seen in the z direction is denoted as Γ, while the interior region is denoted as R.

As a first step, we consider movements parallel to the midsurface of the plate in the undeformed geometry (xy plane). In this regard, we can consider two contributions. The first are the *stretching actions* due to loads at the edge of the plate, the loads being parallel to the midsurface of the plate. For these displacement components, denoted as $(u_1)_s$ and $(u_2)_s$, we assume that a point at position (x_1, x_2, x_3) has identically the same displacement components as the corresponding points (above or below) in the midsurface of the plate. Thus we can say

$$[u_1(x_1, x_2, x_3)]_s = [u(x, y)]_s \tag{6.1a}$$

$$[u_2(x_1, x_2, x_3)]_s = [v(x, y)]_s \tag{6.1b}$$

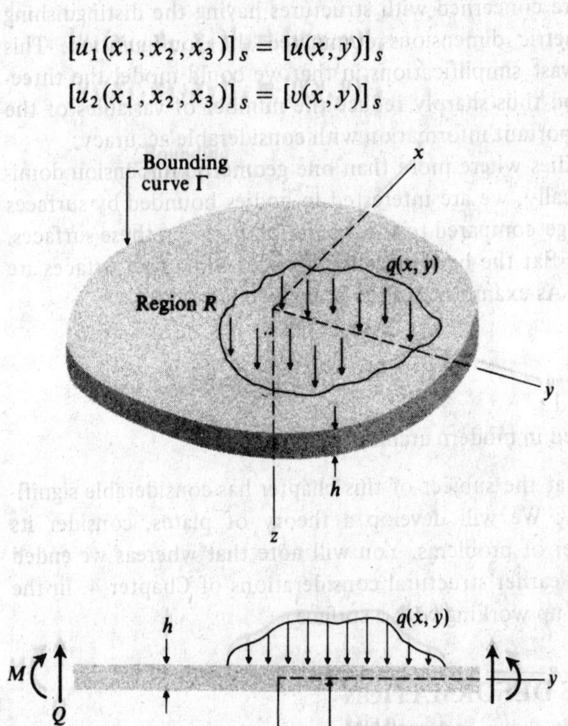

FIGURE 6.1
Plate showing external loads.

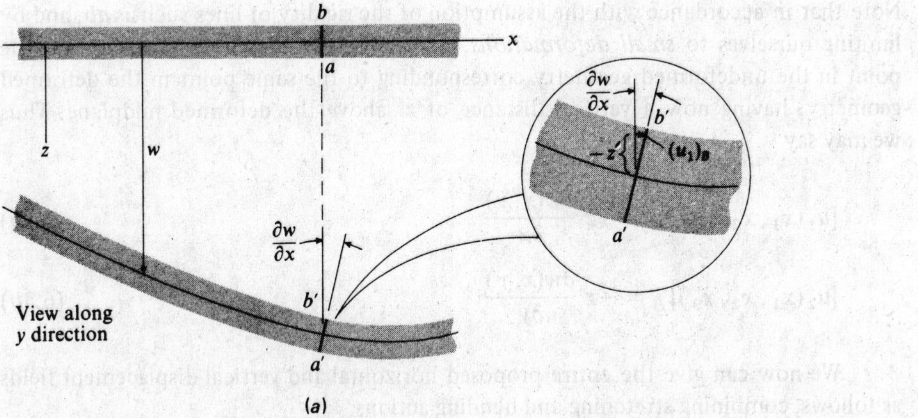

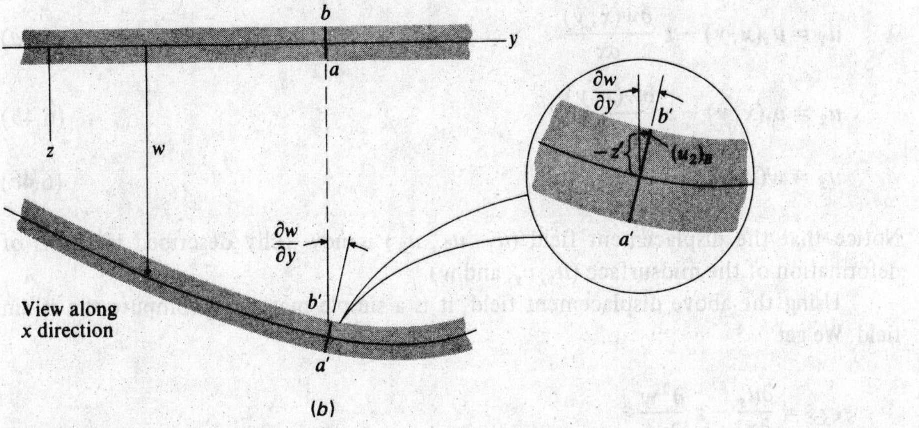

FIGURE 6.2
Views of deflected midsurface of plate.

where $(u)_s$ and $(v)_s$ refer to stretching action of the midsurface. Thus lines connecting the surfaces of the plate and normal to the xy plane in the undeformed geometry ***translate horizontally*** as a result of stretching action. The second contribution is attributed to *bending.* For this action, lines normal to the midsurface in the undeformed geometry remain normal to this surface in the deformed geometry. More specifically, lines, such as ab in Fig. 6.2, connecting the surfaces of the plate and normal to the xy plane in the undeformed geometry, translate vertically as rigid lines while maintaining the same coordinate values x and y. In addition, these lines rotate as rigid elements as a result of bending (see $a'b'$ in Fig. 6.2). The displacements in the x and y directions as a result of the bending action we denote, respectively, as $(u_1)_B$ and $(u_2)_B$. We show these quantities in the inscribed enlargements of Fig. 6.2. It should be clear that

$$[u_1(x_1, x_2, x_3)]_B = -z' \frac{\partial w(x,y)}{\partial x} \tag{6.2a}$$

$$[u_2(x_1, x_2, x_3)]_B = -z' \frac{\partial w(x,y)}{\partial y} \tag{6.2b}$$

Note that in accordance with the assumption of the rigidity of lines such as ab, and by limiting ourselves to *small deformations*, we may take z' as the coordinate z of the point in the undeformed geometry corresponding to the same point in the deformed geometry, having now a vertical distance of z' above the deformed midplane. Thus we may say

$$[u_1(x_1, x_2, x_3)]_B = -z\,\frac{\partial w(x,y)}{\partial x} \tag{6.3a}$$

$$[u_2(x_1, x_2, x_3)]_B = -z\,\frac{\partial w(x,y)}{\partial y} \tag{6.3b}$$

We now can give the entire proposed horizontal and vertical displacement fields as follows, combining stretching and bending actions,

$$u_1 = u_s(x,y) - z\,\frac{\partial w(x,y)}{\partial x} \tag{6.4a}$$

$$u_2 = v_s(x,y) - z\,\frac{\partial w(x,y)}{\partial y} \tag{6.4b}$$

$$u_3 = w(x,y) \tag{6.4c}$$

Notice that the displacement field (u_1, u_2, u_3) is now fully described in terms of deformation of the midsurface (u_s, v_s, and w).†

Using the above displacement field, it is a simple matter to compute the strain field. We get

$$\epsilon_{xx} = \frac{\partial u_s}{\partial x} - z\,\frac{\partial^2 w}{\partial x^2}$$

$$\epsilon_{yy} = \frac{\partial v_s}{\partial y} - z\,\frac{\partial^2 w}{\partial y^2}$$

$$\epsilon_{xy} = \frac{1}{2}\left(\frac{\partial u_s}{\partial y} + \frac{\partial v_s}{\partial x}\right) - z\,\frac{\partial^2 w}{\partial x\,\partial y} \tag{6.5}$$

All other strains are zero. We note immediately an obvious difficulty in that the transverse shear stresses (τ_{xz}, τ_{yz}) will be zero for the proposed displacement field. That these quantities cannot always be so is clear from simple equilibrium considerations. (Recall that we had the same kind of difficulty in the technical theory of beams.) We shall accept this discrepancy for now but will give it proper attention later when discussing "improved theories" of plates.

†It is interesting to note that Eqs. (6.4*a*) and (6.4*b*) can be considered as the first two terms of a Taylor series in powers of the thickness coordinate z. In fact, an alternative derivation of plate theory consists of expanding all quantities (stresses and displacements) into such Taylor series and then matching the coefficients of the corresponding powers of z. It is well to remember here that in retaining only the first two terms of the series we limit ourselves to plates of small thickness h compared to the other lateral dimensions of the plate surfaces.

6.3 STRESS RESULTANT INTENSITY FUNCTIONS AND THE EQUATIONS OF EQUILIBRIUM

In the study of beams we employed such resultant forces at a section as the shear force V, the bending moment M, and the axial force N. For the study of plates we find it useful to introduce *distributions* per unit length of these and other quantities.

For this purpose we have shown in Fig. 6.3 an element of a plate with stresses at the midplane of the plate (these stresses vary in the z direction over the thickness h of the plate). We now define ***shear force intensities*** Q_x and Q_y as follows:

$$Q_x = \int_{-h/2}^{h/2} \tau_{xz}\, dz \tag{6.6a}$$

$$Q_y = \int_{-h/2}^{h/2} \tau_{yz}\, dz \tag{6.6b}$$

It is clear on considering Fig. 6.3 that Q_x is the shear force distribution on a face with a normal in the x direction given per unit length in the y direction, while Q_y is the shear force distribution on a face with a normal in the y direction given per unit length in the x direction. These force intensities per unit length have been shown in Fig. 6.4. We may next introduce ***bending moment intensities*** per unit length as follows:

$$M_x = \int_{-h/2}^{h/2} \tau_{xx} z\, dz \tag{6.7a}$$

$$M_y = \int_{-h/2}^{h/2} \tau_{yy} z\, dz \tag{6.7b}$$

These quantities are shown in Fig. 6.4. It is clear by considering the definition in conjunction with Fig. 6.3 that M_x is the bending moment distribution about the y axis

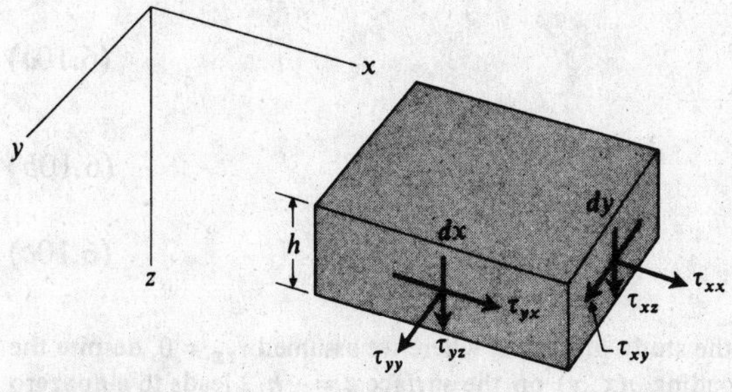

FIGURE 6.3
Plate element showing stresses.

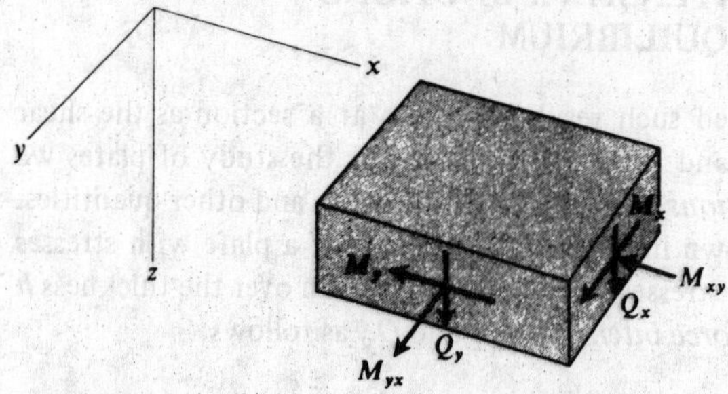

FIGURE 6.4
Plate element showing stress resultant intensities.

on a section having a normal in the x direction per unit length in the y direction, etc. Finally we introduce two ***twisting moment intensities*** per unit length as follows:

$$M_{xy} = \int_{-h/2}^{h/2} \tau_{xy} z \, dz \tag{6.8a}$$

$$M_{yx} = \int_{-h/2}^{h/2} \tau_{yx} z \, dz \tag{6.8b}$$

Again considering the definition of M_{xy} we see that it represents the twisting moment distribution about the x axis at a section whose normal is the x axis, per unit length in the y direction, etc. It is immediately apparent from the complementary property of shear stress that

$$M_{xy} = M_{yx} \tag{6.9}$$

We may readily evaluate the various moment intensity functions in terms of w for the special case of linear elastic behavior by using Hooke's law for *plane stress*† to replace stresses in the defining equations. Thus we have from Hooke's law [see Eq. (1.117)]

$$\tau_{xx} = \frac{E}{1-\nu^2}(\epsilon_{xx} + \nu\epsilon_{yy}) \tag{6.10a}$$

$$\tau_{yy} = \frac{E}{1-\nu^2}(\epsilon_{yy} + \nu\epsilon_{xx}) \tag{6.10b}$$

$$\tau_{xy} = 2G\epsilon_{xy} \tag{6.10c}$$

†This simplification is made as in the study of beams, where we assumed $\tau_{zz} = 0$, despite the fact that the application of transverse loading $q(x, y)$ on the surface $z = -h/2$ leads to a nonzero stress τ_{zz}, and also in spite of the fact that the assumed displacement field with $\epsilon_{zz} = 0$ is not a case of plane stress.

We then have

$$M_x = \int_{-h/2}^{h/2} \frac{E}{1-\nu^2}(\epsilon_{xx} + \nu\epsilon_{yy})z\,dz \tag{6.11a}$$

$$M_y = \int_{-h/2}^{h/2} \frac{E}{1-\nu^2}(\epsilon_{yy} + \nu\epsilon_{xx})z\,dz \tag{6.11b}$$

$$M_{xy} = \int_{-h/2}^{h/2} 2G\epsilon_{xy}z\,dz \tag{6.11c}$$

Now substituting for the strains from Eq. (6.5), we get the following results on carrying out the integration

$$M_x = -D\left(\frac{\partial^2 w}{\partial x^2} + \nu\frac{\partial^2 w}{\partial y^2}\right) \tag{6.12a}$$

$$M_y = -D\left(\frac{\partial^2 w}{\partial y^2} + \nu\frac{\partial^2 w}{\partial x^2}\right) \tag{6.12b}$$

$$M_{xy} = -(1-\nu)D\frac{\partial^2 w}{\partial x\,\partial y} \tag{6.12c}$$

where D, called the ***bending rigidity***, is a constant given as

$$D = \frac{Eh^3}{12(1-\nu^2)} \tag{6.13}$$

As a next step, we will formulate the stress resultant intensity functions along a section of the plane inclined arbitrarily in the ν direction relative to the x, y directions. We shall accomplish this through the stress transformation equations for the stresses illustrated in Fig. 6.5. From a two-dimensional subset of the stress transformation equations [Eqs. (1.11)] it can be seen that

$$\begin{aligned} \tau_{\nu\nu} &= a_{\nu x}^2\tau_{xx} + 2a_{\nu x}a_{\nu y}\tau_{xy} + a_{\nu y}^2\tau_{yy} \\ \tau_{\nu s} &= a_{\nu x}a_{\nu y}(\tau_{yy} - \tau_{xx}) + (a_{\nu x}^2 - a_{\nu y}^2)\tau_{xy} \end{aligned} \tag{6.14}$$

If moment resultants M_ν and $M_{\nu s}$ are defined in a manner analogous to Eqs. (6.7) and (6.8), Eqs. (6.14) may be multiplied by z and integrated over the plate thickness to yield the following moment transformation equations:

$$\begin{aligned} M_\nu &= a_{\nu x}^2 M_x + 2a_{\nu x}a_{\nu y}M_{xy} + a_{\nu y}^2 M_y \\ M_{\nu s} &= a_{\nu x}a_{\nu y}(M_y - M_x) + (a_{\nu x}^2 - a_{\nu y}^2)M_{xy} \end{aligned} \tag{6.15}$$

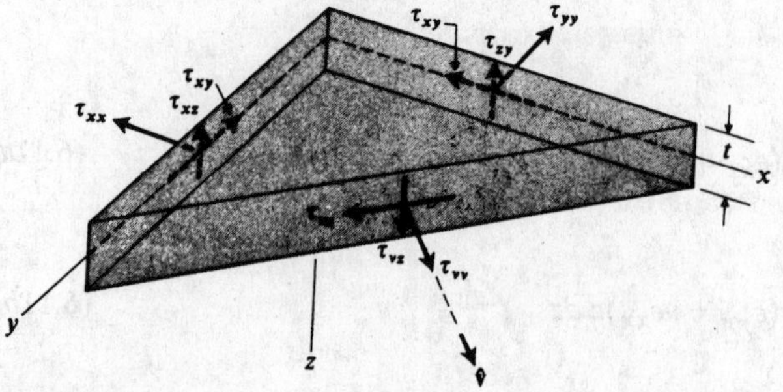

FIGURE 6.5
Stresses on an arbitrary section of a plate element.

Now consider shear stress. From simple equilibrium considerations in the vertical direction for the element in Fig. 6.5

$$\tau_{vz}\, ds = \tau_{yz}\, dx + \tau_{xz}\, dy$$

We may note that†

$$a_{vx} = \frac{dy}{ds} \qquad a_{vy} = \frac{dx}{ds}$$

Hence

$$\tau_{vz} = a_{vy}\tau_{yz} + a_{vx}\tau_{xz}$$

After defining a shear resultant Q_v as in Eqs. (6.6) and integrating over the thickness of the plate, we find

$$Q_v = a_{vy} Q_y + a_{vx} Q_x \tag{6.16}$$

Quantities Q_x, Q_y, M_x, M_y, and M_{xy} can be related by considering equilibrium of the plate element shown in Fig. 6.6, or by integrating the three-dimensional equilibrium equations of elasticity, as we did for the beam in Section 4.3. Thus for equilibrium in the x direction, in the absence of body forces,

$$\frac{\partial \tau_{xx}}{\partial x} + \frac{\partial \tau_{xy}}{\partial y} + \frac{\partial \tau_{xz}}{\partial z} = 0$$

If we multiply this equation by z and integrate over the plate thickness, noting that the operations $\partial/\partial x$ and $\partial/\partial y$ can be interchanged with the z integration, we find

$$\frac{\partial M_x}{\partial x} + \frac{\partial M_{xy}}{\partial y} + \int_{-h/2}^{h/2} z\, \frac{\partial \tau_{xz}}{\partial z}\, dz = 0$$

†Note that ds here is simply a distance increment. Later it will represent the differential of a coordinate s and then we will see that a_{vy} will be $-dx/ds$.

If the remaining thickness integral is integrated by parts,

$$\int_{-h/2}^{h/2} z \frac{\partial \tau_{xz}}{\partial z}\, dz = (z\tau_{xz})\Big|_{-h/2}^{h/2} - \int_{-h/2}^{h/2} \tau_{xz}\, dz = -Q_x$$

We have noted here the definition (6.6*a*), and that no shear stresses are applied at the plate surfaces $z = \pm h/2$. Then we find that

$$Q_x = \frac{\partial M_x}{\partial x} + \frac{\partial M_{xy}}{\partial y} \tag{6.17}$$

In exactly the same way we may integrate the equation of equilibrium in the y direction to obtain

$$Q_y = \frac{\partial M_{xy}}{\partial x} + \frac{\partial M_y}{\partial y} \tag{6.18}$$

Finally, we consider the integration over the thickness of the last equilibrium equation

$$\int_{-h/2}^{h/2} \left(\frac{\partial \tau_{xz}}{\partial x} + \frac{\partial \tau_{yz}}{\partial y} + \frac{\partial \tau_{zz}}{\partial z} \right) dz = 0$$

In view of the definition of the shear resultants, we find from above that

$$\frac{\partial Q_x}{\partial x} + \frac{\partial Q_y}{\partial y} = \tau_{zz}(z)_{z=-h/2} - \tau_{zz}(z)_{z=h/2}$$

Noting that $\tau_{zz}(z)_{z=h/2} = 0$ and that $\tau_{zz}(z)_{z=-h/2} = -q(x, y)$, we obtain finally

$$\frac{\partial Q_x}{\partial x} + \frac{\partial Q_y}{\partial y} + q(x, y) = 0 \tag{6.19}$$

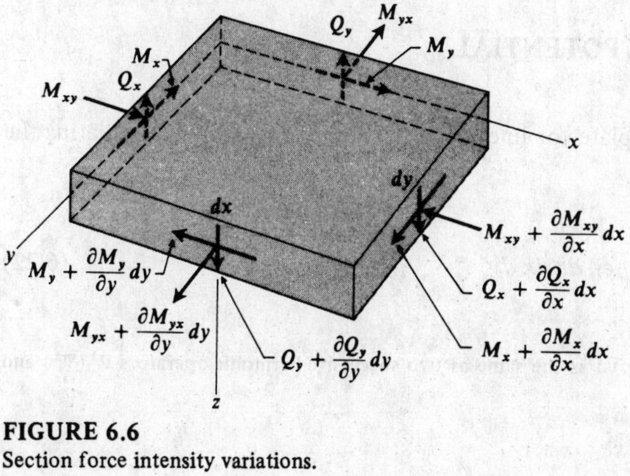

FIGURE 6.6
Section force intensity variations.

You may note that the operations carried out to obtain Eqs. (6.17)-(6.19) correspond to taking moments about the y and x axes and considering force equilibrium in the z direction. In Section 6.6 we shall return to the equilibrium equations of elasticity and integrate them in a different fashion to obtain the details of the distribution of the shear stresses through the thickness.

We may now reduce the three equations of equilibrium [Eqs. (6.17)-(6.19)] to a single equation by eliminating Q_x and Q_y to form the following equation:

$$\frac{\partial^2 M_x}{\partial x^2} + 2\frac{\partial^2 M_{xy}}{\partial x\,\partial y} + \frac{\partial^2 M_y}{\partial y^2} + q = 0 \tag{6.20}$$

where we have used the fact that $M_{xy} = M_{yx}$. Finally, using Eqs. (6.12) we may introduce the function w into the above formulation as the dependent variable. We then get the following equation:

$$\nabla^4 w = \frac{q}{D} \tag{6.21}$$

This is the nonhomogeneous *biharmonic* equation† first obtained for plate theory by Sophie Germain in 1815.

It should be apparent from the above that the results involving the stress resultant intensity function alone [Eqs. (6.15)-(6.20)] are valid for all materials. The results involving w are valid only for linear elastic materials.

Equation (6.21) will form the governing equation of the classical plate theory. Needed in addition are specifications of appropriate boundary conditions (edge conditions) for classical plate theory. In the next section, using the variational approach, we shall arrive at these boundary conditions in a straightforward manner. Sophie Germain's equation will also be derived simultaneously.

6.4 MINIMUM TOTAL POTENTIAL ENERGY APPROACH

The strain energy U of the plate for linear elastic behavior is found by evaluating the following integral:

$$U = \frac{1}{2}\iint_R \int_{-h/2}^{h/2} \tau_{ij}\epsilon_{ij}\,dz\,dx\,dy \tag{6.22}$$

†The biharmonic operator ∇^4 is the same as two successive harmonic operators $\nabla^2(\nabla^2)$ and for rectangular coordinates is

$$\frac{\partial^4}{\partial x^4} + 2\frac{\partial^2}{\partial x^2}\frac{\partial^2}{\partial y^2} + \frac{\partial^4}{\partial y^4}$$

In spite of our comments earlier concerning the transverse strains, we shall employ the state of plane stress in evaluating the above expression. Thus, replacing τ_{ij} in terms of strains [see Eq. (6.10)], we get

$$U = \frac{E}{2(1-\nu^2)} \iint_R \int_{-h/2}^{h/2} [\epsilon_{xx}^2 + 2\nu\epsilon_{xx}\epsilon_{yy} + \epsilon_{yy}^2 + 2(1-\nu)\epsilon_{xy}^2]\, dz\, dx\, dy \tag{6.23}$$

The potential energy for the external loads, meanwhile, is given as

$$V = -\iint_R q(x,y) w(x,y)\, dx\, dy \tag{6.24}$$

where the loads q are assumed to act on the midplane surface of the plate. Considering Eqs. (6.23) and (6.24) and expressing the strains in terms of the displacement field of the midsurface [Eqs. (6.5)], we get the following expression for the total potential energy:

$$\begin{aligned}\pi = \frac{E}{2(1-\nu^2)} \iint_R \int_{-h/2}^{h/2} \Bigg\{ &\left(\frac{\partial u_s}{\partial x} - z\frac{\partial^2 w}{\partial x^2}\right)^2 + \left(\frac{\partial v_s}{\partial y} - z\frac{\partial^2 w}{\partial y^2}\right)^2 \\ &+ 2\nu\left(\frac{\partial u_s}{\partial x} - z\frac{\partial^2 w}{\partial x^2}\right)\left(\frac{\partial v_s}{\partial y} - z\frac{\partial^2 w}{\partial y^2}\right) \\ &+ 2(1-\nu)\left[\frac{1}{2}\left(\frac{\partial u_s}{\partial y} + \frac{\partial v_s}{\partial x}\right) - z\frac{\partial^2 w}{\partial x\,\partial y}\right]^2 \Bigg\}\, dz\, dx\, dy \\ &- \iint_R q(x,y)w(x,y)\, dx\, dy\end{aligned} \tag{6.25}$$

Carrying out the squaring operation and integrating through the thickness while noting that

$$\int_{-h/2}^{h/2} (1, z, z^2)\, dz = \left(h, 0, \frac{h^3}{12}\right) \tag{6.26}$$

we may obtain the total potential energy in the form

$$\begin{aligned}\pi = \frac{C}{2} \iint_R &\left[\left(\frac{\partial u_s}{\partial x}\right)^2 + \left(\frac{\partial v_s}{\partial y}\right)^2 + 2\nu\frac{\partial u_s}{\partial x}\frac{\partial v_s}{\partial y} + \frac{1-\nu}{2}\left(\frac{\partial u_s}{\partial y} + \frac{\partial v_s}{\partial x}\right)^2\right] dx\, dy \\ + \frac{D}{2} \iint_R &\left[\left(\frac{\partial^2 w}{\partial x^2}\right)^2 + \left(\frac{\partial^2 w}{\partial y^2}\right)^2 + 2\nu\frac{\partial^2 w}{\partial x^2}\frac{\partial^2 w}{\partial y^2}\right. \\ &\left. + 2(1-\nu)\left(\frac{\partial^2 w}{\partial x\,\partial y}\right)^2\right] dx\, dy - \iint_R qw\, dx\, dy\end{aligned} \tag{6.27}$$

where C is a new constant called the *extensional stiffness* and is given as

$$C = \frac{Eh}{1-\nu^2} \tag{6.28}$$

The total potential energy functional has three dependent variables—the stretching components u_s and v_s and the vertical displacement variable w. We may use Eq. (2.70) limited to first-order derivatives and extended to apply to several functions. Thus we have for the Euler-Lagrange equations for u_s and v_s:

$$\frac{\partial F}{\partial u_s} - \frac{\partial}{\partial x}\frac{\partial F}{\partial(\partial u_s/\partial x)} - \frac{\partial}{\partial y}\frac{\partial F}{\partial(\partial u_s/\partial y)} = 0$$

$$\frac{\partial F}{\partial v_s} - \frac{\partial}{\partial x}\frac{\partial F}{\partial(\partial v_s/\partial x)} - \frac{\partial}{\partial y}\frac{\partial F}{\partial(\partial v_s/\partial y)} = 0$$

Substituting for F from the first integrand of Eq. (6.27), we get

$$\frac{\partial^2 u_s}{\partial x^2} + \frac{1-\nu}{2}\frac{\partial^2 u_s}{\partial y^2} + \frac{1+\nu}{2}\frac{\partial^2 v_s}{\partial x\,\partial y} = 0 \tag{6.29a}$$

$$\frac{\partial^2 v_s}{\partial y^2} + \frac{1-\nu}{2}\frac{\partial^2 v_s}{\partial x^2} + \frac{1+\nu}{2}\frac{\partial^2 u_s}{\partial x\,\partial y} = 0 \tag{6.29b}$$

Note the equations for u_s and v_s are *uncoupled* from the equation for w. This is exactly as was the case for bending of beams. We shall be concerned here with the stresses and moments in plates as a result of transverse loads. Since the strains due to this action will primarily stem from bending effects rather than stretching effects, we shall only consider the former. Accordingly, we return now to the total potential energy functional and, deleting the u_s and v_s variables, rewrite the expression in the following form:

$$\pi = \frac{D}{2}\iint_R \left\{(\nabla^2 w)^2 + 2(1-\nu)\left[\left(\frac{\partial^2 w}{\partial x\,\partial y}\right)^2 - \frac{\partial^2 w}{\partial x^2}\frac{\partial^2 w}{\partial y^2}\right]\right\} dx\,dy$$
$$- \iint_R qw\,dx\,dy \tag{6.30}$$

We now extremize the above total potential energy functional as follows, adding and subtracting $(\partial^2 w/\partial x^2)(\partial^2 w/\partial y^2)$:

$$\delta^{(1)}\pi = 0 = \frac{D}{2}\iint_R \left[2(\nabla^2 w)\left(\frac{\partial^2 \delta w}{\partial x^2} + \frac{\partial^2 \delta w}{\partial y^2}\right)\right.$$
$$+ (1-\nu)\left(2\frac{\partial^2 w}{\partial x\,\partial y}\frac{\partial^2 \delta w}{\partial x\,\partial y} + 2\frac{\partial^2 w}{\partial y\,\partial x}\frac{\partial^2 \delta w}{\partial y\,\partial x}\right.$$
$$\left.\left. - 2\frac{\partial^2 w}{\partial x^2}\frac{\partial^2 \delta w}{\partial y^2} - 2\frac{\partial^2 w}{\partial y^2}\frac{\partial^2 \delta w}{\partial x^2}\right)\right] dx\,dy - \iint_R q\,\delta w\,dx\,dy \tag{6.31}$$

where we have split up the expression $2(\partial^2 w/\partial x\,\partial y)^2$ into

$$\left(\frac{\partial^2 w}{\partial x\,\partial y}\right)^2 + \left(\frac{\partial^2 w}{\partial y\,\partial x}\right)^2$$

in carrying out the above formulation. We now employ Green's theorem successively. You may wish to take the time to justify that the following result is reached:

$$\begin{aligned}
&\iint_R (D\,\nabla^4 w - q)\,\delta w\,dx\,dy + D\oint_\Gamma \left(\frac{\partial^2 w}{\partial x^2} + \nu\frac{\partial^2 w}{\partial y^2}\right)\frac{\partial\,\delta w}{\partial x}\,dy \\
&\quad - D\oint_\Gamma \left(\frac{\partial^2 w}{\partial y^2} + \nu\frac{\partial^2 w}{\partial x^2}\right)\frac{\partial\,\delta w}{\partial y}\,dx + D\oint_\Gamma (1-\nu)\frac{\partial^2 w}{\partial x\,\partial y}\frac{\partial\,\delta w}{\partial y}\,dy \\
&\quad - D\oint_\Gamma (1-\nu)\frac{\partial^2 w}{\partial x\,\partial y}\frac{\partial\,\delta w}{\partial x}\,dx + D\oint_\Gamma \left(\frac{\partial^3 w}{\partial y^3} + \nu\frac{\partial^3 w}{\partial x^2\,\partial y}\right)\delta w\,dx \\
&\quad - D\oint_\Gamma \left(\frac{\partial^3 w}{\partial x^3} + \nu\frac{\partial^3 w}{\partial x\,\partial y^2}\right)\delta w\,dy + D\oint_\Gamma (1-\nu)\frac{\partial^3 w}{\partial x^2\,\partial y}\,\delta w\,dx \\
&\quad - D\oint_\Gamma (1-\nu)\frac{\partial^3 w}{\partial x\,\partial y^2}\,\delta w\,dy = 0
\end{aligned} \tag{6.32}$$

Now, considering Eqs. (6.12), we see that the integrands of the line integrals can be expressed in terms of the stress resultant intensity functions. Thus we may rewrite the above equation as follows:

$$\begin{aligned}
&\iint_R (D\,\nabla^4 w - q)\,\delta w\,dx\,dy - \oint_\Gamma M_x\frac{\partial\,\delta w}{\partial x}\,dy + \oint_\Gamma M_y\frac{\partial\,\delta w}{\partial y}\,dx \\
&\quad - \oint_\Gamma M_{xy}\frac{\partial\,\delta w}{\partial y}\,dy + \oint_\Gamma M_{xy}\frac{\partial\,\delta w}{\partial x}\,dx - \oint_\Gamma \left(\frac{\partial M_y}{\partial y} + \frac{\partial M_{xy}}{\partial x}\right)\delta w\,dx \\
&\quad + \oint_\Gamma \left(\frac{\partial M_x}{\partial x} + \frac{\partial M_{xy}}{\partial y}\right)\delta w\,dy = 0
\end{aligned} \tag{6.33}$$

Now, using Eqs. (6.17) and (6.18), we may rewrite the last two integrals in terms of shear force intensities. We thus have

$$\iint_R (D\nabla^4 w - q)\delta w\, dx\, dy - \oint_\Gamma M_x \frac{\partial\, \delta w}{\partial x}\, dy + \oint_\Gamma M_y \frac{\partial\, \delta w}{\partial y}\, dx$$

$$- \oint_\Gamma M_{xy} \frac{\partial\, \delta w}{\partial y}\, dy + \oint_\Gamma M_{xy} \frac{\partial\, \delta w}{\partial x}\, dx$$

$$- \oint_\Gamma Q_y\, \delta w\, dx + \oint_\Gamma Q_x\, \delta w\, dy = 0 \tag{6.34}$$

To simplify the functional further, examine a portion of the path Γ, as shown in Fig. 6.7. Considering ν and s to be a rectangular set of coordinates at a point on the boundary, we may say

$$\frac{\partial}{\partial x} = \nabla \cdot \mathbf{i} = \left(\frac{\partial}{\partial \nu}\boldsymbol{\nu} + \frac{\partial}{\partial s}\mathbf{s}\right)\cdot \mathbf{i} = (\boldsymbol{\nu}\cdot\mathbf{i})\frac{\partial}{\partial \nu} + (\mathbf{s}\cdot\mathbf{i})\frac{\partial}{\partial s}$$

$$= \cos\phi \frac{\partial}{\partial \nu} - \cos\left(\frac{\pi}{2} - \phi\right)\frac{\partial}{\partial s} = \cos\phi\frac{\partial}{\partial \nu} - \sin\phi \frac{\partial}{\partial s} \tag{6.35}$$

Hence

$$\frac{\partial}{\partial x} = a_{\nu x}\frac{\partial}{\partial \nu} - a_{\nu y}\frac{\partial}{\partial s} \tag{6.36}$$

Similarly, noting that $\partial/\partial y = \nabla \cdot \mathbf{j}$, we get

$$\frac{\partial}{\partial y} = a_{\nu y}\frac{\partial}{\partial \nu} + a_{\nu x}\frac{\partial}{\partial s} \tag{6.37}$$

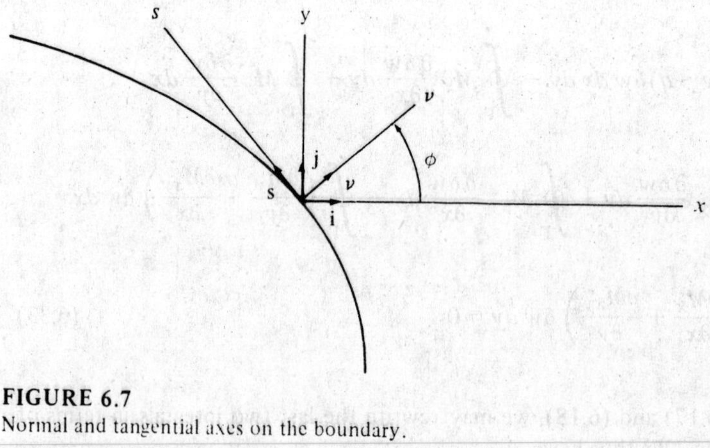

FIGURE 6.7
Normal and tangential axes on the boundary.

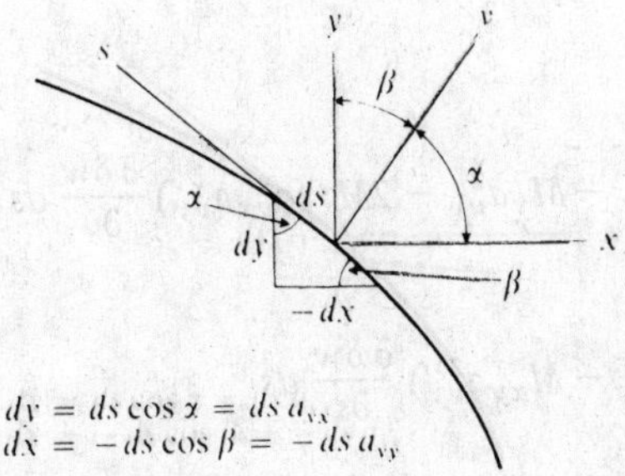

$dy = ds \cos \alpha = ds\, a_{vx}$
$dx = -ds \cos \beta = -ds\, a_{vy}$

FIGURE 6.8
Relation of differentials at boundary.

Now substitute for $\partial/\partial x$ and $\partial/\partial y$ in Eq. (6.34) in accordance with the above equations. We get, on collecting terms,

$$\iint_R (D\nabla^4 w - q)\delta w\, dx\, dy - \oint_\Gamma M_x \left(a_{\nu x} \frac{\partial\, \delta w}{\partial \nu} - a_{\nu y} \frac{\partial\, \delta w}{\partial s} \right) dy$$

$$+ \oint_\Gamma M_y \left(a_{\nu y} \frac{\partial\, \delta w}{\partial \nu} + a_{\nu x} \frac{\partial\, \delta w}{\partial s} \right) dx - \oint_\Gamma M_{xy} \left(a_{\nu y} \frac{\partial\, \delta w}{\partial \nu} + a_{\nu x} \frac{\partial\, \delta w}{\partial s} \right) dy$$

$$+ \oint_\Gamma M_{xy} \left(a_{\nu x} \frac{\partial\, \delta w}{\partial \nu} - a_{\nu y} \frac{\partial\, \delta w}{\partial s} \right) dx - \oint_\Gamma Q_y\, \delta w\, dx + \oint_\Gamma Q_x\, \delta w\, dy = 0 \tag{6.38}$$

We may introduce the differential along a boundary ds in place of dx and dy in the above formulation by noting from Fig. 6.8 that

$$dx = -ds\, a_{\nu y} \tag{6.39a}$$

$$dy = ds\, a_{\nu x} \tag{6.39b}$$

We thus have

$$\iint_R (D\nabla^4 w - q)\delta w\, dx\, dy - \oint_\Gamma M_x \left(a_{\nu x} \frac{\partial\, \delta w}{\partial \nu} - a_{\nu y} \frac{\partial\, \delta w}{\partial s} \right) a_{\nu x}\, ds$$

$$+ \oint_\Gamma M_y \left(a_{\nu y} \frac{\partial\, \delta w}{\partial \nu} + a_{\nu x} \frac{\partial\, \delta w}{\partial s} \right) (-ds\, a_{\nu y}) - \oint_\Gamma M_{xy} \left(a_{\nu y} \frac{\partial\, \delta w}{\partial \nu} + a_{\nu x} \frac{\partial\, \delta w}{\partial s} \right) ds\, a_{\nu x}$$

$$+ \oint_\Gamma M_{yx} \left(a_{\nu x} \frac{\partial\, \delta w}{\partial \nu} - a_{\nu y} \frac{\partial\, \delta w}{\partial s} \right) (-ds\, a_{\nu y}) - \oint_\Gamma Q_y\, \delta w(-ds\, a_{\nu y}) + \oint_\Gamma Q_x\, \delta w(ds\, a_{\nu x}$$

$$= 0$$

Now, collecting terms, we get

$$\iint_R (D\,\nabla^4 w - q)\,\delta w\,dx\,dy + \oint_\Gamma (-M_x a_{\nu x}^2 - M_y a_{\nu y}^2 - 2M_{xy} a_{\nu x} a_{\nu y})\,\frac{\partial\,\delta w}{\partial \nu}\,ds$$

$$+ \oint_\Gamma (M_x a_{\nu x} a_{\nu y} - M_y a_{\nu x} a_{\nu y} - M_{xy} a_{\nu x}^2 + M_{xy} a_{\nu y}^2)\,\frac{\partial\,\delta w}{\partial s}\,ds$$

$$+ \oint_\Gamma (Q_y a_{\nu y} + Q_x a_{\nu x})\,\delta w\,ds = 0$$

By employing Eqs. (6.15) and (6.16) to replace parenthetic expressions in the integrands of the line integrals, we may rewrite the above equation as follows:

$$\iint_R (D\,\nabla^4 w - q)\,\delta w\,dx\,dy - \oint_\Gamma M_\nu\,\frac{\partial\,\delta w}{\partial \nu}\,ds - \oint_\Gamma M_{\nu s}\,\frac{\partial\,\delta w}{\partial s}\,ds$$

$$+ \oint_\Gamma Q_\nu\,\delta w\,ds = 0 \tag{6.40}$$

Let us examine the third integral. Consider it to be for a moment a line integral that is not closed. Then, if $M_{\nu s}$ is continuous and the curve is smooth we may integrate by parts as follows:

$$\int_1^2 M_{\nu s}\,\frac{\partial\,\delta w}{\partial s}\,ds = (M_{\nu s}\,\delta w)|_1^2 - \int_1^2 \frac{\partial M_{\nu s}}{\partial s}\,\delta w\,ds$$

For a closed smooth curve† the expression in parentheses is clearly zero, and so we have for this case

$$\oint_\Gamma M_{\nu s}\,\frac{\partial\,\delta w}{\partial s}\,ds = -\oint_\Gamma \frac{\partial M_{\nu s}}{\partial s}\,\delta w\,ds \tag{6.41}$$

We then get for the variation of the total potential energy:

$$\iint_R (D\,\nabla^4 w - q)\,\delta w\,dx\,dy - \oint_\Gamma M_\nu\,\delta\left(\frac{\partial w}{\partial \nu}\right) ds + \oint_\Gamma \left(Q_\nu + \frac{\partial M_{\nu s}}{\partial s}\right) \delta w\,ds = 0 \tag{6.42}$$

†We shall examine later a rectangular plate with four sections of smooth curves terminating in corners. In that case the integration may be carried out over each edge of the plate and so the closed integral would have contributions from the parenthetic expression at each corner. These are the so-called corner conditions. The parenthetic expression leads to these corner conditions whenever the bounding curve of the plate is only piecewise smooth, i.e., rectangular, triangular, polygonal, etc.

It is then apparent that the Euler-Lagrange equation for this problem is

$$\nabla^4 w = \frac{q}{D} \tag{6.43}$$

the equation presented earlier. Now we get, as a result of the variational process, two sets of boundary conditions, namely the natural and the kinematic boundary conditions. Thus on Γ we require that

$$\text{EITHER } M_\nu = 0 \text{ OR } \frac{\partial w}{\partial \nu} \text{ IS PRESCRIBED}$$

$$\text{EITHER } Q_\nu + \frac{\partial M_{\nu s}}{\partial s} = 0 \text{ OR } w \text{ IS PRESCRIBED} \tag{6.44}$$

That there are only two conditions, despite the fact that there are three variables M_ν, Q_ν, and $M_{\nu s}$ present, may come as a surprise. The first condition is acceptable by physical considerations and needs no further comment. We shall now examine the second condition with a view toward reaching some physical explanation. We present for this purpose the explanation set forth by Thomson and Tait (1880) in their classical treatise *Natural Philosophy*. Accordingly, we show in Fig. 6.9 part of the edge of a plate in which two "panels" of length Δs are identified. The twisting moment is expressed in the second panel as a Taylor expansion with two terms in terms of the twisting moment in the first panel. A third panel of length Δs may be imagined at the center of the aforementioned panels. This is shown in Fig. 6.9 as *AB*. The shear force for this panel is shown as $Q_\nu \Delta s$. Now we make use of the St. Venant principle by replacing the twisting moment distribution in the original two panels by two couples (see Fig. 6.10) having forces of value $M_{\nu s}$ and $[M_{\nu s} + (\partial M_{\nu s}/\partial s)(\Delta s)]$, respectively, with a separation of Δs between the forces. Clearly, any conclusion arising from the new arrangement is valid "away" from the edge. Now we focus our attention on the center panel (between *A* and *B*). The "effective" shear force intensity Q_{eff} for this panel is then seen to be

$$Q_{\text{eff}} = \frac{1}{\Delta s}\left[Q_\nu \Delta s + \left(M_{\nu s} + \frac{\partial M_{\nu s}}{\partial s}\Delta s\right) - M_{\nu s}\right] = Q_\nu + \frac{\partial M_{\nu s}}{\partial s} \tag{6.45}$$

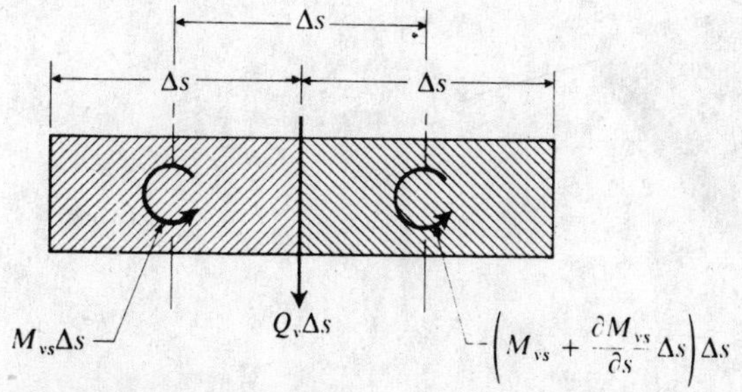

FIGURE 6.9
Edge of plate showing two panels.

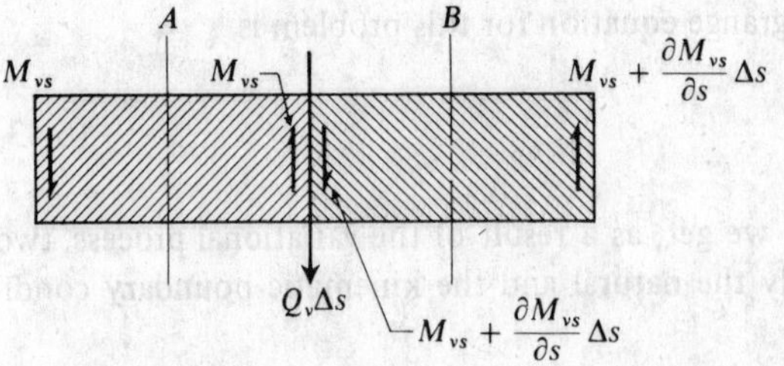

FIGURE 6.10
Replacement of twisting moment distributions by couples.

We may now conclude that the second of our natural boundary conditions renders the *effective shear force intensity equal to zero.*

Thus by the method of minimum total potential energy we have generated the entire boundary-value problem for the classical thin plate theory. Because we employed a constitutive law (Hooke's law) in the formulations, it might seem that the natural boundary conditions are restricted to Hookean materials. In the next section we shall present an alternative derivation of the boundary-value problem by the method of virtual work. No constitutive law is used and we will generate the general moment intensity and shear force intensity equations presented in Section 6.3.† Furthermore, we shall verify the fact that the natural boundary conditions presented in this section are indeed valid for all structural materials.

6.5 PRINCIPLE OF VIRTUAL WORK; RECTANGULAR PLATES

In using the principle of virtual work we shall consider a rectangular plate. In particular, we shall examine closely the corner conditions for such a problem.

We now apply the principle of virtual work, under the assumption of plane stress, to a rectangular plate (see Fig. 6.11) having dimensions $a \times b \times h$ and loaded

†In other fields of study we would not have a virtual work equation, and we would then rely on extremizing the appropriate quadratic functional to find the proper boundary conditions, as we did in Section 6.4.

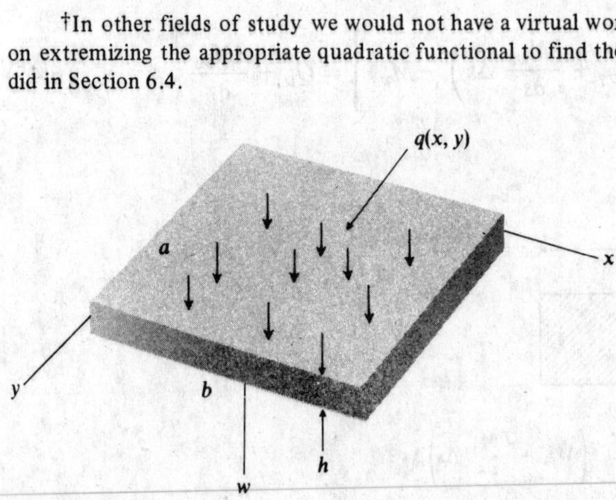

FIGURE 6.11
Rectangular plate.

normal to its centerplane by a loading intensity $q(x, y)$. Thus we have for zero body forces:

$$\int_0^a \int_0^b q(x,y)\,\delta[w(x,y)]\,dx\,dy = \int_0^a \int_0^b \int_{-h/2}^{h/2} (\tau_{xx}\,\delta\epsilon_{xx} + \tau_{yy}\,\delta\epsilon_{yy} + 2\tau_{xy}\,\delta\epsilon_{xy})\,dx\,dy\,dz \tag{6.46}$$

Now replace the strains using only the bending contributions of the displacement field [see Eq. (6.5)]. We thus get, on rearranging the equation,

$$\int_0^a \int_0^b \int_{-h/2}^{h/2} z\left(\tau_{xx}\frac{\partial^2 \delta w}{\partial x^2} + \tau_{yy}\frac{\partial^2 \delta w}{\partial y^2} + 2\tau_{xy}\frac{\partial^2 \delta w}{\partial x\,\partial y}\right) dx\,dy\,dz + \int_0^a \int_0^b q\,\delta w\,dx\,dy = 0$$

Now integrate in the first integral with respect to z and use the stress resultant intensity functions presented in Section 6.3 to arrive at the following equation:

$$\int_0^a \int_0^b \left(M_x \frac{\partial^2 \delta w}{\partial x^2} + 2M_{xy}\frac{\partial^2 \delta w}{\partial x\,\partial y} + M_y \frac{\partial^2 \delta w}{\partial y^2}\right) dx\,dy + \int_0^a \int_0^b q(x,y)\,\delta w\,dx\,dy = 0 \tag{6.47}$$

As in the previous section, we decompose $2\,\partial^2 \delta w/\partial x\,\partial y$ into $(\partial^2 \delta w/\partial x\,\partial y + \partial^2 \delta w/\partial y\,\partial x)$ and employ Green's theorem. We arrive at the following result:

$$\begin{aligned}
&\int_0^a \int_0^b \left(\frac{\partial^2 M_x}{\partial x^2} + 2\frac{\partial^2 M_{xy}}{\partial x\,\partial y} + \frac{\partial^2 M_y}{\partial y^2} + q\right)\delta w\,dx\,dy \\
&+ \int_0^b \left(M_x \frac{\partial \delta w}{\partial x}\right)\bigg|_{x=0}^{x=a} dy + \int_0^a \left(M_y \frac{\partial \delta w}{\partial y}\right)\bigg|_{y=0}^{y=b} dx \\
&- \int_0^b \left[\left(\frac{\partial M_x}{\partial x} + \frac{\partial M_{xy}}{\partial y}\right)\delta w\right]\bigg|_{x=0}^{x=a} dy \\
&- \int_0^a \left[\left(\frac{\partial M_y}{\partial y} + \frac{\partial M_{xy}}{\partial x}\right)\delta w\right]\bigg|_{y=0}^{y=b} dx \\
&+ \int_0^a \left(M_{xy}\frac{\partial \delta w}{\partial x}\right)\bigg|_{y=0}^{y=b} dx + \int_0^b \left(M_{xy}\frac{\partial \delta w}{\partial y}\right)\bigg|_{x=0}^{x=a} dy = 0
\end{aligned} \tag{6.48}$$

Now integrate the last two expressions by parts. We get

$$\int_0^a \left(M_{xy} \frac{\partial\, \delta w}{\partial x} \right) \Bigg|_{y=0}^{y=b} dx = \left[(M_{xy}\, \delta w) \Bigg|_{y=0}^{y=b} \right]_{x=0}^{x=a}$$

$$- \int_0^a \left(\frac{\partial M_{xy}}{\partial x}\, \delta w \right) \Bigg|_{y=0}^{y=b} dx = (M_{xy}\, \delta w)_{(a,b)} - (M_{xy}\, \delta w)_{(a,0)}$$

$$- (M_{xy}\, \delta w)_{(0,b)} + (M_{xy}\, \delta w)_{(0,0)} - \int_0^a \left(\frac{\partial M_{xy}}{\partial x}\, \delta w \right) \Bigg|_{y=0}^{y=b} dx$$

$$\int_0^b \left(M_{xy} \frac{\partial\, \delta w}{\partial y} \right) \Bigg|_{x=0}^{x=a} dy = \left[(M_{xy}\, \delta w) \Bigg|_{x=0}^{x=a} \right]_{y=0}^{y=b}$$

$$- \int_0^b \left(\frac{\partial M_{xy}}{\partial y}\, \delta w \right) \Bigg|_{x=0}^{x=a} dy = (M_{xy}\, \delta w)_{(a,b)} - (M_{xy}\, \delta w)_{(a,0)}$$

$$- (M_{xy}\, \delta w)_{(0,b)} + (M_{xy}\, \delta w)_{(0,0)} - \int_0^b \left(\frac{\partial M_{xy}}{\partial y}\, \delta w \right) \Bigg|_{x=0}^{x=a} dy$$

Inserting these results and noting Eqs. (6.17) and (6.18), we get

$$\int_0^a \int_0^b \left(\frac{\partial^2 M_x}{\partial x^2} + 2 \frac{\partial^2 M_{xy}}{\partial x\, \partial y} + \frac{\partial^2 M_y}{\partial y^2} + q \right) \delta w\, dx\, dy$$

$$+ \int_0^b \left[M_x\, \delta \left(\frac{\partial w}{\partial x} \right) \right]_{x=0}^{x=a} dy + \int_0^a \left[M_y\, \delta \left(\frac{\partial w}{\partial y} \right) \right]_{y=0}^{y=b} dx$$

$$- \int_0^b \left[\left(Q_x + \frac{\partial M_{xy}}{\partial y} \right) \delta w \right]_{x=0}^{x=a} dy - \int_0^a \left[\left(Q_y + \frac{\partial M_{xy}}{\partial x} \right) \delta w \right]_{y=0}^{y=b} dx$$

$$+ 2M_{xy}\, \delta w|_{(a,b)} - 2M_{xy}\, \delta w|_{(0,b)} - 2M_{xy}\, \delta w|_{(a,0)} + 2M_{xy}\, \delta w|_{(0,0)} = 0 \tag{6.49}$$

Now we can see from the above equation that the following equation must be satisfied:

$$\frac{\partial^2 M_x}{\partial x^2} + 2 \frac{\partial^2 M_{xy}}{\partial x\, \partial y} + \frac{\partial^2 M_y}{\partial y^2} + q = 0 \tag{6.50}$$

This is identical to Eq. (6.20) developed directly from equilibrium requirements and valid, consequently, for all materials. And on the boundaries of the plate we see that the following conditions must be met:

Along $x = 0$ *and* $x = a$:

$$\text{EITHER} \quad M_x = 0 \quad \text{OR} \quad \frac{\partial w}{\partial x} \quad \text{IS PRESCRIBED}$$

and

$$\text{EITHER} \quad Q_x + \frac{\partial M_{xy}}{\partial y} = 0 \quad \text{OR} \quad w \quad \text{IS PRESCRIBED}$$

Along $y = 0$ *and* $y = b$:

$$\text{EITHER} \quad M_y = 0 \quad \text{OR} \quad \frac{\partial w}{\partial y} \quad \text{IS PRESCRIBED}$$

and

$$\text{EITHER} \quad Q_y + \frac{\partial M_{xy}}{\partial x} = 0 \quad \text{OR} \quad w \quad \text{IS PRESCRIBED} \tag{6.51}$$

Thus we see that the same edge conditions result as from the analysis with minimum total potential energy. We thus fully verify in the more general undertaking the results of the more limited analysis of the previous section.

Finally, we have to consider the so-called *corner conditions.* Clearly we can either specify w at the corners, thus rendering $\delta w = 0$, or set $M_{xy} = 0$ at the corners. Beyond this, we can conclude from Eq. (6.49) that there actually exist corner forces given by the value $2M_{xy}$ at the corner.† We shall demonstrate this directly in Section 6.7. To get the proper direction of a corner force once M_{xy} has been determined, we may use the scheme of Thomson and Tait presented earlier to set forth the concept of the effective shear. Thus at a corner we use couples giving the proper directions of M_{xy} and M_{yx} on the orthogonal interfaces (see Fig. 6.12*a*). The forces coinciding at

†Expressions $2M_{xy}\,\delta w|_{a,\,b}$ etc., are just the virtual work expressions of these forces.

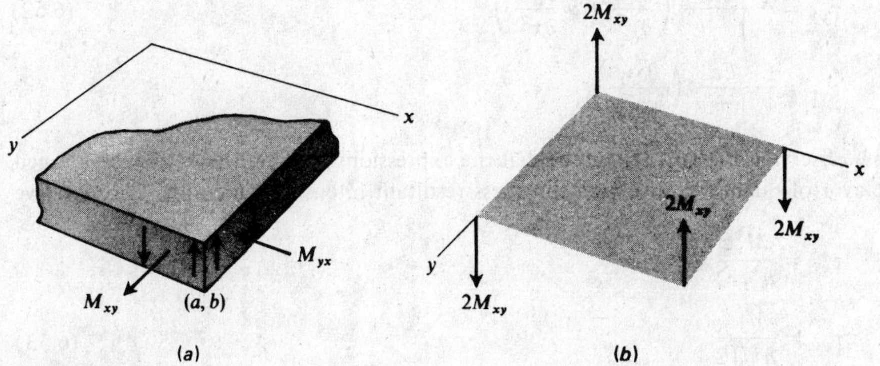

FIGURE 6.12
Corner force development.

the corner edge then give the direction of the corner force. In Fig. 6.12*b* we have shown corner forces for positive values of M_{xy} at all corners.

It is to be pointed out that all these conclusions for the corner conditions could have been reached from the formulations in the preceding section by not limiting ourselves as we did to smooth boundaries.

Now that the theory has been fully set forth we shall briefly examine some aspects of the validity of this theory in the following section.

6.6 A NOTE ON THE VALIDITY OF CLASSICAL PLATE THEORY

As in the case of the technical theory of beams, there are obvious discrepancies in the classical theory of plates that has been presented. Most notably, we have employed the assumption of plane stress in the plate, thus rendering equal to zero transverse shears τ_{zx} and τ_{zy} as well as the transverse normal stress τ_{zz}. It is obvious from equilibrium considerations that such stresses will seldom be zero. Accordingly, we need assurance that the classical plate theory can, under proper conditions, yield meaningful results.

In this regard we can follow one of two approaches. We might test our approximate results against corresponding results stemming from a more exact theory. (This was the procedure we took when we were at this juncture in the study of beams.) Or we might attempt to determine the bounds on the magnitudes of the stresses (given above) that have been deleted from the theory to show that these stresses, although not zero, are nevertheless small compared to the other stresses. In view of the scarcity of exact solutions for plate bending problems from the three-dimensional theory of elasticity, we shall follow the latter procedure.

For this purpose we substitute the assumed form for the strain components as given by Eq. (6.5) into Hooke's law for plane stress as given by Eq. (6.10). The results are stated as follows in Cartesian coordinates:

$$\begin{aligned}
\tau_{xx} &= -\frac{Ez}{1-\nu^2}\left(\frac{\partial^2 w}{\partial x^2} + \nu\frac{\partial^2 w}{\partial y^2}\right)\\
\tau_{yy} &= -\frac{Ez}{1-\nu^2}\left(\frac{\partial^2 w}{\partial y^2} + \nu\frac{\partial^2 w}{\partial x^2}\right)\\
\tau_{xy} &= -\frac{Ez}{1+\nu}\left(\frac{\partial^2 w}{\partial x\,\partial y}\right)
\end{aligned} \tag{6.52}$$

Now observing Eq. (6.12), we see that the expressions in parentheses may be replaced in favor of quantities involving the stress resultant intensity functions. Thus we have

$$\begin{aligned}
\tau_{xx} &= \frac{M_x z}{h^3/12}\\
\tau_{yy} &= \frac{M_y z}{h^3/12}\\
\tau_{xy} &= \frac{M_{xy} z}{h^3/12}
\end{aligned} \tag{6.53}$$

These results are the analogs to the flexure formula of beams and contain the key simplifications made in the classical theory of plates.

Now let us use these stresses in the equations of equilibrium with a view toward determining τ_{zx}, τ_{zy}, and τ_{zz}. Thus in the x direction we have for the case of no body forces present:

$$\frac{\partial \tau_{xx}}{\partial x} + \frac{\partial \tau_{xy}}{\partial y} + \frac{\partial \tau_{xz}}{\partial z} = 0$$

therefore

$$\frac{z}{h^3/12}\frac{\partial M_x}{\partial x} + \frac{z}{h^3/12}\frac{\partial M_{xy}}{\partial y} + \frac{\partial \tau_{xz}}{\partial z} = 0$$

Collecting terms and then using Eq. (6.17), we have

$$\frac{\partial \tau_{xz}}{\partial z} = -\frac{z}{h^3/12}\left(\frac{\partial M_x}{\partial x} + \frac{\partial M_{xy}}{\partial y}\right)$$

$$= -\frac{z}{h^3/12} Q_x$$

Now integrating with respect to z we get

$$\tau_{xz} = -\frac{z^2/2}{h^3/12} Q_x + f(x,y)$$

To determine the arbitrary function $f(x, y)$ note that when $z = \pm h/2$, $\tau_{xz} = 0$.† Hence

$$0 = -\frac{h^2/8}{h^3/12} Q_x + f$$

therefore

$$f = \tfrac{3}{2} h Q_x$$

We may thus give τ_{xz} as follows:

$$\tau_{xz} = \frac{3}{2}\frac{Q_x}{h}\left(1 - 4\frac{z^2}{h^2}\right) \tag{6.54}$$

Next, using the equilibrium equation in the y direction, we may show in a similar fashion that

$$\tau_{yz} = \frac{3}{2}\frac{Q_y}{h}\left(1 - 4\frac{z^2}{h^2}\right) \tag{6.55}$$

†We assume that no shears are applied at the top and bottom of the plate; only transverse loading $q(x, y)$ is given.

Finally, we consider the z direction and note that

$$\frac{\partial \tau_{zz}}{\partial z} = -\frac{\partial \tau_{xz}}{\partial x} - \frac{\partial \tau_{yz}}{\partial y}$$

$$= -\frac{3}{2h}\left(1 - \frac{4z^2}{h^2}\right)\left(\frac{\partial Q_x}{\partial x} + \frac{\partial Q_y}{\partial y}\right)$$

where we have used Eqs. (6.54) and (6.55) to replace the stresses on the right side of the equation. Now noting Eq. (6.19) we may replace the last quantity in parentheses by $-q$ to get the following result:

$$\frac{\partial \tau_{zz}}{\partial z} = \frac{3q}{2h}\left(1 - 4\frac{z^2}{h^2}\right) \tag{6.56}$$

Integrating, we get for τ_{zz}:

$$\tau_{zz} = \frac{3q}{2h}\left(z - \frac{4z^3}{3h^2}\right) + g(x, y)$$

The boundary conditions for τ_{zz} are

$$(a) \text{ when } z = -\frac{h}{2} \qquad \tau_{zz} = -q$$

$$(b) \text{ when } z = \frac{h}{2} \qquad \tau_{zz} = 0$$

Hence we have from condition (a):

$$-q = \frac{3q}{2h}\left(-\frac{h}{2} + \frac{4h^3}{24h^2}\right) + g(x, y) \tag{6.57}$$

therefore

$$g = -\tfrac{1}{2}q$$

Using this result for g, it is immediately clear that condition (b) is satisfied. We thus have for τ_{zz}:

$$\tau_{zz} = -q\left(\frac{1}{2} - \frac{3}{2}\frac{z}{h} + 2\frac{z^3}{h^3}\right) \tag{6.58}$$

We thus have a set of stresses τ_{xz}, τ_{yz}, and τ_{zz} [Eqs. (6.54), (6.55), and (6.58)] that are computed from rigorous equilibrium considerations, using results from a theory that neglected these very stresses. If it turns out that the stresses computed above are small, we have inner consistency in our theory. To do this, we shall next make an order-of-magnitude study of all six stresses for comparison. For this purpose consider then a portion of a plate A (see Fig. 6.13) having some dimension L that characterizes

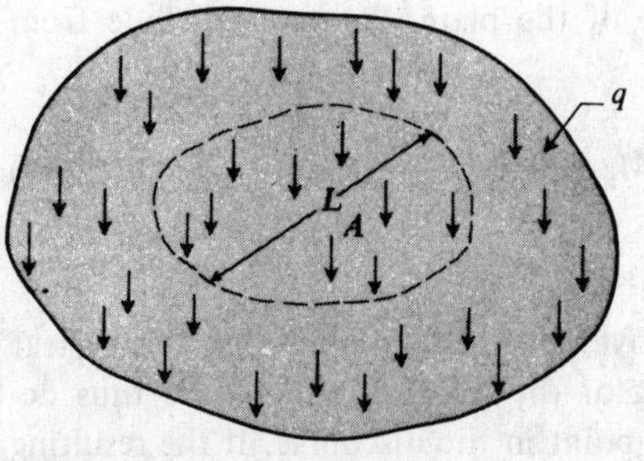

FIGURE 6.13
Plate showing inner portion for study.

lateral distances for this portion of the plate. Then we can say, taking q as some average loading intensity for the portion of the plate shown, that the order of magnitude of the total external force is

$$O(F) = O(q)O(L^2)$$

Considering equilibrium of the plate section A in the vertical direction, we can equate the transverse load given above with the resultant of shear force distribution. Thus, using the order of magnitude of the average value of Q_ν, we may write the following order-of-magnitude equation:

$$O(Q_\nu)\,O(L) = O(q)\,O(L^2)$$

therefore

$$O(Q_\nu) = O(q)\,O(L) = O(qL) \tag{6.59}$$

We have thus established the order of magnitude of an average value of Q_ν for A in terms of the order of magnitude of qL. Now consider an order-of-magnitude study of moments acting on the element. Equate moments about an axis C–C going through the line of action of the resultant of the force distribution q (see Fig. 6.14)

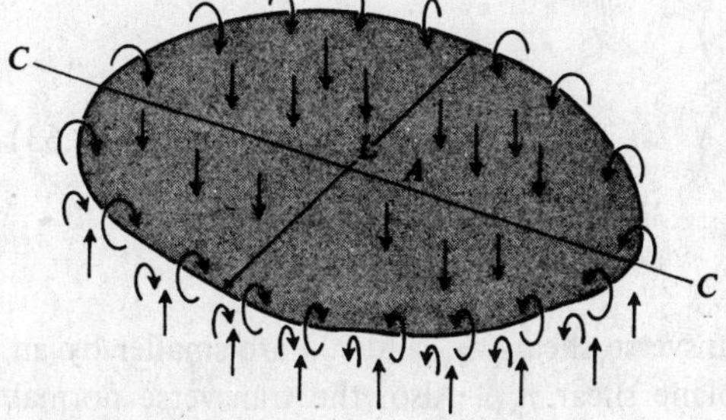

FIGURE 6.14
Inner portion of plate of dimension L.

with those of force distributions on the edge of the plate. We can conclude from equilibrium:

$$0 = \underbrace{O(Q_\nu)O(L)}_{\text{shear force}}\underbrace{O(L)}_{\text{arm}} + O(M_\nu)O(L) + O(M_{\nu s})O(L) \tag{6.60}$$

Replace $O(Q_\nu)$ by using Eq. (6.59) in the above equation. Assume next that shear effects are *as significant* as either the bending or the twisting effects. We thus deliberately make shear effects significant at this point in the discourse. If the resulting shear stresses turn out *still* to be small in our order-of-magnitude studies, then we can rest assured that this did not arise from our prematurely diminishing their significance somewhere in the development. We then equate the order of magnitude of bending and twisting moments, respectively, with that of the shear effects [using Eq. (6.59)] as they appear in the above equation:

$$O(M_\nu) = O(qL^2) \tag{6.61a}$$

$$O(M_{\nu s}) = O(qL^2) \tag{6.61b}$$

We thus have order-of-magnitude formulations for the stress resultant intensity functions Q_ν, M_ν, and $M_{\nu s}$, which may now be used in conjunction with stresses for comparison. First go back to Eqs. (6.53) to consider stresses τ_{xx}, τ_{yy}, and τ_{xy}. Since z will be of the order of magnitude of h, we can say, using Eq. (6.61),

$$\begin{aligned} O(\tau_{xx}) &= O\left(\frac{qL^2}{h^2}\right) \\ O(\tau_{yy}) &= O\left(\frac{qL^2}{h^2}\right) \\ O(\tau_{xy}) &= O\left(\frac{qL^2}{h^2}\right) \end{aligned} \tag{6.62}$$

Now considering τ_{zx}, τ_{zy}, and τ_{zz} from Eqs. (6.54), (6.55), and (6.58) in a similar fashion, we conclude, using Eqs. (6.59) and (6.61),

$$\begin{aligned} O(\tau_{zx}) &= O(qL)O\left(\frac{1}{h}\right) = O\left(q\,\frac{L}{h}\right) \\ O(\tau_{zy}) &= O(qL)O\left(\frac{1}{h}\right) = O\left(q\,\frac{L}{h}\right) \\ O(\tau_{zz}) &= O(q) \end{aligned} \tag{6.63}$$

We can now make comparisons. The transverse shears τ_{zx} and τ_{zy} are smaller by an order of magnitude h/L than the midplane shear τ_{xy}. Also, the transverse normal stress τ_{zz} is smaller by an order h^2/L^2 than the midplane normal stresses τ_{xx} and τ_{yy}. Thus for a very thin plate where $L/h \gg 10$ we have inner consistency, and the classical theory may be expected to give good results. For a moderately thin plate, especially in the vicinity of concentrated loads, it might be well to account for the effects of transverse shear deformation as we did earlier for beam theory. Improved theories derived for this purpose by Mindlin and Reissner will be presented in a later section of the chapter.

We now examine solutions of simple problems with the classical theory of plates.

6.7 EXAMPLES FROM CLASSICAL PLATE THEORY; SIMPLY SUPPORTED RECTANGULAR PLATES

In this and the following two sections we shall consider examples illustrating certain aspects of the classical theory of plates. We are by no means presenting a comprehensive discussion of the work done in applying plate theory;† rather we shall examine only salient features of the classical applications.

In this section we consider rectangular plates having as domains $0 \leqslant x \leqslant a$, $0 \leqslant y \leqslant b$ (see Fig. 6.11). We consider first the case where the edges are simply supported, which means that $w = 0$ on all edges [this is the kinematic condition of Eq. (6.51)]. Also, from Eq. (6.51) we use the natural boundary condition $M_x = 0$ for $x = 0$ and $x = a$ and $M_y = 0$ for $y = 0$ and $y = b$. But from Eq. (6.12) this means that

$$\frac{\partial^2 w}{\partial x^2} + \nu \frac{\partial^2 w}{\partial y^2} = 0 \qquad \text{for } x = 0 \text{ and } x = a \tag{6.64a}$$

$$\frac{\partial^2 w}{\partial y^2} + \nu \frac{\partial^2 w}{\partial x^2} = 0 \qquad \text{for } y = 0 \text{ and } y = b \tag{6.64b}$$

Since $w = 0$ along $x = 0$ and $x = a$ it means that $\partial^2 w/\partial y^2 = 0$ along these edges. Similarly, $\partial^2 w/\partial x^2 = 0$ along edges $y = 0$ and $y = b$. The boundary condition for the plate can then be given as follows:

Along $x = 0, x = a$:

$$w = \frac{\partial^2 w}{\partial y^2} = 0 \tag{6.65a}$$

Along $y = 0, y = b$:

$$w = \frac{\partial^2 w}{\partial x^2} = 0 \tag{6.65b}$$

Now, using as coordinate functions in the Ritz method the eigenfunctions‡ for this problem, $\sin(m\pi x/a)\sin(n\pi y/b)$, we form the following infinite sequence for w:

†For a more complete compendium of solutions to plate problems the reader is referred to Timoshenko and Woinowsky-Krieger (1959).

‡We shall discuss eigenfunctions in Chapter 7.

$$w = \sum_{m=1}^{\infty} \sum_{n=1}^{\infty} w_{mn} \sin \frac{m\pi x}{a} \sin \frac{n\pi y}{b} \tag{6.66}$$

where w_{mn} are undetermined coefficients. Now we can determine w_{mn} by minimizing the total potential energy functions in accordance with the Ritz method (Chapter 3). Since the eigenfunctions are a complete set here, and the eigenfunctions satisfy a full set of boundary conditions, we know from our earlier work that a solution so found must be an exact solution to the problem.

A simpler way to get the coefficients w_{mn}, rather than by formally using the approach of Ritz, is to proceed by expanding the loading $q(x,y)$ into a double Fourier series, using the same eigenfunctions as in Eq. (6.66). Thus, using i and j as dummy indices in place of m and n, we have

$$q(x,y) = \sum_{i=1}^{\infty} \sum_{j=1}^{\infty} q_{ij} \sin \frac{i\pi x}{a} \sin \frac{j\pi y}{b} \tag{6.67}$$

where the coefficients q_{mn} are determined by the familiar technique of multiplying both sides of the equation by $\sin(m\pi x/a) \sin(n\pi y/b)$ and integrating over the domain. Thus:

$$\int_0^a \int_0^b q(x,y) \sin \frac{m\pi x}{a} \sin \frac{n\pi y}{b}\, dx\, dy$$

$$= \int_0^a \int_0^b \sum_{i=1}^{\infty} \sum_{j=1}^{\infty} q_{ij} \sin \frac{i\pi x}{a} \sin \frac{m\pi x}{a} \sin \frac{j\pi y}{b} \sin \frac{n\pi y}{b}\, dx\, dy$$

Making use of the orthogonality properties of eigenfunctions,† we can solve for q_{mn} from the above equations to get

$$q_{mn} = \frac{4}{ab} \int_0^a \int_0^b q(x,y) \sin \frac{m\pi x}{a} \sin \frac{n\pi y}{b}\, dx\, dy \tag{6.68}$$

Therefore, we know the values of q_{ij} in Eq. (6.67). Now substituting $q(x,y)$ as given by Eq. (6.67) and w as given by Eq. (6.66) into the basic equation (6.43), we get the following result on equating the coefficients of $\sin(m\pi x/a) \sin(n\pi y/b)$:

$$w_{mn} = \frac{1}{\pi^4 D} \frac{q_{mn}}{[(m/a)^2 + (n/b)^2]^2} \tag{6.69}$$

These are the Ritz coefficients, as you may readily verify.

†The functions $f_1, f_2, \ldots, f_n$ are orthogonal in a domain if $\iint f_i f_j \, dx\, dy = 0$ for $i \neq j$ for the domain. We shall discuss this property in more detail in Chapter 7.

As a specific illustration, consider the rectangular plate of the previous discussion to have an applied loading distribution given as

$$q(x,y) = q_{11} \sin \frac{\pi x}{a} \sin \frac{\pi y}{b}$$

We see from Eqs. (6.68) and (6.69) that there is only one nonzero Ritz coefficient w_{11}, given as

$$w_{11} = \frac{1}{\pi^4 D} \frac{q_{11}}{[(1/a)^2 + (1/b)^2]^2} \tag{6.70}$$

and so the exact solution for this problem simply is

$$w = \frac{q_{11}}{\pi^4 D[(1/a)^2 + (1/b)^2]^2} \sin \frac{\pi x}{a} \sin \frac{\pi y}{b} \tag{6.71}$$

We shall now illustrate the existence of the corner forces discussed earlier. First, note that the total downward force F_D from the loading is

$$F_D = \int_0^a \int_0^b q(x,y)\, dx\, dy = \frac{4q_{11}ab}{\pi^2} \tag{6.72}$$

Next, consider the effective shear forces at the edges of the plate, as shown in Fig. 6.15. The effective shear forces $(Q_x)_{\text{eff}}$ and $(Q_y)_{\text{eff}}$ can be given as follows, by first using Eqs. (6.17) and (6.18) and then using Eqs. (6.12):

$$(Q_x)_{\text{eff}} = Q_x + \frac{\partial M_{xy}}{\partial y} = \frac{\partial M_x}{\partial x} + 2\frac{\partial M_{xy}}{\partial y}$$

$$= -D\left[\frac{\partial^3 w}{\partial x^3} + (2-\nu)\frac{\partial^3 w}{\partial x\, \partial y^2}\right]$$

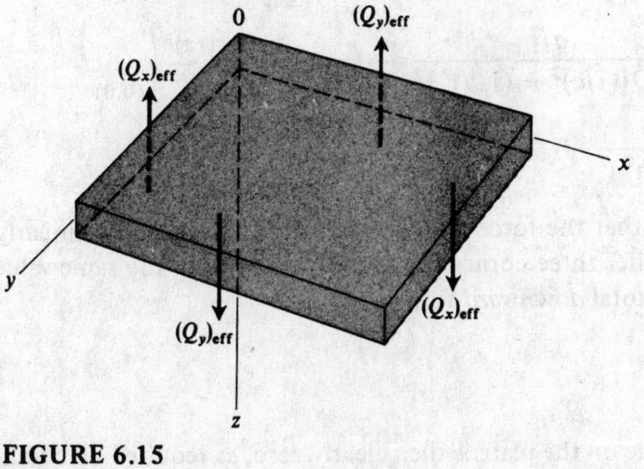

FIGURE 6.15
Rectangular plate showing effective shear.

$$(Q_y)_{\text{eff}} = Q_y + \frac{\partial M_{xy}}{\partial x} = \frac{\partial M_y}{\partial y} + 2\,\frac{\partial M_{xy}}{\partial x}$$

$$= -D\left[\frac{\partial^3 w}{\partial y^3} + (2-\nu)\,\frac{\partial^3 w}{\partial x^2\,\partial y}\right]$$

Now employing Eq. (6.71) and Eq. (6.70) to replace w for the specific problem at hand, we get

$$\begin{aligned} (Q_x)_{\text{eff}} &= Dw_{11}\left[\left(\frac{\pi}{a}\right)^3 + (2-\nu)\,\frac{\pi^3}{ab^2}\right]\cos\frac{\pi x}{a}\,\sin\frac{\pi y}{b} \\ (Q_y)_{\text{eff}} &= Dw_{11}\left[\left(\frac{\pi}{b}\right)^3 + (2-\nu)\,\frac{\pi^3}{a^2 b}\right]\sin\frac{\pi x}{a}\,\cos\frac{\pi y}{b} \end{aligned} \tag{6.73}$$

By integrating the above expressions along the edges of the plate we may determine a net vertical force F_u from the effective shear (actually, the forces on all four edges turn out to be in the upward direction). Thus, using Eq. (6.70) to replace w_{11}, we get

$$F_u = \frac{4q_{11}ab}{\pi^2} + \frac{8q_{11}(1-\nu)}{\pi^2 ab(1/a^2 + 1/b^2)^2}$$

Comparing this upward force with the downward force from the loading we see that there is a net *upward* force of

$$\frac{8q_{11}(1-\nu)}{\pi^2 ab(1/a^2 + 1/b^2)^2}$$

Clearly, the corner forces must be taken into account. These you will recall are given as $2M_{xy}$ with a sign convention given by Fig. 6.12*b*. Thus at the origin we have, using Eqs. (6.12*c*) and (6.71),

$$(F_{\text{corner}})_{(0,0)} = 2(M_{xy})_{\substack{x=0\\y=0}} = -2(1-\nu)D\left(\frac{\partial^2 w}{\partial x\,\partial y}\right)_{\substack{x=0\\y=0}}$$

$$= \left\{-2(1-\nu)D\,\frac{q_{11}}{\pi^4 D[(1/a)^2 + (1/b)^2]^2}\,\frac{\pi}{a}\,\frac{\pi}{b}\cos\frac{\pi x}{a}\cos\frac{\pi y}{b}\right\}_{(0,0)}$$

$$= -\frac{2q_{11}(1-\nu)}{\pi^2 ab(1/a^2 + 1/b^2)^2}$$

Consulting Fig. 6.12*b*, we see that the force at (0, 0) is downward. We may similarly show that the forces at the other three corners are downward, having the same value as given above. Thus there is a total *downward* force of

$$\frac{8q_{11}(1-\nu)}{\pi^2 ab(1/a^2 + 1/b^2)^2}$$

from the corners. The net force on the plate is then clearly zero, as required by equilibrium.

---

*6.8 RECTANGULAR PLATES; LÉVY'S METHOD

We now examine a more general boundary condition for the rectangular plate, namely a plate that is simply supported on two opposite edges (say at $x = 0$ and at $x = a$) while the other edges ($y = 0$ and $y = b$) are supported in an arbitrary manner. We follow the procedure set forth by the French mathematician Lévy. He assumed a solution of the form

$$w(x, y) = \sum_{m=1}^{\infty} Y_m(y) \sin \frac{m\pi x}{a} \tag{6.74}$$

where $Y_m(y)$ is yet to be determined and must satisfy appropriate boundary conditions at $y = 0$ and $y = b$. We substitute for w, as given above, into the total potential energy as given by Eq. (6.30). Thus:

$$\begin{aligned}\pi = \frac{D}{2} \int_0^b \int_0^a \Bigg\{ &\left(\sum_{m=1}^{\infty} \left[-Y_m \left(\frac{m\pi}{a}\right)^2 \sin \frac{m\pi x}{a} + Y_m'' \sin \frac{m\pi x}{a} \right] \right)^2 \\ &+ 2(1-\nu) \left[\sum_{m=1}^{\infty} Y_m' \frac{m\pi}{a} \cos \frac{m\pi x}{a} \right]^2 \\ &- 2(1-\nu) \left[\sum_{m=1}^{\infty} -Y_m \left(\frac{m\pi}{a}\right)^2 \sin \frac{m\pi x}{a} \right] \sum_{p=1}^{\infty} Y_p'' \sin \frac{p\pi x}{a} \Bigg\} dx\, dy \\ &- \int_0^b \int_0^a q \left(\sum_{m=1}^{\infty} Y_m \sin \frac{m\pi x}{a} \right) dx\, dy \end{aligned} \tag{6.75}$$

Note that in the first integration each term, after squaring operations are carried out, contains either $\sin^2 (m\pi x/a)$, $\cos^2 (m\pi x/a)$, or some product of $\sin (p\pi x/a) \sin (m\pi x/a)$ or $\cos (p\pi x/a) \cos (m\pi x/a)$ with $p \neq m$. The integration with respect to x yields $a/2$ for the squares of the sine and the cosine and yields zero for the sine and cosine products with different indices. The resulting equation can then be given as follows:

$$\begin{aligned}\pi = \frac{D}{2} \frac{a}{2} \int_0^b \Bigg\{ &\sum_{m=1}^{\infty} \left[-Y_m \left(\frac{m\pi}{a}\right)^2 + (Y_m'') \right]^2 \\ &+ 2(1-\nu) \sum_{m=1}^{\infty} (Y_m')^2 \left(\frac{m\pi}{a}\right)^2 - 2(1-\nu) \sum_{m=1}^{\infty} (-Y_m)(Y_m'') \left(\frac{m\pi}{a}\right)^2 \\ &- \sum_{m=1}^{\infty} 2q_m Y_m \Bigg\} dy \end{aligned} \tag{6.76}$$

where

$$q_m = \frac{2}{aD} \int_0^a q \sin \frac{m\pi x}{a}\, dx \tag{6.77}$$

We have here a functional with one independent variable y and with infinitely many functions Y_m of this variable having as the highest-order derivative the value two. From Eq. (2,64) we get as the appropriate Euler-Lagrange equations for extremizing the above functional

$$-\frac{d^2}{dy^2}\frac{\partial F}{\partial Y_m''}+\frac{d}{dy}\frac{\partial F}{\partial Y_m'}-\frac{\partial F}{\partial Y_m}=0 \qquad m=1,2,\ldots$$

We then get for Y_m, on substituting for F, the following ordinary differential equation:

$$Y_m^{\mathrm{IV}}-2\left(\frac{m\pi}{a}\right)^2 Y_m''+\left(\frac{m\pi}{a}\right)^4 Y_m=q_m \tag{6.78}$$

We also have from the variational process the following boundary conditions, as can readily be seen by consulting Eq. (2.63) and extrapolating the results to apply to a problem with many functions Y_m. Thus we have at the ends $y=0$ and $y=b$:

$$\begin{aligned}\frac{\partial F}{\partial Y_m''}&=0 \quad \text{OR} \quad Y_m' \quad \text{PRESCRIBED}\\ -\frac{d}{dy}\frac{\partial F}{\partial Y_m''}+\frac{\partial F}{\partial Y_m''}&=0 \quad \text{OR} \quad Y_m \quad \text{PRESCRIBED}\end{aligned} \tag{6.79}$$

Using F from Eq. (6.76), we may write these boundary conditions as follows:

$$\begin{aligned}Y_m''-\nu\left(\frac{m\pi}{a}\right)^2 Y_m&=0 \quad \text{OR} \quad Y_m' \quad \text{PRESCRIBED}\\ Y_m'''-(2-\nu)\left(\frac{m\pi}{a}\right)^2 Y_m'&=0 \quad \text{OR} \quad Y_m \quad \text{PRESCRIBED}\end{aligned} \tag{6.80}$$

The first of the above conditions represents the moment-slope duality, while the second equation gives the effective shear-displacement pair.

We thus see that the problem is reduced to solving ordinary differential equations [Eqs. (6.78)]. We illustrate this in the following example:

EXAMPLE 6.1 As a simple example to illustrate the method of Lévy, consider a rectangular plate of dimensions a $\times$ b simply supported at edges $x=0$ and $x=a$ and fixed at edges $y=0$ and $y=b$. This is shown in Fig. 6.16. A triangular load is shown with a maximum intensity of q_0 at $x=a$.

For the problem at hand we can immediately give the boundary conditions for Y_m as follows [see Eqs. (6.80)]:

$$Y_m=Y_m'=0 \quad \text{at } y=0 \text{ and } y=b \tag{a}$$

The term q_m can next readily be determined. Thus, employing $q=(x/a)q_0$ in Eq. (6.77), we get

$$\begin{aligned}q_m&=\frac{2}{aD}\int_0^a \frac{x}{a}q_0\sin\frac{m\pi x}{a}\,dx=\frac{2q_0}{a^2D}\int_0^a x\sin\frac{m\pi x}{a}\,dx\\ &=-\frac{2q_0}{m\pi D}\cos m\pi=(-1)^{m+1}\frac{2q_0}{m\pi D}\end{aligned} \tag{b}$$

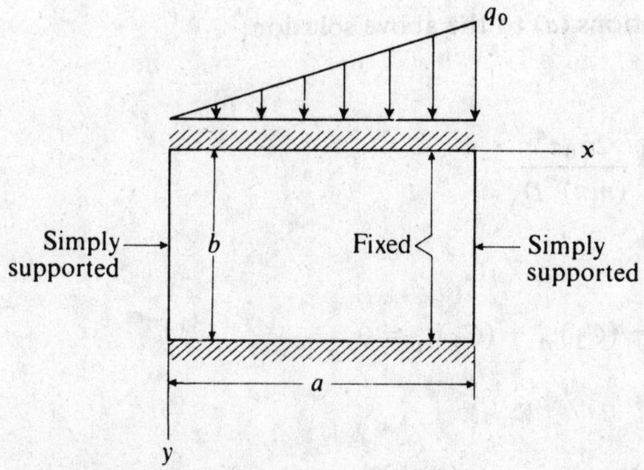

FIGURE 6.16
Rectangular plate problem.

The differential equation to be solved is then [see Eq. (6.78)]

$$Y_m^{IV} - 2\left(\frac{m\pi}{a}\right)^2 Y_m'' + \left(\frac{m\pi}{a}\right)^4 Y_m = (-1)^{m+1}\frac{2q_0}{m\pi D} \qquad (c)$$

The complementary solution is readily determined by employing the trial solution e^{py}. The characteristic equation becomes

$$p^4 - 2\left(\frac{m\pi}{a}\right)^2 p^2 + \left(\frac{m\pi}{a}\right)^4 = 0$$

therefore

$$\left[p^2 - \left(\frac{m\pi}{a}\right)^2\right]^2 = 0$$

Clearly $p = \pm m\pi/a$ has double roots. Hence we can give the complementary solution as follows:

$$(Y_m)_{\text{comp}} = (C_1)_m e^{(m\pi/a)y} + (C_2)_m e^{(-m\pi/a)y} + (C_3)_m y e^{(m\pi/a)y} + (C_4)_m y e^{(-m\pi/a)y}$$

Noting by inspection that a particular solution is

$$(Y_m)_{\text{part}} = (-1)^{m+1}\frac{2q_0 a^4}{(m\pi)^5 D}$$

we may give the general solution as follows:

$$Y_m = (C_1)_m e^{(m\pi/a)y} + (C_2)_m e^{(-m\pi/a)y} + (C_3)_m y e^{(m\pi/a)y} + (C_4)_m y e^{(-m\pi/a)y} + (-1)^{m+1}\frac{2q_0 a^4}{(m\pi)^5 D} \qquad (d)$$

Now apply the boundary conditions (*a*) to the above solution:

$\underline{Y_m = 0 \text{ at } y = 0}$

$$(C_1)_m + (C_2)_m = (-1)^m \frac{2q_0a^4}{(m\pi)^5 D}$$

$\underline{Y'_m = 0 \text{ at } y = 0}$

$$\frac{m\pi}{a}(C_1)_m - \frac{m\pi}{a}(C_2)_m + (C_3)_m + (C_4)_m = 0$$

$\underline{Y_m = 0 \text{ at } y = b}$

$$e^{(m\pi/a)b}(C_1)_m + e^{(-m\pi/a)b}(C_2)_m + be^{(m\pi/a)b}(C_3)_m + be^{(-m\pi/a)b}(C_4)_m = \frac{(-1)^m 2q_0a^4}{(m\pi)^5 D}$$

$\underline{Y'_m = 0 \text{ at } y = b}$

$$\frac{m\pi}{a}e^{(m\pi/a)b}(C_1)_m - \frac{m\pi}{a}e^{(-m\pi/a)b}(C_2)_m + e^{(m\pi/a)b}\left(1 + \frac{m\pi b}{a}\right)(C_3)_m + e^{(-m\pi/a)b}\left(1 - \frac{m\pi b}{a}\right)(C_4)_m = 0 \tag{e}$$

Let us introduce the notation next

$$\beta_m = m\pi\frac{b}{a}$$
$$\alpha_m = \frac{(-1)^m 2q_0a^4}{(m\pi)^5 D} \tag{f}$$

In matrix form Eqs. (*e*) then become

$$\begin{bmatrix} 1 & 1 & 0 & 0 \\ \frac{\beta_m}{b} & -\frac{\beta_m}{b} & 1 & 1 \\ e^{\beta_m} & e^{-\beta_m} & be^{\beta_m} & be^{-\beta_m} \\ \frac{\beta_m}{b}e^{\beta_m} & -\frac{\beta_m}{b}e^{-\beta_m} & e^{\beta_m}(1+\beta_m) & e^{-\beta_m}(1-\beta_m) \end{bmatrix} \begin{bmatrix} (C_1)_m \\ (C_2)_m \\ (C_3)_m \\ (C_4)_m \end{bmatrix} = \alpha_m \begin{bmatrix} 1 \\ 0 \\ 1 \\ 0 \end{bmatrix}$$

We get the following results for the coefficients:

$$(C_1)_m = \frac{\alpha_m[1 + 2\beta_m + 2\beta_m^2 - e^{\beta_m}(1+\beta_m) + e^{-\beta_m}(1-\beta_m) - e^{-2\beta_m}]}{2 + 4\beta_m^2 - e^{2\beta_m} - e^{-2\beta_m}}$$

$$(C_2)_m = \frac{\alpha_m[-e^{-\beta_m}(1-\beta_m) + e^{\beta_m}(1+\beta_m) - e^{2\beta_m} + 1 - 2\beta_m + 2\beta_m^2]}{2 + 4\beta_m^2 - e^{2\beta_m} - e^{-2\beta_m}}$$

$$(C_3)_m = \frac{(\alpha_m\beta_m/b)[e^{-\beta_m}(2\beta_m - 1) + e^{-2\beta_m} + e^{\beta_m} - 1 - 2\beta_m]}{2 + 4\beta_m^2 - e^{2\beta_m} - e^{-2\beta_m}}$$

$$(C_4)_m = \frac{(\alpha_m\beta_m/b)[e^{\beta_m}(1+2\beta_m) - e^{-\beta_m} - e^{2\beta_m} + 1 - 2\beta_m]}{2 + 4\beta_m^2 - e^{2\beta_m} - e^{-2\beta_m}}$$

We can then give the solution as follows:

$$w(x,y) = \sum_{m=1}^{\infty} [(C_1)_m e^{(m\pi/a)y} + (C_2)_m e^{(-m\pi/a)y} + (C_3)_m y e^{(m\pi/a)y}$$
$$+ (C_4)_m y e^{(-m\pi/a)y} + \alpha_m] \sin\frac{m\pi x}{a} \qquad (g)$$

If we wish to compute M_x and M_y we may use Eqs. (6.12*a*) and (6.12*b*). Thus

$$M_x = -D\left(\frac{\partial^2 w}{\partial x^2} + \nu\frac{\partial^2 w}{\partial y^2}\right)$$
$$M_y = -D\left(\frac{\partial^2 w}{\partial y^2} + \nu\frac{\partial^2 w}{\partial x^2}\right) \qquad (h)$$

where we may use Eq. (*g*) for *w*. In Table 6.1 we give results from the above calculations for various values of *m* in the case of a square plate ($a/b = 1$). All results apply to the center of the plate.

Timoshenko and Woinowsky-Kreiger (1959, p. 190) have given an infinite series solution to the above problem by superposing the solution for a simply supported rectangular plate under the triangular load $(x/a)q_0$ with the solution to the same problem under an applied torque distribu-

TABLE 6.1
Results from Lévy's Method

Highest value of *m*	wD/q_0a^4	M_x/q_0a^2	M_y/q_0a^2
1	9.8093×10^{-4}	1.4002×10^{-2}	1.7307×10^{-2}
2	9.8093×10^{-4}	1.4002×10^{-2}	1.7307×10^{-2}
3	9.5679×10^{-4}	1.1810×10^{-2}	1.6505×10^{-2}
4	9.5679×10^{-4}	1.1810×10^{-2}	1.6505×10^{-2}
5	9.5887×10^{-4}	1.2324×10^{-2}	1.6661×10^{-2}
6	9.5887×10^{-4}	1.2324×10^{-2}	1.6661×10^{-2}
7	9.5848×10^{-4}	1.2136×10^{-2}	1.6605×10^{-2}
8	9.5848×10^{-4}	1.2136×10^{-2}	1.6605×10^{-2}
9	9.5859×10^{-4}	1.2224×10^{-2}	1.6632×10^{-2}
10	9.5859×10^{-4}	1.2224×10^{-2}	1.6632×10^{-2}
11	9.5855×10^{-4}	1.2176×10^{-2}	1.6617×10^{-2}

TABLE 6.2.
Deflection Data for Rectangular Plate

b/a	Lévy (m goes to 11)	Finite difference
0.5	$8.1576 \times 10^{-5} q_0 a^4/D$	$10.813 \times 10^{-5} q_0 a^4/D$ (8 × 4 grid)
1	$9.5855 \times 10^{-4} q_0 a^4/D$	$11.014 \times 10^{-4} q_0 a^4/D$ (6 × 4 grid)
2	$4.2224 \times 10^{-3} q_0 a^4/D$	$4.4547 \times 10^{-3} q_0 a^4/D$ (4 × 8 grid)

tion along edges $y = 0$ and $y = a$, where the torque is of such a value as to render $\partial w/\partial y = 0$ at these edges. The results are identical to those found here since basically the same approach (sinusoidal expansions in the direction x) underlies both computations.

It may be of interest to compare results for deflection at the center of the plate obtained by the Lévy method with those stemming from a finite difference calculation for the same problem. Such results are shown in Table 6.2.

6.9 THE CLAMPED RECTANGULAR PLATE; APPROXIMATE SOLUTIONS

In this section we shall examine the case of the clamped plate having dimensions $2a \times 2b$ and subject to a uniform load distribution q_0 (see Fig. 6.17). We shall present approximate solutions via several methods and compare certain results of such computations with those from exact solutions.

A. Ritz Method

As a first step we shall use the Ritz method in conjunction with the total potential energy π [see Eq. (6.30)], which we now rewrite as

$$\pi = \frac{D}{2} \iint_R \left\{ (\nabla^2 w)^2 + 2(1-\nu)\left[\left(\frac{\partial^2 w}{\partial x\, \partial y}\right)^2 - \frac{\partial^2 w}{\partial x^2}\frac{\partial^2 w}{\partial y^2}\right] - \frac{2q_0 w}{D} \right\} dA \tag{6.81}$$

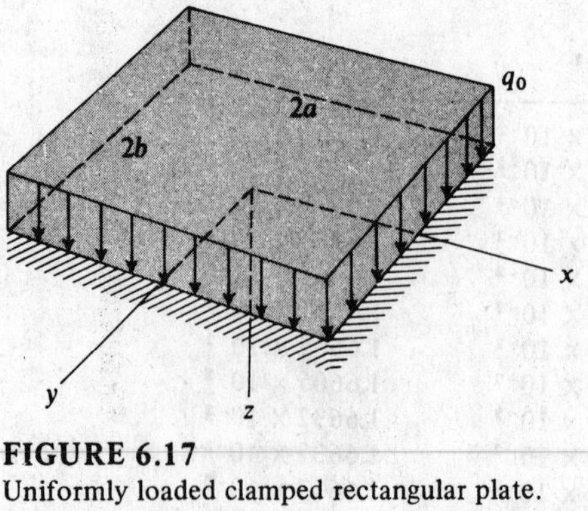

FIGURE 6.17
Uniformly loaded clamped rectangular plate.

We will first show that this integral can be considerably simplified as a result of the boundary conditions of the problem, namely:

$$\begin{aligned} w &= \frac{\partial w}{\partial x} = 0 \qquad x = \pm a \\ w &= \frac{\partial w}{\partial y} = 0 \qquad y = \pm b \end{aligned} \tag{6.82}$$

Thus consider the integral† from Eq. (6.81)

$$\iint_R \left[\left(\frac{\partial^2 w}{\partial x\, \partial y} \right)^2 - \frac{\partial^2 w}{\partial x^2} \frac{\partial^2 w}{\partial y^2} \right] dA$$

We shall first integrate by parts each expression in the integrand. Thus

$$\begin{aligned} \iint_R \left(\frac{\partial^2 w}{\partial x\, \partial y} \right)^2 dA &= \iint_R \frac{\partial^2 w}{\partial x\, \partial y} \frac{\partial}{\partial x} \frac{\partial w}{\partial y}\, dA \\ &= - \iint_R \frac{\partial^3 w}{\partial x^2\, \partial y} \frac{\partial w}{\partial y}\, dA + \oint_\Gamma \frac{\partial^2 w}{\partial x\, \partial y} \frac{\partial w}{\partial y} a_{\nu x}\, ds \\ &= - \iint_R \frac{\partial^3 w}{\partial x^2\, \partial y} \frac{\partial w}{\partial y}\, dA + \oint_\Gamma \frac{\partial^2 w}{\partial x\, \partial y} \frac{\partial w}{\partial y}\, dy \end{aligned} \tag{6.83}$$

where we have used Eq. (6.39*b*) to replace $a_{\nu x}\, ds$. Also,

$$\begin{aligned} \iint_R \frac{\partial^2 w}{\partial x^2} \frac{\partial^2 w}{\partial y^2}\, dA &= \iint_R \frac{\partial^2 w}{\partial x^2} \frac{\partial}{\partial y} \frac{\partial w}{\partial y}\, dA \\ &= - \iint_R \frac{\partial^3 w}{\partial y\, \partial x^2} \frac{\partial w}{\partial y}\, dA + \oint \frac{\partial^2 w}{\partial x^2} \frac{\partial w}{\partial y} a_{\nu y}\, ds \\ &= - \iint_R \frac{\partial^3 w}{\partial y\, \partial x^2} \frac{\partial w}{\partial y}\, dA - \oint \frac{\partial^2 w}{\partial x^2} \frac{\partial w}{\partial y}\, dx \end{aligned} \tag{6.84}$$

†The expression

$$\left(\frac{\partial^2 w}{\partial x\, \partial y} \right)^2 - \frac{\partial^2 w}{\partial x^2} \frac{\partial^2 w}{\partial y^2}$$

is called the *Gaussian curvature.* It plays an important role in the differential geometry of curved surfaces.

where we have used Eq. (6.39*a*) to replace $a_{\nu y}\, ds$. We now may form the integral of the Gaussian curvature in the following way:

$$\iint_R \left[\left(\frac{\partial^2 w}{\partial x\,\partial y}\right)^2 - \frac{\partial^2 w}{\partial x^2}\frac{\partial^2 w}{\partial y^2}\right] dA = \oint_\Gamma \frac{\partial w}{\partial y}\left[\frac{\partial}{\partial y}\left(\frac{\partial w}{\partial x}\right) dy + \frac{\partial}{\partial x}\left(\frac{\partial w}{\partial x}\right) dx\right]$$

$$= \oint_\Gamma \frac{\partial w}{\partial y}\, d\left(\frac{\partial w}{\partial x}\right) = \oint \frac{\partial w}{\partial y}\frac{d}{ds}\left(\frac{\partial w}{\partial x}\right) ds \qquad (6.85)$$

If the boundary is piecewise smooth, such as the rectangular plate we will be considering, then the line integral on the right side of the above equation is actually a sum of line integrals between the corners of the boundary. It is immediately apparent that for a clamped rectangular plate $\partial w/\partial y = 0$ on the boundaries $x = \pm a$ and $\partial w/\partial x = 0$ on the boundaries $y = \pm b$ (see Fig. 6.17). For the case at hand then we can conclude that the Gaussian curvature integral is zero and can be deleted from Eq. (6.81).

Let us consider the case of a plate with a *smooth boundary* such as an elliptic or circular plate. For this purpose we integrate the right side of Eq. (6.85) by parts

$$\iint_R \left[\left(\frac{\partial^2 w}{\partial x\,\partial y}\right)^2 - \frac{\partial^2 w}{\partial x^2}\frac{\partial^2 w}{\partial y^2}\right] dA = -\oint_\Gamma \frac{\partial w}{\partial x}\frac{d}{ds}\left(\frac{\partial w}{\partial y}\right) ds + \left(\frac{\partial w}{\partial y}\frac{\partial w}{\partial x}\right)_{\text{end pts.}}$$

It is apparent that for the case of a continuous curve the last expression is zero. Also, by expressing $(\partial^2 w/\partial x\,\partial y)^2$ in Eq. (6.83) and $(\partial^2 w/\partial x^2)(\partial^2 w/\partial y^2)$ in Eq. (6.84), respectively, as

$$\frac{\partial^2 w}{\partial x\,\partial y}\frac{\partial}{\partial y}\left(\frac{\partial w}{\partial x}\right) \qquad \text{and} \qquad \frac{\partial^2 w}{\partial y^2}\frac{\partial}{\partial x}\left(\frac{\partial w}{\partial x}\right)$$

rather than as we did, we may conclude, by proceeding as above from Eq. (6.83) to Eq. (6.85), that

$$\iint_R \left[\left(\frac{\partial^2 w}{\partial x\,\partial y}\right)^2 - \frac{\partial^2 w}{\partial x^2}\frac{\partial^2 w}{\partial y^2}\right] dA = -\oint \frac{\partial w}{\partial x}\frac{d}{ds}\left(\frac{\partial w}{\partial y}\right) ds \qquad (6.86)$$

We can then conclude from Eqs. (6.85) and (6.86) that

$$\iint_R \left[\left(\frac{\partial^2 w}{\partial x\,\partial y}\right)^2 - \frac{\partial^2 w}{\partial x^2}\frac{\partial^2 w}{\partial y^2}\right] dA = \frac{1}{2}\oint_\Gamma \left[\frac{\partial w}{\partial y}\frac{d}{ds}\left(\frac{\partial w}{\partial x}\right) - \frac{\partial w}{\partial x}\frac{d}{ds}\left(\frac{\partial w}{\partial y}\right)\right] ds \qquad (6.87)$$

We see that if

$$\frac{\partial w}{\partial x} = \frac{\partial w}{\partial y} = 0$$

at all points along the boundary then again we can delete the integral of the Gaussian curvature in the total potential energy expression. But if the above conditions hold we can also conclude

$$\frac{\partial w}{\partial \nu} = \frac{\partial w}{\partial x}\frac{dx}{d\nu} + \frac{\partial w}{\partial y}\frac{dy}{d\nu} = 0 \qquad (a)$$

$$\frac{\partial w}{\partial s} = \frac{\partial w}{\partial x}\frac{dx}{ds} + \frac{\partial w}{\partial y}\frac{dy}{ds} = 0 \qquad (b)$$

where ν and s are, respectively, normal and tangential to the boundary. The last condition (b) is satisfied if $w = 0$ on the boundary and so we can conclude from above that for a *clamped plate* ($\partial w/\partial \nu = w = 0$) having a continuous boundary we can drop the integral of the Gaussian curvature from the total potential energy expression.†

Returning to the case of the clamped rectangular plate, we then have for π

$$\pi = \frac{D}{2}\int_{-a}^{+a}\int_{-b}^{+b}\left[(\nabla^2 w)^2 - \frac{2q_0 w}{D}\right] dx\, dy \qquad (6.88)$$

For a one-parameter approach with the Ritz method we may use the following function:

$$w = C_1(x^2 - a^2)^2(y^2 - b^2)^2 \qquad (6.89)$$

Note that the boundary conditions are satisfied and the function is even in the x and y coordinates, as is to be expected of the solution. Substituting the above function into Eq. (6.88) and extremizing with respect to C_1, you may readily verify that we get

$$C_1 = 0.383\,\frac{q_0/D}{7a^4 + 4a^2b^2 + 7b^4}$$

The approximate solution is then

$$w_1 = \frac{0.383(q_0/D)}{7a^4 + 4a^2b^2 + 7b^4}(x^2 - a^2)^2(y^2 - b^2)^2 \qquad (6.90)$$

The maximum deflection occurs at the center of the plate $x = 0, y = 0$ and for a square plate $a = b$ we get from Eq. (6.90)

$$w_1 = 0.02127\,\frac{q_0 a^4}{D} \qquad (6.91)$$

†We shall make use of this in Section 6.10 when we consider clamped elliptic plates. Also, in Problem 6.20 we will ask you to prove that for *polygonal plates* the first variation of the integral of the Gaussian curvature vanishes if simply $w = 0$ on the boundary. This means we can delete the Gaussian curvature expression in π for such situations (a simply supported rectangular plate is an example) when we anticipate taking the first variation of the total potential energy.

A series solution† for the same problem gives the following result:

$$w_{\max} = 0.0202 \frac{q_0 a^4}{D} \tag{6.92}$$

We can improve the result by employing three coordinate functions as follows:

$$\begin{aligned} w_3 &= C_1 (x^2 - a^2)^2 (y^2 - b^2)^2 + C_2 x^2 (x^2 - a^2)^2 (y^2 - b^2)^2 \\ &\quad + C_3 y^2 (x^2 - a^2)^2 (y^2 - b^2)^2 \end{aligned}$$

The result of such a calculation gives

$$w_{\max} = 0.02067 \frac{q_0 a^4}{D} \tag{6.93}$$

which is closer to the series solution.

B. Galerkin's Method

We now consider the same problem using Galerkin's method. Accordingly, we go to Eq. (3.98) and apply it to the plate equation:

$$\nabla^4 w = \frac{q_0}{D} \tag{6.94}$$

to get the following equation:

$$\int_{-a}^{+a} \int_{-b}^{+b} \left(\nabla^4 \tilde{w} - \frac{q_0}{D} \right) \Phi_i \, dA = 0 \qquad i = 1, 2, \ldots \tag{6.95}$$

The boundary conditions for the problem are given as

$$\begin{aligned} w = \frac{\partial w}{\partial x} = 0 \quad & \text{at } x = -a \text{ and } x = +a \\ w = \frac{\partial w}{\partial y} = 0 \quad & \text{at } y = -b \text{ and } y = +b \end{aligned} \tag{6.96}$$

We will now introduce the nondimensional variables η and ζ defined as follows:

$$\begin{aligned} \zeta &= \frac{x}{a} \\ \eta &= \frac{y}{b} \end{aligned} \tag{6.97}$$

†See Table 35 in Timoshenko and Woinowsky-Krieger (1959). (Note we must adjust this result since our "a" is twice the "a" of Table 35.)

This means that the boundary conditions are given as

$$w = \frac{\partial w}{\partial \zeta} = 0 \quad \text{on } \zeta = \pm 1$$
$$w = \frac{\partial w}{\partial \eta} = 0 \quad \text{on } \eta = \pm 1 \tag{6.98}$$

With this in mind we now set forth the following functions:

$$\begin{aligned} g_1 &= (\zeta^2 - 1)^2 \\ g_2 &= (\zeta^3 - \zeta)^2 \\ h_1 &= (\eta^2 - 1)^2 \\ h_2 &= (\eta^3 - \eta)^2 \end{aligned} \tag{6.99}$$

We can then formulate a set of four coordinate functions Φ_i that satisfy the boundary conditions given above

$$\begin{aligned} \Phi_1 &= g_1 h_1 = (\zeta^2 - 1)^2 (\eta^2 - 1)^2 \\ \Phi_2 &= g_1 h_2 = (\zeta^2 - 1)^2 (\eta^3 - \eta)^2 \\ \Phi_3 &= g_2 h_1 = (\zeta^3 - \zeta)^2 (\eta^2 - 1)^2 \\ \Phi_4 &= g_2 h_2 = (\zeta^3 - \zeta)^2 (\eta^3 - \eta)^2 \end{aligned} \tag{6.100}$$

We can then give $\tilde{w}$ as follows:

$$\tilde{w} = C_{11} g_1 h_1 + C_{12} g_1 h_2 + C_{21} g_2 h_1 + C_{22} g_2 h_2 \tag{6.101}$$

where the C's are to be determined by the Galerkin method.

Now going back to Eq. (6.95), we need only integrate over one quadrant as a result of symmetry. (The resulting multiplier, 4, can then be discarded.) Also note that

$$\nabla^2 \equiv \frac{\partial^2}{\partial x^2} + \frac{\partial^2}{\partial y^2} = \frac{\partial^2}{a^2\, \partial \zeta^2} + \frac{\partial^2}{b^2\, \partial \eta^2}$$

We thus have the following four equations on multiplying through by b^3/a:

$$\int_0^1 \int_0^1 \left[\left(\frac{b^2}{a^2} \frac{\partial^2}{\partial \zeta^2} + \frac{\partial^2}{\partial \eta^2} \right)^2 (C_{11} g_1 h_1 + C_{12} g_1 h_2 + C_{21} g_2 h_1 + C_{22} g_2 h_2) - \frac{b^4 q_0}{D} \right] g_1 h_1 \, d\zeta \, d\eta = 0$$

$$\int_0^1 \int_0^1 \left[\left(\frac{b^2}{a^2} \frac{\partial^2}{\partial \zeta^2} + \frac{\partial^2}{\partial \eta^2} \right)^2 (C_{11} g_1 h_1 + C_{12} g_1 h_2 + C_{21} g_2 h_1 + C_{22} g_2 h_2) - \frac{b^4 q_0}{D} \right] g_1 h_2 \, d\zeta \, d\eta = 0$$

$$\int_0^1\int_0^1\left[\left(\frac{b^2}{a^2}\frac{\partial^2}{\partial\zeta^2}+\frac{\partial^2}{\partial\eta^2}\right)^2(C_{11}g_1h_1+C_{12}g_1h_2+C_{21}g_2h_1+C_{22}g_2h_2)-\frac{b^4q_0}{D}\right]g_2h_1\,d\zeta\,d\eta=0$$

$$\int_0^1\int_0^1\left[\left(\frac{b^2}{a^2}\frac{\partial^2}{\partial\zeta^2}+\frac{\partial^2}{\partial\eta^2}\right)^2(C_{11}g_1h_1+C_{12}g_1h_2+C_{21}g_2h_1+C_{22}g_2h_2)-\frac{b^4q_0}{D}\right]g_2h_2\,d\zeta\,d\eta=0$$

For a given ratio b^2/a^2 we may solve the preceding equations for coefficients C_{ij} in terms of D, b, and q_0. In the case of a *square plate* we have $a = b$ and $b^2/a^2 = 1$. The above equations become

$$\begin{bmatrix}13.3747 & 1.21588 & 1.21588 & 0.135098\\ 1.21588 & 2.60843 & 0.135098 & 0.218235\\ 1.21588 & 0.135098 & 2.60843 & 0.218235\\ 0.135088 & 0.218235 & 0.218235 & 0.118993\end{bmatrix}\begin{bmatrix}C_{11}\\ C_{12}\\ C_{21}\\ C_{22}\end{bmatrix}=\frac{a^4q_0}{D}\begin{bmatrix}0.284444\\ 0.0406349\\ 0.0406349\\ 0.00580499\end{bmatrix}$$

Solving with the aid of a computer we get

$$\begin{bmatrix}C_{11}\\ C_{12}\\ C_{21}\\ C_{22}\end{bmatrix}=\frac{a^4q_0}{D}\begin{bmatrix}0.0202319\\ 0.00534806\\ 0.00534806\\ 0.00624451\end{bmatrix}$$

The approximate solution to the problem is then given as follows:

$$\begin{aligned}\tilde{w}=\Bigg\{&(0.0202319)\left[\left(\frac{x}{a}\right)^2-1\right]^2\left[\left(\frac{y}{b}\right)^2-1\right]^2\\ &+(0.00534806)\left[\left(\frac{x}{a}\right)^2-1\right]^2\left[\left(\frac{y}{b}\right)^3-\frac{y}{b}\right]^2\\ &+(0.00534806)\left[\left(\frac{x}{b}\right)^3-\frac{x}{b}\right]^2\left[\left(\frac{y}{b}\right)^2-1\right]^2\\ &+(0.00624451)\left[\left(\frac{x}{b}\right)^3-\frac{x}{b}\right]^2\left[\left(\frac{y}{b}\right)^3-\frac{y}{b}\right]^2\Bigg\}\frac{a^4q_0}{D}\end{aligned} \quad (6.102)$$

As a check on the results we consider the deflection at the center of the plate ($x = y = 0$). We get

$$\tilde{w}_{\max} = 0.0202319 \frac{a^4 q_0}{D} \tag{6.103}$$

From an analytic series solution we have the following corresponding result:

$$w_{\max} = 0.0202 \frac{a^4 q_0}{D}$$

The results are clearly very close to each other.

C. Kantorovich's Method

As a final step in this section we turn to the method of Kantorovich for the clamped plate under a uniform load q_0. As we showed in Section 5.6, the Kantorovich method (see Kantorovich and Krylov, 1958) can be reformulated to be the solution of a Galerkin integral [see Eq. (5.56)]. Accordingly, we can use Eq. (6.95) presented for the Galerkin method in the previous section. We shall consider $(y^2 - b^2)^2$ to be the coordinate function $\Phi_1(y)$ and use for w the function w_1 given by $C_1(x)(y^2 - b^2)^2$ with the unknown function as $C_1(x)$. Thus we have initially

$$\int_{-a}^{+a} \left[\int_{-b}^{+b} \left(\nabla^4 w_1 - \frac{q_0}{D} \right) \Phi_1(y)\, dy \right] dx = 0$$

This equation will be satisfied if we set the expression in brackets equal to zero. Thus

$$\int_{-b}^{+b} \left(\nabla^4 w_1 - \frac{q_0}{D} \right) \Phi_1(y)\, dy = 0$$

Substituting for w_1 and Φ_1 we have

$$\int_{-b}^{+b} \left\} \nabla^4 [C_1(x)(y^2 - b^2)^2] - \frac{q_0}{D} \right\{ (b^2 - y^2)^2\, dy = 0$$

therefore

$$\int_{-b}^{+b} \left[24C_1 + 2(12y^2 - 4b^2) \frac{d^2 C_1}{dx^2} + (y^2 - b^2)^2 \frac{d^4 C_1}{dx^4} - \frac{q_0}{D} \right] (y^2 - b^2)^2\, dy = 0$$

Integrating and multiplying through by $315/(256b^5)$, we then get

$$b^4 \frac{d^4 C_1}{dx^4} - 6b^2 \frac{d^2 C_1}{dx^2} + \frac{63}{2} C_1 = \frac{21}{16} \frac{q_0}{D}$$

The characteristic equation for the above differential equation is given as follows:

$$b^4 p^4 - 6b^2 p^2 + \frac{63}{2} = 0$$

Solving for $p^2 b^2$ with the quadratic formula, we get

$$p^2 b^2 = 3 \pm i(4.75)$$

Hence, taking roots we get four values of p

$$p_{1,2,3,4} = \frac{1}{b}(\pm\alpha \pm i\beta) \qquad \text{where} \left\{ \begin{aligned} \alpha &= 2.075 \\ \beta &= 1.143 \end{aligned} \right.$$

The complementary solution then is

$$(C_1)_c = B_1 e^{(\alpha+i\beta)x/b} + B_2 e^{(\alpha-i\beta)x/b} + B_3 e^{(-\alpha-i\beta)x/b} + B_4 e^{(-\alpha+i\beta)x/b}$$

This may be written as follows, on expanding $e^{\pm\alpha}$ in terms of hyperbolic functions and expanding $e^{\pm i\beta}$ in terms of sines and cosines:

$$(C_1)_c = A_1 \cosh\alpha\frac{x}{b}\cos\beta\frac{x}{b} + A_2\cosh\alpha\frac{x}{b}\sin\beta\frac{x}{b} + A_3\sinh\alpha\frac{x}{b}\sin\beta\frac{x}{b}$$
$$+ A_4\sinh\alpha\frac{x}{b}\cos\beta\frac{x}{b}$$

The particular solution is

$$(C_1)_p = \frac{2}{63}\,\frac{21}{16}\,\frac{q_0}{D} = \frac{1}{24}\,\frac{q_0}{D}$$

Since we expect that $C_1(x)$ will be an even function of x, we can immediately set $A_2 = A_4 = 0$ so that the general solution for $C_1(x)$ is as follows:

$$C_1(x) = A_1\cosh\alpha\frac{x}{b}\cos\beta\frac{x}{b} + A_3\sinh\alpha\frac{x}{b}\sin\beta\frac{x}{b} + \frac{1}{24}\,\frac{q_0}{D}$$

We next use the condition $C(a) = C'(a) = 0$ to satisfy the boundary condition for w_1 on edges $x = \pm a$. In this way we get the remaining constants A_1 and A_3. Thus

$$A_1\cosh\alpha\frac{a}{b}\cos\beta\frac{a}{b} + A_3\sinh\alpha\frac{a}{b}\sin\beta\frac{a}{b} + \frac{1}{24}\,\frac{q_0}{D} = 0$$

$$A_1\frac{\alpha}{b}\sinh\alpha\frac{a}{b}\cos\beta\frac{a}{b} - A_1\frac{\alpha}{b}\cosh\alpha\frac{a}{b}\sin\beta\frac{a}{b}$$
$$+ A_3\left(-\frac{\beta}{b}\cosh\alpha\frac{a}{b}\sin\beta\frac{a}{b} + \frac{\beta}{b}\sinh\alpha\frac{a}{b}\cos\beta\frac{a}{b}\right) = 0$$

We get

$$A_1 = \frac{\gamma_1}{\gamma_0} \frac{1}{24} \frac{q_0}{D}$$

$$A_3 = \frac{\gamma_2}{\gamma_0} \frac{1}{24} \frac{q_0}{D}$$

where, taking $a/b = \mu$, we have

$$\gamma_0 = \beta \sinh \alpha\mu \cosh \alpha\mu + \alpha \sin \beta\mu \cos \beta\mu$$

$$\gamma_1 = -(\alpha \cosh \alpha\mu \sin \beta\mu + \beta \sinh \alpha\mu \cos \beta\mu)$$

$$\gamma_2 = \alpha \sinh \alpha\mu \cos \beta\mu - \beta \cosh \alpha\mu \sin \beta\mu$$

The solution stemming from this analysis is then

$$w = \frac{1}{24} \frac{q_0}{D} \left[\left(\frac{\gamma_1}{\gamma_0} \cosh \alpha \frac{x}{b} \cos \beta \frac{x}{b} + \frac{\gamma_2}{\gamma_0} \sinh \alpha \frac{x}{b} \sin \beta \frac{x}{b} \right) + 1 \right] (y^2 - b^2)^2$$

To find the maximum deflection set $x = y = 0$ above and find w. We get for a square plate, where $a = b$:

$$w_{\max} = \frac{1}{24} \frac{q_0}{D} \left(\frac{\gamma_1}{\gamma_0} + 1 \right) a^4$$

$$= \frac{1}{24} \frac{q_0 a^4}{D} (0.479) = 0.01992 \frac{q_0 a^4}{D}$$

Note that by using a single coordinate function we have achieved an accuracy with the Kantorovich method involving a variable coefficient comparable to that achieved with four coordinate functions having constant coefficients.

6.10 ELLIPTIC AND CIRCULAR PLATES

Consider an *elliptic* plate, as shown in Fig. 6.18, where the boundary is expressed as

$$\left(\frac{x}{a}\right)^2 + \left(\frac{y}{b}\right)^2 - 1 = 0$$

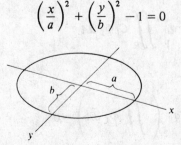

FIGURE 6.18
Elliptic plate.

We shall examine the case where the edge is clamped and a uniform load q_0 acts over the plate surface. The appropriate boundary conditions are then

$$w = \frac{\partial w}{\partial \nu} = 0 \tag{6.104}$$

Because we have a smooth boundary that is clamped, we know from the previous section that we can delete the Gaussian curvature expression from the total potential energy functional. We then have for π in this case:

$$\pi = \frac{D}{2} \iint_R (\nabla^2 w)^2 \, dx\, dy - \iint_R q_0 w \, dx\, dy \tag{6.105}$$

By using Green's theorem and the simple vector identity

$$\nabla \cdot (\Phi \mathbf{F}) = \Phi\, \nabla \cdot \mathbf{F} + \mathbf{F} \cdot \nabla \Phi$$

we ask you to demonstrate in Problem 6.13 that the above form of the total potential energy can be expressed as follows, in order to simplify later calculations:

$$\pi = \frac{D}{2} \iint_R w\, \nabla^4 w \, dx\, dy - \iint_R q_0 w \, dx\, dy \tag{6.106}$$

For the Ritz method we shall consider as a coordinate function the following function:

$$\Phi_1 = \left[\left(\frac{x}{a}\right)^2 + \left(\frac{y}{b}\right)^2 - 1 \right]^2$$

Clearly such a function satisfies the boundary conditions of the problem. We consider a one-term sequence given as follows:

$$w_1 = C_1 \left[\left(\frac{x}{a}\right)^2 + \left(\frac{y}{b}\right)^2 - 1 \right]^2 \tag{6.107}$$

Substituting into Eq. (6.106) we get

$$\pi = \frac{D}{2} \iint_R C_1^2 \Phi_1 \left(\frac{24}{a^4} + \frac{24}{b^4} + \frac{16}{a^2 b^2} \right) dA - q_0 \iint_R C_1 \Phi_1 \, dA$$

Hence, setting $\partial \pi / \partial C_1 = 0$, we have

$$DC_1 \left(\frac{24}{a^4} + \frac{24}{b^4} + \frac{16}{a^2 b^2} \right) \left(\iint_R \Phi_1 \, dA \right) - q_0 \left(\iint_R \Phi_1 \, dA \right) = 0$$

therefore

$$C_1 = \frac{q_0}{D(24/a^4 + 24/b^4 + 16/a^2b^2)}$$

Thus we get for w_1

$$w_1 = \frac{q_0}{D(24/a^4 + 24/b^4 + 16/a^2b^2)} \left[\left(\frac{x}{a}\right)^2 + \left(\frac{y}{b}\right)^2 - 1 \right]^2 \tag{6.108}$$

Actually, the above solution is an exact solution to the problem, as can readily be demonstrated by direct substitution into the differential equation

$$\nabla^4 w = \frac{q_0}{D}$$

We can then compute strains and stresses from Eq. (6.108) by employing Eqs. (6.5) and (6.10).

To get the solution for a clamped *circular* plate uniformly loaded by q_0 is now a simple matter. Say the radius of the plate is a. We merely replace "b" by "a" in Eq. (6.108) and let $x^2 + y^2 = r^2$. The result becomes

$$w = \frac{q_0}{D(64/a^4)} \left(1 - \frac{r^2}{a^2}\right)^2 \tag{6.109}$$

Other coordinate functions that may be used in the Ritz method for loadings symmetric about the major and minor diameters of the elliptic plate are

$$\Phi_n = \left[\left(\frac{x}{a}\right)^2 + \left(\frac{y}{b}\right)^2 - 1 \right]^n \tag{6.110}$$

where $n > 2$. In the case of a clamped circular plate with loadings symmetric about the origin we have as possible coordinate functions

$$\Phi_n = \left(1 - \frac{r^2}{a^2}\right)^n \tag{6.111}$$

where $n > 2$.

We now consider a circular plate that is *free* at the edges but is supported by an *elastic foundation* (see Fig. 6.19). A concentrated load P acts at the center of the plate. To account for the elastic support a potential energy term

$$U_{\text{found}} = \frac{1}{2} \iint_R kw^2 \, dA \tag{6.112}$$

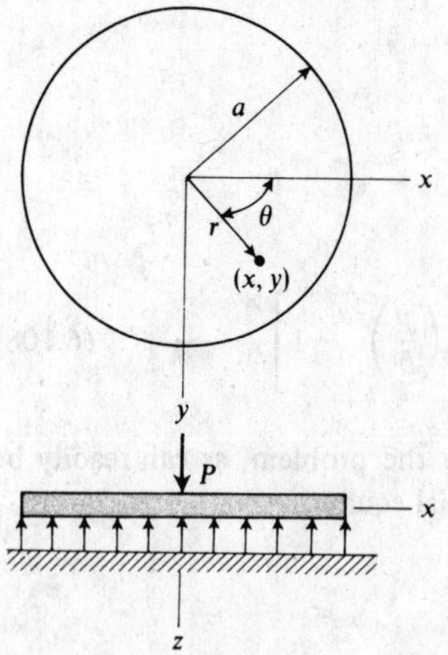

FIGURE 6.19
Circular plate on an elastic foundation.

must be added to the total potential energy, where the term k is the modulus of the foundation. Accordingly, the total potential energy functional is

$$\pi = \frac{D}{2} \iint_R \left\{ (\nabla^2 w)^2 + 2(1-\nu) \left[\left(\frac{\partial^2 w}{\partial x \, \partial y} \right)^2 - \frac{\partial^2 w}{\partial x^2} \frac{\partial^2 w}{\partial y^2} \right] \right\} dx\, dy$$

$$- \iint_R \left(qw - \frac{1}{2} kw^2 \right) dA \tag{6.113}$$

In the development of Section 6.4 the inclusion of the elastic support may be accomplished by replacing q in Eq. (6.31) by $(q - kw)$. The new Euler-Lagrange equation then becomes

$$D \nabla^4 w + kw = q$$

Solutions to the above equation can be given in terms of so-called modified Kelvin functions (see Timoshenko and Woinowsky-Krieger, 1959, chap. 8). These are rather difficult functions to use for numerical calculations. For this reason we shall consider obtaining an approximate solution by working with the total potential energy functional π. It is best to employ polar coordinates here and so we transform the coordinates from rectangular to polar in Eq. (6.113) to form the following functional:

$$\pi = \int_0^{2\pi} \int_0^a \left(\frac{D}{2} \left\{ \left(\frac{\partial^2 w}{\partial r^2} + \frac{1}{r} \frac{\partial w}{\partial r} + \frac{1}{r^2} \frac{\partial^2 w}{\partial \theta^2} \right)^2 \right. \right.$$

$$\left. -2(1-\nu) \left[\frac{\partial^2 w}{\partial r^2} \left(\frac{1}{r} \frac{\partial w}{\partial r} + \frac{1}{r^2} \frac{\partial^2 w}{\partial \theta^2} \right) - \left(\frac{1}{r} \frac{\partial^2 w}{\partial r \, \partial \theta} - \frac{1}{r^2} \frac{\partial w}{\partial \theta} \right)^2 \right] \right\}$$

$$\left. - q(r, \theta) w + k \frac{w^2}{2} \right) r\, dr\, d\theta \tag{6.114}$$

We shall consider an approximate solution of the form

$$w = A + B\left(\frac{r}{a}\right)^2 \tag{6.115}$$

where A and B are to be determined. Substituting into Eq. (6.114) we get

$$\pi = \frac{4\pi D}{a^2}(1+\nu)B^2 + \pi k\left(\frac{1}{2}A^2a^2 + \frac{1}{2}ABa^2 + \frac{1}{6}B^2a^2\right) - PA$$

Minimizing with respect to A and B yields the following pair of simultaneous algebraic equations:

$$A + B\left[\frac{2}{3} + \frac{16D(1+\nu)}{ka^4}\right] = 0 \tag{6.116a}$$

$$A + \frac{1}{2}B = \frac{P}{\pi ka^2} \tag{6.116b}$$

For a particular numerical example where

$P = 45{,}000$ N $\qquad h = 15$ mm

$k = .015$ N/mm^3 $\qquad E = 2.1 \times 10^{11}$ Pa

$a = 1.5$ m $\qquad \nu = 0.3$

we find at the center that

$$w_{r=0} = 0.00158 \text{ m}$$

A more exact solution is given by Timoshenko as

$$w_{\text{exact}} = 0.0019 \text{ m}$$

The difference is less than 17% of the exact value.

We might point out here that if we obtained moments (and stresses) from the above approximate deflection function we would be considerably in error. The reason is that, for circular plates, concentrated loads introduce singularities in the moments and shears at the position of application of the loads. Such singularities require w to have the form $r^2 \log r$. Such a form appears in the aforementioned Kelvin function but not in our approximate function.

*6.11 IMPROVED THEORY—AXISYMMETRIC CIRCULAR PLATES

In the classical theory of plates presented thus far in this chapter we have neglected transverse shear, i.e., set it equal to zero, despite the fact that we knew from equilibrium considerations that it was not zero. We then justified this step in Section 6.6

by showing that such stresses are small for thin plates ($L/h \gg 10$). At this time, we shall present an improved theory which takes into account transverse shear. This addition in our treatment of plates is analogous to the Timoshenko beam theory presented in Chapter 4.

We shall consider here the case of the axisymmetric circular plate.† The theory is due to Hencky. We proceed by introducing a function $\psi(r)$, to be determined later, such that

$$\begin{aligned} u_r &= -z\psi(r) \\ u_\theta &= 0 \\ u_z &= w(r) \end{aligned} \tag{6.117}$$

Examining Fig. 6.20 in conjunction with the above equations, we see that the assumed displacement field permits transverse area elements normal to r, such as areas ΔA, to displace vertically a distance w and at the same time, since $\partial u_r/\partial z = -\psi(r)$, to rotate about an axis in the transverse direction, θ, an amount dependent only on r (through the function ψ). This is still a simple displacement field. The strains for this displacement field (see Problem 1.17) can then be given as follows:

$$\begin{aligned} \epsilon_{rr} &= \frac{\partial u_r}{\partial r} = -z\,\frac{d\psi}{dr} = -z\psi' \\ \epsilon_{\theta\theta} &= \frac{1}{r}\,\frac{\partial u_\theta}{\partial \theta} + \frac{u_r}{r} = -\frac{z}{r}\,\psi \\ \epsilon_{zz} &= \frac{\partial u_z}{\partial z} = 0 \\ \epsilon_{r\theta} &= \frac{1}{2}\left(\frac{1}{r}\,\frac{\partial u_r}{\partial \theta} + \frac{\partial u_\theta}{\partial r} - \frac{u_\theta}{r}\right) = 0 \\ \epsilon_{rz} &= \frac{1}{2}\left(\frac{\partial u_z}{\partial r} + \frac{\partial u_r}{\partial z}\right) = \frac{1}{2}\,(w' - \psi) \\ \epsilon_{\theta z} &= \frac{1}{2}\left(\frac{\partial u_\theta}{\partial z} + \frac{1}{r}\,\frac{\partial u_z}{\partial \theta}\right) = 0 \end{aligned} \tag{6.118}$$

For the plate as follows for linear elastic behavior, we now formulate the strain energy:

$$U = \frac{1}{2}\int_0^{2\pi}\int_{r_0}^{r_1}\int_{-h/2}^{h/2} (\tau_{\theta\theta}\,\epsilon_{\theta\theta} + \tau_{rr}\epsilon_{rr} + \tau_{zz}\,\epsilon_{zz} + 2\tau_{r\theta}\,\epsilon_{r\theta} + 2\tau_{rz}\,\epsilon_{rz} + 2\tau_{\theta z}\,\epsilon_{\theta z})\,d\theta\; r\,dr\,dz$$

where r_0 and r_1 are the inner and outer radii of the plate. (We assume for now that there may be a hole at the center of the plate of radius r_0.) We thus have, after retaining only nonzero terms and integrating with respect to θ,

$$U = \pi \int_{r_0}^{r_1}\int_{-h/2}^{h/2} (\tau_{rr}\epsilon_{rr} + \tau_{\theta\theta}\,\epsilon_{\theta\theta} + 2\tau_{rz}\,\epsilon_{rz})r\,dr\,dz \tag{6.119}$$

†We shall consider the rectangular plate in this regard in Chapter 7, when we consider the vibration of plates.

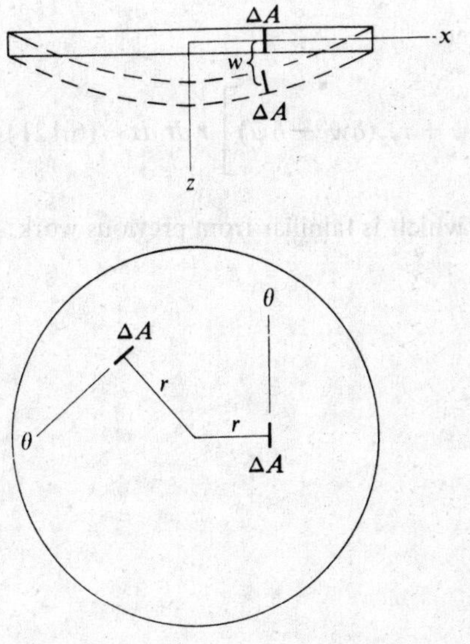

FIGURE 6.20
Axisymmetric circular plate.

We are thus retaining transverse shear effects. Now employ Hooke's law, given as

$$\tau_{rr} = \frac{E}{1-\nu^2}(\epsilon_{rr} + \nu\epsilon_{\theta\theta}) \tag{6.120a}$$

$$\tau_{\theta\theta} = \frac{E}{1-\nu^2}(\epsilon_{\theta\theta} + \nu\epsilon_{rr}) \tag{6.120b}$$

$$\tau_{rz} = 2G\epsilon_{rz} \tag{6.120c}$$

Note that the transverse shear stress, as a result of our assumptions, is uniform with respect to thickness. We shall later take into account the fact that the transverse shear stress and strain do vary over the thickness. Substituting into Eq. (6.119), we have

$$U = \pi \int_{r_0}^{r_1} \int_{-h/2}^{h/2} \left\{ \frac{E}{1-\nu^2} [(\epsilon_{rr}^2 + \nu\epsilon_{\theta\theta}\epsilon_{rr}) + (\epsilon_{\theta\theta}^2 + \nu\epsilon_{rr}\epsilon_{\theta\theta})] + 4G\epsilon_{rz}^2 \right\} r\,dr\,dz$$

Taking the first variation, we get

$$\delta^{(1)}U = \pi \int_{r_0}^{r_1} \int_{-h/2}^{h/2} \left\{ \frac{2E}{1-\nu^2} [(\epsilon_{rr} + \nu\epsilon_{\theta\theta})\delta\epsilon_{rr} + (\epsilon_{\theta\theta} + \nu\epsilon_{rr})\delta\epsilon_{\theta\theta}] + 8G\epsilon_{rz}\,\delta\epsilon_{rz} \right\} r\,dr\,dz$$

Noting Eqs. (6.120), we can replace the coefficients of the strain variations with stresses to form

$$\delta^{(1)}U = 2\pi \int_{r_0}^{r_1} \int_{-h/2}^{h/2} (\tau_{rr}\,\delta\epsilon_{rr} + \tau_{\theta\theta}\,\delta\epsilon_{\theta\theta} + 2\tau_{rz}\,\delta\epsilon_{rz}) r\,dr\,dz$$

Now replacing the strains with Eqs. (6.118) we get

$$\delta^{(1)}U = 2\pi \int_{r_0}^{r_1} \int_{-h/2}^{h/2} \left[-z\tau_{rr}\,\delta\psi' - \frac{z}{r}\,\tau_{\theta\theta}\,\delta\psi + \tau_{rz}(\delta w' - \delta\psi) \right] r\,dr\,dz \quad (6.121)$$

We now introduce the following nomenclature, which is familiar from previous work:

$$M_r = \int_{-h/2}^{h/2} z\tau_{rr}\,dz$$

$$M_\theta = \int_{-h/2}^{h/2} z\tau_{\theta\theta}\,dz$$

$$Q = \int_{-h/2}^{h/2} \tau_{rz}\,dz$$

Integrating with respect to z in Eq. (6.121) while noting that ψ and w are not functions of z and using the above results, we get

$$\delta^{(1)}U = 2\pi \int_{r_0}^{r_1} \left[-M_r\,\delta\psi' - \frac{M_\theta}{r}\,\delta\psi + Q\,\delta(w' - \psi) \right] r\,dr$$

Note next that

$$\delta^{(1)}V = -2\pi \int_{r_0}^{r_1} q\,\delta w\,r\,dr$$

For the first variation of total potential energy $(U + V)$ we then have

$$\delta^{(1)}(U + V) = 2\pi \int_{r_0}^{r_1} \left[-M_r\,\delta\psi' - \frac{M_\theta}{r}\,\delta\psi + Q\,\delta(w' - \psi) - q\,\delta w \right] r\,dr = 0$$

This may be written as

$$\int_{r_0}^{r_1} \left[-rM_r\,\frac{d}{dr}(\delta\psi) - M_\theta\,\delta\psi + rQ\,\frac{d}{dr}(\delta w) - rQ\,\delta\psi - rq\,\delta w \right] dr = 0$$

Integrating the first and third terms by parts and collecting terms we get

$$\int_{r_0}^{r_1} [(rM_r)' - M_\theta - rQ]\,\delta\psi\,dr + \int_{r_0}^{r_1} [-(rQ)' - rq]\,\delta w\,dr + (-rM_r)\,\delta\psi \Big|_{r_0}^{r_1}$$

$$+ rQ\,\delta w \Big|_{r_0}^{r_1} = 0$$

The following, then, are the Euler-Lagrange equations:

$$rM_r' + M_r - M_\theta - rQ = 0 \tag{6.122a}$$

or

$$M_r' + \frac{M_r - M_\theta}{r} = Q \tag{6.122b}$$

And

$$[-(rQ)' - rq] = 0 \tag{6.123a}$$

or

$$\frac{1}{r}\frac{d}{dr}(rQ) = -q \tag{6.123b}$$

The boundary conditions are at $r = r_0, r_1$:

$$\begin{aligned} rM_r &= 0 \quad \text{OR} \quad \psi \text{ IS PRESCRIBED} \\ rQ &= 0 \quad \text{OR} \quad w \text{ IS PRESCRIBED} \end{aligned} \tag{6.124}$$

In order to express the plate equations in terms of displacements, note from Eqs. (6.120*a*) and (6.118) that

$$M_r = \int_{-h/2}^{h/2} z\tau_{rr}\, dz = \int_{-h/2}^{h/2} \frac{E}{1-\nu^2}(\epsilon_{rr} + \nu\epsilon_{\theta\theta})z\, dz$$

$$= \frac{E}{1-\nu^2}\left(-\psi' - \nu\frac{\psi}{r}\right)\int_{-h/2}^{h/2} z^2\, dz = -D\left(\psi' + \nu\frac{\psi}{r}\right)$$

Similarly,

$$M_\theta = -D\left(\frac{\psi}{r} + \nu\psi'\right)$$

As for Q, we shall now introduce the correction mentioned earlier to account for variation of transverse shear over the thickness of the plate. Thus, as we did in Timoshenko's improved theory, we proceed as follows:

$$Q = \int_{-h/2}^{h/2} \tau_{rz}\, dz = k(\tau_{rz})_{z=0}h$$

We are accordingly assuming here that a constant factor k, properly chosen, can account for the nonuniformity of shear across the thickness everywhere in the plate. Now using Hooke's law and Eq. (6.118), we have for the above equation

$$Q = kG(w' - \psi)h \tag{6.125}$$

We can now express Eqs. (6.122) and (6.123) as follows, using preceding results for M_r, M_θ, and Q:

$$-D\frac{d}{dr}\left(\psi' + \nu\frac{\psi}{r}\right) + \frac{D}{r}\left(-\psi' - \nu\frac{\psi}{r} + \nu\psi' + \frac{\psi}{r}\right) = hkG(w' - \psi)$$

$$\frac{1}{r}\frac{d}{dr}[hkGr(w' - \psi)] = -q$$

The first of the above equations may be simplified in form to

$$-D\left(\psi'' + \frac{\psi'}{r} - \frac{\psi}{r^2}\right) = kG(w' - \psi)h \tag{6.126}$$

As for the second equation, we have on integrating:

$$hkGr(w' - \psi) = -\int_{r_0}^{r} q(\xi)\xi\, d\xi + C_1$$

To evaluate C_1, we restrict the discussion at this time to a whole plate (no hole)—hence $r_0 = 0$. Accordingly, setting $r_0 = 0$ in the above equation and excluding a point load at the center, we see that $C_1 = 0$. We can solve for ψ from the above equation as follows:

$$\psi = w' + \frac{1}{hkGr}\int_0^r q(\xi)\xi\, d\xi \tag{6.127}$$

Hence, once we know w, we can readily determine ψ from the above equation. We shall accordingly concentrate on Eq. (6.126) with a view toward eliminating ψ in an expeditious manner. For this purpose we return to Eq. (6.125) and solve for ψ as follows:

$$\psi = w' - \frac{Q}{kGh} \tag{6.128}$$

Substituting into Eq. (6.126), we get the following result:

$$-D\left(w''' - \frac{Q''}{hkG} + \frac{w''}{r} - \frac{Q'}{rkGh} - \frac{w'}{r^2} + \frac{Q}{hkGr^2}\right) = Q$$

Rearranging the terms we get

$$-D\left(w''' + \frac{w''}{r} - \frac{w'}{r^2}\right) = Q - \frac{D}{hkG}\left(Q'' + \frac{Q'}{r} - \frac{Q}{r^2}\right) \tag{6.129}$$

Now multiply the above equation by r and then apply the operator $-(1/r)\,d/dr$ as follows:

$$D\frac{1}{r}\frac{d}{dr}\left(rw''' + w'' - \frac{w'}{r}\right) \equiv D\left(\frac{1}{r}\frac{d}{dr} + \frac{d^2}{dr^2}\right)\left(\frac{1}{r}\frac{d}{dr} + \frac{d^2}{dr^2}\right)w$$

$$\equiv D\nabla^2\nabla^2 w = D\nabla^4 w$$

$$= -\frac{1}{r}\frac{d}{dr}(rQ) + \frac{D}{hkG}\frac{1}{r}\frac{d}{dr}\left(rQ'' + Q' - \frac{Q}{r}\right) \tag{6.130}$$

But from Eq. (6.123*b*), $-(1/r)\,d(rQ)/dr = q$. Also note from Eq. (6.123*b*) that

$$\nabla^2 q = \frac{1}{r}\frac{d}{dr}\left(r\frac{dq}{dr}\right) = \frac{1}{r}\frac{d}{dr}\left[r\frac{d}{dr}\left(-\frac{1}{r}\frac{d}{dr}(rQ)\right)\right]$$

$$= -\frac{1}{r}\frac{d}{dr}\left[r\frac{d}{dr}\left(Q' + \frac{Q}{r}\right)\right] = -\frac{1}{r}\frac{d}{dr}\left(rQ'' + Q' - \frac{Q}{r}\right)$$

We can then write Eq. (6.130) as follows, using the above results:

$$D\nabla^4 w = q - \frac{D}{hkG}\nabla^2 q$$

therefore

$$\boxed{D\nabla^4 w + \frac{D}{hkG}\nabla^2 q - q = 0} \tag{6.131}$$

We thus have an equation in terms of w only, including now the simplification due to the k factor.

To illustrate the use of the theory presented, we consider the case of uniform loading q_0 on the *clamped* circular plate. On consulting Eq. (6.131) it is immediately clear that the differential equation in this case is exactly the same for the improved theory as for the classical theory. However, the *boundary conditions are different.* For comparison purposes we set forth the boundary conditions for the classical and improved theories.† Thus at $r = a$ we require in each case

Classical theory $\quad w = \dfrac{dw}{dv} = 0$ (6.132)

Improved theory $\quad w = \psi = 0$

†For the case of the *simply supported* circular plate we have at the edge:

Classical theory $\quad w = w'' + \nu\dfrac{w'}{r} = 0$

Improved theory $\quad w = \psi' + \nu\dfrac{\psi}{r} = 0$

where ψ, you will recall from Section 6.12, is the rotation about a circumferential line of area elements having normals in the radial direction. Furthermore, ψ is given in terms of w and q by Eq. (6.127).

We may consider the solution for the improved theory in two parts owing to the linearity of the equations. First, there is the bending deflection (no shear effects) alone, which has been solved in Section 6.10

$$w_B = \frac{q_0 a^4}{64D}\left[1 - \left(\frac{r}{a}\right)^2\right]^2$$

For the deflection w_S due to transverse shear only we will first heuristically propose such a solution and then demonstrate that the total deflection function $w_B + w_S$ solves the boundary-value problem for the improved theory. Consider accordingly a portion of the plate having radius r. From equilibrium considerations we know that

$$Q = \frac{q_0 r}{2} \tag{6.133}$$

But recall that

$$Q = \int_{-h/2}^{h/2} \tau_{rz}\, dz = k(\tau_{rz})_{z=0} h$$

Hence, employing Eq. (6.133),

$$(\tau_{rz})_{z=0} = \frac{Q}{kh} = \frac{q_0 r}{2hk} \tag{6.134}$$

Let us consider shear deformation at the centerplane of the plate. An element is shown in Fig. 6.21*a* at the centerplane in the undeformed geometry. In the deformed geometry we assume, as in our study of beams, that the vertical sides of the element

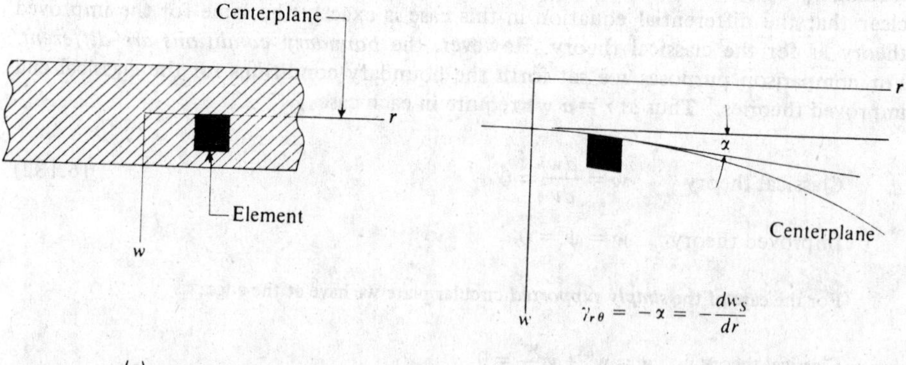

(*a*) (*b*)

FIGURE 6.21
Deformation of an element in centerplane.

remain vertical while the horizontal sides stay parallel to the centerplane in the deformed geometry, as shown in Fig. 6.21*b*. From this diagram we conclude that

$$\gamma_{rz} = -\alpha = -\frac{dw_S}{dr}$$

Hence, using Hooke's law,

$$(\tau_{rz})_{z=0} = -G\,\frac{dw_S}{dr} \tag{6.135}$$

Substituting the above result into Eq. (6.134), we get

$$\frac{dw_S}{dr} = -\frac{q_0 r}{2kGh}$$

Integrating, we get

$$w_S = \frac{q_0}{4kGh}(-r^2 + C_1') = \frac{q_0 a^2}{4kGh}\left(C_1 - \frac{r^2}{a^2}\right) \tag{6.136}$$

where C_1 is the constant of integration. It is now a simple matter to show that w_S satisfies the homogeneous part of the differential equation, which, for the axisymmetric case involving cylindrical coordinates, is given as follows:

$$\nabla^4 w = \left(\frac{\partial^2}{\partial r^2} + \frac{1}{r}\frac{\partial}{\partial r}\right)\left(\frac{\partial^2}{\partial r^2} + \frac{1}{r}\frac{\partial}{\partial r}\right) w = 0$$

The proposed solution

$$w = w_B + w_S = \frac{q_0 a^4}{64D}\left[1 - \left(\frac{r}{a}\right)^2\right]^2 + \frac{a^2 q_0}{4kGh}\left(C_1 - \frac{r^2}{a^2}\right)$$

then satisfies the entire differential equation

$$\nabla^4 w = q_0$$

for the problem at hand.†

We now proceed to the boundary conditions. Clearly, to have $w_S = 0$ when $r = a$ it is necessary that $C_1 = 1$. As for the second condition, we require that [see Eqs. (6.132) and (6.127)]:

$$\psi(a) = w'(a) + \frac{1}{hkG}\,\frac{1}{a}\int_0^a q_0 \xi \, d\xi = 0$$

Since $w_B'(a) = 0$ (the classical condition for bending) we need only include $w_S(a)$ in the above formulation. We may directly see on substitution and on carrying out the

†Note that the equation is actually an ordinary differential equation for the problem at hand.

integration that the above condition is also satisfied. Thus the proposed solution leads to an exact solution of the boundary-value problem here. It will be useful to introduce $D\ [= Eh^3/12(1-\nu^2)]$ and to replace G by $E/2(1+\nu)$ in w_S as follows:

$$w_S = \frac{q_0 a^2}{\{4kE/[2(1+\nu)]\}D[12(1-\nu^2)/Eh^3]h}\left[1-\left(\frac{r}{a}\right)^2\right]$$

$$= \frac{q_0 a^2 h^2}{24k(1-\nu)D}\left[1-\left(\frac{r}{a}\right)^2\right]$$

The solution then is

$$w = \frac{q_0 a^4}{64D}\left[1-\left(\frac{r}{a}\right)^2\right]^2 + \frac{q_0 a^2 h^2}{24k(1-\nu)D}\left[1-\left(\frac{r}{a}\right)^2\right]$$

As for k, we know from the exact solution of plates that τ_{rz} varies parabolically across the thickness of the plate (Timoshenko and Goodier, 1951) and so we take $k = \frac{4}{3}$ to correspond to this result. One can show that this displacement then corresponds to an exact solution from the theory of elasticity (see Love, 1944).

6.12 CLOSURE

In this chapter we used the variational approach to formulate a useful boundary-value problem for plates. We then used variational methods to develop approximate solutions to these boundary-value problems. Much of the work paralleled what was done in Chapter 4 on beams and Chapter 5 on torsion. You will recall in particular that we introduced in the discussion of torsion a method for finding the lower limit of the quadratic functional I for the so-called von Neumann problem. This was the Trefftz method. From this we can get a sequence of functions v_n that converges to the solution under certain circumstances. However, we used the method to find an upper bound to the torsional rigidity, thereby affording us, with the aid of the Ritz method, a means of bracketing the correct value of this quantity. We thus entered into the difficult area of error analysis. Now for the case of clamped plates and simply supported plates there is an extension to the Trefftz method discussed earlier, namely the method of Rafalson. Here we can similarly find a lower bound of the quadratic functional to go with an upper bound associated with the Ritz method. In this approach a sequence w_n can be found that converges to the correct solution under certain appropriate conditions. We shall not get into this development in this text other than to point out its existence since this takes us beyond the level we have prescribed for the book.

We have restricted ourselves entirely to static problems of beams and plates up to this time. In the following chapter we shall investigate simple features of the vibration of beams and plates and thereby introduce the eigenvalue problem. You will see that variational methods are again extremely useful for such problems.

REFERENCES

Kantorovich, L. V., and Krylov, V. I., "Approximate Methods of Higher Analysis," Interscience, New York, 1958, p. 322.

Love, A. E. H., "Mathematical Theory of Elasticity," Dover, New York, 1944, p. 485.
Timoshenko, S., and Goodier, J. N., "Theory of Elasticity," McGraw-Hill, New York, 1951, p. 351.
Timoshenko, S., and Woinowsky-Kreiger, S., "Theory of Plates," McGraw-Hill, New York, 1959.

READING

Langhaar, H. L., "Energy Methods in Applied Mechanics," Wiley, New York, 1962.
Mansfield, E. H., "The Bending and Stretching of Plates," Macmillan, New York, 1964.
Morley, L., "Skew Plates and Structures," Macmillan, New York, 1963.
Rivello, R. M., "Theory and Analysis of Flight Structures," McGraw-Hill, New York, 1969.
Timoshenko, S., and Woinowsky-Kreiger, S., "Theory of Plates," McGraw-Hill, New York, 1959.
Way, S., "Plates," in Flügge, W., "Handbook of Engineering Mechanics," McGraw-Hill, New York, 1962.

PROBLEMS

6.1 Derive Eq. (6.32) using Green's theorem and starting from Eq. (6.31).

6.2 Derive the following expression for the complementary strain energy of a rectangular plate $a \times b \times h$. The material is linear elastic. Hint: see Eq. (6.53).

$$U^* = \frac{1}{2}\int_0^a\int_0^b [M_x^2 - 2\nu M_x M_y + M_y^2 + 2(1+\nu)M_{xy}^2]\, dx\, dy$$

6.3 An *orthotropic continuum* has three planes of symmetry with respect to elastic properties. For plane stress, the generalized Hooke's law for such a material is given as

$$\tau_{xx} = C_{11}\epsilon_{xx} + C_{12}\epsilon_{yy}$$
$$\tau_{yy} = C_{12}\epsilon_{xx} + C_{22}\epsilon_{yy}$$
$$\tau_{xy} = 2G_{12}\epsilon_{xy}$$

(Note that there are four elastic moduli.) Formulate the equations of equilibrium as well as the boundary conditions for a rectangular *orthotropic plate* in terms of w. Demonstrate that your equation degenerates to the case of the isotropic plate. Use whatever results of Sections 6.3 and 6.4 that are valid for this case.

6.4 Develop a reciprocal theorem between the bending deflection and the transverse load on a plate. That is, show that

$$\iint [q^{(1)}(x,y)w^{(2)}(x,y) - q^{(2)}(x,y)w^{(1)}(x,y)]\, dx\, dy = 0$$

where superscripts (1) and (2) refer to different loadings and corresponding deflection conditions on the same plate.

Hint: Start with the reciprocal theorem (Problem 1.22)

$$\iiint \tau_{ij}^{(1)}\epsilon_{ij}^{(2)}\, dv = \iiint \tau_{ij}^{(2)}\epsilon_{ij}^{(1)}\, dv \qquad (a)$$

Employ for plane stress in Eq. (*a*) the strain displacement relations, the stress intensity resultant function, and finally Eq. (6.50).

6.5 Consider an orthotropic plate (see Problem 6.3) where the planes of symmetry are the coordinate planes of a cylindrical coordinate system. Express Eqs. (6.4) for the case of deformation of an axisymmetric plate. Using the strain displacement relations for polar coordinates (see Problem 1.17), show that the strain energy for an orthotropic, axisymmetric, circular plate is given, using data from the previous problem, as

$$U=\pi\int_{r_0}^{r_1}\left[D_{11}\left(\frac{d^2w}{dr^2}\right)^2+2D_{12}\frac{1}{r}\frac{dw}{dr}\frac{d^2w}{dr^2}+D_{22}\left(\frac{1}{r}\frac{dw}{dr}\right)^2\right]r\,dr$$

where

$$D_{ij}=\tfrac{1}{12}C_{ij}h^3$$

Show that the equation of equilibrium for such a plate is

$$D_{11}\left(r\frac{d^2w}{dr^2}\right)''-D_{22}\left(\frac{1}{r}\frac{dw}{dr}\right)'=rq(r)$$

Note that $\epsilon_{\theta\theta}=(z/r)\,\partial w/\partial r$.

6.6 Derive an exact series solution for a simply supported rectangular plate (see Fig. 6.22) loaded uniformly with loading q_0. Show that the solution can be given as follows:

$$w(x,y)=\frac{16q_0}{\pi^6}\frac{12(1-\nu^2)}{Eh^3}\sum_{m(\text{odd})}^{\infty}\sum_{n(\text{odd})}^{\infty}\frac{\sin(m\pi x/a)\sin(n\pi y/b)}{(mn^5/b^4)\{1+[(m/a)/(n/b)]^2\}^2}$$

Show that for a plate where $b/a\to 0$, the above solution degenerates to that of a beam uniformly loaded by loading p_o and given as

$$w(y)=\frac{4p_0b^4}{\pi^5EI}\sum_{n(\text{odd})}^{\infty}\frac{\sin(n\pi y/b)}{n^5}$$

where I is the moment of inertia per unit width. Also, $(q_o)(1)=p_o$ and $\nu=0$ in the development. Finally, take $a/m=\infty$.

Hint:

$$\sum_{m(\text{odd})}^{\infty}\frac{\sin m\pi x}{m}=\frac{\pi}{4}$$

Thus we have derived the deflection of a beam by considering that of a rectangular plate and letting one dimension become very large compared to the other dimension.

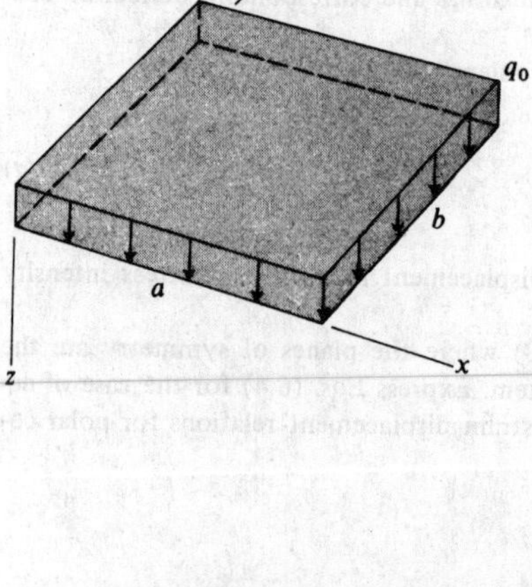

FIGURE 6.22

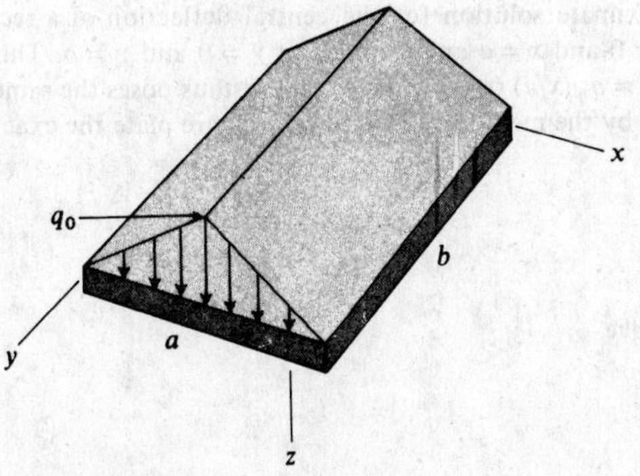

FIGURE 6.23

*6.7 Consider the rectangular piate shown in Fig. 6.23 with a triangular load. This plate is simply supported at $x = 0$ and $x = a$ and is fixed at $y = 0$ and $y = b$. Using the Lévy method, show that

$$q_m = \frac{8q_0}{Dm^2\pi^2} \sin \frac{m\pi}{2}$$

Next determine the constants $(C_1)_m$, $(C_2)_m$, $(C_3)_m$, and $(C_4)_m$ and show that for the following data

$$h = .03 \text{ m}$$
$$a = .6 \text{ m}$$
$$b = .6 \text{ m}$$
$$E = 2.1 \times 10^{11} \text{ Pa}$$
$$\nu = 0.3$$
$$q_0 = 7 \times 10^5 \text{ Pa}$$

we get the following results:

Constant	$m = 1$	$m = 3$	$m = 5$	$m = 7$
$(C_1)_m$	-2.028×10^{-4}	1.6753×10^{-9}	0	0
$(C_2)_m$	-1.2511×10^{-3}	1.9927×10^{-6}	-9.3049×10^{-8}	1.2358×10^{-8}
$(C_3)_m$	7.4372×10^{-5}	-7.5723×10^{-10}	0	0
$(C_4)_m$	-1.721×10^{-3}	9.3833×10^{-6}	-7.3081×10^{-7}	1.3588×10^{-7}

Finally, determine the deflection at the center of the plate to be as follows:

	$m = 1$	$m = 3$	$m = 5$	$m = 7$
w	2.6896×10^{-3}	2.6873×10^{-3}	2.6875×10^{-3}	2.6875×10^{-3}

Use Eq. (g) of Example 6.1.

6.8 Using the Ritz method, find an approximate solution for the central deflection of a rectangular plate simply supported at $x = 0$ and $x = a$ and clamped at $y = 0$ and $y = b$. This plate is loaded by a triangular loading $q = q_0(x/a)$ (see Fig. 6.16), and it thus poses the same problem that was solved in Example 6.1 by the method of Lévy. For a square plate the exact answer is

$$w\left(\frac{a}{2}, \frac{b}{2}\right) = 0.0009586\,\frac{q_0 a^4}{D}$$

and for a rectangular plate $b/a = 2$ we have

$$w\left(\frac{a}{2}, \frac{b}{2}\right) = 0.004222\,\frac{q_0 a^4}{D}$$

Compare your results with the above.

6.9 Find an approximate solution for the center point deflection of a simply supported square plate under the triangular loading (see Fig. 6.24)

$$q(x, y) = q_0\,\frac{x}{a}$$

Compare your result with the exact answer:

$$w\left(\frac{a}{2}, \frac{a}{2}\right) = 0.00203\,\frac{q_0 a^4}{D}$$

6.10 Find an approximate Ritz solution for the deflection at the center of a simply supported, square plate loaded by moment intensities (moment per unit length) M_0 along the edges $y = 0$ and $y = a$ so as to cause downward deflection of the plate. Use a function that satisfies the kinematic boundary conditions of the problem. Compare your result with the exact solution for this problem:

$$w\left(\frac{a}{2}, \frac{a}{2}\right) = 0.0368\,\frac{M_0 a^2}{D}$$

6.11 Consider a rectangular plate simply supported at the edges $x = 0$ and $y = 0$ and freely supported at the edges $x = a$ and $y = b$ (see Fig. 6.25). The load is a force at $x = a, y = b$. Find an approximate solution that satisfies the boundary conditions. What corner force does this solution generate at $x = a, y = b$? Comment on the result.

6.12 Consider a clamped rectangular plate ($0 \leqslant x \leqslant a$, $0 \leqslant y \leqslant b$) with a concentrated load at the center. Find an approximate solution for the deflection and compare it with the following accurate results:

Square plate ($a = b$) $\quad w_{\max} = 0.00560\,\dfrac{Pa^2}{D}$

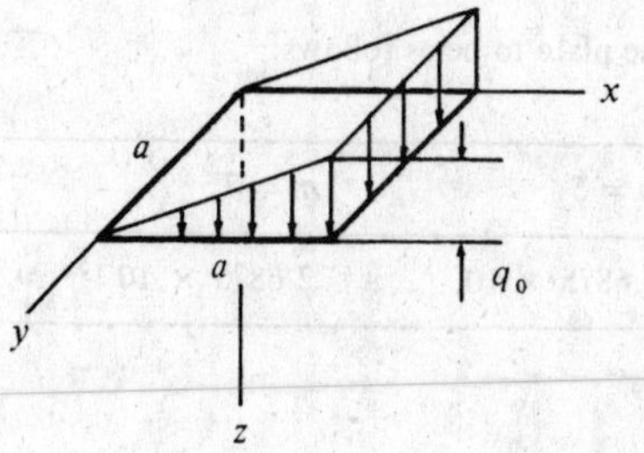

FIGURE 6.24

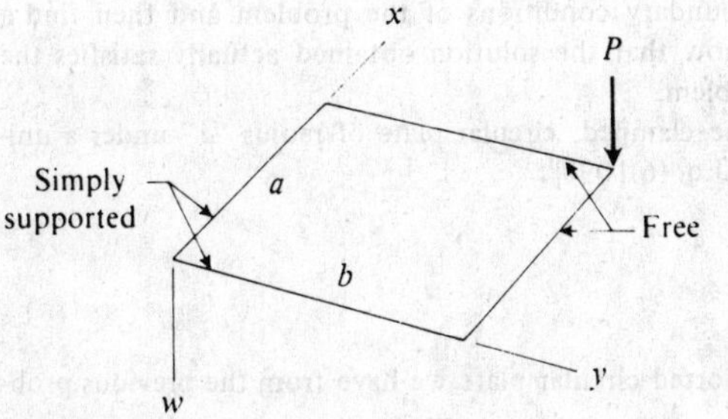

FIGURE 6.25

$$\text{Rectangular plate } \left(a = \frac{b}{2}\right) \qquad w_{\max} = 0.00722\,\frac{Pa^2}{D}$$

*6.13 Using Green's theorem with $\phi = \nabla^2 w$ and $\psi = w$, the boundary condition (6.104), and the vector identity

$$\nabla \cdot (\Phi \mathbf{F}) = \Phi\, \nabla \cdot \mathbf{F} + \mathbf{F} \cdot \nabla \Phi$$

show that Eq. (6.105) can be rewritten as Eq. (6.106). Show that Eq. (6.108) is an exact solution to the clamped elliptic plate problem.

6.14 Show that the Gaussian curvature for an axisymmetric, isotropic simply supported circular plate is

$$\frac{1}{r}\, w'(r) w''(r) \tag{a}$$

Now show that the total potential energy is given for such a plate (radius "a") as follows:

$$U + V = 2\pi \int_0^a \left\{ \frac{D}{2} \left[\left(w''(r) + \frac{1}{r}\, w'(r) \right)^2 - \frac{2(1-\nu)}{r}\, w'(r) w''(r) \right] - wq \right\} r\, dr \tag{b}$$

The boundary conditions for this problem may be given as

$$\text{at } r = a \qquad w = 0 \qquad M_r = 0$$

Noting that when $\theta = 0$ we have

$$M_r = M_x$$

Show that [see Eq. (6.12)]

$$M_r = -D \left(\frac{\partial^2 w}{\partial r^2} + \frac{\nu}{r}\, \frac{\partial w}{\partial r} \right)$$

for the axisymmetric case. Hence the boundary conditions at $r = a$ are

$$w = w'' + \frac{\nu}{r}\, w' = 0$$

6.15 Consider a simply supported circular plate loaded by a moment intensity M_0 uniformly around the edge. Referring to Problem 6.14 and, starting with an approximate function

$w = A + Br^2$, satisfy the kinematic boundary conditions of the problem and then find a Ritz approximate solution. Finally, show that the solution obtained actually satisfies the natural boundary conditions of the problem.

6.16 The solutions for the deflection of the clamped, circular plate of radius "a" under a uniform loading q_0 has been shown to be [Eq. (6.109)]

$$w_q = \frac{q_0 a^4}{64D}\left[1 - \left(\frac{r}{a}\right)^2\right]^2$$

For the pure bending of a simply supported circular plate we have from the previous problem the result

$$w_{M_0} = \frac{M_0 a^2}{2(1+\nu)D}\left[1 - \left(\frac{r}{a}\right)^2\right]$$

Obtain from these the solution for a simply supported plate under a uniform loading q_0. (See Problem 6.14 for information as to π and the boundary conditions for a circular plate.)

6.17 Find an approximate solution for the clamped circular orthotropic plate of radius "a" (see Problems 6.3 and 6.5) for a centrally located concentrated load. What is the ratio of maximum deflections between the orthotropic plate and the isotropic plate for the same coordinate function you have used?

*6.18 Using the results of Problem 6.14, obtain an approximate solution for the deflection of a simply supported circular plate of radius "a" loaded by a ring loading of total value P at radius b where $0 \leqslant b \leqslant a$. The exact solution for $r = 0$ is

$$w(0) = \frac{P}{8\pi D}\left[\frac{3+\nu}{2(1+\nu)}(a^2 - b^2) - b^2 \ln \frac{a}{b}\right]$$

For $b/a = \frac{1}{2}$ and $\nu = 0.3$ compare solutions at $r = 0$. Use a polynomial $A + Br^2 + Cr^4$ for the coordinate function that satisfies the boundary conditions of the problem. Specifically, show that

$$w(r) = C\left(\frac{5+\nu}{1+\nu}a^4 - \frac{6+2\nu}{1+\nu}a^2 r^2 + r^4\right)$$

6.19 Using Eqs. (6.53), (6.54), and (6.58), we assume the following stress distribution for axisymmetric deformation of a Hookean material:

$$\tau_{rr} = \frac{M_r z}{h^3/12}$$

$$\tau_{\theta\theta} = \frac{M_\theta z}{h^3/12}$$

$$\tau_{rz} = \frac{3}{2}\frac{Q_r}{h}\left[1 - \left(\frac{z}{h/2}\right)^2\right]$$

$$\tau_{zz} = \frac{3}{4} q(r)\left[\frac{2}{3} - \frac{z}{h/2} + \frac{1}{3}\left(\frac{z}{h/2}\right)^3\right]$$

In addition we assume a strain displacement field as given by Eq. (6.118) in the Hencky theory of plates. Using the Reissner functional given as

$$R = \iiint (\tau_{ij}\epsilon_{ij} - \mathfrak{U}^*)\, dv - \int_{r_0}^{r_1}\int_0^{2\pi} q(r,\theta)\, w(r,\theta)\, r\, dr\, d\theta$$

where $\mathcal{U}^*$ is the complementary strain energy density function, derive a plate theory including both shear *and* normal stress effects.

Hint: Neglecting τ_{zz}^2, show first that

$$R = 2\pi \int_{r_0}^{r_1} \left[-M_r\psi'(r) + Q_r(w' - \psi) - \frac{1}{r} M_\theta \psi - \frac{M_r^2}{2D_1} - \frac{M_\theta^2}{2D_1} + \nu \frac{M_r M_\theta}{D_1} \right.$$
$$\left. - \frac{\nu q}{D_1} \frac{h^2}{10} (M_r + M_\theta) - \frac{3}{5} \frac{Q_r^2}{Gh} - qw \right] r\, dr$$

where $D_1 = Eh^3/12$. Next consider ψ, w, M_r, M_θ, and Q_r as the extremization variables to reach the following results:

$$\text{Equilibrium} \begin{cases} rQ_r = -M_\theta + \dfrac{d}{dr}(rM_r) \\ \dfrac{d}{dr}(rQ_r) = -qr \end{cases}$$

$$\text{Stress displacement relations} \begin{cases} M_r - \nu M_\theta = -D_1 \psi'(r) - \dfrac{\nu q h^2}{10} \\ M_\theta - \nu M_r = -D_1 \dfrac{\psi}{r} - \dfrac{\nu q h^2}{10} \\ Q_r = \frac{5}{6} Gh(w' - \psi) \end{cases}$$

The boundary conditions are

$rM_r = 0$ or ψ specified

$rQ_r = 0$ or w specified

6.20 Show that the contribution to the first variation of the bending energy of plates of the Gaussian curvature vanishes (1) for smooth-edged plates that are clamped and (2) for polygonal plates with only zero displacement required at the edges.

Hint: Start with Eq. (6.87) and then use Eqs. (6.36) and (6.37) before integrating by parts to get the line integrals

$$\oint \frac{\partial^2 w}{\partial s^2} \frac{\partial \delta w}{\partial \nu}\, ds \quad \text{and} \quad \oint \frac{\partial^3 w}{\partial \nu\, \partial s^2}\, ds$$

*6.21 Use the Reissner principle to derive an improved plate theory for axisymmetric deformation of a circular plate loaded by surface loading $q(r)$, edge moments per unit length M_0, and edge loading per unit length Q_0, downward. Obtain all appropriate differential equations, stress-displacement relations, and all boundary conditions. Proceed by first writing Reissner's functional R for the case where $\tau_{zz} = \tau_{r\theta} = \tau_{\theta z} = 0$. Next, using Eqs. (6.124) and the first set of equations of Problem 6.19, while using $\tau_{rz} = Q/kh$ in $\mathcal{U}^*$, show that R is given as

$$R = 2\pi \int_{r_0}^{r_1} \left[-r\psi' M_r + r(w' - \psi)Q - M_\theta \psi - \frac{1}{E} \left(\frac{rM_r^2}{h^3/12} + \frac{rM_\theta^2}{h^3/12} \right. \right.$$
$$\left. \left. - 2\nu \frac{rM_r M_\theta}{h^3/12} - \frac{1}{2G} \frac{Q^2 r}{k^2 h} \right) - 2\pi q(r) w(r) r \right] dr - 2\pi r_1 Q_0 w(r_1) + 2\pi r_1 M_0 \psi(r_1)$$

6.22 Solve the pure bending problem for a circular plate by using the Hencky theory, i.e., a simply supported circular plate loaded by uniform edge moment per unit length M_0 at the edge $r = a$. Proceed by showing, on using Eqs. (6.122*a*) and (6.123*a*) for $q = 0$ and then using ensuing formulas for M_r and M_θ, that the governing equation is

$$r\psi'' + \psi' - \frac{\psi}{r} = -\frac{C_1}{D}$$

with C_1 an arbitrary constant. Show that

$$\psi = Ar + \frac{B}{r} - \frac{C_1}{2D} r \ln r$$

is the general solution. [Why is $-(C_1/2D)r$ not used as a particular solution here?] Using Eq. (6.128) and eliminating singularities at $r = 0$, show that

$$w = \frac{Ar^2}{2} + C_2$$

Finally, using boundary conditions at $r = a$ for w and M_r, show that the deflection curve is

$$w = \frac{M_0}{2D(1+\nu)}(a^2 - r^2)$$

7

DYNAMICS OF BEAMS AND PLATES

7.1 INTRODUCTION

Up to this time we have considered only the case of structural bodies in static equilibrium. We now examine variational aspects of the dynamics of beams and plates. Our procedure will be first to present Hamilton's principle, since this principle underlies much of what we do in this chapter. In Part A we shall center our attention on beams, first deriving the equations of motion from Hamilton's principle and then considering both exact and approximate solutions for free vibrations. For the latter calculations we shall feature the Rayleigh and Rayleigh-Ritz methods. The same pattern is then followed in Part B for the study of plates. In Part C of this chapter we examine the Rayleigh quotient, used in earlier parts of the chapter, in a more general manner and develop strong supporting arguments for some physically inspired assertions made earlier concerning the Rayleigh and the Rayleigh-Ritz methods. The very powerful maximum theorem and "min-max" theorem from the calculus of variations are presented. Thus this section is a more theoretical exposition. It is supportive of earlier work as well as basic to new work on stability in Chapter 8.

7.2 HAMILTON'S PRINCIPLE

Those who have studied dynamics of particles and rigid bodies at the intermediate level should recall Hamilton's principle as employed for discrete systems. The functional of interest is

$$\int_{t_1}^{t_2} (T - V)\, dt$$

where T is the total kinetic energy of the bodies and V is the potential energy of the forces. The variable t is, of course, the time. Using $q_1, q_2, \ldots, q_n$ as the generalized coordinates and assuming they are independent, the Euler-Lagrange equations then yield the well-known *Lagrange* equations of motion, given as follows in terms of the Lagrangian $L = T - V$:

$$\frac{d}{dt}\left(\frac{\partial L}{\partial \dot{q}_i}\right) - \frac{\partial L}{\partial q_i} = 0 \qquad i = 1, 2, \ldots, n$$

These equations are then the starting point for Lagrangian mechanics.

To develop Hamilton's principle for continua from the method of virtual work, we employ the D'Alembert principle, which, you will recall, sets forth Newton's law in the following form:

$$\mathbf{F} - m\mathbf{a} = \mathbf{0}$$

The vector $-m\mathbf{a}$ is considered a force, with the result that the above equation has a form corresponding to the equations of statics. For a continuous deformable body the inertia force for the D'Alembert principle is given as $-\rho\, d^2\mathbf{u}/dt^2$ and is a body force distribution playing the same role as $\mathbf{B}$ in the discussion of Section 3.2. Accordingly, Eq. (3.3), which is valid for static conditions, may be generalized to include dynamic conditions if the integral

$$-\iiint_V \rho \frac{d^2 u_i}{dt^2}\, \delta u_i\, dv$$

is added to the left side of the equation. We thus have at time t for a virtual displacement field δu_i consistent with the constraints

$$-\iiint_V \rho \frac{d^2 u_i}{dt^2}\, \delta u_i\, dv + \iiint_V B_i\, \delta u_i\, dv + \iint_{S_1} T_i^{(\nu)}\, \delta u_i\, dS = \iiint_V \tau_{ij}\, \delta \epsilon_{ij}\, dv \tag{7.1}$$

We may introduce the force potential V such that

$$V = -\iiint_V B_i u_i\, dv - \iint_{S_1} T_i^{(\nu)} u_i\, dS$$

The second and third terms on the right side of Eq. (7.1) can then be given as

$$\iiint_V B_i\, \delta u_i\, dv + \iint_{S_1} T_i^{(\nu)}\, \delta u_i\, dS = -\delta^{(1)} V$$

Similarly, we assume the existence of a strain energy density function $\mathcal{U}$ for the body such that $\tau_{ij} = \partial \mathcal{U}/\partial \epsilon_{ij}$. The right side of Eq. (7.1) becomes $\delta^{(1)}U$ and we may give this equation in the following form:

$$-\iiint_V \rho \frac{d^2 u_i}{dt^2} \delta u_i \, dv - \delta^{(1)}V - \delta^{(1)}U = 0$$

The variables u_i, V, and U in the above equation may be functions of time and so the statement applies for time t. We now integrate with respect to time from the limit t_1 (denoting the onset of the motion) to t_2. We thus have

$$-\int_{t_1}^{t_2} \iiint_V \rho \frac{d^2 u_i}{dt^2} \delta u_i \, dv \, dt - \int_{t_1}^{t_2} \delta^{(1)}V \, dt - \int_{t_1}^{t_2} \delta^{(1)}U \, dt = 0$$

We now interchange integration operations in the first expression and the variation and integration operations in the remaining terms as follows:

$$-\iiint_V \left(\rho \int_{t_1}^{t_2} \frac{d^2 u_i}{dt^2} \delta u_i \, dt \right) dv - \delta^{(1)} \left(\int_{t_1}^{t_2} V \, dt \right) - \delta^{(1)} \left(\int_{t_1}^{t_2} U \, dt \right) = 0 \tag{7.2}$$

Next we integrate by parts the integral in parentheses in the first term. Thus

$$\int_{t_1}^{t_2} \frac{d^2 u_i}{dt^2} \delta u_i \, dt = \frac{du_i}{dt} \delta u_i \bigg|_{t_1}^{t_2} - \int_{t_1}^{t_2} \frac{du_i}{dt} \frac{d}{dt} (\delta u_i) \, dt$$

We adopt the rule here that $\delta u_i = 0$ at $t = t_1$ and $t = t_2$. That is, we take *the virtual displacement fields to be zero at the time limits t_1 and t_2*. The above statement may then be written as follows:

$$\int_{t_1}^{t_2} \frac{d^2 u_i}{dt^2} \delta u_i \, dt = -\int_{t_1}^{t_2} \dot{u}_i \, \delta \dot{u}_i \, dt = -\int_{t_1}^{t_2} \delta^{(1)} \left(\frac{\dot{u}_i^2}{2} \right) dt$$

Substituting back into Eq. (7.2) we have

$$\iiint_V \frac{\rho}{2} \int_{t_1}^{t_2} \delta^{(1)} \dot{u}_i^2 \, dt \, dv - \delta^{(1)} \left(\int_{t_1}^{t_2} V \, dt \right) - \delta^{(1)} \left(\int_{t_1}^{t_2} U \, dt \right) = 0$$

Again considering the first expression, we have for constant density

$$\iiint_V \frac{\rho}{2} \int_{t_1}^{t_2} \delta^{(1)} \dot{u}_i^2 \, dt \, dv = \int_{t_1}^{t_2} \iiint_V \frac{\rho}{2} \delta^{(1)} \dot{u}_i^2 \, dv \, dt$$

$$= \int_{t_1}^{t_2} \delta^{(1)} \left(\iiint_V \frac{\rho \dot{u}_i^2}{2} \, dv \right) dt = \int_{t_1}^{t_2} \delta^{(1)} T \, dt = \delta^{(1)} \left(\int_{t_1}^{t_2} T \, dt \right)$$

where T is the kinetic energy of the body at time t. We now may give Hamilton's principle in the following manner:

$$\delta^{(1)}\left(\int_{t_1}^{t_2} (T - V - U)\,dt\right) = 0$$

Noting that $(V + U)$ is the *total potential energy* π, we then get the following familiar form for the Hamilton principle:

$$\boxed{\delta^{(1)}\left(\int_{t_1}^{t_2} (T - \pi)\,dt\right) = \delta^{(1)}\left(\int_{t_1}^{t_2} L\,dt\right) = 0} \tag{7.3}$$

where L, still called the Lagrangian, is now $T - \pi$. *Hamilton's principle* states that, *of all the paths of admissible configurations that the body can take as it goes from configuration 1 at time t_1 to configuration 2 at time t_2, the path that satisfies Newton's law at each instant during the interval (and is thus the actual locus of configurations) is the path that extremizes the time integral of the Lagrangian during the interval.*

We shall directly illustrate the use of Hamilton's principle in the following section, where we consider the vibrating beam.

Part A
Beams

7.3 EQUATIONS OF MOTION FOR VIBRATING BEAMS

We shall return to Chapter 4 to generalize the assumed kinematics of a simple beam to include time variation. Omitting axially applied external forces, we thus have from Eq. (4.4)

$$\begin{aligned} u_1 &= -z\,\frac{\partial w(x,t)}{\partial x} \\ u_2 &= 0 \\ u_3 &= w(x,t) \end{aligned} \tag{7.4}$$

We will need the Lagrangian for use in Hamilton's principle and so we proceed by considering the kinetic energy of the beam. Thus we have, using the preceding results,

$$T = \frac{1}{2}\int_0^L \iint_A \rho \dot{u}_i \dot{u}_i \, dy\, dz\, dx = \frac{1}{2}\int_0^L \iint_A \rho \left(\frac{\partial w}{\partial t}\right)^2 dx\, dy\, dz$$

$$+ \frac{1}{2}\int_0^L \iint_A \rho \left(\frac{\partial^2 w}{\partial t\, \partial x}\right)^2 z^2 \, dx\, dy\, dz$$

$$= \int_0^L \frac{\rho A}{2}\left(\frac{\partial w}{\partial t}\right)^2 dx + \int_0^L \frac{\rho I}{2}\left(\frac{\partial^2 w}{\partial t\, \partial x}\right)^2 dx \tag{7.5}$$

The kinetic energy is thus seen to be composed of two parts. The first represents the kinetic energy due to translatory motion in the vertical direction z. The second expression involves half the product of an angular velocity $\partial(\partial w/\partial x)/\partial t$, squared, of a beam element dx about the y axis times the mass moment of inertia about the y axis for this element, $(\rho I\, dx)$. Thus the second expression represents the kinetic energy due to rotation of the beam elements. For most problems, *particularly those involving thin beams*, this contribution to T may be neglected since it is usually very small for long, slender beams compared to the other term. In those cases we say that we are neglecting the *rotatory inertia* for the problem.† We shall here neglect this contribution and so we shall use for T the following result:

$$T = \int_0^L \frac{\rho A}{2}\left(\frac{\partial w}{\partial t}\right)^2 dx \tag{7.6}$$

As for the strain energy of the beam, we include at this time only that energy which is due to bending. You will recall from Chapter 4 [see development of Eq. (4.8)] that

$$U = \int_0^L \frac{EI}{2}\left(\frac{\partial^2 w}{\partial x^2}\right)^2 dx \tag{7.7}$$

The Lagrangian that we shall employ is then

$$L = \int_0^L \frac{\rho A}{2}\left(\frac{\partial w}{\partial t}\right)^2 dx - \int_0^L \frac{EI}{2}\left(\frac{\partial^2 w}{\partial x^2}\right)^2 dx + \int_0^L q(x,t)w(x,t)\, dx$$

$$= \int_0^L \left[\frac{\rho A}{2}(\dot{w})^2 - \frac{EI}{2}(w_{xx})^2 + qw\right] dx$$

†This should be easily seen by noting that ρA is proportional to the depth of the beam while ρI is proportional to the depth cubed. However, it should be noted that the rotary inertia may not be negligible even for thin beams when complex mode shapes (to be described soon) at high frequencies are involved.

where $\dot{w} = \partial w/\partial t$ and $w_{xx} = \partial^2 w/\partial x^2$. Hamilton's principle then requires that

$$\delta^{(1)} \int_{t_1}^{t_2} \int_0^L \left(\frac{\rho A}{2} \dot{w}^2 - \frac{EI}{2} w_{xx}^2 + qw \right) dx\, dt = 0 \tag{7.8}$$

Denoting the integrand as F and choosing a one-parameter family of admissible functions $\tilde{w}(x, t) = w(x, t) + \epsilon\eta(x, t)$, where $\eta(x, t_1) = \eta(x, t_2) = 0$ as required by Hamilton's principle, we have for Eq. (7.8)†

$$\left[\frac{d}{d\epsilon} \left(\int_{t_1}^{t_2} \int_0^L \tilde{F}\, dx\, dt \right) \right]_{\epsilon=0} = 0$$

therefore

$$\int_{t_1}^{t_2} \int_0^L \left(\frac{\partial F}{\partial \dot{w}} \dot{\eta} + \frac{\partial F}{\partial w_{xx}} \eta_{xx} + \frac{\partial F}{\partial w} \eta \right) dx\, dt = 0 \tag{7.9}$$

Integrate the first expression by parts with respect to time. Thus

$$\int_{t_1}^{t_2} \int_0^L \frac{\partial F}{\partial \dot{w}} \frac{\partial \eta}{\partial t}\, dx\, dt = \int_0^L \left(\eta \frac{\partial F}{\partial \dot{w}} \right)_{t_1}^{t_2} dx - \int_{t_1}^{t_2} \int_0^L \eta \frac{\partial}{\partial t} \frac{\partial F}{\partial \dot{w}}\, dx\, dt$$

Since $\eta = 0$ at t_1 and t_2 we may delete the first expression on the right side of the above equation. Now integrate the second expression in Eq. (7.9) twice by parts with respect to x as follows:

$$\int_{t_1}^{t_2} \int_0^L \frac{\partial F}{\partial w_{xx}} \eta_{xx}\, dx\, dt = \int_{t_1}^{t_2} \left(\frac{\partial F}{\partial w_{xx}} \eta_x \right)_0^L dt - \int_{t_1}^{t_2} \int_0^L \left(\frac{\partial}{\partial x} \frac{\partial F}{\partial w_{xx}} \right) \eta_x\, dx\, dt$$

$$= \int_{t_1}^{t_2} \left(\frac{\partial F}{\partial w_{xx}} \eta_x \right)_0^L dt - \int_{t_1}^{t_2} \left[\frac{\partial}{\partial x} \left(\frac{\partial F}{\partial w_{xx}} \right) \eta \right]_0^L dt$$

$$+ \int_{t_1}^{t_2} \int_0^L \left(\frac{\partial^2}{\partial x^2} \frac{\partial F}{\partial w_{xx}} \right) \eta\, dx\, dt$$

Substituting these results back into Eq. (7.9) we get

$$\int_{t_1}^{t_2} \int_0^L \left(-\frac{\partial}{\partial t} \frac{\partial F}{\partial \dot{w}} + \frac{\partial^2}{\partial x^2} \frac{\partial F}{\partial w_{xx}} + \frac{\partial F}{\partial w} \right) \eta\, dx\, dt + \int_{t_1}^{t_2} \left(\frac{\partial F}{\partial w_{xx}} \eta_x \right)_0^L dt$$

$$- \int_{t_1}^{t_2} \left[\frac{\partial}{\partial x} \left(\frac{\partial F}{\partial w_{xx}} \right) \eta \right]_0^L dt = 0$$

†Since we derived Hamilton's principle using the delta operator we shall work here with a single parameter family representation and will ask you to develop the same results using the operator approach.

Since η and η_x could possibly be zero at $x = 0$ and $x = L$, we conclude from the above formulation that the following equation must necessarily be satisfied along the beam:

$$\frac{\partial}{\partial t}\frac{\partial F}{\partial \dot{w}} - \frac{\partial^2}{\partial x^2}\frac{\partial F}{\partial w_{xx}} - \frac{\partial F}{\partial w} = 0 \tag{7.10}$$

Inserting the expression $[(\rho A/2)\dot{w}^2 - (EI/2)(w_{xx})^2 + qw]$ for F, we then get the basic equation of motion for the beam:

$$\boxed{\rho A \frac{\partial^2 w}{\partial t^2} + \frac{\partial^2}{\partial x^2}\left(EI\frac{\partial^2 w}{\partial x^2}\right) = q} \tag{7.11}$$

The boundary conditions at $x = 0$ and $x = L$ clearly are

$$\begin{aligned} &\text{EITHER} \quad \frac{\partial F}{\partial w_{xx}} = 0 \quad \text{OR} \quad w_x \quad \text{IS PRESCRIBED} \\ &\text{EITHER} \quad \frac{\partial}{\partial x}\frac{\partial F}{\partial w_{xx}} = 0 \quad \text{OR} \quad w \quad \text{IS PRESCRIBED} \end{aligned} \tag{7.12}$$

Inserting the function F and going back to the usual notation for derivatives, we get for these end conditions

$$\begin{aligned} &\text{EITHER} \quad \frac{\partial^2 w}{\partial x^2} = 0 \quad \text{OR} \quad \frac{\partial w}{\partial x} \quad \text{IS PRESCRIBED} \\ &\text{EITHER} \quad \frac{\partial}{\partial x}\left(EI\frac{\partial^2 w}{\partial x^2}\right) = 0 \quad \text{OR} \quad w \quad \text{IS PRESCRIBED} \end{aligned} \tag{7.13}$$

It is to be noted that the boundary conditions can be time-dependent.

We shall now investigate the free vibration of beams, using the simplified equations presented in this section. Later we shall present an improved theory accounting for transverse shear and rotatory inertia when we consider the vibrations of the so-called Timoshenko beam.

7.4 FREE VIBRATIONS OF A SIMPLY SUPPORTED BEAM

With no external loads the differential equation of motion for the beam is given as

$$\frac{\partial^2}{\partial x^2}\left(EI\frac{\partial^2 w}{\partial x^2}\right) + \rho A\frac{\partial^2 w}{\partial t^2} = 0 \tag{7.14}$$

We employ a *separation of variables* approach by expressing w as the product of a function W of x and a function T of t. Thus

$$w = W(x)T(t) \tag{7.15}$$

Substituting into Eq. (7.14) and dividing by WT gives the following result:

$$-\frac{1}{W}\frac{1}{\rho A}\frac{d^2}{dx^2}\left(EI\frac{d^2W}{dx^2}\right)=\frac{d^2T/dt^2}{T} \tag{7.16}$$

Since each side is separately a function of a different variable, we set each side equal to a constant $-\omega^2$, in the familiar manner for this technique. We get two ordinary differential equations as a result:

$$\frac{d^2T}{dt^2}+\omega^2T=0 \tag{7.17a}$$

$$\frac{d^2}{dx^2}\left(EI\frac{d^2W}{dx^2}\right)-\rho A\omega^2W=0 \tag{7.17b}$$

For the case where EI and ρA are constants, the general solution to the equations is given as follows:

$$W=G\cosh kx+B\sinh kx+H\cos kx+D\sin kx \tag{7.18a}$$

$$T=E\sin\omega t+F\cos\omega t \tag{7.18b}$$

where

$$k=\left(\frac{\rho A\omega^2}{EI}\right)^{1/4} \tag{7.19}$$

and where G, B, H, D, E, and F are integration constants.

We may satisfy the end conditions for this problem by using at each end at all times a natural boundary condition $\partial^2w/\partial x^2=0$ to give a zero moment there [see Eq. (4.16)], and a kinematic condition $w=0$. We thus have the following equations as a result of imposing these conditions on the solution W as given by Eq. (7.18a):

$$G+H=0$$

$$k^2(G-H)=0$$

$$G\cosh kL+B\sinh kL+H\cos kL+D\sin kL=0$$

$$k^2(G\cosh kL+B\sinh kL-H\cos kL-D\sin kL)=0$$

Inspection of the first two equations immediately indicates that $G=H=0$. We thus have for the remaining equations

$$\begin{aligned}B\sinh kL+D\sin kL&=0\\ B\sinh kL-D\sin kL&=0\end{aligned} \tag{7.20}$$

For a nontrivial solution the determinant formed by the coefficients of B and D in the above equations must be zero. This results in the following equation, called the *frequency or characteristic equation:*

$$\sin kL \sinh kL = 0 \tag{7.21}$$

We shall rule out the possibility that $k = 0$ to satisfy this equation. Our reason for this is that $k = 0$ implies $\omega = 0$ from Eq. (7.19), and a zero value of ω requires from Eq. (7.18*b*) that T is a constant, so that $w = WT$ is then independent of time. This can only mean here that the bar is at rest† and we have a trivial result. The only other possibility then is that

$$kL = n\pi \qquad n = 1, 2, \ldots \tag{7.22}$$

so that the sine function will be equal to zero. We thus have an infinite discrete set of possible values for ω^2, which from Eqs. (7.22) and (7.19) are given as follows:

$$\omega_n = \left(\frac{n\pi}{L}\right)^2 \left(\frac{EI}{\rho A}\right)^{1/2} \qquad n = 1, 2, \ldots \tag{7.23}$$

These are the ***natural frequencies*** or ***eigenvalues*** of the beam. Substituting the allowed values of kL, namely $n\pi$, into Eqs. (7.20), we see that the constant B must now be zero to satisfy these equations. We then have as possible solutions for W

$$W_n = D_n \sin \frac{n\pi x}{L} \qquad n = 1, 2, \ldots \tag{7.24}$$

The function $\sin n\pi x/L$ is termed the nth ***mode shape*** or the nth ***eigenfunction*** corresponding ***respectively*** to the nth ***natural frequency*** or the nth ***eigenvalue*** of the beam. A possible solution w_n for the problem is accordingly

$$w_n(x, t) = \sin \frac{n\pi x}{L} \,(A_n \cos \omega_n t + B_n \sin \omega_n t) \tag{7.25}$$

where D_n has been incorporated with the arbitrary constants in Eq. (7.18*b*) to form A_n and B_n. Note that with the aid of a phasor diagram we may replace the expression in parentheses with a single harmonic function as follows:

$$w_n(x, t) = \sin \frac{n\pi x}{L} \,[C_n \cos (\omega_n t + \alpha_n)] \tag{7.26}$$

where C_n and α_n are arbitrary constants replacing A_n and B_n.‡ From this we can see that the centerline deformation of the beam has the shape of the sinusoid $\sin n\pi x/L$ with an amplitude q_n, which is a harmonic function of time given as

$$q_n = C_n \cos (\omega_n t + \alpha_n)$$

†Since $w = WT =$ const is also a possible solution, the condition $k = 0$ can then represent a rigid-body movement. This movement can be of interest in the beam with free end conditions—the so-called free-free beam. However, in this case we have simple pin supports at the ends and rigid-body movement must be ruled out.

‡Note that $C_n = \sqrt{A_n^2 + B_n^2}$ and $\alpha_n = \tan^{-1} A_n/B_n$.

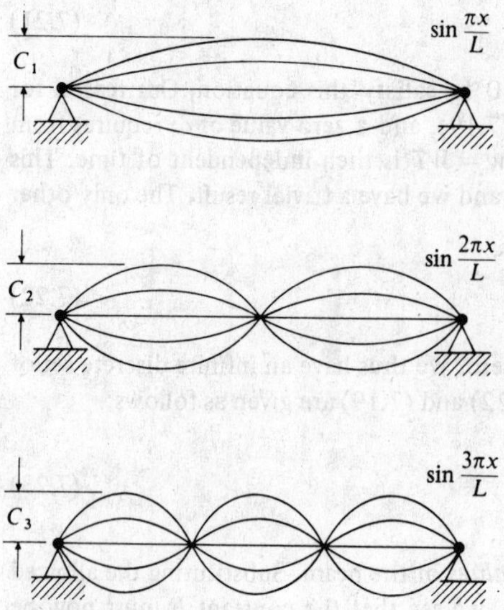

FIGURE 7.1
First three vibration modes.

This motion is shown for $n = 1, 2, 3$ in Fig. 7.1. The motion depicted by Eqs. (7.25) and (7.26) is that of the beam moving from one extreme configuration for the nth mode to the other (as denoted in the diagram for the first three modes) with a frequency ω_n. The general vibration of the beam is a superposition of all such modes of vibration, each having constants C_n and α_n (or A_n or B_n). That is,

$$w(x, t) = \sum_{n=1}^{\infty} \sin \frac{n\pi x}{L} C_n \cos (\omega_n t + \alpha_n) \tag{7.27a}$$

$$w(x, t) = \sum_{n=1}^{\infty} \sin \frac{n\pi x}{L} [A_n \cos \omega_n t + B_n \sin \omega_n t] \tag{7.27b}$$

The infinite set of constants C_n and α_n, or A_n and B_n, must be determined so that the initial conditions of the problem are satisfied, namely

$$\begin{aligned} w(x, 0) &= \phi(x) \\ \frac{\partial w(x, 0)}{\partial t} &= \psi(x) \end{aligned} \tag{7.28}$$

where ϕ and ψ are given functions. Thus, using Eq. (7.27b), we have

$$\phi = \sum_{n=1}^{\infty} \sin \frac{n\pi x}{L} A_n$$

$$\psi = \sum_{n=1}^{\infty} \sin \frac{n\pi x}{L} \omega_n B_n$$

The technique, you will recall, is to multiply in each case by sin $(n\pi x/L)$ and integrate from 0 to L. Noting that†

$$\int_0^L \sin \frac{n\pi x}{L} \sin \frac{m\pi x}{L} dx = \frac{L}{2} \delta_{mn}$$

where δ_{mn} is the Kronecker delta, we may readily solve for A_m and B_m as follows:

$$A_m = \frac{2}{L} \int_0^L \phi(x) \sin \frac{m\pi x}{L} dx \tag{7.29a}$$

$$B_m = \frac{2}{L\omega_n} \int_0^L \psi(x) \sin \frac{m\pi x}{L} dx \tag{7.29b}$$

The free vibrations of the simply supported beam are now fully determined.

We may readily solve for free vibrations of beams under different end conditions. As exercises you will be asked to verify the following results.

(a) Cantilevered Beam (Fig. 7.2*a*)

$$\omega_n = k_n^2 \left(\frac{EI}{\rho A} \right)^{1/2} \tag{7.30a}$$

where k_n is determined by the following frequency equation:

$$\cos k_n L \cosh k_n L = -1 \tag{7.30b}$$

The eigenfunctions are

$$W_n(x) = \cosh k_n x - \cos k_n x - \frac{\cos k_n L + \cosh k_n L}{\sin k_n L + \sinh k_n L} (\sinh k_n x - \sin k_n x) \tag{7.30c}$$

†We will discuss orthogonality of eigenfunctions in greater generality later in the chapter.

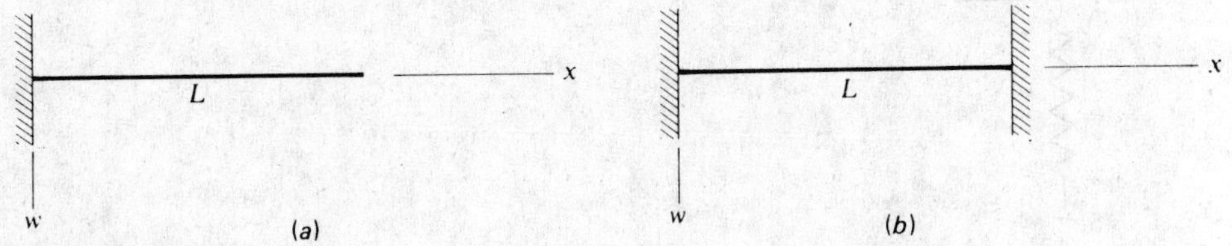

FIGURE 7.2
Cantilevered and clamped beams.

(b) Beam Clamped at Both Ends (Fig. 7.2*b*)†

$$\omega_n^2 = k_n^4 \frac{EI}{\rho A} \tag{7.31a}$$

where k_n is determined by the following frequency equation:

$$\cos kL \cosh kL = 1 \tag{7.31b}$$

The eigenfunctions are

$$W_n(x) = \cosh k_n x - \cos k_n x - \frac{\cos k_n L - \cosh k_n L}{\sin k_n L - \sinh k_n L} (\sinh k_n x - \sin k_n x)$$

Since the knowledge of the lowest natural frequency of a beam is of importance in engineering applications, we shall now present a powerful technique for approximating this frequency for beams. This technique will be particularly useful when the beams do not have uniform geometry and mechanical properties (i.e., I and/or E are not constant but are functions of position), so that analytical approaches of the type presented above are not straightforward and simple.

7.5 RAYLEIGH'S METHOD FOR BEAMS

We shall now present an approximate method for determining the natural frequency ω_1 for beams. Recall from the free vibration of an elementary spring-mass system (see Fig. 7.3) that the potential energy of the spring has its greatest value when the kinetic energy of the mass is zero and vice versa. Because of conservation of mechanical energy one can then say‡

$$T_{\max} = U_{\max}$$

therefore

$$\tfrac{1}{2} M \dot{x}_{\max}^2 = \tfrac{1}{2} k x_{\max}^2 \tag{7.32}$$

†For a free-free beam the frequency equation is the same as that for a beam clamped at both ends (i.e., the clamped-clamped beam). See exercises.

‡The same results are obtainable by equating mean kinematic and potential energies.

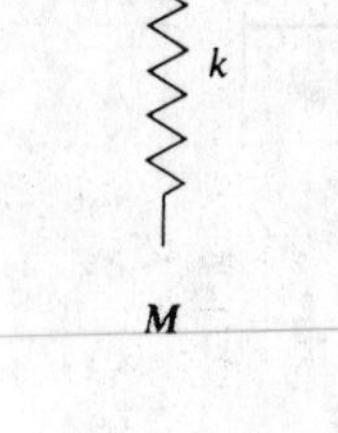

FIGURE 7.3
Spring mass system.

Since the system is vibrating sinusoidally, we can furthermore say

$$x = A \sin \omega t$$

therefore

$$\dot{x} = A\omega \cos \omega t$$

Accordingly, we conclude

$$x_{\max} = A, \quad \dot{x}_{\max} = A\omega$$

Now going back to Eq. (7.32), we have

$$\tfrac{1}{2}M(A\omega)^2 = \tfrac{1}{2}kA^2$$

therefore

$$\omega = \sqrt{\frac{k}{M}}$$

We are thus able to determine the natural frequency ω.

For a beam vibrating in a natural mode $W_n(x)$ with frequency ω_n we have similarly a kinetic energy

$$T = \int_0^L \frac{\rho A}{2} \dot{w}_n^2 \, dx$$

and a strain energy

$$U = \int_0^L \frac{EI}{2} (w_n'')^2 \, dx$$

Since $w_n = W_n C_n \sin(\omega_n t + \alpha_n)$, we have

$$T = \frac{1}{2} \int_0^L W_n^2 \rho A C_n^2 \omega_n^2 \cos^2(\omega_n t + \alpha_n) \, dx$$

$$U = \frac{1}{2} \int_0^L EI\, C_n^2 (W_n'')^2 \sin^2(\omega_n t + \alpha_n) \, dx$$

For reasons of conservation of mechanical energy, we equate the maximum values of T and U, as in the earlier case, to form

$$T_{\max} = \frac{1}{2} \omega_n^2 C_n^2 \int_0^L \rho A W_n^2 \, dx = U_{\max} = \frac{1}{2} C_n^2 \int_0^L EI (W_n'')^2 \, dx$$

From this we may solve for ω_n^2 as follows:

$$\omega_n^2 = \frac{\int_0^L EI(W_n'')^2\,dx}{\int_0^L \rho A W_n^2\,dx} \tag{7.33}$$

The expression on the right side is the so-called ***Rayleigh quotient*** applied to beams. With the correct eigenfunction W_n in the quotient, clearly we get the proper eigenvalue ω_n^2. We now demonstrate that we may get good approximate results for an eigenvalue ω_1^2 if, in place of W_1 for the first mode, we use a function $\tilde{W}_1$ that on the one hand satisfies the boundary conditions of the problem and on the other hand reasonably resembles the first mode function W_1. Since the mode shape W_1 corresponding to the lowest eigenvalue is usually the only one that can readily be approximated as described above (one can use the *deflection under its own weight* for such purposes), we have accordingly restricted the approach to this case. Thus by using $\tilde{W}_1$ in the Rayleigh quotient, we obtain an approximate eigenvalue $\tilde{\omega}_1^2$. We will show (Part C) that Rayleigh's quotient is a functional that has an *extremum* with respect to admissible functions $\tilde{W}_1$ satisfying the boundary conditions, the extremal function being the eigenfunction W_1. This means that using a function $\tilde{W}_1$ in the Rayleigh quotient such that $(\tilde{W}_1 - W_1)$ has a small average value over the beam will result in an even *smaller* value for $(\tilde{\omega}_1^2 - \omega_1^2)$. This is the reason for the success of the Rayleigh method.† Furthermore, we shall now demonstrate and later prove (Part C) that the approximation of ω_1 is always *from above*–i.e., $(\tilde{\omega}_1^2 - \omega_1^2) \geqslant 0$.

EXAMPLE 7.1 We now illustrate the use of the Rayleigh quotient by estimating the fundamental frequency of a cantilever beam (see Fig. 7.4*a*). We choose the following function to approximate the fundamental mode:

$$W_{\text{app}} = \frac{\rho A}{24EI}(x^4 - 4Lx^3 + 6L^2x^2) \tag{7.34}$$

†This is analogous to an extremal condition for a function $y(x)$ at $x = a$. A small departure from the extremal position "a" [the departure being analogous to $(\tilde{W}_1 - W_1)_{\text{av}}$] results in an even smaller change in $y(x)$ [the change being analogous to $(\tilde{\omega}_1^2 - \omega_1^2)$].

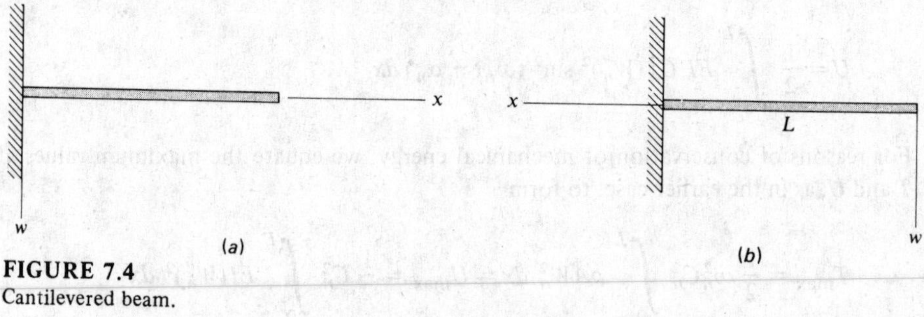

FIGURE 7.4
Cantilevered beam.

You may readily show that this is the static deflection equation for the cantilever beam and accordingly satisfies most boundary conditions of the problem. Substituting this function into the Rayleigh quotient gives us

$$\omega = \frac{3.53}{L^2}\sqrt{\frac{EI}{\rho A}} \tag{7.35}$$

Now going back to the exact solution of the cantilever problem we may show by a process of trial and error that the smallest root of the frequency equation [Eq. (7.30*b*)] is

$$k_1 L = 1.875$$

Hence from Eq. (7.30*a*) we have

$$\omega_1 = \frac{(1.875)^2}{L^2}\sqrt{\frac{EI}{\rho A}} = \frac{3.52}{L^2}\sqrt{\frac{EI}{\rho A}} \tag{7.36}$$

The closeness of the result speaks for itself.

As a second approximation we assume

$$W_{\text{app}} = a\left(1 - \cos\frac{\pi x}{2L}\right) \tag{7.37}$$

where a is an undetermined constant. This function satisfies both the geometric boundary condition at the clamped end at $x = 0$ [i.e., $W(0, t) = \partial W(0, t)/\partial x = 0$] and also satisfies one of the two natural boundary conditions at the free end [$\partial^2 W(L, t)/\partial x^2 = 0$]. Upon substituting this approximation into the Rayleigh quotient we get the following result:†

$$\omega = \frac{(1.915)^2}{L^2}\sqrt{\frac{EI}{\rho A}} = \frac{3.667}{L^2}\sqrt{\frac{EI}{\rho A}}$$

While still a reasonably good approximation, it is not as good as the one found by using the static deflection curve. Upon that in both cases we have approximations from above, as noted (without proof) earlier.

As a final step in this section we now investigate the significance of the rotatory inertia term neglected in the development of the equations of motion of the beam in Section 7.3. We consider for this purpose the simply supported beam whose natural frequencies we have already determined. The Rayleigh quotient with the rotatory inertia included becomes (as you may demonstrate) for a uniform beam

$$\omega^2 = \frac{EI\int_0^L [W''(x)]^2\,dx}{\rho A\int_0^L [W(x)]^2\,dx + \rho I\int_0^L [W'(x)]^2\,dx} \tag{7.38}$$

†This example demonstrates that we can still get acceptable results if not all the natural boundary conditions are satisfied at an endpoint of the beam.

We take as the approximate function sin $\pi x/L$, which is the exact fundamental mode shape of the solution with no rotatory inertia.† By a simple calculation we get

$$\omega^2 = \frac{EI\pi^4/\rho AL^4}{1 + (\pi^2/12)(h/L)^2}$$

The numerator represents the fundamental frequency from the elementary theory. We then conclude that for long thin beams ($h/L \ll 1$) the effect of rotatory inertia is only a slight reduction of the fundamental frequency from the elementary result. For such beams and for low-order mode shapes, we are then justified in neglecting rotatory inertia. (In a later section we shall consider rotatory inertia and transverse shear in the Timoshenko beam.)

7.6 RAYLEIGH–RITZ METHOD FOR BEAMS

In the preceding section we used the Rayleigh quotient to give an upper bound to ω_1, the fundamental frequency of vibration of a beam. Although we presented the Rayleigh quotient in terms of beams, the principles set forth apply to any free vibration of a linear elastic body. Indeed, we shall use the Rayleigh quotient later in this chapter for estimating the fundamental natural frequency of certain plate problems, and in Part C of the chapter we shall present a rather general discussion of the Rayleigh quotient with supporting arguments for the conclusions of the previous section, substantiated up to now only by examples.

We shall now employ the Ritz approach of Chapter 3 in conjunction with the Rayleigh quotient to form the very valuable Rayleigh-Ritz method. Instead of employing a single function ϕ for the Rayleigh quotient, we shall now select a set of n linearly independent functions‡ ϕ_i, each satisfying the boundary conditions of the problem, to represent W_{app} as a parameter-laden sum as follows

$$W_{\text{app}} = A_1\phi_1 + \cdots + A_n\phi_n \tag{7.39}$$

where the A_i are undetermined constants. Now substituting this approximate function into the Rayleigh quotient and denoting it as Λ^2, we get

$$\Lambda^2 = \frac{\displaystyle\int_0^L EI\left(\sum_{i=1}^n A_i\phi_i''\right)^2 dx}{\displaystyle\int_0^L \rho A\left(\sum_{i=1}^n A_i\phi_i\right)^2 dx} \tag{7.40}$$

Introducing the following notation

$$a_{ij} = \int_0^L EI\phi_i''\phi_j''\,dx \qquad D_{ij} = \int_0^L \rho A\phi_i\phi_j\,dx \tag{7.41}$$

†Actually, the exact solution for the fundamental mode with rotatory inertia is the same as the exact solution for the simpler theory.

‡A linearly independent set of functions ϕ_i can sum up to zero identically ($\Sigma_i A_i\phi_i = 0$) only if all the coefficients A_i are zero.

we get for Λ^2

$$\Lambda^2 = \frac{\sum_{i=1}^{n} \sum_{j=1}^{n} a_{ij} A_i A_j}{\sum_{i=1}^{n} \sum_{j=1}^{n} D_{ij} A_i A_j} \tag{7.42}$$

Now we extremize Λ^2 with respect to the coefficients A_i as follows:

$$\frac{\partial(\Lambda^2)}{\partial A_i} = 0$$

therefore

$$\frac{2 \sum_{j=1}^{n} a_{ij} A_j}{\sum_{i=1}^{n} \sum_{j=1}^{n} D_{ij} A_i A_j} - \frac{2\left(\sum_{i=1}^{n} \sum_{j=1}^{n} a_{ij} A_i A_j\right)\left(\sum_{j=1}^{n} D_{ij} A_j\right)}{\left(\sum_{j=1}^{n} \sum_{i=1}^{n} D_{ij} A_i A_j\right)^2} = 0 \qquad i = 1, 2, \ldots, n$$

Canceling $2/\Sigma_{i=1}^{n} \Sigma_{j=1}^{n} D_{ij} A_i A_j$ and employing Eq. (7.42) in the second term, we get

$$\sum_{j=1}^{n} a_{ij} A_j - \Lambda^2 \sum_{j=1}^{n} D_{ij} A_j = 0$$

therefore

$$\sum_{j=1}^{n} (a_{ij} - \Lambda^2 D_{ij}) A_j = 0 \qquad i = 1, 2, \ldots, n \tag{7.43}$$

We thus have a homogeneous system of n equations for the n constants A_j. For a non-trivial solution the determinant of the coefficients must be zero. We thus get the frequency equation familiar from the vibration of discrete systems:

$$|a_{ij} - \Lambda^2 D_{ij}| = 0 \tag{7.44}$$

Satisfaction of the above determinant will yield an nth-order equation in Λ^2 having n roots $\Lambda_1^2, \Lambda_2^2, \ldots, \Lambda_n^2$, which we arrange in ascending order of magnitude. For each root Λ_j^2 we can determine from Eq. (7.43) the ratios of a set of constants $A_i^{(j)}$ so that we form a function $W_{\text{app}}^{(j)}$ as follows:

$$W_{\text{app}}^{(j)} = \sum_{i=1}^{n} A_i^{(j)} \phi_i \qquad j = 1, 2, \ldots, n \tag{7.45}$$

We will demonstrate now by example and will later prove that the roots Λ_j^2 *are approximations from above of the first n natural frequencies of the system.* And Eq. (7.45) represents *approximations of the first n mode shapes of the system.* (The latter will not be nearly as good approximations as the frequencies.)

What we are doing here is representing a beam with an infinite number of degrees of freedom by one with n degrees of freedom. Naturally, the more functions ϕ_i that are employed, the better the approximation Λ_j^2 should be for the natural frequencies. The method is most successful when we have a general notion of the shape of the modes so that the functions ϕ_i can be chosen to reasonably resemble them.

EXAMPLE 7.2 We go back to the cantilever beam of Example 7.1 (see Fig. 7.4*b*) and employ for the Rayleigh-Ritz method two functions:†

$$\phi_1 = \left(1 - \frac{x}{L}\right)^2$$

$$\phi_2 = \frac{x}{L}\left(1 - \frac{x}{L}\right)^2$$

These functions both satisfy the kinematic boundary conditions of the problem at the support and have a resemblance to what can be expected for the first and second mode shapes. (See Fig. 7.5*a*.) We form the approximate function W_{app} as follows:

$$W_{\text{app}} = A_1\left(1 - \frac{x}{L}\right)^2 + A_2 \frac{x}{L}\left(1 - \frac{x}{L}\right)^2$$

The constants a_{ij} and D_{ij} are then evaluated as follows:

$$a_{11} = EI \int_0^L (\phi_1'')^2 \, dx = EI \int_0^L \left(\frac{2}{L^2}\right)^2 dx = \frac{4EI}{L^3}$$

$$a_{21} = a_{12} = EI \int_0^L \phi_1'' \phi_2'' \, dx = EI \int_0^L \frac{2}{L^2}\frac{2}{L^2}\left(3\frac{x}{L} - 2\right) dx = -\frac{2EI}{L^3}$$

†As in the Rayleigh method, this example shows that good results can often be found by using functions ϕ_i that do not satisfy natural boundary conditions at one endpoint of the beam.

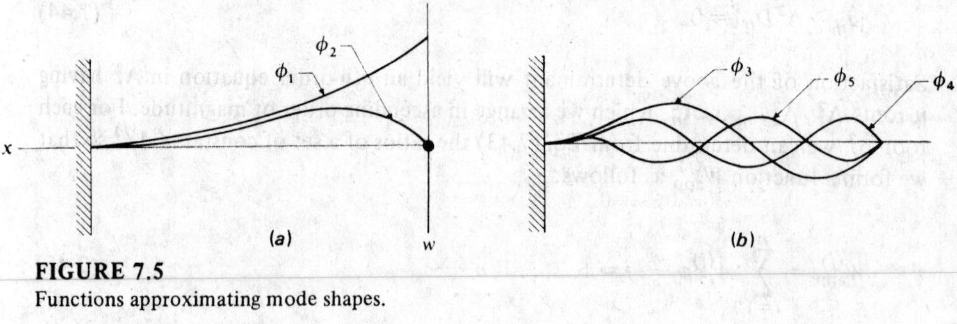

FIGURE 7.5
Functions approximating mode shapes.

$$a_{22} = EI \int_0^L (\phi_2'')^2 \, dx = EI \int_0^L \left(\frac{2}{L}\right)^2 \left(3\,\frac{x}{L} - 2\right)^2 dx = \frac{4EI}{L^3}$$

$$D_{11} = \rho A \int_0^L \left(1 - \frac{x}{L}\right)^4 dx = \frac{\rho LA}{5}$$

$$D_{21} = D_{12} = \rho A \int_0^L \left(1 - \frac{x}{L}\right)^2 \frac{x}{L}\left(1 - \frac{x}{L}\right)^2 dx = \frac{\rho AL}{30}$$

$$D_{22} = \rho A \int_0^L \left(1 - \frac{x}{L}\right)^4 \left(\frac{x}{L}\right)^2 dx = \frac{\rho AL}{105}$$

The frequency determinant then becomes

$$\begin{vmatrix} \dfrac{4EI}{L^3} - \Lambda^2\left(\dfrac{\rho LA}{5}\right) & \dfrac{-2EI}{L^3} - \Lambda^2\left(\dfrac{\rho AL}{30}\right) \\ \dfrac{-2EI}{L^3} - \Lambda^2\left(\dfrac{\rho AL}{30}\right) & \dfrac{4EI}{L^3} - \Lambda^2\left(\dfrac{\rho AL}{105}\right) \end{vmatrix} = 0$$

We then get, on carrying out the determinant,

$$\left[\frac{4EI}{L^3} - \Lambda^2\left(\frac{\rho LA}{5}\right)\right]\left[\frac{4EI}{L^3} - \Lambda^2\left(\frac{\rho AL}{105}\right)\right] - \left[\frac{2EI}{L^3} + \Lambda^2\left(\frac{\rho AL}{30}\right)\right]^2 = 0$$

Letting $4EI/L^3 = \alpha$ and $\rho LA = \beta$, we have

$$\left(\alpha - \frac{\beta}{5}\Lambda^2\right)\left(\alpha - \frac{\beta}{105}\Lambda^2\right) - \left(\frac{\alpha}{2} + \Lambda^2\,\frac{\beta}{30}\right)^2 = 0$$

We get the following equation in Λ from the above:

$$\beta^2(0.000794)\Lambda^4 - \alpha\beta(0.243)\Lambda^2 + 0.75\alpha^2 = 0$$

The roots are

$$\Lambda_1^2 = \frac{\alpha}{\beta}(3.15) = \frac{4EI}{\rho L^4 A}(3.15) = 12.60\,\frac{EI}{\rho L^4 A}$$

$$\Lambda_2^2 = \frac{\alpha}{\beta}(303) = \frac{4EI}{\rho L^4 A}(303) = 1212\,\frac{EI}{\rho L^4 A}$$

The results for the first two modes as developed exactly are

$$\omega_1^2 = k_1^4\,\frac{EI}{\rho AL^4} = (1.875)^4\,\frac{EI}{\rho L^4 A} = 12.30\,\frac{EI}{\rho L^4 A}$$

$$\omega_2^2 = k_2^4\,\frac{EI}{\rho AL^4} = (4.694)^4\,\frac{EI}{\rho L^4 A} = 483\,\frac{EI}{\rho L^4 A}$$

While there is a very close approximation from above for the first mode, we get a very poor result for the natural frequency of the second mode.

We now shall attempt an improvement in the evaluation of ω_2 by employing six functions having the following form:

$$\phi_1 = \left(1 - \frac{x}{L}\right)^2$$

$$\phi_2 = \frac{x}{L}\left(1 - \frac{x}{L}\right)^2$$

$$\phi_3 = \left(\frac{x}{L} - 0.5\right)\frac{x}{L}\left(1 - \frac{x}{L}\right)^2$$

$$\phi_4 = \left(\frac{x}{L} - 0.75\right)\left(\frac{x}{L} - 0.25\right)\frac{x}{L}\left(1 - \frac{x}{L}\right)^2$$

$$\phi_5 = \left(\frac{x}{L} - 0.2\right)\left(\frac{x}{L} - 0.5\right)\left(\frac{x}{L} - 0.8\right)\frac{x}{L}\left(1 - \frac{x}{L}\right)^2$$

$$\phi_6 = \left(\frac{x}{L} - 0.18\right)\left(\frac{x}{L} - 0.34\right)\left(\frac{x}{L} - 0.6\right)\left(\frac{x}{L} - 0.84\right)\frac{x}{L}\left(1 - \frac{x}{L}\right)^2$$

Figure 7.5*b* illustrates some of the new functions chosen. Notice that they have increasing numbers of nodal points, as is to be expected for ascending modes. The results are given in Table 7.1, where the first column indicates the number of functions used from the above list and where $\gamma = EI/\rho AL^4$. Notice the convergence from above of the results particularly for the first four modes. In Table 7.2 we show a comparison of the results with the exact results. We can get a reasonably good result up to the fourth eigenvalue. It is clear from the computation here that it requires many functions and much work to get higher eigenvalues.

*7.7 THE TIMOSHENKO BEAM

As a final step in our study of the dynamics of beams we go to the Timoshenko beam, where you will recall we include the effects of both transverse shear and rotatory inertia. From Eq. (4.30) for no axial force, we assume the following displacement field for such beams:

$$\begin{aligned} u_1(x,y,z,t) &= -z\psi(x,t) \\ u_2(x,y,z,t) &= 0 \\ u_3(x,y,z,t) &= w(x,t) \end{aligned} \tag{7.46}$$

TABLE 7.1
Natural Frequency Results Using Rayleigh-Ritz

n	Λ_1^2/γ	Λ_2^2/γ	Λ_3^2/γ	Λ_4^2/γ	Λ_5^2/γ	Λ_6^2/γ
2	12.480	1211.5				
3	12.400	494.3	13,958.1			
4	12.362	491.0	4,012.8	79,296.4		
5	12.362	485.5	3,999.3	16,517.2	316,640	
6	12.362	485.5	3,880.7	16,507.3	51,790	1,030,864

TABLE 7.2
Rayleigh-Ritz and Exact Results

Approximate solutions	Exact solutions
$\Lambda_1^2/\gamma = 12.36$	$\omega_1^2/\gamma = 12.36$
$\Lambda_2^2/\gamma = 485.5$	$\omega_2^2/\gamma = 485.48$
$\Lambda_3^2/\gamma = 3{,}807$	$\omega_3^2/\gamma = 3{,}807$
$\Lambda_4^2/\gamma = 16{,}507$	$\omega_4^2/\gamma = 14{,}617$
$\Lambda_5^2/\gamma = 51{,}790$	$\omega_5^2/\gamma = 39{,}944$
$\Lambda_6^2/\gamma = 1{,}030{,}684$	$\omega_6^2/\gamma = 173{,}881$

where the functions are now dependent on time. The nonzero strains are then

$$\epsilon_{xx} = -z\,\frac{\partial \psi}{\partial x} \tag{7.47a}$$

$$\epsilon_{xz} = \frac{1}{2}\left(\frac{\partial w}{\partial x} - \psi\right) \tag{7.47b}$$

Note that ϵ_{xz} is thus taken as constant over a section in the above approximation. Considering the cross section of the beam to be rectangular $h \times b$ (see Fig. 7.6) as in the development of Chapter 4 and using the same approximation $\tau_{xx} = E\epsilon_{xx}$ as was employed earlier, we get for the bending moment at a section the following result with the aid of Eq. (7.47*a*):

$$M = \int_{-h/2}^{h/2} \tau_{xx}\, zb\, dz = -EI\,\frac{\partial \psi}{\partial x} \tag{7.48}$$

We can give the shear force V_S by using Hooke's law $\tau_{xz} = 2\epsilon_{xz}G$ and Eq. (7.47*b*) as follows:

$$V_S = \int_{-h/2}^{h/2} \tau_{xz}\, b\, dz = kGA\left(\frac{\partial w}{\partial x} - \psi\right) \tag{7.49}$$

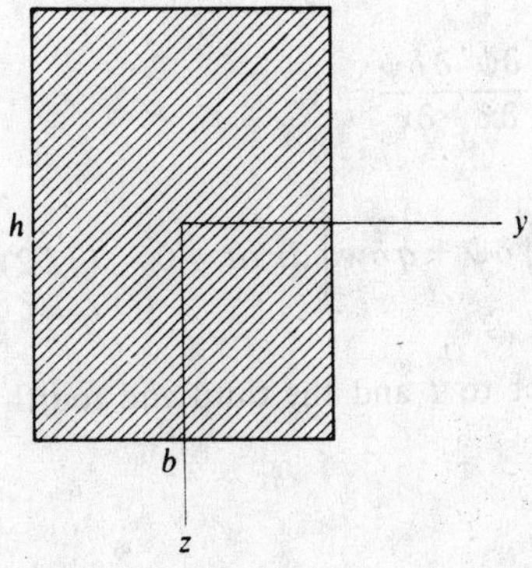

FIGURE 7.6
Beam with rectangular cross section.

where k is the shear constant which we insert as a correction factor to account for the fact that τ_{xz} is not (in reality) uniform over the height of the section.

We may now express the energies that will be needed for Hamilton's principle. For the kinetic energy we have

$$T = \frac{1}{2}\int_{-h/2}^{h/2}\int_{-b/2}^{b/2}\int_{0}^{L} \rho \dot{u}_i \dot{u}_i \, dx\, dy\, dz$$

$$= \frac{1}{2}\int_{-h/2}^{h/2}\int_{-b/2}^{b/2}\int_{0}^{L} \rho \left[z^2 \left(\frac{\partial \psi}{\partial t}\right)^2 + \left(\frac{\partial w}{\partial t}\right)^2 \right] dx\, dy\, dz$$

$$= \frac{1}{2}\int_{0}^{L} \left[\rho I \left(\frac{\partial \psi}{\partial t}\right)^2 + \rho A \left(\frac{\partial w}{\partial t}\right)^2 \right] dx \tag{7.50}$$

Note: we now retain the kinetic energy due to rotation of elements of the beam. As for the strain energy plus the potential energy of the loading q we employ Eq. (4.37) as follows:

$$\pi = U + V = \int_{0}^{L} \left[\frac{EI}{2}\left(\frac{\partial \psi}{\partial x}\right)^2 + \frac{kGA}{2}\left(\frac{\partial w}{\partial x} - \psi\right)^2 - qw \right] dx \tag{7.51}$$

Accordingly, Hamilton's principle is given as

$$\delta^{(1)} \int_{t_1}^{t_2} L\, dt = \delta^{(1)} \int_{t_1}^{t_2}\int_{0}^{L} \left\{ \frac{1}{2}\left[\rho I \left(\frac{\partial \psi}{\partial t}\right)^2 + \rho A \left(\frac{\partial w}{\partial t}\right)^2 \right] - \frac{EI}{2}\left(\frac{\partial \psi}{\partial x}\right)^2 - \frac{kGA}{2}\left(\frac{\partial w}{\partial x} - \psi\right)^2 + qw \right\} dx\, dt = 0$$

The integrand here is composed of two functions to be varied, w and ψ. Carrying out the first variation, we get

$$\int_{t_1}^{t_2}\int_{0}^{L} \left[\rho I \frac{\partial \psi}{\partial t}\frac{\partial\, \delta\psi}{\partial t} + \rho A \frac{\partial w}{\partial t}\frac{\partial\, \delta w}{\partial t} - EI \frac{\partial \psi}{\partial x}\frac{\partial\, \delta\psi}{\partial x} - kGA\left(\frac{\partial w}{\partial x} - \psi\right)\frac{\partial\, \delta w}{\partial x} + kGA\left(\frac{\partial w}{\partial x} - \psi\right)\delta\psi + q\,\delta w \right] dx\, dt = 0 \tag{7.52}$$

Integrating the first two terms by parts with respect to t and the third and fourth terms with respect to x, we get

$$\int_{t_1}^{t_2}\int_0^L\left\{-\frac{\partial}{\partial t}(\rho I\dot{\psi})\,\delta\psi-\frac{\partial}{\partial t}(\rho A\dot{w})\,\delta w+\frac{\partial}{\partial x}\left(EI\frac{\partial\psi}{\partial x}\right)\delta\psi\right.$$

$$\left.+\frac{\partial}{\partial x}\left[kGA\left(\frac{\partial w}{\partial x}-\psi\right)\right]\delta w+kGA\left(\frac{\partial w}{\partial x}-\psi\right)\delta\psi+q\,\delta w\right\}dx\,dt$$

$$+\int_0^L\rho I\frac{\partial\psi}{\partial t}\,\delta\psi\Big|_{t_1}^{t_2}dx+\int_0^L\rho A\frac{\partial w}{\partial t}\,\delta w\Big|_{t_1}^{t_2}dx$$

$$-\int_{t_1}^{t_2}EI\frac{\partial\psi}{\partial x}\,\delta\psi\Big|_{x_1}^{x_2}dt-\int_{t_1}^{t_2}kGA\left(\frac{\partial w}{\partial x}-\psi\right)\delta w\Big|_{x_1}^{x_2}dt=0 \quad (7.53)$$

Noticing that $\delta\psi=\delta w=0$ at times t_1 and t_2, we then obtain on grouping terms in the above equation:

$$\int_{t_1}^{t_2}\int_0^L\left[\left\{-\frac{\partial}{\partial t}(\rho I\dot{\psi})+\frac{\partial}{\partial x}\left(EI\frac{\partial\psi}{\partial x}\right)+kGA\left(\frac{\partial w}{\partial x}-\psi\right)\right\}\delta\psi\right.$$

$$\left.+\left\{-\frac{\partial}{\partial t}(\rho A\dot{w})+\frac{\partial}{\partial x}\left[kGA\left(\frac{\partial w}{\partial x}-\psi\right)\right]+q\right\}\delta w\right]dx$$

$$-\int_{t_1}^{t_2}EI\frac{\partial\psi}{\partial x}\,\delta\psi\Big|_0^L dt-\int_{t_1}^{t_2}kGA\left(\frac{\partial w}{\partial x}-\psi\right)\delta w\Big|_0^L dt=0$$

The Euler-Lagrange equations for the Timoshenko beam are accordingly:

$$\boxed{\begin{aligned}&-\frac{\partial}{\partial t}(\rho A\dot{w})+\frac{\partial}{\partial x}[kGA(w_x-\psi)]+q=0\\&-\frac{\partial}{\partial t}(\rho I\dot{\psi})+\frac{\partial}{\partial x}(EI\psi_x)+kGA(w_x-\psi)=0\end{aligned}} \quad (7.54)$$

For constant values of E, I, G, A, etc., we then have

$$\rho A\ddot{w}-kGA(w_{xx}-\psi_x)-q=0 \quad (7.55a)$$

$$\rho I\ddot{\psi}-EI\psi_{xx}-kGA(w_x-\psi)=0 \quad (7.55b)$$

The boundary conditions at the ends then become:

$$\begin{aligned}&\text{EITHER } w_x-\psi=0 \quad \text{OR} \quad w \text{ IS SPECIFIED}\\&\text{EITHER } \psi_x=0 \quad \text{OR} \quad \psi \text{ IS SPECIFIED}\end{aligned} \quad (7.56)$$

Noting Eqs. (7.48) and (7.49), we may express Eqs. (7.55) in terms of V_s and M as follows:

$$\begin{aligned} \frac{\partial V_s}{\partial x} &= \rho A\ddot{w} - q \\ V_s - \frac{\partial M}{\partial x} &= \rho I\ddot{\psi} \end{aligned} \tag{7.57}$$

where at the ends

$$\begin{aligned} &\text{EITHER} \quad V_s = 0 \quad \text{OR} \quad w \text{ IS SPECIFIED} \\ &\text{EITHER} \quad M = 0 \quad \text{OR} \quad \psi \text{ IS SPECIFIED} \end{aligned} \tag{7.58}$$

We can write Eqs. (7.55) in terms of the transverse displacement w alone by solving for ψ in Eq. (7.55*a*) and substituting into Eq. (7.55*b*). We get, on rearranging terms, the single Timoshenko beam equation:

$$\begin{aligned} EI\frac{\partial^4 w}{\partial x^4} + \rho A\frac{\partial^2 w}{\partial t^2} - \rho I\left(1 + \frac{E}{kG}\right)\frac{\partial^4 w}{\partial t^2\,\partial x^2} + \frac{\rho^2 I}{kG}\frac{\partial^4 w}{\partial t^4} \\ = q + \frac{\rho I}{kGA}\frac{\partial^2 q}{\partial t^2} - \frac{EI}{kGA}\frac{\partial^2 q}{\partial x^2} \end{aligned} \tag{7.59}$$

We shall now consider the free vibrations of a beam, using Timoshenko beam theory. The appropriate form of Eq. (7.59) with $q = 0$ is restated as follows:

$$EI\frac{\partial^4 w}{\partial x^4} + \rho A\frac{\partial^2 w}{\partial t^2} - \rho I\left(1 + \frac{E}{kG}\right)\frac{\partial w^4}{\partial t^2\,\partial x^2} + \frac{\rho^2 I}{kG}\frac{\partial^4 w}{\partial t^4} = 0 \tag{7.60}$$

We assume a separation of variables for the above equation in the form

$$w(x, t) = W(x)\cos\omega t \tag{7.61}$$

That is, we assume the bar is oscillating at a frequency ω with a mode shape $W(x)$. Substituting the above solution into Eq. (7.60), we get for W:

$$EIW^{\text{IV}} - \rho A\omega^2 W + \rho I\left(1 + \frac{E}{kG}\right)\omega^2 W'' + \frac{\rho^2 I}{kG}\omega^4 W = 0 \tag{7.62}$$

The general solution to this ordinary differential equation has the form

$$W = A\sin\lambda x + B\cos\lambda x + C\sinh\lambda x + D\cosh\lambda x$$

We now take the case where the beam is simply supported at both ends. We shall employ as boundary conditions for w the correct kinematic condition $w = 0$ at the ends and as an approximation to zero moment will use here the classical natural boundary condition $w'' = 0$ at the ends. These boundary conditions can be satisfied by taking $B = C = D = 0$ and having $\lambda = n\pi/L$ for $n = 1, 2, \ldots$. Solutions from these steps can then be given as follows:

$$W_n = A_n \sin \frac{n\pi x}{L} \tag{7.63}$$

By substituting the above result back into Eq. (7.62) we may then determine the allowable values of ω_n to be associated with the eigenfunctions $\sin n\pi x/L$. Thus we get (on canceling $A_n \sin n\pi x/L$):

$$EI\left(\frac{n\pi}{L}\right)^4 - \rho A\omega_n^2 - \rho I\left(1 + \frac{E}{kG}\right)\omega_n^2\left(\frac{n\pi}{L}\right)^2 + \frac{\rho^2 I}{kG}\omega_n^4 = 0 \tag{7.64}$$

Dividing through by $-EI/L^4$ and rearranging terms, we get

$$-\frac{\rho^2 L^4}{EkG}\omega_n^4 + \left[\frac{\rho A L^4}{EI} + \frac{\rho L^2}{E}\left(1 + \frac{E}{kG}\right)(n\pi)^2\right]\omega_n^2 - (n\pi)^4 = 0 \tag{7.65}$$

Now introduce the following terms

$$\Omega_n^2 = (\text{dimensionless frequency})^2 = \frac{\rho I}{EA}\omega_n^2 \tag{7.66a}$$

$$r^2 = (\text{radius of gyration})^2 = \frac{I}{A} \tag{7.66b}$$

into Eq. (7.65). We get

$$-\frac{E}{kG}\left(\frac{L}{r}\right)^4\Omega_n^4 + \left[\left(\frac{L}{r}\right)^4 + \left(\frac{L}{r}\right)^2\left(1 + \frac{E}{kG}\right)(n\pi)^2\right]\Omega_n^2 - (n\pi)^4 = 0 \tag{7.67}$$

We would next like to show how our results can be altered so that we can omit shear and rotatory inertia and thus get back to earlier results for the case of the simply supported beam according to classical beam theory. If we go back to Eqs. (7.50) and (7.51), we see that by letting $k = 0$ and $\rho I = 0$ we get the energy quantities used for the simple beam theory. We cannot use this procedure in Eq. (7.67) because of the various operations performed in arriving at the equation (such as dividing through by k). We can proceed *formally*, however, to get to the simple beam theory. If we set equal to zero in Eq. (7.67) the expression $-(E/kG)(L/r)^4$ as well as $(L/r)^2 \times (1 + E/kG)$, we arrive at the following result:

$$\left(\frac{L}{r}\right)^4 \Omega_n^2 - (n\pi)^4 = 0$$

therefore

$$\Omega_n^2 = \frac{n^4\pi^4 r^4}{L^4} \tag{7.68}$$

From Eq. (7.66*a*) we can solve for ω_n, using the above result. Replacing r^2 by I/A we then have

$$\omega_n^2 = \frac{(n\pi)^4}{L^4}\frac{EI}{\rho A} \tag{7.69}$$

We recognize these frequencies as the natural frequencies given by the simple beam theory found in Section 7.4 [see Eq. (7.23)]. It should be clear that, of the expressions deleted, the ones containing the term k relate to transverse shear, and so if we wish formally to include *only rotatory inertia* we set equal to zero only the expressions

$$-\frac{E}{kG}\left(\frac{L}{r}\right)^4 \quad \text{and} \quad \left(\frac{L}{r}\right)^2\frac{E}{kG}$$

We then have from Eq. (7.67) the result

$$\left[\left(\frac{L}{r}\right)^4 + \left(\frac{L}{r}\right)^2 (n\pi)^2\right]\Omega_n^2 - (n\pi)^4 = 0$$

Hence the dimensionless frequency involving only rotatory inertia becomes

$$(\Omega_n^2)_{\text{rot inertia}} = \frac{(n\pi)^4}{(L/r)^4[1+(n\pi)^2(r/L)^2]} = \frac{(\Omega_n^2)_{\text{class}}}{[1+(n\pi)^2(r/L)^2]} \tag{7.70}$$

where $(\Omega_n^2)_{\text{class}}$ represents the dimensionless frequency for the classical beam in accordance with Eq. (7.68). To indicate the effects of rotatory inertia we have plotted $(\Omega_n^2)_{\text{rot inertia}}/(\Omega_n^2)_{\text{class}}$ as a function of (L/r) for various values of n as shown in Fig. 7.7. For small values of n and for large slenderness ratio L/r there is little difference between the classical frequency and that which includes rotatory inertia. However, for short stubby beams or for higher modes. the rotatory inertia can be significant.

Now if wish only to include *shear effects*, we again go back to Eq. (7.67). This time we formally delete the term $(L/r)^2(n\pi)^2$ as well as the term $-(E/kG) \times (L/r)^4\,\Omega_n^4$, since the latter contains *both* shear effects and rotatory inertia, as can be seen on tracing steps back to the original energy expressions for T and $U+V$. The frequency equation now becomes

$$\left[\left(\frac{L}{r}\right)^4 + \frac{E}{kG}\left(\frac{L}{r}\right)^2 (n\pi)^2\right](\Omega_n^2)_{\text{shear}} - (n\pi)^4 = 0$$

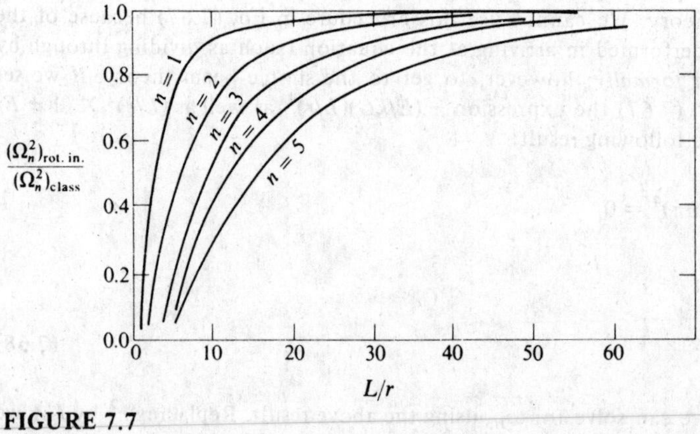

FIGURE 7.7

Curves showing effects of rotatory inertia on natural frequency of vibration.

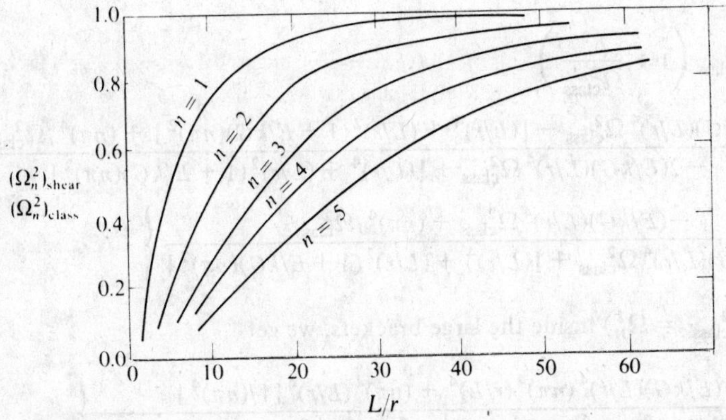

FIGURE 7.8
Curves showing effects of shear on the natural frequency of vibration.

therefore

$$(\Omega_n^2)_{\text{shear}} = \frac{(n\pi)^4}{(L/r)^4\,[1 + (E/kG)(r/L)^2\,(n\pi)^2\,]} = \frac{(\Omega_n^2)_{\text{class}}}{1 + (E/kG)(r/L)^2\,(n\pi)^2} \tag{7.71}$$

A plot of $(\Omega_n^2)_{\text{shear}}/(\Omega_n^2)_{\text{class}}$ versus L/r for $E/kG = 3.06$ and for various values of n has been included in Fig. 7.8. Note that for large values of n and for small values of L/r we get pronounced shear effects that must be taken into account.

We now turn to the full equation for Ω^4 including both shear and rotatory inertia effects. The frequency equation is too complicated to give useful information as it stands, so we will proceed by giving Ω^2 as follows:

$$\Omega^2 = \Omega_{\text{class}}^2 + \delta^2 \tag{7.72}$$

where we will in subsequent calculations neglect higher-order terms in δ^2. This approach will be valid when there is not a large effect from rotatory inertia and shear. Accordingly, substituting Eq. (7.72) into the frequency equation (7.67) we get

$$-\left(\frac{E}{kG}\right)\left(\frac{L}{r}\right)^4 (\Omega_{\text{class}}^4 + 2\delta^2\,\Omega_{\text{class}}^2) + \left[\left(\frac{L}{r}\right)^4 + \left(\frac{L}{r}\right)^2\left(1 + \frac{E}{kG}\right)(n\pi)^2\right](\Omega_{\text{class}}^2 + \delta^2) - (n\pi)^4 = 0$$

Solving for δ^2 from the above equation we have

$$\delta^2 = \frac{(E/kG)(L/r)^4\,\Omega_{\text{class}}^4 - [(L/r)^4 + (L/r)^2(1 + E/kG)(n\pi)^2\,]\,\Omega_{\text{class}}^2 + (n\pi)^4}{-2(E/kG)(L/r)^4\,\Omega_{\text{class}}^2 + [(L/r)^4 + (L/r)^2(1 + E/kG)(n\pi)^2]} \tag{7.73}$$

Hence going back to Eq. (7.72) and using Eq. (7.73) we have

$$\Omega^2 = \Omega^2_{\text{class}} + \delta^2 = \Omega^2_{\text{class}}\left(1 + \frac{\delta^2}{\Omega^2_{\text{class}}}\right)$$

$$= \Omega^2_{\text{class}}\left\{1 + \frac{(E/kG)(L/r)^4\Omega^2_{\text{class}} - [(L/r)^4 + (L/r)^2(1+E/kG)(n\pi)^2] + (n\pi)^4/\Omega^2_{\text{class}}}{-2(E/kG)(L/r)^4\Omega^2_{\text{class}} + [(L/r)^4 + (L/r)^2(1+E/kG)(n\pi)^2]}\right\}$$

$$= \Omega^2_{\text{class}}\left\{\frac{-(E/kG)(L/r)^4\Omega^2_{\text{class}} + (n\pi)^4/\Omega^2_{\text{class}}}{-2(E/kG)(L/r)^4\Omega^2_{\text{class}} + [(L/r)^4 + (L/r)^2(1+E/kG)(n\pi)^2]}\right\}$$

Using Eq. (7.68) for Ω^2_{class} $(= \Omega_n^2)$ inside the large brackets, we get

$$\Omega^2 = \Omega^2_{\text{class}}\left\{\frac{-(E/kG)(L/r)^4(n\pi)^4(r/L)^4 + (n\pi)^4(L/r)^4[1/(n\pi)^4]}{-2(E/kG)(L/r)^4(n\pi)^4(r/L)^4 + [(L/r)^4 + (L/r)^2(1+E/kG)(n\pi)^2]}\right\}$$

Canceling terms and multiplying numerator and denominator by $(r/L)^4$, we get

$$\Omega^2 = \Omega^2_{\text{class}}\left[\frac{1 - (E/kG)(n\pi)^4(r/L)^4}{1 + (1+E/kG)(n\pi)^2(r/L)^2 - 2(E/kG)(n\pi)^4(r/L)^4}\right] \tag{7.74}$$

For the case where $r/L < 1$ we may give the following simplification by dropping terms with $(r/L)^4$:

$$\Omega^2 \approx \Omega^2_{\text{class}}\left[\frac{1}{1 + (1+E/kG)(n\pi)^2(r/L)^2}\right]$$

$$\approx \Omega^2_{\text{class}}\left[1 - \left(1 + \frac{E}{kG}\right)\left(\frac{r}{L}\right)^2(n\pi)^2\right] \tag{7.75}$$

Figure 7.9 shows $\Omega^2/\Omega^2_{\text{class}}$ plotted against L/r for various values of n for a value of E/kG given as 3.06.

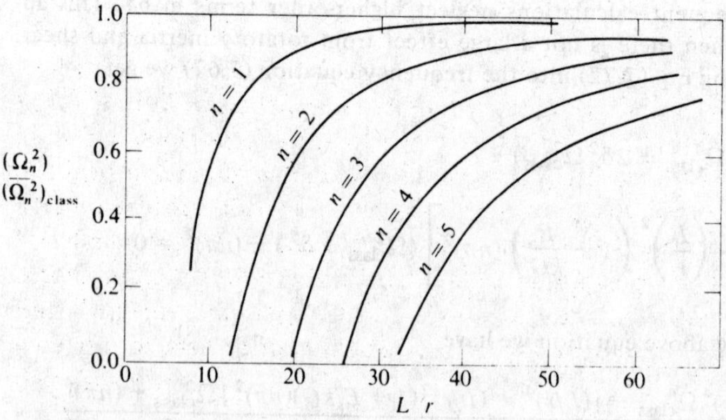

FIGURE 7.9
Combined effects of rotatory inertia and shear.

Part B
Plates

7.8 EQUATIONS OF MOTION FOR PLATES

We now proceed with a discussion for plates that parallels the discussion on beams. Accordingly, we shall first develop the equations of motion and boundary conditions for classical plate theory, using Hamilton's principle.

The kinetic energy of the plate is needed. For the classical case this results from vertical movement of elements of the plate only (rotatory inertia will be discussed later). Thus we have

$$T = \frac{1}{2} \iint_R h\rho \dot{w}^2 \, dx \, dy \tag{7.76}$$

where R denotes the transverse area of the plate. The total potential energy has been given by Eq. (6.30) as

$$U + V = \frac{D}{2} \iint_R \left\{ (\nabla^2 w)^2 + 2(1-\nu) \left[\left(\frac{\partial^2 w}{\partial x \, \partial y} \right)^2 - \frac{\partial^2 w}{\partial x^2} \frac{\partial^2 w}{\partial y^2} \right] \right\} dx \, dy$$
$$- \iint_R qw \, dx \, dy \tag{7.77}$$

Hamilton's principle then requires that

$$\delta^{(1)} \left[\int_{t_1}^{t_2} \iint_R \left(\frac{h\rho}{2} \dot{w}^2 - \frac{D}{2} \left\{ (\nabla^2 w)^2 + 2(1-\nu) \left[\left(\frac{\partial^2 w}{\partial x \, \partial y} \right)^2 - \frac{\partial^2 w}{\partial x^2} \frac{\partial^2 w}{\partial y^2} \right] \right\} \right.\right.$$
$$\left.\left. - qw \right) dx \, dy \, dt \right] = 0 \tag{7.78}$$

Carrying out the first variation by using the operator approach, we get

$$\int_{t_1}^{t_2} \iint_R \left\{ \rho h \frac{\partial w}{\partial t} \frac{\partial \delta w}{\partial t} - D \left[\nabla^2 w \, \nabla^2 \delta w + 2(1-\nu) \left(\frac{\partial^2 w}{\partial x \, \partial y} \frac{\partial^2 \delta w}{\partial x \, \partial y} \right.\right.\right.$$
$$\left.\left.\left. - \frac{1}{2} \frac{\partial^2 w}{\partial x^2} \frac{\partial^2 \delta w}{\partial y^2} - \frac{1}{2} \frac{\partial^2 w}{\partial y^2} \frac{\partial^2 \delta w}{\partial x^2} \right) \right] - q \, \delta w \right\} dx \, dy \, dt = 0 \tag{7.79}$$

Noting that $\nabla^2 \delta w = \partial^2 \delta w/\partial x^2 + \partial^2 \delta w/\partial y^2$, we next integrate by parts with respect to time for the first term in the integrand, and then successively with respect to the spatial coordinates for all but the last term. We get

$$\int_{t_1}^{t_2} \iint_R \left\{ -h\rho \frac{\partial^2 w}{\partial t^2} \delta w - D\left[\frac{\partial^2}{\partial x^2} \nabla^2 w + \frac{\partial^2}{\partial y^2} \nabla^2 w + 2(1-\nu)\left(\frac{\partial^4 w}{\partial x^2 \partial y^2} - \frac{1}{2}\frac{\partial^4 w}{\partial x^2 \partial y^2} - \frac{1}{2}\frac{\partial^4 w}{\partial x^2 \partial y^2} \right)\right] \delta w - q\,\delta w \right\} dx\,dy\,dt$$
$$+ \int_{t_1}^{t_2} \left[\oint_\Gamma (\cdots)\,dx + \oint_\Gamma (\cdots)\,dy \right] dt = 0 \qquad (7.80)$$

The line integrals themselves are identical to those developed in the static development of plates in Chapter 6 since the dynamic term from the kinetic energy makes no contribution to the line integrals. Accordingly, we shall not have to dwell on the line integrals here but will be able to use the results of Section 6.4. Collecting terms in the surface integrals, we may write the above equation as follows:

$$\int_{t_1}^{t_2} \left[\iint_R \left\{ -h\rho \ddot{w} + D\left[-\frac{\partial^2}{\partial x^2}(\nabla^2 w) - \frac{\partial^2}{\partial y^2}(\nabla^2 w) + q \right] \right\} \delta w\,dx\,dy \right.$$
$$\left. + \oint_\Gamma (\cdots)\,dx + \oint_\Gamma (\cdots)\,dy \right] dt = 0 \qquad (7.81)$$

This becomes

$$\int_{t_1}^{t_2} \left[\iint_R (-h\rho\ddot{w} - D\nabla^4 w + q)\,\delta w\,dx\,dy + \oint_\Gamma (\cdots)\,dx \right.$$
$$\left. + \oint_\Gamma (\cdots)\,dy \right] dt = 0 \qquad (7.82)$$

We can conclude from the above that the differential equation of motion is

$$\boxed{D\nabla^4 w + \rho h \ddot{w} = q} \qquad (7.83)$$

The boundary conditions stemming from the line integrals are given by Eq. (6.44) and are rewritten below. On the boundary

$$\text{EITHER } M_\nu = 0 \quad \text{OR} \quad \frac{\partial w}{\partial \nu} \text{ IS PRESCRIBED} \qquad (7.84a)$$

$$\text{EITHER} \quad Q_\nu + \frac{\partial M_{\nu s}}{\partial s} = 0 \quad \text{OR} \quad w \quad \text{IS PRESCRIBED} \tag{7.84b}$$

where ν is the outward normal direction from the boundary. The natural boundary conditions can be readily given in terms of w by employing Eqs. (6.15) and (6.12) in Eq. (7.84a) and Eqs. (6.15)-(6.18) and Eq. (6.12) in Eq. (7.84b).

7.9 FREE VIBRATIONS OF A SIMPLY SUPPORTED PLATE

As a way of introducing certain concepts concerning the vibration of plates and providing "exact" data to be used later in comparing results from approximate methods, we now consider the free vibrations of a thin plate measuring $a \times b \times h$ and simply supported on all edges (see Fig. 7.10).

The differential equation for w in this case is Eq. (7.83) with $q = 0$. We shall express this equation in the following simple form:

$$\beta^2 \nabla^4 w + \ddot{w} = 0 \tag{7.85}$$

where

$$\beta^2 = \frac{D}{\rho h} \tag{7.86}$$

As is the usual procedure, we attempt a separation of variables as follows:

$$w = W(x, y)T(t) \tag{7.87}$$

This results in the usual way in an ordinary differential equation in time and a partial differential equation in spatial coordinates (x, y). Thus

$$\frac{\ddot{T}}{T} = -\omega^2 \qquad \text{therefore } \ddot{T} + \omega^2 T = 0 \tag{7.88a}$$

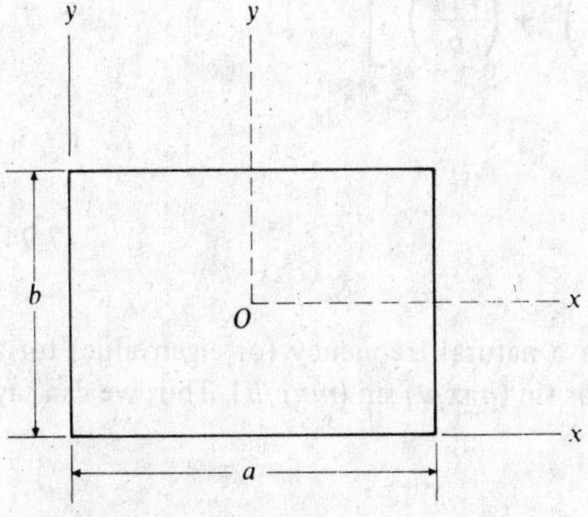

FIGURE 7.10
Simply supported plate.

$$\frac{\beta^2 \nabla^4 W}{W} = \omega^2 \qquad \text{therefore } \nabla^4 W - \frac{\omega^2}{\beta^2} W = 0 \tag{7.88b}$$

where ω^2 is the separation parameter. The solution to Eq. (7.88*a*) is immediately seen to be

$$T = B \cos \omega t + A \sin \omega t \tag{7.89}$$

As for Eq. (7.88*b*), we again proceed by a separation of variables. However, because of the simplicity of the problem here we shall be able a priori to choose functions $X(x)$ and $Y(y)$ for a product solution $W(x, y)$. We shall consider the product $X(x)Y(y)$ as follows:

$$W(x,y) = X(x)Y(y) = \sin \frac{n\pi x}{a} \sin \frac{m\pi y}{b} \tag{7.90}$$

Recall from Section 6.4 (static bending of simply supported rectangular plates) that this product satisfies the complete boundary conditions for a simply supported plate—i.e.,

$$W = M_x = 0 \quad \text{at } x = 0, a \tag{7.91}$$

$$W = M_y = 0 \quad \text{at } y = 0, b \tag{7.92}$$

Now substitute this product [Eq. (7.90)] into Eq. (7.88*b*) to find

$$\left[\left(\frac{n\pi}{a}\right)^4 + 2\left(\frac{n\pi}{a}\right)^2\left(\frac{m\pi}{b}\right)^2 + \left(\frac{m\pi}{b}\right)^4 - \frac{\omega_{nm}^2}{\beta^2}\right] \sin \frac{n\pi x}{a} \sin \frac{m\pi y}{b} = 0 \tag{7.93}$$

where the subscripts n, m have been used for the separation variable ω^2 to associate it with the particular product having integers n and m. We then conclude that

$$\omega_{nm}^2 = \beta^2 \left[\left(\frac{n\pi}{a}\right)^4 + 2\left(\frac{n\pi}{a}\right)^2\left(\frac{m\pi}{b}\right)^2 + \left(\frac{m\pi}{b}\right)^4\right]$$

therefore

$$\omega_{nm}^2 = \beta^2 \pi^4 \left[\left(\frac{n}{a}\right)^2 + \left(\frac{m}{b}\right)^2\right]^2 \tag{7.94}$$

From the above steps we see that ω_{nm} is a natural frequency (or eigenvalue) for a mode shape (or eigenfunction) W_{nm} given as $\sin(n\pi x/a)\sin(m\pi y/b)$. Thus we can say for the (n, m)th mode:

$$\omega_{nm} = \beta\pi^2 \left[\left(\frac{n}{a}\right)^2 + \left(\frac{m}{b}\right)^2\right] \tag{7.95a}$$

$$W_{nm} = \sin \frac{n\pi x}{a} \sin \frac{m\pi y}{b} \tag{7.95b}$$

The free vibration of the plate is a superposition of all the modes with proper amplitudes and properly phased together to satisfy the *initial conditions* of the problem. Thus we can say

$$w = \sum_{m=1}^{\infty} \sum_{n=1}^{\infty} \sin \frac{n\pi x}{a} \sin \frac{m\pi y}{b} (A_{nm} \sin \omega_{nm} t + B_{nm} \cos \omega_{nm} t) \tag{7.96}$$

where the double infinity of constants A_{nm} and B_{nm} are determined to satisfy the conditions

$$\begin{aligned} w(x, y, 0) &= \phi(x, y) \\ \frac{\partial w(x, y, 0)}{\partial t} &= \psi(x, y) \end{aligned} \tag{7.97}$$

with ϕ and ψ as known functions. Accordingly, submitting Eq. (7.96) to the above conditions we get

$$\sum_{n=1}^{\infty} \sum_{m=1}^{\infty} \sin \frac{n\pi x}{a} \sin \frac{m\pi y}{b} B_{nm} = \phi(x, y) \tag{7.98a}$$

$$\sum_{n=1}^{\infty} \sum_{m=1}^{\infty} \sin \frac{n\pi x}{a} \sin \frac{m\pi y}{b} \omega_{nm} A_{nm} = \psi(x, y) \tag{7.98b}$$

Now making use of the fact that

$$\begin{aligned} \int_0^a \sin \frac{r\pi x}{a} \sin \frac{s\pi x}{a} \, dx &= \frac{a}{2} \delta_{rs} \\ \int_0^b \sin \frac{p\pi y}{b} \sin \frac{q\pi y}{b} \, dy &= \frac{b}{2} \delta_{pq} \end{aligned} \tag{7.99}$$

we note that

$$\iint_R W_{rs} W_{pq} \, dx \, dy = \frac{ab}{4} \delta_{rs} \delta_{pq} \tag{7.100}$$

This is the orthogonality property of the eigenfunction for this problem. Now multiplying Eq. (7.98*a*) by $\sin (p\pi x/a) \sin (q\pi y/b)$ and integrating over R:

$$\iint_R \left(\sum_{m=1}^{\infty} \sum_{n=1}^{\infty} \sin \frac{n\pi x}{a} \sin \frac{m\pi y}{b} \sin \frac{p\pi x}{a} \sin \frac{q\pi y}{b} B_{nm} \right) dx\, dy$$

$$= \iint_R \phi(x,y) \sin \frac{p\pi x}{a} \sin \frac{q\pi y}{b}\, dx\, dy$$

Using the orthogonality property given by Eq. (7.100) we may solve for B_{pq} from the above formulation in a direct manner. Thus:

$$B_{pq} = \frac{4}{ab} \iint_R \phi(x,y) \sin \frac{p\pi x}{a} \sin \frac{q\pi y}{b}\, dx\, dy \tag{7.101a}$$

Similarly, we have for A_{pq}:

$$A_{pq} = \frac{4}{ab\omega_{pq}} \iint_R \psi(x,y) \sin \frac{p\pi x}{a} \sin \frac{q\pi y}{b}\, dx\, dy \tag{7.101b}$$

We have thus solved the problem of the rectangular simply supported freely vibrating plate. It is to be pointed out that there are no such simple solutions available for the clamped rectangular plate and the free rectangular plate. This is partly due to the fact that the governing equation [Eq. (7.88)] for such plates is not separable in rectangular coordinates. In the problems we shall be able to investigate solutions of the circular plates, since for cylindrical coordinates the basic plate equation is separable.

7.10 RAYLEIGH'S METHOD FOR PLATES

We now present Rayleigh's method for determining an approximation from above for the fundamental frequency of a plate. The basis for the method is as given in Section 7.5, where we discussed beams. We assume the plate is vibrating freely with frequency ω_1 in the *fundamental mode.* We can then represent the motion as follows:

$$w = W_1(x,y) \cos \omega_1 t \tag{7.102}$$

Since we do not have dissipation, we equate the maximum kinetic energy of the system to the maximum strain energy of the system. That is,

$$T_{\max} = U_{\max} \tag{7.103}$$

To find $T_{\max}$ note that we can give T as follows, using Eq. (7.102):

$$T = \frac{1}{2} \iint_R h\rho \dot{w}^2\, dA = \frac{\omega_1^2 \sin^2 \omega_1 t}{2} \iint_R h\rho W_1^2\, dA$$

Clearly, for T_{max} we take $\sin^2 \omega_1 t = 1$ to get

$$T_{max} = \frac{1}{2} h\omega_1^2 \rho \iint_R W_1^2 \, dA$$

As for U, we use Eq. (7.102) in Eq. (7.77) with q set equal to zero. Thus

$$U = \frac{D}{2} \cos^2 \omega_1 t \iint_R \left\{ (\nabla^2 W_1)^2 + 2(1-\nu) \left[\left(\frac{\partial^2 W_1}{\partial x \, \partial y} \right)^2 - \frac{\partial^2 W_1}{\partial x^2} \frac{\partial^2 W_1}{\partial y^2} \right] \right\} dA$$

For U_{max} we set $\cos^2 \omega_1 t = 1$. Thus

$$U_{max} = \frac{D}{2} \iint_R \left\{ (\nabla^2 W_1)^2 + 2(1-\nu) \left[\left(\frac{\partial^2 W_1}{\partial x \, \partial y} \right)^2 - \frac{\partial^2 W_1}{\partial x^2} \frac{\partial^2 W_1}{\partial y^2} \right] \right\} dA$$

Substituting into Eq. (7.103), we can now form Rayleigh's quotient as follows, by solving for ω_1^2:

$$\omega_1^2 = \frac{D \iint_R \{(\nabla^2 W_1)^2 + 2(1-\nu)[(\partial^2 W_1/\partial x \, \partial y)^2 - (\partial^2 W_1/\partial x^2)(\partial^2 W_1/\partial y^2)]\} \, dA}{h\rho \iint_R W_1^2 \, dA} \tag{7.104}$$

It will be shown that Rayleigh's quotient given above is a functional that, like the corresponding functional for beams, is a minimum for the lowest eigenfunction with respect to functions $\tilde{W}_1$ that satisfy the boundary conditions of the problem. Hence a function $\tilde{W}_1$ satisfying boundary conditions and a good approximation to the fundamental mode shape will yield, on substitution into Eq. (7.104), a value of $\tilde{\omega}_1^2$ that will be an even better approximation of ω_1^2. Furthermore, this approximation will be *from above.* We have here a powerful method of approximating the very important fundamental frequency of a plate—a tool that is particularly valuable considering the paucity of solutions available for the plate.

We now illustrate this approach for the case of the simply supported rectangular plate, which was solved in the preceding section, in order to illustrate the method and to compare approximate results with the exact results for this case.

EXAMPLE 7.3 We again examine the rectangular simply supported flat plate having dimensions $a \times b \times h$. It will be simplest here to place the reference at the center of the plate, as shown in Fig. 7.10 (see dashed reference coordinates).

To approximate the first mode we use the following even function:

$$w_{app} = \left[1 - \left(\frac{x}{a/2} \right)^2 \right] \left[1 - \left(\frac{y}{b/2} \right)^2 \right] = \left(1 - \frac{4x^2}{a^2} \right) \left(1 - \frac{4y^2}{b^2} \right) \tag{a}$$

This function satisfies the kinematic boundary condition $w = 0$ but not the natural boundary condition $\partial^2 w/\partial \nu^2 = 0$. We now substitute the above function into Eq. (7.104) to find ω_1. After straightforward computations we get

$$\omega_1^2 = \frac{40D}{h\rho}\left(\frac{3}{a^4} + \frac{200}{a^2 b^2} + \frac{3}{b^4}\right) \tag{b}$$

We compute ω_1^2 for various values of a/b and compare with the exact solution [Eq. (7.94)]. The results are given in Table 7.3, where we use the frequency parameter $(h\rho a^4/D)\omega^2 = k^2$. We see that for $a/b = 1$ the ratio of frequencies $\omega_1/\omega_{\text{exact}}$ is

$$\frac{\omega_1}{\omega_{\text{exact}}} = \sqrt{\frac{440}{389}} = 1.06 \tag{c}$$

In an effort to improve the accuracy of the computation we now employ a function that satisfies both the kinematic boundary condition $w = 0$ and the natural boundary condition $\partial^2 w/\partial \nu^2 = 0$. Thus consider the following even function:

$$W_1 = \left(x^4 - \frac{3a^2}{2}x^2 + \frac{5a^4}{16}\right)\left(y^4 - \frac{3b^2}{2}y^2 + \frac{5}{16}b^4\right) \tag{d}$$

You may readily demonstrate that this function satisfies the boundary conditions. Substituting into Eq. (7.104) we get, on solving for ω_1^2,

$$\omega_1^2 = 98.7\,\frac{D}{h\rho}\left(\frac{1}{a^2} + \frac{1}{b^2}\right)^2 \tag{e}$$

This result now coincides, within the computational accuracy, with the exact solution presented earlier.

7.11 RAYLEIGH-RITZ METHOD FOR PLATES

We now set forth the Rayleigh-Ritz method as applied to plates. We will be able thereby to improve the estimate of the fundamental frequency, and if we can estimate higher mode shapes we will also be able to get approximations from above of higher natural frequencies. As was indicated in Section 7.6 on beams, we proceed by giving W_{app} in terms of a set of linearly independent functions ϕ_i which satisfy the boundary conditions of the problem. That is, let

$$W_{\text{app}}(x, y) = A_1\phi_1(x, y) + A_2\phi_2(x, y) + \cdots + A_n\phi_n(x, y) \tag{7.105}$$

TABLE 7.3
Frequency Parameter for Rayleigh's Method

a/b	1	2	3
k_1^2	440	2,890	11,640
k_{exact}^2	389	2,435	9,740

where A_i are the undetermined coefficients. Now substitute this result into the Rayleigh quotient for plates as follows:

$$\Lambda^2 = \frac{1}{h\rho \iint_R \left(\sum_{i=1}^{n} A_i \phi_i\right)^2 dA} \left(D \iint_R \left\{ \left(\sum_{i=1}^{n} A_i \nabla^2 \phi_i \right)^2 + 2(1-\nu) \left[\left(\sum_{i=1}^{n} A_i \frac{\partial^2 \phi_i}{\partial x \, \partial y} \right)^2 - \left(\sum_{i=1}^{n} A_i \frac{\partial^2 \phi_i}{\partial x^2} \right) \left(\sum_{j=1}^{n} A_j \frac{\partial^2 \phi_j}{\partial y^2} \right) \right] \right\} dA \right) \tag{7.106}$$

We next introduce the following notation, replacing x and y by x_1 and x_2, respectively:

$$c_{ij} = h\rho \iint_R \phi_i \phi_j \, dA$$

$$a_{ijpq} = D \iint_R \frac{\partial^2 \phi_i}{\partial x_p^2} \frac{\partial^2 \phi_j}{\partial x_q^2} \, dA \quad \begin{cases} i, j = 1, 2, \ldots, n \\ p, q = 1, 2 \end{cases} \tag{7.107}$$

$$b_{ij} = D \iint_R \frac{\partial^2 \phi_i}{\partial x_1 \, \partial x_2} \frac{\partial^2 \phi_j}{\partial x_1 \, \partial x_2} \, dA$$

Then we can give the Rayleigh quotient as†

$$\Lambda^2 = \frac{\sum_{j=1}^{n} \sum_{i=1}^{n} \left\{ A_i A_j \left[\sum_{q=1}^{2} \sum_{p=1}^{2} a_{ijpq} + 2(1-\nu)(b_{ij} - a_{ij12}) \right] \right\}}{\sum_{i=1}^{n} \sum_{j=1}^{n} A_i A_j c_{ij}} \tag{7.108}$$

Now extremizing with respect to A_i we have n equations of the form

$$\frac{\partial \Lambda^2}{\partial A_i} = 0 = \frac{2 \sum_{j=1}^{n} A_j \left[\sum_{q=1}^{2} \sum_{p=1}^{2} a_{ijpq} + 2(1-\nu)(b_{ij} - a_{ij12}) \right]}{\sum_{i=1}^{n} \sum_{j=1}^{n} A_i A_j c_{ij}} - \frac{\left\{ \sum_{k=1}^{n} \sum_{j=1}^{n} A_k A_j \left[\sum_{q=1}^{2} \sum_{p=1}^{2} a_{kjpq} + 2(1-\nu)(b_{kj} - a_{kj12}) \right] \right\} \left(2 \sum_{j=1}^{n} A_j c_{ij} \right)}{\left(\sum_{k=1}^{n} \sum_{j=1}^{n} A_k A_j c_{kj} \right)^2}$$

$$i = 1, 2, \ldots, n$$

†If any expressions in this equation are not clear, you may justify them in your mind by using function $A_1\phi_1 + A_2\phi_2$ to test the formulations.

Note that i has become a free index in the first fraction above and in the last bracketed quantity in the numerator of the second fraction. To avoid confusion we have, as a result, switched from i to k in the remaining expressions of the second fraction, where i has played and would continue to play the role of a dummy index. Using Eq. (7.108) and canceling

$$\frac{2}{\sum_{k=1}^{n}\sum_{j=1}^{n} A_k A_j c_{kj}}$$

we get

$$\sum_{j=1}^{n} A_j\left[\sum_{q=1}^{2}\sum_{p=1}^{2} a_{ijpq} + 2(1-\nu)(b_{ij} - a_{ij12})\right] - \Lambda^2 \sum_{j=1}^{n} A_j c_{ij} = 0 \qquad i = 1, 2, \ldots, n$$

Hence

$$\sum_{j=1}^{n}\left[\sum_{q=1}^{2}\sum_{p=1}^{2} a_{ijpq} + 2(1-\nu)(b_{ij} - a_{ij12}) - \Lambda^2 c_{ij}\right] A_j = 0 \qquad i = 1, 2, \ldots, n$$

Using the following notation:

$$\alpha_{ij} = \sum_{q=1}^{2}\sum_{p=1}^{2} a_{ijpq} + 2(1-\nu)(b_{ij} - a_{ij12}) \tag{7.109}$$

we get

$$\sum_{i=1}^{n} (\alpha_{ij} - \Lambda^2 c_{ij})A_j = 0 \qquad i = 1, 2, \ldots, n \tag{7.110}$$

We thus get a system of equations of the form given by Eq. (7.43) for beams. We know that a necessary condition for a nontrivial result is that

$$|\alpha_{ij} - \Lambda^2 c_{ij}| = 0 \tag{7.111}$$

This gives us an algebraic equation of degree n in Λ^2. The n roots will then be approximations from above of the first n natural frequencies of the problem. Also, the ratios of the A_i for a given Λ^2 can be determined from Eq. (7.110), thus establishing the approximate eigenfunctions. In Part C of the chapter we shall justify this statement.

We will illustrate the Rayleigh-Ritz method in the following example.

EXAMPLE 7.4 We now consider the calculation of higher-order eigenvalues for the vibration of a simply supported rectangular plate shown in Fig. 7.10, with the origin at the center.

As a first step, we shall choose a system of n coordinate functions that satisfies both the kinematic boundary condition $w = 0$ and the natural boundary condition $\partial^2 w/\partial \nu^2 = 0$ of the problem and that reasonably resembles what may be expected to be the first n even eigenfunctions. We shall formulate the coordinate functions ϕ_i as products of two functions—one of which is a function of x alone and the other a function of y alone. That is,

$$\phi_i(x, y) = g(x)h(y) \tag{a}$$

We will require accordingly for the boundary conditions that

$$\begin{aligned}
&\phi_i\left(\frac{a}{2}, y\right) = \left(\frac{\partial^2 \phi_i}{\partial x^2}\right)_{(a/2, y)} = 0 \quad \text{therefore } g\left(\frac{a}{2}\right) = g''\left(\frac{a}{2}\right) = 0\\
&\phi_i\left(-\frac{a}{2}, y\right) = \left(\frac{\partial^2 \phi_i}{\partial x^2}\right)_{(-a/2, y)} = 0 \quad \text{therefore } g\left(-\frac{a}{2}\right) = g''\left(-\frac{a}{2}\right) = 0\\
&\phi_i\left(x, \frac{b}{2}\right) = \left(\frac{\partial^2 \phi_i}{\partial y^2}\right)_{(x, b/2)} = 0 \quad \text{therefore } h\left(\frac{b}{2}\right) = h''\left(\frac{b}{2}\right) = 0\\
&\phi_i\left(x, -\frac{b}{2}\right) = \left(\frac{\partial^2 \phi_i}{\partial y^2}\right)_{(x, -b/2)} = 0 \quad \text{therefore } h\left(-\frac{b}{2}\right) = h''\left(-\frac{b}{2}\right) = 0
\end{aligned} \tag{b}$$

Since we know the mode shapes of the rectangular simply supported plate, we shall set forth approximations for the first five even mode shapes to show how we can approximate the eigenvalues for these modes.

For instance, consider the fundamental mode shape of the plate. If we start with the even function

$$g_1 = C_1 + C_2 x^2 + C_3 x^4$$

we may choose one constant arbitrarily and then choose the other two constants to satisfy the boundary conditions on g at $x = -a/2$ given by Eqs. (b). Because the function is even, we then also satisfy the boundary conditions at $x = +a/2$. Since we can here exclude nodal points (see Fig. 7.11) over the interval $-a/2 < x < a/2$, we can set forth a function in the x direction approximating the fundamental mode in that direction. We have for these requirements for $g_1(x)$:

$$g_1(x) = -5 + \frac{24x^2}{a^2} - \frac{16x^4}{a^4} \tag{c}$$

Similarly, for approximating the fundamental mode shape in the y direction we express $h_1(y)$ as

$$h_1(y) = -5 + \frac{24y^2}{b^2} - \frac{16y^4}{b^4} \tag{d}$$

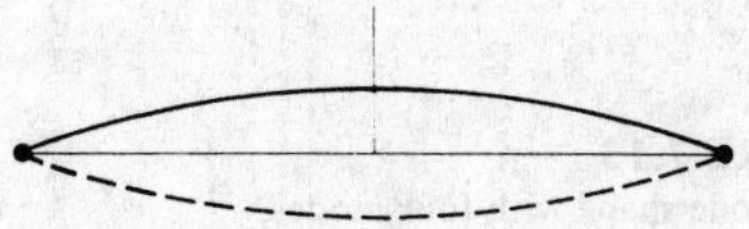

FIGURE 7.11
Fundamental mode shape.

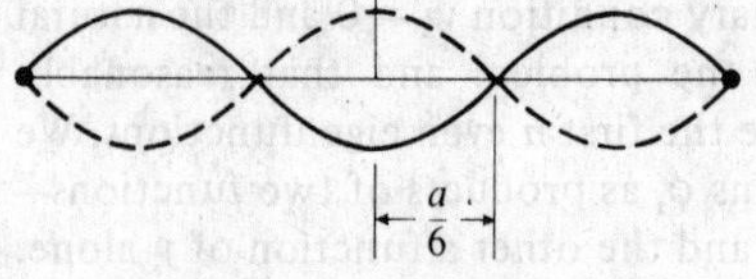

FIGURE 7.12
Even mode shape with two modes.

Hence we can give ϕ_1 as

$$\phi_1 = g_1(x)h_1(y)$$

We know that the next even mode shape will have two nodal points in either the x or the y direction (see Fig. 7.12) and so we consider the even function:

$$C_1 + C_2x^2 + C_3x^4 + C_4x^6$$

Choose one constant arbitrarily and then compute the other constants to satisfy the boundary conditions [Eqs. (b)] at $x = -a/2$ plus the requirement of having a nodal point at $x = a/6$ (and hence also at $-a/6$). The following function denoted as $g_3(x)$ satisfies the above requirements:

$$g_3(x) = 19 - \frac{804x^3}{a^3} + \frac{4496x^4}{a^4} - \frac{6336x^6}{a^6} \tag{e}$$

Similarly, for the y direction we can form $h_3(y)$ as follows:

$$h_3(y) = 19 - \frac{804y^2}{b^2} + \frac{4496y^4}{b^4} - \frac{6336y^6}{b^6} \tag{f}$$

As for the mode shape Fig. 7.13, we proceed by considering a function of the form:

$$C_1 + C_2x^2 + C_3x^4 + C_4x^6 + C_5x^8$$

Choosing one constant arbitrarily, then satisfying the boundary conditions at $x = a/2$, and finally setting the function equal to zero at $x = a/10$ and $x = 3a/10$ (the nodal points), we may establish a set of C's. Thus, denoting the function as $g_5(x)$, we have

$$g_5(x) = 10.4 - \frac{1228.7x^2}{a^2} + \frac{19735.8x^4}{a^4} - \frac{94200x^6}{a^6} + \frac{137000x^8}{a^8} \tag{g}$$

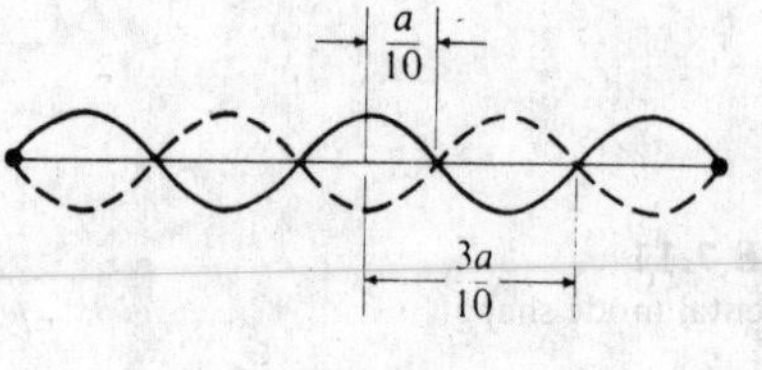

FIGURE 7.13
Even mode shape with four modes.

Similarly, we may form a function $h_5(y)$ as follows:

$$h_5(y) = 10.4 - \frac{1228.7y^2}{b^2} + \frac{19735.8y^4}{b^4} - \frac{94200y^6}{b^6} + \frac{137000y^8}{b^8} \qquad (h)$$

We may now approximate the first five even mode shapes in the following manner:

$$\phi_1 = g_1 h_1$$
$$\phi_2 = g_1 h_3$$
$$\phi_3 = g_3 h_3$$
$$\phi_4 = g_1 h_5$$
$$\phi_5 = g_3 h_5$$

Now compute α_{ij} [see Eq. (7.109)] and c_{ij} [see Eq. (7.107)] using the above functions. We may solve Eq. (7.111) successively by using ϕ_1 (i.e., take $n = 1$), then using ϕ_1 and ϕ_2 ($n = 2$), and so forth until all five functions are employed ($n = 5$). We then get successively approximations from above of the eigenvalue corresponding to the first symmetric mode (for $n = 1$), then the eigenvalues of the first two symmetric modes (for $n = 2$), etc. These results are shown in Table 7.4 for $a/b = 1$ and $a/b = 2$. A Poisson ratio of 0.5 has been used for simplicity.

We thus see that when *we can approximate the eigenfunctions reasonably closely* we may reach very good approximations from above of the corresponding eigenvalues by the Rayleigh-Ritz method. In an exercise you will be asked to make calculations such as those above for the odd modes of the simply supported rectangular plate.

TABLE 7.4
Eigenvalues Using the Rayleigh-Ritz Method

n	Λ_1^2	Λ_2^2	Λ_3^2	Λ_4^2	Λ_5^2
			$a/b = 1$		
1	389.969				
2	389.803	10,213		Multiply each term by $(D/\rho h a^4)$	
3	389.802	10,213	32,787		
4	389.802	9,747	32,787	80,713	
5	389.802	9,747	32,136	80,711	131,154
Exact	389.635	9,741	31,560	65,848	112,604
			$a/b = 2$		
1	2,437.811				
2	2,435.475	140,616		Multiply each term by $(D/\rho h a^4)$	
3	2,435.475	140,615	205,723		
4	2,435.474	133,444	205,723	1,227,430	
5	2,435.474	133,444	197,865	1,227,427	1,403,654
Exact	2,435.219	133,352	197,257	993,667	1,157,313

*7.12 TRANSVERSE SHEAR AND ROTATORY INERTIA–REISSNER-MINDLIN PLATE THEORY

In Section 6.11 we examined an improved theory for the axisymmetric circular plate wherein the effect of transverse shear was taken into account. We now consider the rectangular plate and take into account the effects of transverse shear and rotatory inertia while formulating the free vibration problem for this case. The theory was given for statics by Reissner and extended to dynamics by Mindlin.

We start in the usual way by proposing a displacement field. Deleting stretching effects, we have

$$
\begin{aligned}
u_1 &= -z\psi_x(x, y, t) = -z\psi_x \\
u_2 &= -z\psi_y(x, y, t) = -z\psi_y \qquad (7.112) \\
u_3 &= w(x, y, t) = w
\end{aligned}
$$

where ψ_x and ψ_y are components of the vector field ψ. The deformation differs from the classical case [Eqs. (6.4)] in that functions ψ_i, yet to be determined, replace $w_{,i}$ in u_1 and u_2. This means we are maintaining the assumption that line elements originally normal to the midplane remain straight on deformation (no warping), but we are abandoning the assumption that such line elements remain normal to the midplane after deformation. The assumption of no curvature is not correct and we shall later introduce means of correcting for it. The strain field for the assumed displacement field follows immediately as

$$
\begin{aligned}
\epsilon_{xx} &= -z\frac{\partial \psi_x}{\partial x} \\
\epsilon_{yy} &= -z\frac{\partial \psi_y}{\partial y} \\
\epsilon_{xy} &= -\frac{z}{2}\left(\frac{\partial \psi_x}{\partial y} + \frac{\partial \psi_y}{\partial x}\right) \qquad (7.113) \\
\epsilon_{xz} &= \frac{1}{2}\left(\frac{\partial w}{\partial x} - \psi_x\right) \\
\epsilon_{yz} &= \frac{1}{2}\left(\frac{\partial w}{\partial y} - \psi_y\right)
\end{aligned}
$$

We shall consider the terms ψ_x and ψ_y as variables in the ensuing discussion. Next we shall ascertain the resultant intensity functions. Thus for M_x we have

$$M_x = \int_{-h/2}^{h/2} \tau_{xx} z\, dz = \int_{-h/2}^{h/2} \frac{E}{1-\nu^2}(\epsilon_{xx} + \nu\epsilon_{yy}) z\, dz$$

where we have assumed a plane stress distribution in employing Hooke's law [Eq. (1.118)]. Now, using Eq. (7.113) in the above formulation, we get

$$M_x = \frac{E}{1-\nu^2}\int_{-h/2}^{h/2} z^2\left(-\frac{\partial\psi_x}{\partial x} - \nu\frac{\partial\psi_y}{\partial y}\right) dz$$

$$= -\frac{Eh^3}{12(1-\nu^2)}\left(\frac{\partial\psi_x}{\partial x} + \nu\frac{\partial\psi_y}{\partial y}\right) = -D\left(\frac{\partial\psi_x}{\partial x} + \nu\frac{\partial\psi_y}{\partial y}\right) \tag{7.114}$$

Similarly, we have for the other moment resultant functions

$$M_y = -D\left(\frac{\partial\psi_y}{\partial y} + \nu\frac{\partial\psi_x}{\partial x}\right) \tag{7.115a}$$

$$M_{xy} = -(1-\nu)\frac{D}{2}\left(\frac{\partial\psi_x}{\partial y} + \frac{\partial\psi_y}{\partial x}\right) \tag{7.115b}$$

We now consider the resultant shear force intensity function. For Q_x we have

$$Q_x = \int_{-h/2}^{h/2} \tau_{xz}\, dz = k\tau_{xz} h \tag{7.116}$$

where we have introduced a shear factor k to correct for the error stemming from Eq. (7.113) that τ_{xz} is a constant over the thickness of the plate (i.e., is not a function of z). Here is where warping effects, mentioned earlier, are taken into account. Now, using Hooke's law and Eq. (7.113), we have

$$Q_x = khG\left(\frac{\partial w}{\partial x} - \psi_x\right) \tag{7.117}$$

Similarly, we get for Q_y

$$Q_y = khG\left(\frac{\partial w}{\partial y} - \psi_y\right) \tag{7.118}$$

Since we shall employ Hamilton's principle we next turn toward evaluation of the various energies involved in order to be able to express the Lagrangian function in terms of our plate variables. First we consider the kinetic energy T. Thus:

$$T = \frac{1}{2}\iint_R \int_{-h/2}^{h/2} \rho \dot{u}_i \dot{u}_i\, dz\, dA$$

$$= \frac{1}{2}\iint_R \int_{-h/2}^{h/2} \rho\left[z^2\left(\frac{\partial\psi_x}{\partial t}\right)^2 + z^2\left(\frac{\partial\psi_y}{\partial t}\right)^2 + \left(\frac{\partial w}{\partial t}\right)^2\right] dz\, dA$$

Note we are now retaining rotatory inertia effects. Integrating with respect to z we get

$$T = \frac{1}{2} \iint_R \left[\frac{\rho h^3}{12} \left(\frac{\partial \psi_x}{\partial t} \right)^2 + \frac{\rho h^3}{12} \left(\frac{\partial \psi_y}{\partial t} \right)^2 + \rho h \left(\frac{\partial w}{\partial t} \right)^2 \right] dA \tag{7.119}$$

As for the strain energy U, we have

$$U = \frac{1}{2} \iint_R \int_{-h/2}^{h/2} (\tau_{xx}\epsilon_{xx} + \tau_{yy}\epsilon_{yy} + \tau_{zz}\epsilon_{zz} + 2\tau_{xy}\epsilon_{xy} + 2\tau_{xz}\epsilon_{xz}$$
$$+ 2\tau_{yz}\epsilon_{yz})\, dz\, dA$$

We now assume a plane stress state for τ_{xx} and τ_{yy} (i.e., we drop τ_{zz}) and employ Hooke's law [Eq. (1.118)] in the above equation† for τ_{xx}, τ_{yy}, and τ_{xy}:

$$U = \frac{1}{2} \iint_R \int_{-h/2}^{h/2} \left[\frac{E}{1-\nu^2} (\epsilon_{xx} + \nu\epsilon_{yy})\epsilon_{xx} + \frac{E}{1-\nu^2} (\epsilon_{yy} + \nu\epsilon_{xx})\epsilon_{yy} \right.$$
$$\left. + 4G\epsilon_{xy}^2 + 2\tau_{xz}\epsilon_{xz} + 2\tau_{yz}\epsilon_{yz} \right] dA\, dz$$

Next use Eq. (7.113) to replace the strains in the above equation. We get

$$U = \frac{1}{2} \iint_R \int_{-h/2}^{h/2} \left\{ \frac{E}{1-\nu^2} z^2 \left[\left(\frac{\partial \psi_x}{\partial x} \right)^2 + \left(\frac{\partial \psi_y}{\partial y} \right)^2 + 2\nu \frac{\partial \psi_x}{\partial x} \frac{\partial \psi_y}{\partial y} \right] \right.$$
$$\left. + Gz^2 \left(\frac{\partial \psi_x}{\partial y} + \frac{\partial \psi_y}{\partial x} \right)^2 + \tau_{xz} \left(\frac{\partial w}{\partial x} - \psi_x \right) + \tau_{yz} \left(\frac{\partial w}{\partial y} - \psi_y \right) \right\} dA\, dz$$

Integrate with respect to z over the thickness of the plate:

$$U = \frac{1}{2} \iint_R \left\{ \frac{Eh^3}{12(1-\nu)} \left[\left(\frac{\partial \psi_x}{\partial x} \right)^2 + \left(\frac{\partial \psi_y}{\partial y} \right)^2 + 2\nu \frac{\partial \psi_x}{\partial x} \frac{\partial \psi_y}{\partial y} \right] \right.$$
$$\left. + \frac{Gh^3}{12} \left(\frac{\partial \psi_x}{\partial y} + \frac{\partial \psi_y}{\partial x} \right)^2 + kh\tau_{xz} \left(\frac{\partial w}{\partial x} - \psi_x \right) + kh\tau_{yz} \left(\frac{\partial w}{\partial y} - \psi_y \right) \right\} dA$$

where we have used Eq. (7.116) and the corresponding equation for τ_{yz}, keeping in mind that $(\partial w/\partial y - \psi_y)$ and $(\partial w/\partial x - \psi_x)$ are not functions of z. Introducing D and replacing τ_{xz} and τ_{yz} by using Hooke's law and Eq. (7.113), we get

†We are thus using a simplified constitutive law at this juncture. We have alluded to this kind of simplification procedure on a number of earlier occasions.

$$U = \frac{1}{2}\iint_R \left\{ D\left[\left(\frac{\partial\psi_x}{\partial x}\right)^2 + \left(\frac{\partial\psi_y}{\partial y}\right)^2 + 2\nu\frac{\partial\psi_x}{\partial x}\frac{\partial\psi_y}{\partial y}\right] + \frac{Gh^3}{12}\left(\frac{\partial\psi_x}{\partial y} + \frac{\partial\psi_y}{\partial x}\right)^2 \right.$$
$$\left. + khG\left[\left(\frac{\partial w}{\partial x} - \psi_x\right)^2 + \left(\frac{\partial w}{\partial y} - \psi_y\right)^2\right]\right\} dA \qquad (7.120)$$

Finally, noting that $V = -\iint_R wq\, dA$, we may state Hamilton's principle as follows, on using Eqs. (7.119) and (7.120):

$$\delta^{(1)}\int_{t_1}^{t_2} L\, dt = 0 = \delta^{(1)}\int_{t_1}^{t_2} \frac{1}{2}\iint_R \left\{ \frac{\rho h^3}{12}\left(\frac{\partial\psi_x}{\partial t}\right)^2 + \frac{\rho h^3}{12}\left(\frac{\partial\psi_y}{\partial t}\right)^2 + \rho h\left(\frac{\partial w}{\partial t}\right)^2 \right.$$
$$- D\left[\left(\frac{\partial\psi_x}{\partial x}\right)^2 + \left(\frac{\partial\psi_y}{\partial y}\right)^2 + 2\nu\frac{\partial\psi_x}{\partial x}\frac{\partial\psi_y}{\partial y}\right]$$
$$- \frac{Gh^3}{12}\left(\frac{\partial\psi_x}{\partial y} + \frac{\partial\psi_y}{\partial x}\right)^2 - kGh\left[\left(\frac{\partial w}{\partial x} - \psi_x\right)^2\right.$$
$$\left.\left. + \left(\frac{\partial w}{\partial y} - \psi_y\right)^2\right] + 2wq\right\} dA\, dt = 0$$

Carrying out the variation operation using the δ operator, we then get

$$\int_{t_1}^{t_2}\iint_R \left\{ \frac{\rho h^3}{12}\frac{\partial\psi_x}{\partial t}\frac{\partial\,\delta\psi_x}{\partial t} + \frac{\rho h^3}{12}\frac{\partial\psi_y}{\partial t}\frac{\partial\,\delta\psi_y}{\partial t} + \rho h\frac{\partial w}{\partial t}\frac{\partial\,\delta w}{\partial t} \right.$$
$$- D\left(\frac{\partial\psi_x}{\partial x}\frac{\partial\,\delta\psi_x}{\partial x} + \frac{\partial\psi_y}{\partial y}\frac{\partial\,\delta\psi_y}{\partial y} + \nu\frac{\partial\psi_x}{\partial x}\frac{\partial\,\delta\psi_y}{\partial y} + \nu\frac{\partial\psi_y}{\partial y}\frac{\partial\,\delta\psi_x}{\partial x}\right)$$
$$- \frac{Gh^3}{12}\left(\frac{\partial\psi_x}{\partial y} + \frac{\partial\psi_y}{\partial x}\right)\left(\frac{\partial\,\delta\psi_x}{\partial y} + \frac{\partial\,\delta\psi_y}{\partial x}\right)$$
$$- kGh\left[\left(\frac{\partial w}{\partial x} - \psi_x\right)\left(\frac{\partial\,\delta w}{\partial x} - \delta\psi_x\right) + \left(\frac{\partial w}{\partial y} - \psi_y\right)\left(\frac{\partial\,\delta w}{\partial y} - \delta\psi_y\right)\right]$$
$$\left. + q\,\delta w\right\} dA\, dt = 0 \qquad (7.121)$$

We shall next integrate by parts with respect to time for the first three terms, keeping in mind that all variations at the limits t_1 and t_2 are zero. And for all but the last of the other terms we employ Green's theorem in the following form [see Eq. (I.32), Appendix I] with $a_{\nu x}\, ds = dy$ and $a_{\nu y}\, ds = -dx$ in accordance with Eqs. (6.39) and Fig. 6.8:

$$\iint_R G \frac{\partial H}{\partial x}\, dx\, dy = -\iint_R H \frac{\partial G}{\partial x}\, dx\, dy + \oint_\Gamma GH\, dy$$
$$\iint_R G \frac{\partial H}{\partial y}\, dx\, dy = -\iint_R H \frac{\partial G}{\partial y}\, dx\, dy - \oint_\Gamma GH\, dx \qquad (7.122)$$

We then get the following result:

$$\int_{t_1}^{t_2} \iint_R \left[-\frac{\rho h^3}{12} \frac{\partial^2 \psi_x}{\partial t^2} \delta\psi_x - \frac{\rho h^3}{12} \frac{\partial^2 \psi_y}{\partial t^2} \delta\psi_y - \rho h \frac{\partial^2 w}{\partial t^2} \delta w \right.$$
$$+ D \left(\frac{\partial^2 \psi_x}{\partial x^2} \delta\psi_x + \frac{\partial^2 \psi_y}{\partial y^2} \delta\psi_y + \nu \frac{\partial^2 \psi_x}{\partial x\, \partial y} \delta\psi_y + \nu \frac{\partial^2 \psi_y}{\partial x\, \partial y} \delta\psi_x \right)$$
$$+ \frac{Gh^3}{12} \left(\frac{\partial^2 \psi_x}{\partial y^2} \delta\psi_x + \frac{\partial^2 \psi_x}{\partial y\, \partial x} \delta\psi_y + \frac{\partial^2 \psi_y}{\partial x\, \partial y} \delta\psi_x + \frac{\partial^2 \psi_y}{\partial x^2} \delta\psi_y \right)$$
$$+ kGh \left(\frac{\partial^2 w}{\partial x^2} \delta w + \frac{\partial w}{\partial x} \delta\psi_x - \frac{\partial \psi_x}{\partial x} \delta w - \psi_x \delta\psi_x + \frac{\partial^2 w}{\partial y^2} \delta w + \frac{\partial w}{\partial y} \delta\psi_y \right.$$
$$\left. \left. - \frac{\partial \psi_y}{\partial y} \delta w - \psi_y \delta\psi_y \right) + q\, \delta w \right] dA\, dt + \int_{t_1}^{t_2} \oint_\Gamma \left[-D \left(\frac{\partial \psi_x}{\partial x} \delta\psi_x\, dy \right. \right.$$
$$\left. - \frac{\partial \psi_y}{\partial y} \delta\psi_y\, dx - \nu \frac{\partial \psi_x}{\partial x} \delta\psi_y\, dx + \nu \frac{\partial \psi_y}{\partial y} \delta\psi_x\, dy \right)$$
$$- \frac{Gh^3}{12} \left(- \frac{\partial \psi_x}{\partial y} \delta\psi_x\, dx + \frac{\partial \psi_x}{\partial y} \delta\psi_y\, dy - \frac{\partial \psi_y}{\partial x} \delta\psi_x\, dx + \frac{\partial \psi_y}{\partial x} \delta\psi_y\, dy \right)$$
$$\left. - kGh \left(\frac{\partial w}{\partial x} \delta w\, dy - \psi_x \delta w\, dy - \frac{\partial w}{\partial y} \delta w\, dx + \psi_y \delta w\, dx \right) \right] dt = 0$$

We now collect terms in the above formulation with respect to the variation terms:

$$\int_{t_1}^{t_2} \iint_R \left\{ \left[-\frac{\rho h^3}{12} \frac{\partial^2 \psi_x}{\partial t^2} + D \left(\frac{\partial^2 \psi_x}{\partial x^2} + \nu \frac{\partial^2 \psi_y}{\partial x\, \partial y} \right) + \frac{Gh^3}{12} \left(\frac{\partial^2 \psi_x}{\partial y^2} + \frac{\partial^2 \psi_y}{\partial x\, \partial y} \right) \right. \right.$$
$$\left. + kGh \left(\frac{\partial w}{\partial x} - \psi_x \right) \right] \delta\psi_x + \left[-\frac{\rho h^3}{12} \frac{\partial^2 \psi_y}{\partial t^2} + D \left(\frac{\partial^2 \psi_y}{\partial y^2} + \nu \frac{\partial^2 \psi_x}{\partial x\, \partial y} \right) \right.$$
$$\left. + \frac{Gh^3}{12} \left(\frac{\partial^2 \psi_y}{\partial x^2} + \frac{\partial^2 \psi_x}{\partial y\, \partial x} \right) + kGh \left(\frac{\partial w}{\partial y} - \psi_y \right) \right] \delta\psi_y$$
$$\left. + \left[-\rho h \frac{\partial^2 w}{\partial t^2} + kGh \left(\frac{\partial^2 w}{\partial x^2} - \frac{\partial \psi_x}{\partial x} + \frac{\partial^2 w}{\partial y^2} - \frac{\partial \psi_y}{\partial y} + q \right) \right] \delta w \right\} dA\, dt \qquad (7.123)$$

$$+\int_{t_1}^{t_2}\oint_\Gamma\left\{\left[-D\left(\frac{\partial\psi_x}{\partial x}\,dy+\nu\,\frac{\partial\psi_y}{\partial y}\,dy\right)+\frac{Gh^3}{12}\left(\frac{\partial\psi_x}{\partial y}\,dx+\frac{\partial\psi_y}{\partial x}\,dx\right)\right]\delta\psi_x\right.$$
$$+\left[D\left(\frac{\partial\psi_y}{\partial y}\,dx+\nu\,\frac{\partial\psi_x}{\partial x}\,dx\right)-\frac{Gh^3}{12}\left(\frac{\partial\psi_x}{\partial y}\,dy+\frac{\partial\psi_y}{\partial x}\,dy\right)\right]\delta\psi_y$$
$$\left.-\left[kGh\left(\frac{\partial w}{\partial x}\,dy-\psi_x\,dy-\frac{\partial w}{\partial y}\,dx+\psi_y\,dx\right)\right]\delta w\right\}dt=0 \tag{7.123 (Cont.)}$$

In accordance with previous remarks it is clear that each of the coefficients of the varied terms in the integral $\int_{t_1}^{t_2}\iint_R$ for the variations must be zero. We consider the coefficient of $\delta\psi_x$ now. We have then

$$-\frac{\rho h^3}{12}\,\frac{\partial^2\psi_x}{\partial t^2}+D\,\frac{\partial}{\partial x}\left(\frac{\partial\psi_x}{\partial x}+\nu\,\frac{\partial\psi_y}{\partial y}\right)+\frac{Gh^3}{12}\,\frac{\partial}{\partial y}\left(\frac{\partial\psi_x}{\partial y}+\frac{\partial\psi_y}{\partial x}\right)$$
$$+\,kGh\left(\frac{\partial w}{\partial x}-\psi_x\right)=0$$

Noting that

$$D=\frac{Eh^3}{12(1-\nu^2)}$$

and that

$$G=\frac{E}{2(1+\nu)}$$

we may replace the term $Gh^3/12$ by $(D/2)(1-\nu)$ in the third expression above. Now observing Eqs. (7.114), (7.115*b*), and (7.117), we get the following result:

$$-\frac{\rho h^3}{12}\,\frac{\partial^2\psi_x}{\partial t^2}-\frac{\partial M_x}{\partial x}-\frac{\partial M_{xy}}{\partial y}+Q_x=0$$

We may carry out similar operations for the coefficients of $\delta\psi_y$ and δw in the integral $\int_{t_1}^{t_2}\iint_R$ of Eq. (7.123). The resulting equations, including the above equation, then may be stated as follows:

$$Q_x=\frac{\partial M_x}{\partial x}+\frac{\partial M_{xy}}{\partial y}+\frac{\rho h^3}{12}\,\frac{\partial^2\psi_x}{\partial t^2}$$
$$Q_y=\frac{\partial M_{xy}}{\partial x}+\frac{\partial M_y}{\partial y}+\frac{\rho h^3}{12}\,\frac{\partial^2\psi_y}{\partial t^2} \tag{7.124}$$
$$\frac{\partial Q_x}{\partial x}+\frac{\partial Q_y}{\partial y}+q=\rho h\,\frac{\partial^2 w}{\partial t^2}$$

If we delete the time derivatives we get a set having a familiar form from our work in static classical plate theory (Section 6.3). Keep in mind, however, that the shear effects are built into the system by the way we have formulated the resultant intensity functions, Q_x and Q_y. It will be more useful at this time to use equations of motion in terms of displacement functions, and so we go to Eq. (7.123) and set the coefficients of the variations to zero in the first integral and replace $Gh^3/12$ by $(D/2)(1-\nu)$ to obtain

$$D\left(\frac{\partial^2\psi_x}{\partial x^2}+\frac{1-\nu}{2}\frac{\partial^2\psi_x}{\partial y^2}+\frac{1+\nu}{2}\frac{\partial^2\psi_y}{\partial x\,\partial y}\right)+kGh\left(\frac{\partial w}{\partial x}-\psi_x\right)-\frac{\rho h^3}{12}\frac{\partial^2\psi_x}{\partial t^2}=0$$

$$D\left(\frac{1-\nu}{2}\frac{\partial^2\psi_y}{\partial x^2}+\frac{\partial^2\psi_y}{\partial y^2}+\frac{1+\nu}{2}\frac{\partial^2\psi_x}{\partial x\,\partial y}\right)+kGh\left(\frac{\partial w}{\partial y}-\psi_y\right)-\frac{\rho h^3}{12}\frac{\partial^2\psi_y}{\partial t^2}=0$$

$$-kGh\left(\frac{\partial^2 w}{\partial x^2}+\frac{\partial^2 w}{\partial y^2}-\frac{\partial\psi_x}{\partial x}-\frac{\partial\psi_y}{\partial y}\right)+\rho h\frac{\partial^2 w}{\partial t^2}=q \tag{7.125}$$

We may eliminate the functions ψ_i from the above equations by tedious but straightforward means to arrive at the following equation for w:

$$\left(\nabla^2-\frac{\rho}{kG}\frac{\partial^2}{\partial t^2}\right)\left(D\nabla^2-\frac{\rho h^3}{12}\frac{\partial^2}{\partial t^2}\right)w+\rho h\frac{\partial^2 w}{\partial t^2}$$
$$=\left(1-\frac{D}{kGh}\nabla^2+\frac{\rho h^2}{12kG}\frac{\partial^2}{\partial t^2}\right)q \tag{7.126}$$

To delete rotatory inertia in the above equation, we simply set the expressions

$$\frac{\rho h^3}{12}\frac{\partial^2 w}{\partial t^2} \quad \text{and} \quad \frac{\rho h^2}{12kG}\frac{\partial^2 q}{\partial t^2}$$

equal to zero. We then have

$$D\left(\nabla^2-\frac{\rho}{kG}\frac{\partial^2}{\partial t^2}\right)\nabla^2 w+\rho h\frac{\partial^2 w}{\partial t^2}=\left(1-\frac{D}{kGh}\nabla^2\right)q \tag{7.127}$$

And if we wish to delete only the effects of shear deformation we let $1/kGh$ go to zero. We then get

$$\left(D\nabla^2-\frac{\rho h^3}{12}\frac{\partial^2}{\partial t^2}\right)\nabla^2 w+\rho h\frac{\partial^2 w}{\partial t^2}=q \tag{7.128}$$

If both modifications are made simultaneously we get the classical plate equation:

$$D\nabla^4 w+\rho h\frac{\partial^2 w}{\partial t^2}=q \tag{7.129}$$

Now we get back to the line integral of Eq. (7.123). We collect terms as follows in setting this line integral equal to zero:

$$\int_{t_1}^{t_2} \oint_\Gamma \left\{ \delta\psi_x\, dy \left[-D\left(\frac{\partial \psi_x}{\partial x} + \nu \frac{\partial \psi_y}{\partial y} \right) \right] - \delta\psi_x\, dx \left[-\frac{Gh^3}{12} \left(\frac{\partial \psi_x}{\partial y} + \frac{\partial \psi_y}{\partial x} \right) \right] \right.$$

$$+ \delta\psi_y\, dy \left[-\frac{Gh^3}{12} \left(\frac{\partial \psi_x}{\partial y} + \frac{\partial \psi_y}{\partial x} \right) \right] - \delta\psi_y\, dx \left[-D\left(\frac{\partial \psi_y}{\partial y} + \nu \frac{\partial \psi_x}{\partial x} \right) \right]$$

$$\left. + \delta w\, dy \left[-kGh \left(\frac{\partial w}{\partial x} - \psi_x \right) \right] - \delta w\, dx \left[-kGh \left(\frac{\partial w}{\partial y} - \psi_y \right) \right] \right\} dt = 0 \tag{7.130}$$

Using Eqs. (7.114), (7.115), (7.117), and (7.118), this becomes

$$\int_{t_1}^{t_2} \oint_\Gamma [M_x\, \delta\psi_x\, dy - M_y\, \delta\psi_y\, dx + M_{xy}\, \delta\psi_y\, dy - M_{xy}\, \delta\psi_x\, dx$$

$$- Q_x\, \delta w\, dy + Q_y\, \delta w\, dx]\; dt = 0$$

Again in accordance with Eqs. (6.39) (see also Fig. 6.8) we replace dx by $-a_{\nu y}\, ds$ and dy by $a_{\nu x}\, ds$ to get

$$\int_{t_1}^{t_2} \oint_\Gamma [M_x a_{\nu x}\, \delta\psi_x + M_y a_{\nu y}\, \delta\psi_y + M_{xy} a_{\nu x}\, \delta\psi_y + M_{xy} a_{\nu y}\, \delta\psi_x$$

$$- Q_x a_{\nu x}\, \delta w - Q_y a_{\nu y}\, \delta w]\; ds\, dt = 0 \tag{7.131}$$

Using ν and s as coordinates we can say

$$\begin{aligned} \psi_x &= a_{\nu x}\psi_\nu - a_{\nu y}\psi_s \\ \psi_y &= a_{\nu y}\psi_\nu + a_{\nu x}\psi_s \end{aligned} \tag{7.132}$$

where ψ_ν and ψ_s are components of ψ in the ν and s directions, respectively. Solving for ψ_ν and ψ_s we get

$$\begin{aligned} \psi_\nu &= a_{\nu x}\psi_x + a_{\nu y}\psi_y \\ \psi_s &= -a_{\nu y}\psi_x + a_{\nu x}\psi_y \end{aligned} \tag{7.133}$$

Now going back to Eq. (7.131) we have

$$\int_{t_1}^{t_2} \oint_\Gamma \left\{ M_x a_{\nu x}(a_{\nu x}\, \delta\psi_\nu - a_{\nu y}\, \delta\psi_s) + M_y a_{\nu y}(a_{\nu y}\, \delta\psi_\nu + a_{\nu x}\, \delta\psi_s) \right.$$

$$+ M_{xy}\,[a_{\nu x}(a_{\nu y}\, \delta\psi_\nu + a_{\nu x}\, \delta\psi_s) + a_{\nu y}(a_{\nu x}\, \delta\psi_\nu - a_{\nu y}\, \delta\psi_s)]$$

$$\left. - (Q_x a_{\nu x} + Q_y a_{\nu y})\, \delta w \right\} ds\, dt = 0$$

Collecting terms,

$$\int_{t_1}^{t_2}\oint_{\Gamma}\{(M_x a_{\nu x}^2 + M_y a_{\nu y}^2 + 2a_{\nu x}a_{\nu y}M_{xy})\delta\psi_\nu + [-M_x a_{\nu x}a_{\nu y} + M_y a_{\nu x}a_{\nu y}$$
$$+ M_{xy}(a_{\nu x}^2 - a_{\nu y}^2)]\,\delta\psi_s - (Q_x a_{\nu x} + Q_y a_{\nu y})\delta w\}\,ds\,dt = 0$$

Going back to Eqs. (6.15) and (6.16) we see that the above equation can be given as follows:

$$\int_{t_1}^{t_2}\oint (M_\nu\,\delta\psi_\nu + M_{\nu s}\,\delta\psi_s - Q_\nu\,\delta w)\,ds\,dt = 0 \tag{7.134}$$

We conclude from the above statement that along the boundary of the plate

EITHER $M_\nu = 0$	OR	ψ_ν	IS SPECIFIED	(7.135*a*)
EITHER $M_{\nu s} = 0$	OR	ψ_s	IS SPECIFIED	(7.135*b*)
EITHER $Q_\nu = 0$	OR	w	IS SPECIFIED	(7.135*c*)

We get here three conditions on the edge of the plate as opposed to the two conditions in classical plate theory that required considerable discussion.

We have thus posed the entire boundary-value problem. We can solve Eq. (7.126) for w and then, going back to Eqs. (7.125), we can get ψ_x and ψ_y. The natural boundary conditions have to do with the resultant force, moment, and torque intensities imposed at the edges. These can then be written as conditions on w, ψ_x, and ψ_y on the boundaries by employing Eqs. (6.15) and (6.16) as well as Eqs. (7.114), (7.115), (7.117), and (7.118). As for the factor k, we have as yet not had to specify it in this theory. Actually for isotropic materials we find that the factor $\frac{5}{6}$ gives good results in many situations.

We shall not attempt full solutions of the boundary-value problem that has been developed. Instead, we shall consider the case of a simply supported rectangularplate and examine the natural frequencies of vibration for this case when rotatory inertia and shear are included. Accordingly, for a plate $a \times b \times h$ (see Fig. 7.10) we assume that there is a free vibration mode. We assume w to have the following form for this case:

$$w = A_{mn}\sin\frac{m\pi x}{a}\sin\frac{n\pi y}{b}\cos\omega_{mn}t \tag{7.136}$$

Note that the function specifies that w is zero on the boundaries for integer values of n and m and so we satisfy the kinematic conditions for w given by Eq. (7.135*c*). Since we shall only seek ω_{mn}, we shall not need to solve the entire boundary-value problem. Hence we shall not get involved with ψ_x and ψ_y, or the boundary conditions involving these functions. Accordingly, we substitute Eq. (7.136) into Eq. (7.126) with q set equal to zero to get the following result, on canceling A_{mn} sin $(m\pi x/a)$ sin $(n\pi y/b)$ cos $\omega_{mn}t$:

$$\left[-\left(\frac{m\pi}{a}\right)^2-\left(\frac{n\pi}{b}\right)^2+\frac{\rho\omega_{mn}^2}{kG}\right]\Big\}-D\left[\left(\frac{m\pi}{a}\right)^2+\left(\frac{n\pi}{a}\right)^2\right]+\frac{\rho h^3\omega_{mn}^2}{12}\Big\}$$
$$-\rho h\omega_{mn}^2=0 \tag{7.137}$$

To simplify to the classical theory we can show that $\rho h^3/12$ and $1/kGh$ must be set equal to zero. When this is done above we get

$$\frac{\rho h\omega_{mn}^2}{D}=\left[\left(\frac{m\pi}{a}\right)^2+\left(\frac{n\pi}{b}\right)^2\right]^2 \tag{7.138}$$

which is precisely the result reached earlier in Section 7.9 for the classical theory. We now introduce the following dimensionless parameters:

$$\Omega_{mn}^2=\frac{\rho h^3(1-\nu^2)}{12E}\,\omega_{mn}^2 \tag{7.139a}$$

$$S^2=\frac{1}{12}\left(\frac{h}{a}\right)^2 \tag{7.139b}$$

$$\lambda_{mn}=\frac{a/m}{b/n}\qquad\text{(aspect ratio)} \tag{7.139c}$$

Substituting for ω_{mn} in Eq. (7.138) using Eq. (7.139a) and replacing D by $Eh^3/[12(1-\nu^2)]$ we get, on using Eqs. (7.139b) and (7.139c), the classical value of Ω_{mn}^2:

$$(\Omega_{mn}^2)_{\text{class}}=(m\pi)^4S^4(1+\lambda_{mn}^2)^2 \tag{7.140}$$

And in going back to the improved theory [Eq. (7.137)] the aforementioned substitutions lead to the following result:

$$\Omega_{mn}^2-[\Omega_{mn}^2-\Gamma_{mn}S^2(m\pi)^2]\left[\frac{E'}{kG}\Omega_{mn}^2-(m\pi)^2S^2\Gamma_{mn}\right]=0 \tag{7.141a}$$

where

$$\Gamma_{mn}=1+\lambda_{mn}^2\qquad\text{and}\qquad E'=\frac{E}{1-\nu^2} \tag{7.141b}$$

To obtain a simplified solution to Eq. (7.141a) for Ω_{mn}^2 we let

$$\Omega_{mn}^2=(\Omega_{mn}^2)_{\text{class}}+\delta^2 \tag{7.142}$$

The success of this procedure rests on the condition that δ^4 is much smaller than $(\Omega_{mn}^2)_{\text{class}}$ or δ^2. Substituting Eq. (7.142) into Eq. (7.141a) we obtain the result:

$$(\Omega_{mn}^2)_{\text{class}} + \delta^2 - \left[(\Omega_{mn}^2)_{\text{class}} + \delta^2 - \Gamma_{mn} S^2 (m\pi)^2\right] \left\{ \frac{E'}{kG} \left[(\Omega_{mn}^2)_{\text{class}} + \delta^2\right] - (m\pi)^2 S^2 \Gamma_{mn} \right\} = 0$$

We get, on carrying out the multiplication and dropping terms involving δ^4,

$$\begin{aligned}(\Omega_{mn}^2)_{\text{class}} &+ \delta^2 - (\Omega_{mn}^4)_{\text{class}} \frac{E'}{kG} - (\Omega_{mn}^2)_{\text{class}} \frac{E'}{kG} \delta^2 - \delta^2 (\Omega_{mn}^2)_{\text{class}} \frac{E'}{kG} \\ &+ (m\pi)^2 S^2 \Gamma_{mn} (\Omega_{mn}^2)_{\text{class}} + (m\pi)^2 S^2 (\Gamma_{mn}) \delta^2 \\ &+ \Gamma_{mn} S^2 (m\pi)^2 \frac{E'}{kG} (\Omega_{mn}^2)_{\text{class}} + \Gamma_{mn} S^2 (m\pi)^2 \frac{E'}{kG} \delta^2 \\ &- (m\pi)^4 S^4 \Gamma_{mn}^2 = 0\end{aligned}$$

Solving for δ^2 we have

$$\delta^2 = \frac{-(\Omega_{mn}^2)_{\text{class}} + (E'/kG)(\Omega_{mn}^4)_{\text{class}} - (m\pi)^2 S^2 \Gamma_{mn} (1 + E'/kG)(\Omega_{mn}^2)_{\text{class}}}{1 + (m\pi)^2 S^2 \Gamma_{mn} (1 + E'/kG) - 2(\Omega_{mn}^2)_{\text{class}} E'/kG} + \frac{(m\pi)^4 S^4 \Gamma_{mn}}{1 + (m\pi)^2 S^2 \Gamma_{mn} (1 + E'/kG) - 2(\Omega_{mn}^2)_{\text{class}} E'/kG}$$

We may now give Ω_{mn}^2 as follows:

$$\begin{aligned}\Omega_{mn}^2 &= (\Omega_{mn}^2)_{\text{class}} + \delta^2 \\ &= (\Omega_{mn}^2)_{\text{class}} \left[\frac{-(\Omega_{mn}^2)_{\text{class}} E'/kG + (m\pi)^4 S^4 \pi^2 / (\Omega_{mn}^2)_{\text{class}}}{1 + (m\pi)^2 S^2 \Gamma_{mn} (1 + E'/kG) - 2(\Omega_{mn}^2)_{\text{class}} E'/kG} \right]\end{aligned}$$

Replace $(\Omega_{mn}^2)_{\text{class}}$ in the bracketed expression above by $(m\pi)^4 S^4 \Gamma_{mn}^2$ from Eqs. (7.140) and (7.141*b*):

$$\Omega_{mn}^2 = (\Omega_{mn}^2)_{\text{class}} \frac{1 - (m\pi)^4 S^4 \Gamma_{mn}^2 E'/kG}{1 + (m\pi)^2 S^2 \Gamma_{mn} (1 + E'/kG) - 2(m\pi)^4 S^4 \Gamma_{mn}^2 E'/kG}$$

For thin plates $S^2 \ll 1$. We can then drop the last terms in the numerator and denominator and, expanding what is left as a power series, we can then say as an approximation:

$$\Omega_{mn}^2 \cong (\Omega_{mn}^2)_{\text{class}} \left[1 - \left(1 + \frac{E'}{kG}\right) S^2 (m\pi)^2 \Gamma_{mn} \right] \tag{7.143}$$

We now have the means of ascertaining natural frequencies of the simply supported plate including rotatory inertia and shear.

Part C
General Considerations

7.13 THE EIGENFUNCTION-EIGENVALUE PROBLEM RESTATED

In the consideration of the vibrations of beams and plates undertaken in Parts A and B of this chapter we proceeded by the familiar separation of variables technique to arrive at an equation of the form

$$L(W) = \omega^2 M(W) \tag{7.144}$$

where L and M are differential operators acting on the dependent variable $W(x, y)$. Also, there were linear homogeneous boundary conditions of the form

$$N(W) = 0 \tag{7.145}$$

where N is a linear differential operator. Thus in the case of the beam we see from Eq. (7.17*b*) that

$$L = \frac{d^2}{dx^2}(EI)\frac{d^2}{dx^2} \tag{7.146a}$$

$$M = \rho A \tag{7.146b}$$

And in the case of classical plates we see from Eq. (7.88*b*) that

$$L = \nabla^4 \tag{7.147a}$$

$$M = \frac{1}{\beta^2} \tag{7.147b}$$

These are examples of eigenvalue-eigenfunction problems. The eigenvalue-eigenfunction problem then asks: *What function W, satisfying the given boundary conditions of the problem and acted on by the operator L, gives a result that is equal to some number* ω^2 *times the result of the M operator acting on this function W?* The function W that satisfies the aforestated condition is called the *eigenfunction*, while the accompanying value ω^2 is called the *eigenvalue.* In the problems examined in Parts A and B of the chapter we found that the ω^2, for the problems considered, formed an infinite discrete set of numbers giving the natural frequencies of the vibration problem. The eigenfunctions for the problem were the so-called mode shapes for free vibration. (It is possible that several functions W may correspond to a single eigenvalue ω^2 and we then have a *degeneracy* present. We shall not get involved here in such considerations.)

In the two problems undertaken (the simply supported beam and the simply supported plate) we found that integrals of the product of two eigenfunctions over the domain of the problem were zero when these eigenfunctions were different from each other. This was a useful property in establishing the constants of integration to satisfy the initial conditions of the problem. Actually, you will soon see that this condition is a consequence of a property of the function called *orthogonality*. We now define orthogonal eigenfunctions as those satisfying the equation $L(W) = \omega^2 M(W)$ for which the following conditions hold:

$$\iiint W_i L(W_j)\, dv = 0 \quad \text{if } i \neq j \tag{7.148a}$$

$$\iiint W_i M(W_j)\, dv = 0 \quad \text{if } i \neq j \tag{7.148b}$$

We will now show that the eigenfunctions W_i are orthogonal in the above sense if the operators L and M are *self-adjoint.*† Using the more universal notation λ_i for ω_i^2, we consider eigenfunctions W_i and W_j satisfying the differential equation as follows:

$$L(W_i) = \lambda_i M(W_i) \tag{7.149a}$$

$$L(W_j) = \lambda_j M(W_j) \tag{7.149b}$$

Hence

$$W_j L(W_i) = \lambda_i W_j M(W_i) \tag{7.150a}$$

$$W_i L(W_j) = \lambda_j W_i M(W_j) \tag{7.150b}$$

where we have multiplied Eq. (7.149*a*) by W_j and Eq. (7.149*b*) by W_i. Integrating over the domain of the problem and subtracting, we then get

$$\iiint_D [W_j L(W_i) - W_i L(W_j)]\, dv = \iiint_D [\lambda_i W_j M(W_i) - \lambda_j W_i M(W_j)]\, dv$$

Now employing the self-adjoint property of the operators we get

$$0 = (\lambda_i - \lambda_j) \iiint_D W_i M(W_j)\, dv$$

Clearly, if $\lambda_i \neq \lambda_j$—i.e., if the eigenvalues are different, we conclude that

$$\iiint_D W_i M(W_j)\, dv = 0 \tag{7.151}$$

†A symmetric or self-adjoint operator S, you will recall from Chapter 3, is one for which $\langle u, Sv \rangle = \langle v, Su \rangle$.

Now, going back to Eq. (7.150*a*), we integrate over the domain and note the above conclusion. We then must conclude that

$$\iiint_D W_j L(W_i)\, dv = 0 \tag{7.152}$$

thus completing the proof. Now we have already shown in Chapter 3 that the operators L and M are self-adjoint for beams and plates under certain boundary conditions (see Section 3.11). We leave it for you to show accordingly that the simply supported beam and the simply supported plate examined earlier have boundary conditions that satisfy the conditions presented in Chapter 3 and thereby have orthogonal eigenfunctions.

7.14 THE RAYLEIGH QUOTIENT IN TERMS OF OPERATORS

We will now develop Rayleigh's quotient directly from the differential equation for W in the case of a beam. Thus we have for the beam from Eq. (7.17*b*):

$$\frac{d^2}{dx^2}\left(EI\frac{d^2W}{dx^2}\right) = \rho A\omega^2 W$$

Now multiply both sides of the equation by W. Integrating over the domain of the beam we get

$$\int_0^L W\frac{d^2}{dx^2}\left(EI\frac{d^2W}{dx^2}\right)dx = \omega^2\int_0^L \rho A W^2\, dx \tag{7.153}$$

Next integrate the left side of the equation by parts twice as follows:

$$\int_0^L W\frac{d^2}{dx^2}\left(EI\frac{d^2W}{dx^2}\right)dx = \int_0^L EI\left(\frac{d^2W}{dx^2}\right)^2 dx + W\frac{d}{dx}\left(EI\frac{d^2W}{dx^2}\right)\Bigg|_0^L$$

$$-EI\frac{dW}{dx}\frac{d^2W}{dx^2}\Bigg|_0^L \quad .$$

For simply supported or cantilevered conditions it is clear that the last two expressions on the right side of the above equation are zero. We use this result in Eq. (7.153) to solve for ω^2 as follows:

$$\omega^2 = \frac{\displaystyle\int_0^L EI(d^2W/dx^2)^2\, dx}{\displaystyle\int_0^L \rho A W^2\, dx} \tag{7.154}$$

But this is *Rayleigh's quotient* for the problem [see Eq. (7.33)]. Let us next express the above equation with the differential operators. Thus, going back to Eq. (7.153), we can conclude that

$$\omega^2 = \frac{\int_0^L WL(W)\,dx}{\int_0^L WM(W)\,dx} \tag{7.155a}$$

Indeed, if we go back to Eq. (7.144), multiply by W, and integrate over the domain we get the above result directly.

In the general case, we can give the Rayleigh quotient as follows, using the notation of Chapter 3:

$$\boxed{\omega^2 = \frac{\langle W, LW\rangle}{\langle W, MW\rangle}} \tag{7.155b}$$

Thus, using L and M for a plate in Eq. (7.155) and employing Green's theorem plus the boundary conditions which we formulated for plates, we can establish Eq. (7.104), which was developed earlier by using conservation of energy considerations. We shall see in the following sections that the Rayleigh quotient as given by Eq. (7.155) [we shall denote it at times as $R(W)$] has very useful mathematical properties that permit the approximation of the eigenvalues.

7.15 STATIONARY VALUES OF THE RAYLEIGH QUOTIENT

We now prove a very important theorem concerning stationary properties of the Rayleigh quotient. This theorem will form the basis for supporting the statement made earlier concerning the direction of approximation for the Rayleigh-Ritz method and will lend credence to the process.

Let $(\phi_1, \phi_2, \ldots, \phi_{k-1})$ be the first nonzero $(k-1)$ eigenfunctions for the eigenvalue problem

$$L(W) = \omega^2 M(W) \tag{7.156}$$

where the boundary conditions are such that L and M are self-adjoint. We shall extremize the Rayleigh quotient for the above case with respect to functions ψ that satisfy the boundary conditions of the problem and that are *orthogonal* to the aforestated $(k-1)$ eigenfunctions. We will prove that the *minimum of the Rayleigh quotient with respect to* ψ *is the* kth *eigenvalue and the extremizing function* ψ *is simply* ϕ_k, *the* kth *eigenfunction.*

As a first step we express ψ as an infinite series expansion in the nonzero eigenfunctions:

$$\psi = \sum_{n=1}^{\infty} C_n \phi_n \tag{7.157}$$

where C_n are computed in the familiar manner by making use of the orthogonality property of ϕ_n. The $(k-1)$ orthogonality conditions then may be expressed as follows:

$$\iiint_D \phi_i L(\psi)\, dv = \iiint_D \phi_i L\left(\sum_{n=1}^{\infty} C_n \phi_n\right) dv = \sum_{n=1}^{\infty} C_n \iiint_D \phi_i L(\phi_n)\, dv = 0$$

$$\iiint_D \phi_i M(\psi)\, dv = \iiint_D \phi_i M\left(\sum_{n=1}^{\infty} C_n \phi_n\right) dv = \sum_{n=1}^{\infty} C_n \iiint_D \phi_i M(\phi_n)\, dv = 0$$

$$i = 1, 2, \ldots, k-1$$

Now consider $i = 1$. We have from above, making use of the orthogonal property of the eigenfunctions,

$$C_1 \iiint_D \phi_1 M(\phi_1)\, dv = 0$$

But $\iiint \phi_1 M(\phi_1)\, dv$ represents a kinetic energy in this discussion and so it must be positive definite. Assuming as we did that ϕ_1 is not everywhere zero, then clearly $C_1 = 0$. In a similar manner we can conclude that $C_2 = C_3 = \cdots C_{k-1} = 0$. Accordingly, we have from the orthogonality requirements on ψ the result that

$$C_1 = C_2 = \cdots C_{k-1} = 0$$

We now examine the Rayleigh quotient. For this purpose we note on using Eq. (7.157) for ψ

$$\iiint_D \psi L(\psi)\, dv = \iiint_D \sum_{n=k}^{\infty} C_n \phi_n \sum_{j=k}^{\infty} C_j L(\phi_j)\, dv$$

$$= \sum_{n=k}^{\infty} \sum_{j=k}^{\infty} C_n C_j \iiint_D \phi_n L(\phi_j)\, dv$$

$$= \sum_{n=k}^{\infty} C_n^2 \iiint_D \phi_n L(\phi_n)\, dv$$

$$= \sum_{n=k}^{\infty} C_n^2 \omega_n^2 \iiint_D \phi_n M(\phi_n)\, dv \tag{7.158a}$$

where we have used the orthogonality property of the eigenfunctions in the next to last expression and Eq. (7.156) in the last expression. Also, using Eq. (7.157) for ψ and again using the orthogonality property of the eigenfunctions, we get

$$\iiint_D \psi M(\psi)\, dv = \iiint_D \sum_{n=k}^{\infty} C_n\phi_n \sum_{j=k}^{\infty} C_j M(\phi_j)\, dv$$

$$= \sum_{n=k}^{\infty} \sum_{j=k}^{\infty} C_n C_j \iiint_D \phi_n M(\phi_j)\, dv$$

$$= \sum_{n=k}^{\infty} C_n^2 \iiint_D \phi_n M(\phi_n)\, dv \tag{7.158b}$$

We now introduce the notation

$$\kappa_n = \iiint_D \phi_n M(\phi_n)\, dv \tag{7.159}$$

so that the Rayleigh quotient R [see Eqs. (7.155)] can be given as follows, making use of Eqs. (7.158*a*) and (7.158*b*):

$$R = \frac{\sum_{n=k}^{\infty} C_n^2 \kappa_n \omega_n^2}{\sum_{n=k}^{\infty} C_n^2 \kappa_n} \tag{7.160}$$

We can rewrite the above equation as follows:†

$$R = \omega_k^2 + \frac{\kappa_{k+1} C_{k+1}^2 (\omega_{k+1}^2 - \omega_k^2) + \kappa_{k+2} C_{k+2}^2 (\omega_{k+2}^2 - \omega_k^2) + \cdots}{\sum_{n=k}^{\infty} C_n^2 \kappa_n} \tag{7.161}$$

Since $\omega_{k+1} > \omega_k$, etc., and since κ must be positive definite (kinetic energy), we may conclude that the minimum value for R is ω_k^2–the kth eigenvalue. This occurs when $C_{k+1}, C_{k+2}, \ldots$ are all zero. Thus the only nonzero coefficient for this condition is C_k, and so going back to Eq. (7.157) we see that the function ψ is simply ϕ_k up to a proportionality constant. *Thus the minimizing function ψ for the Rayleigh quotient is the* k*th eigenfunction.*

Of course, if $k = 1$, the function ψ does not require orthogonality restrictions vis-à-vis eigenfunctions; it must merely satisfy the boundary conditions of the problem.

†You may easily justify the formulation by recombining the terms under a common denominator and carrying out necessary cancellation of terms in the numerator.

Thus we see here that *any* such function when substituted into the Rayleigh quotient will yield a result λ_1 that will be greater than or equal to ω_1^2, the lowest eigenvalue. And by using a parameter-laden set of functions to express ψ we can adjust the constants to give the smallest Rayleigh quotient for that family of functions and thereby come closer to the lowest eigenvalue ω_1^2 *from above.* Finally, since the eigenfunction ϕ_1 is an extremal function for the Rayleigh quotient, clearly a function ψ ***reasonably close*** to ϕ_1 should give a λ *very close* to ω_1^2. These remarks then give the basis for Rayleigh's method of finding the lowest eigenvalue.

The question now arises as to the basis of the Rayleigh-Ritz method for finding eigenvalues beyond the lowest eigenvalue. We now re-examine the Rayleigh-Ritz method itself in a more formal manner in order to prepare the groundwork for assessing its validity.

7.16 RAYLEIGH–RITZ METHOD RE–EXAMINED

We shall now develop the Rayleigh-Ritz method for finding eigenvalues beyond the lowest eigenvalue by a method that is closely tied to the theorem presented in the previous section.

In the usual manner we consider the linear combination of n functions ψ_i that satisfy the boundary conditions of the problem to form W as follows:

$$W = C_1\psi_1 + \cdots + C_n\psi_n \tag{7.162}$$

By limiting the possible deflections of the body to what can be described by the expression given above, we have tacitly built in hidden constraints such that the body, instead of having an infinite number of degrees of freedom, will soon be seen to have only n degrees of freedom. We are, in effect, promulgating a new system to represent the original system. We can find the lowest eigenvalue Λ_1^2 of this hypothetical system by extremizing the Rayleigh quotient. Clearly Λ_1^2 will be some sort of approximation of the eigenvalue ω_1^2 of the actual problem. We can determine the C's† and thus establish the corresponding eigenfunction for our hypothetical system. Thus:

$$W^{(1)} = C_1^{(1)}\psi_1 + \cdots + C_n^{(1)}\psi_n \tag{7.163}$$

The superscript indicates that we are dealing with the first mode. Presumably $W^{(1)}$ must then be some approximation of the first eigenfunction of the actual problem. To get the second eigenvalue of the hypothetical system, we employ the minimum theorem of the previous section by minimizing the Rayleigh quotient using admissible functions W given by Eq. (7.162) but now subject to the requirement that they be orthogonal to $W^{(1)}$. This generates Λ_2^2 and a second eigenfunction for the hypothetical system given in the form:

$$W^{(2)} = C_1^{(2)}\psi_1 + \cdots + C_n^{(2)}\psi_n$$

†This can be done up to a multiplicative constant.

The procedure is continued in this manner. Thus for the kth eigenvalue we adjust the C's to extremize the Rayleigh quotient and also to make W orthogonal to the $k-1$ lower eigenfunctions of the hypothetical system. Let us now do this more formally. Accordingly, for the process to approximate the kth eigenvalue from above we have

$$\Lambda_k^2 = \frac{\iiint_D WL(W)\,dv}{\iiint_D WM(W)\,dv} = \frac{\iiint_D \left(\sum_{i=1}^{n} C_i\psi_i\right) L\left(\sum_{j=1}^{n} C_j\psi_j\right) dv}{\iiint_D \left(\sum_{i=1}^{n} C_i\psi_i\right) M\left(\sum_{j=1}^{n} C_j\psi_j\right) dv}$$

$$= \frac{\sum_{i=1}^{n}\sum_{j=1}^{n} C_iC_jG_{ij}}{\sum_{i=1}^{n}\sum_{j=1}^{n} C_iC_jE_{ij}} \tag{7.164}$$

where

$$G_{ij} = \iiint_D \psi_i L(\psi_j)\,dv$$
$$E_{ij} = \iiint_D \psi_i M(\psi_j)\,dv \tag{7.165}$$

We have the following orthogonality requirement:

$$\iiint_D W^{(S)} M\left(\sum_{i=1}^{n} C_i\psi_i\right) dv = 0 \qquad S = 1, 2, \ldots, k-1$$

Replacing $W^{(S)}$ by $\Sigma_{j=1}^{n} C_j^{(S)}\psi_j$, we may rewrite the above equation as follows:

$$\iiint_D W^{(S)} M\left(\sum_{i=1}^{n} C_i\psi_i\right) dv = \iiint_D \left(\sum_{j=1}^{n} C_j^{(S)}\psi_j\right)\left(\sum_{i=1}^{n} C_i M(\psi_i)\right) dv = 0$$

Therefore

$$\sum_{j=1}^{n}\sum_{i=1}^{n} C_j^{(S)} C_i E_{ij} = 0 \qquad S = 1, 2, \ldots, k-1 \tag{7.166}$$

The $k-1$ equations above are *constraining equations* for the extremization process, and so we extremize the function Λ_k^{2*} given as

$$\Lambda_k^{2*} = \frac{\sum_{i=1}^{n} \sum_{j=1}^{n} C_i C_j G_{ij}}{\sum_{i=1}^{n} \sum_{j=1}^{n} C_i C_j E_{ij}} - \sum_{S=1}^{k-1} \lambda^{(S)} \sum_{j=1}^{n} \sum_{i=1}^{n} C_j^{(S)} C_i E_{ij}$$

where $\lambda^{(S)}$ are Lagrange multipliers. To extremize Λ_k^{2*} we proceed as follows:

$$\frac{\partial \Lambda_k^{2*}}{\partial C_i} = 0 \qquad i = 1, 2, \ldots, n$$

therefore

$$\frac{2 \sum_{j=1}^{n} C_j G_{ij}}{\sum_{i=1}^{n} \sum_{j=1}^{n} C_i C_j E_{ij}} - \frac{\left(\sum_{i=1}^{n} \sum_{j=1}^{n} C_i C_j G_{ij}\right)\left(2 \sum_{j=1}^{n} C_j E_{ij}\right)}{\left(\sum_{j=1}^{n} \sum_{i=1}^{n} C_i C_j E_{ij}\right)^2} - \sum_{S=1}^{k-1} \lambda^{(S)} \sum_{j=1}^{n} C_j^{(S)} E_{ij} = 0$$

$$i = 1, 2, \ldots, n$$

Next multiply through by $\frac{1}{2} \Sigma_{i=1}^{n} \Sigma_{j=1}^{n} C_i C_j E_{ij}$ to get

$$\sum_{j=1}^{n} C_j G_{ij} - \frac{\sum_{i=1}^{n} \sum_{j=1}^{n} C_i C_j G_{ij}}{\sum_{i=1}^{n} \sum_{j=1}^{n} C_i C_j E_{ij}} \sum_{j=1}^{n} C_j E_{ij}$$

$$-\left(\sum_{S=1}^{k-1} \lambda^{(S)} \sum_{j=1}^{n} C_j^{(S)} E_{ij}\right)\left(\frac{1}{2} \sum_{v=1}^{n} \sum_{p=1}^{n} C_v C_p E_{vp}\right) = 0 \qquad i = 1, 2, \ldots, n$$

Note that we have changed dummy indices in the last term for convenience. Using Eq. (7.164) to bring Λ_k^2 into the second expression of the above equation, we get

$$\sum_{j=1}^{n} C_j G_{ij} - \Lambda_k^2 \sum_{j=1}^{n} C_j E_{ij}$$

$$-\frac{1}{2}\left(\sum_{S=1}^{k-1} \lambda^{(S)} \sum_{j=1}^{n} C_j^{(S)} E_{ij}\right)\left(\sum_{v=1}^{n} \sum_{p=1}^{n} C_v C_p E_{vp}\right) = 0 \qquad i = 1, \ldots, n \tag{7.167}$$

Note first that *if the* $k-1$ *Lagrange multipliers are zero in value*, we then have

$$\sum_{j=1}^{n} (G_{ij} - \Lambda_k^2 E_{ij}) C_j = 0 \qquad i = 1, 2, \ldots, n \tag{7.168}$$

We will prove in Appendix II that this is indeed the situation (the proof, while straightforward, is somewhat tedious and offers no new insight). Considering the C's as unknowns we then have n simultaneous homogeneous equations for the unknowns. For a nontrivial solution it is necessary that the determinant of the C's be zero. That is, dropping the subscript k for reasons soon to be apparent, we have

$$|G_{ij} - \Lambda^2 E_{ij}| = 0 \tag{7.169}$$

We will get n roots for Λ^2, which we order in magnitude as $\Lambda^{2(1)}, \Lambda^{2(2)}, \ldots, \Lambda^{2(n)}$. Note in particular that no matter what k is for the problem (assuming it is no greater than n) we arrive at the *same* equation (7.169) to be solved. That is, this equation applies for *all* the eigenvalues of the hypothetical system from 1 to n that we might be seeking to establish by this method. And so we interpret the n roots Λ^2 as the eigenvalues of the hypothetical system and hence as the sought-for approximations from above of the n eigenvalues of the actual problem. By using the proper operators L and M for the beam and the plate we can also see that the formulation of the Λ's above is *exactly the same formulation* that was used in Parts A and B, respectively, for the Rayleigh-Ritz method, thus justifying our use here of the same notation Λ that was used earlier. Note also that we can get a set of constants $C_i^{(k)}$ for each root Λ_k^2 by substituting Λ_k^2 into Eq. (7.168) and solving for the constants $C_i^{(k)}$. However, because the equations are homogeneous, we can solve uniquely only for the *ratios* of these constants. This gives the *shape* of the eigenfunctions of the hypothetical system and hence approximations of the actual eigenfunctions but not their amplitudes.

Now that we have presented the Rayleigh-Ritz method in a framework directly from the vital minimum theorem of Section 7.15; we will be able soon to show that the n values Λ^2 in ascending order of size as computed by the Rayleigh-Ritz method are approximations from above of the first n eigenvalues of the problem. For this step we need the use of the maximum-minimum principle, which we now present.

7.17 MAXIMUM-MINIMUM PRINCIPLE

We now consider $k-1$ functions $g_1, g_2, \ldots, g_{k-1}$, which are perfectly arbitrary except for simple integrability requirements in the domain of interest. Now as admissible functions ψ to extremize the Rayleigh quotient for a given eigenvalue-eigenfunction problem, we shall consider those that satisfy the boundary conditions of the problem and satisfy the following $k-1$ orthogonality conditions:

$$\iiint_D g_i M(\psi)\, dv = 0 \qquad i = 1, 2, \ldots, k-1 \tag{7.170}$$

We will now show that the *minimum value of the Rayleigh quotient with respect to ψ for the given set of functions g_i will be less than the* kth *eigenvalue of the problem and will equal the* kth *eigenvalue of the problem if the g's are taken as the first $k-1$ eigenfunctions ϕ_i of the problem*—thus getting back to the case presented in Section 7.15.

To prove this, we first show that there always exists a function ψ_0 that is a linear combination of the first k eigenfunctions

$$\psi_0 = C_1\phi_1 + \cdots + C_k\phi_k \tag{7.171}$$

and that satisfies Eq. (7.170) for any set of functions g_i. To show this, substitute Eq. (7.171) into Eq. (7.170) as follows:

$$\iiint_D g_i M\left(\sum_{j=1}^{k} C_j\phi_j\right) dv = \sum_{j=1}^{k} \iiint_D g_i C_j M(\phi_j)\, dv = 0 \qquad i = 1, 2, \ldots, k-1 \tag{7.172}$$

Since i goes from 1 to $k-1$ there are $k-1$ equations. Considering the C's as k unknowns, it is clear that we can always find, although not uniquely, a set of values for these constants that will satisfy Eq. (7.172). That is, because we have more unknowns than equations here, we can always determine a set of values for the unknowns that satisfy the equations.

Now employ the function ψ_0 in the Rayleigh quotient R. We may in this regard use the results stemming from Eqs. (7.158*a*) and (7.158*b*) (by summing now from 1 to k) as well as the definition given by Eq. (7.159). This leads to Eq. (7.160), which is made valid for our case here by using $n = 1$ to k, giving the result

$$R = \frac{C_1^2\kappa_1\omega_1^2 + C_2^2\kappa_2\omega_2^2 + \cdots + C_k^2\kappa_k\omega_k^2}{C_1^2\kappa_1 + C_2^2\kappa_2 + \cdots + C_k^2\kappa_k} \tag{7.173}$$

Remembering that $\omega_1^2 \leqslant \omega_2^2 \leqslant \cdots \leqslant \omega_k^2$ and noting that the κ's are positive definite, we can conclude on replacing $\omega_1^2, \ldots, \omega_{k-1}^2$ by ω_k^2 in the above equation that

$$R \leqslant \frac{C_1^2\kappa_1\omega_k^2 + \cdots + C_k^2\kappa_k\omega_k^2}{C_1^2\kappa_1 + \cdots + C_k^2\kappa_k} = \omega_k^2 \tag{7.174}$$

Thus we see that the Rayleigh quotient for ψ_0, which is a member of the family of admissible functions ψ, *is less than or equal to* the kth eigenvalue of the problem. Certainly the *minimum* of R with respect to the *whole* admissible class of functions ψ must then be less than or equal to ω_k^2. The equality is achieved according to the minimum theorem of Section 7.15 when the functions g_i used for orthogonalization are the first $k-1$ eigenfunctions of the problem. The theorem is now proved.

What is the relation between the minimum principle presented in Section 7.15 (and used to justify Rayleigh's method for the first mode) and the present maximum-minimum principle? We shall use for this explanation some plots whose detailed meaning is not of concern here. Thus in Fig. 7.14 we have shown a region I which is meant to give all the values of the Rayleigh quotient when the admissible functions are made orthogonal to the first $k-1$ *eigenfunctions* of the problem. We identify the family of functions involved in the orthogonality conditions involved in the orthogonality conditions as $\Theta_{(\,)}$. Thus for region I we have indicated $\Theta_{(\phi_i)}$ in the diagram.

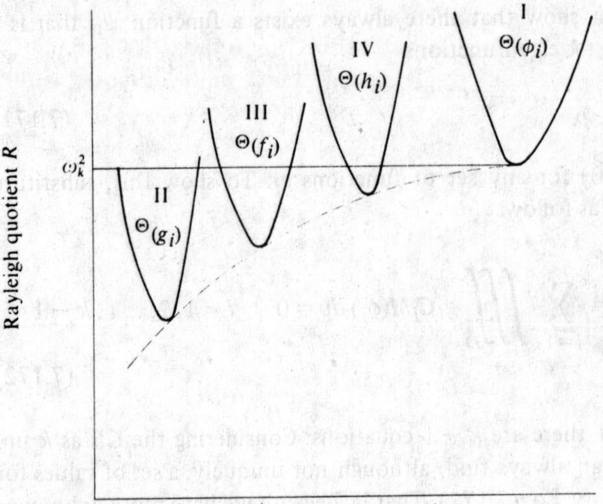

FIGURE 7.14
Idealized plots showing minimum-maximum principle.

Notice that the *minimum* value for this family of Rayleigh quotients is ω_k^2, in accordance with the *minimum principle.* Next consider a system of Rayleigh quotients for which the admissible functions have been made orthogonal to a set of $k-1$ functions g_i (not equal to ϕ_i). This set of functions is denoted as $\Theta_{(g_i)}$ and the system of Rayleigh quotients is shown in the diagram as region II. Notice now that the minimum value of R is less than ω_k^2, in accordance with the *maximum-minimum* principle of this section. By changing the functions g_i we may form other regions in our plot such as regions III and IV and thereby increase the minimum value of the Rayleigh quotient as shown in the diagram. One could then conceivably continually increase the minimum values of R for each family Θ by continuously changing the family of functions g_i (see dashed line on the diagram). The maximum value of such minimums is attained when Θ consists of the eigenfunctions ϕ_i.

We shall next employ the maximum-minimum principle to show that $\Lambda_k^2 \geqslant \omega_k^2$, thus justifying the earlier assertions of estimating from above by the Rayleigh-Ritz method.

7.18 JUSTIFICATION OF THE ESTIMATION FROM ABOVE ASSERTION OF THE RAYLEIGH-RITZ METHOD

We would like to show here that $\Lambda_k^2 \geqslant \omega_k^2$, where Λ_k^2 is calculated from the Rayleigh-Ritz procedure as used in Parts A and B and as presented more formally in Part C of this chapter.

Consider two systems A and B identical in every way except that system B has additional hidden constraints, making it a "stiffer" system. Such constraints, we have pointed out, occur as a result of simplifications made in the manner in which system B deforms. Such simplifications occur when we replace a body having an infinite number of degrees of freedom by a substitute having a finite number of degrees of freedom, as we did in the Rayleigh-Ritz method described in Section 7.16. The admissible functions $\bar{\psi}_i$ for minimizing the Rayleigh quotient for B will be a ***subset*** of the set ψ_i for system A. Taking ϕ_i to be the ***eigenfunctions*** of system A, we can say, on minimizing the Rayleigh quotient with respect to the complete set ψ_i while maintaining orthogonality with respect to the first $k-1$ eigenfunctions ϕ_i of system A,

$$\omega_k^2 = [R_{\min}(\psi_i)]_{\Theta(\phi_i)} \tag{7.175}$$

This result is directly from the minimum theorem. If we now use ***subset*** $\bar{\psi}_i$ to extremize the Rayleigh quotient above, we clearly get a result that is ***greater than or equal to*** the value of the minimum Rayleigh quotient with the complete set ψ_i. Thus we can say

$$\omega_k^2 \leqslant [R_{\min}(\bar{\psi}_i)]_{\Theta(\phi_i)} \tag{7.176}$$

Now in the Rayleigh-Ritz method, as we pointed out, we effectively replace the system with infinite degrees of freedom by one with as many degrees of freedom as the number of coordinate functions used. We shall consider B to be such a system formed from system A. The eigenfunctions of system B are the so-called ***approximate eigenfunctions*** of system A coming out of the Rayleigh-Ritz method. We denote them as $(\phi_i)_{\text{app}}$. Thus using the minimum theorem again for system B and denoting Λ_k^2 as the kth eigenvalue for system B:

$$\Lambda_k^2 = [R_{\min}(\bar{\psi}_i)]_{\Theta(\phi_i)_{\text{app}}} \tag{7.177}$$

Now we compare the right sides of Eq. (7.176) and (7.177). We can use the maximum-minimum principle, noting that $(\phi_i)_{\text{app}}$ and ***not*** ϕ_i are the ***eigenfunctions*** for system B,† to conclude that

$$[R_{\min}(\bar{\psi}_i)]_{\Theta(\phi_i)} \leqslant [R_{\min}(\bar{\psi}_i)]_{\Theta(\phi_i)_{\text{app}}} \tag{7.178}$$

Accordingly, using Eq. (7.176) to replace the left side of the above inequality and Eq. (7.177) to replace the right side, we have

$$\omega_k^2 \leqslant \Lambda_k^2$$

We have thus accomplished our goal, since ω_k^2 may be considered the eigenvalues of the problem and Λ_k^2 clearly are the approximate eigenvalues of the problem found by the Rayleigh-Ritz method.

7.19 CLOSURE

We began this chapter by presenting Hamilton's principle, which permits us to set forth differential equations of motion. In particular, we formulated the equations of

†ϕ_i correspond to the functions g_i of the maximum-minimum principle.

motion for the simple beam and the Timoshenko beam in Part A and then, in Part B, we did the same for simple plates and the Mindlin plate. Once these equations were developed, we concentrated on finding approximate solutions for natural frequencies of these bodies via Rayleigh's method and the Rayleigh-Ritz method. At first we employed conservation of energy as the underlying basis for the method. We found in our examples that the approximations for the natural frequencies obtained by these methods were always from above. In Part C we generalized the arguments of the preceding sections to encompass the eigenvalue-eigenfunction problem, thereby presenting the Rayleigh quotient and its properties. With the aid of the minimum theorem and the minimum-maximum theorem we were able to justify that the approximations for eigenfunctions from the Rayleigh and Rayleigh-Ritz methods had to be from above. The generalizations of Part C will permit us to employ the Rayleigh and Rayleigh-Ritz methods for the stability problems of columns and plates in Chapter 8.

You will recall that in the chapter on torsion we obtained approximations from above for the torsional rigidity via the Ritz method, and in addition obtained approximations from below via the Trefftz method, thus bracketing the correct torsional rigidity. We wish to point out here that it is also possible to find approximations from below of eigenvalues, thus permitting a bracketing of the correct eigenvalues when results from the Rayleigh and Rayleigh-Ritz methods are used. The method is due to A. Weinstein and involves the use of weakened or relaxed boundary conditions. Detailed discussion of the method, however, is beyond the scope of this text.†

READING

Den Hartog, J. P.: "Mechanical Vibrations," McGraw-Hill, New York, 1947.
Gould, S. H.: "Variational Methods for Eigenvalue Problems," Univ. of Toronto Press, Toronto, 1966.
Meirovitch, L.: "Analytical Methods in Vibrations," Macmillan, New York, 1970.
Mikhlin, S. G.: "Variational Methods of Mathematical Physics," Macmillan, New York, 1964.
Nowacki, W.: "Dynamics of Elastic Systems," Chapman and Hall, London, 1963.
Rayleigh, J. W. S.: "The Theory of Sound," Dover, New York, 1954.
Temple, G., and Bickley, W. G.: "Rayleigh's Principle," Dover, New York, 1956.
Timoshenko, S.: "Vibration Problems in Engineering," Van Nostrand, New York, 1955.
Tong, K. N.: "Theory of Mechanical Vibration," Wiley, New York, 1960.
Volterra, E., and Zachmanoglou, E. C.: "Dynamics of Vibrations," Merrill, Columbus, Ohio, 1965.

PROBLEMS

7.1 Derive Hamilton's principle for a system by the δ-operator approach rather than by the single-parameter family approach as done in Section 7.3.

7.2 Verify Eqs. (7.30*a*) and (7.30*b*).

7.3 Derive Eq. (7.38).

7.4 Consider the nonuniform cantilevered beam shown in Fig. 7.15. Using a simple parabolic shape for the fundamental mode, show that Rayleigh's quotient gives as the fundamental natural frequency:

$$\omega = \sqrt{\frac{30}{12}\,\frac{(h_0 + h_1)^4 - h_0^4}{h_1^2 + 6h_0 h_1}\,\frac{E}{\rho L^4}}$$

†For a detailed exposition of Weinstein's method see Gould (1966).

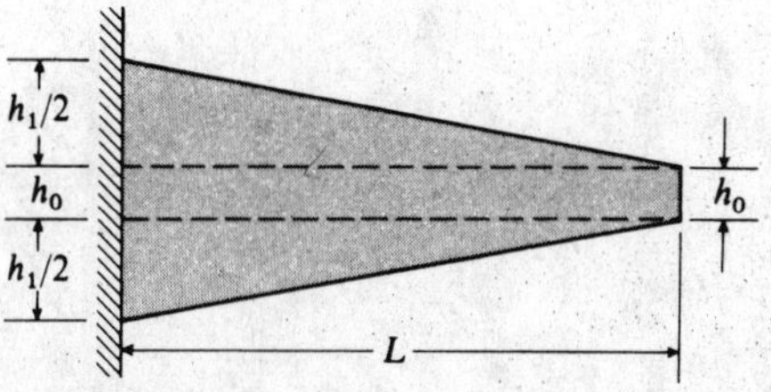

FIGURE 7.15

Compare your result (after suitable adjustments) with the corresponding result for a uniform cantilevered beam:

$$\omega = 1.015 \frac{h_0}{L^2}\sqrt{\frac{E}{\rho}}$$

and that for a pure triangle:

$$\omega = 1.535 \frac{h_1}{L^2}\sqrt{\frac{E}{\rho}}$$

(This result is due to Kirchhoff.)

7.5 Consider a cantilevered beam with a mass M attached firmly at a position x^* (see Fig. 7.16) from the base ($x = 0$). What is the Rayleigh quotient for transverse vibration? Find an approximation to the fundamental natural frequency. Is it possible that the added mass could not affect natural frequencies of higher modes? Explain. Where should the mass be placed to decrease as much as possible the fundamental frequency?

7.6 Two identical cantilevered beams A and B are pinned together at what would be their free ends (see Fig. 7.17). Using functions W_A and W_B for each cantilever, what are the proper boundary conditions these functions must satisfy? Find two functions that satisfy all boundary conditions except one condition at the pin. Now with the aid of the Rayleigh quotient find the fundamental natural frequency approximately.

7.7 Derive the differential equation and boundary conditions for the torsional oscillation of a bar in the absence of warping in terms of the rotation angle $\theta(z, t)$ and using the polar moment of inertia of the cross section $I_p = \iint (x^2 + y^2)\, dA$. Also, discuss the conditions under which the effect of the geometric cross section of the bar will not be a factor in the solution.

7.8 Do the first part of Problem 7.7, this time including the effects of two rigid cylinders, respectively of uniform mass per unit length ρ_1 and ρ_2 and polar moments of inertia of the cross section I_1 and I_2, placed at the ends $z = 0$ and $z = L$ of the bar.

7.9 Derive the differential equations and boundary conditions for longitudinal vibration of a uniform bar. Also obtain the appropriate Rayleigh quotient for such a problem.

7.10 In the preceding problem devise approximate eigenfunctions that satisfy the end conditions of

(*a*) Both ends free–i.e., $\tau_{xx} = E(\partial u/\partial x) = 0$.

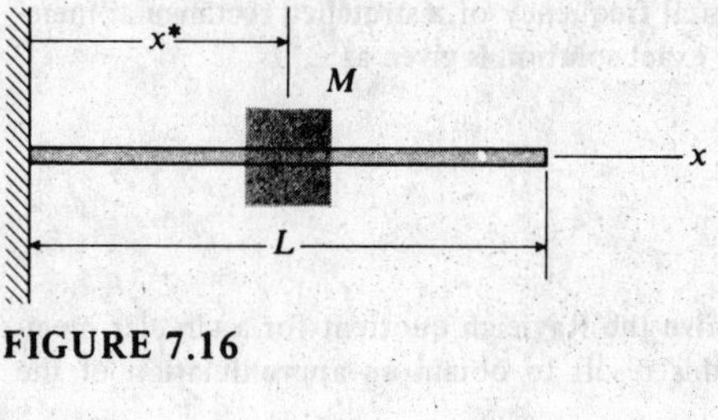

FIGURE 7.16

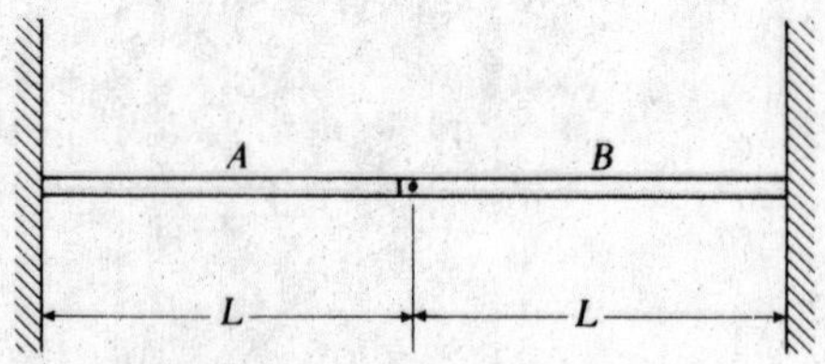

FIGURE 7.17

(*b*) Fixed at both ends–i.e., $u = 0$.
(*c*) Fixed at $x = 0$ and free at $x = L$.
Using the Rayleigh quotient, find the lowest eigenvalues for these three cases.

7.11 Find the Rayleigh quotient for torsional vibration of Problem 7.8. What is the effect of the added masses on the free vibration frequency?

7.12 Using a cosine function of z to approximate a mode shape with a single nodal point along the shaft, approximate from the Rayleigh quotient the fundamental frequency of the system presented in Problems 7.8 and 7.11. Show that if $I_1, I_2, < I_p$ (recall that I_1 and I_2 are polar moments of inertia of the added rigid cylinders and I_p is the polar moment of inertia of the connecting shaft) that

$$\omega \approx \frac{\pi}{L}\sqrt{\frac{G}{\rho}}$$

The shaft and masses have the same density. Note that this corresponds to weak (or no) coupling with the added masses at the boundary conditions, which are ordinarily inhomogeneous.

7.13 We derived in an earlier problem (Chapter 3) the strain energy of a stretched membrane:

$$U = \frac{S}{2}\iint_A \left[\left(\frac{\partial w}{\partial x}\right)^2 + \left(\frac{\partial w}{\partial y}\right)^2\right] dx\, dy = \frac{S}{2}\iint_A (\nabla w)^2\, dA \qquad (a)$$

where S is the force per unit length along the membrane. Show from Hamilton's principle that the governing equation for the motion of the membrane is

$$\nabla^2 w = \frac{\gamma}{S}\frac{\partial^2 w}{\partial t^2}$$

where $\gamma = \rho t$, t being the membrane thickness. Determine the boundary conditions for the problem. Show that the Rayleigh quotient for the dynamic behavior of the stretched membrane is

$$\omega^2 = \frac{S\iint [(\partial W/\partial x)^2 + (\partial W/\partial y)^2]\, dx\, dy}{\gamma \iint W^2\, dx\, dy}$$

7.14 Using Rayleigh's quotient, find the fundamental frequency of a stretched rectangular membrane supported at $x = 0, a$ and $y = 0, b$. The exact solution is given as

$$\omega_{11}^2 = \frac{S}{\gamma}\pi^2\left(\frac{1}{a^2} + \frac{1}{b^2}\right)$$

See Problem 7.13.

7.15 With the aid of Eq. (*a*) of Problem 7.13 derive the Rayleigh quotient for a circular membrane for axisymmetric deformation. Use this result to obtain an approximation of the fundamental natural frequency whose exact value is $2.404(\sqrt{S/\gamma})/a$, where a is the radius, S is the tensile force per unit length, and $\gamma = \rho t$.

7.16 With the aid of Eq. (*b*) of Problem 6.14 formulate the Rayleigh quotient for axisymmetric circular plates. Now obtain an approximation of the lowest frequency of vibration of a clamped circular plate. The exact result is

$$\omega = 2.955 \sqrt{\frac{Eh^2}{\rho a^4(1-\nu^2)}}$$

7.17 Obtain an approximate result for the lowest natural frequency of axisymmetric vibration of an annulus clamped at the inner edge ($r = b$) and free at the outer edge ($r = a$). Southwell has given the following exact result

$$\text{for} \qquad \frac{b}{a} = 0.276 \qquad \omega^* = 1.81$$

where

$$\omega^* = \omega \sqrt{\frac{(1-\nu^2)\rho}{E}} \frac{a^2}{h}$$

where $\nu = 0.30$ and h is the thickness.

7.18 Using the operator approach, give the Rayleigh quotient for torsional oscillation of a uniform shaft if the equation of torsional motion is

$$\frac{\partial^2 u}{\partial t^2} = \frac{G}{\rho} \frac{\partial^2 \theta}{\partial z^2}$$

7.19 Using the operator approach, what is the Rayleigh quotient for longitudinal vibration of a uniform bar if the equation of longitudinal motion is

$$\frac{\partial^2 u}{\partial t^2} = \frac{E}{\rho} \frac{\partial^2 u}{\partial x^2}$$

8

ELASTIC STABILITY

8.1 INTRODUCTION

This chapter is devoted to consideration of some of the foundations of the theory of elastic stability. We will examine the meaning of stability, especially in the context of variational methods, and we will discuss methods of obtaining stability bounds for various problems. We will use the simple Euler column problem to indicate a variety of approaches. And we will use variational methods to set forth approximate techniques for solving stability problems involving columns and plates.

In all of our work in this area we will restrict ourselves to linear elastic materials while including the geometrically nonlinear terms that are required to examine the stability of a state of equilibrium.

Part A
Stability of Rigid-Body Systems

8.2 STABILITY

We say that a configuration in a state of equilibrium is *stable* if, after some slight disturbance causing a small change in the configuration, there follows a return to the original configuration. Thus consider Fig. 8.1*a*, which shows a cylinder in a circular well. Clearly, this equilibrium configuration is a stable one. In Fig. 8.1*b* we have shown an *unstable* equilibrium configuration. If the cylinder is perfectly balanced, no change in configuration occurs. However, an infinitesimal disturbance causes finite movement of the cylinder. Figure 8.1*c* represents *neutral stability*. All positions of the cylinder

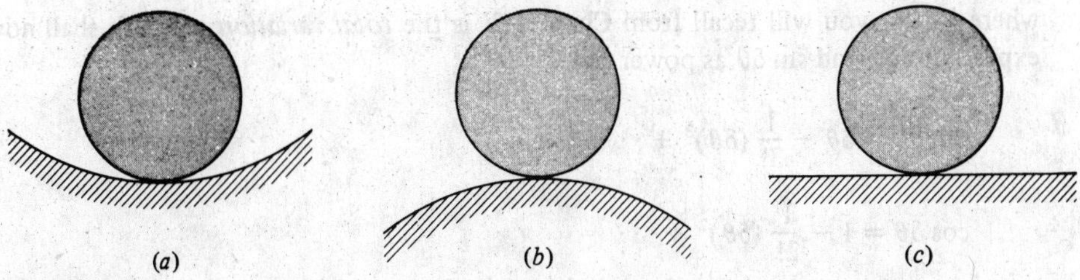

FIGURE 8.1
Three kinds of equilibrium.

are equilibrium positions. Furthermore, there is no change in the total potential energy as the position is varied.

Note that the stable equilibrium illustrated above corresponds to a position of a local *minimum* for the potential energy of the cylinder, while the unstable equilibrium corresponds to a position of a local *maximum* for the total potential energy of the cylinder. We may say for conservative systems that ***a local minimum for the total potential energy corresponds to a stable equilibrium configuration.***

Let us now consider the planar cylinder section shown in Fig. 8.2. Point C is the center of gravity of the cylinder section and R is the radius. In Fig. 8.2*a* we show an equilibrium position for the cylinder section ($\theta = 0$).

The total potential energy may be given (see Fig. 8.2*b*) as a function of θ as

$$\pi = W[(R-h)-(R-h)\cos\theta] \tag{8.1}$$

where W is the weight of the cylinder and h is the distance from the center of gravity to the ground in the equilibrium configuration. If the angle θ is now given a variation $\delta\theta$, the new potential energy is

$$\pi(\theta+\delta\theta) = W[(R-h)-(R-h)\cos(\theta+\delta\theta)]$$

We can rewrite both sides of this equation as†

$$\pi(\theta)+\delta^{(T)}\pi(\theta) = W[(R-h)-(R-h)(\cos\theta\cos\delta\theta-\sin\theta\sin\delta\theta)] \tag{8.2}$$

†Recall from Section 2.4 that $F(\theta+\delta\theta)-F(\theta)=\delta^{(T)}F$.

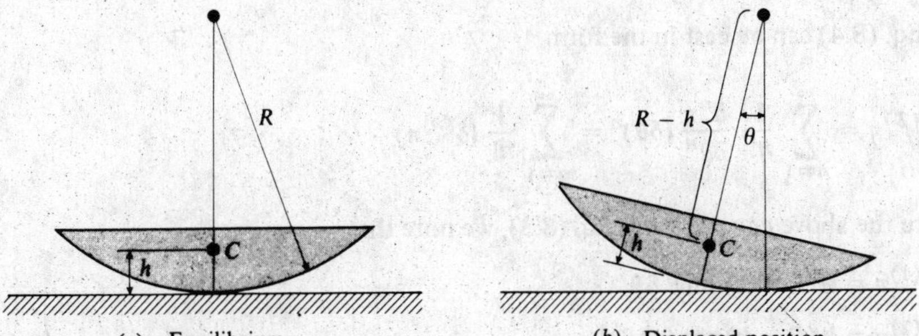

FIGURE 8.2
Stability of rigid-body disk.

where $\delta^{(T)}\pi$, you will recall from Chapter 2, is the *total variation* of π. We shall now expand $\cos\delta\theta$ and $\sin\delta\theta$ as power series:

$$\sin\delta\theta = \delta\theta - \frac{1}{3!}(\delta\theta)^3 + \cdots$$

$$\cos\delta\theta = 1 - \frac{1}{2!}(\delta\theta)^2 + \cdots$$

Equation (8.2) can then be written as

$$\pi + \delta^{(T)}\pi = W[(R-h)-(R-h)\cos\theta] + W(R-h)(\sin\theta)\delta\theta$$
$$+ \frac{1}{2!}W(R-h)(\cos\theta)(\delta\theta)^2 - \frac{1}{3!}W(R-h)(\sin\theta)(\delta\theta)^3 + \cdots$$

Noting Eq. (8.1), we can give $\delta^{(T)}\pi$ as

$$\delta^{(T)}\pi = W(R-h)\sin\theta\,\delta\theta + \frac{1}{2!}W(R-h)\cos\theta(\delta\theta)^2$$
$$- \frac{1}{3!}W(R-h)\sin\theta(\delta\theta)^3 + \cdots \tag{8.3}$$

We can also express $\delta^{(T)}\pi$ as a Taylor series expansion in the following way:

$$\delta^{(T)}\pi = \pi(\theta+\delta\theta) - \pi(\theta) = \frac{\partial\pi}{\partial\theta}\delta\theta + \frac{1}{2!}\frac{\partial^2\pi}{\partial\theta^2}(\delta\theta)^2 + \frac{1}{3!}\frac{\partial^3\pi}{\partial\theta^3}(\delta\theta)^3 + \cdots \tag{8.4}$$

Note that the first expression on the extreme right side of Eq. (8.4) is the *first variation* of π, i.e., $\delta^{(1)}\pi$, and we state this formally:

$$\delta^{(1)}\pi = \frac{\partial\pi}{\partial\theta}\delta\theta$$

Also, higher-order variations of π such as the nth order variation can be written as

$$\delta^{(n)}\pi = \frac{\partial^n\pi}{\partial\theta^n}(\delta\theta)^n \tag{8.5}$$

Hence Eq. (8.4) can be cast in the form

$$\delta^{(T)}\pi = \sum_{n=1}^{\infty}\frac{1}{n!}\frac{\partial^n\pi}{\partial\theta^n}(\delta\theta)^n = \sum_{n=1}^{\infty}\frac{1}{n!}(\delta^{(n)}\pi)$$

Compare the above equation and Eq. (8.3), we note that

$$\delta^{(1)}\pi = W(R-h)\sin\theta\,\delta\theta$$
$$\delta^{(2)}\pi = W(R-h)\cos\theta(\delta\theta)^2 \tag{8.6}$$
$$\delta^{(3)}\pi = -W(R-h)\sin\theta(\delta\theta)^3$$

and so on.

To establish *equilibrium configurations* for the problem at hand, *we set the first variation equal to zero*, as we have done in the past. This requires that

$$W(R-h)\sin\theta = 0$$

We see that $\theta = 0$ is the expected position of equilibrium.

We now consider Eq. (8.3) for the case $\theta = 0$. We get

$$\delta^{(T)}\pi = \frac{1}{2!}W(R-h)(\delta\theta)^2 + \cdots$$

For π to be a *minimum* at $\theta = 0$ for *stable equilibrium*, the expression $\delta^{(T)}\pi$ must be positive for $\pm\delta\theta$, as explained in Section 2.1. We accordingly require for stability that the dominant term in the above expansion be positive, i.e.,

$$W(R-h)(\delta\theta)^2 = (\delta^{(2)}\pi)_{eq} > 0$$

This condition is satisfied for $R > h$. Now decrease R until it reaches the value h. Clearly, the second variation $(\delta^{(2)}\pi)_{eq}$ becomes zero. Furthermore, it is easily seen from Eq. (8.6) that *all variations are zero for this condition.* This means that there is no change in the total potential energy for all configurations in the neighborhood of the equilibrium configuration. We have thus reached a case of *neutral stability.* We shall denote the value of the adjustable parameter that first renders the system *not stable* as the *critical value* of the parameter. Thus for this problem, R is the adjustable parameter chosen and its critical value is h. We thus reach the conclusion that when the adjustable parameter R reaches its critical value h as it is decreased in value from $R > h$, we *cease* to have stable equilibrium. We shall next establish a criterion for finding the critical value. This will be useful in later computations involving arbitrary loads.

Accordingly, we consider the second variation $\delta^{(2)}\pi$:

$$\delta^{(2)}\pi = W(R-h)\cos\theta(\delta\theta)^2$$

Note that at the equilibrium position, $\theta = 0$, we pointed out that we had stable equilibrium for $R > h$. Also note that the second variation of the total potential energy is then positive-definite† *for a range of values of θ in the neighborhood of the equilibrium configuration.* For decreasing R, note that the second variation *ceases* to be positive definite exactly when $R = h$. We may make this observation. The body reaches a "critical" configuration and thus *ceases to have stability when the second variation of the total potential energy ceases to be positive definite.* We may use this criterion for the second variation of the total potential energy to find the critical value of the adjustable parameter.

We now consider a second rigid-body problem so that we can introduce both the concept of the *buckling load* and a second criterion for stability. Accordingly, we show (in Fig. 8.3*a*) a vertical rigid rod acted on by a force P. Two identical linear

†In this discussion a function $F(x_i)$ is said to be *positive definite* if the function is greater than zero in a neighborhood about $x_i = 0$ and can equal zero only at the position $x_i = 0$.

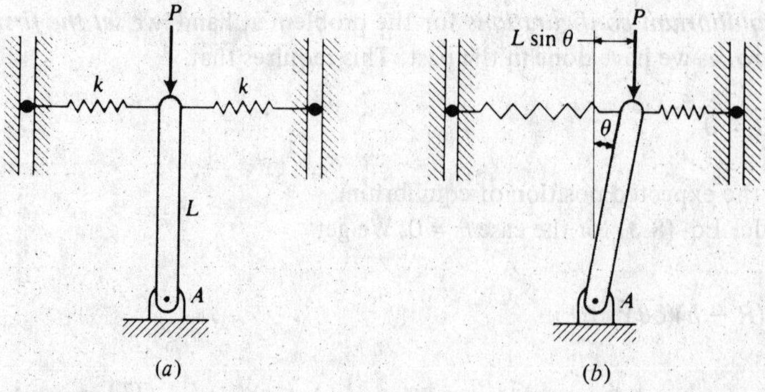

FIGURE 8.3
Elastically constrained rigid body.

springs having spring constant k connect the rod to pins in frictionless slots held by immovable walls. In Fig. 8.3*b* we show the rod in a deflected position measured by the angle θ. We first set forth the total potential energy of the system and then examine the first variation for possible equilibrium configurations. Thus we have for π

$$\pi = 2(\tfrac{1}{2})(k)(L \sin \theta)^2 - PL(1 - \cos \theta) \tag{8.7}$$

The first variation then becomes

$$\delta^{(1)}\pi = \frac{\partial \pi}{\partial \theta}\,\delta\theta = (2kL^2 \sin \theta \cos \theta - PL \sin \theta)\,\delta\theta$$

Replacing $2 \sin \theta \cos \theta$ by $\sin 2\theta$ in the first expression on the right side of the equation and setting the coefficient of $\delta\theta$ equal to zero for equilibrium, we get

$$(kL^2 \sin 2\theta - PL \sin \theta) = 0$$

For small values of δ we can express this equation as

$$(2kL^2 - PL)\theta = 0 \tag{8.8}$$

Clearly, $\theta = 0$ corresponds to a state of equilibrium for *any load P*.

Let us now discuss the stability of this configuration. We consider P to be the adjustable parameter just as R was in the last example. The second variation for this problem becomes

$$\delta^{(2)}\pi = \frac{\partial^2 \pi}{\partial \theta^2}\,(\delta\theta)^2 = (2kL^2 \cos 2\theta - PL \cos \theta)(\delta\theta)^2$$

For $\theta = 0$ we see that $\delta^{(2)}\pi$ is positive if the load P is less than $2kL$. Also, for a given load $P < 2kL$ there is a *range* of values of θ about $\theta = 0$ for which $\delta^{(2)}\pi$ is positive. Thus if $P < 2kL$, the second variation is *positive definite* at the equilibrium configuration $\theta = 0$. When $P = 2kL$ we have for the above equation

$$\delta^{(2)}\pi = PL(\cos 2\theta - \cos\theta)(\delta\theta)^2 \tag{8.9}$$

and it becomes clear that the second variation is negative in the neighborhood of $\theta = 0$. The second variation has thus *ceased* to be positive definite. This value of P clearly is the *critical* one or, we say, the *buckling load* for this system. It is denoted as P_{cr}.

Note that we have arrived at the buckling load by considering the positive definiteness of the second variation. This is called an *energy approach* toward ascertaining the buckling load. We shall now present yet another way of arriving at the buckling load by considering the first variation of the total potential energy. Thus by reexamining Eq. (8.8) we see that $\delta^{(1)}\pi = 0$ not only for $\theta = 0$ (which information we have used) but also for $P = P_{cr}$. This means that under the buckling load any value of θ can be, theoretically at least, an equilibrium position. A plot of θ versus P for equilibrium is shown in Fig. 8.4. Note that the locus of such states includes the axis $\theta = 0$ and a horizontal line $P = P_{cr}$. The point A where the path OA diverges into two paths is called a *bifurcation point*. Thus we see that the critical load occurs in this problem at the bifurcation point in such an equilibrium plot. (We will see in a later section that for elastic bodies there may be an infinity of bifurcation points corresponding to an infinite set of discrete values of load. Each such load is called a critical load; the *lowest critical* load is also called the *buckling load*.) We shall refer to an equilibrium configuration corresponding to *zero external load* as a *trivial equilibrium configuration*. The preceding discussion permits us to conclude that in the *presence of the critical loads we can have nontrivial equilibrium configurations*. Indeed, the search for loads by which nontrivial equilibrium configurations are permitted is called the *equilibrium method* for finding critical loads.

In closing this section we may draw the following conclusions from the examples considered. First, the buckling load signals the upper limit of loading for stable equilibrium. Further, this load can be found by determining the value of loading at which the second variation of the total potential energy ceases to be positive definite at an equilibrium configuration (the energy method). Or we can seek those values of load that permit nontrivial equilibrium configurations (equilibrium method).

In Part C we shall consider the elastic stability of columns, through which we can extend the conclusions reached above and present again, by example, other notions

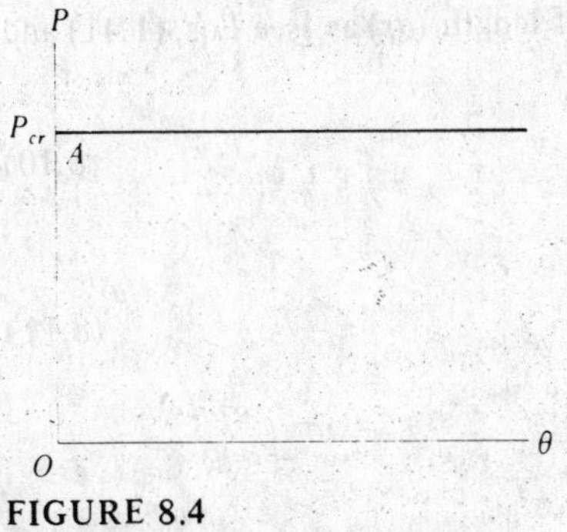

FIGURE 8.4
Buckling load and bifurcation point.

pertinent to the understanding of elastic stability. First, however, we shall introduce some elementary concepts of nonlinear elastic theory.

Part B
Some Results from Nonlinear Elasticity Theory

8.3 COMMENT

We wish to present here some results from the nonlinear theory of elasticity, particularly for large deformations of elastic columns and plates. The principal complexity in dealing with large deformations is the nonlinearity of the equations. Another major problem is that in the general theory the local stresses are difficult to identify because the deformations cause changes in the areas with respect to which the stresses are measured. Thus it is of paramount importance to specify in the general case whether one is operating in terms of the original undeformed coordinates or the deformed set of coordinates that move with the deforming body.

However, for practical problems of elastic stability such precision is not required. What we do require is the ability to allow for multiple solutions (remember that uniqueness is a principal feature of classical elasticity) and sufficiently large nonlinearity to allow projections of planar forces to have a significant component out of their initial plane of action. In the rigid-body problems of the preceding section, the nonlinearities are reflected in the terms $\sin\theta$ and $\cos\theta$.

In Part B we shall devote ourselves to the development of two useful theories that concern large deflections of columns and large deflections of plates. The theories are fundamentally different, as the first is based on the incompressible (inextensible) behavior of the column centerline, while the latter is based on the coupling between the out-of-plane bending deflection and the (nonlinear) strains in the middle surface of the structure.

8.4 LARGE BENDING DEFLECTIONS OF COLUMNS

From the nonlinear theory of elasticity, we may use the strains ϵ_{ij} to find the deformed length (ds^*) of a line element originally of length (ds) as [see Eqs. (1.41) and (1.45) in Section 1.7]:

$$(ds^*)^2 = (ds)^2 + 2\epsilon_{ij}\,dx_i\,dx_j \tag{8.10}$$

where

$$(ds)^2 = dx_i\,dx_i \tag{8.11}$$

and

$$2\epsilon_{ij} = \frac{\partial u_i}{\partial x_j} + \frac{\partial u_j}{\partial x_i} + \frac{\partial u_k}{\partial x_i}\frac{\partial u_k}{\partial x_j} \tag{8.12}$$

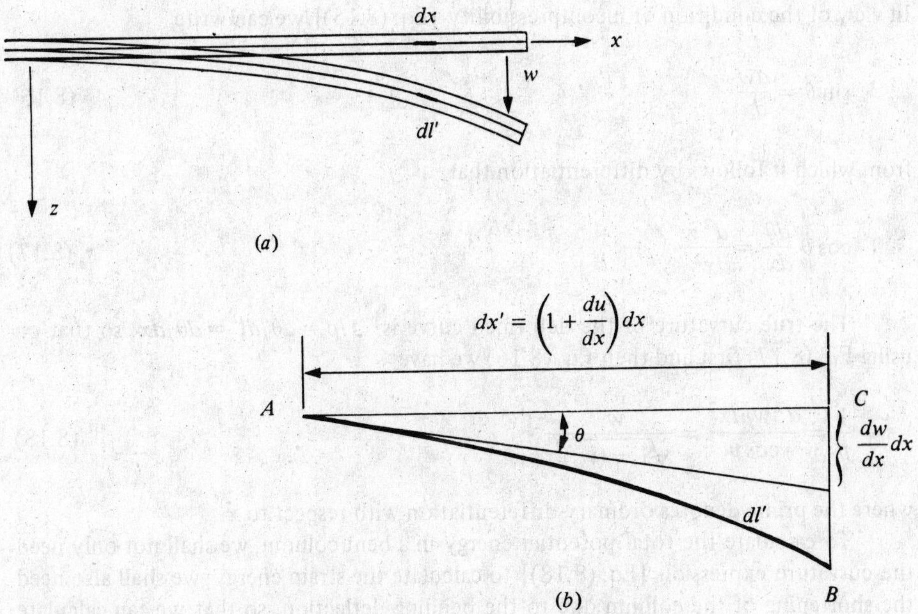

FIGURE 8.5
Deformed beam.

In writing these results in this form, we are using a Lagrangian approach as we are identifying in Eq. (8.12) the displacement components u_i as functions of the undeformed coordinates x_j.

Let $u_1 = u(x)$ and $u_3 = w(x)$ be, respectively, axial and transverse displacements measured from the undeformed axis of the column, which lies on the x axis (see Fig. 8.5*a*). Using Eq. (8.10) we can easily calculate the length of an element (dl') that originally was of length dx, i.e.,

$$(dl')^2 = (dx)^2(1 + 2\epsilon_{xx}) \tag{8.13}$$

where, from Eq. (8.12),

$$\epsilon_{xx} = \frac{du}{dx} + \frac{1}{2}\left(\frac{du}{dx}\right)^2 + \frac{1}{2}\left(\frac{dw}{dx}\right)^2 \tag{8.14}$$

If the column is incompressible during its bending from the straight-line configuration, then $dl' = dx$. Substituting this and Eq. (8.14) into Eq. (8.13) forces us to conclude that

$$\left(1 + \frac{du}{dx}\right)^2 + \left(\frac{dw}{dx}\right)^2 = 1 \tag{8.15}$$

From Fig. 8.5*b* we can see that

$$\sin\theta = \frac{(dw/dx)\,dx}{AB} = \frac{(dw/dx)\,dx}{\sqrt{(1 + du/dx)^2 + (dw/dx)^2}\,(dx)}$$

In view of the condition of incompressibility [Eq. (8.15)] we can write

$$\sin\theta = \frac{dw}{dx} \tag{8.16}$$

from which it follows by differentiation that

$$\cos\theta \frac{d\theta}{dx} = \frac{d^2w}{dx^2} \tag{8.17}$$

The true curvature of the deformed curve is† $1/\rho = d\theta/dl' = d\theta/dx$, so that on using Eq. (8.17) first and then Eq. (8.16) we have

$$\frac{1}{\rho} = \frac{d^2w/dx^2}{\cos\theta} = \frac{w''}{\sqrt{1-(w')^2}} \tag{8.18}$$

where the prime denotes ordinary differentiation with respect to x.

To calculate the total potential energy in a bent column, we shall not only need the curvature expression [Eq. (8.18)] to calculate the strain energy; we shall also need the shortening of the column due to the bending deflection, so that we can calculate the energy input by the applied axial load. The total shortening would simply be

$$\int_0^L \frac{du}{dx}\,dx = \int_0^L [\sqrt{1-(w')^2} - 1]\,dx \tag{8.19}$$

by virtue of the incompressibility condition [Eq. (8.15)]. Remember that this shortening is due to the bending only and that the column neutral axis is unstrained.

For the familiar elementary theory we would assume the column deformation to be small enough so that $w'(x) < 1$, in which case the curvature [see Eq. (8.18)] simplifies to

$$\frac{1}{\rho} \cong \frac{d^2w}{dx^2} \tag{8.20a}$$

while the shortening is, after expanding the radical in Eq. (8.19) as a power series,

$$\int_0^L \frac{du}{dx}\,dx \cong -\frac{1}{2}\int_0^L (w')^2\,dx \tag{8.20b}$$

†We may also obtain this result by considering the deformed coordinates $x_1' = x + u, x_3' = w$, and noting a useful result from differential geometry

$$\frac{1}{\rho} = \pm\frac{(dx_1'/dx)(d^2x_3'/dx^2) - (dx_3'/dx)(d^2x_1'/dx^2)}{[(dx_1'/dx)^2 + (dx_3'/dx)^2]^{3/2}}$$

If we let $x_1' = x, x_3' = y(x)$, we could also obtain another familiar result.

Now we determine the strain energy due to bending for this simple one-dimensional problem. Using the Euler-Bernoulli hypothesis for bending of beams in the form $\epsilon_{xx}(x, z) = z/\rho$, we have

$$U = \frac{E}{2} \iiint_V [\epsilon_{xx}(x, z)]^2 \, dv = \frac{E}{2} \iiint_V \left(\frac{z}{\rho}\right)^2 dv$$

Hence

$$U = \frac{EI}{2} \int_0^L \left(\frac{1}{\rho}\right)^2 dx$$

This equation reflects the fact that the beam deformation is due to bending and that the neutral axis ($z = 0$) is incompressible.

The potential of the applied compressive axial load P is

$$V_P = P[u(L) - u(0)] = P \int_0^L \frac{du}{dx} \, dx$$

which is thus simply the product of the load and the shortening. The total potential energy is

$$\pi = \frac{EI}{2} \int_0^L \left(\frac{1}{\rho}\right)^2 dx - P \int_0^L \frac{du}{dx} \, dx \tag{8.21a}$$

Using the approximation embodied in Eqs. (8.20), we can give π for the elementary theory as

$$\pi = \frac{EI}{2} \int_0^L \left(\frac{d^2 w}{dx^2}\right)^2 dx - \frac{P}{2} \int_0^L (w')^2 \, dx \tag{8.21b}$$

8.5 THE VON KÁRMÁN THEORY OF PLATES

In this section we shall derive the von Kármán theory of plates. This is a nonlinear theory that allows for comparatively large rotations of line elements originally normal to the x, y axes in the midplane of the plate (Fig. 8.6). These rotation terms allow projections of the in-plane forces $\bar{N}_\nu$ and $\bar{N}_{\nu s}$ to be felt in the transverse direction, normal to the plane of the plate.

This plate theory is derived assuming that the strains and rotations are both small compared to unity, so that we can ignore changes in geometry in the definition of stress components and in the limits of integration needed for work and energy con-

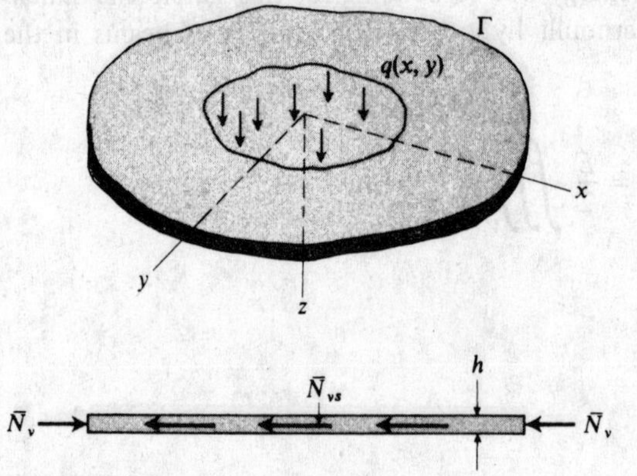

FIGURE 8.6
Plate with transverse and in-plane loads.

siderations. We further stipulate that the strains will be smaller than the rotations, in the sense described below.

We introduce the linear strain parameters e_{ij} and the rotation parameters ω_{ij} defined as

$$\begin{aligned} 2e_{ij} &= u_{i,j} + u_{j,i} \\ 2\omega_{ij} &= u_{i,j} - u_{j,i} \end{aligned} \tag{8.22}$$

The strain-displacement Eq. (8.12) can then be written as [see Eq. (1.52)]

$$2\epsilon_{ij} = 2e_{ij} + (e_{ki} + \omega_{ki})(e_{kj} + \omega_{kj}) \tag{8.23}$$

For the simplifications discussed, namely, $e_{ki} < \omega_{ki}$, Eq. (8.23) reduces to

$$\epsilon_{ij} = e_{ij} + \tfrac{1}{2}\omega_{ki}\omega_{kj} \tag{8.24}$$

Finally, we use the Kirchhoff assumption† that lines normal to the undeformed middle surface remain normal to this surface in the deformed geometry and are unextended after deformation. This means that

$$u_1(x_1, x_2, x_3) = u(x, y) - z\,\frac{\partial w(x, y)}{\partial x} \tag{8.25a}$$

$$u_2(x_1, x_2, x_3) = v(x, y) - z\,\frac{\partial w(x, y)}{\partial y} \tag{8.25b}$$

$$u_3(x_1, x_2, x_3) = w(x, y) \tag{8.25c}$$

where u, v, and w are the displacement components of the middle surface of the plate. We may now give the strain parameters and rotation parameters as follows:

†The two-dimensional extension of the Euler-Bernoulli hypothesis.

$$e_{11} = \frac{\partial u}{\partial x} - z\,\frac{\partial^2 w}{\partial x^2} \qquad e_{12} = \frac{1}{2}\left(\frac{\partial u}{\partial y} + \frac{\partial v}{\partial x} - 2z\,\frac{\partial^2 w}{\partial x\,\partial y}\right)$$

$$e_{22} = \frac{\partial v}{\partial y} - z\,\frac{\partial^2 w}{\partial y^2} \qquad e_{13} = e_{23} = 0 \tag{8.26a}$$

$$e_{33} = \frac{\partial w}{\partial z} = 0$$

$$\omega_{12} = \frac{1}{2}\left(\frac{\partial u}{\partial y} - \frac{\partial v}{\partial x}\right)$$

$$\omega_{13} = -\frac{\partial w}{\partial x} \tag{8.26b}$$

$$\omega_{23} = -\frac{\partial w}{\partial y}$$

We now observe tne following. The rotation parameter ω_{12} approximates a rotation component about the z axis, while ω_{23} and ω_{13} approximate rotation components about axes parallel to the x and y axes, respectively, in the midplane of the plate. For a thin, hence flexible, plate we can reasonably expect that

$$\omega_{12} \ll \omega_{23}, \omega_{13} \tag{8.27}$$

Neglecting ω_{12}, we can now employ Eqs. (8.24) and (8.26) to find the ϵ_{ij}:

$$\epsilon_{11} = \frac{\partial u}{\partial x} + \frac{1}{2}\left(\frac{\partial w}{\partial x}\right)^2 - z\,\frac{\partial^2 w}{\partial x^2}$$

$$\epsilon_{22} = \frac{\partial v}{\partial y} + \frac{1}{2}\left(\frac{\partial w}{\partial y}\right)^2 - z\,\frac{\partial^2 w}{\partial y^2}$$

$$\epsilon_{33} = \frac{1}{2}\left(\frac{\partial w}{\partial x}\right)^2 + \frac{1}{2}\left(\frac{\partial w}{\partial y}\right)^2 \tag{8.28}$$

$$\epsilon_{12} = \frac{1}{2}\left(\frac{\partial u}{\partial y} + \frac{\partial v}{\partial x} - 2z\,\frac{\partial^2 w}{\partial x\,\partial y}\right) + \frac{1}{2}\,\frac{\partial w}{\partial x}\,\frac{\partial w}{\partial y}$$

$$\epsilon_{13} \cong \epsilon_{23} \cong 0$$

For a constitutive law we will employ Hooke's law for plane stress over the thickness of the plate. Thus we shall be concerned here only with ϵ_{11}, ϵ_{22}, and ϵ_{12}. Accordingly, for the assumptions presented here, we can say† that the first variation of the strain energy U is

$$\delta^{(1)}U = \iiint_V \tau_{ij}\,\delta\epsilon_{ij}\,dv = \iint_R \int_{-h/2}^{h/2} (\tau_{11}\,\delta\epsilon_{11} + 2\tau_{12}\,\delta\epsilon_{12} + \tau_{22}\,\delta\epsilon_{22})\,dz\,dA \tag{8.29}$$

†The discussion here closely parallels the derivations in Sections 6.4 and 6.5.

Now we replace the strain terms in the above expression and commute the delta operator with the derivative operators.

$$\delta^{(1)}U = \iint_R \int_{-h/2}^{h/2} \left[\tau_{11}\left(\frac{\partial\,\delta u}{\partial x} + \frac{\partial w}{\partial x}\frac{\partial\,\delta w}{\partial x} - z\frac{\partial^2\,\delta w}{\partial x^2}\right) + \tau_{12}\left(\frac{\partial\,\delta u}{\partial y} + \frac{\partial\,\delta v}{\partial x} - 2z\frac{\partial^2\,\delta w}{\partial x\,\partial y} + \frac{\partial w}{\partial x}\frac{\partial\,\delta w}{\partial y} + \frac{\partial w}{\partial y}\frac{\partial\,\delta w}{\partial x}\right) + \tau_{22}\left(\frac{\partial\,\delta v}{\partial y} + \frac{\partial w}{\partial y}\frac{\partial\,\delta w}{\partial y} - z\frac{\partial^2\,\delta w}{\partial y^2}\right)\right] dA\,dz \tag{8.30}$$

Next we integrate with respect to z and introduce resultant stress and moment intensity functions N_x, N_y, and N_{xy}, and M_x, M_y, and M_{xy}, respectively:

$$M_x = \int_{-h/2}^{h/2} \tau_{xx} z\,dz \qquad N_x = \int_{-h/2}^{h/2} \tau_{xx}\,dz$$

$$M_y = \int_{-h/2}^{h/2} \tau_{yy} z\,dz \qquad N_y = \int_{-h/2}^{h/2} \tau_{yy}\,dz \tag{8.31}$$

$$M_{xy} = \int_{-h/2}^{h/2} \tau_{xy} z\,dz = M_{yx} \qquad N_{xy} = \int_{-h/2}^{h/2} \tau_{xy}\,dz = N_{yx}$$

where N_x (N_y), as shown in Fig. 8.7, is a force in the x direction (y direction) measured per unit length in the y direction (x direction) and where N_{xy} (N_{yx}) is a force

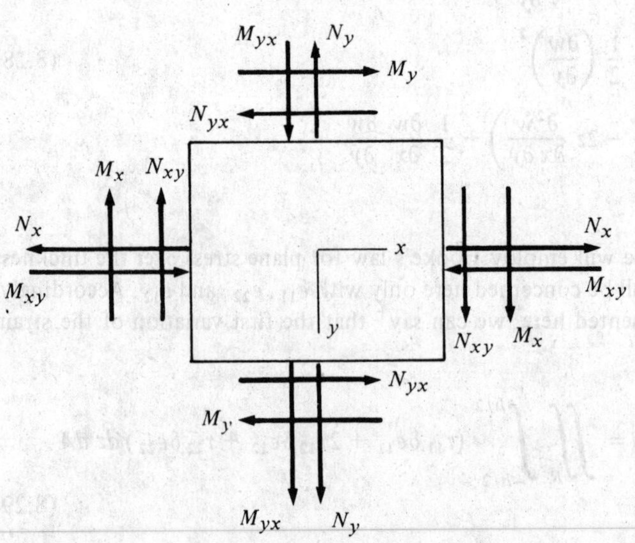

FIGURE 8.7
Plate element showing stress and moment intensity functions.

in the $x(y)$ direction measured per unit length in the $y(x)$ direction. Similarly, the quantity M_x represents a moment per unit length in the y direction, with its vector in the y direction. Finally, M_{xy} is the twisting moment per unit length in the y direction with its vector in the x direction. The terms N_x, N_y, N_{xy}, and N_{yx} may for practical purposes be compared with normal and shear stresses and we may conclude that $N_{xy} = N_{yx}$. We then get for Eq. (8.30) the following result:

$$\delta^{(1)}U = \iint_R \left[N_x \left(\frac{\partial\, \delta u}{\partial x} + \frac{\partial w}{\partial x}\frac{\partial\, \delta w}{\partial x} \right) - M_x \frac{\partial^2 \delta w}{\partial x^2} \right.$$
$$+ N_{xy} \left(\frac{\partial\, \delta u}{\partial y} + \frac{\partial\, \delta v}{\partial x} + \frac{\partial w}{\partial x}\frac{\partial\, \delta w}{\partial y} + \frac{\partial w}{\partial y}\frac{\partial\, \delta w}{\partial x} \right)$$
$$\left. - 2M_{xy} \frac{\partial^2 \delta w}{\partial x\, \partial y} + N_y \left(\frac{\partial\, \delta v}{\partial y} + \frac{\partial w}{\partial y}\frac{\partial\, \delta w}{\partial y} \right) - M_y \frac{\partial^2 \delta w}{\partial y^2} \right] dx\, dy \qquad (8.32)$$

The first variation of the potential of the applied forces meanwhile takes the form (noting that $\bar{N}_\nu$ is taken as positive in compression as shown in Fig. 8.6)

$$\delta^{(1)}V = -\iint_R q\, \delta w\, dx\, dy + \oint_\Gamma \bar{N}_\nu\, \delta u_\nu\, ds + \oint_\Gamma \bar{N}_{\nu s}\, \delta u_s\, ds \qquad (8.33)$$

where u_ν and u_s are the in-plane displacements of the boundary of the plate in directions normal and tangential to the boundary, respectively. We are using the undeformed geometry for the applied loads above rather than the deformed geometry—thereby restricting the result to reasonably small deformations. By using the above result for $\delta^{(1)}V$ and using Eq. (8.32) for $\delta^{(1)}U$, we may form $\delta^{(1)}\pi$. The total potential energy so formed approximates the actual total potential energy for the kind of deformation restrictions embodied in Kirchhoff's assumptions. And, since we have used undeformed geometry for stresses and external loads, we are limited to small deformations in employing this functional. Finally, because we used Eq. (8.24) for strain, we are assuming that strains are much smaller than rotations. Thus we have for the total potential energy principle:

$$\delta^{(1)}\pi = 0 = \iint_R \left[N_x \left(\frac{\partial\, \delta u}{\partial x} + \frac{\partial w}{\partial x}\frac{\partial\, \delta w}{\partial x} \right) - M_x \frac{\partial^2 \delta w}{\partial x^2} \right.$$
$$+ N_{xy} \left(\frac{\partial\, \delta u}{\partial y} + \frac{\partial\, \delta v}{\partial x} + \frac{\partial w}{\partial x}\frac{\partial\, \delta w}{\partial y} + \frac{\partial w}{\partial y}\frac{\partial\, \delta w}{\partial x} \right)$$
$$\left. - 2M_{xy} \frac{\partial^2 \delta w}{\partial x\, \partial y} + N_y \left(\frac{\partial\, \delta v}{\partial y} + \frac{\partial w}{\partial y}\frac{\partial\, \delta w}{\partial y} \right) - M_y \frac{\partial^2 \delta w}{\partial y^2} \right] dx\, dy$$
$$- \iint_R q\, \delta w\, dx\, dy + \oint_\Gamma \bar{N}_\nu\, \delta u_\nu\, ds + \oint_\Gamma \bar{N}_{\nu s}\, \delta u_s\, ds \qquad (8.34)$$

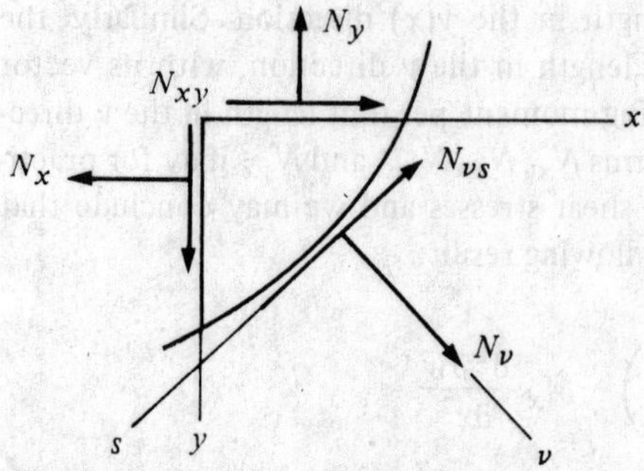

FIGURE 8.8
Plate element at the boundary.

We may proceed to carry out the extremization process. We employ Green's theorem one or more times to get the δu's and δv's out from the partial derivatives. Then we proceed to simplify the expressions in the line integrals by noting from equilibrium (see Fig. 8.8) that

$$N_\nu = N_x a_{\nu x}^2 + 2N_{xy} a_{\nu x} a_{\nu y} + N_y a_{\nu y}^2 \tag{8.35a}$$

$$N_{\nu s} = (N_y - N_x) a_{\nu x} a_{\nu y} + N_{xy}(a_{\nu x}^2 - a_{\nu y}^2) \tag{8.35b}$$

where $a_{\nu x}$ and $a_{\nu y}$ are the direction cosines of the outward normal of the boundary. Furthermore, simple vector projections permit us to say

$$\begin{aligned} u_\nu &= a_{\nu x} u + a_{\nu y} v \\ u_s &= -a_{\nu y} u + a_{\nu x} v \end{aligned} \tag{8.36}$$

And we also note that [see Eqs. (6.37)]

$$\begin{aligned} \frac{\partial}{\partial x} &= a_{\nu x} \frac{\partial}{\partial \nu} - a_{\nu y} \frac{\partial}{\partial s} \\ \frac{\partial}{\partial y} &= a_{\nu y} \frac{\partial}{\partial \nu} + a_{\nu x} \frac{\partial}{\partial s} \end{aligned} \tag{8.37}$$

Finally, we introduce the transverse shear forces of plate theory [see the development of Eqs. (6.16)-(6.18)]:

$$\begin{aligned} Q_\nu &= Q_x a_{\nu x} + Q_y a_{\nu y} \\ Q_x &= \frac{\partial M_x}{\partial x} + \frac{\partial M_{xy}}{\partial y} \\ Q_y &= \frac{\partial M_y}{\partial y} + \frac{\partial M_{xy}}{\partial x} \end{aligned} \tag{8.38}$$

Now using Eqs. (8.35)-(8.38) we may rewrite Eq. (8.34) as†

$$\delta^{(1)}\pi = -\iint_R \Bigg\{ \left(\frac{\partial N_x}{\partial x} + \frac{\partial N_{xy}}{\partial y} \right) \delta u + \left(\frac{\partial N_{xy}}{\partial x} + \frac{\partial N_y}{\partial y} \right) \delta v$$

$$+ \left[\frac{\partial^2 M_x}{\partial x^2} + 2 \frac{\partial^2 M_{xy}}{\partial x\, \partial y} + \frac{\partial^2 M_y}{\partial y^2} + \frac{\partial}{\partial x}\left(N_x \frac{\partial w}{\partial x} \right) + \frac{\partial}{\partial y}\left(N_{xy} \frac{\partial w}{\partial x} \right) \right.$$

$$\left. + \frac{\partial}{\partial x}\left(N_{xy} \frac{\partial w}{\partial y} \right) + \frac{\partial}{\partial y}\left(N_y \frac{\partial w}{\partial y} \right) + q \right] \delta w \Bigg\} \, dx\, dy$$

$$+ \oint_\Gamma (N_\nu + \bar{N}_\nu)\, \delta u_\nu \, ds + \oint_\Gamma (N_{\nu s} + \bar{N}_{\nu s})\, \delta u_s \, ds - \oint_\Gamma M_\nu \frac{\partial\, \delta w}{\partial \nu} \, ds$$

$$+ \oint_\Gamma \left(Q_\nu + \frac{\partial M_{\nu s}}{\partial s} + N_\nu \frac{\partial w}{\partial \nu} + N_{\nu s} \frac{\partial w}{\partial s} \right) \delta w \, ds - [M_{\nu s}\, \delta w]_\Gamma = 0 \qquad (8.39)$$

The last expression accounts for "corners" in the boundary. From the above equations we may now make a series of deductions. First, in region R we can conclude that

$$\frac{\partial N_x}{\partial x} + \frac{\partial N_{xy}}{\partial y} = 0 \qquad (8.40a)$$

$$\frac{\partial N_{xy}}{\partial x} + \frac{\partial N_y}{\partial y} = 0 \qquad (8.40b)$$

$$\frac{\partial^2 M_x}{\partial x^2} + 2 \frac{\partial^2 M_{xy}}{\partial x\, \partial y} + \frac{\partial^2 M_y}{\partial y^2} + \frac{\partial}{\partial x}\left(N_x \frac{\partial w}{\partial x} \right) + \frac{\partial}{\partial y}\left(N_{xy} \frac{\partial w}{\partial x} \right)$$

$$+ \frac{\partial}{\partial x}\left(N_{xy} \frac{\partial w}{\partial y} \right) + \frac{\partial}{\partial y}\left(N_y \frac{\partial w}{\partial y} \right) + q = 0 \qquad (8.40c)$$

The first two equations above clearly are identical to the equations of equilibrium for plane stress, as is to be expected. We shall use these equations now to simplify the third equation after we use the differentiation operators on the expressions involving products. We are thus able to eliminate expressions involving the first partial of w. We get

$$\frac{\partial^2 M_x}{\partial x^2} + 2 \frac{\partial^2 M_{xy}}{\partial x\, \partial y} + \frac{\partial^2 M_y}{\partial y^2} + N_x \frac{\partial^2 w}{\partial x^2} + 2N_{xy} \frac{\partial^2 w}{\partial x\, \partial y} + N_y \frac{\partial^2 w}{\partial y^2} + q = 0 \qquad (8.41)$$

†This closely parallels the development of Section 6.4. The details are left as an exercise for the reader.

Now comparing Eq. (8.41) with the classical case we see that we have here introduced *nonlinear terms* $[N_x(\partial^2 w/\partial x^2) + 2N_{xy}(\partial^2 w/\partial x\,\partial y) + N_y(\partial^2 w/\partial y^2)]$ involving the in-plane force intensities as additional "transverse loading" terms.†

Considering next the remainder of Eq. (8.39) we can stipulate the following boundary conditions along Γ:

$$\text{EITHER } N_\nu = -\bar{N}_\nu \quad \text{OR} \quad u_\nu \text{ IS SPECIFIED} \tag{8.42a}$$

$$\text{EITHER } N_{\nu s} = -\bar{N}_{\nu s} \quad \text{OR} \quad u_s \text{ IS SPECIFIED} \tag{8.42b}$$

$$\text{EITHER } M_\nu = 0 \quad \text{OR} \quad \frac{\partial w}{\partial \nu} \text{ IS SPECIFIED} \tag{8.42c}$$

$$\text{EITHER } Q_\nu + \frac{\partial M_{\nu s}}{\partial s} + N_\nu \frac{\partial w}{\partial \nu} + N_{\nu s}\frac{\partial w}{\partial \nu} = 0 \quad \text{OR} \quad w \text{ IS SPECIFIED} \tag{8.42d}$$

$$\text{At discontinuities} \quad [M_{\nu s}\,\delta w] = 0 \tag{8.42e}$$

The last three conditions are familiar from work on plates except that the effective shear force $(Q_\nu + \partial M_{\nu s}/\partial s)$ is now augmented by projections of the in-plane plate forces at the plate edges.

The equations of equilibrium may be solved if a constitutive law is used. We will employ here (as pointed out earlier) the familiar Hooke's law for plane stress. We will use this constituve law to replace the resultant intensity functions by appropriate derivatives of the displacement field of the midplane of the plate. Consider, for example, the quantity M_x. Using Hooke's law and Eqs. (8.28), we have

$$M_x = \int_{-h/2}^{h/2} z\tau_{xx}\,dz = \int_{-h/2}^{h/2} z\,\frac{E}{1-\nu^2}(\epsilon_{xx} + \nu\epsilon_{yy})\,dz$$

$$= \int_{-h/2}^{h/2} z\,\frac{E}{1-\nu^2}\left[\frac{\partial u}{\partial x} + \frac{1}{2}\left(\frac{\partial w}{\partial x}\right)^2 - z\frac{\partial^2 w}{\partial x^2} + \nu\frac{\partial v}{\partial y} + \frac{\nu}{2}\left(\frac{\partial w}{\partial y}\right)^2 - \nu z\frac{\partial^2 w}{\partial y^2}\right] dz$$

Integrating and inserting limits, we get

$$M_x = \frac{E}{1-\nu^2}\,\frac{h^3}{12}\left(-\frac{\partial^2 w}{\partial x^2} - \nu\frac{\partial^2 w}{\partial y^2}\right) = -D\left(\frac{\partial^2 w}{\partial x^2} + \nu\frac{\partial^2 w}{\partial y^2}\right) \tag{8.43}$$

where D is the familiar bending rigidity, given as $D = Eh^3/12(1-\nu^2)$. Similarly, we have

†In the Russian literature these terms are often referred to as "reduced loads" of the in-plane forces or "reduced forces."

$$M_y = -D\left(\frac{\partial^2 w}{\partial y^2} + \nu\,\frac{\partial^2 w}{\partial x^2}\right) \tag{8.44a}$$

$$M_{xy} = -(1-\nu)D\,\frac{\partial^2 w}{\partial x\,\partial y} \tag{8.44b}$$

$$N_x = C\left\{\left[\frac{\partial u}{\partial x} + \frac{1}{2}\left(\frac{\partial w}{\partial x}\right)^2\right] + \nu\left[\frac{\partial v}{\partial y} + \frac{1}{2}\left(\frac{\partial w}{\partial y}\right)^2\right]\right\} \tag{8.44c}$$

$$N_y = C\left\{\left[\frac{\partial v}{\partial y} + \frac{1}{2}\left(\frac{\partial w}{\partial y}\right)^2\right] + \nu\left[\frac{\partial u}{\partial x} + \frac{1}{2}\left(\frac{\partial w}{\partial x}\right)^2\right]\right\} \tag{8.44d}$$

$$N_{xy} = C\left(\frac{1-\nu}{2}\right)\left[\frac{\partial u}{\partial y} + \frac{\partial v}{\partial x} + \frac{\partial w}{\partial x}\,\frac{\partial w}{\partial y}\right] \tag{8.44e}$$

where C is the extensional rigidity, given as

$$C = \frac{Eh}{1-\nu^2} \tag{8.45}$$

We could now substitute in Eq. (8.41) for the resultant intensity functions, using the above relations to get the equilibrium equations in terms of the displacement components of the midplane plate. However, we shall follow another route that leads to a somewhat less complicated system of equations.

Note accordingly that Eqs. (8.40*a*) and (8.40*b*) will be individually satisfied if we define an Airy stress function F as follows:

$$N_x = \frac{\partial^2 F}{\partial y^2} \qquad N_y = \frac{\partial^2 F}{\partial x^2} \qquad N_{xy} = -\frac{\partial^2 F}{\partial x\,\partial y} \tag{8.46}$$

Then, replacing M_x, M_y, and M_{xy}, it is a simple matter to show that the Eq. (8.40*c*) can be written as

$$D\,\nabla^4 w = \frac{\partial^2 F}{\partial y^2}\,\frac{\partial^2 w}{\partial x^2} - 2\,\frac{\partial^2 F}{\partial x\,\partial y}\,\frac{\partial^2 w}{\partial x\,\partial y} + \frac{\partial^2 F}{\partial x^2}\,\frac{\partial^2 w}{\partial y^2} + q \tag{8.47}$$

We now have a single partial differential equation with two dependent variables, w and F. Since we are now studying in-plane effects by a stress approach, we must ensure the compatibility of the in-plane displacements. This will give us a second companion equation to go with Eq. (8.47). To do this, we shall seek to relate the strains ϵ_{xx}, ϵ_{yy}, and ϵ_{xy} at the midplane surface in such a way that when we employ Eq. (8.28) to replace the strains we end up with a result that does not contain the in-plane displacement components u and v. Thus you may readily demonstrate by substituting from Eq. (8.28) that

$$\left(\frac{\partial^2 \epsilon_{xx}}{\partial y^2} + \frac{\partial^2 \epsilon_{yy}}{\partial x^2} - 2\,\frac{\partial^2 \epsilon_{xy}}{\partial x\,\partial y}\right)_{z=0} = \left(\frac{\partial^2 w}{\partial x\,\partial y}\right)^2 - \frac{\partial^2 w}{\partial x^2}\,\frac{\partial^2 w}{\partial y^2}$$

Since this equation ensures the proper relation of strains at the midplane surface to the midplane displacement component w without explicitly involving in-plane displacement components u and v, it serves as the desired compatibility equation for the strains at the midplane surface. We next express the compatibility equation in terms of the stress resultant intensity function. To do this, substitute for strains, using Eq. (8.28), and then note Eqs. (8.44) stemming from Hooke's law:

$$\frac{1}{Eh}\left[\frac{\partial^2 N_x}{\partial y^2} - \nu\frac{\partial^2 N_y}{\partial y^2} + \frac{\partial^2 N_y}{\partial x^2} - \nu\frac{\partial^2 N_x}{\partial x^2} + 2(1+\nu)\frac{\partial^2 N_{xy}}{\partial x\,\partial y}\right]$$
$$= \left(\frac{\partial^2 w}{\partial x\,\partial y}\right)^2 - \frac{\partial^2 w}{\partial x^2}\frac{\partial^2 w}{\partial y^2}$$

Finally, replacing the resultant intensity functions in terms of the Airy function [see Eq. (8.46)], we get

$$\nabla^4 F = Eh\left[\left(\frac{\partial^2 w}{\partial x\,\partial y}\right)^2 - \frac{\partial^2 w}{\partial x^2}\frac{\partial^2 w}{\partial y^2}\right] \tag{8.48}$$

The above equation and Eq. (8.47), which we now rewrite as

$$D\nabla^4 w = \frac{\partial^2 F}{\partial y^2}\frac{\partial^2 w}{\partial x^2} - 2\frac{\partial^2 F}{\partial x\,\partial y}\frac{\partial^2 w}{\partial x\,\partial y} + \frac{\partial^2 F}{\partial x^2}\frac{\partial^2 w}{\partial y^2} + q \tag{8.49}$$

are the celebrated *von Kármán plate equations.* Note that they are still highly nonlinear. The equations do have considerable mutual symmetry. This is brought out by defining a nonlinear operator L:

$$L(p,q) = \frac{\partial^2 p}{\partial y^2}\frac{\partial^2 q}{\partial x^2} - 2\frac{\partial^2 p}{\partial x\,\partial y}\frac{\partial^2 q}{\partial x\,\partial y} + \frac{\partial^2 p}{\partial x^2}\frac{\partial^2 q}{\partial y^2} \tag{8.50}$$

Then the von Kármán plate equations can be given as

$$\nabla^4 F = -\frac{Eh}{2}L(w,w) \tag{8.51a}$$

$$D\nabla^4 w = L(F,w) + q \tag{8.51b}$$

The deletion of the nonlinear operator leaves us with the (uncoupled) plane-stress problem of two-dimensional elasticity theory and the classic plate bending equation.

Part C
Elastic Stability of Columns

8.6 THE EULER LOAD; EQUILIBRIUM METHOD

We consider now the classic problem of the theory of elastic stability, dating back to the work of Euler, who in his studies on variational calculus founded, as it were, the

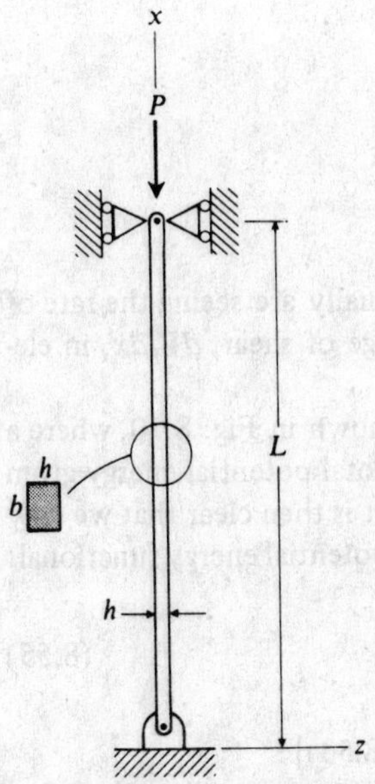

FIGURE 8.9
Pin-ended column.

theory of elastic stability. This problem consists of an initially straight pin-ended, long, slender column under the action of a compressive force P as shown in Fig. 8.9.

We shall first compute the total potential energy. We use the "small deflection" approximations to the curvature [Eq. (8.20*a*)] and the shortening [Eq. (8.20*b*)] to get Eq. (8.21*b*):

$$\pi = \frac{EI}{2}\int_0^L \left(\frac{d^2w}{dx^2}\right)^2 dx - \frac{P}{2}\int_0^L \left(\frac{dw}{dx}\right)^2 dx$$

$$= \int_0^L \left[\frac{EI}{2}\left(\frac{d^2w}{dx^2}\right)^2 - \frac{P}{2}\left(\frac{dw}{dx}\right)^2\right] dx \tag{8.52}$$

By using Eq. (2.62), we can state the Euler-Lagrange equation of the above functional

$$EI\frac{d^4w}{dx^4} + P\frac{d^2w}{dx^2} = 0 \tag{8.53}$$

subject to the boundary conditions at $x = 0, L$ that [see Eq. (2.63)]

$$EI\frac{d^3w}{dx^3} + P\frac{dw}{dx} = 0 \quad \text{or} \quad w = 0 \tag{8.54a}$$

$$EI\frac{d^2w}{dx^2} = 0 \quad \text{or} \quad \frac{dw}{dx} = 0 \tag{8.54b}$$

We thus see that in both the equilibrium equation [Eq. (8.53)] and the shear boundary condition [Eq. (8.54*a*)], a transverse projection $P\,dw/dx$ of the axial load P has

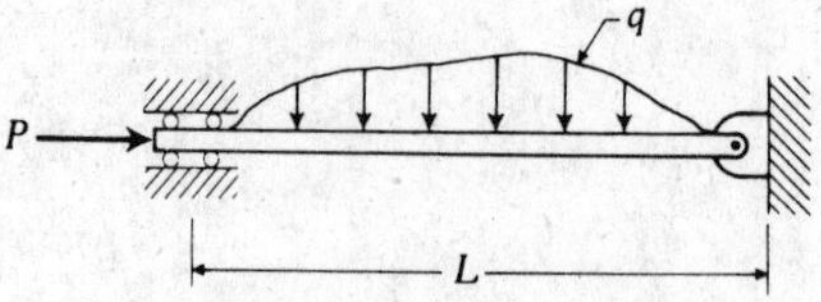

FIGURE 8.10
Beam column.

been "introduced." (In the equation of equilibrium we actually are seeing the rate of change of that projected load, exactly as we see the change of shear, dV/dx, in elementary beam theory.)

Suppose next we have a *beam column* problem, as shown in Fig. 8.10, where a loading $q(x)$ acts in addition to the axial load P. Thus the total potential energy given by Eq. (8.52) must be augmented by the term $-\int_0^L qw\,dx$. It is then clear that we may reach the following equation on extremizing this new total potential energy functional:

$$EI\frac{d^4w}{dx^4} + P\frac{d^2w}{dx^2} = q(x) \tag{8.55}$$

We thus have a generalization here of the earlier form [Eq. (8.53)].

We point out that although Eq. (8.55) is linear, it reflects the nonlinearity of its derivation since certain types of superpositions are not possible. In particular, we could not superpose the effect of P and the effects of $q(x)$ separately; that is, we cannot superpose the solution of

$$EI\frac{d^4w}{dx^4} = q(x)$$

and

$$EI\frac{d^4w}{dx^4} + P\frac{d^2w}{dx^2} = 0$$

Further, while we can superpose solutions of Eq. (8.55) for superposed values of $q(x)$, we cannot do this for superposed values of P.

Recall that the vanishing of the first variation for the rigid-body problems produced *algebraic equations* from which we could deduce that for a given load, namely the critical load, equilibrium configurations other than the so-called trivial one were possible equilibrium states. In the case here we have a *differential equation* resulting from the vanishing of the first variation. However, for the column we may now ask for essentially the same information as we asked earlier; namely, for what loads P are there nontrivial configurations $w(x)$ that satisfy equilibrium conditions (i.e., our differential equation and boundary conditions)? To pose such a question mathematically is to set forth an *eigenvalue problem* where the admissible values of P, namely the critical loads P_{cr}, are the *eigenvalues.* Accordingly, the various techniques discussed in Chapter 7 relative to the Rayleigh quotient are applicable here for determining, at least approximately, certain critical loads. We shall investigate this aspect of the problem later. Now we shall present an exact solution for the simple Euler column (Fig. 8.9).

You may readily verify that the solution to Eq. (8.53) can be given as follows:

$$w = A \sin kx + B \cos kx + Cx + D \tag{8.56a}$$

where

$$k^2 = \frac{P}{EI} \tag{8.56b}$$

Satisfaction of the boundary conditions, namely $w = d^2w/dx^2 = 0$ at $x = 0, L$, gives us the following system of equations:

$$B + D = 0$$

$$A \sin kL + B \cos kL + CL + D = 0$$

$$B = 0$$

$$A \sin kL + B \cos kL = 0$$

From these equations we find that

$$B = C = D = 0 \tag{8.57a}$$

$$A \sin kL = 0 \tag{8.57b}$$

For a *nontrivial* solution the following conditions must be met:

$$k = \frac{n\pi}{L}, \qquad n = 1, 2, \ldots \tag{8.58}$$

This generates a series of critical loads (the eigenvalues) given as

$$P_{cr} = \frac{n^2\pi^2 EI}{L^2} \qquad n = 1, 2, \ldots \tag{8.59}$$

The smallest of these critical loads, the buckling load ($n = 1$), is called the *Euler load* and will be denoted as P_E. The mode shapes for the Euler load as well as higher critical loads are those of sinusoids and are shown to Fig. 8.11.

To understand the physical aspects of the Euler load and the other critical loads, let us imagine that we are loading the column starting from $P = 0$ and plotting the ratio P/P_E against the amplitude of the deflection curve, which we denote as w_0. Below the value $P/P_E = 1$, we know from our analysis that to satisfy all the boundary conditions [see Eq. (8.57*b*)] $A = 0$ and we have a trivial solution. This means that $w_0 = 0$; the curve P/P_E versus w_0 coincides with the P/P_E axis, as shown in Fig. 8.12. At point (1), where $P/P_E = 1$, we may have *any value* of A and hence any value of w_0. Above this point we are again restricted to the trivial solution $A = 0$ and hence $w_0 = 0$. Point (1) is a bifurcation point like that described earlier for the rigid-body

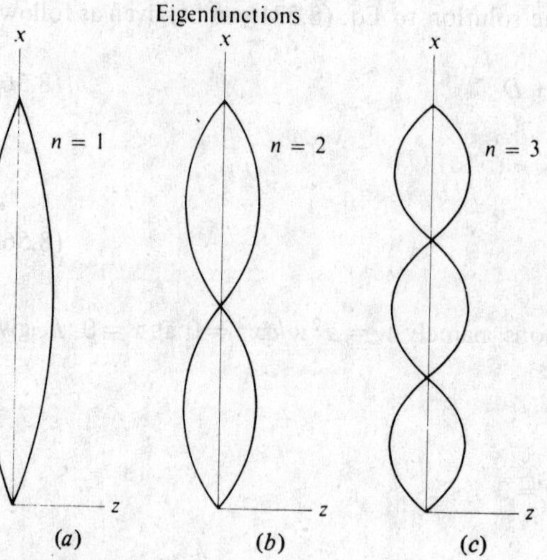

FIGURE 8.11
Eigenfunctions.

problem. In this case we have an infinity of such points as P/P_E becomes, successively, $2^2, 3^2$, etc. These clearly correspond to the series of critical loads.

In ascertaining the critical loads, as in the rigid-body examples, we asked the question: Are there loads for which there are nontrivial equilibrium configurations satisfying the boundary conditions of the problem? For the rigid-body example undertaken earlier the question was answered by simple algebra; in the elastic problem we dealt with an eigenvalue problem. In later sections we will see that one can ask other questions of the system to generate information about critical loads. However, the amount of information found is not always the same; it depends on the question asked and the system considered.

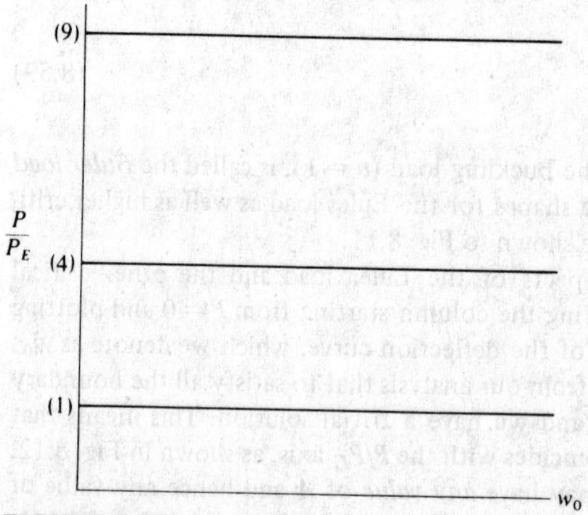

FIGURE 8.12
Bifurcation points.

8.7 ENERGY METHODS

Up to this time we have employed the first variation of the total potential energy to set forth the eigenvalue problem, which gave us the critical loads. Thus in the preceding section we considered the Euler column and, by seeking nontrivial solutions to the equilibrium equation, were able to ascertain the critical loads. We shall now demonstrate that we can generate the Euler load for this problem by other methods—the so-called energy methods, one of which was presented earlier for rigid bodies.

The second variation of the total potential energy can play a significant role in the study of elastic stability, as was the case for rigid-body stability. You will recall from the latter considerations that $\delta^{(2)}\pi$ ceased to be positive definite at the buckling load. This is ordinarily the case for elastic bodies on which conservative forces act.

As a preliminary note in this regard, let us recall that we can find all orders of variations of π through Eq. (8.5) when we have a single independent variable present. For s independent variables x_i, we can find formulations for all orders of variations by first carrying out a Taylor series expansion for π, as we did in Eq. (8.4) for one variable. Thus:†

$$\pi(x_1 + \delta x_1, x_2 + \delta x_2, \ldots, x_s + \delta x_s) = \pi(x_1, x_2, \ldots, x_s)$$

$$+ \sum_{p=1}^{s} \frac{\partial \pi}{\partial x_p} \delta x_p + \frac{1}{2!} \sum_{p=1}^{s} \sum_{q=1}^{s} \frac{\partial^2 \pi}{\partial x_p \, \partial x_q} \delta x_p \, \delta x_q$$

$$+ \frac{1}{3!} \sum_{p=1}^{s} \sum_{q=1}^{s} \sum_{r=1}^{s} \frac{\partial^3 \pi}{\partial x_p \, \partial x_q \, \partial x_r} \delta x_p \, \delta x_q \, \delta x_r + \cdots$$

Hence,

$$\pi(x_1 + \delta x_1, x_2 + \delta x_2, \ldots, x_s + \delta x_s) - \pi(x_1, x_2, \ldots, x_s)$$

$$= \delta^{(T)}\pi = \sum_{p=1}^{s} \frac{\partial \pi}{\partial x_p} \delta x_p + \frac{1}{2!} \sum_{p=1}^{s} \sum_{q=1}^{s} \frac{\partial^2 \pi}{\partial x_p \, \partial x_q} \delta x_p \, \delta x_q$$

$$+ \frac{1}{3!} \sum_{p=1}^{s} \sum_{q=1}^{s} \sum_{r=1}^{s} \frac{\partial^3 \pi}{\partial x_p \, \partial x_q \, \partial x_r} \delta x_p \, \delta x_q \, \delta x_r + \cdots \tag{8.60}$$

We can then say

$$\delta^{(1)}\pi = \sum_{p=1}^{s} \frac{\partial \pi}{\partial x_p} \delta x_p$$

$$\delta^{(2)}\pi = \sum_{p=1}^{s} \sum_{q=1}^{s} \frac{\partial^2 \pi}{\partial x_p \, \partial x_q} \delta x_p \, \delta x_q \qquad \text{etc.}$$

†Consult any standard advanced calculus book for Taylor series expansions in n variables.

With this background, let us now reexamine the pin-ended column. We note that the general solution obtained for the Euler column is

$$w = \sum_{n=1}^{\infty} A_n \sin \frac{n\pi x}{L}$$

The total potential energy, given by Eq. (8.52) for this case, is then

$$\pi = \frac{EIL}{4} \sum_{n=1}^{\infty} A_n^2 \left(\frac{n\pi}{L}\right)^4 - \frac{PL}{4} \sum_{n=1}^{\infty} A_n^2 \left(\frac{n\pi}{L}\right)^2$$

We now find the total variation of π, considering the A_n to be independent variables. For this purpose we may employ Eq. (8.60) with s going to ∞. Since A_n appears to the power 2, only the first two expressions on the right side of Eq. (8.60) give nonzero results. We then have by such a computation:

$$\delta^{(T)}\pi = \sum_{n=1}^{\infty} \left[\frac{EIL}{2}\left(\frac{n\pi}{L}\right)^4 - \frac{PL}{2}\left(\frac{n\pi}{L}\right)^2\right] A_n \, \delta A_n$$

$$+ \frac{1}{2!} \sum_{n=1}^{\infty} \left[\frac{EIL}{2}\left(\frac{n\pi}{L}\right)^4 - \frac{PL}{2}\left(\frac{n\pi}{L}\right)^2\right] (\delta A_n)^2$$

The second variation of π then may be seen to be

$$\delta^{(2)}\pi = \sum_{n=1}^{\infty} \left[\frac{EIL}{2}\left(\frac{n\pi}{L}\right)^4 - \frac{PL}{2}\left(\frac{n\pi}{L}\right)^2\right] (\delta A_n)^2$$

$$= \sum_{n=1}^{\infty} \left[\left(\frac{n\pi}{L}\right)^2 \frac{L}{2}\left(EI\frac{n^2\pi^2}{L^2} - P\right)\right] (\delta A_n)^2$$

$$= \sum_{n=1}^{\infty} \left[\left(\frac{n\pi}{L}\right)^2 \frac{L}{2}(n^2 P_E - P)\right] (\delta A_n)^2 \tag{8.61}$$

It is clear from above that when $P < P_E$ the second variation is positive definite. And, using a theorem of Sylvester (see LaSalle and Lefschetz, 1961, p. 36), we may conclude that $\delta^{(2)}\pi$ *ceases* to be positive definite when the load P equals or exceeds the Euler load. *We can then say here, as in the case of the rigid-body discussion, that the buckling load is the one for which the second variation of the total potential energy ceases to be positive definite.* We may extend this conclusion to other linear elastic stability problems wherein we have constant (not time-varying) applied loads.

To get *another* energy point of view of the buckling load, consider the total variation of a functional in the following way:

$$\tilde{I} - I = \delta^{(T)} I = \int_{x_1}^{x_2} F(x, y + \epsilon\eta, y' + \epsilon\eta')\, dx - \int_{x_1}^{x_2} F(x, y, y')\, dx$$

Now expand $F(x, y + \epsilon\eta, y' + \epsilon\eta')$ as a power series in the parameter ϵ about the value $\epsilon = 0$. We get, after cancellation of terms,

$$\delta^{(T)} I = \left[\frac{\partial}{\partial \epsilon} \int_{x_1}^{x_2} F(x, y + \epsilon\eta, y' + \epsilon\eta')\, dx \right]_{\epsilon=0} \epsilon$$

$$+ \frac{1}{2!} \left[\frac{\partial^2}{\partial \epsilon^2} \int_{x_1}^{x_2} F(x, y + \epsilon\eta, y' + \epsilon\eta')\, dx \right]_{\epsilon=0} \epsilon^2$$

$$+ \frac{1}{3!} \left[\frac{\partial^3}{\partial \epsilon^3} \int_{x_1}^{x_2} F(x, y + \epsilon\eta, y' + \epsilon\eta')\, dx \right]_{\epsilon=0} \epsilon^3 + \cdots$$

The first expression on the right side of the above equation is the well-known first variation [see Eq. (2.21)]. The second variation may be taken as the coefficient of 1/2!, the third variation as the coefficient of 1/3!, etc. With this background, let us now consider again the total potential energy π for the Euler column. We have from Eq. (8.52):

$$\pi = \frac{EI}{2} \int_0^L (w'')^2\, dx - \frac{P}{2} \int_0^L (w')^2\, dx \tag{8.62}$$

To get $\tilde{\pi}$ use $\tilde{w} = w + \epsilon\eta$. Then we have

$$\tilde{\pi} = \frac{EI}{2} \int_0^L (w'' + \epsilon\eta'')^2\, dx - \frac{P}{2} \int_0^L (w' + \epsilon\eta')^2\, dx$$

$$= \frac{EI}{2} \int_0^L (w'')^2\, dx - \frac{P}{2} \int_0^L (w')^2\, dx + \left(EI \int_0^L w''\eta''\, dx - P \int_0^L w'\eta'\, dx \right) \epsilon$$

$$+ \left[\frac{EI}{2} \int_0^L (\eta'')^2\, dx - \frac{P}{2} \int_0^L (\eta')^2\, dx \right] \epsilon^2 \tag{8.63}$$

Noting Eq. (8.62), we may give $\delta^{(T)}\pi$ as follows:

$$\tilde{\pi} - \pi = \delta^{(T)} \pi = \left[EI \int_0^L w''\eta''\, dx - P \int_0^L w'\eta'\, dx \right] \epsilon$$

$$+ \left[EI \int_0^L (\eta'')^2\, dx - P \int_0^L (\eta')^2\, dx \right] \frac{\epsilon^2}{2!}$$

Thus we can conclude that

$$\delta^{(2)}\pi = \left[EI \int_0^L (\eta'')^2 \, dx - P \int_0^L (\eta')^2 \, dx \right] \epsilon^2 \tag{8.64}$$

We shall now *extremize* the second variation with respect to functions that satisfy the end conditions of the problem. The corresponding Euler-Lagrange equation for such an extremization is then

$$EI \frac{d^4 \eta}{dx^4} + P \frac{d^2 \eta}{dx^2} = 0$$

The determination of the existence of appropriate functions η then leads us to the same problem set forth earlier for determining the critical loads. That is, *setting the first variation of the second variation of the total potential energy equal to zero leads to the same eigenvalue problem presented earlier for establishing the critical loads.* Accordingly, the critical loads may be characterized as those loads permitting the establishment of the extremal function with appropriate boundary conditions for the following condition:

$$\delta^{(1)}(\delta^{(2)}\pi) = 0 \tag{8.65}$$

This is the criterion of Trefftz (1933, pp. 160–165) and constitutes the second of our energy methods.

We shall make use of this criterion in the study of plates.

8.8 IMPERFECTION ANALYSIS

In the preceding sections we examined a *perfectly straight* column to find that the equilibrium method, stemming from setting the first variation of π equal to zero, gave us critical loads; the extremization of the second variation, or noting when π ceases to be positive definite, gave us the buckling load. We shall now set forth another criterion for the buckling load by considering a column that is initially *not* straight, such as the one shown in Fig. 8.13*a*. The initial deflection is given by the function

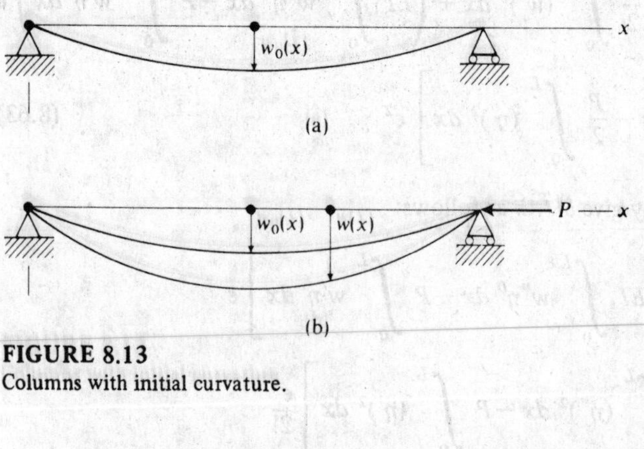

FIGURE 8.13
Columns with initial curvature.

$w_0(x)$. The total deflection with the presence of axial load P then is given by $w(x)$, as shown in Fig. 8.13b.

We will give $w_0(x)$ as a Fourier series expansion in the following way:

$$w_0(x) = \sum_{n=1}^{\infty} a_n \sin \frac{n\pi x}{L} \tag{8.66}$$

so that

$$a_n = \frac{2}{L} \int_0^L w_0(x) \sin \frac{n\pi x}{L}\, dx$$

The function $w(x)$ can in turn be given as

$$w(x) = \sum_{n=1}^{\infty} b_n \sin \frac{n\pi x}{L} \tag{8.67}$$

where b_n is a set of constants. Note that the end conditions are satisfied by each term in the above expansion. The total potential energy π for this case is then

$$\pi = \frac{1}{2} EI \int_0^L \left[\frac{d^2}{dx^2}(w - w_0) \right]^2 dx - P\Delta \tag{8.68}$$

We may give Δ, the contraction of the column along the x axis, as follows:

$$\Delta = \frac{1}{2} \int_0^L \left(\frac{dw}{dx} \right)^2 dx - \frac{1}{2} \int_0^L \left(\frac{dw_0}{dx} \right)^2 dx \tag{8.69}$$

Accordingly, we have for π

$$\pi = \frac{1}{2} EI \int_0^L \left[\frac{d^2}{dx^2}(w - w_0) \right]^2 dx - \frac{P}{2} \int_0^L \left[\left(\frac{dw}{dx} \right)^2 - \left(\frac{dw_0}{dx} \right)^2 \right] dx \tag{8.70}$$

By using Eqs. (8.66) and (8.67) to replace w and w_0, respectively, π becomes a function of the terms b_n. Thus

$$\pi = \frac{1}{2} EI \int_0^L \left[\sum_1^{\infty} (a_n - b_n) \left(\frac{n\pi}{L} \right)^2 \sin \frac{n\pi x}{L} \right]^2 dx$$

$$- \frac{P}{2} \int_0^L \left[\left(\sum_1^{\infty} b_n \frac{n\pi}{L} \cos \frac{n\pi x}{L} \right)^2 - \left(\sum_1^{\infty} a_n \frac{n\pi}{L} \cos \frac{n\pi x}{L} \right)^2 \right] dx$$

To extremize π we require that†

$$\frac{\partial \pi}{\partial b_m} = 0 \qquad m = 1, 2, \ldots$$

Hence:

$$-\frac{EI}{2}\int_0^L 2\frac{\pi^2}{L^2}\left[\sum_1^\infty n^2(a_n - b_n)\sin\frac{n\pi x}{L}\right]\left[\left(\frac{m\pi}{L}\right)^2 \sin\frac{m\pi x}{L}\right]dx$$

$$-\frac{P}{2}\int_0^L 2\left(\sum_1^\infty b_n \frac{n\pi}{L}\cos\frac{n\pi x}{L}\right)\frac{m\pi}{L}\cos\frac{m\pi x}{L}\,dx = 0 \qquad m = 1, 2, \ldots$$

Due to orthogonality, the terms in each integral vanish except when $n = m$. We then get from the above equation

$$EI\frac{\pi^4 m^4}{L^4}(a_m - b_m) + \frac{P}{L^2}m^2\pi^2 b_m = 0 \qquad m = 1, 2, \ldots \tag{8.71}$$

We now solve for the b_m:

$$b_m = \frac{a_m}{1 - PL^2/\pi^2 m^2 EI} = \frac{a_m}{1 - P/m^2 P_E} \qquad m = 1, 2, \ldots$$

We can then give $w(x)$ as follows:

$$w(x) = \sum_{n=1}^\infty \frac{a_n}{1 - (1/n^2)P/P_E}\sin\frac{n\pi x}{L} \tag{8.72}$$

Unlike the results obtained for the straight beam stemming from the extremization of the total potential energy, we see here that a *nontrivial deflection exists from the very onset of the axial loading.* Note that as the load approaches the buckling load the first term in the series *becomes indefinitely large.*‡ We may accordingly denote here another criterion for the buckling load: that it represents *the load for which the initial deviations from straightness become amplified beyond any finite amount.* Note also that when $P = n^2 P_E$ we get singular behavior in w, and so the imperfection analysis, by this action, yields in addition the other critical loads.

It is worthwhile to note here why knowing the buckling load is vital in design. Every column will have some initial curvature. As a consequence, large deflections will occur as the axial load approaches its critical value. Also, the same effect occurs

†Remember that π is a function of the b's and is not a functional here.

‡Note that if a_1 is large, the assumptions regarding linear elasticity with small rotations may be violated long before P reaches the buckling load P_E.

when the line of action of the axial load is ***not colinear*** with the centerline of the column. The loading is then said to be *eccentric.* Since there will always be some loading eccentricity present, we see again the need to keep the axial load sufficiently lower than the buckling load.

8.9 THE KINETIC METHOD

We now present another procedure for establishing the buckling load. It is called the *kinetic method.* In the analysis we shall pose the following question: Is there a value of the load P for which the most general free vibration of the system, however small initially, becomes unbounded with time?

We once again consider the pin-ended column of the previous investigations. To derive the equation of motion from Hamilton's principle, we employ for the total potential energy the form given by Eq. (8.52), while for the kinetic energy we employ the expression

$$\frac{\rho A}{2}\int_0^L \dot{w}^2\,dx$$

as explained in Section 7.3. Hence the Lagrangian that we shall employ in Hamilton's principle is

$$L = \int_0^L \frac{\rho A}{2}\left(\frac{\partial w}{\partial t}\right)^2 dx - \int_0^L \frac{EI}{2}\left(\frac{\partial^2 w}{\partial x^2}\right)^2 dx + \int_0^L \frac{P}{2}\left(\frac{\partial w}{\partial x}\right)^2 dx$$

$$= \int_0^L \left(\frac{\rho A}{2}\dot{w}^2 - \frac{EI}{2}w_{xx}^2 + \frac{P}{2}w_x^2\right) dx$$

Hamilton's principle now requires that

$$\delta^{(1)}\int_{t_1}^{t_2}\int_0^L \left(\frac{\rho A}{2}\dot{w}^2 - \frac{EI}{2}w_{xx}^2 + \frac{P}{2}w_x^2\right) dx\,dt = 0$$

Using the familiar one-parameter family of varied functions $\tilde{w}(x,t) = w(x,t) + \epsilon\eta(x,t)$, where $\eta(x,t_1) = \eta(x,t_2) = 0$, we have

$$\frac{d}{d\epsilon}\left(\int_{t_1}^{t_2}\int_0^L \tilde{F}\,dx\,dt\right)_{\epsilon=0}$$

$$= \int_{t_1}^{t_2}\int_0^L \left(\frac{\partial F}{\partial \dot{w}}\dot{\eta} + \frac{\partial F}{\partial w_{xx}}\eta_{xx} + \frac{\partial F}{\partial w_x}\eta_x\right) dx\,dt = 0 \tag{8.73}$$

We integrate the first expression in the integrand by parts with respect to time. Thus:

$$\int_{t_1}^{t_2}\int_0^L \frac{\partial F}{\partial \dot{w}}\,\dot{\eta}\,dx\,dt = \int_0^L \left(\eta\,\frac{\partial F}{\partial \dot{w}}\right)_{t_1}^{t_2} dx - \int_{t_1}^{t_2}\int_0^L \eta\,\frac{\partial}{\partial t}\,\frac{\partial F}{\partial \dot{w}}\,dx\,dt$$

Note that the first term on the right side is zero because of the initial and final time conditions on η. Next integrate the last expressions in Eq. (8.73) by parts with respect to x. Thus:

$$\int_0^L\int_{t_1}^{t_2} \frac{\partial F}{\partial w_x}\,\eta_x\,dt\,dx = \int_{t_1}^{t_2}\left(\frac{\partial F}{\partial w_x}\,\eta\right)_0^L dt - \int_{t_1}^{t_2}\int_0^L\left(\frac{\partial}{\partial x}\,\frac{\partial F}{\partial w_x}\right)\eta\,dx\,dt$$

Since we require that η satisfy the end conditions of the problem, the first term on the right side of this equation is zero. Finally, we integrate by parts the second expression in the integrand of Eq. (8.73) twice with respect to x. Thus:

$$\int_0^L\int_{t_1}^{t_2} \frac{\partial F}{\partial w_{xx}}\,\eta_{xx}\,dt\,dx = \int_{t_1}^{t_2}\left(\frac{\partial F}{\partial w_{xx}}\,\eta_x\right)_0^L dt - \int_{t_1}^{t_2}\left[\left(\frac{\partial}{\partial x}\,\frac{\partial F}{\partial w_{xx}}\right)\eta\right]_0^L dt$$

$$+ \int_0^L\int_{t_1}^{t_2}\left(\frac{\partial^2}{\partial x^2}\,\frac{\partial F}{\partial w_{xx}}\right)\eta\,dx\,dt$$

The second integral on the right side of the equation clearly is zero. We then have, on substituting the above results into Eq. (8.73),

$$-\int_{t_1}^{t_2}\int_0^L\left(\frac{\partial}{\partial t}\,\frac{\partial F}{\partial \dot{w}} - \frac{\partial^2}{\partial x^2}\,\frac{\partial F}{\partial w_{xx}} + \frac{\partial}{\partial x}\,\frac{\partial F}{\partial w_x}\right)\eta\,dx\,dt + \int_{t_1}^{t_2}\left(\frac{\partial F}{\partial w_{xx}}\,\eta_x\right)_0^L dt = 0$$

Inserting for F the expression

$$\frac{\rho A}{2}\,\dot{w}^2 - \frac{EI}{2}\,w_{xx}^2 + \frac{P}{2}\,w_x^2$$

we may then give the equation of motion in the following way:

$$\rho A \ddot{w} + \frac{\partial^2}{\partial x^2}\left(EI\,\frac{\partial^2 w}{\partial x^2}\right) + P\,\frac{\partial^2 w}{\partial x^2} = 0 \tag{8.74}$$

Also, we may require at the ends that $\partial F/\partial w_{xx} = EIw_{xx} = 0$ and $w = 0$.

We shall take the solution for the *freely vibrating* simply supported column to have the following form (for uniform EI):

$$w(x,t) = \sum_{n=1}^{\infty} A_n \sin\frac{n\pi x}{L}\,e^{i\omega_n t} \tag{8.75}$$

where A_n are complex constants that are determined from the initial conditions and ω_n are frequency terms that are chosen to render each term in the series a solution to the differential equation. Note that the boundary conditions $w = w_{xx} = 0$ at the ends of the column are satisfied. On substitution of the expansion into the differential equation we find that

$$\rho A\omega_n^2 = \left(\frac{n\pi}{L}\right)^2 (n^2 P_E - P) \tag{8.76}$$

Notice from the above result that when $P > P_E$, ω_1 becomes imaginary and, as a result, the first term ($n = 1$) in the series expansion for $w(x, t)$ above diverges as time becomes large. We can conclude from this that the ***buckling load is one that, when exceeded, results in unbounded behavior of the vibratory motion.*** This in turn means that when P is above the buckling load the equilibrium configuration is unstable since any lateral disturbance, however slight, will, within the assumptions of the theory, result in ever growing lateral motion.

8.10 GENERAL REMARKS

In reviewing the various approaches to the buckling problem of the pin-ended column, we note that the ***equilibrium method*** gave an infinite set of eigenvalues or critical loads where nontrivial configurations could satisfy the requirements of equilibrium. The lowest critical value we called the buckling load. The ***imperfection analysis*** showed that at the critical loads we got infinite deflections but that between these loads, theoretically, finite deflection could be maintained. Using one of the so-called ***energy methods*** (the energy method of Trefftz), we found that the extremization of $\delta^{(2)}\pi$ again led to the set of critical loads. Furthermore, with the other energy method we found that the buckling load P_E marked the load where, on increasing the value of P from zero, the second variation of π ceased to be positive definite. The implication here, from the extrapolation of the discussions on rigid-body systems, is that up to the buckling load the equilibrium configurations are stable, but at and beyond the buckling load the equilibrium configurations are not stable. This was again verified when we considered the ***kinetic approach*** for loads that exceed the critical load.

All these results stem from consideration of a particular simple problem, namely the pin-ended column. What can we say about other ***linearized*** problems? ***All these conclusions can generally be extrapolated intact.***

8.11 THE ELASTICA

Up to this time we have used simplified formulations for the study of the elastic stability of columns. From these formulations we mathematically produced such results as bifurcation points, nontrivial equilibrium configurations, and so on, at the so-called critical loads. In this section we examine a more exact formulation for the column problem, and we shall then be able to relate the results of the simplified analysis to the dictates of the more exact theory. Specifically, in previous sections, we have, in effect, employed the following simplified functional for the total potential energy:

$$\pi = \frac{EI}{2}\int_0^L \left(\frac{d^2w}{dx^2}\right)^2 dx - \frac{P}{2}\int_0^L \left(\frac{dw}{dx}\right)^2 dx \tag{8.77}$$

The expression d^2w/dx^2 used in the first integral is actually an approximation of the curvature κ of the member. You will recall from Section 8.4 that κ is given exactly as

$$\kappa = \frac{d^2w/dx^2}{[1 + (dw/dx)^2]^{3/2}}$$

Or, if we use the variables θ and s at the neutral axis (Figs. 8.5 and 8.14), we may also give κ exactly as

$$\kappa \equiv \frac{1}{\rho} = \frac{d\theta}{dl'} = \frac{d\theta}{ds} \tag{8.78}$$

In the second integral of Eq. (8.77), we have, in effect, approximated the shortening of the column Δ by the expression

$$\frac{1}{2}\int_0^L \left(\frac{dw}{dx}\right)^2 dx$$

We developed this approximation in Example 3.3, and you will recall that a small slope dw/dx was a key assumption in that development. An exact evaluation of Δ is given by Eq. (8.19) with the substitution of Eq. (8.16):

$$\Delta = L - \int_0^L \cos\theta\, ds = -\left(\int_0^L \cos\theta\, ds - L\right) \tag{8.79}$$

Using Eqs. (8.78) and (8.79), we can give the total potential energy in more precise form as

$$\pi = \frac{EI}{2}\int_0^L \left(\frac{d\theta}{ds}\right)^2 ds + P\left(\int_0^L \cos\theta\, ds - L\right) \tag{8.80}$$

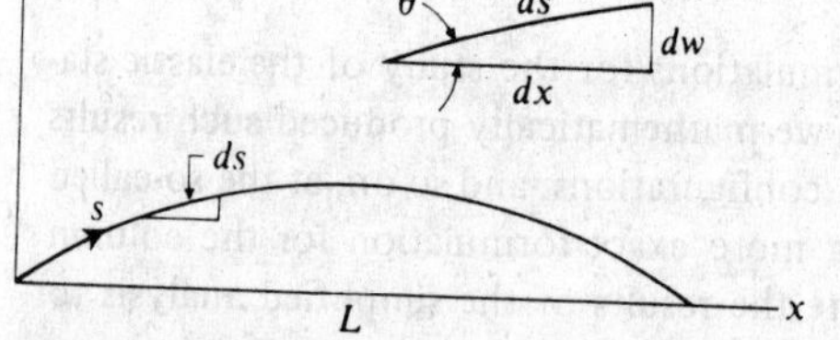

FIGURE 8.14
Deflected neutral axis of column.

We now compute the first variation of this functional:

$$\delta^{(1)}\pi = EI\int_0^L \frac{d\theta}{ds}\frac{d}{ds}(\delta\theta)\,ds - P\int_0^L \sin\theta\,\delta\theta\,ds$$

Integrating the first integral by parts, we get

$$\delta^{(1)}\pi = \left[EI\frac{d\theta}{ds}\,\delta\theta\right]_0^L - \int_0^L\left(EI\frac{d^2\theta}{ds^2} + P\sin\theta\right)\delta\theta\,ds$$

Setting the first variation equal to zero gives us the following Euler-Lagrange equation:

$$EI\frac{d^2\theta}{ds^2} + P\sin\theta = 0 \tag{8.81}$$

while the boundary conditions at $s = 0, L$ are

$$\text{EITHER}\quad EI\frac{d\theta}{ds} = 0 \qquad \text{OR} \qquad \theta \quad \text{IS SPECIFIED} \tag{8.82}$$

Since θ represents the local slope of the column, it is clear that the above boundary conditions represent the "*moment-slope duality.*" The differential equation is nonlinear.† However, we will see that it is integrable in terms of *elliptic integrals.* To carry out an immediate quadrature, multiply the equation by $d\theta/ds$. Thus:

$$EI\frac{d^2\theta}{ds^2}\frac{d\theta}{ds} + P\sin\theta\frac{d\theta}{ds} = 0$$

This result may be written in the following form:

$$\frac{d}{ds}\left[\frac{EI}{2}\left(\frac{d\theta}{ds}\right)^2 - P\cos\theta\right] = 0$$

We may integrate to find

$$\frac{EI}{2}\left(\frac{d\theta}{ds}\right)^2 = P\cos\theta + C_1 \tag{8.83}$$

For a simply supported column, θ is not specified at the ends and so we use the natural boundary condition $d\theta/ds = 0$ at the ends. Denoting θ at $x = 0$ as α, and θ at $x = L$ as $-\alpha$, we can apply the natural boundary conditions to Eq. (8.83) to calculate:

$$C_1 = -P\cos\alpha$$

†It actually has the same form as the exact equation for the simple pendulum.

so that

$$\frac{d\theta}{ds} = \left[\frac{2P}{EI}(\cos\theta - \cos\alpha)\right]^{1/2} = \left[\frac{4P}{EI}\left(\sin^2\frac{\alpha}{2} - \sin^2\frac{\theta}{2}\right)\right]^{1/2} \tag{8.84}$$

where we have used familiar trignonometric identities to rewrite the right side of the equation. We may next set up the following quadrature by separating variables:

$$\int_0^L ds = L = \sqrt{\frac{EI}{4P}} \int_{-\alpha}^{+\alpha} \frac{d\theta}{(\sin^2\alpha/2 - \sin^2\theta/2)^{1/2}}$$

$$= \sqrt{\frac{EI}{4P}} \int_0^{\alpha} \frac{2\,d\theta}{(\sin^2\alpha/2 - \sin^2\theta/2)^{1/2}} \tag{8.85}$$

(In the last step above we made use of the symmetry of the deflection curve about the line $x = L/2$.) We now institute a change of variable as follows:

$$\sin\frac{\theta}{2} = p\sin\phi \tag{8.86a}$$

where

$$p = \sin\frac{\alpha}{2} \tag{8.86b}$$

With some algebraic manipulation, we may rewrite Eq. (8.85) as

$$L = 2\sqrt{\frac{EI}{P}} \int_0^{\pi/2} \frac{d\phi}{(1 - p^2\sin^2\phi)^{1/2}} \tag{8.87}$$

Upon introduction of the Euler load P_E, this equation becomes

$$\frac{P}{P_E} = \frac{2}{\pi} \int_0^{\pi/2} \frac{d\phi}{(1 - p^2\sin^2\phi)^{1/2}} \tag{8.88}$$

Using tables of elliptic integrals of the first kind (Abramowitz and Stegun, 1965), we can find P/P_E as a function of p, and hence of α, the slope at the ends. This is done by choosing a value of P/P_E and then finding from the tables the proper value of p to satisfy the above equation.

We shall next relate p to the maximum deflection w_{max} of the column so that we will be able to plot P/P_E against w_{max}. For this purpose, note from Fig. 8.14 that

$$\sin\theta = \frac{dw}{ds} \tag{8.89}$$

We may immediately perform a quadrature to solve for w. Thus, replacing $\sin\theta$ by using Eq. (8.81):

$$w = -\frac{EI}{P}\frac{d\theta}{ds} + C_2 \tag{8.90}$$

Clearly, $w = 0$ at the ends, where, as a result of the natural boundary condition, $d\theta/ds$ must also be zero, rendering the constant of integration equal to zero. Employing Eq. (8.84) to replace $d\theta/ds$ above, we then have for w:

$$w = -\sqrt{\frac{2EI}{P}}(\cos\theta - \cos\alpha)^{1/2} \tag{8.91}$$

The maximum displacement occurs at $\theta = 0$ and so we have for w_{max}:

$$w_{max} = -\sqrt{\frac{2EI}{P}}(1 - \cos\alpha)^{1/2} \tag{8.92}$$

This may be rewritten in the following form, on introducing P_E and employing a trigonometric identity for the bracketed expression:

$$\frac{w_{max}}{L} = \frac{2}{\pi}\frac{\sin\alpha/2}{\sqrt{P/P_E}} = \frac{2}{\pi}\frac{p}{\sqrt{P/P_E}} \tag{8.93}$$

Thus with p and P/P_E appropriately related by Eq. (8.88), we can determine the corresponding value of w_{max}/L from the preceding equation. A plot of P/P_E versus w_{max}/L is shown in Fig. 8.15. At $P/P_E = 1$ the exact curve is tangent to the line

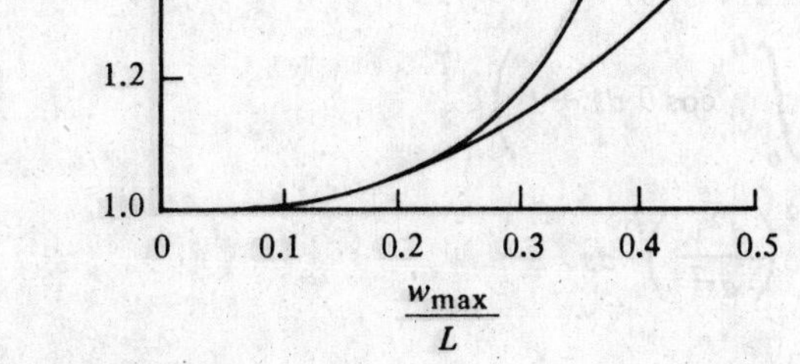

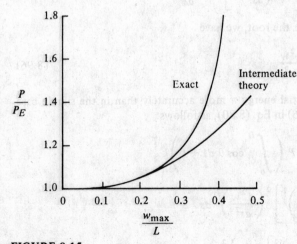

FIGURE 8.15
Postbuckling deflection curves.

$P/P_E = 1$. We see from the exact analysis that we still get the bifurcation point, as we did in the linearized theory, but we no longer have the indeterminacy of load versus deflection at the buckling load. We call the regime of deformation beyond the bifurcation point the *postbuckling regime.* Since the column can withstand even higher loads in the vicinity of the bifurcation point, we say that we have postbuckling *stability*. (If after the bifurcation point the structure could withstand only decreased loads, obviously we would have a situation of postbuckling *instability*.) Clearly, in practical problems the question of postbuckling behavior is of significance.

In the following section we shall consider an "intermediate" theory that will reproduce the exact results up to about $w_{max}/L = 0.3$ and that involves a considerably simpler computation than that of the elastica.

8.12 AN INTERMEDIATE THEORY†

We again consider the pin-connected column of the previous sections. Our procedure will be to use certain power series expansions to form a more accurate expression for π than was used for the elementary case. We now express $\cos\theta$ as follows, making use of Eq. (8.89):

$$\cos\theta = \sqrt{1-\sin^2\theta} = \sqrt{1-\left(\frac{dw}{ds}\right)^2} \tag{8.94}$$

Next we use the binomial expansion to obtain the following approximation:

$$\cos\theta \cong 1 - \frac{1}{2}\left(\frac{dw}{ds}\right)^2 - \frac{1}{8}\left(\frac{dw}{ds}\right)^4 \tag{8.95}$$

Similarly, to obtain an approximate curvature expression, noting that $\theta = \sin^{-1}(dw/ds)$, we have

$$\frac{d\theta}{ds} = \frac{d}{ds}\left(\sin^{-1}\frac{dw}{ds}\right) = \left[1-\left(\frac{dw}{ds}\right)^2\right]^{-1/2}\frac{d^2w}{ds^2}$$

Using the binomial expansion for the root, we have

$$\frac{d\theta}{ds} \cong \left[1+\frac{1}{2}\left(\frac{dw}{ds}\right)^2\right]\frac{d^2w}{ds^2} \tag{8.96}$$

We may express the total potential energy π more accurately than in the simple case by utilizing Eqs. (8.95) and (8.96) in Eq. (8.80), as follows:

$$\begin{aligned}\pi &= \frac{EI}{2}\int_0^L\left(\frac{d\theta}{ds}\right)^2 ds + P\left(\int_0^L \cos\theta\, ds - L\right)\\ &= \frac{EI}{2}\int_0^L\left[1+\frac{1}{2}\left(\frac{dw}{ds}\right)^2\right]^2\left(\frac{d^2w}{ds^2}\right)^2 ds\\ &\quad -\frac{P}{2}\int_0^L\left[1+\frac{1}{4}\left(\frac{dw}{ds}\right)^2\right]\left(\frac{dw}{ds}\right)^2 ds\end{aligned} \tag{8.97}$$

†Adapted from Mayers (1964).

[Note that if we delete $(dw/ds)^2$ in the brackets as being small compared with unity and replace s by x as the integration variable, we revert to the classical theory presented at the outset.] We shall proceed by using the Rayleigh-Ritz method here, employing as a single coordinate function the eigenfunction of the classical simple theory with s as the variable. That is:

$$w = A \sin \frac{\pi s}{L} \tag{8.98}$$

Clearly such a function satisfies the geometric and natural boundary conditions of the simply supported column. Substituting Eq. (8.98) into Eq. (8.97) and using $a = A\pi/L$, we get after integration and algebraic manipulations:

$$\pi = \frac{EI\pi^2}{4L}\left[a^2 + \frac{1}{4}a^4 + \frac{1}{32}a^6 - \frac{P}{P_E}\left(a^2 + \frac{3}{16}a^4\right)\right] \tag{8.99}$$

Extremizing π with respect to a yields the following equation:

$$\frac{3}{16}a^4 + \left(1 - \frac{3}{4}\frac{P}{P_E}\right)a^2 + 2\left(1 - \frac{P}{P_E}\right) = 0 \tag{8.100}$$

We may solve for a for each value of P/P_E. Noting that w_{max}/L equals a/π, we may then plot P/P_E against w_{max}/L, forming the so-called intermediate theory in Fig. 8.15. Note that the agreement between exact and intermediate theories is quite good for $w_{max}/L < 0.3$. (In most situations of practical interest this agreement is more than satisfactory, for it is not likely that the assumption of linear elasticity would be valid much beyond that large a deflection.)

By the technique presented here for the pin-ended column, we can often obtain reasonably accurate load-deflection curves without an inordinate amount of work.

Most of the buckling analyses we have considered thus far have been linearized eigenvalue problems that define the buckling load. In Sections 8.11 and 8.12 we extended the results to include deflections beyond the bifurcation point and we were able to discuss *postbuckling stability*. We wish to call the reader's attention to the work of Koiter to predict postbuckling stability. This theory was developed in 1946 but has received increased attention in recent years (see Dym, 1974, Chapters 5 and 6).

Part D
Elastic Stability of Plates

8.13 THE BUCKLING EQUATION FOR RECTANGULAR PLATES

In this section we shall present a linearized equation leading to the eigenvalue problem for the study of buckling of plates. Specifically, we shall present means of determining critical values of constant applied edge loads $\bar{N}_x, \bar{N}_y, \bar{N}_{xy}$, and $\bar{N}_{yx}$ for the rectangular plate (see Fig. 8.16). (We assume here that $\bar{N}_{xy} = \bar{N}_{yx}$.) In doing so, we follow a procedure analogous to that taken in Section 8.6, where we first found the equation

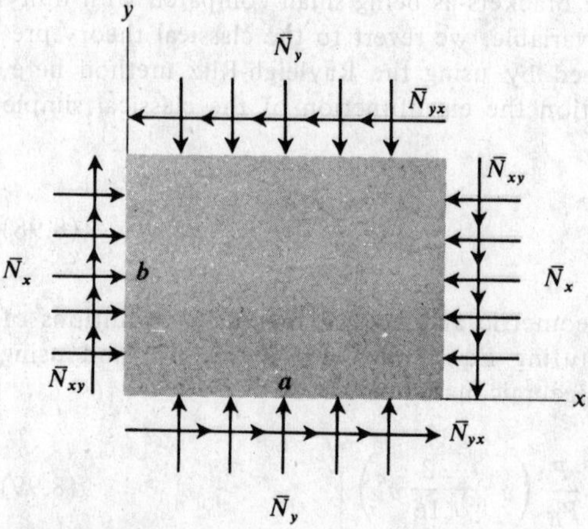

FIGURE 8.16
Rectangular plate.

of equilibrium of a column by extremizing the total potential energy. We shall then linearize the equation while noting the restrictions required by this process. We already have equations of equilibrium of plates with given edge loads from our work in Section 8.5. We accordingly now rewrite the von Kármán plate equations in terms of rectangular components:

$$D\,\nabla^4 w = \frac{\partial^2 F}{\partial y^2}\frac{\partial^2 w}{\partial x^2} - 2\,\frac{\partial^2 F}{\partial x\,\partial y}\,\frac{\partial^2 w}{\partial x\,\partial y} + \frac{\partial^2 F}{\partial x^2}\,\frac{\partial^2 w}{\partial y^2} \tag{8.101a}$$

$$\nabla^4 F = Eh\left[\left(\frac{\partial^2 w}{\partial x\,\partial y}\right)^2 - \frac{\partial^2 w}{\partial x^2}\,\frac{\partial^2 w}{\partial y^2}\right] \tag{8.101b}$$

where

$$\frac{\partial^2 F}{\partial y^2} = N_x$$
$$\frac{\partial^2 F}{\partial x^2} = N_y \tag{8.102}$$
$$-\frac{\partial^2 F}{\partial x\,\partial y} = N_{xy}$$

In the column analysis we could neglect extensions along the neutral axis of the column. Here we shall first set the terms N_x, N_y, and N_{xy} equal to the corresponding edge loads, plus a small perturbation quantity. That is, we take[†]

$$\frac{\partial^2 F}{\partial y^2} = -\bar{N}_x + \frac{\partial^2 \widetilde{F}}{\partial y^2} \tag{8.103}$$

[†]Note that we have taken compression as positive for the applied loads and have reversed the direction of $\bar{N}_{xy}$ as positive.

$$\frac{\partial^2 F}{\partial x^2} = -\bar{N}_y + \frac{\partial^2 \tilde{F}}{\partial x^2}$$

$$\frac{\partial^2 F}{\partial x\, \partial y} = \bar{N}_{xy} + \frac{\partial^2 \tilde{F}}{\partial x\, \partial y} \qquad (8.103)\ (Cont.)$$

where the tildes indicate the perturbation quantity. We shall agree to keep only linear terms in the perturbation quantities and in w and its derivatives; that is, we shall neglect all products and squares of these quantities. Accordingly, when we substitute Eqs. (8.103) into Eqs. (8.101), we obtain

$$D\,\nabla^4 \tilde{w} = -\bar{N}_x \frac{\partial^2 \tilde{w}}{\partial x^2} - 2\bar{N}_{xy} \frac{\partial^2 \tilde{w}}{\partial x\, \partial y} - \bar{N}_y \frac{\partial^2 \tilde{w}}{\partial y^2} \qquad (8.104a)$$

$$\nabla^4 \tilde{F} = 0 \qquad (8.104b)$$

Note that the *equations above are now uncoupled*, and we need now only concern ourselves with the first of the pair.

To gain insight as to what simplifications have effectively been incorporated in the study by the previous (rather formal) mathematical steps, we turn (as we did for the column) to considerations of the total potential energy. In particular, let us consider again the potential of the loading for a plate of general shape, as shown in Fig. 8.17. We can say, using negative edge loads as positive,

$$V = -\left[\oint_\Gamma (-\bar{N}_\nu) u_\nu\, ds + \oint_\Gamma (-\bar{N}_{\nu s}) u_s\, ds\right]$$

In the case of a rectangular plate (Fig. 8.16), this equation becomes

$$V = -\int_0^a [\bar{N}_y u_y + \bar{N}_{yx} u_x]_{y=0}\, dx - \int_0^b [(-\bar{N}_x) u_x + (-\bar{N}_{xy}) u_y]_{x=a}\, dy$$

$$- \int_a^0 [(-\bar{N}_y) u_y + (-\bar{N}_{yx}) u_x]_{y=b} (-dx) - \int_b^0 [\bar{N}_x u_x + \bar{N}_{xy} u_y]_{x=0} (-dy)$$

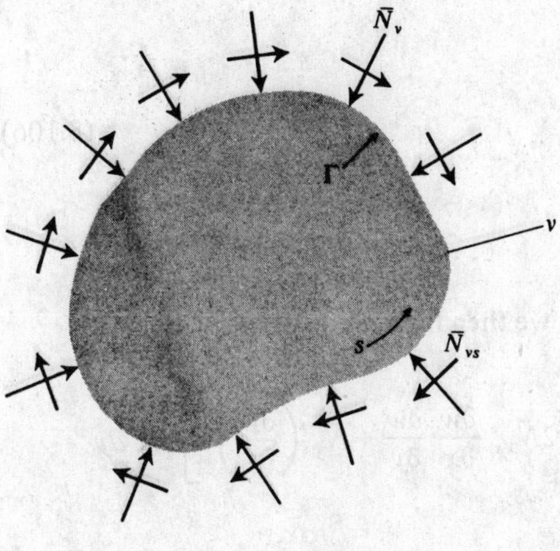

FIGURE 8.17
Plate of general shape.

Collecting terms, we get

$$V = \int_0^a \bar{N}_y [(u_y)_{y=b} - (u_y)_{y=0}]\, dx + \int_0^b \bar{N}_x [(u_x)_{x=a} - (u_x)_{x=0}]\, dy$$

$$+ \int_0^a \bar{N}_{yx} [(u_x)_{y=b} - (u_x)_{y=0}]\, dx + \int_0^b \bar{N}_{xy} [(u_y)_{x=a} - (u_y)_{x=0}]\, dy$$

Now such expressions as $(u_y)_{y=b} - (u_y)_{y=0}$ can be given as $\int_0^b (\partial u_y/\partial y)\, dy$, etc., and accordingly noting that $N_{xy} = N_{yx}$, the above equation can be given as follows:

$$V = \int_0^a \int_0^b \left[\bar{N}_y \frac{\partial u_y}{\partial y} + \bar{N}_{xy} \left(\frac{\partial u_x}{\partial y} + \frac{\partial u_y}{\partial x} \right) + \bar{N}_x \frac{\partial u_x}{\partial x} \right] dx\, dy \tag{8.105}$$

As a next step let us go back to Eq. (8.28), which gives strains in terms of displacements for the von Kármán plate theory. Note from this listing that in the midplane of the plate

$$\epsilon_{xx} = \frac{\partial u_x}{\partial x} + \frac{1}{2} \left(\frac{\partial w}{\partial x} \right)^2$$

$$\epsilon_{yy} = \frac{\partial u_y}{\partial y} + \frac{1}{2} \left(\frac{\partial w}{\partial y} \right)^2$$

$$\epsilon_{xy} = \frac{1}{2} \left(\frac{\partial u_y}{\partial x} + \frac{\partial u_x}{\partial y} \right) + \frac{1}{2} \frac{\partial w}{\partial x} \frac{\partial w}{\partial y}$$

If for the ensuing analysis we set the strains equal to zero *in the midplane of the plate*, we can say from above

$$\begin{aligned} \frac{\partial u_x}{\partial x} &= -\frac{1}{2} \left(\frac{\partial w}{\partial x} \right)^2 \\ \frac{\partial u_y}{\partial y} &= -\frac{1}{2} \left(\frac{\partial w}{\partial y} \right)^2 \\ \frac{\partial u_y}{\partial x} + \frac{\partial u_x}{\partial y} &= -\frac{\partial w}{\partial x} \frac{\partial w}{\partial y} \end{aligned} \tag{8.106}$$

Substituting these results into Eq. (8.105), we then have for V:

$$V = -\frac{1}{2} \int_0^a \int_0^b \left[\bar{N}_x \left(\frac{\partial w}{\partial x} \right)^2 + 2\bar{N}_{xy} \frac{\partial w}{\partial x} \frac{\partial w}{\partial y} + \bar{N}_y \left(\frac{\partial w}{\partial y} \right)^2 \right] dx\, dy$$

We may now give a total potential energy functional, using the above result for the external loads and using Eq. (6.27) with $u_s = v_s = q = 0$ for the strain energy of bending. We then get

$$\pi = \frac{D}{2}\int_0^a\int_0^b\left\{(\nabla^2 w)^2 + 2(1-\nu)\left[\left(\frac{\partial^2 w}{\partial x\,\partial y}\right)^2 - \frac{\partial^2 w}{\partial x^2}\frac{\partial^2 w}{\partial y^2}\right]\right\}dx\,dy$$
$$-\frac{1}{2}\int_0^a\int_0^b\left[\bar{N}_x\left(\frac{\partial w}{\partial x}\right)^2 + 2\bar{N}_{xy}\frac{\partial w}{\partial x}\frac{\partial w}{\partial y} + \bar{N}_y\left(\frac{\partial w}{\partial y}\right)^2\right]dx\,dy \tag{8.107}$$

It will now be left as an exercise to show that the extremization of the above functional yields the Euler-Lagrange equation, Eq. (8.104*a*), the very equation reached by the linearization process. And so we confirm that *this equation disregards the extensions of the midsurface.*

We can now determine the boundary conditions for the linearized buckling equation from the extremization of the functional (8.107). Or, more easily, they may be deduced from the von Kármán plate theory [see Eq. (8.42)] by replacing N_ν and $N_{\nu s}$, respectively, by the applied loads $-\bar{N}_\nu$ and $-\bar{N}_{\nu s}$. Thus, we have

$$\text{EITHER } M_\nu = 0 \quad \text{OR} \quad \frac{\partial w}{\partial \nu} \text{ IS SPECIFIED} \tag{8.108a}$$

$$\text{EITHER } Q_\nu + \frac{\partial M_{\nu s}}{\partial s} - \bar{N}_\nu\frac{\partial w}{\partial \nu} - \bar{N}_{\nu s}\frac{\partial w}{\partial s} = 0 \quad \text{OR} \quad w \text{ IS SPECIFIED} \tag{8.108b}$$

and at discontinuities on the boundary

$$M_{\nu s}\,\delta w = 0 \tag{8.108c}$$

This completes the formulation of the linearized buckling equation for rectangular plates, which allows the computation of loads that permit nontrivial solutions $w \neq 0$ of the buckling equation. This is the so-called *equilibrium method* set forth in Part C. Later we shall consider the *energy method* for plates.

8.14 THE EQUILIBRIUM METHOD–AN EXAMPLE

We shall now consider the case of a simply supported rectangular plate, $a \times b \times h$, in uniaxial uniform compression (see Fig. 8.18). This means that $\bar{N}_{xy} = \bar{N}_y = 0$, and the governing equation is given as

$$D\,\nabla^4 w + \bar{N}_x\frac{\partial^2 w}{\partial x^2} = 0 \tag{8.109}$$

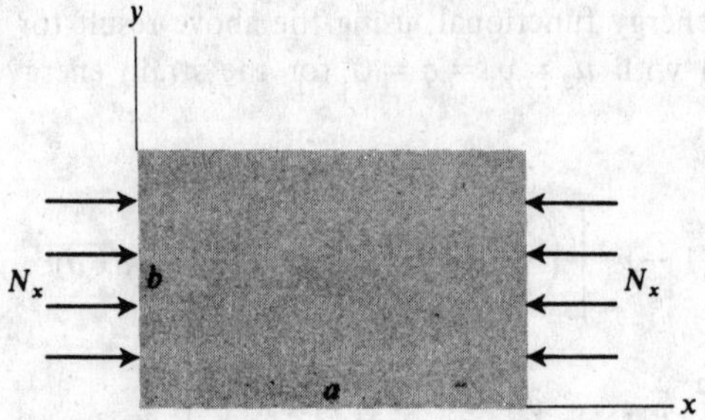

FIGURE 8.18
Uniaxial compression.

The boundary conditions for this problem are†

$$\text{At } x = 0, a \qquad w = \frac{\partial^2 w}{\partial x^2} = 0$$

$$\text{At } y = 0, b \qquad w = \frac{\partial^2 w}{\partial y^2} = 0 \tag{8.110}$$

From our work in linear plate theory we shall propose the following expression as a solution to Eq. (8.109):

$$w = A_{mn} \sin \frac{m\pi x}{a} \sin \frac{n\pi y}{b} \tag{8.111}$$

To satisfy the boundary conditions (8.110), m and n must clearly be integers. To avoid a trivial solution, we take m and n as nonzero. Substituting into Eq. (8.109), we have

$$\left\{ D\left[\left(\frac{m\pi}{a}\right)^2 + \left(\frac{n\pi}{b}\right)^2\right]^2 - \bar{N}_x \left(\frac{m\pi}{a}\right)^2 \right\} A_{mn} \sin \frac{m\pi x}{a} \sin \frac{n\pi y}{b} = 0$$

To get a nontrivial solution for the equilibrium equation, it is clear from the above that critical loads $\bar{N}_x$ exist and are given as

$$(\bar{N}_x)_{cr} = D\left(\frac{a}{m\pi}\right)^2 \left[\left(\frac{m\pi}{a}\right)^2 + \left(\frac{n\pi}{b}\right)^2\right]^2 \tag{8.112}$$

We may introduce the critical stress τ_{cr} by dividing through by h in the above equation. Replacing D by its basic definition [see Eq. (6.13)], we have

$$\tau_{cr} = \frac{\pi^2 E}{12(1-\nu^2)} \left(\frac{h}{b}\right)^2 \left[\left(\frac{mb}{a}\right)^2 + 2n^2 + n^4 \left(\frac{a}{mb}\right)^2\right] \tag{8.113}$$

†Note that for $x = a$, and $x = 0$, M_ν becomes M_x, and from Eq. (6.12a) we see that for a simply supported straight edge the requirement for $M_x = 0$ leads to the requirement that $\partial^2 w/\partial x^2 = 0$.

We have here a double infinity of discrete values for τ_{cr}. Whereas for columns it was easy to ascertain by inspection which critical load would be lowest (and thus the buckling load), this is not the case here. It is clear on inspection of Eq. (8.113) that τ_{cr} increases with n for any value of a/b, which we shall call the *aspect ratio*. This is not so for m. Accordingly, as a first step in our efforts to find the minimum buckling stress, we set n equal to its smallest value (i.e., unity). We can then give Eq. (8.113) in the following form:

$$\tau_{cr} = \frac{Ek_{cr}\pi^2}{12(1-\nu^2)}\left(\frac{h}{b}\right)^2 \tag{8.114}$$

where k_{cr}, termed the ***buckling coefficient***, has the form

$$k_{cr} = \left(m\frac{b}{a} + \frac{1}{m}\frac{a}{b}\right)^2 \tag{8.115}$$

We show k_{cr} plotted against the aspect ratio a/b for different values of m in Fig. 8.19. It is clear that the value of m giving the smallest buckling coefficient depends on the aspect ratio. Thus for $a/b = 1$, we see that $m = 1$, but for $a/b = 2$, m must be 2. As a/b gets larger, the critical buckling coefficient approaches the value 4. Establishing m and n for the lowest critical loads for a given ratio a/b thus establishes the buckling mode for the plate. For instance, with $a/b = 2$, $n = 1$, and $m = 2$ there must be a nodal line at $x = a/2$ [see Fig. 8.20 and Eq. (8.111)] and we say that the plate has two buckles. The curved lines in the diagram may be thought of as contour lines of the buckled shape, the full lines indicating an upward deflection and the dashed lines indicating a downward deflection. For $m = 3$ there will be three such buckles, and so on. Finally, notice from Fig. 8.19 that when a/b is an integer, then $m = a/b$ to

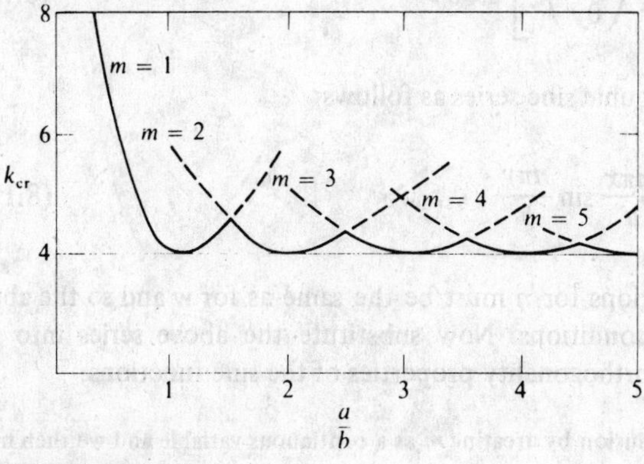

FIGURE 8.19
Plots of k_{cr} for different values of m.

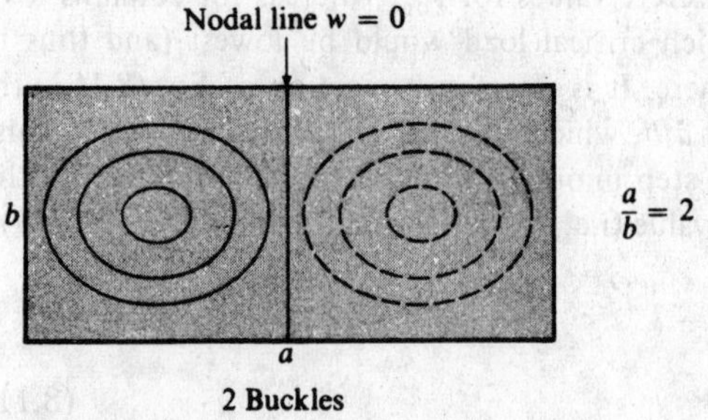

FIGURE 8.20
Plate with two buckles.

achieve the minimum value of k_{cr}.† For such cases the length of the plate (x direction) is divided into m buckles, i.e., m half-waves of length b.

8.15 THE RECTANGULAR PLATE VIA THE ENERGY METHOD

We now consider again the same simply supported rectangular plate examined in the previous section. We will be concerned with the positive definiteness of the *second variation of the total potential energy* $\delta^{(2)}\pi$ in this method.

As shown in Section 8.7 (dealing with a single-parameter approach), the second variation of a functional of η may be considered to have the same form as the functional itself, and we may state for $\delta^{(2)}\pi$:

$$\delta^{(2)}\pi = \int_0^a \int_0^b D\left\{(\nabla^2\eta)^2 + 2(1-\nu)\left[\left(\frac{\partial^2\eta}{\partial x\,\partial y}\right)^2 - \frac{\partial^2\eta}{\partial x^2}\frac{\partial^2\eta}{\partial y^2}\right]\right\} dx\,dy - \int_0^a \int_0^b \left[\bar{N}_x\left(\frac{\partial\eta}{\partial y}\right)^2\right] dx\,dy \tag{8.116}$$

We may now represent η as a double sine series as follows:

$$\eta = \sum_{n=1}^{\infty}\sum_{m=1}^{\infty} A_{mn} \sin\frac{m\pi x}{a}\sin\frac{n\pi y}{b} \tag{8.117}$$

Note that the boundary conditions for η must be the same as for w and so the above series satisfies the boundary conditions. Now substitute the above series into Eq. (8.116). We get, on noting the orthogonality properties of the sine functions,

†We can reach the same conclusion by treating m as a continuous variable and we then minimize k_{cr} with respect to m. Thus

$$\frac{\partial k_{cr}}{\partial m} = 0$$

This gives us $m = a/b$.

$$\delta^{(2)}\pi = \frac{\pi^2}{4} \sum_{m=1}^{\infty} \sum_{n=1}^{\infty} \left\{ \left[D\pi^2 ab \left(\frac{m^2}{a^2} + \frac{n^2}{b^2} \right)^2 - \frac{b\bar{N}_x}{a} m^2 \right] A_{mn}^2 \right\} \tag{8.118}$$

Since the constants A_{mn} are independent of each other, it should be clear that in order for $\delta^{(2)}\pi$ to be positive definite *each coefficient of the A_{mn} must be positive.* Accordingly, $\delta^{(2)}\pi$ is no longer positive definite, *when any coefficient becomes zero.* This leads us to the same result arrived at in Eq. (8.112).

8.16 THE CIRCULAR PLATE VIA THE ENERGY METHOD

We now consider the axisymmetric buckling of a circular plate (see Fig. 8.21). A uniform loading intensity $\bar{N}_\nu = N$ is applied normal to the periphery, as shown in the diagram.

We may express the total potential energy with the aid of Eq. (6.114) as follows:

$$\pi = U + V = \int_0^{2\pi} \int_0^a \frac{D}{2} \left[\left(\frac{\partial^2 w}{\partial r^2} + \frac{1}{r} \frac{\partial w}{\partial r} \right)^2 - 2(1-\nu) \frac{\partial^2 w}{\partial r^2} \frac{1}{r} \frac{\partial w}{\partial r} \right] dr\, r\, d\theta$$

$$+ \int_0^{2\pi} N(u_r r)_{r=a}\, d\theta \tag{8.119}$$

On integrating with respect to θ, we may reformulate this expression as

$$U + V = \pi D \int_0^a \left[r \frac{\partial^2 w}{\partial r^2} + \frac{1}{r} \left(\frac{\partial w}{\partial r} \right)^2 + 2\nu \frac{\partial w}{\partial r} \frac{\partial^2 w}{\partial r^2} \right] dr + 2\pi D\alpha^2 (ru_r)_{r=a} \tag{8.120}$$

where $\alpha^2 = N/D$. We now examine the expression $(u_r r)_{r=a}$. This may be written as

$$(ru_r)_{r=a} = \int_0^a \frac{d}{dr}(ru_r)\, dr = \int_0^a r \frac{du_r}{dr}\, dr + \int_0^a u_r\, dr$$

$$= \int_0^a r \frac{du_r}{dr}\, dr + \int_0^a r \frac{u_r}{r}\, dr \tag{8.121}$$

Note next that the in-plane strain-displacement relations for cylindrical coordinates comparable to the relations used here for rectangular coordinates are

$$\epsilon_{rr} = \frac{du_r}{dr} + \frac{1}{2} \left(\frac{dw}{dr} \right)^2$$
$$\epsilon_{\theta\theta} = \frac{u_r}{r} \tag{8.122}$$

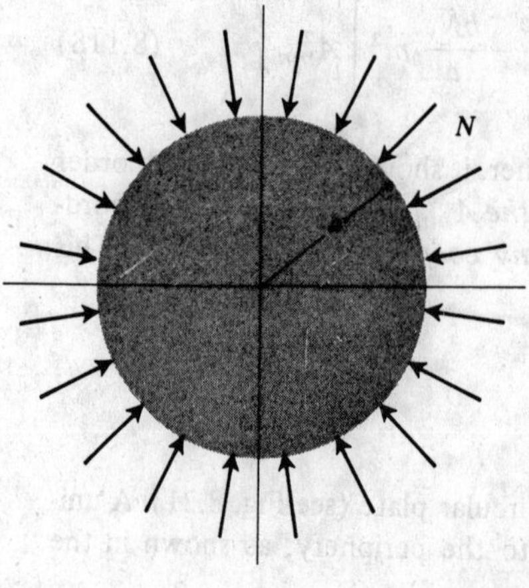

FIGURE 8.21
Circular plate.

If we set the in-plane strains equal to zero, we see that

$$\frac{du_r}{dr} = -\frac{1}{2}\left(\frac{dw}{dr}\right)^2$$
$$\frac{u_r}{r} = 0 \tag{8.123}$$

Substituting these results into Eq. (8.121), we get

$$(ru_r)_{r=a} = -\frac{1}{2}\int_0^a r\left(\frac{dw}{dr}\right)^2 dr$$

The total potential energy then becomes

$$U + V = \pi D \int_0^a \left(rw_{rr}^2 + \frac{1}{r}w_r^2 + 2\nu w_r w_{rr} - \alpha^2 r w_r^2\right) dr$$

The second variation then can be given as

$$\delta^{(2)}\pi = 2\pi D \int_0^a \left(r\eta_{rr}^2 + \frac{1}{r}\eta_r^2 + 2\nu\eta_r\eta_{rr} - \alpha^2 r\eta_r^2\right) dr$$

Employing the *Trefftz criterion*, we extremize the above functional. The Euler-Lagrange equation for this case will be established with the aid of Eq. (2.62). Thus

$$-\frac{d^2}{dr^2}\frac{\partial F}{\partial \eta_{rr}} + \frac{d}{dr}\frac{\partial F}{\partial \eta_r} = 0 \tag{8.124}$$

where

$$F = r\eta_{rr}^2 + \frac{1}{r}\eta_r^2 + 2\nu\eta_r\eta_{rr} - \alpha^2 r\eta_r^2 \tag{8.125}$$

We may perform one quadrature on Eq. (8.124) immediately to get

$$-\frac{d}{dr}\frac{\partial F}{\partial \eta_{rr}} + \frac{\partial F}{\partial \eta_r} = C_1 \tag{8.126}$$

Now substituting for F from Eq. (8.125), we get for the above equation:

$$r\eta_{rrr} + \eta_{rr} + \left(\alpha^2 r - \frac{1}{r}\right)\eta_r = C_1 \tag{8.127}$$

The boundary conditions on η may be deduced from Eq. (2.63) to be, at $r = 0$ and $r = a$,

$$\frac{\partial F}{\partial \eta_{rr}} = 0 \qquad \text{OR} \qquad \eta_r \ \text{ IS SPECIFIED} \tag{8.128a}$$

$$-\frac{d}{dr}\left(\frac{\partial F}{\partial \eta_{rr}}\right) + \frac{\partial F}{\partial \eta_r} = 0 \qquad \text{OR} \qquad \eta \ \text{ IS SPECIFIED} \tag{8.128b}$$

The natural boundary condition of Eq. (8.128*b*) is, using Eq. (8.125) for F

$$r\eta_{rrr} + \eta_{rr} + \left(\alpha^2 r - \frac{1}{r}\right)\eta_r = 0$$

at the endpoints. We see immediately from Eq. (8.127) that the constant C_1 must be zero as a result of the above boundary condition applied at $r = a$. Now consider the resulting differential Eq. (8.127) further. Denoting η_r as G and multiplying through by r, we find

$$r^2 G_{rr} + rG_r + (\alpha^2 r^2 - 1)G = 0 \tag{8.129}$$

This is Bessel's equation of the first order. The solution for this equation which is regular at the origin is

$$G = C_2 J_1(\alpha r) \tag{8.130}$$

Since $G(=\eta_r) = 0$ at $r = 0$, we see that the kinematic boundary condition, Eq. (8.128*a*), is now satisfied at $r = 0$.† To get η itself, we now consider the remaining boundary conditions.

†This results from axial symmetry and the exclusion of discontinuities of slope.

For a *clamped plate* we know that $\eta = \eta_r = 0$ at $r = a$. We may readily ensure that $\eta = 0$ at $r = a$ by choosing the integration constant properly when we integrate Eq. (8.129). As for $\eta_r = 0$ at $r = a$, we require that

$$G(a) = J_1(\alpha a) = 0$$

Choosing the first positive root of J_1, we find

$$\alpha a = 3.832$$

Accordingly, the buckling load N_{cr} is, on recalling from Eq. (8.120) that $\alpha^2 = N/D$,

$$N_{cr} = D\alpha^2 = \frac{14.65D}{a^2} \tag{8.131}$$

Now consider the *simply supported* plate. The condition that $\eta = 0$ at $r = a$ can again be satisfied by choosing the integration constant properly on integrating Eq. (8.129) for η. The other condition that must be satisfied is the natural boundary condition, Eq. (8.128*a*). That is, η must be chosen such that

$$\frac{\partial F}{\partial \eta_{rr}} = 0 \qquad \text{at } r = a$$

Substituting for F in this equation, we get

$$2r\eta_{rr} + 2\nu\eta_r = 0 \qquad \text{at } r = a$$

Substituting for η_r $(= G)$ using Eq. (8.130),

$$r\frac{dJ_1(\alpha r)}{dr} + \nu J_1(\alpha r) = 0 \qquad \text{at } r = a$$

Differentiating the Bessel function† yields at $r = a$:

$$\alpha a J_0(\alpha a) - J_1(\alpha a) + \nu J_1(\alpha a) = 0$$

so that

$$J_1(\alpha a)(1 - \nu) = \alpha a J_0(\alpha a)$$

with $\nu = 0.3$, the smallest root of the above equation is $\alpha a = 2.049$. Hence, the buckling load in this situation is

$$N_{cr} = 4.20\,\frac{D}{a^2} \tag{8.132}$$

†See any standard advanced mathematics text such as Wylie (1951, chapter 8) for a discussion of Bessel functions, including differentiation properties.

Part E
Approximation Methods

8.17 COMMENT

We shall now consider approximate solutions for buckling loads and buckling modes via variational methods. Primarily, we shall extend the techniques formulated in Chapter 7 for finding approximate natural frequencies (eigenvalues for the vibration problem) to that of finding approximate critical loads (eigenvalues of the elastic stability problem). We shall first establish the Rayleigh quotient for columns and plates. With the background material of Chapter 7 as a basis, we shall then employ the Rayleigh and Rayleigh-Ritz methods of approximation. After this we shall consider the Kantorovich method.

8.18 THE RAYLEIGH QUOTIENT FOR BEAM-COLUMNS

We shall first formulate the Rayleigh quotient for the column, following procedures analogous to those set forth in Part C of Chapter 7. Accordingly, we begin with the differential equation [Eq. (8.53)] for the column:

$$\frac{d^2}{dx^2}\left(EI\frac{d^2w}{dx^2}\right) = -P\frac{d^2w}{dx^2} \tag{8.133}$$

Using the form for a differential equation given by Eq. (7.144), namely

$$L(w) = \omega^2 M(w) \tag{8.134}$$

we find that for the column

$$\begin{aligned} L &\equiv \frac{d^2}{dx^2}EI\frac{d^2}{dx^2} \\ M &\equiv -\frac{d^2}{dx^2} \\ \omega^2 &\equiv P \end{aligned} \tag{8.135}$$

We have already shown in Section 3.11 that $(d^2/dx^2)EI(d^2/dx^2)$ must be a self-adjoint, positive definite operator for the usual end supports of beams (and hence columns), and we leave it as an exercise to show that $-d^2/dx^2$ is also a self-adjoint, positive definite operator. We can then conclude (see Section 7.13) that the eigenfunctions of the boundary-value problem involving Eq. (8.133) and the usual end conditions must be orthogonal. The Rayleigh quotient R for a problem having *constant bending stiffness EI* is then given as

$$R = \frac{\int_0^L wL(w)\,dx}{\int_0^L wM(w)\,dx} = -\frac{EI\int_0^L w(d^4w/dx^4)\,dx}{\int_0^L w(d^2w/dx^2)\,dx} \tag{8.136}$$

We may rewrite the Rayleigh quotient in a "better" form if we integrate the expression in the denominator by parts. Thus:

$$-\int_0^L w \frac{d^2 w}{dx^2}\, dx = \int_0^L \left(\frac{dw}{dx}\right)^2 dx - w\frac{dw}{dx}\bigg|_0^L \tag{8.137}$$

The last expression vanishes for fixed or simple supports at the ends of the columns. Also, we can integrate the numerator of Eq. (8.136) by parts twice. Thus:

$$\int_0^L w\frac{d^4 w}{dx^4}\, dx = \int_0^L \left(\frac{d^2 w}{dx^2}\right)^2 dx + w\frac{d^3 w}{dx^3}\bigg|_0^L - \frac{dw}{dx}\frac{d^2 w}{dx^2}\bigg|_0^L \tag{8.138}$$

Again, for fixed or simply supported end conditions, the last two expressions on the right side of the above equation vanish. We thus have, on using Eqs. (8.137) and (8.138) to replace the denominator and numerator respectively, for the Rayleigh quotient of Eq. (8.136),

$$R = \frac{EI\displaystyle\int_0^L (d^2 w/dx^2)^2\, dx}{\displaystyle\int_0^L (dw/dx)^2\, dx} \tag{8.139}$$

We can then say that P is found from

$$P = \frac{EI\displaystyle\int_0^L (d^2 w/dx^2)^2\, dx}{\displaystyle\int_0^L (dw/dx)^2\, dx} \tag{8.140}$$

8.19 THE RAYLEIGH AND RAYLEIGH–RITZ METHODS APPLIED TO COLUMNS

In accordance with Section 7.15 we may find approximations from above of the buckling load P_{cr} by using the Rayleigh method on the Rayleigh quotient functional. As a first step we illustrate the use of the Rayleigh method for the case of the column clamped at $x = 0$ and pinned at $x = L$.

We choose as an approximate function

$$w_{app} = a_1\left[\left(\frac{x}{L}\right)^2 - \left(\frac{x}{L}\right)^3\right] \tag{8.141}$$

Note that this function satisfies the boundary conditions $w(0) = w'(0) = w(L) = 0$ *but not the natural boundary condition* $w''(L) = 0$. Substituting the above approximation into Eq. (8.140), we get

$$P_1 = \frac{EI\,4a^2/L^3}{0.133a^2/L} = 30\,\frac{EI}{L^2} \tag{8.142}$$

This exact buckling load is $20.19\,EI/L^2$ and we have relatively poor agreement. We can improve the accuracy of the result by choosing a function that satisfies the complete set of boundary conditions of the problem. (Recall that we reached the same conclusion in Chapter 4 in our discussion of the Ritz method applied to beams.) We now choose for w_1 the function:

$$w_1 = a_1\left[\left(\frac{x}{L}\right)^2 - \frac{5}{3}\left(\frac{x}{L}\right)^3 + \frac{2}{3}\left(\frac{x}{L}\right)^4\right] \tag{8.143}$$

The complete set of boundary conditions [including $w''(L) = 0$] is now satisfied. Substituting the above function into Eq. (8.140), we find

$$P_{cr} = 21.0\,\frac{EI}{L^2}$$

The agreement is now good.

Next we employ the Rayleigh-Ritz method to find a particularly good approximation of the buckling load and to find approximations *from above* of critical loads beyond the first. The basis for this procedure has been established in Section 7.6, where, for an approximate eigenfunction w_{app}, we employed a parameter-laden sum of coordinate functions:

$$w_{app} = A_1\phi_1 + A_2\phi_2 + \cdots + A_n\phi_n$$

We thus have for the approximate Rayleigh quotient:

$$R_{app} = \Lambda^2 = \frac{EI\int_0^L \left(\sum\limits_{i=1}^{n} A_i\phi_i''\right)^2 dx}{\int_0^L \left(\sum\limits_{i=1}^{n} A_i\phi_i'\right)^2 dx}$$

We introduce the notation

$$a_{ij} = \int_0^L EI\phi_i''\phi_j''\,dx$$

$$b_{ij} = \int_0^L \phi_i'\phi_j'\,dx \tag{8.144}$$

so that Λ^2 becomes

$$\Lambda^2 = \frac{\sum\limits_{i=1}^{n}\sum\limits_{j=1}^{n} a_{ij}A_iA_j}{\sum\limits_{i=1}^{n}\sum\limits_{j=1}^{n} b_{ij}A_iA_j}$$

In extremizing Λ^2 with respect to the constants A_i, we have shown in Section 7.6 that we reach the following requirement:

$$|a_{ij} - \Lambda^2 b_{ij}| = 0 \tag{8.145}$$

The roots $\Lambda_1^2, \ldots, \Lambda_n^2$ are the approximations from above of the first n critical loads.

We illustrate this calculation for a simply supported column by considering for w_n the functions

$$\begin{aligned}
w_1 &= A_1(\cos\xi - \cos 3\xi) \\
w_2 &= A_1(\cos\xi - \cos 3\xi) + A_2(\cos 3\xi - \cos 5\xi) \\
w_3 &= A_1(\cos\xi - \cos 3\xi) + A_2(\cos 3\xi - \cos 5\xi) + A_3(\cos 5\xi - \cos 7\xi) \\
&\vdots \\
w_n &\doteq A_1(\cos\xi - \cos 3\xi) + \cdots + A_n[\cos(2n-1)\xi - \cos(2n+1)\xi]
\end{aligned} \tag{8.146}$$

where $\xi = \pi x/2L$. Note that these functions satisfy only the kinematic boundary conditions $w = 0$ at the ends. Substituting the above functions into Eq. (8.144) and then using these results in Eq. (8.145), we get the results given in Table 8.1. Note that as n goes up, the approximation for the lower critical values improves. Note also that *the approximations are from above.*

We now turn our attention to plates, in particular rectangular plates.

8.20 RAYLEIGH QUOTIENT FOR RECTANGULAR PLATES

The differential equation for buckling of a plate was shown to be (Section 8.13):

$$D\nabla^4 w = -\bar{N}_x \frac{\partial^2 w}{\partial x^2} - 2\bar{N}_{xy}\frac{\partial^2 w}{\partial x\,\partial y} - \bar{N}_y \frac{\partial^2 w}{\partial y^2} \tag{8.147}$$

where the barred quantities are the edge loads, which ultimately cause buckling. If these loads are developed so as to have at all times *fixed ratios between each other*, and if the loads are constant on each edge, we can give the above equation in the following form:

$$D\nabla^4 w = -P^*\left(\alpha\frac{\partial^2 w}{\partial x^2} + 2\beta\frac{\partial^2 w}{\partial x\,\partial y} + \gamma\frac{\partial^2 w}{\partial y^2}\right) \tag{8.148}$$

TABLE 8.1
Critical Loads via Rayleigh-Ritz

n	P_1L^2/EI	P_2L^2/EI	P_3L^2/EI
1	20.22		
2	20.202	59.726	
3	20.195	59.695	118.95
Exact	20.19	59.67	118.90

where P^* will be called the loading factor, a *variable* parameter, and α, β, and γ are constants. Then we can say

$$L \equiv D\nabla^4$$
$$M \equiv -\left(\alpha\frac{\partial^2}{\partial x^2} + 2\beta\frac{\partial^2}{\partial x\,\partial y} + \gamma\frac{\partial^2}{\partial y^2}\right) \tag{8.149}$$
$$\omega^2 \equiv P^*$$

We may then state that, as a result of our discussion in Section 7.14,

$$P^* = \frac{D\iint w\nabla^4 w\,dx\,dy}{-\iint(\alpha\,\partial^2 w/\partial x^2 + 2\beta\,\partial^2 w/\partial x\,\partial y + \gamma\,\partial^2 w/\partial y^2)w\,dx\,dy} \tag{8.150}$$

We shall now consider the use of Rayleigh's method for solving the buckling problem of a simply supported plate, $a \times b$, in uniaxial compression (see Fig. 8.18). The Rayleigh quotient for this case may be given as, noting that $P^* = \bar{N}$, $\alpha = 1$, and $\beta = \gamma = 0$,

$$\bar{N} = \frac{D\iint w\nabla^4 w\,dx\,dy}{-\iint w(\partial^2 w/\partial x^2)\,dx\,dy}$$

By using Green's theorem to integrate both integrals by parts in the above equation (see Problem 6.13) and considering the boundary conditions, we find that

$$\bar{N} = \frac{D\iint(\nabla^2 w)^2\,dx\,dy}{\iint(\partial w/\partial x)^2\,dx\,dy} \tag{8.151}$$

Consider the case where $a/b = 1$, i.e., the square plate. We shall use the following function as an approximate buckling mode:

$$w = C_1 xy(x-a)(y-b)$$

so the *kinematic boundary condition* $w = 0$ is satisfied. Substituting into Eq. (8.151), we have for $\bar{N}_{cr}$

$$\bar{N}_{cr} = \frac{44.0D}{a^2}$$

The exact value for $\bar{N}_{cr}$ developed in Section 8.14 is

$$(\bar{N}_{cr})_{ex} = \frac{39.5D}{a^2}$$

The chosen function does not satisfy the natural boundary conditions, and an error of 11% results.

8.21 THE KANTOROVICH METHOD

We next consider a rectangular plate of dimensions $a \times b \times h$, clamped on all edges, and loaded by a uniform force distribution N normal to the edges (see Fig. 8.22). For the study of the elastic stability of such a problem we may take P^* in Eq. (8.148) as N with $\alpha = \gamma = 1$ and $\beta = 0$. We then have for the appropriate differential equation the following result, with x and y as the independent variables,

$$\nabla^4 w + \lambda^2 \nabla^2 w = 0 \tag{8.152}$$

where $\lambda^2 = N/D$. The boundary conditions are easily seen to be

$$w = \frac{\partial w}{\partial x} = 0 \quad \text{at } x = \pm a \tag{8.153a}$$

$$w = \frac{\partial w}{\partial y} = 0 \quad \text{at } y = \pm b \tag{8.153b}$$

As pointed out in Section 6.9 on the static deflection of a rectangular plate, the *Kantorovich method is equivalent to rendering the Galerkin integral equal to zero.* Thus for a one-term approximation with $g_0(y)$ as a known function satisfying boundary conditions (8.153*b*), and $f_1(x)$ as the unknown function, we have

$$w_{1,0}(x,y) = f_1(x) g_0(y) \tag{8.154}$$

We then have, on setting the Galerkin integral equal to zero,

$$\int_{-a}^{a} \int_{-b}^{b} (\nabla^4 w_{1,0} + \lambda^2 \nabla^2 w_{1,0}) g_0 \, dx \, dy = 0 \tag{8.155}$$

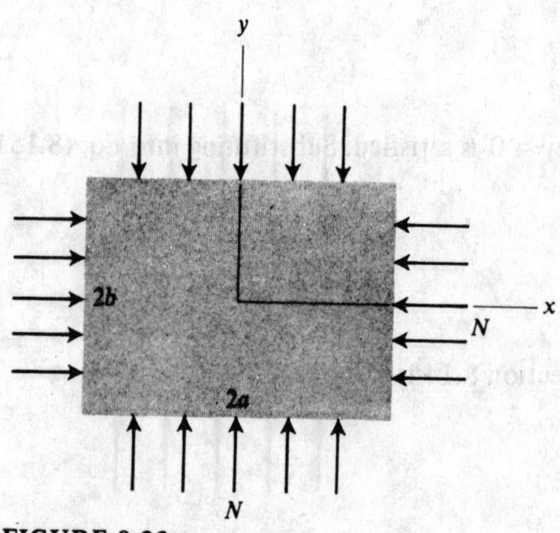

FIGURE 8.22
Rectangular plate.

As a choice for $w_{1,0}$ we take

$$w_{1,0} = f_1(x)(y^2 - b^2)^2 \tag{8.156}$$

Substituting Eq. (8.156) into the Galerkin integral, we get, after integrating and dividing through by b^5,

$$\left(b^4 \frac{256}{315}\right)\frac{d^4 f_1}{dx^4} + b^2\left[(\lambda b)^2 \frac{256}{315} - \frac{512}{105}\right]\frac{d^2 f_1}{dx^2} - \left[(\lambda b)^2 \frac{256}{105} - \frac{128}{5}\right] f_1 = 0 \tag{8.157}$$

We have here an ordinary fourth-order differential equation with constant coefficients. The procedure is to consider a trial solution e^{mx} to form a characteristic equation for m. By substituting, we get

$$b^4 m^4 + b^2[(\lambda b)^2 - 6]m^2 - \left[3(\lambda b)^2 - \frac{63}{2}\right] = 0$$

Since the magnitude of λb is not known, the roots m^2 may be real, imaginary, or complex. Consider the roots m^2, using the quadratic formula. We get

$$m^2 = \frac{-[(\lambda b)^2 - 6] \pm \sqrt{[(\lambda b)^2 - 6]^2 + [3(\lambda b)^2 - 63/2]4}}{2b^2} \tag{8.158}$$

If $(\lambda b)^2 = 21/2$, the second bracketed expression in the above root is zero and the magnitude of the square root equals $(\lambda b)^2 - 6$. We then have as roots for m^2

$$m^2 = 0 \qquad m^2 = -\frac{(\lambda b)^2 - 6}{b^2}$$

The values of m are then either zero or pure imaginary numbers. If now

$$(\lambda b)^2 > \frac{21}{2} \tag{8.159}$$

the square root in Eq. (8.158) exceeds the first term $[(\lambda b)^2 - 6]$ in the numerator and we have for possible roots m two real numbers and two imaginary numbers. For cases where $(\lambda b)^2 < 21/2$ the magnitude of the root, if it is real, is less than $(\lambda b)^2 - 6$, and so we get four pure imaginary values of m. Also, there is the possibility of getting four complex values of m if the root is imaginary. Now the only kinds of roots that yield critical values of N are those corresponding to conditions (8.159)–that is, two real roots and two imaginary roots for m. The roots for m will accordingly be denoted as follows:

$$m_1 = m$$

$$m_2 = -m$$

$$m_3 = i\Gamma$$

$$m_4 = -i\Gamma$$

where we have from Eq. (8.158)

$$m = \frac{1}{b}\left(\left\{\left[\frac{(\lambda b)^2}{2} - 3\right]^2 + \left[3(\lambda b)^2 - \frac{63}{2}\right]\right\}^{1/2} - \left[\frac{(\lambda b)^2}{2} - 3\right]\right)^{1/2}$$
$$\Gamma = \frac{1}{b}\left(\left\{\left[\frac{(\lambda b)^2}{2} - 3\right]^2 + \left[3(\lambda b)^2 - \frac{63}{2}\right]\right\}^{1/2} + \left[\frac{(\lambda b)^2}{2} - 3\right]\right)^{1/2} \tag{8.160}$$

We use only the positive roots in the above equation. We thus have a solution to the differential Eq. (8.157):

$$f_1(x) = A_1 \sinh mx + A_2 \cosh mx + A_3 \sin \Gamma x + A_4 \cos \Gamma x$$

To nondimensionalize the above result, replace mx by $am(x/a)$ and denote am as ρ_1; replace Γx by $a\Gamma(x/a)$ and denote $a\Gamma$ simply as κ_1. Thus we have for $f_1(x)$

$$f_1(x) = A_1 \sinh \rho_1 \frac{x}{a} + A_2 \cosh \rho_1 \frac{x}{a} + A_3 \sin \kappa_1 \frac{x}{a} + A_4 \cos \kappa_1 \frac{x}{a}$$

where, on considering Eq. (8.160), we have

$$\left\{\begin{matrix}\rho_1\\ \kappa_1\end{matrix}\right\} = \frac{a}{b}\left(\left\{\left[\frac{(\lambda b)^2}{2} - 3\right]^2 + \left[3(\lambda b)^2 - \frac{63}{2}\right]\right\}^{1/2} \mp \left[\frac{(\lambda b)^2}{2} - 3\right]\right)^{1/2} \tag{8.161}$$

For critical values of N corresponding to mode shapes that are symmetric with respect to both the x and y axes, we set A_1 and A_3 equal to zero. Thus we have

$$f_1(x) = A_2 \cosh \rho_1 \frac{x}{a} + A_4 \cos \kappa_1 \frac{x}{a} \tag{8.162}$$

The boundary conditions in the x direction now require that at $x = \pm a$

$$f_1 = 0 \tag{8.163a}$$
$$\frac{df_1}{dx} = 0 \tag{8.163b}$$

We thus obtain

$$A_2 \cosh \rho_1 + A_4 \cos \kappa_1 = 0 \tag{8.164a}$$
$$A_2 \rho_1 \sinh \rho_1 - A_4 \kappa_1 \sin \kappa_1 = 0 \tag{8.164b}$$

For a nontrivial solution the determinant of the constants must be zero. This results in the following requirement:

$$\kappa_1 \cosh \rho_1 \sin \kappa_1 = -\rho_1 \sinh \rho_1 \cos \kappa_1$$

or

$$\kappa_1 \tan \kappa_1 = -\rho_1 \tanh \rho_1 \tag{8.165}$$

For the case of a square plate ($a = b$), we find from Eqs. (8.161) the values of $(\lambda b)^2$ that give values of κ_1 and ρ_1 that permit satisfaction of the above equation. With the aid of a computer or by graphic representation, we may show that

$$(\lambda b)^2 = 13.29, 40.00, 88.70, \ldots$$

The buckling load N_{cr} then becomes

$$N_{cr} = \lambda^2 D = \frac{13.29}{b^2} D \tag{8.166}$$

As for obtaining an approximation of the ***buckling mode shape*** corresponding to this result, note that the approximate values of ρ_1 and κ_1 for $(\lambda b)^2 = 13.29$ are

$$\begin{aligned} \rho_1 &= 1.0046 \\ \kappa_1 &= 2.8814 \end{aligned} \tag{8.167}$$

Now by substituting these values into Eq. (8.164*a*), we can find A_4 in terms of A_2. We can then say for $w_{1,0}$ that

$$w_{1,0} = A_2 \left[\cos \kappa_1 \cosh \left(\rho_1 \frac{x}{a} \right) - \cosh \rho_1 \cos \left(\kappa_1 \frac{x}{a} \right) \right] (y^2 - b^2)^2$$

8.22 CLOSURE

We thus come to the end of Part II of the text wherein we used the variational methods of Part I to derive properly posed boundary-value problems for important areas of structural mechanics. This includes the appropriate differential equation along with kinematic and natural boundary conditions. Also, we presented a number of approximation techniques stemming from variational considerations. These techniques gave us approximate solutions covering the entire spatial domain of the problem.

In Part III of the text we will again use variational considerations to present powerful approximating techniques giving us discretized approximate solutions to important classes of boundary-value problems. This is the method of finite elements. We shall consider in Part III the various structural problems studied in Part II, and in the last chapter we shall extend the methodology to other areas of study. Modern design of complex structures makes much use of the finite element procedure, as do other areas of study of fluid mechanics, heat transfer, electromagnetic theory, and so on, where difficult problems must be solved. We assume no prior knowledge of finite elements on the part of the reader.

REFERENCES

Abramowitz, M., and Stegun, I. A. (eds.), "Handbook of Mathematical Functions," National Bureau of Standards, U.S. Department of Commerce, Applied Mathematics Series, vol. 55, 1965.

Dym, C. L., "Stability Theory and Its Applications to Structural Mechanics," Noordhoof International, Leyden, The Netherlands, 1974.

LaSalle, J., and Lefschetz, S., "Stability by Liapounov's Direct Method," Academic, New York, 1961.

Mayers, J., "Aircraft and Missile Structures," unpublished lecture notes, Stanford University, Stanford, Calif., 1964.

Trefftz, E., "Zur Theorie der Stabilitat des Elastischen Gleichgewichts," *Z. Angew. Math. Mech.*, vol. 13, pp. 160–165, 1933.

Wylie, C. R., "Advanced Engineering Mathematics," McGraw-Hill, New York, 1951.

READING

Bleich, F., "Buckling Strength of Metal Structures," McGraw-Hill, New York, 1952.

Bolotin, V. V., "Dynamic Stability of Elastic Systems," Holden-Day, San Francisco, 1964.

Bolotin, V. V., "Nonconservative Problems of the Theory of Elastic Systems," Pergamon, New York, 1963.

Cox, H. L., "The Buckling of Plates and Shells," Pergamon, New York, 1961.

Dym, C. L., "Stability Theory and Its Applications to Structural Mechanics," Noordhoff-International, Leyden, The Netherlands, 1974.

Funge, Y. C., "Foundations of Solid Mechanics," Prentice-Hall, Englewood Cliffs, N.J., 1965.

Gerard, G., "Introduction to Structural Stability Theory," McGraw-Hill, New York, 1962.

Herrmann, G. (ed.), "Dynamic Stability of Structures," Pergamon, New York, 1963.

Koiter, W. T., "The Stability of Elastic Equilibrium," doctoral dissertation, Technische Hooge School, Delft, 1945; Technical Rept. AFFDL TR-70-25, Wright-Patterson Air Force Base, February 1970.

Langhaar, H. L., "Energy Methods in Applied Mechanics," Wiley, New York, 1962.

Libove, C., Elastic Stability, p. 44-1 in Flugge, W., "Handbook of Engineering Mechanics," McGraw-Hill, 1962.

Mayers, J., "Aircraft and Missile Structures," unpublished lecture notes, Stanford University, Stanford, Calif., 1964.

Timoshenko, S. P., and Gere, J. M., "Theory of Elastic Stability," McGraw-Hill, New York, 1961.

Ziegler, H., "Principles of Structural Stability," Blaisdell, Waltham, Mass., 1968.

PROBLEMS

8.1 Show that the buckling load of a column clamped at both ends is $P = 4P_E = 4\pi^2 EI/L^2$.

8.2 Obtain an approximate (Rayleigh-Ritz) solution for the buckling load of a clamped-clamped column.

8.3 Show that the buckling load of a column clamped at $x = 0$ and free at $x = L$ is $P = P_E/4 = \pi^2 EI/4L^2$.

8.4 Obtain an approximate (Rayleigh-Ritz) solution for the buckling load of a clamped-free column.

8.5 Consider the clamped-free column loaded as shown in Fig. 8.23. Here we take the load P not as the usual axial force but directed along a line through $x = 0$. Explain why the buckling load of this problem is equal to that of a pinned-pinned column.

8.6 Consider now a clamped-free column subjected to a load P constrained to remain tangential to the free end of the column as it deforms (Fig. 8.24). You may verify that this is a nonconservative problem, as the loading is path-dependent. What are the boundary conditions

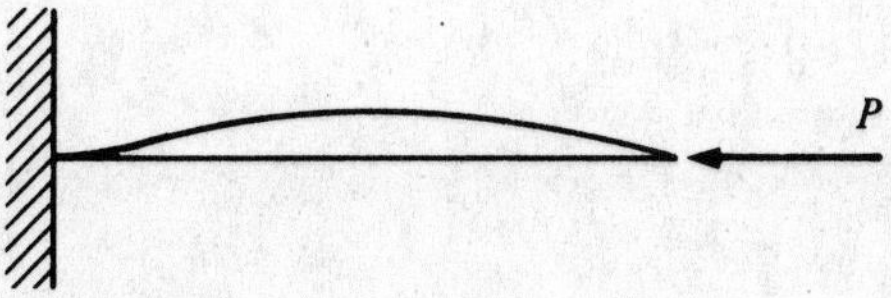

FIGURE 8.23

at $x = L$? Show that the solution for the Euler equation subject to the boundary conditions for this problem leads to the characteristic equation

$$\begin{vmatrix} k & 0 & 1 \\ \sin kl & \cos kl & 0 \\ \cos kl & -\sin kl & 0 \end{vmatrix} = 0$$

What does this imply about the buckling load? Why do we obtain this result?

8.7 Solve Problem 8.6 by the kinetic method. Show that this leads to the characteristic equation

$$(\lambda_1^4 + \lambda_2^4) + 2\lambda_1^2\lambda_2^2 \cos \lambda_1 L \cosh \lambda_2 L + \lambda_1\lambda_2(\lambda_1^2 - \lambda_2^2) \sin \lambda_1 L \sinh \lambda_2 L = 0$$

where

$$\lambda_{1,2}^2 = \pm\frac{k^2}{2} + \sqrt{\left(\frac{k^2}{2}\right)^2 + \omega^2}$$

with

$$k^2 = \frac{P}{EI} \quad \text{and} \quad \omega^2 = \frac{\rho A \Omega^2}{EI}$$

Ω being the actual frequency of the vibrating column. Substituting for the λ's, the characteristic equation is then

$$0 = k^4 + 2\omega^2 + 2\omega^2 \cos \lambda_1 L \cosh \lambda_2 L + \omega k^2 \sin \lambda_1 L \sinh \lambda_2 L$$

The lowest root may be obtained as $P_{cr} = 2\pi^2 EI/L^2$. Thus the usefulness of the kinetic method for nonconservative problems has been demonstrated. Also note that for $P \rightarrow 0$, i.e.,

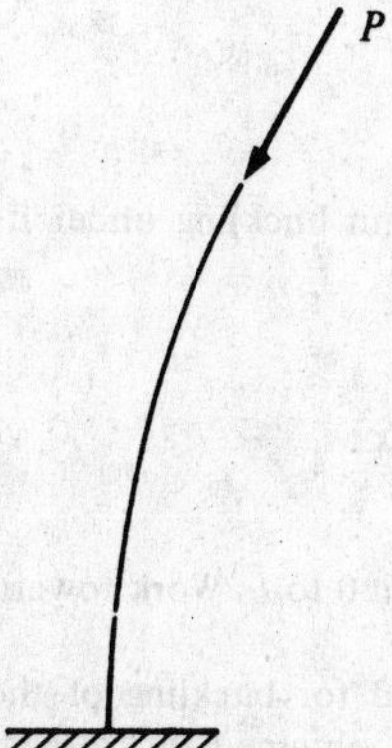

FIGURE 8.24

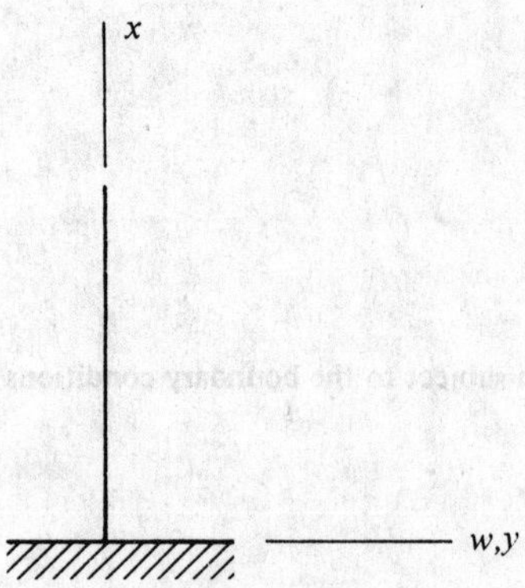

FIGURE 8.25

$k^2 \to 0$, the characteristic equation is that for the frequency of an unloaded clamped-free bar:

$$\cos \sqrt{\omega L} \cosh \sqrt{\omega L} = -1$$

*8.8 Find the buckling load of a clamped-free column (standing vertically!) acted upon by its own weight (see Fig. 8.25). Considering shear force, show that $V \equiv EIw''' = -q(L-x)w'$. Then show that the differential equation can be reduced to

$$EI \frac{d^3 w}{dx^3} = -q(L-x) \frac{dw}{dx} \qquad (a)$$

where q is a uniformly distributed load. The calculation can be simplified by introducing the variables

$$\zeta = \frac{2}{3} \sqrt{\frac{q(L-x)^3}{EI}} \qquad (b)$$

and $u = dw/d\zeta$ so that the governing equation becomes a Bessel equation,

$$\frac{d^2 u}{d\zeta^2} + \frac{1}{\zeta} \frac{du}{d\zeta} + \left(1 - \frac{1}{9\zeta^2}\right) u = 0 \qquad (c)$$

Then using appropriate tables of Bessel functions, and satisfying appropriate boundary conditions, find the critical load

$$(qL)_{cr} = 7.837 \frac{EI}{L^2}$$

8.9 Show that the energy functional for the previous problem (a column buckling under its own weight) is

$$U + V = \frac{EI}{2} \int_0^L \left(\frac{d^2 w}{dx^2}\right)^2 dx - \frac{q}{2} \int_0^L (L-x) \left(\frac{dw}{dx}\right)^2 dx$$

Hint: Multiply Eq. (*a*) of Problem 8.8 by $\delta(w')\, dx$ and integrate from 0 to L. Work toward forming $\delta^{(1)}[\ ] = 0$. Then [ ] is the desired functional.

8.10 Obtain an approximate buckling load by the Rayleigh-Ritz method for buckling of the clamped-free column under its own weight. (The exact answer was obtained in Problem

8.8.) Use as a function the static deflection of a cantilevered beam under a triangular load that is zero at the tip. That is,

$$w = A\left[\left(\frac{x}{L}\right)^5 - 5\left(\frac{x}{L}\right)^4 + 10\left(\frac{x}{L}\right)^3 - 10\left(\frac{x}{L}\right)^2\right]$$

8.11 Obtain the buckling load for a pinned-pinned column supported by a uniform elastic foundation with modulus K. The critical load is given by

$$P_{\text{cr}} = P_E\left(m^2 + \frac{KL^4}{m^2\pi^4 EI}\right)$$

where m is the axial wave number. Show that for $KL^4/\pi^4 EI < 4$ the column buckles with only one wave ($m = 1$) and $P = P_E$; for $KL^4/\pi^4 EI > 4$, show that two waves occur. *Hint*: Set $(P_{\text{cr}})_{m=1} = (P_{\text{cr}})_{m=2}$ and determine $KL^4/\pi^4 EI$ for this condition.

8.12 Use a Galerkin integral to obtain the buckling load for a simply supported rectangular plate (see Fig. 8.26) under loading $\bar{N}x = N_x^0 y/b$, buckling into m waves in the x direction and one wave in the y direction. What is the minimum value of the buckling coefficient

$$k_{x\Delta} = \frac{N_x^0 b^2}{\pi^2 D}$$

8.13 For a square plate, simply supported, extend the above results (Problem 8.12) to get a two-term solution for $m = n = 1$ and $m = 2$, $n = 1$. Does this improve the accuracy of the results or point to another possible buckling load and mode?

8.14 Derive an interaction formula for biaxial buckling under uniform loads N_x^0 and N_y^0 for a rectangular plate that is simply supported. Show that for buckling

$$m^2 k_x + m^2 k_y = \left(m^2 \frac{b}{a} + m^2 \frac{a}{b}\right)^2$$

where k_x $(= N_x^0 b^2/\pi^2 D)$ and k_y $(= N_y^0 a^2/\pi^2 D)$ are buckling coefficients. Plot k_x against k_y for a square plate for $m = 1$, $n = 1$, for $m = 1$, $n = 2$, and for $m = 2$, $n = 1$. For these plots identify line segments for lowest buckling loads. These segments are stability boundaries.

8.15 Compute the shear buckling coefficient $N_{xy}^0 b^2/\pi^2 D = k_{xy}$ for a long, simply supported, rectangular plate. As an approximation use

$$w = w_{m1} \sin\frac{m\pi x}{a} \sin\frac{\pi y}{b} + w_{m2} \cos\frac{m\pi x}{a} \sin\frac{2\pi y}{b}$$

in conjunction with the Galerkin integral approach. Why must such a pattern be chosen instead of the simpler forms we use for compressive loading? Why do we stipulate a long plate? Show that

$$k_{xy} = \frac{3\pi}{16} \frac{\Lambda^4 + 5\Lambda^2 + 4}{\Lambda}$$

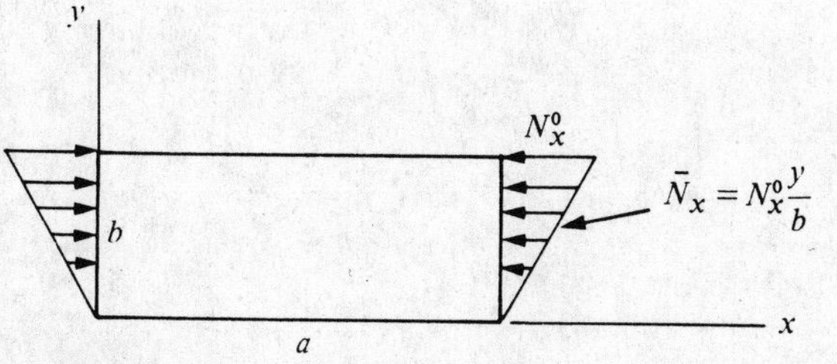

FIGURE 8.26

where $\Lambda = b/(a/m)$ = (plate width)(axial buckle length) is the aspect ratio for a single axial buckle. Find minimum $(k_{xy})_{cr}$. An exact answer for the infinite plate, minimized with respect to an aspect ratio $\Lambda = b/(a/m)$, is $k_{cr} = 5.35$.

8.16 Obtain the buckling coefficient for a rectangular plate, under loading $\bar{N}_x = N_x^0$, that is *clamped* at the loaded edges. The exact answer, for a square plate, is $(k_x)_{cr} = 6.7432$. Note that a function of the form $\sin^2 \pi x/a$ satisfies useful boundary conditions (for this problem) at $x = 0, a$. Do we need to consider the Gaussian curvature contribution in π? Note that $\sin^2 \theta = \frac{1}{2}(1 - \cos 2\theta)$.

III

FINITE ELEMENTS

We now present an introductory study of finite elements. We assume no prior knowledge in this area on the part of the reader. Consequently, the early chapters of Part III are elementary and written with pedagogical considerations uppermost in mind. The consideration of plane trusses in Chapter 9 is our vehicle to introduce the reader to finite element concepts and methodology, beginning where we left off in Chapter 3. Having thus eased into the subject, we then present in Part B of Chapter 9 some useful preliminary generalizations and notation for finite elements by using triangular elements while considering plane stress. The triangular element is chosen here because it is complex enough to bring out certain salient features and simple enough to avoid excessive complications. Those relations reached that are general are boxed in. We thus achieve a modest base to work from as we finish Chapter 9, and in Chapter 10 we immediately apply these results to study beams, where we can most easily apply the theory while dealing with one of the simplest of elements.

In Chapter 11 we are ready for a more exacting study of the fundamentals of finite element methodology. We consider various kinds of elements and different interpolation functions; we delve into convergence criteria and so on. From this wider and higher base, we can then study plane stress, plates, torsion, elastic stability, and dynamics in Chapters 12-16 For the needed functionals we use results of Part II, where we considered all the areas covered

in Part III, thus eliminating the need to seek outside information before embarking on a finite element procedure.

In these undertakings, we have deemed it best to work with rough grids of six or seven elements and to work out details for a single element, and then, finally, to present the global results. In many of the problems, we ask the student to work such problems with a hand calculator, using the computer only to invert a matrix at the final phase of computations. Also, we have presented problems with fine grids for which an entire program must be used. The program may be developed by the student as a project, or he or she may use programs made available to the student, such as SAP IV. In Chapter 17 we then make much use of the quadratic functional material of Chapter 3 to consider the finite element solution of ordinary differential equations, heat conduction, and potential fluid flow–thus opening up our study to new areas. The Galerkin method is also studied so that readers will not be totally dependent on the existence of a quadratic functional for their problems.

9

INTRODUCTION TO FINITE ELEMENTS

Part A Trusses

9.1 FINITE ELEMENTS AND TRUSSES

In Chapter 3 we considered simple plane trusses. Actually, that study can be considered a simple form of the finite element procedure. At this time, we shall reiterate some of the earlier work and expand it so as to give a simple foundation for the study of finite elements.

Accordingly, consider the simple, plane truss in Fig. 9.1. The members of the truss are the *finite elements* of the system and are numbered separately from the joints. We shall call the joints *nodal points.* At each nodal point i (Fig. 9.2), we have indicated two displacement components (u_i, v_i). These are called the *nodal displacement* components for the truss. Corresponding to each pair of nodal displacement components is a set of forces (U_i, V_i) that represent the total force components *on the members at joint i from node i.* A *double* arrow has been used, you will notice, for a nodal displacement component such as u_i and the corresponding nodal force component U_i, and so on.

In Section 3.8 we showed that the forces U_i, V_i are related to the nodal displacement components as follows:†

$$\begin{Bmatrix} U_1 \\ V_1 \\ \vdots \\ U_6 \\ V_6 \end{Bmatrix} = [K] \begin{Bmatrix} u_1 \\ v_1 \\ \vdots \\ u_6 \\ v_6 \end{Bmatrix} \tag{9.1}$$

where $[K]$ is the *stiffness* matrix for the *entire* truss or, as we say in finite elements, $[K]$ is the *global stiffness matrix.* Let us expand the above equation to include the elements of $[K]$, which you will recall from Chapter 3 must be symmetric. Thus we have

†We will use the convention that { } represents a *column* matrix while ⌊ ⌋ represents a *row* matrix.

$$
\begin{Bmatrix} U_1 \\ V_1 \\ U_2 \\ V_2 \\ U_3 \\ V_3 \\ U_4 \\ V_4 \\ U_5 \\ V_5 \\ U_6 \\ V_6 \end{Bmatrix}
=
\begin{bmatrix}
K_{U_1u_1} & K_{U_1v_1} & K_{U_1u_2} & K_{U_1v_2} & K_{U_1u_3} & K_{U_1v_3} & K_{U_1u_4} & K_{U_1v_4} & K_{U_1u_5} & K_{U_1v_5} & K_{U_1u_6} & K_{U_1v_6} \\
K_{V_1u_1} & K_{V_1v_1} & K_{V_1u_2} & K_{V_1v_2} & K_{V_1u_3} & K_{V_1v_3} & K_{V_1u_4} & K_{V_1v_4} & K_{V_1u_5} & K_{V_1v_5} & K_{V_1u_6} & K_{V_1v_6} \\
K_{U_2u_1} & K_{U_2v_1} & K_{U_2u_2} & K_{U_2v_2} & K_{U_2u_3} & K_{U_2v_3} & K_{U_2u_4} & K_{U_2v_4} & K_{U_2u_5} & K_{U_2v_5} & K_{U_2u_6} & K_{U_2v_6} \\
K_{V_2u_1} & K_{V_2v_1} & K_{V_2u_2} & K_{V_2v_2} & K_{V_2u_3} & K_{V_2v_3} & K_{V_2u_4} & K_{V_2v_4} & K_{V_2u_5} & K_{V_2v_5} & K_{V_2u_6} & K_{V_2v_6} \\
K_{U_3u_1} & K_{U_3v_1} & K_{U_3u_2} & K_{U_3v_2} & K_{U_3u_3} & K_{U_3v_3} & K_{U_3u_4} & K_{U_3v_4} & K_{U_3u_5} & K_{U_3v_5} & K_{U_3u_6} & K_{U_3v_6} \\
K_{V_3u_1} & K_{V_3v_1} & K_{V_3u_2} & K_{V_3v_2} & K_{V_3u_3} & K_{V_3v_3} & K_{V_3u_4} & K_{V_3v_4} & K_{V_3u_5} & K_{V_3v_5} & K_{V_3u_6} & K_{V_3v_6} \\
K_{U_4u_1} & K_{U_4v_1} & K_{U_4u_2} & K_{U_4v_2} & K_{U_4u_3} & K_{U_4v_3} & K_{U_4u_4} & K_{U_4v_4} & K_{U_4u_5} & K_{U_4v_5} & K_{U_4u_6} & K_{U_4v_6} \\
K_{V_4u_1} & K_{V_4v_1} & K_{V_4u_2} & K_{V_4v_2} & K_{V_4u_3} & K_{V_4v_3} & K_{V_4u_4} & K_{V_4v_4} & K_{V_4u_5} & K_{V_4v_5} & K_{V_4u_6} & K_{V_4v_6} \\
K_{U_5u_1} & K_{U_5v_1} & K_{U_5u_2} & K_{U_5v_2} & K_{U_5u_3} & K_{U_5v_3} & K_{U_5u_4} & K_{U_5v_4} & K_{U_5u_5} & K_{U_5v_5} & K_{U_5u_6} & K_{U_5v_6} \\
K_{V_5u_1} & K_{V_5v_1} & K_{V_5u_2} & K_{V_5v_2} & K_{V_5u_3} & K_{V_5v_3} & K_{V_5u_4} & K_{V_5v_4} & K_{V_5u_5} & K_{V_5v_5} & K_{V_5u_6} & K_{V_5v_6} \\
K_{U_6u_1} & K_{U_6v_1} & K_{U_6u_2} & K_{U_6v_2} & K_{U_6u_3} & K_{U_6v_3} & K_{U_6u_4} & K_{U_6v_4} & K_{U_6u_5} & K_{U_6v_5} & K_{U_6u_6} & K_{U_6v_6} \\
K_{V_6u_1} & K_{V_6v_1} & K_{V_6u_2} & K_{V_6v_2} & K_{V_6u_3} & K_{V_6v_3} & K_{V_6u_4} & K_{V_6v_4} & K_{V_6u_5} & K_{V_6v_5} & K_{V_6u_6} & K_{V_6v_6}
\end{bmatrix}
\begin{Bmatrix} u_1 \\ v_1 \\ u_2 \\ v_2 \\ u_3 \\ v_3 \\ u_4 \\ v_4 \\ u_5 \\ v_5 \\ u_6 \\ v_6 \end{Bmatrix}
\tag{9.2}
$$

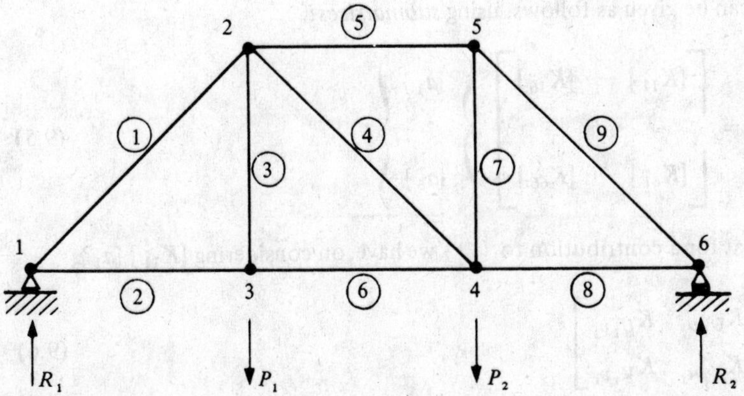

FIGURE 9.1
Truss with members and joints separately labeled.

We see from the above equation that $K_{U_i u_i}$ is the horizontal force component from pin i needed for a unit horizontal deflection of pin i with all other nodal displacement components kept at zero. Similarly, $K_{V_p v_q}$ is the vertical force component from pin p needed for a unit vertical deflection of pin q with all other nodal displacement components held at zero.

To simplify the notation, we now define $\{q_i\}$ as the force vector from node i onto the members of joint i. That is

$$\{q_i\} = \begin{Bmatrix} U_i \\ V_i \end{Bmatrix} \tag{9.3}$$

Also, we define $\{a_i\}$ as the *nodal displacement vector* at node i. That is

$$\{a_i\} = \begin{Bmatrix} u_i \\ v_i \end{Bmatrix} \tag{9.4}$$

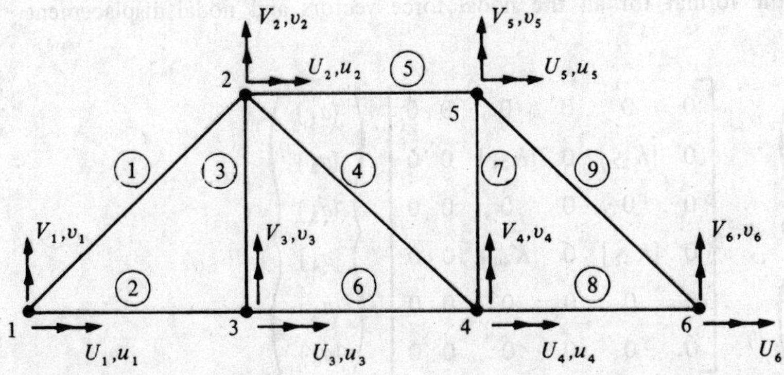

FIGURE 9.2
Truss showing nodal displacement components and forces from joints onto truss.

Then, Eq. (9.2) can be given as follows, using *submatrices*:

$$\left\{\begin{matrix}\{q_1\}\\ \vdots\\ \vdots\\ \{q_6\}\end{matrix}\right\} = \begin{bmatrix}[K_{11}] & \cdots & [K_{16}]\\ \vdots & & \vdots\\ [K_{61}] & \cdots & [K_{66}]\end{bmatrix}\left\{\begin{matrix}\{a_1\}\\ \vdots\\ \vdots\\ \{a_6\}\end{matrix}\right\} \tag{9.5}$$

Clearly, we see that for a contribution to $\{{U_1 \atop V_1}\}$ we have, on considering $[K_{11}]\{a_1\}$,

$$[K_{11}] = \begin{bmatrix} K_{U_1u_1} & K_{U_1v_1}\\ K_{V_1u_1} & K_{V_1v_1}\end{bmatrix} \tag{9.6}$$

Also,

$$[K_{ij}] = \begin{bmatrix} K_{U_iu_j} & K_{U_iv_j}\\ K_{V_iu_j} & K_{V_iv_j}\end{bmatrix} \tag{9.7}$$

Let us next consider an *element* of the truss, say the element 4. Then using Eq. (9.5) applied to this element, we can say for the relation between forces and displacements of nodes 2 and 4:

$$\left\{\begin{matrix}\{q_2\}\\ \{q_4\}\end{matrix}\right\} = \begin{bmatrix}[K_{22}] & [K_{24}]\\ [K_{42}] & [K_{44}]\end{bmatrix}\left\{\begin{matrix}\{a_2\}\\ \{a_4\}\end{matrix}\right\}$$

or

$$\{q\}^{e=4} = [K]^{e=4}\{a\}^{e=4} \tag{9.8}$$

where $[K]^{e=4}$ is the stiffness matrix for element 4. This formulation can be given as follows, using the format for all the nodal force vectors and nodal displacement vectors:

$$\left\{\begin{matrix}\{q_1\}\\ \{q_2\}\\ \{q_3\}\\ \{q_4\}\\ \{q_5\}\\ \{q_6\}\end{matrix}\right\} = \begin{bmatrix} 0 & 0 & 0 & 0 & 0 & 0\\ 0 & [K_{22}] & 0 & [K_{24}] & 0 & 0\\ 0 & 0 & 0 & 0 & 0 & 0\\ 0 & [K_{42}] & 0 & [K_{44}] & 0 & 0\\ 0 & 0 & 0 & 0 & 0 & 0\\ 0 & 0 & 0 & 0 & 0 & 0\end{bmatrix}\left\{\begin{matrix}\{a_1\}\\ \{a_2\}\\ \{a_3\}\\ \{a_4\}\\ \{a_5\}\\ \{a_6\}\end{matrix}\right\} \tag{9.9}$$

In expanded form, we can say for the preceding result:

$$\begin{Bmatrix} U_1 \\ V_1 \\ U_2 \\ V_2 \\ U_3 \\ V_3 \\ U_4 \\ V_4 \\ U_5 \\ V_5 \\ U_6 \\ V_6 \end{Bmatrix} = \begin{bmatrix} 0 & 0 & 0 & 0 & 0 & 0 & 0 & 0 & 0 & 0 & 0 & 0 \\ 0 & 0 & 0 & 0 & 0 & 0 & 0 & 0 & 0 & 0 & 0 & 0 \\ 0 & 0 & K_{U_2u_2} & K_{U_2v_2} & 0 & 0 & K_{U_2u_4} & K_{U_2v_4} & 0 & 0 & 0 & 0 \\ 0 & 0 & K_{V_2u_2} & K_{V_2v_2} & 0 & 0 & K_{V_2u_4} & K_{V_2v_4} & 0 & 0 & 0 & 0 \\ 0 & 0 & 0 & 0 & 0 & 0 & 0 & 0 & 0 & 0 & 0 & 0 \\ 0 & 0 & 0 & 0 & 0 & 0 & 0 & 0 & 0 & 0 & 0 & 0 \\ 0 & 0 & K_{U_4u_2} & K_{U_4v_2} & 0 & 0 & K_{U_4u_4} & K_{U_4v_4} & 0 & 0 & 0 & 0 \\ 0 & 0 & K_{V_4u_2} & K_{V_4v_2} & 0 & 0 & K_{V_4u_4} & K_{V_4v_4} & 0 & 0 & 0 & 0 \\ 0 & 0 & 0 & 0 & 0 & 0 & 0 & 0 & 0 & 0 & 0 & 0 \\ 0 & 0 & 0 & 0 & 0 & 0 & 0 & 0 & 0 & 0 & 0 & 0 \\ 0 & 0 & 0 & 0 & 0 & 0 & 0 & 0 & 0 & 0 & 0 & 0 \\ 0 & 0 & 0 & 0 & 0 & 0 & 0 & 0 & 0 & 0 & 0 & 0 \end{bmatrix} \begin{Bmatrix} u_1 \\ v_1 \\ u_2 \\ v_2 \\ u_3 \\ v_3 \\ u_4 \\ v_4 \\ u_5 \\ v_5 \\ u_6 \\ v_6 \end{Bmatrix} \tag{9.10}$$

By first evaluating each element stiffness matrix and then inserting the results into the above format, we can thereby get the global stiffness matrix for the whole truss.

We shall illustrate in the following section how we can easily formulate the stiffness matrix for any element of the truss.

9.2 STIFFNESS MATRIX FOR AN ELEMENT

We now find the stiffness matrix for an element e of the truss having nodes i and j as shown in Fig. 9.3. To get $[K]^e$, we form the following matrix equation for member ij.

$$\begin{Bmatrix} U_i \\ V_i \\ U_j \\ V_j \end{Bmatrix} = \begin{bmatrix} K_{U_iu_i} & K_{U_iv_i} & K_{U_iu_j} & K_{U_iv_j} \\ K_{V_iu_i} & K_{V_iv_i} & K_{V_iu_j} & K_{V_iv_j} \\ K_{U_ju_i} & K_{U_jv_i} & K_{U_ju_j} & K_{U_jv_j} \\ K_{V_ju_i} & K_{V_jv_i} & K_{V_ju_j} & K_{V_jv_j} \end{bmatrix} \begin{Bmatrix} u_i \\ v_i \\ u_j \\ v_j \end{Bmatrix} = [K]^e \begin{Bmatrix} \{a_i\} \\ \{a_j\} \end{Bmatrix} \tag{9.11}$$

In Fig. 9.4 we have instituted a small displacement u_j, keeping other nodal displacement components equal to zero. The elongation δl of the member is then

$$\delta l = u_j \cos \beta$$

For a small displacement u_j, we can take angles $\alpha \approx \beta$. Hence

$$\delta l \approx u_j \cos \alpha$$

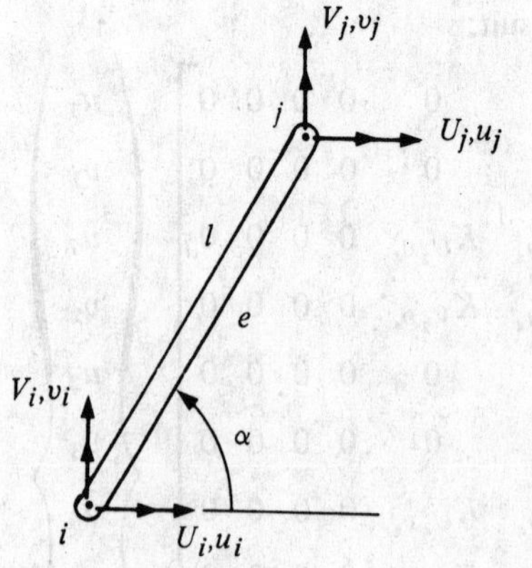

FIGURE 9.3
Element e from a truss.

With modulus of elasticity E, the force P in the member is easily determined as

$$P = \frac{\delta l}{l} EA = \frac{u_j \cos\alpha\, EA}{l} \qquad \text{(tension)} \tag{9.12}$$

where A is the cross-sectional area of the member. We can now give $K_{U_i u_j}$ as†

$$K_{U_i u_j} = -\frac{P\cos\alpha}{u_j} = -\frac{\cos^2\alpha\, EA}{l}$$

Similarly,

$$K_{V_i u_j} = -\frac{P\sin\alpha}{u_j} = -\frac{\sin\alpha\cos\alpha\, EA}{l}$$

$$= -\frac{\sin 2\alpha\, EA}{2l}$$

†Recall that U_i is the force on the member from pin i to be associated with unit nodal displacement component u_j at pin j. Since member ij will be in tension for u_j, U_i must be in the negative x direction so that $K_{U_i u_j}$ is negative.

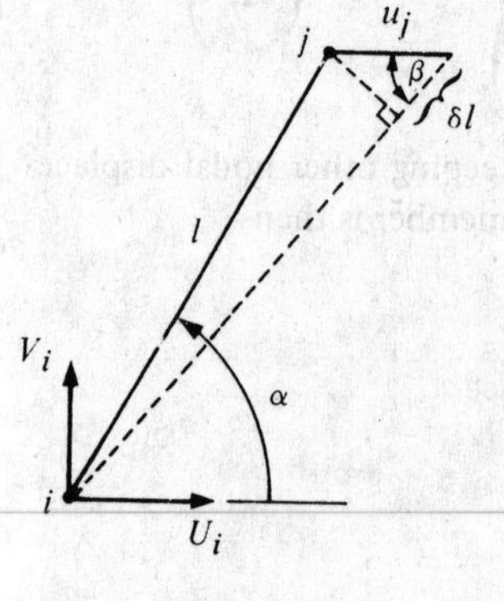

FIGURE 9.4
Element with displacement u_j.

Now if we consider the forces U_j and V_j from u_j at the other nodal point j, we get the same results as for U_i and V_i except for sign, as can readily be deduced from ***equilibrium***. We can accordingly say for u_j that

$$K_{U_i u_j} = -\frac{\cos^2 \alpha\, EA}{l}$$

$$K_{V_i u_j} = -\frac{\sin 2\alpha\, EA}{2l}$$

$$K_{U_j u_j} = +\frac{\cos^2 \alpha\, EA}{l} \tag{9.13}$$

$$K_{V_j u_j} = +\frac{\sin 2\alpha\, EA}{2l}$$

If we now consider u_i (i.e., u at the other nodal point), we get compression of the member. It is easily seen that we get the above results with changed signs and with u_j replaced by u_i. Thus:

$$K_{U_i u_i} = +\frac{\cos^2 \alpha\, EA}{l}$$

$$K_{V_i u_i} = +\frac{\sin 2\alpha\, EA}{2l}$$

$$K_{U_j u_i} = -\frac{\cos^2 \alpha\, EA}{l} \tag{9.14}$$

$$K_{V_j u_i} = -\frac{\sin 2\alpha\, EA}{2l}$$

Now consider v_j in Fig. 9.5. We can then say for δl:

$$\delta l = v_j \cos \beta \approx v_j \sin \alpha$$

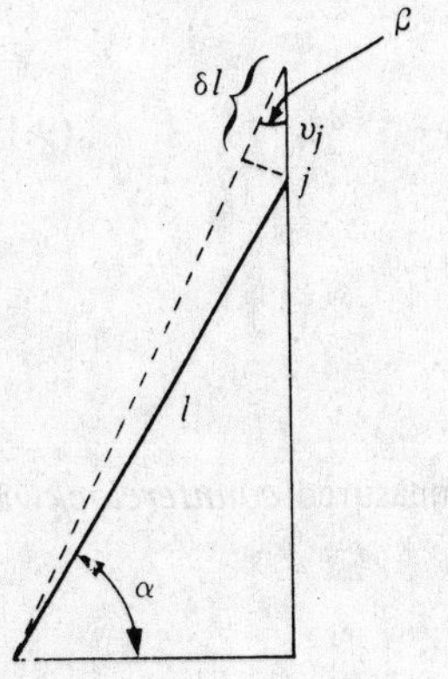

FIGURE 9.5
Element with displacement v_j.

Hence, for the force P in the member we have

$$P = \frac{v_j \sin \alpha}{l} EA \qquad \text{(tension)}$$

We can now write

$$K_{U_i v_j} = -\frac{P \cos \alpha}{v_j} = -\frac{\sin 2\alpha\, EA}{2l}$$

$$K_{V_i v_j} = -\frac{P \sin \alpha}{v_j} = -\frac{\sin^2 \alpha\, EA}{l}$$

For U_j and V_j, respectively, we get the same results as above except for sign (as a result of equilibrium), so that for v_j we have in total:

$$\begin{aligned}
K_{U_i v_j} &= -\frac{\sin 2\alpha\, EA}{2l} \\
K_{V_i v_j} &= -\frac{\sin^2 \alpha\, EA}{l} \\
K_{U_j v_j} &= +\frac{\sin 2\alpha\, EA}{2l} \\
K_{V_j v_j} &= +\frac{\sin^2 \alpha\, EA}{l}
\end{aligned} \qquad (9.15)$$

And for v_i we get compression in the member. It is easily seen again that we get the same result as above but with opposite signs and with v_i replacing v_j. Thus

$$\begin{aligned}
K_{U_i v_i} &= +\frac{\sin 2\alpha\, EA}{2l} \\
K_{V_i v_i} &= +\frac{\sin^2 \alpha\, EA}{l} \\
K_{U_j v_i} &= -\frac{\sin 2\alpha\, EA}{2l} \\
K_{V_j v_i} &= -\frac{\sin^2 \alpha\, EA}{l}
\end{aligned} \qquad (9.16)$$

The stiffness matrix for any element ij, with the angle α measured *counterclockwise* from the x axis to ij, can be given as follows:

$$[K]^e = \frac{EA}{l}\begin{bmatrix} \cos^2\alpha & \frac{\sin 2\alpha}{2} & -\cos^2\alpha & -\frac{\sin 2\alpha}{2} \\ \frac{\sin 2\alpha}{2} & \sin^2\alpha & -\frac{\sin 2\alpha}{2} & -\sin^2\alpha \\ -\cos^2\alpha & -\frac{\sin 2\alpha}{2} & \cos^2\alpha & \frac{\sin 2\alpha}{2} \\ -\frac{\sin 2\alpha}{2} & -\sin^2\alpha & \frac{\sin 2\alpha}{2} & \sin^2\alpha \end{bmatrix} \tag{9.17}$$

We thus have available to us a simple stiffness matrix for an element that must be properly placed in a truss matrix [such as is shown in Eq. (9.10)] to get the global matrix.

We now discuss the building of the global matrix. Examine for this purpose the 12 × 12 matrix in Eq. (9.10) for the truss in Fig. 9.2. The element stiffness matrix shown in the array corresponds to an element having *nodal points 2 and 4*. Hence, the array shown above for this element with $\alpha = -45°$ would be placed in the third and fourth rows (for nodal point 2) and the seventh and eighth rows (for nodal point 4). Furthermore, the columns also have the same numbering as the rows, namely, third and fourth and seventh and eighth. In addition, element 9 having nodal points 5 and 6 would occupy rows and columns 9 and 10 and rows and columns 11 and 12. The angle α would then be $-45°$.

We shall now illustrate this (as well as other considerations) in the following simple example.

EXAMPLE 9.1 Find the stiffness matrix for the truss shown in Fig. 9.6.

There are five members and eight nodal displacement components. We will set up the format for the entire truss analogous to Eq. (9.10).

$$\begin{Bmatrix} U_1 \\ V_1 \\ U_2 \\ V_2 \\ U_3 \\ V_3 \\ U_4 \\ V_4 \end{Bmatrix} = \begin{array}{c} \begin{matrix} \quad 1 & \quad\;\; 1 & \quad\;\; 2 & \quad\;\; 2 & \quad\;\; 3 & \quad\;\; 3 & \quad\;\; 4 & \quad\;\; 4 \end{matrix} \\ \begin{bmatrix} K_{U_1u_1} & K_{U_1v_1} & K_{U_1u_2} & K_{U_1v_2} & K_{U_1u_3} & K_{U_1v_3} & K_{U_1u_4} & K_{U_1v_4} \\ K_{V_1u_1} & K_{V_1v_1} & K_{V_1u_2} & K_{V_1v_2} & K_{V_1u_3} & K_{V_1v_3} & K_{V_1u_4} & K_{V_1v_4} \\ K_{U_2u_1} & K_{U_2v_1} & K_{U_2u_2} & K_{U_2v_2} & K_{U_2u_3} & K_{U_2v_3} & K_{U_2u_4} & K_{U_2v_4} \\ K_{V_2u_1} & K_{V_2v_1} & K_{V_2u_2} & K_{V_2v_2} & K_{V_2u_3} & K_{V_2v_3} & K_{V_2u_4} & K_{V_2v_4} \\ K_{U_3u_1} & K_{U_3v_1} & K_{U_3u_2} & K_{U_3v_2} & K_{U_3u_3} & K_{U_3v_3} & K_{U_3u_4} & K_{U_3v_4} \\ K_{V_3u_1} & K_{V_3v_1} & K_{V_3u_2} & K_{V_3v_2} & K_{V_3u_3} & K_{V_3v_3} & K_{V_3u_4} & K_{V_3v_4} \\ K_{U_4u_1} & K_{U_4v_1} & K_{U_4u_2} & K_{U_4v_2} & K_{U_4u_3} & K_{U_4v_3} & K_{U_4u_4} & K_{U_4v_4} \\ K_{V_4u_1} & K_{V_4v_1} & K_{V_4u_2} & K_{V_4v_2} & K_{V_4u_3} & K_{V_4v_3} & K_{V_4u_4} & K_{V_4v_4} \end{bmatrix} \end{array} \begin{Bmatrix} u_1 \\ v_1 \\ u_2 \\ v_2 \\ u_3 \\ v_3 \\ u_4 \\ v_4 \end{Bmatrix} \tag{a}$$

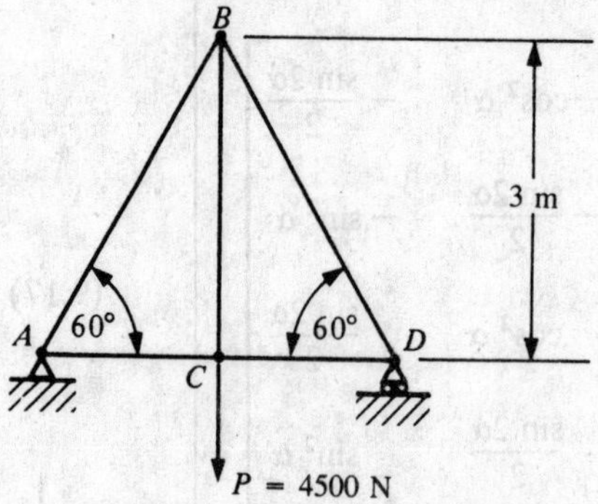

FIGURE 9.6
Simple truss.

As you may readily demonstrate, the stiffness matrix as indicated earlier will be symmetric and so only half the terms need be calculated.

The procedure is to get the stiffness matrix for each element (see Fig. 9.7) and then assemble the global stiffness matrix.

Element ①, $\alpha = 60^\circ$, $l = 3.464$ m, nodal points 1-2

$$[K]^1 = \frac{EA}{3.464}\begin{bmatrix} 0.250 & 0.433 & -0.250 & -0.433 \\ 0.433 & 0.750 & -0.433 & -0.750 \\ -0.250 & -0.433 & 0.250 & 0.433 \\ -0.433 & -0.750 & 0.433 & 0.750 \end{bmatrix} \qquad (b)$$

Element ②, $\alpha = 0$, $l = 1.732$ m, nodal points 1-3

$$[K]^2 = \frac{EA}{1.732}\begin{bmatrix} 1 & 0 & -1 & 0 \\ 0 & 0 & 0 & 0 \\ -1 & 0 & 1 & 0 \\ 0 & 0 & 0 & 0 \end{bmatrix} \qquad (c)$$

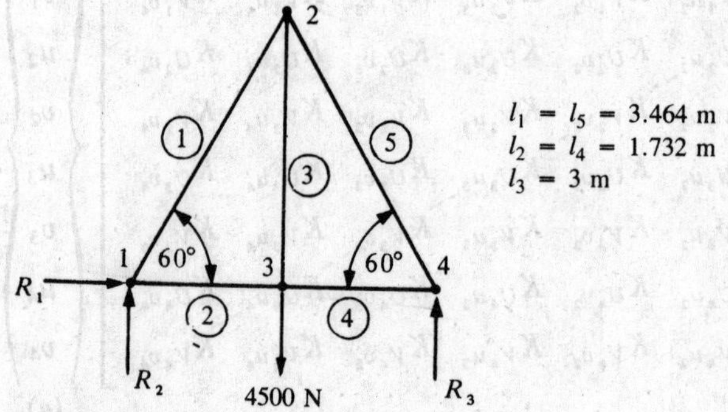

FIGURE 9.7
Truss with members and nodes numbered.

Element ③, $\alpha = 90^\circ$, $l = 3$ m, nodal points 3-2

$$[K]^3 = \frac{EA}{3}\begin{bmatrix} 0 & 0 & 0 & 0 \\ 0 & 1 & 0 & -1 \\ 0 & 0 & 0 & 0 \\ 0 & -1 & 0 & 1 \end{bmatrix} \qquad (d)$$

Element ④, $\alpha = 0$, $l = 1.732$ m, nodal points 3-4

$$[K]^4 = \frac{EA}{1.732}\begin{bmatrix} 1 & 0 & -1 & 0 \\ 0 & 0 & 0 & 0 \\ -1 & 0 & 1 & 0 \\ 0 & 0 & 0 & 0 \end{bmatrix} \qquad (e)$$

Element ⑤, $\alpha = 120^\circ$, $l = 3.464$ m, nodal points 4-2

$$[K]^5 = \frac{EA}{3.464}\begin{bmatrix} 0.250 & -0.433 & -0.250 & 0.433 \\ -0.433 & 0.750 & 0.433 & -0.750 \\ -0.250 & 0.433 & 0.250 & -0.433 \\ 0.433 & -0.750 & -0.433 & 0.750 \end{bmatrix} \qquad (f)$$

We next assemble the global stiffness matrix by noting the nodal points and inserting terms to correspond to these nodal points. For instance, element ① involves nodal points 1 and 2. Merely lay terms in Eq. (*b*), after multiplying by the coefficient 1/3.464, onto the grid shown in Eq. (*a*) to cover nodal points 1 and 2.

	1	1	2	2	3	3	4	4
U_1	①	①	①	①				
V_1	①	①	①	①				
U_2	①	①	①	①				
V_2	①	①	①	①				
U_3								
V_3								
U_4								
V_4								

Next go to element ② involving nodal points 1 and 3. We insert Eq. (*c*), after multiplying by the coefficient 1/1.732, into the global array as shown below.

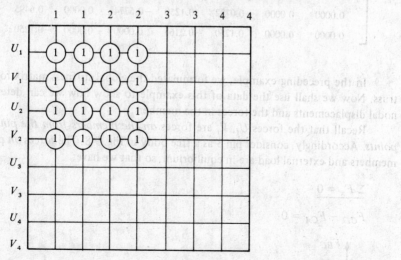

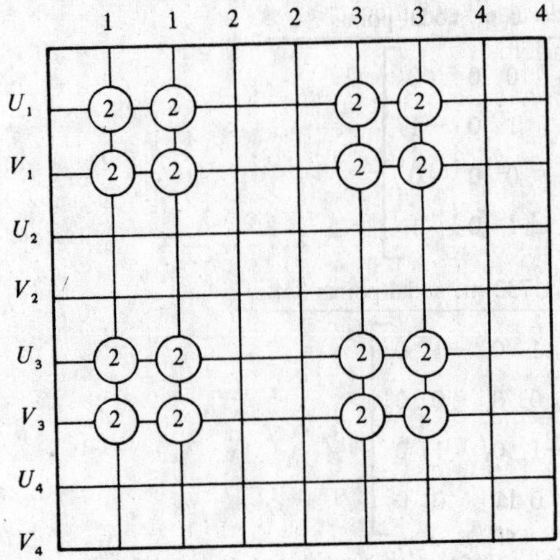

We continue to insert the stiffness coefficients, *adding* those that appear at the same grid points. Of course, when we insert a stiffness coefficient, we use numerical values from Eqs. (b) through (f). The final result is presented next.

$$[K] = EA\begin{bmatrix} 0.6495 & 0.1250 & -0.0722 & -0.1250 & -0.5774 & 0.0000 & 0.0000 & 0.0000 \\ 0.1250 & 0.2165 & -0.1250 & -0.2165 & 0.0000 & 0.0000 & 0.0000 & 0.0000 \\ -0.0722 & -0.1250 & 0.1443 & 0.0000 & 0.0000 & 0.0000 & -0.0722 & 0.1250 \\ -0.1250 & -0.2165 & 0.0000 & 0.7664 & 0.0000 & -0.3333 & 0.1250 & -0.2165 \\ -0.5774 & 0.0000 & 0.0000 & 0.0000 & 1.1547 & 0.0000 & -0.5774 & 0.0000 \\ 0.0000 & 0.0000 & 0.0000 & -0.3333 & 0.0000 & 0.3333 & 0.0000 & 0.0000 \\ 0.0000 & 0.0000 & -0.0722 & 0.1250 & -0.5774 & 0.0000 & 0.6495 & -0.1250 \\ 0.0000 & 0.0000 & 0.1250 & -0.2165 & 0.0000 & 0.0000 & -0.1250 & 0.2165 \end{bmatrix} \qquad (h)$$

In the preceding example, we formulated the global stiffness matrix for a simple truss. Now we shall use the data of this example to show how we can determine the nodal displacements and the forces in the members.

Recall that the forces U_i, V_i are forces *on the members from the pins or nodal points.* Accordingly, consider pin 3 as a free body in Fig. 9.8. The forces *on* pin 3 *from* members and external load are in equilibrium, so that we have:

$$\underline{\Sigma F_x = 0}$$

$$F_{CD} - F_{CA} = 0$$

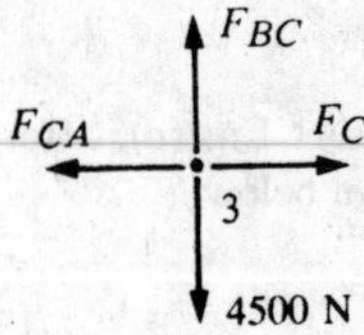

FIGURE 9.8
Pin 3 as a free body.

But $-(F_{CD}-F_{CA})$ is the horizontal force *from* the pin 3 *onto* the members at 3, and so $-(F_{CD}-F_{CA})=U_3$. We can conclude from above that $U_3=0$. Now consider the vertical direction for equilibrium.

$$\underline{\Sigma F_y = 0}$$

$$F_{BC} - 4500 = 0$$

Therefore

$$F_{BC} = 4500 \text{ N}$$

But $(-F_{BC})$ is the force in the vertical direction *from* pin 3 *onto* the members, and hence $(-F_{BC})=V_3$. Using the above equation, we can readily see that

$$V_3 = -4500 \text{ N}$$

From the consideration at nodal point 3, we conclude that we can replace U_3 and V_3, respectively, by the known *external loads* in the horizontal and vertical directions, using the regular signs for these loads. At nodal point 1, we replace U_1, V_1 by R_1, R_2 and at nodal point 4 we replace U_4, V_4 by 0, R_3. The supporting forces R_i are taken as unknown and are determined in the process. We can now write

$$\begin{Bmatrix} U_1 \\ V_1 \\ U_2 \\ V_2 \\ U_3 \\ V_3 \\ U_4 \\ V_4 \end{Bmatrix} = \begin{Bmatrix} R_1 \\ R_2 \\ 0 \\ 0 \\ 0 \\ -4500 \\ 0 \\ R_3 \end{Bmatrix}$$

We next go back to Eq. (9.1). Putting the force matrix on the right side of the equation as is customary, we have

$$\begin{bmatrix} & & \\ & K & \\ & & \end{bmatrix} \begin{Bmatrix} u_1 \\ v_1 \\ u_2 \\ v_2 \\ u_3 \\ v_3 \\ u_4 \\ v_4 \end{Bmatrix} = \begin{Bmatrix} R_1 \\ R_2 \\ 0 \\ 0 \\ 0 \\ -4500 \\ 0 \\ R_3 \end{Bmatrix} \tag{9.18}$$

The idea would now be to solve for $\{a\}$ in terms of the loads. We cannot do this yet because the stiffness matrix $[K]$ in the above equation is *singular*, that is, the determinant of $[K]$ equals zero.†

To remedy this difficulty, we must insert the *boundary conditions.* Let us for a moment consider the following matrix equation:

$$\begin{bmatrix} K_{11} & K_{12} & K_{13} & K_{14} \\ K_{21} & K_{22} & K_{23} & K_{24} \\ K_{31} & K_{32} & K_{33} & K_{34} \\ K_{41} & K_{42} & K_{43} & K_{44} \end{bmatrix} \begin{Bmatrix} u_1 \\ v_1 \\ u_2 \\ v_2 \end{Bmatrix} = \begin{Bmatrix} U_1 \\ V_1 \\ U_2 \\ V_2 \end{Bmatrix} \tag{9.19}$$

where one of the nodal displacements, say u_2, is known to be zero. The above matrix equation can be written as follows:

$$\begin{bmatrix} K_{11} & K_{12} & 0 & K_{14} \\ K_{21} & K_{22} & 0 & K_{24} \\ 0 & 0 & 0 & 0 \\ K_{41} & K_{42} & 0 & K_{44} \end{bmatrix} \begin{Bmatrix} u_1 \\ v_1 \\ 0 \\ v_2 \end{Bmatrix} = \begin{Bmatrix} U_1 \\ V_1 \\ 0 \\ V_2 \end{Bmatrix}$$

You may directly verify that when you multiply out the first, second, and fourth rows above you get the correct equations for Eq. (9.19) for these rows with $u_2 = 0$. If we strike out the third row and the third column, we will not affect the results for U_1, V_1, and V_2. Thus, we can work with the following matrix equation:

$$\begin{bmatrix} K_{11} & K_{12} & K_{14} \\ K_{21} & K_{22} & K_{24} \\ K_{41} & K_{42} & K_{44} \end{bmatrix} \begin{Bmatrix} u_1 \\ v_1 \\ v_2 \end{Bmatrix} = \begin{Bmatrix} U_1 \\ V_1 \\ V_2 \end{Bmatrix}$$

The stiffness matrix so reduced is called the ***reduced stiffness matrix.*** If next both u_2 and v_2 are prescribed as equal to zero, we would delete the third and fourth rows and columns of Eq. (9.19) to form the reduced matrix. We would then have

$$\begin{bmatrix} K_{11} & K_{12} \\ K_{21} & K_{22} \end{bmatrix} \begin{Bmatrix} u_1 \\ v_1 \end{Bmatrix} = \begin{Bmatrix} U_1 \\ V_1 \end{Bmatrix}$$

As for the problem at hand, we form the reduced matrix by eliminating the first, second, and last rows and columns in Eq. (9.18) since $u_1 = v_1 = v_4 = 0$.

†This results from the fact that the boundary conditions have not as yet been applied, thus rigid-body movement has not yet been excluded.

Suppose next that in Eq. (9.19) u_2 is specified as a *nonzero* value "a". Then we can rewrite Eq. (9.19) as follows:

$$\begin{bmatrix} K_{11} & K_{12} & 0 & K_{14} \\ K_{21} & K_{22} & 0 & K_{24} \\ 0 & 0 & 1 & 0 \\ K_{41} & K_{42} & 0 & K_{44} \end{bmatrix} \begin{Bmatrix} u_1 \\ v_1 \\ u_2 \\ v_2 \end{Bmatrix} = \begin{Bmatrix} (U_1 - K_{13}a) \\ (V_1 - K_{23}a) \\ a \\ (V_2 - K_{43}a) \end{Bmatrix} \tag{9.20}$$

When we multiply the third row out, we get $u_2 = a$ as required. When we multiply out the other rows, we get the correct results for Eq. (9.19) for these rows if we set $u_2 = a$. We can again delete row 3 and column 3 to form the following result:

$$\begin{bmatrix} K_{11} & K_{12} & K_{14} \\ K_{21} & K_{22} & K_{24} \\ K_{41} & K_{42} & K_{44} \end{bmatrix} \begin{Bmatrix} u_1 \\ v_1 \\ v_2 \end{Bmatrix} = \begin{Bmatrix} (U_1 - K_{13}a) \\ (V_1 - K_{23}a) \\ (V_2 - K_{43}a) \end{Bmatrix}$$

If $v_1 = a$ and $u_2 = b$, we extend the method as follows:

$$\begin{bmatrix} K_{11} & 0 & 0 & K_{14} \\ 0 & 1 & 0 & 0 \\ 0 & 0 & 1 & 0 \\ K_{41} & 0 & 0 & K_{44} \end{bmatrix} \begin{Bmatrix} u_1 \\ v_1 \\ u_2 \\ v_2 \end{Bmatrix} = \begin{Bmatrix} (U_1 - K_{12}a - K_{13}b) \\ a \\ b \\ (V_2 - K_{42}a - K_{43}b) \end{Bmatrix} \tag{9.21}$$

Clearly, if you evaluate the equations from Eq. (9.19) for the first and fourth rows and put in $v_1 = a$ and $u_2 = b$ after carrying out the steps, you will see that the above is a correct formulation. Now we can delete the second and third rows and columns above without invalidating the remaining formulation. Thus we have

$$\begin{bmatrix} K_{11} & K_{14} \\ K_{41} & K_{44} \end{bmatrix} \begin{Bmatrix} u_1 \\ v_2 \end{Bmatrix} = \begin{Bmatrix} (U_1 - K_{12}a - K_{13}b) \\ (V_2 - K_{42}a - K_{43}b) \end{Bmatrix}$$

Note in Eq. (9.21) before reduction that, for those rows for which the displacement components are *not* specified constants, we must subtract from the forces K values corresponding to the position of the zeros in *these particular rows.* Each such K value, furthermore, must be multiplied by the specific *known* displacement in vector $\{a\}$ that would normally be multiplied by the K value when carrying out the matrix product $[K]\{a\}$. Thus, for the first row in Eq. (9.21) u_1 is unspecified. The zeros in this row correspond to K_{12} and K_{13}. Now K_{12} normally multiplies v_1, which is known to have value a, and K_{13} multiplies u_2, which is known to have value b. Thus we subtract from U_1 the value $K_{12}a$ and $K_{12}b$, as shown in the equation. It is thus a simple process to reduce the matrix equation. Clearly, the procedure described earlier with zero values of prescribed nodal displacements is but a special case of the procedure just described.

EXAMPLE 9.2 Determine the nodal deflections and the supporting forces for the truss in Example 9.1

On deleting the first two rows and columns as well as the last row and column, the basic matrix equation becomes

$$EA\begin{bmatrix} 0.1443 & 0 & 0 & 0 & -0.0722 \\ 0 & 0.7664 & 0 & -0.3333 & 0.1250 \\ 0 & 0 & 1.1547 & 0 & -0.5774 \\ 0 & -0.3333 & 0 & 0.3333 & 0 \\ -0.0722 & 0.1250 & -0.5774 & 0 & 0.6495 \end{bmatrix}\begin{Bmatrix} u_2 \\ v_2 \\ u_3 \\ v_3 \\ u_4 \end{Bmatrix} = \begin{Bmatrix} 0 \\ 0 \\ 0 \\ -4500 \\ 0 \end{Bmatrix}$$

The reduced matrix is no longer singular, and we can now find the inverse of the reduced stiffness matrix. Inverting the above matrix equation, we have

$$\begin{Bmatrix} u_2 \\ v_2 \\ u_3 \\ v_3 \\ u_4 \end{Bmatrix} = \frac{1}{EA}\begin{bmatrix} 7.798 & -0.5 & 0.867 & -0.5 & 1.733 \\ -0.5 & 2.598 & -0.5 & 2.598 & -1 \\ 0.867 & -0.5 & 1.733 & -0.5 & 1.733 \\ -0.5 & 2.598 & -0.5 & 5.598 & -1 \\ 1.733 & -1 & 1.733 & -1 & 3.465 \end{bmatrix}\begin{Bmatrix} 0 \\ 0 \\ 0 \\ -4500 \\ 0 \end{Bmatrix}$$

We thus have

$$u_2 = \frac{1}{EA}\,2{,}251\text{ m}$$

$$v_2 = -\frac{1}{EA}\,11{,}689\text{ m}$$

$$u_3 = \frac{1}{EA}\,2{,}251\text{ m}$$

$$v_3 = -\frac{1}{EA}\,25{,}190\text{ m}$$

$$u_4 = \frac{1}{EA}\,4{,}500\text{ m}$$

where the units of EA are pounds force.

To get the supporting forces, go back to Eq. (9.18) and set up the equation for the R_1, R_2, and R_3 rows. Thus

$$EA\begin{bmatrix} 0.6495 & 0.1625 & -0.0722 & -0.1250 & -0.5774 & 0 & 0 & 0 \\ 0.1625 & 0.2165 & -0.125 & -0.2165 & 0 & 0 & 0 & 0 \\ 0 & 0 & 0.125 & -0.2165 & 0 & 0 & -0.125 & 0.2165 \end{bmatrix}\begin{Bmatrix} 0 \\ 0 \\ u_2 \\ v_2 \\ u_3 \\ v_3 \\ u_4 \\ 0 \end{Bmatrix} = \begin{Bmatrix} R_1 \\ R_2 \\ R_3 \end{Bmatrix}$$

Substituting the computed values for the nodal displacements, we now get the supporting forces from calculations involving 8 decimal place accuracy:

$$R_1 = 0 \text{ N}$$

$$R_2 = 2{,}250 \text{ N}$$

$$R_3 = 2{,}250 \text{ N}$$

These clearly are the expected results and act as a partial check on the work done up to this point. Finally, to get the forces in the members, we proceed as follows (see Fig. 9.9):

$$P_e = \frac{A_e E_e}{L_e}\left[(u_j \cos\phi_e + v_j \sin\phi_e) - (u_i \cos\phi_e + v_i \sin\phi_e)\right] \tag{9.22}$$

In matrix notation this becomes

$$P_e = \left(\frac{AE}{L}\right)_e [G]_e \begin{Bmatrix} \{a_i\} \\ \{a_j\} \end{Bmatrix}$$

where

$$[G] = \lfloor -\cos\phi_e, -\sin\phi_e, \cos\phi_e, \sin\phi_e \rfloor \tag{9.23}$$

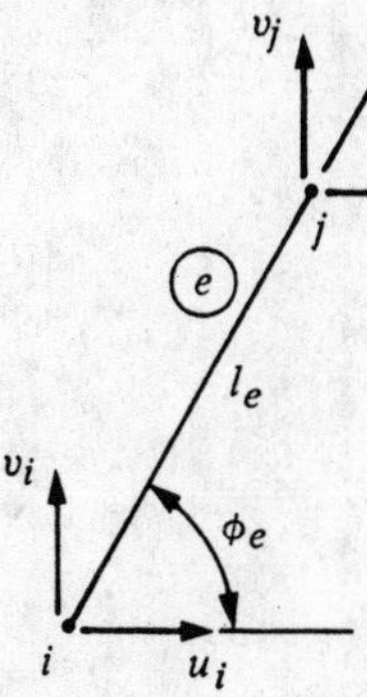

FIGURE 9.9
Element shown for finding force P_e in the element.

The forces in the members are then

$$P_1 = 2{,}597 \text{ N compression}$$
$$P_2 = 1{,}300 \text{ N tension}$$
$$P_3 = 4{,}500 \text{ N tension}$$
$$P_4 = 1{,}300 \text{ N tension}$$
$$P_5 = 2{,}597 \text{ N compression}$$

These results may easily be checked by the method of joints.

In this section we have first shown how one forms the element stiffness matrix, and in the example we have assembled a global stiffness matrix for a simple truss. We have also shown how one changes $[K]$ from a singular matrix to one permitting the solution of the unknown nodal displacements and forces.

In the next section, with this introduction as a background, we will give a more general approach to finite elements.

Part B
More General Preliminary Considerations

9.3 BASIC CONSIDERATIONS FOR FINITE ELEMENTS

We will set up basic ideas and the accompanying notation for finite elements in this section. We will work with plane stress in formulating many of these ideas but will indicate as we go along when the results are more generally valid by boxing in the pertinent results of general use. Accordingly, let us consider a triangular element (Fig. 9.10), which we denote as element e, with three nodes i, j, and p. The *nodal displacement components* for these three nodes are, respectively, (u_i, v_i), (u_j, v_j),

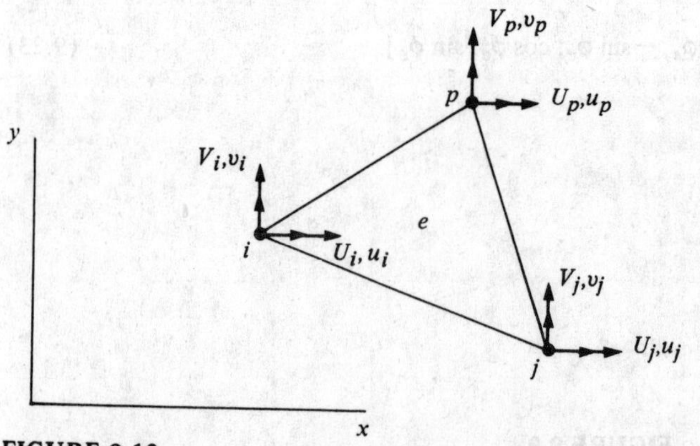

FIGURE 9.10
Triangular element.

and (u_p, v_p). The corresponding nodal forces on the element needed for this set of nodal displacements are (U_i, V_i), (U_j, V_j), and (U_p, V_p). We make the following additional definitions for the nodal displacement matrices for element e:

$$\begin{Bmatrix} u_i \\ v_i \end{Bmatrix} = \{a_i\}$$

$$\begin{Bmatrix} u_j \\ v_j \end{Bmatrix} = \{a_j\} \tag{9.24a}$$

$$\begin{Bmatrix} u_p \\ v_p \end{Bmatrix} = \{a_p\}$$

$$\begin{Bmatrix} \{a_i\} \\ \{a_j\} \\ \{a_p\} \end{Bmatrix} = \{a\}^e \tag{9.24b}$$

We next give the *displacement field* in terms of the nodal displacement vector $\{a\}^e$ for an element. The displacement function $\{\vec{u}(x, y)\}$ is defined as follows for plane strain:

$$\{\vec{u}(x,y)\} = \begin{Bmatrix} u(x,y) \\ v(x,y) \end{Bmatrix} \tag{9.25}$$

Note that this function gives the displacement components of any point in the element. Next we wish to relate the displacement field for an element to the nodal displacements for the element. For this purpose we introduce the *interpolation* functions or, as they are often called, the *shape* functions. Thus, we state for our triangular element

$$\begin{Bmatrix} u(x,y) \\ v(x,y) \end{Bmatrix} = [N] \begin{Bmatrix} a_1 \\ a_2 \\ a_3 \\ a_4 \\ a_5 \\ a_6 \end{Bmatrix}$$

where $[N]$ is the interpolation function matrix, which here must have 12 elements forming 6 columns and 2 rows. Each member of $[N]$ is a function of position. We may write the above equation identifying the various terms with nodes as follows:

$$\{\vec{u}(x,y)\} = \lfloor [N_i]\,[N_j]\,[N_p] \rfloor \begin{Bmatrix} \{a_i\} \\ \{a_j\} \\ \{a_p\} \end{Bmatrix} \tag{9.26}$$

where for node i we have for the interpolation functions

$$[N_i] = \begin{bmatrix} (N_i)_{11} & (N_i)_{12} \\ (N_i)_{21} & (N_i)_{22} \end{bmatrix}$$

Associations with the other nodes may be similarly formed. Thus for our triangular element we have 12 interpolation functions. In the general case, we have

$$\boxed{\{\vec{u}(x,y,z)\} = [N]\,\{a\}} \tag{9.27}$$

Now the interpolation functions must be such that the displacement field $\{\vec{u}\}$ at the *position* of a nodal point gives the proper nodal *displacements* at that nodal point. Thus for nodal point i we have

$$\{\vec{u}(x_i,y_i)\} = [N_i(x_i,y_i)]\,\{a_i\} + [N_j(x_i,y_i)]\,\{a_j\} + [N_p(x_i,y_i)]\,\{a_p\} \tag{9.28}$$

and we require that $[N_j(x_i,y_i)] = 0$, $[N_p(x_i,y_i)] = 0$, and $[N_i(x_i,y_i)] = [I]$, where $[I]$ is the unitary matrix. Similarly,

$$\begin{cases} [N_i(x_j,y_j)] = 0 \\ [N_j(x_j,y_j)] = [I] \\ [N_p(x_j,y_j)] = 0 \end{cases} \qquad \begin{cases} [N_i(x_p,y_p)] = 0 \\ [N_j(x_p,y_p)] = 0 \\ [N_p(x_p,y_p)] = [I] \end{cases} \tag{9.29}$$

Now we relate the *strain field* matrix $\{\epsilon\}$ to the *displacement* field. Thus, for small strain we have in general

$$\boxed{\{\epsilon\} = [L]\,\{\vec{u}(x,y,z)\}} \tag{9.30}$$

where $[L]$ is a suitable differential operator developed by geometric considerations. As an example, for *plane stress* we would have

$$\{\epsilon\} \equiv \begin{Bmatrix} \epsilon_{xx} \\ \epsilon_{yy} \\ \epsilon_{xy} \end{Bmatrix} = \begin{Bmatrix} \dfrac{\partial u}{\partial x} \\ \dfrac{\partial v}{\partial y} \\ \dfrac{1}{2}\left(\dfrac{\partial u}{\partial y} + \dfrac{\partial v}{\partial x}\right) \end{Bmatrix} = \begin{bmatrix} \dfrac{\partial}{\partial x} & 0 \\ 0 & \dfrac{\partial}{\partial y} \\ \dfrac{1}{2}\dfrac{\partial}{\partial y} & \dfrac{1}{2}\dfrac{\partial}{\partial x} \end{bmatrix} \begin{Bmatrix} u(x,y) \\ v(x,y) \end{Bmatrix} \tag{9.31}$$

where the operator $[L]$ acts on $\left\{ {u(x,y) \atop v(x,y)} \right\}$ in the above equation.

To relate $\{\epsilon\}$ to the *nodal displacement vectors*, we introduce the *nodal displacement–strain* matrix $[B]$ by replacing $\{\vec{u}(x, y, z)\}$ in Eq. (9.30), using Eq. (9.27). We then get

$$\boxed{\{\epsilon\} = [L][N]\{a\} = [B]\{a\}} \tag{9.32}$$

where

$$\boxed{[B] = [L][N]} \tag{9.33}$$

As a final step in this section, we consider the *constitutive law* matrix for linear elastic materials. Let $[\sigma]$ be the stress matrix. We then have the following relation:

$$\boxed{[\sigma] = [D]([\epsilon] - [\epsilon_0]) + [\sigma_0]} \tag{9.34}$$

where $[\epsilon]$ is the total strain matrix, $[\epsilon_0]$ represents the *initial strain* matrix and $[\sigma_0]$ represents the *residual* stress matrix. The matrix $[D]$ represents the *constitutive law* matrix. In the case of plane stress

$$\{\sigma\} = \begin{Bmatrix} \tau_{xx} \\ \tau_{yy} \\ \tau_{xy} \end{Bmatrix} \tag{9.35}$$

and you may verify from Eqs. (6.10) and Eq. (1.98*a*) that for an isotropic, Hookean material

$$[D] = \frac{E}{1-\nu^2}\begin{bmatrix} 1 & \nu & 0 \\ \nu & 1 & 0 \\ 0 & 0 & \dfrac{1-\nu}{2} \end{bmatrix} \tag{9.36}$$

We shall now consider the general theory for the *displacement method*.

9.4 GENERAL THEORY FOR THE DISPLACEMENT METHOD

We will now employ the principle of total potential energy for an element to formulate the equation of equilibrium in terms of nodal displacements. That is, we shall extremize the total potential energy functional for an element with respect to the nodal displacement components of the element.† We shall consider that initial strain and residual stress are present, both indicated by the subscript zero. In the ensuing discussion, we will consider the nodes of an element as separate entities, leaving the element minus the nodes as a separate entity. This is very similar to the way in which we separately treated the pins of a truss and the members of the truss as discussed in Part A of this chapter.

Acting on an element from outside, we may have body force distributions b_i and traction force distributions T_i. We shall consider these forces to be replaced by an equivalent system of point forces acting only *on the nodes.* The forces *from* the nodes acting *on* the element are denoted as $\{q\}$. The method of total potential energy for the element then states that

$$\delta^{(1)}\pi = 0 = \iiint_V (\tau_{xx}\,\delta\epsilon_{xx} + \cdots + \tau_{xy}\,\delta\epsilon_{xy})\,dv - \lfloor \delta a \rfloor \{q\} \qquad (9.37)$$

In matrix form we have‡

$$\iiint_V [\delta\epsilon]^T [\sigma]\,dv - \{\delta a\}^T \{q\} = 0 \qquad (9.38)$$

Let us examine the first expression in the above equation. The integrand may be written as follows, using Eq. (9.34) for $[\sigma]$:

$$[\delta\epsilon]^T[\sigma] = [\delta\epsilon]^T([D]([\epsilon] - [\epsilon_0]) + [\sigma_0])$$

Now use Eq. (9.32) for $[\delta\epsilon]^T$ and $[\epsilon]$ on the right side of the above equation. On separating the expressions we get

$$[\delta\epsilon]^T[\sigma] = \{\delta a\}^T[B]^T[D][B]\{a\} - \{\delta a\}^T[B]^T[D][\epsilon_0] + \{\delta a\}^T[B]^T[\sigma_0] \qquad (9.41)$$

Let us next evaluate the forces $\{q\}$, which, you will recall, apply *to* the element *from* the nodes. To do this we consider the nodes for the element as a system of free bodies. Acting on these nodes from the element we have the point force system $\{q'\}$, which

†Recall that we used equilibrium equations at each joint to arrive at the stiffness matrix when we considered trusses in Section 9.2. In more general applications of the finite element approach, we shall use the procedure of extremizing an appropriate quadratic functional separately over each element domain. We shall look into this later in Chapter 17.

‡From the definition of a matrix product we have

$$A_{ij}B_{jk} \equiv [A][B] \qquad (9.39)$$

and

$$A_{ji}B_{jk} = [A]^T[B] \qquad (9.40)$$

where $[A]^T$ is the transpose of $[A]$. Furthermore, note that the transpose of $[A][B]$ is equal to $[B]^T[A]^T$.

is the *reaction* to $\{q\}$. Thus we have $\{q'\} = -\{q\}$. Also acting on the nodes is a loading system, which is statically equivalent to the external loadings on the element consisting for now of a body force distribution $b_i(x, y, z)$ and a traction force distribution $T_i(x, y, z)$. The static equivalent of these external force distributions must be in equilibrium with the force system $\{q'\}$.

Hence, we may use the principle of virtual work for the nodal system of the element. Thus

$$\{\delta a\}^T\{-q\} + \iiint_V \{\delta \vec{u}\}^T\{b\}\, dv + \oint\!\!\oint_A \{\delta \vec{u}\}^T\{T\}\, dA = 0 \tag{9.42}$$

Note that for the virtual work on the nodes, stemming from $\{b\}$ and $\{T\}$, we have used the virtual work of these loadings *directly on* the element itself. Replacing $\{\delta \vec{u}\}^T$ by $\{\delta a\}^T[N]^T$, we have

$$\{\delta a\}^T\{q\} = \{\delta a\}^T \iiint_V [N]^T\{b\}\, dv + \{\delta a\}^T \oint\!\!\oint_A [N]^T\{T\}\, dA$$

where we have extracted $\{\delta a\}^T$ from the integral. Solving for $\{q\}$ in the above matrix equation, we have

$$\boxed{\{q\} = \iiint_V [N]^T\{b\}\, dv + \oint\!\!\oint_A [N]^T\{T\}\, dA} \tag{9.43}$$

Substitute from Eqs. (9.41) and (9.43) into Eq. (9.38) to replace $[\delta\epsilon]^T[\sigma]$ and $\{q\}$. We get

$$\iiint_V \Big[\{\delta a\}^T[B]^T[D][B]\{a\} - \{\delta a\}^T[B]^T[D][\epsilon_0] + \{\delta a\}^T[B]^T[\sigma_0]\Big]\, dv$$

$$- \{\delta a\}^T \iiint_V [N]^T\{b\}\, dv - \{\delta a\}^T \oint\!\!\oint_A [N]^T\{T\}\, dA = 0$$

Finally, extract $\{\delta a\}^T$ and $\{a\}$ from the integrals. We get

$$\{\delta a\}^T\left(\iiint_V [B]^T[D][B]\, dv\right)\{a\} - \{\delta a\}^T \iiint_V [B]^T[D][\epsilon_0]\, dv$$

$$+ \{\delta a\}^T \iiint_V [B]^T[\sigma_0]\, dv - \{\delta a\}^T \iiint_V [N]^T\{b\}\, dv$$

$$- \{\delta a\}^T \oint\!\!\oint_A [N]^T\{T\}\, dA = 0 \tag{9.44}$$

We denote the first integral as $[K]^e$, and we will soon see that it represents the element stiffness matrix. That is,

$$[K]^e = \iiint_V [B]^T [D] [B]\, dv \tag{9.45}$$

Let us denote the effects of initial strain and residual stress as $\{f\}^e$. That is,

$$\{f\}^e = \iiint_V [B]^T [\sigma_0]\, dv - \iiint_V [B]^T [D] [\epsilon_0]\, dv \tag{9.46}$$

With this notation and using Eq. (9.43), Eq. (9.44) becomes

$$\{\delta a\}^T ([K]^e \{a\}^e + \{f\}^e - \{q\}^e) = 0$$

Since $\{\delta a\}^T$ is arbitrary, we conclude from the above equation that for each element

$$[K]^e \{a\}^e = \{q\}^e - \{f\}^e \tag{9.47}$$

We thus arrive at the equation of equilibrium for the finite element methodology. It should be clear now why $[K]^e$ as defined by Eq. (9.45) must be the stiffness matrix for the element.

For the global equation, we assemble the stiffness matrix exactly as was done for the case of trusses. Thus, for 12 nodal displacement components, we form a 12 × 12 grid as shown below for the global stiffness matrix. Now consider an element having nodes 2, 3, and 5. The element stiffness matrix clearly will consist of 36 terms. We enter these terms into the grid as shown. We do the same for all the other elements, adding K-values as they superpose.

1 u_1	2 v_1	3 u_2	4 v_2	5 u_3	6 v_3	7 u_4	8 v_4	9 u_5	10 v_5	11 u_6	12 v_6	
												U_1 1
												V_1 2
		K_{33}	K_{34}	K_{35}	K_{36}			K_{39}	$K_{3,10}$			U_2 3
		K_{43}	K_{44}	K_{45}	K_{46}			K_{49}	$K_{4,10}$			V_2 4
		K_{53}	K_{54}	K_{55}	K_{56}			K_{59}	$K_{5,10}$			U_3 5
		K_{63}	K_{64}	K_{65}	K_{66}			K_{69}	$K_{6,10}$			V_3 6
												U_4 7
												V_4 8
		K_{93}	K_{94}	K_{95}	K_{96}			K_{99}	$K_{9,10}$			U_5 9
		$K_{10,3}$	$K_{10,4}$	$K_{10,5}$	$K_{10,6}$			$K_{10,9}$	$K_{10,10}$			V_5 10
												U_6 11
												V_6 12

Superposing all the inserted elemental stiffness elements, we can arrive at the global stiffness matrix.

9.5 LOCAL AND GLOBAL COORDINATE SYSTEMS

We have been using one reference xy for all the elements. Thus xy is the *global* reference. It is often the case that other axes associated with each of the elements may be used in addition to the global coordinates. Such axes are called *local* axes. These axes may be rotated relative to the global axes as illustrated in Fig. 9.11, where we show a plane element with $\bar{x}\bar{y}$ as local axes and xy as global axes. The force components from the nodes on the element are shown for both local and global axes. The angle θ is measured counterclockwise.

The equation of equilibrium for local coordinates is given as

$$[\bar{K}]\{\bar{a}\} = \{\bar{q}\} \tag{9.48}$$

Next, we can relate the force components shown in Fig. 9.11 for local and global axes in the following way:

$$\begin{aligned} \bar{U}_1 &= U_1 \cos\theta + V_1 \sin\theta \\ \bar{V}_1 &= -U_1 \sin\theta + V_1 \cos\theta \\ \bar{U}_2 &= U_2 \cos\theta + V_2 \sin\theta \\ \bar{V}_2 &= -U_2 \sin\theta + V_2 \cos\theta \end{aligned} \tag{9.49}$$

In matrix form we have

$$\begin{Bmatrix} \bar{U}_1 \\ \bar{V}_1 \\ \bar{U}_2 \\ \bar{V}_2 \end{Bmatrix} = \begin{bmatrix} \cos\theta & \sin\theta & 0 & 0 \\ -\sin\theta & \cos\theta & 0 & 0 \\ 0 & 0 & \cos\theta & \sin\theta \\ 0 & 0 & -\sin\theta & \cos\theta \end{bmatrix} \begin{Bmatrix} U_1 \\ V_1 \\ U_2 \\ V_2 \end{Bmatrix} \tag{9.50}$$

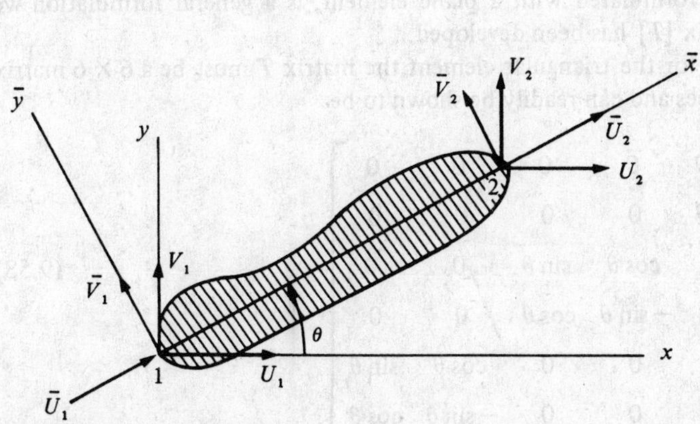

FIGURE 9.11
Plane element with local and global axes.

Hence, in compact form

$$\{\bar{q}\} = [T]\{q\} \tag{9.51}$$

where $[T]$ is called the *transformation* matrix. This matrix has the property that

$$[T]^T = [T]^{-1} \tag{9.52}$$

and for this reason is said to be an *orthogonal* matrix. From Eq. (9.52) we note further that $[T]^T[T] = [I] = [T][T]^T$, where $[I]$ is the unitary matrix. Using the orthogonality property of $[T]$, we can solve for $\{q\}$ in Eq. (9.51) in the following ways:

$$\{q\} = [T]^{-1}\{\bar{q}\} = [T]^T\{\bar{q}\} \tag{9.53}$$

If we consider nodal displacement components for local and global axes, we get the same transformation matrix $[T]$ as given in Eq. (9.50), since this formulation represents the transformation of all vectors in two-dimensional space. Hence we can say

$$\{\bar{a}\} = [T]\{a\} \tag{9.54}$$

Now go back to Eq. (9.48) and replace $\{\bar{q}\}$ from Eq. (9.51) and $\{\bar{a}\}$ from Eq. (9.54). We then get

$$[\bar{K}][T]\{a\} = [T]\{q\} \tag{9.55}$$

Next, premultiply by $[T]^T$. We get, on using the orthogonality property of $[T]$,

$$[T]^T[\bar{K}][T]\{a\} = [T]^T[T]\{q\} = [I]\{q\} = \{q\} \tag{9.56}$$

Comparing this equation with the basic equation $[K]\{a\} = \{q\}$, we must conclude that

$$\boxed{[K] = [T]^T[\bar{K}][T]} \tag{9.57}$$

The transformation formulated on the right side of Eq. (9.57) is called a *similarity* transformation. It generates the global stiffness matrix from the local stiffness matrix. This result, although formulated with a plane element, is a general formulation we can use once the matrix $[T]$ has been developed.

As an example, for the triangular element the matrix T must be a 6×6 matrix to account for six forces and can readily be shown to be

$$[T] = \begin{bmatrix} \cos\theta & \sin\theta & 0 & 0 & 0 & 0 \\ -\sin\theta & \cos\theta & 0 & 0 & 0 & 0 \\ 0 & 0 & \cos\theta & \sin\theta & 0 & 0 \\ 0 & 0 & -\sin\theta & \cos\theta & 0 & 0 \\ 0 & 0 & 0 & 0 & \cos\theta & \sin\theta \\ 0 & 0 & 0 & 0 & -\sin\theta & \cos\theta \end{bmatrix} \tag{9.58}$$

9.6 CLOSURE

In this chapter we first considered simple trusses to introduce the reader to the finite element method. In Part B we presented in a physically meaningful manner a rather general set of procedures and definitions for finite elements that will serve many of our needs as we go on to subsequent chapters. For convenience, we now summarize the key results of Part B of this chapter. Note that for trusses we did not need matrix $[N]$ since the displacement field for the elements is determined simply and uniquely by the nodal displacements.

$$\{\vec{u}\} = [N]\{a\} \tag{9.59a}$$

$$\{\epsilon\} = [L]\{\vec{u}\} \tag{9.59b}$$

$$[B] = [L][N] \tag{9.59c}$$

$$\{\epsilon\} = [B]\{a\} \tag{9.59d}$$

$$[\sigma] = [D]([\epsilon] - [\epsilon_0]) + [\sigma_0] \tag{9.59e}$$

$$[K] = \iiint_V [B]^T [D][B]\, dv \tag{9.59f}$$

$$[K]\{a\} = \{q\} - \{f\} \tag{9.59g}$$

PROBLEMS

9.1 (*a*) What does $K_{V_6 v_4}$ represent? (*b*) What does $K_{U_3 u_3}$ represent?

9.2 What are the elements of the submatrices $[K_{24}]$ and $[K_{16}]$?

9.3 In Eq. (9.10) insert the terms for K_{15} and K_{46} after deleting the terms presently in the K matrix.

9.4 Derive Eq. (9.14) as we derived Eq. (9.13) in the text.

9.5 For the following matrix equation involving $[K]$, $\{a\}$, and $\{q\}$, what is the reduced equation?

$$\begin{bmatrix} 600 & & & & & \\ 300 & 600 & & \text{symmetric} & & \\ 200 & 100 & 200 & & & \\ 400 & 50 & 60 & 90 & & \\ -300 & 30 & 80 & 70 & 500 & \\ 800 & 400 & -400 & 150 & 300 & 800 \end{bmatrix} \begin{Bmatrix} 0 \\ v_1 \\ 0 \\ v_2 \\ u_3 \\ 0 \end{Bmatrix} = \begin{Bmatrix} R_1 \\ 200 \\ R_2 \\ 300 \\ 600 \\ R_3 \end{Bmatrix}$$

9.6 In the preceding problem, u_1 and v_3 are specified constants g and p, respectively. What is the reduced equation? Take $R_2 = 400$ in the preceding problem.

9.7 If in Problem 9.5, $u_1 = 0$, $v_1 = g$, and $u_3 = 0$, what is the reduced equation for the force vector given as

$$\begin{Bmatrix} R_1 \\ V_1 \\ U_2 \\ V_2 \\ R_3 \\ V_3 \end{Bmatrix}$$

9.8 In Problem 9.5 express the matrix equation for determining R_1, R_2, and R_3.

9.9 Express Eq. (9.26) with all terms $(N_i)_{11}$, and so on, shown in the various matrices for the triangular element.

9.10 Find the displacements of the joints and the forces in the members for the truss in Fig. 9.12.

9.11 Find the displacements of the joints and the forces in the members for the truss in Fig. 9.13.

9.12 Consider a one-dimensional element having nodes i and j at the ends with a_i and a_j as vertical nodal displacements (Fig. 9.14). If $w(x)$ for the element is given as

$$w(x) = \lfloor N \rfloor \{a\}$$

show that for

$$\lfloor N \rfloor = \left\lfloor \left(1 - 3\frac{x^2}{L^2} + 2\frac{x^3}{L^3}\right), \left(3\frac{x^2}{L^2} - 2\frac{x^3}{L^3}\right) \right\rfloor$$

we get the proper values of w at nodes i and j in terms of nodal displacements a_i and a_j.

9.13 In the next chapter we shall consider an interpolation matrix for a beam element (Fig. 9.15) having two nodes. There will be four nodal displacements consisting of a vertical deflection and a counterclockwise rotation at each node. The interpolation function matrix is then

$$\lfloor N \rfloor = \left\lfloor \left(1 - 3\frac{x^2}{L^2} + 2\frac{x^3}{L^3}\right), \left(x - 2\frac{x^2}{L} + \frac{x^3}{L^2}\right), \left(3\frac{x^2}{L^2} - 2\frac{x^3}{L^3}\right), \left(-\frac{x^2}{L} + \frac{x^3}{L^2}\right) \right\rfloor$$

Consider $w(x)$ to be the vertical deflection. Then

$$w(0) = a_1$$

$$\frac{dw(0)}{dx} = a_2$$

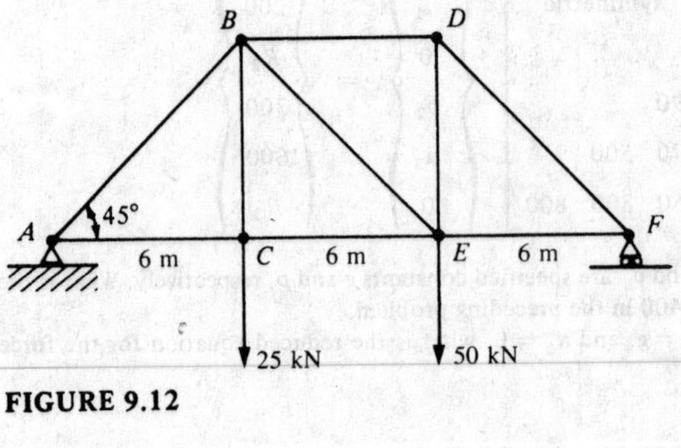

FIGURE 9.12

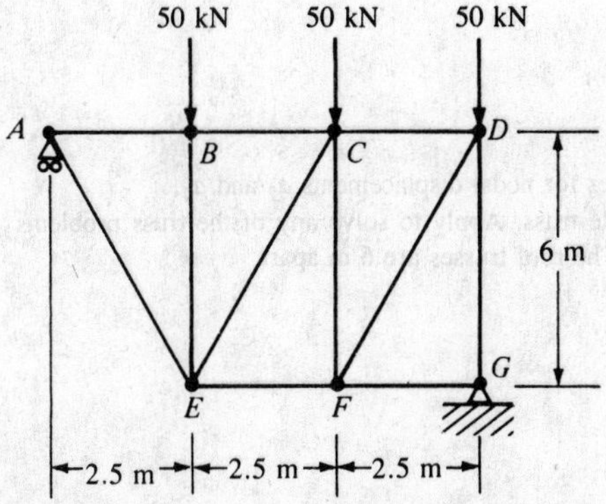

FIGURE 9.13

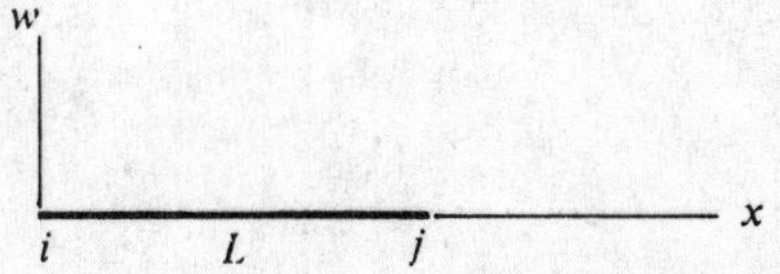

FIGURE 9.14

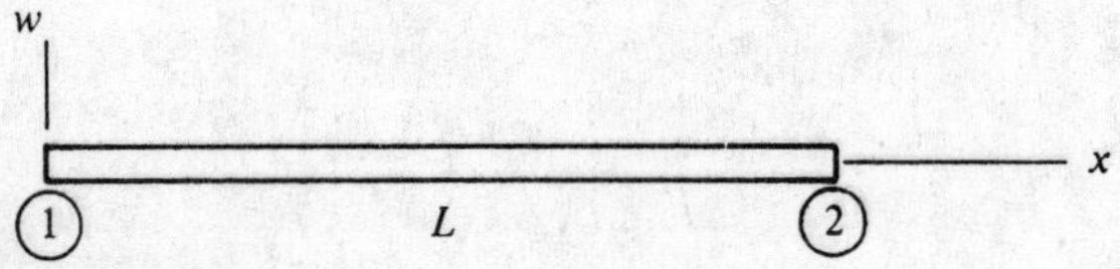

FIGURE 9.15

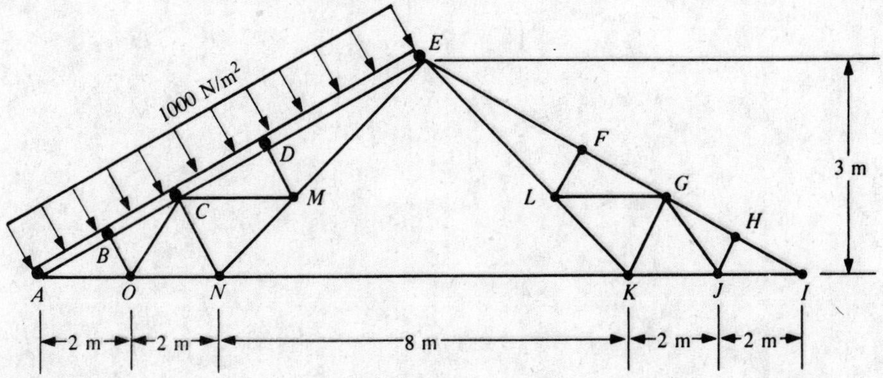

FIGURE 9.16

$$w(L) = a_3$$

$$\frac{dw(L)}{dx} = a_4$$

Show that the matrix N gives the proper values for nodal displacements a_2 and a_4.

*9.14 Formulate a computer program for any simple truss. Apply to solve any of the truss problems 9.10 → 9.11 or the roof truss in Fig. 9.16. The roof trusses are 6 m apart.

10

FINITE ELEMENTS FOR BEAMS

10.1 INTRODUCTION

Beams under vertical loads are almost always encountered in structural work. Furthermore, the computation of shear forces, bending moments, slopes, and deflections is often so tedious that it becomes worthwhile at times to make use of the digital computer via the method of finite elements.

In the method of finite elements for beams, we decompose the beam into small but finite lengths, such as the element in Fig. 10.1, where the beam and an element are shown in the undeformed geometry. We consider the ends of the element as nodal points and have accordingly identified nodes 1 and 2 for the finite element in Fig. 10.1. The method of finite elements for beams computes the desired information for each finite element and then adjusts the results (there will be constants of integration) so that when the elements are put together there results a smooth deformation for the beam that closely resembles the deformation found by the usual analytical methods of integration. The finite element approach gives the desired information at the nodal points of the finite elements comprising the beam, rather than at all points as is the case for analytical solutions.

10.2 DEFINITIONS FOR BEAM ELEMENT DEFORMATION AND LOAD

In our general discussion in Part B of Chapter 9, we considered the nodes to be separate from the elements and we moved all the external forces acting on the element to the nodes while ensuring statical equivalence via the method of virtual work. The generalized forces on the element from the nodes were denoted as $\{q\}$, and it is this force vector that appears in the basic finite element equation $[K]\{a\} = \{q\}$ that we shall use in this chapter. Equation (9.43) gives us a means of determining $\{q\}$ in terms

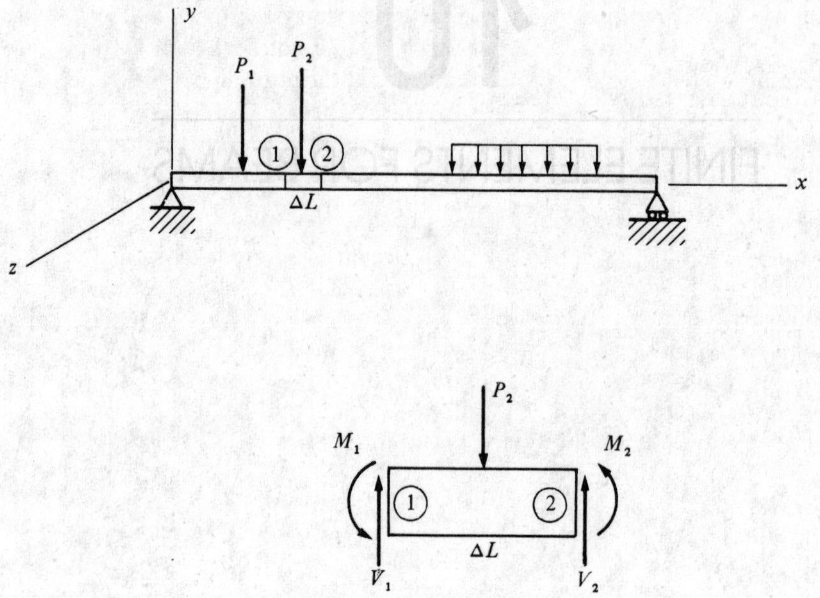

Finite element

FIGURE 10.1
Beam showing a finite element with loads acting on the element.

of the external loads acting on the element. Instead of using this formula, we shall pursue a simpler approach that will enable us to better understand Eq. (9.43) and its development before using it in more complex undertakings of future chapters. First we wish to clear the elements, now separated from their nodes, of the external forces in an appropriate manner. In general, we shall try wherever possible to choose elements so that the external point loads occur on the nodes at the outset. In cases where an applied force occurs between the nodes, such as in the element shown in Fig. 10.1, we can replace the forces by a system of forces and couple moments acting on the element from the nodes. We want two things from this replacement. First, we want *static equivalence* between the initial load and its replacement system. Second, we want the deformations resulting from these two loading systems to be close to each other. We know that the *slopes* of the neutral axis at the *ends* of an element play a predominant role in determining the deformation between the ends. Thus, we ensure that the end slopes resulting from the initial load are equal to the slopes resulting from the replacement system. To better understand this, imagine that the element is simply supported at the nodes as shown in Fig. 10.2*a* with the initial load P applied. The deformed shape is shown dashed. In Fig. 10.2*b* we then show the replacement system of forces and couple moments on the same element so chosen as to give *static equivalence and equal end slopes.* How do we get F_A, M_A, F_B, and M_B? Clearly, if we superpose the initial loading in Fig. 10.2*a* and the *reaction* to the loading in Fig. 10.2*b*, we would achieve zero slopes at the ends of the simply supported element with this system of loads. The simply supported beam would then be, in effect, *fixed* at both ends, as shown in Fig. 10.2*c*. Solving for the supporting forces and couples in Fig. 10.2*c* and then taking the ***reactions*** to this supporting force system gives us the desired F_A, M_A, F_B, and F_c (see Fig. 10.2*d*). Actually, we do not have to go through this kind of procedure. It is simpler to place the nodes at the position of point forces and point couples. We have gone through the procedure for pedagogical reasons.

If we have a uniform loading between two nodes as shown in Fig. 10.3*a*, we may go through the same exercise described above for the load P to get the replace-

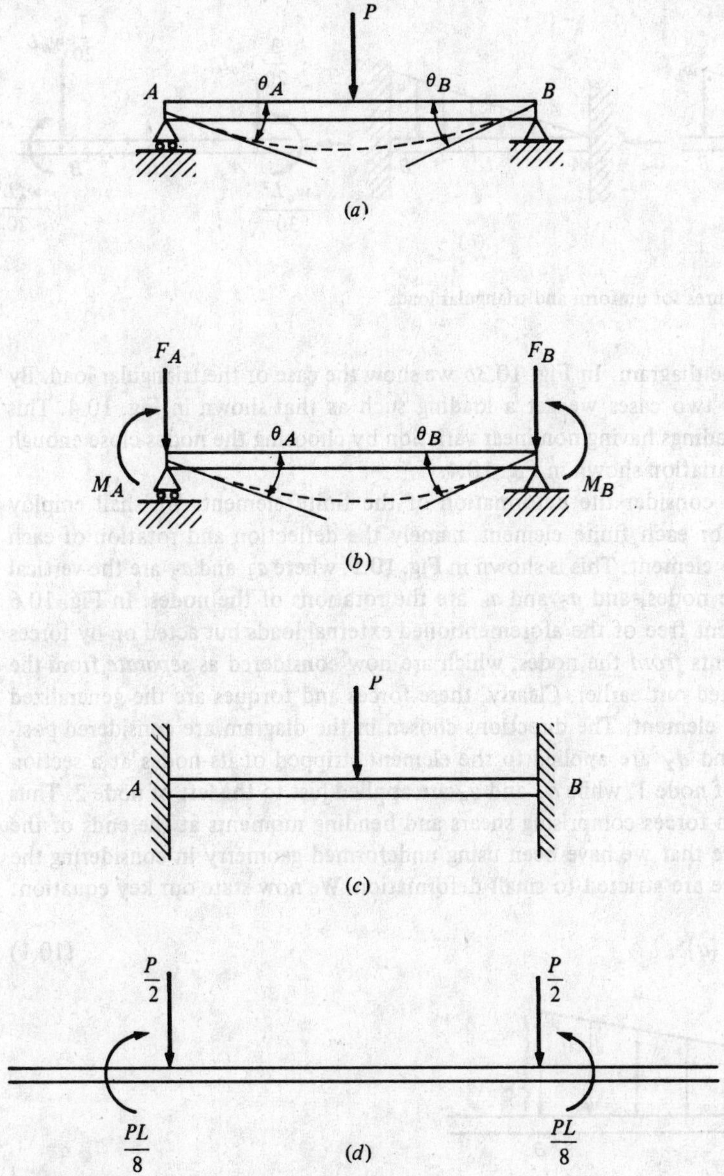

FIGURE 10.2
Replacement of a point force at the center of an element.

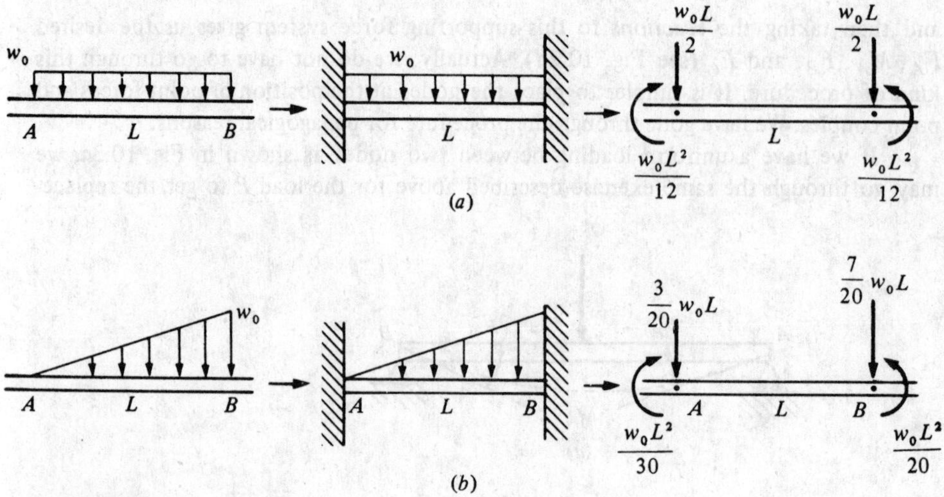

FIGURE 10.3
Replacement procedures for uniform and triangular loads.

ment shown in the diagram. In Fig. 10.3*b* we show the case of the triangular load. By superposing these two cases we get a loading such as that shown in Fig. 10.4. This can be used for loadings having nonlinear variation by choosing the nodes close enough to approach the variation shown in Fig. 10.4.

Now let us consider the deformation of the finite element. We shall employ four parameters for each finite element, namely the deflection and rotation of each node of the finite element. This is shown in Fig. 10.5, where a_1 and a_3 are the vertical deflections of the nodes, and a_2 and a_4 are the rotations of the nodes. In Fig. 10.6 we show an element free of the aforementioned external loads but acted on by forces and couple moments *from* the nodes, which are now considered as *separate* from the elements, as pointed out earlier. Clearly, these forces and torques are the generalized forces $\{q\}$ on the element. The directions chosen in the diagram are considered positive. Forces q_1 and q_2 are applied to the element stripped of its nodes at a section just to the right of node 1, while q_3 and q_4 are applied just to the left of node 2. Thus the q's are section forces comprising shears and bending moments at the ends of the element. Also note that we have been using undeformed geometry in considering the forces and thus we are stricted to small deformation. We now state our key equation:

$$[K]^e\{a\}^e = \{q\}^e \tag{10.1}$$

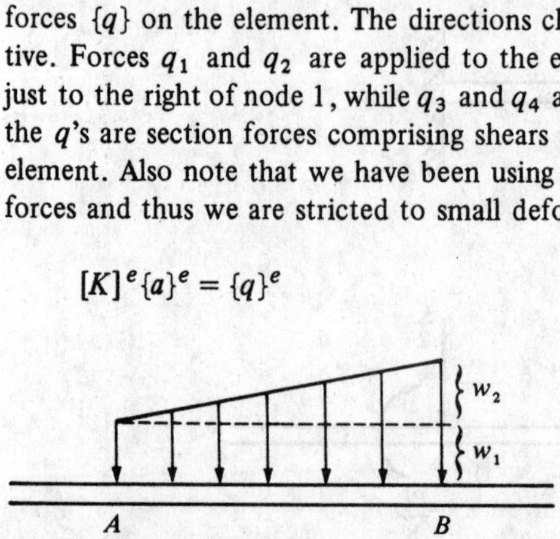

FIGURE 10.4
Loading which may be considered a superposition of uniform loading and triangular loading.

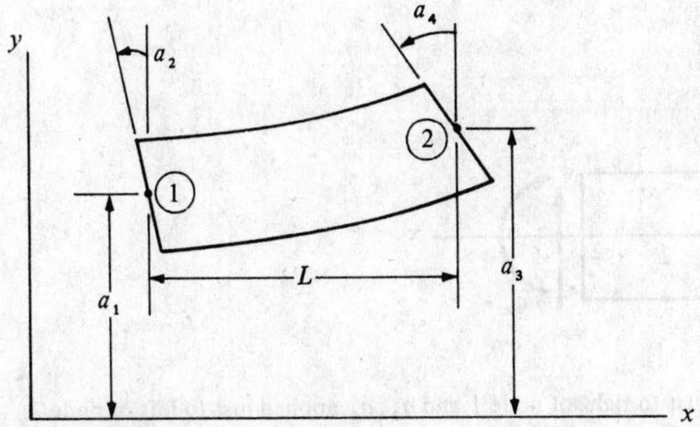

FIGURE 10.5
Element with nodal displacement components.

10.3 KEY MATRICES FOR BEAM ELEMENT AND SYSTEM

Next we wish to derive the interpolation function matrix $[N]$. The differential equation for the finite element in the domain *between* the nodes is, with no transverse loads,

$$\frac{d^4 w}{dx^4} = 0 \tag{10.2}$$

The general solution to the differential equation is easily seen to be

$$w(x) = \alpha_1 + \alpha_2 x + \alpha_3 x^2 + \alpha_4 x^3 \tag{10.3}$$

where the α's are constants.† In considering the nodes, we can relate the nodal displacement components to $w(x)$ as follows:

$$\begin{aligned} w(0) &= a_1 \\ w'(0) &= a_2 \\ w(L) &= a_3 \\ w'(L) &= a_4 \end{aligned} \tag{10.4}$$

Employing Eq. (10.3), we see that

$$\begin{aligned} a_1 &= \alpha_1 \\ a_2 &= \alpha_2 \\ a_3 &= \alpha_1 + \alpha_2 L + \alpha_3 L^2 + \alpha_4 L^3 \\ a_4 &= \alpha_2 + 2\alpha_3 L + 3\alpha_4 L^2 \end{aligned} \tag{10.5}$$

†The number of nodal displacements for an element gives the so-called *degrees of freedom* of the element. Note here that the number of constants α_i equals the number of degrees of freedom. The approach followed here, with a *parameter-laden polynomial* used for the field variable $w(x)$, will be used later in the text in a number of instances to find interpolation functions.

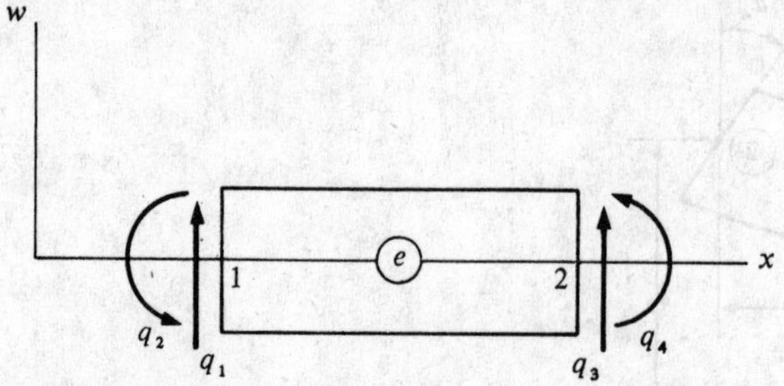

FIGURE 10.6
Finite element with q_1, q_2 applied just to right of node 1 and q_3, q_4 applied just to left of node 2.

In matrix form we have

$$\begin{Bmatrix} a_1 \\ a_2 \\ a_3 \\ a_4 \end{Bmatrix} = \begin{bmatrix} 1 & 0 & 0 & 0 \\ 0 & 1 & 0 & 0 \\ 1 & L & L^2 & L^3 \\ 0 & 1 & 2L & 3L^2 \end{bmatrix} \begin{Bmatrix} \alpha_1 \\ \alpha_2 \\ \alpha_3 \\ \alpha_4 \end{Bmatrix} \tag{10.6}$$

Hence, we have

$$\{a\} = [A]\{\alpha\} \tag{10.7}$$

where $[A]$, *the coefficient matrix*, is given as

$$[A] = \begin{bmatrix} 1 & 0 & 0 & 0 \\ 0 & 1 & 0 & 0 \\ 1 & L & L^2 & L^3 \\ 0 & 1 & 2L & 3L^2 \end{bmatrix} \tag{10.8}$$

Taking the inverse of $[A]$, we can solve for $\{\alpha\}$ in terms of $\{a\}$:

$$\{\alpha\} = [A]^{-1}\{a\} \tag{10.9}$$

where

$$[A]^{-1} = \begin{bmatrix} 1 & 0 & 0 & 0 \\ 0 & 1 & 0 & 0 \\ -\frac{3}{L^2} & -\frac{2}{L} & \frac{3}{L^2} & -\frac{1}{L} \\ \frac{2}{L^3} & \frac{1}{L^2} & -\frac{2}{L^3} & \frac{1}{L^2} \end{bmatrix} \tag{10.10}$$

To get $[N]$, note first from Eq. (10.3) that

$$w = \lfloor 1, x, x^2, x^3 \rfloor \{\alpha\}^e = \lfloor 1, x, x^2, x^3 \rfloor [A]^{-1} \{a\}^e = \lfloor C \rfloor [A]^{-1} \{a\}^e \qquad (10.11)$$

where we have used Eq. (10.9). Matrix $\lfloor C \rfloor$ is called the polynomial matrix. Noting Eq. (9.59*a*), where $\{\vec{u}(x, y, z)\} \equiv w(x)$ for beams, as well as the right side of Eq. (10.11), we have:†

$$[N] = \lfloor C \rfloor [A]^{-1} = \lfloor 1, x, x^2, x^3 \rfloor \begin{bmatrix} 1 & 0 & 0 & 0 \\ 0 & 1 & 0 & 0 \\ -\dfrac{3}{L^2} & -\dfrac{2}{L} & \dfrac{3}{L^2} & -\dfrac{1}{L} \\ \dfrac{2}{L^3} & \dfrac{1}{L^2} & -\dfrac{2}{L^3} & \dfrac{1}{L^2} \end{bmatrix} \qquad (10.12)$$

Thus, we arrive at the interpolation function matrix:

$$[N] = \left\lfloor \left(1 - 3\frac{x^2}{L^2} + 2\frac{x^3}{L^3}\right), \left(x - 2\frac{x^2}{L} + \frac{x^3}{L^2}\right), \left(3\frac{x^2}{L^2} - 2\frac{x^3}{L^3}\right), \left(-\frac{x^2}{L} + \frac{x^3}{L^2}\right) \right\rfloor \qquad (10.13)$$

We note that

$$\lfloor N \rfloor_{x=0} = \lfloor 1,\ 0,\ 0,\ 0 \rfloor$$
$$\lfloor N \rfloor_{x=L} = \lfloor 0,\ 0,\ 1,\ 0 \rfloor$$
$$\lfloor N' \rfloor_{x=0} = \lfloor 0,\ 1,\ 0,\ 0 \rfloor$$
$$\lfloor N' \rfloor_{x=L} = \lfloor 0,\ 0,\ 0,\ 1 \rfloor$$

Hence, the equation $w = \lfloor N \rfloor \{a\}$ gives the proper values of w at the nodes when $[N]$ is given by Eq. (10.13). We also note that for the individual nodes

$$\lfloor N_1 \rfloor = \left\lfloor \left(1 - 3\frac{x^2}{L^2} + 2\frac{x^3}{L^3}\right), \left(x - 2\frac{x^2}{L} + \frac{x^3}{L^2}\right) \right\rfloor$$
$$\lfloor N_2 \rfloor = \left\lfloor \left(3\frac{x^2}{L^2} - 2\frac{x^3}{L^3}\right), \left(-\frac{x^2}{L} + \frac{x^3}{L^2}\right) \right\rfloor \qquad (10.14)$$

†The choice of the shape function is vital in considering the convergence of the finite element method as one increases the number of finite elements in a domain. We shall come back to this consideration later in the chapter. Here the interpolation functions give *exactly* the same deflections between nodal displacements and deflection of the neutral axis as that of strength of materials, as a result of Eqs. (10.2)–(10.3).

Next we define a generalized "strain" ϵ to be

$$\epsilon \equiv -\frac{d^2 w}{dx^2} \tag{10.15}$$

and a generalized "stress" σ to be

$$\sigma \equiv -EI\frac{d^2 w}{dx^2} \tag{10.16}$$

Hence, Eq. (9.59*e*), with no initial strains ϵ_0 or residual stress σ_0, becomes in accordance with the above two equations:

$$\sigma = EI\epsilon$$

Accordingly, the constitutive law matrix is

$$[D] = EI \tag{10.17}$$

Also, consider Eq. (9.59*b*) relating the strain and deformation fields:

$$\{\epsilon\} = [L]\{\vec{u}\} \tag{10.18}$$

With $w \equiv \vec{u}$ and $\epsilon \equiv -d^2w/dx^2$, the strain-displacement field operator matrix $[L]$ is then

$$[L] = \left[-\frac{d^2}{dx^2}\right] \tag{10.19}$$

Now we can go to Eq. (9.59*c*) to find the $[B]$ matrix needed for assembling the stiffness matrix. Thus:

$$[B] = [L][N] = \left(-\frac{d^2}{dx^2}\right)\left\lfloor\left(1 - 3\frac{x^2}{L^2} + 2\frac{x^3}{L^3}\right), \left(x - 2\frac{x^2}{L} + \frac{x^3}{L^2}\right), \left(3\frac{x^2}{L^2} - 2\frac{x^3}{L^3}\right), \left(-\frac{x^2}{L} + \frac{x^3}{L^2}\right)\right\rfloor \tag{10.20a}$$

Therefore

$$\lfloor B\rfloor = \left\lfloor\left(\frac{6}{L^2} - 12\frac{x}{L^3}\right), \left(\frac{4}{L} - \frac{6x}{L^2}\right), \left(-\frac{6}{L^2} + \frac{12x}{L^3}\right), \left(\frac{2}{L} - \frac{6x}{L^2}\right)\right\rfloor \tag{10.20b}$$

Next, we get to the stiffness matrix $[K]^e$. For this purpose we go back briefly to Section 9.4, where we now use the extremization of the total potential energy for beams. The total potential energy for bending is [see Eq. (4.8) for the case where there are only end loads $\{q\}$ on the element†]:

†The external loads and supporting forces will come up later.

$$\pi = \int_0^L \frac{EI}{2}\left(\frac{d^2w}{dx^2}\right)^2 dx - \{a\}^T\{q\}$$

Hence:

$$\delta^{(1)}\pi = 0 = \int_0^L \frac{EI}{2}\, 2\left(\frac{d^2w}{dx^2}\right)\delta\left(\frac{d^2w}{dx^2}\right) dx - \{\delta a\}^T\{q\}$$

Noting that $\delta(d^2w/dx^2) = -\{\delta\epsilon\}$ and that $EI\, d^2w/dx^2 = -\{\sigma\}$, we can say

$$\delta^{(1)}\pi = \int_0^L \{\delta\epsilon\}^T\{\sigma\}\, dx - \{\delta a\}^T\{q\} = 0 \tag{10.21}$$

Let us consider the integrand in the above equation. Using Eq. (9.59*d*) for $\{\delta\epsilon\}^T$, we have

$$\{\delta\epsilon\}^T = [\delta a]^T[B]^T$$

From Eqs. (9.59*e*) and (9.59*d*), furthermore, we have

$$\{\sigma\} = [D]\{\epsilon\} = [D][B]\{a\}$$

Substituting these results for $\{\delta\epsilon\}^T$ and $\{\sigma\}$ into Eq. (10.21), we get

$$\{\delta a\}^T\left[\left(\int_0^L [B]^T[D][B]\, dx\right)\{a\}^e - \{q\}\right] = 0$$

We now see that Eq. (9.59*f*) is still valid for beams, but instead of integrating over a volume, for our "stress" and "strain" we must integrate only along the length L of an element. Using Eqs. (10.20) and (10.17), we accordingly have for $[K]^e$:†

$$[K]^e = \int_0^L \begin{Bmatrix} \left(\dfrac{6}{L^2} - \dfrac{12x}{L^3}\right) \\ \left(\dfrac{4}{L} - \dfrac{6x}{L^2}\right) \\ \left(-\dfrac{6}{L^2} + \dfrac{12x}{L^3}\right) \\ \left(\dfrac{2}{L} - \dfrac{6x}{L^2}\right) \end{Bmatrix} (EI) \left\lfloor \left(\frac{6}{L^2} - \frac{12x}{L^3}\right), \left(\frac{4}{L} - \frac{6x}{L^2}\right), \left(-\frac{6}{L^2} + \frac{12x}{L^3}\right), \left(\frac{2}{L} - \frac{6x}{L^2}\right) \right\rfloor dx$$

†Note that the stiffness matrix $[K]^e$ can be written as

$$[K]^e = \int_0^L EI\{N''\}\lfloor N''\rfloor\, dx$$

This follows from Eq. (10.20*a*), where $\{B\}$ is $\{N''\}$, and also from Eq. (10.17), where $[D] = EI$. Substituting into the formulation $\int [B]^T[D][B]\, dx$ gives us the above result.

Carrying out the integrations, we get

$$[K]^e = \frac{EI}{L^3}\begin{bmatrix} 12 & 6L & -12 & 6L \\ 6L & 4L^2 & -6L & 2L^2 \\ -12 & -6L & 12 & -6L \\ 6L & 2L^2 & -6L & 4L^2 \end{bmatrix} \tag{10.22}$$

We thus have the following equation between the nodal displacement and the forces acting on the element next to the nodal points:

$$[K]^e\{a\}^e = \{q\}^e \tag{10.23}$$

where $[K]^e$ is given by Eq. (10.22). To get the global stiffness matrix, we insert the element stiffness matrices into the grid involving all the nodal displacement components exactly as described in the previous chapter.

10.4 PROCEDURAL DETAILS FOR GLOBAL EQUATIONS

Let us now consider a beam made up of two elements as shown in Figs. 10.7 and 10.8. We wish to find the global stiffness matrix for this system. We form a six by six grid and for element 1 with nodal points 1 and 2 we insert matrix (10.22) to correspond to the first four rows and first four columns. Now go to element 2 having nodal points 2 and 3. Again matrix (10.22) is inserted intact into the last four rows and last four columns. Where there is overlap of the inserted matrices, we merely add the terms at each grid point. The results are shown in the following equation.

$$[K] = \frac{EI}{L^3}\begin{array}{c} \begin{array}{cccccc} 1 & 2 & 3 & 4 & 5 & 6 \end{array} \\ \begin{bmatrix} 12 & 6L & -12 & 6L & 0 & 0 \\ 6L & 4L^2 & -6L & 2L^2 & 0 & 0 \\ -12 & -6L & 24 & 0 & -12 & 6L \\ 6L & 2L^2 & 0 & 8L^2 & -6L & 2L^2 \\ 0 & 0 & -12 & -6L & 12 & -6L \\ 0 & 0 & 6L & 2L^2 & -6L & 4L^2 \end{bmatrix} \end{array} \begin{array}{c} 1 \\ 2 \\ 3 \\ 4 \\ 5 \\ 6 \end{array} \tag{10.24}$$

We now wish to set up the global matrix equation corresponding to Eq. (9.59*g*) for the above two-element beam. We shall take $\{f\}$ to be zero here, and shall now determine what constitutes $\{q\}$. We have shown for the simple truss that this vector represents the external point loads at each nodal point and we shall show this to

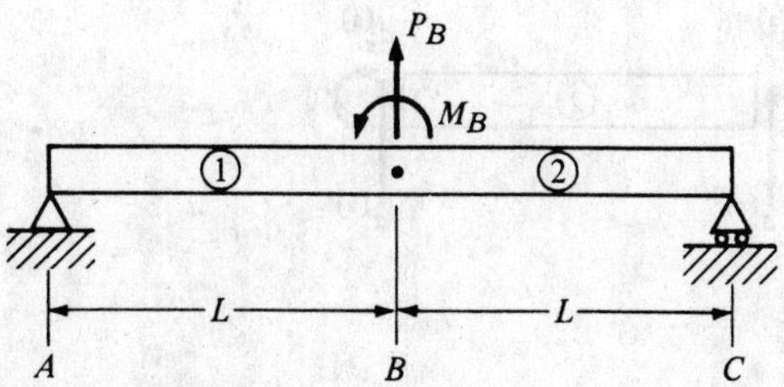

FIGURE 10.7
Beam with two elements.

be the case here. Considering each element separately, as shown in Fig. 10.8, we can say:

Element (1)

$$\begin{Bmatrix} q_1^{(1)} \\ q_2^{(1)} \\ q_3^{(1)} \\ q_4^{(1)} \end{Bmatrix} = \frac{EI}{L^3} \begin{bmatrix} 12 & 6L & -12 & 6L \\ 6L & 4L^2 & -6L & 2L^2 \\ -12 & -6L & 12 & -6L \\ 6L & 2L^2 & -6L & 4L^2 \end{bmatrix} \begin{Bmatrix} a_1^{(1)} \\ a_2^{(1)} \\ a_3^{(1)} \\ a_4^{(1)} \end{Bmatrix} \tag{10.25}$$

Element (2)

$$\begin{Bmatrix} q_3^{(2)} \\ q_4^{(2)} \\ q_5^{(2)} \\ q_6^{(2)} \end{Bmatrix} = \frac{EI}{L^3} \begin{bmatrix} 12 & 6L & -12 & 6L \\ 6L & 4L^2 & -6L & 2L^2 \\ -12 & -6L & 12 & -6L \\ 6L & 2L^2 & -6L & 4L^2 \end{bmatrix} \begin{Bmatrix} a_3^{(2)} \\ a_4^{(2)} \\ a_5^{(2)} \\ a_6^{(2)} \end{Bmatrix} \tag{10.26}$$

We have already shown how, for the global stiffness matrix, columns and rows corresponding to $a_3^{(1)}$ and $a_4^{(1)}$ of element 1 overlap with columns and rows corresponding to $a_3^{(2)}$ and $a_4^{(2)}$. So we know how to form $[K]$ for the entire system. Furthermore, from *compatibility* considerations at nodal point B, we can say (see Fig. 10.9):

$$\begin{aligned} a_3^{(1)} &= a_3^{(2)} \\ a_4^{(1)} &= a_4^{(2)} \end{aligned} \tag{10.27}$$

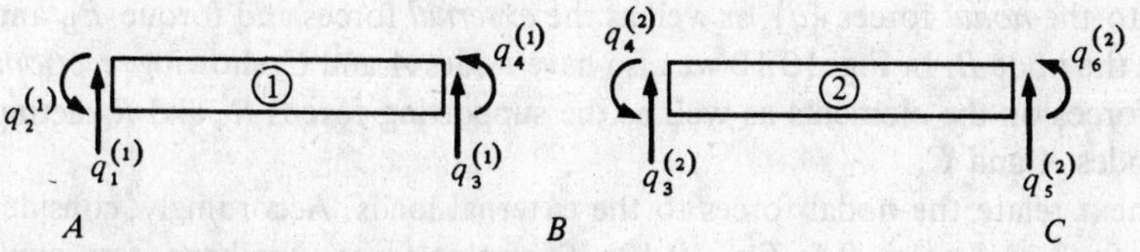

FIGURE 10.8
Elements are separated and stripped of nodes.

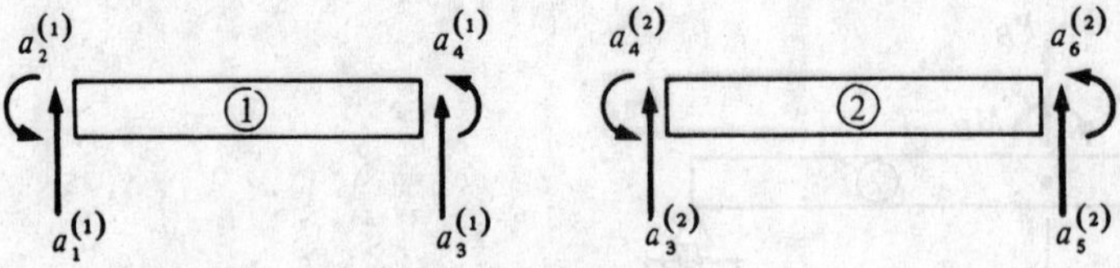

FIGURE 10.9
Elements showing nodal displacements.

We can now give the nodal displacement vector for the entire beam as follows:

$$\begin{Bmatrix} a_1^{(1)} \\ a_2^{(1)} \\ a_3^{(1)} = a_3^{(2)} \\ a_4^{(1)} = a_4^{(2)} \\ a_5^{(2)} \\ a_6^{(2)} \end{Bmatrix} = \begin{Bmatrix} \Delta_A \\ \theta_A \\ \Delta_B \\ \theta_B \\ \Delta_C \\ \theta_C \end{Bmatrix} \tag{10.28}$$

where the new notation is referred to the displacements and rotation of the joints A, B, and C. We then have $[K]$ and $\{a\}$ for the entire beam. Finally, we must determine $\{q\}$ for the entire beam.

In forming the global representation of the forces, we note from Chapter 9 that $\{q\}$ are the total forces *from* the nodes *onto* the continguous elements. To get the *total of the nodal forces acting on the elements from each node*, we can inspect Fig. 10.8. We see that, corresponding to the nodal displacements of Eq. (10.28), we have the following *total* of nodal forces acting *on* the continguous elements *from* the nodes.

$$\begin{Bmatrix} q_1^{(1)} \\ q_2^{(1)} \\ q_3^{(1)} + q_3^{(2)} \\ q_4^{(1)} + q_4^{(2)} \\ q_5^{(2)} \\ q_6^{(2)} \end{Bmatrix} \tag{10.29}$$

We now consider Fig. 10.10*a*, where for node B we have an infinitesimal slice exposing the *reactions* to the *nodal* forces $\{q\}$, as well as the *external* forces and torque, P_B and M_B, acting on the node B. In Fig. 10.10 we also have nodes A and C, showing *reactions* to the *nodal* forces on the elements as well as the supporting forces R_A and R_C acting directly on nodes A and C.

Let us next relate the nodal forces to the external loads. Accordingly, consider the free body for nodal point B in Fig. 10.10*a*. Summing forces, we have from equilibrium

$\underline{\Sigma F_y = 0}$

$$P_B - q_3^{(1)} - q_3^{(2)} = 0$$

Therefore

$$P_B = q_3^{(1)} + q_3^{(2)} \tag{10.30a}$$

Summing moments, we have from equilibrium

$\underline{\Sigma M = 0}$

$$M_B - q_4^{(1)} - q_4^{(2)} = 0$$

Therefore

$$M_B = q_4^{(1)} + q_4^{(2)} \tag{10.30b}$$

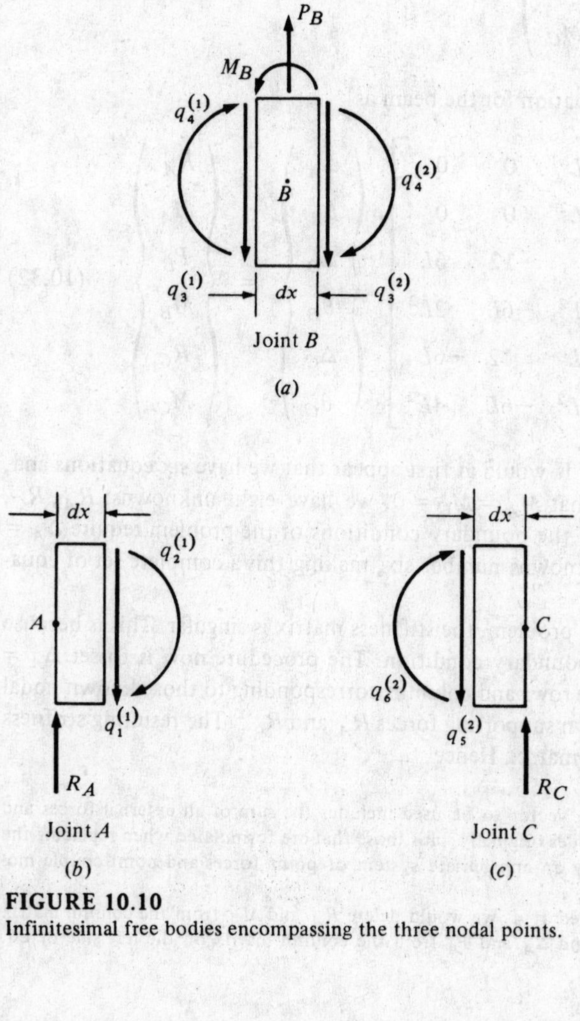

FIGURE 10.10
Infinitesimal free bodies encompassing the three nodal points.

Now $q_4^{(1)}$ and $q_4^{(2)}$ represent positive bending moments in accordance with our convention for the present treatment of finite elements. Also, $q_3^{(1)}$ and $q_3^{(2)}$ represent positive shears. Hence, noting Eqs. (10.30*a*), we can replace $q_3^{(1)} + q_3^{(2)}$ in Eq. (10.29) by P_B. Similarly, noting Eq. (10.30*b*), we can replace $q_4^{(1)} + q_4^{(2)}$ by M_B in Eq. (10.29). From equilibrium considerations of joint A (see Fig. 10.10), meanwhile, we can replace $q_1^{(1)}$ by R_A and $q_2^{(1)}$ by M_A (in this case $M_A = 0$) in Eq. (10.29). Finally, from equilibrium considerations at joint C, we can replace $q_5^{(2)}$ by R_C and $q_6^{(2)}$ by M_C (also equal to zero here) in Eq. (10.29). Thus, the force system that we can use is:

$$\begin{Bmatrix} q_1^{(1)} \\ q_2^{(1)} \\ q_3^{(1)} + q_3^{(2)} \\ q_4^{(1)} + q_4^{(2)} \\ q_5^{(2)} \\ q_6^{(2)} \end{Bmatrix} = \begin{Bmatrix} R_A \\ M_A \\ P_B \\ M_B \\ R_C \\ M_C \end{Bmatrix} \tag{10.31}$$

We now assemble the global equation for the beam as†

$$\frac{EI}{L^3}\begin{bmatrix} 12 & 6L & -12 & 6L & 0 & 0 \\ 6L & 4L^2 & -6L & 2L^2 & 0 & 0 \\ -12 & -6L & 24 & 0 & -12 & 6L \\ 6L & 2L^2 & 0 & 8L^2 & -6L & 2L^2 \\ 0 & 0 & -12 & -6L & 12 & -6L \\ 0 & 0 & 6L & 2L^2 & -6L & 4L^2 \end{bmatrix} \begin{Bmatrix} \Delta_A \\ \theta_A \\ \Delta_B \\ \theta_B \\ \Delta_C \\ \theta_C \end{Bmatrix} = \begin{Bmatrix} R_A \\ M_A \\ P_B \\ M_B \\ R_C \\ M_C \end{Bmatrix} \tag{10.32}$$

Can we solve these equations? It would at first appear that we have six equations and, noting from earlier remarks that $M_A = M_C = 0$, we have eight unknowns: $R_A, R_C, \Delta_A, \theta_A, \Delta_B, \theta_B, \Delta_C, \theta_C$. But the boundary conditions of the problem require $\Delta_A = \Delta_C = 0$, and so the actual unknowns number six, making this a complete set of equations.

However, as in the truss problem, the stiffness matrix is singular. This is because we have not yet put in the boundary condition. The procedure now is to set $\Delta_A = \Delta_C = 0$. Then we eliminate the rows and columns corresponding to those known nodal displacements and the unknown supporting forces R_A and R_c.‡ The resulting stiffness matrix is our *reduced* stiffness matrix. Hence

†Note that the force system vector to be used includes the sum of all external forces and couple moments that are at the nodes originally, plus those that are formulated when replacing the external loads on the elements by an appropriate system of point forces and point couple moments at the nodes.

‡If the beam were cantilevered at A, we would delete R_A and M_A from the column matrix on the right side of Eq. (10.32) and Δ_A and θ_A from the column matrix on the left side of Eq. (10.32) and so on.

$$\frac{EI}{L^3}\begin{bmatrix} 4L^2 & -6L & 2L^2 & 0 \\ -6L & 24 & 0 & 6L \\ 2L^2 & 0 & 8L^2 & 2L^2 \\ 0 & 6L & 2L^2 & 4L^2 \end{bmatrix}\begin{Bmatrix} \theta_A \\ \Delta_B \\ \theta_B \\ \theta_C \end{Bmatrix} = \begin{Bmatrix} M_A \\ P_B \\ M_B \\ M_C \end{Bmatrix} \tag{10.33}$$

Now setting $M_A = M_C = 0$ and considering the case where $P_B = -4500$ N and $M_B = 0$, we invert the stiffness matrix and solve for θ_A, Δ_B, θ_B, and θ_C. We get the following results:

$$\theta_A = -\frac{1{,}125L^2}{EI} \qquad \Delta_B = -\frac{750L^3}{EI}$$
$$\theta_B = 0 \qquad \theta_C = +\frac{1{,}125L^2}{EI} \tag{10.34}$$

Equation (10.34) gives us an approximation of the deformation of the beam. To get the unknown supporting forces (in this case a trivial result), go back to Eq. (10.32) and consider rows R_A and R_C as follows:

$$\frac{EI}{L^3}\begin{bmatrix} 12 & 6L & -12 & 6L & 0 & 0 \\ 0 & 0 & -12 & -6L & 12 & -6L \end{bmatrix}\begin{Bmatrix} \Delta_A \\ \theta_A \\ \Delta_B \\ \theta_B \\ \Delta_C \\ \theta_C \end{Bmatrix} = \begin{Bmatrix} R_A \\ R_C \end{Bmatrix}$$

Inserting known values for the nodal displacements, we get

$$R_A = \frac{EI}{L^3}\left[(12)(0) + 6L\left(-\frac{1{,}125L^2}{EI}\right) - 12\left(-\frac{750L^3}{EI}\right) - 6L(0) + 0 + 0\right]$$
$$= -6(1{,}125) + 12(750) = 2{,}250 \text{ N}$$
$$R_C = \frac{EI}{L^3}\left[0 + 0 - 12\left(-\frac{750L^3}{EI}\right) - 6L(0) + 12(0) - 6L\left(\frac{1{,}125L^2}{EI}\right)\right]$$
$$= 2{,}250 \text{ N}$$

From Eq. (10.31) we can get shear and bending moments just ouside the nodes.

EXAMPLE 10.1 We now consider the statically indeterminate cantilever beam shown in Fig. 10.11. We have chosen three finite elements in order to have the

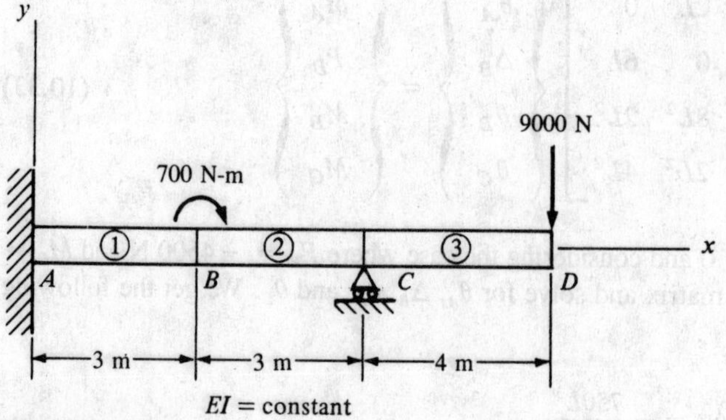

FIGURE 10.11
Statically indeterminate cantilever with three elements.

applied and supporting forces and torques at the nodes. To get the *element stiffness matrices*, we can then use Eq. (10.22). Thus

$$[K]^{(1)} = \frac{EI}{27}\begin{bmatrix} 12 & 18 & -12 & 18 \\ 18 & 36 & -18 & 18 \\ -12 & -18 & 12 & -18 \\ 18 & 18 & -18 & 36 \end{bmatrix} \qquad (a)$$

$$[K]^{(2)} = \frac{EI}{27}\begin{bmatrix} 12 & 18 & -12 & 18 \\ 18 & 36 & -18 & 18 \\ -12 & -18 & 12 & -18 \\ 18 & 18 & -18 & 36 \end{bmatrix} \qquad (b)$$

$$[K]^{(3)} = \frac{EI}{27}\begin{bmatrix} 5.063 & 10.125 & -5.063 & 10.125 \\ 10.125 & 27 & -10.125 & 13.5 \\ -5.063 & -10.125 & 5.063 & -10.125 \\ 10.125 & 13.5 & -10.125 & 27 \end{bmatrix} \qquad (c)$$

Note that a_3 and a_4 of element 1 coincide with a_3 and a_4 of element 2. We recall in forming $[K]$ that the third and fourth rows and columns of $[K]^{(1)}$ overlap the first and second rows and columns of $[K]^{(2)}$. Furthermore, the overlapped elements add. This is clearly also true for the third and fourth rows and columns of element 2 vis-à-vis the first and second rows and columns of element 3. The following is according to the proper beam stiffness matrix $[K]$. Taking blank entries to imply a value of zero, we have

$$[K] = \frac{EI}{27}\begin{bmatrix} 12 & 18 & -12 & 18 & & & & \\ 18 & 36 & -18 & 18 & & & & \\ -12 & -18 & 24 & 0 & -12 & 18 & & \\ 18 & 18 & 0 & 72 & -18 & 18 & & \\ & & -12 & -18 & 17.1 & 7.875 & -5.063 & 10.125 \\ & & 18 & 18 & -7.875 & 63 & -10.125 & 13.5 \\ & & & & -5.063 & -10.125 & 5.063 & -10.125 \\ & & & & 10.125 & 13.5 & -10.125 & 27 \end{bmatrix} \qquad (d)$$

$[K]^{(1)}$, $[K]^{(2)}$, $[K]^{(3)}$

The basic equation approximating beam behavior that corresponds to Eq. (10.31) for two elements is given as follows for this problem:

$$[K]\begin{Bmatrix} \Delta_A \\ \theta_A \\ \Delta_B \\ \theta_B \\ \Delta_C \\ \theta_C \\ \Delta_D \\ \theta_D \end{Bmatrix} = \begin{Bmatrix} R_A \\ M_A \\ P_B \\ M_B \\ R_C \\ M_C \\ P_D \\ M_D \end{Bmatrix} \qquad (e)$$

Our next step is to simplify the above system. Note first that $\Delta_A = \theta_A = 0$—that is, we have two constraints at the first node— and $\Delta_C = 0$ at the third node. This means that we delete R_A, M_A, and R_C from the column matrix on the right side and Δ_A, θ_A, and Δ_C from the column matrix on the left side of the above equation. In addition, we delete rows and columns 1, 2, and 5 from the beam stiffness matrix $[K]$. We are then left with the following equation to work with:

$$\frac{EI}{27}\begin{bmatrix} 24 & 0 & 18 & 0 & 0 \\ 0 & 72 & 18 & 0 & 0 \\ 18 & 18 & 63 & -10.125 & 13.5 \\ 0 & 0 & -10.125 & 5.063 & -10.125 \\ 0 & 0 & 13.5 & -10.125 & 27 \end{bmatrix}\begin{Bmatrix} \Delta_B \\ \theta_B \\ \theta_C \\ \Delta_D \\ \theta_D \end{Bmatrix} = \begin{Bmatrix} P_B \\ M_B \\ M_C \\ P_D \\ M_D \end{Bmatrix} \qquad (f)$$

Noting that $M_B = -700$ N-m and $P_D = -9000$ N, we get, on finding the inverse matrix for $[K]$, the following equations:

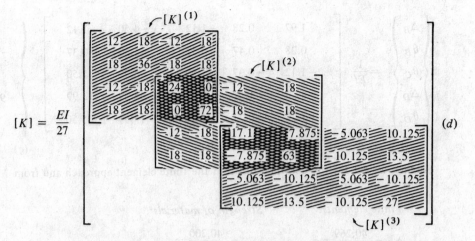

$$\begin{Bmatrix} \Delta_B \\ \theta_B \\ \theta_C \\ \Delta_D \\ \theta_D \end{Bmatrix} = \frac{1}{EI} \begin{bmatrix} 1.97 & 0.28 & -1.12 & -4.50 & -1.12 \\ 0.28 & 0.47 & -0.37 & -1.50 & -0.37 \\ -1.12 & -0.37 & 1.50 & 5.99 & 1.50 \\ -4.50 & -1.5 & 5.99 & 45.29 & 13.99 \\ -1.12 & -0.37 & 1.50 & 13.99 & 5.50 \end{bmatrix} \begin{Bmatrix} \\ -70 \\ \\ -900 \\ \\ \end{Bmatrix} \qquad (g)$$

Solving, we get for the deformation from the finite element approach and from strength of materials calculations:†

Finite elements	*Strength of materials*
$\Delta_B = \dfrac{40,269}{EI}$ m	$\Delta_B = \dfrac{40,300}{EI}$ m
$\theta_B = \dfrac{13,161}{EI}$ rad	$\theta_B = \dfrac{12,123}{EI}$ rad
$\theta_C = -\dfrac{53,692}{EI}$ rad	$\theta_C = -\dfrac{53,736}{EI}$ rad
$\Delta_D = -\dfrac{406,594}{EI}$ m	$\Delta_D = -\dfrac{406,850}{EI}$ m
$\theta_D = -\dfrac{125,627}{EI}$ rad	$\theta_D = -\dfrac{125,737}{EI}$ rad

(h)

We now have an approximation for the deformation of the beam. To get the shear and bending moment approximation, go back to Eq. (*e*) for each element. Thus, for the first element we have, using Eq. (*a*),

$$\begin{Bmatrix} q_1^{(1)} \\ q_2^{(1)} \\ q_3^{(1)} \\ q_4^{(1)} \end{Bmatrix} = \frac{EI}{27} \begin{bmatrix} 12 & 18 & -12 & 18 \\ 18 & 36 & -18 & 18 \\ -12 & -18 & 12 & -18 \\ 18 & 18 & -18 & 36 \end{bmatrix} \begin{Bmatrix} 0 \\ 0 \\ \dfrac{40,269}{EI} \\ \dfrac{13,161}{EI} \end{Bmatrix} \qquad (i)$$

We have then from finite elements and from strength of materials

Finite elements	*Strength of materials*
$q_1^{(1)} = -9,123$ N	$V(0^+) = 9,131$ N
$q_2^{(1)} = -18,072$ N-m	$M(0^+) = 18,088$ N-m
$q_3^{(1)} = 9,123$ N	$V(0^-) = 9,131$ N
$q_4^{(1)} = -9,298$ N-m	$V(0^-) = -9,305$ N-m

(j)

†Should the closeness of the finite element and strength of materials results surprise you even for such large elements? See comments on this point in the closure section of this chapter.

We can operate similarly on the other two elements, thereby obtaining an approximation to the shear force and bending moment distribution in the beam.

To obtain the supporting forces we go back to Eqs. (*e*) and (*d*) and use the rows that were crossed out when we reduced the stiffness matrix. Thus we have

$$\begin{Bmatrix} R_A \\ M_A \\ R_C \end{Bmatrix} = \frac{EI}{27}\begin{bmatrix} 12 & 18 & -12 & 18 & 0 & 0 & 0 & 0 \\ 18 & 36 & -18 & 18 & 0 & 0 & 0 & 0 \\ 0 & 0 & -12 & -18 & 17.06 & -7.875 & -5.063 & 10.125 \end{bmatrix}\begin{pmatrix} \Delta_A \\ \theta_A \\ \Delta_B \\ \theta_B \\ \Delta_C \\ \theta_C \\ \Delta_D \\ \theta_D \end{pmatrix} \tag{k}$$

If we substitute for the nodal deflections that are known and those that have been solved we can get the unknown supporting forces. Thus we have for both finite elements and strength of materials:

Finite elements	*Strength of materials*	
$R_A = -9{,}123$ N	$R_A = -9{,}131$ N	
$M_A = -18{,}072$ N-m	$M_A = -18{,}088$ N-m	(*l*)
$R_C = 18{,}123$ N	$R_C = 18{,}131$ N	

10.5 CLOSURE

In Chapter 9 we considered trusses to introduce the finite element concept in the most simple manner. We then were able to set forth the general displacement method for finite elements. In this chapter we were able to use this method to solve beam problems. We could expand the procedures considered here to solve frame problems.

You will note that for both trusses and beams we got excellent results even though in the latter case very few elements were used. These good results were not fortuitous. For trusses the nodal displacements *completely* and *exactly* determine the deformation of all the members (or elements) of the truss. Thus we have *continuity* of deformation at the nodes between concurrent members. In the case of beams the relation between the nodal displacements for an element and the deformation of the element itself (as determined via the interpolation function) is *exactly* that of *strength of materials.* Thus nodal displacements of an element determine exactly the deformation of the element itself from the viewpoint of strength of materials (i.e., the deflection and slope of the neutral axis). Furthermore, we achieve in the finite element process *continuity* of deflection and slope between elements at the nodes.†

When such continuity of deformation between elements is possible, we say that we have *conformable* elements. In more complex structures such as plates and torsion rods we may not choose to use comformable elements to work with, for reasons to be discussed later. In such cases convergence criteria for the finite element process as the size of the element is decreased are very important. That is, achieving good results requires additional considerations beyond those needed for truss and beam problems. Hence, before we can proceed to more complex problems, we must pause to present in some detail such basic considerations as the choice of elements, consideration of the appropriate number of degrees of freedom for an element, finding of proper interpolation functions, and the important questions of the convergence requirements for the finite element process. We consider these items and more in the following chapter. We can then move to plane stress, the bending of plates, torsion, etc., as well as non-solid-mechanical applications of the finite element process.

PROBLEMS

10.1 Starting with the interpolation function matrix $[N]$ as given by Eq. (10.13) and the $[D]$ matrix as given by Eq. (10.17), compute K_{11} and K_{12} of the element stiffness matrix, given as

$$\int_0^L [B]^T[D][B]\,dx$$

Check your results with Eq. (10.22).

10.2 In the preceding problem, get K_{11} and K_{12} of the element stiffness matrix for an element with the formulation

$$[K]^e = \int_0^L EI\{N''\}\lfloor N''\rfloor\,dx$$

10.3 Consider an axial stiffness element as shown in Fig. 10.12. This element has two nodal displacements and hence has two degrees of freedom. Find the following matrices for this element:

$[N]$
$[D]$
$[K]$

In Eq. (9.48*f*), use for dv the product $A\,dx$, where A is the cross-sectional area.

†Consider the case where only point forces and point couples act on the beam and choose the elements so that these point loads occur at nodes. This was the case in Example 10.1. In such instances the number of finite elements is not significant for beam calculations–the results are the same as those from strength of materials, where only one element effectively is used, normally the whole beam itself. For *distributed* loads we get only approximate solutions by the finite element process. Here, size of the elements is an important consideration.

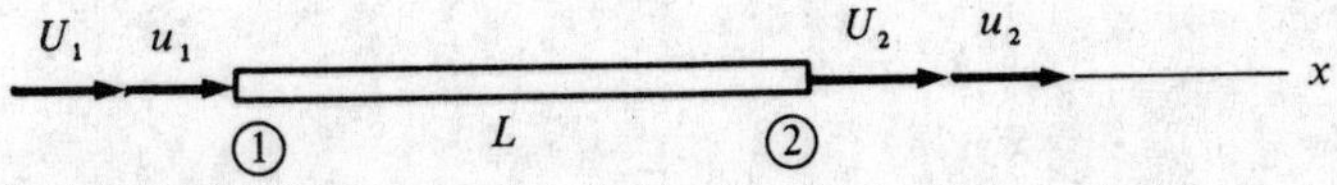

FIGURE 10.12

10.4 By the method of finite elements, find axial displacements at points a, b, and c. Use the data given in Fig. 10.13. Also use the result of the preceding problem, where we got

$$[K]^e = \frac{A_e E_e}{L_e} \begin{bmatrix} 1 & -1 \\ -1 & 1 \end{bmatrix}$$

10.5 For an axial element (see Problem 10.3) at an angle θ (see Fig. 10.14), we have shown that the local stiffness matrix is given as

$$[\bar{K}] = \frac{AE}{L} \begin{bmatrix} 1 & -1 \\ -1 & 1 \end{bmatrix}$$

Let us imagine that we have $\bar{v}_2$ and $\bar{v}_1$ but that for local coordinates they are zero. What is the stiffness matrix $[\bar{K}]$, including the u's? Now determine the stiffness matrix $[K]$ for global coordinates, using the results of Section 9.5.

10.6 If we have a triangular element in plane stress with 6 degrees of freedom (see Fig. 10.15), what is the transformation matrix $[T]$ relating nodal displacements between local and global coordinates?

10.7 Using the three elements indicated in Fig. 10.16, find the nodal displacements and rotations for the cantilever beam. Take $EI =$ const.

$A_1 = 600 \text{ mm}^2$
$A_2 = 1,200 \text{ mm}^2$
$A_3 = 1,800 \text{ mm}^2$

$F_1 = 500 \text{ N}$
$F_2 = 300 \text{ N}$
$F_3 = 400 \text{ N}$

$E_1 = 2.1 \times 10^{11} \text{ Pa}$
$E_2 = 0.7 \times 10^{11} \text{ Pa}$
$E_3 = 1.4 \times 10^{11} \text{ Pa}$

$\leftarrow$150 mm$\rightarrow\leftarrow$175 mm$\rightarrow$100 mm

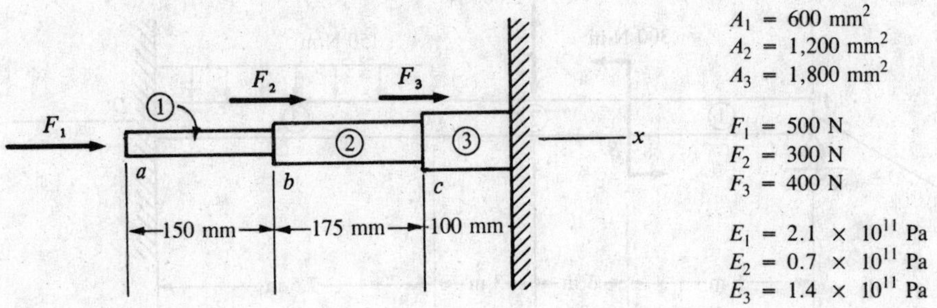

FIGURE 10.13

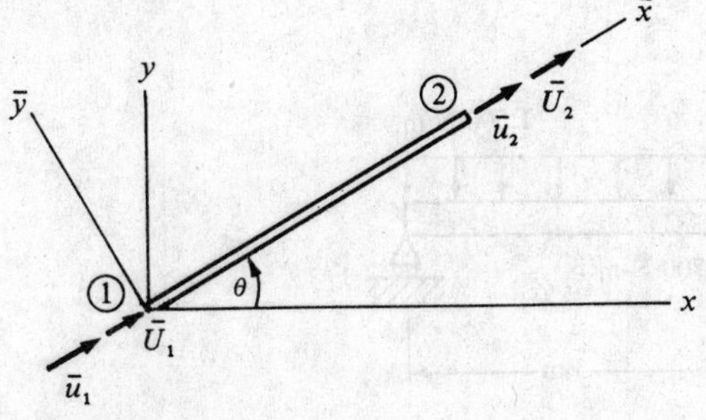

FIGURE 10.14

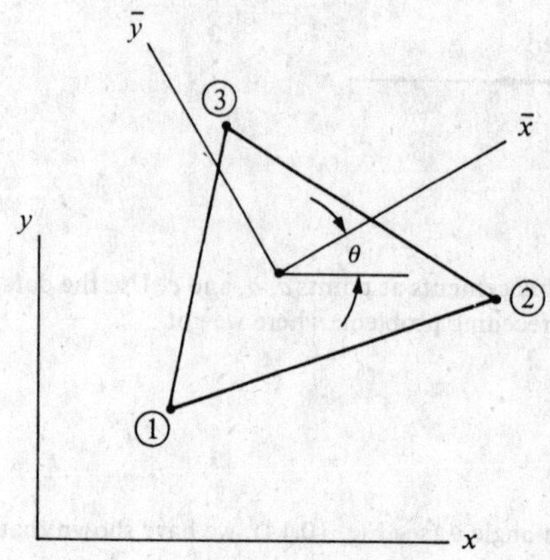

FIGURE 10.15

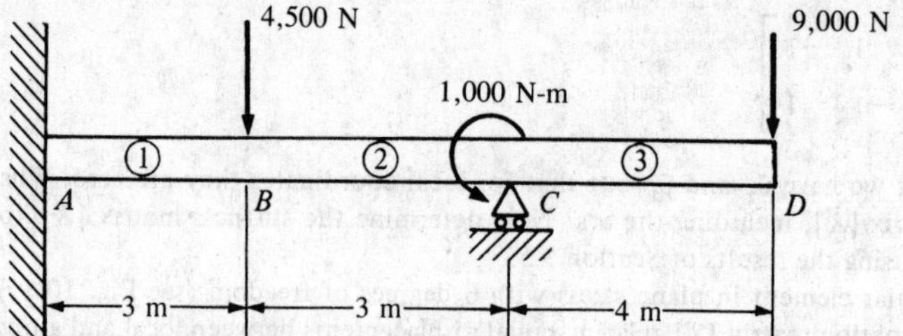

FIGURE 10.16

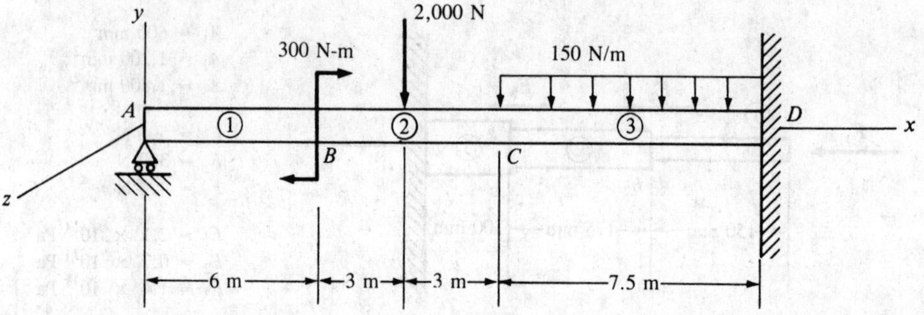

FIGURE 10.17

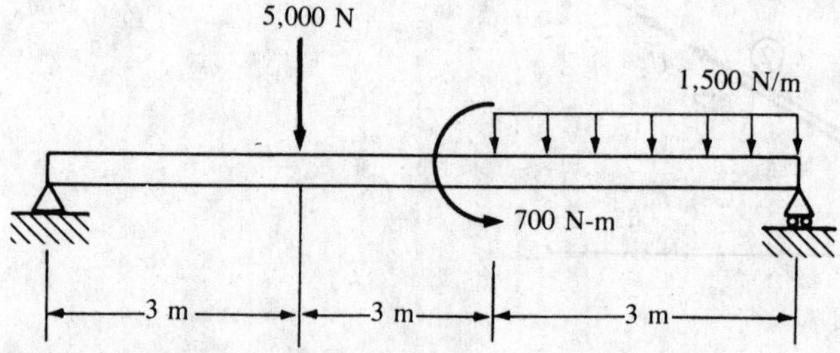

FIGURE 10.18

10.8 Using three elements AB, BC, and CD (Fig. 10.17), find the nodal displacements and rotations as well as the supporting forces. Take $EI =$ const.

10.9 Using three small elements under the uniform load and two large elements for the rest of the beam (Fig. 10.18), find the nodal displacements and internal forces at nodes.

*10.10 Write a program for any beam problem and apply this program to Problems 10.7 to 10.9.

11

ELEMENTS AND INTERPOLATION FUNCTIONS

11.1 INTRODUCTION

In Chapter 9 we considered trusses, where the members of the truss were themselves the finite elements. We then considered triangular elements to help illustrate the methodology of finite elements in a fairly general manner. In Chapter 10 we used this general approach to study beams, where the finite elements were finite lengths of beams. In Chapter 11 we shall introduce other kinds of elements along with pertinent definitions in anticipation of dealing with more complex finite element problems. We shall discuss requirements for convergence of the finite element process as the elements become smaller and more numerous. We shall also consider interpolation functions.

11.2 GENERAL ELEMENT CONSIDERATIONS

Let us first present some elements in which the nodes are at the vertices of the elements. We shall call these elements simple elements. In Fig. 11.1 we show some commonly used simple elements.

You will recall that in discussing beam deflection we expressed the deflection $w(x)$ as a parameter-laden polynomial. This will often be the case as we proceed to study other problems. If we use polynomials for the case of one nodal variable per node, then for *linear* variation of the variable in an element, we use nodes only at the *corners*, as shown in Fig. 11.1. If a quadratic variation of the variable for this case is to be used, we place an *additional* node at approximately the middle of each edge. This is shown in Fig. 11.2. For higher degrees of variation we may include more nodes for the elements. We point out here that nodes may be added *inside* the elements as well as along the edges. However, interior nodes are generally to be avoided because of computational complexity.

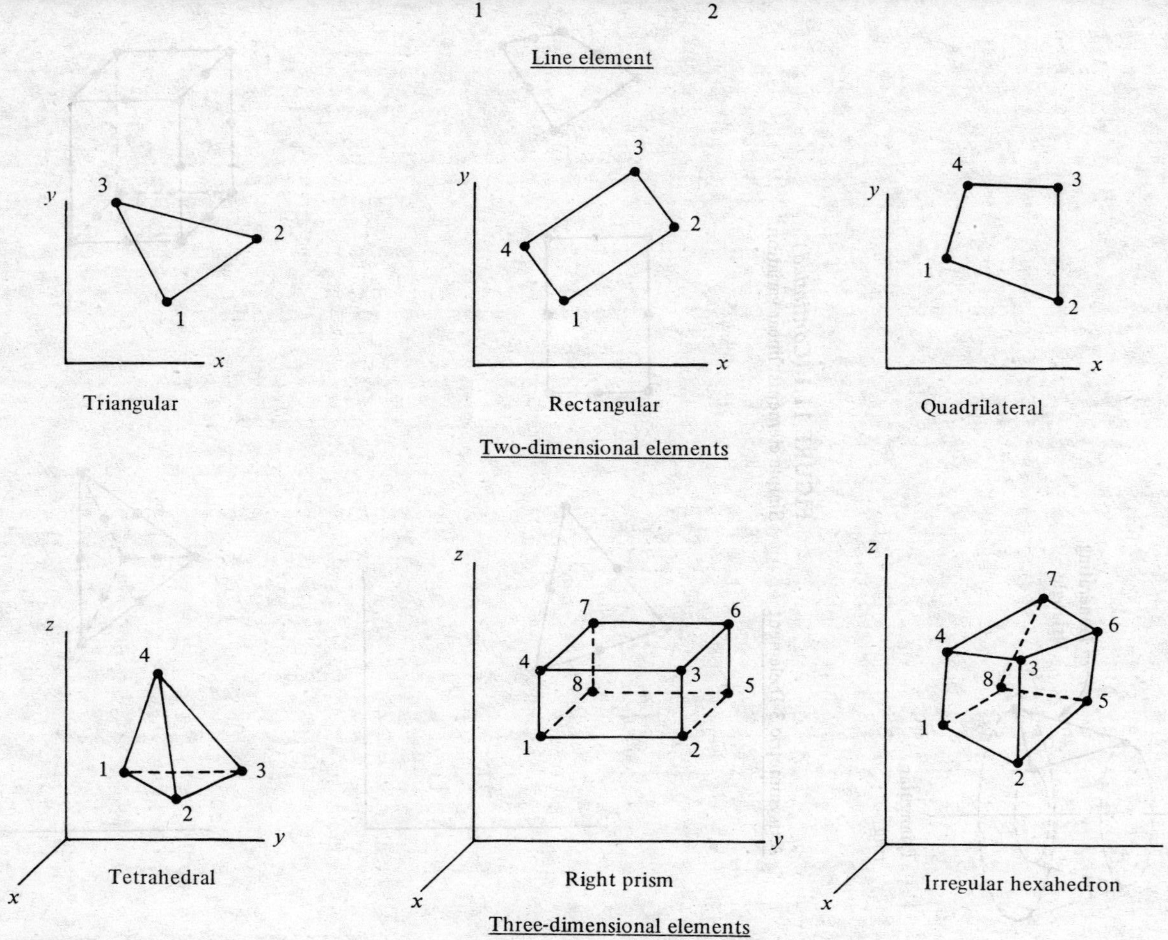

FIGURE 11.1
Simple elements; linear variation.

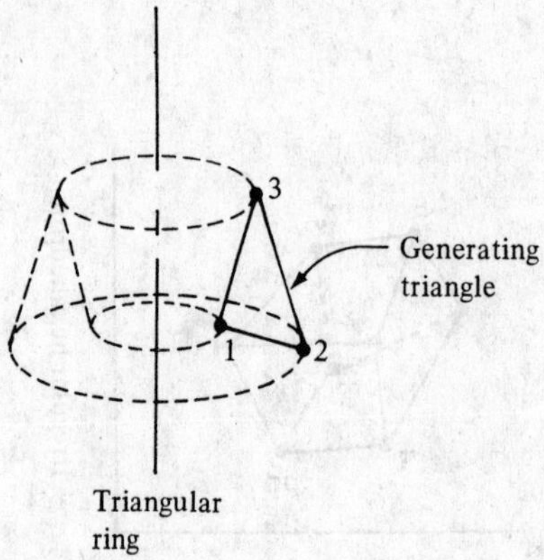

Axisymmetric 3-D element

FIGURE 11.1 (*Continued*)
Simple elements; linear variation.

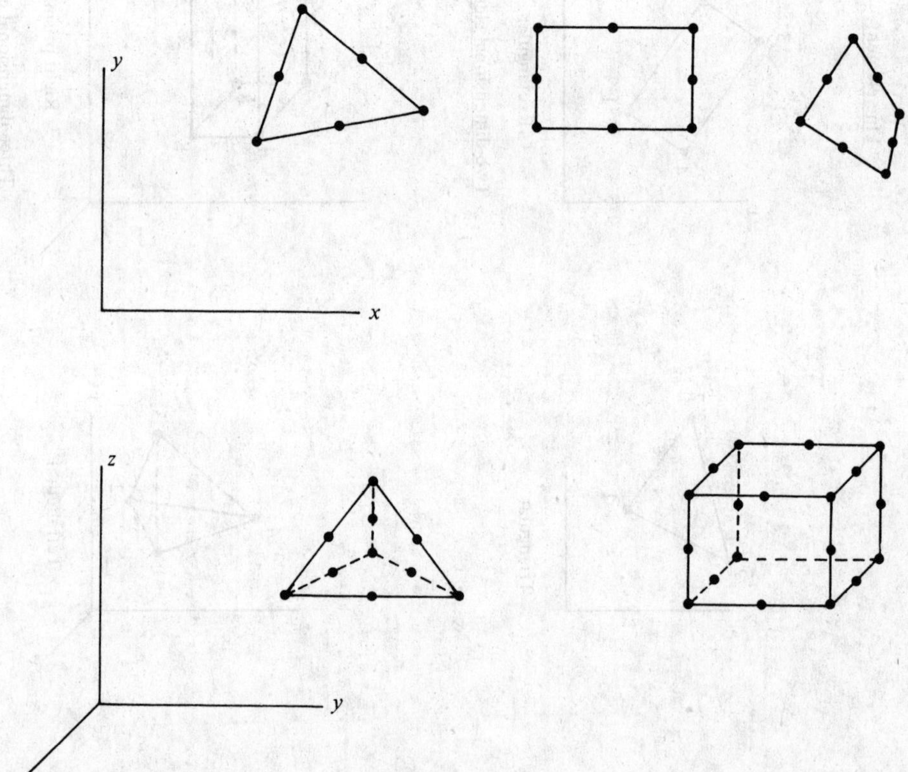

FIGURE 11.2
Quadratic elements for one nodal variable per node.

Let us go back again to the previous work on beams. Recall that for each node we had two nodal variables, namely the displacement and rotation. Thus, an element has four nodal variables. Each of these variables is called a *degree of freedom* (DOF), so that for the beam element we had four degrees of freedom. We will use this terminology for other elements. For instance, the linear triangle with one nodal variable per node has three degrees of freedom.

11.3 CONVERGENCE REQUIREMENTS FOR INTERPOLATION FUNCTIONS

There are certain requirements for the interpolation functions of an element that will now be presented. These requirements, given below, must be met if we are to be assured of convergence of the process as the elements are increased in number and decreased in size. Thus:

1. The interpolation functions must not allow a strain to appear if the nodal displacements are such as to be compatible with a rigid-body displacement.
2. If the nodal displacements are compatible with a uniform strain in the elements, then the interpolation functions must yield this strain for the nodal displacements.

When dealing with fields of study other than solid mechanics, condition 2 can be stated in the following, more general manner:

2A. If the nodal variables are compatible with uniform states of the variable ϕ and any of its derivatives up to the highest in the appropriate quadratic functional $I(\phi)$, then these uniform states must be preserved in the element as it shrinks to zero size. This condition is called the *completeness* condition.

We present yet another condition for convergence for solid mechanics.

3. The interpolation functions should be chosen so that the strain remains finite at the boundary. Note, however, that the strain can be indeterminate at a boundary. If the strain involves only first derivatives, this means that the displacement field must be continuous across a boundary. If, on the other hand, we are dealing with a generalized "strain" involving second derivatives of displacement components, such as the beam problem of Chapter 10, then the first partials of the displacement field components must be continuous across a boundary.

Again, if we generalize condition 3 to other fields of study, we have the following condition:

3A. The dependent variable ϕ and all its derivatives up to *one less* than the highest-order derivative in the appropriate quadratic functional $I(\phi)$ should be continuous at the element interfaces.

The latter conditions, 3 and 3A, are called *compatibility* conditions. Elements satisfying these conditions are called *conformable* elements.

To describe the degree of compatibility achieved by interpolation functions at the element interfaces, we have a specific notation in common use. We say that we have C^0 continuity when the field variable only (and none of its derivatives) maintains continuity at an interelement interface. We say that we have C^1 continuity when the field variable and its first derivatives are continuous at element interfaces. C^2 continuity indicates that the field variable, its first derivatives, and its second derivatives are all continuous at boundary interfaces, and so forth. Now we can say that for *assurance of finite element convergence for a functional having p as the highest order of a derivative, completeness requires C^p continuity while compatibility requires C^{p-1} continuity.*

Now that we have described certain elements in terms of general shape and have considered certain requirements on interpolation functions for convergence, we turn to ways of finding interpolation functions.

Part A
One-Dimensional Elements

11.4 INTERPOLATION FUNCTIONS FOR ONE-DIMENSIONAL ELEMENTS

In Chapter 10 we considered the beam problem and we started by expressing the field variable $w(x)$ in terms of a third-order, parameter-laden polynomial [see Eq. (10:3)]. Going back to Section 10.3, you will note that we related the nodal displacement vector $\{a\}$ to the parameter vector $\{\alpha\}$ [see Eq. (10.6)] to form the coefficient matrix $[A]$. Noting that the field variable w is expressible as [see Eq. (10.11)]

$$w = \lfloor C \rfloor \{\alpha\}^e = \lfloor C \rfloor [A]^{-1} \{a\}^e \tag{11.1}$$

and also that w is expressed as [see Eq. (9.59*a*)]

$$w = [N]\{a\} \tag{11.2}$$

we then have for $[N]$ the result

$$[N] = \lfloor C \rfloor [A^{-1}] \tag{11.3}$$

We shall have occasion to use this approach again as we move along in the text.

In the above discussions we worked with a polynomial with one independent variable [see Eq. (10.3)]. We can express this one-dimensional approach in general for any field variable ϕ as

$$\phi(x) = \sum_{i=0}^{n} \alpha_i x^i \tag{11.4}$$

With $n = 1$ we have a linear variation, with $n = 2$ we have a quadratic variation, and so on. Note that for an nth-order polynomial there will be $n + 1$ terms. We now illustrate the use of the above formulations.

EXAMPLE 11.1 Consider a one-dimensional element with two degrees of freedom and length L. Find the interpolation function for this element.

We consider an element with nodes i and j. From Eq. (11.4) we have for the field variable ϕ:

$$\phi(x) = \alpha_0 + \alpha_1 x$$

Then

$$\phi(x_i) \equiv a_i = \alpha_0 + \alpha_1 x_i$$

$$\phi(x_j) \equiv a_j = \alpha_0 + \alpha_1 x_j$$

In matrix form we have

$$\{a\} = \begin{bmatrix} 1 & x_i \\ 1 & x_j \end{bmatrix} \begin{Bmatrix} \alpha_0 \\ \alpha_1 \end{Bmatrix}$$

Clearly,

$$[A] = \begin{bmatrix} 1 & x_i \\ 1 & x_j \end{bmatrix}$$

and we then have for $[A]^{-1}$

$$[A]^{-1} = \frac{1}{x_j - x_i} \begin{bmatrix} x_j & -x_i \\ -1 & 1 \end{bmatrix}$$

Then from Eq. (11.3)

$$[N] = \lfloor C \rfloor [A]^{-1} = \lfloor 1, x \rfloor \frac{1}{x_j - x_i} \begin{bmatrix} x_j & -x_i \\ -1 & 1 \end{bmatrix}$$

$$= \frac{1}{x_j - x_i} \lfloor (x_j - x), (-x_i + x) \rfloor$$

Therefore,

$$[N] = \left\lfloor \frac{x_j - x}{x_j - x_i}, \frac{x - x_i}{x_j - x_i} \right\rfloor$$

In the previous discussion for one-dimensional elements, using polynomials of various degrees to express the field variable ϕ, we indicated how we could obtain the interpolation function. An inverse of the coefficient matrix $[A]$ was needed. To avoid inversions for higher-order approximations, we now present the *Lagrange interpolation* function N_p for one-dimensional elements with n nodes. Using the notation $\mathcal{L}_p$, we have for this interpolation function with one D.O.F. per node:

$$N_p \equiv \mathcal{L}_p = \prod_{\substack{i=1 \\ i \neq p}}^{n} \frac{x - x_i}{x_p - x_i}$$

$$= \frac{(x - x_1)(x - x_2) \cdots (x - x_{p-1})(x - x_{p+1}) \cdots (x - x_n)}{(x_p - x_1)(x_p - x_2) \cdots (x_p - x_{p-1})(x_p - x_{p+1}) \cdots (x_p - x_n)} \tag{11.5}$$

where Π indicates the product of the parenthetic binomial expressions. There are $n - 1$ binomial expressions in the numerator and denominator. Note that in the one-dimensional element there are n nodes, and that when $x = x_p$, that is at the pth node, we have unity for $(N_p)_p$ since numerator and denominator become identical, whereas at any other node where $x \neq x_p$ we have zero, as required for the pth interpolation function. To illustrate Eq. (11.5), we have for a three-node, three DOF element the following three interpolation functions:

$$\begin{aligned} N_1 &= \frac{(x - x_2)(x - x_3)}{(x_1 - x_2)(x_1 - x_3)} \\ N_2 &= \frac{(x - x_1)(x - x_3)}{(x_2 - x_1)(x_2 - x_3)} \\ N_3 &= \frac{(x - x_1)(x - x_2)}{(x_3 - x_1)(x_3 - x_2)} \end{aligned} \tag{11.6}$$

It is to be noted that the continuity of the field variable will be preserved with the Lagrange interpolation function and so we have C^0 continuity.

If we wish C^1 or higher continuity between elements for a one-dimensional element, we may use the *Hermite polynomial* element, where ϕ as well as derivatives $d\phi/dx$, $d^2\phi/dx^2, \ldots, d^p\phi/dx^p$ at the nodes can be considered *degrees of freedom*, and not just the field variable as was the case for Lagrange interpolation functions. We will first consider a linear element with ends 1 and 2 of length L. Let us suppose that the field variable ϕ and its derivations up to order $m - 1$ are to be degrees of freedom at the ends of the element. Then we can say

$$\begin{aligned} \phi = N_1(\phi)_1 + N_2\left(\frac{d\phi}{dx}\right)_1 + \cdots + N_m\left(\frac{d^{m-1}\phi}{dx^{m-1}}\right)_1 \\ + N_{m+1}(\phi)_2 + N_{m+2}\left(\frac{d\phi}{dx}\right)_2 + \cdots + N_{2m}\left(\frac{d^{m-1}\phi}{dx^{m-1}}\right)_2 \end{aligned} \tag{11.7}$$

where the N's are Hermite polynomials. To construct the proper interpolation function, we use a polynomial representation for each N_i of order $2m - 1$ and having $2m$ constants. That is, using normalized coordinates $\xi = x/L$, we have

$$N_i = (C_1)_i + (C_2)_i\xi + (C_3)_i\xi^2 + \cdots + (C_{2m})_i\xi^{(2m-1)} \tag{11.8}$$

We must now choose the $2m$ constants for interpolation function N_i so that the appropriate derivative of N_i is unity for the ith degree of freedom at the appropriate node and is zero for all the other interpolation function derivatives evaluated at this node. From Eq. (11.7) we can say

$$\phi = N_1(\phi)_1 + N_2\left(\frac{d\phi}{dx}\right)_1 + \cdots + N_m\left(\frac{d^{m-1}\phi}{dx^{m-1}}\right)_1 + N_{m+1}(\phi)_2 + \cdots + N_{2m}\left(\frac{d^{m-1}\phi}{dx^{m-1}}\right)_2$$

$$\frac{d\phi}{dx} = \frac{dN_1}{dx}(\phi)_1 + \frac{dN_2}{dx}\left(\frac{d\phi}{dx}\right)_1 + \cdots + \frac{dN_m}{dx}\left(\frac{d^{m-1}\phi}{dx^{m-1}}\right)_1 + \frac{dN_{m+1}}{dx}(\phi)_2 + \cdots + \frac{dN_{2m}}{dx}\left(\frac{d^{m-1}\phi}{dx^{m-1}}\right)_2 \tag{11.9}$$

$$\vdots$$

$$\frac{d^{m-1}\phi}{dx^{m-1}} = \frac{d^{m-1}N_1}{dx^{m-1}}(\phi)_1 + \frac{d^{m-1}N_2}{dx^{m-1}}\left(\frac{d\phi}{dx}\right)_1 + \cdots + \frac{d^{m-1}N_m}{dx^{m-1}}\left(\frac{d^{m-1}\phi}{dx^{m-1}}\right)_1 + \frac{d^{m-1}N_{m+1}}{dx^{m-1}}(\phi)_2 + \cdots + \frac{d^{m-1}N_{2m}}{dx^{m-1}}\left(\frac{d^{m-1}\phi}{dx^{m-1}}\right)_2$$

where the notation $(\;)_s$ indicates "at node s." To achieve the proper property of N_1 at node 1 for DOF 1, we require from the system of Eqs. (11.9) (on considering N_1 and all of its derivatives for the two nodes):

$$\begin{aligned} &(N_1)_1 = 1 && (N_1)_2 = 0 \\ &\left(\frac{dN_1}{dx}\right)_1 = 0 && \left(\frac{dN_1}{dx}\right)_2 = 0 \\ &\quad\vdots && \quad\vdots \\ &\left(\frac{d^{m-1}N_1}{dx^{m-1}}\right)_1 = 0 && \left(\frac{d^{m-1}N_1}{dx^{m-1}}\right)_2 = 0 \end{aligned} \tag{11.10}$$

where only one equation is not zero on the right-hand side. Substituting for N_1 using Eq. (11.8), we can determine the $2m$ constants for N_1. Now we require from Eqs. (11.9) for node 1, where we have DOF 2, the following conditions (on considering the *second column of terms* on the right sides of these equations for the two nodes)

$$\begin{aligned} &(N_2)_1 = 0 && (N_2)_2 = 0 \\ &\left(\frac{dN_2}{dx}\right)_1 = 1 && \left(\frac{dN_2}{dx}\right)_2 = 0 \\ &\quad\vdots && \quad\vdots \\ &\left(\frac{d^{m-1}N_2}{dx^{m-1}}\right)_1 = 0 && \left(\frac{d^{m-1}N_2}{dx^{m-1}}\right)_2 = 0 \end{aligned} \tag{11.11}$$

where only one equation is not zero on the right-hand side. Again, substituting Eq. (11.8) for $i = 2$, we get a set of constants for N_2. We do this for all the degrees of freedom and thus get all the desired interpolation functions. We now illustrate the process in detail.

EXAMPLE 11.2 Compute the *Hermitian* interpolation functions for the simple flexure element used in the preceding chapter for four degrees of freedom (see Fig. 11.3).

The field variable ϕ is now w, the deflection of the centerline, and so we have w_1, $(dw/dx)_1$, w_2, and $(dw/dx)_2$ as nodal displacements. We will have four interpolation functions, each a cubic polynomial in ξ. We first consider N_1.

$$N_1 = C_1 + C_2\xi + C_3\xi^2 + C_4\xi^3 \tag{a}$$

The basic equations corresponding to Eq. (11.9) are

$$w = N_1 w_1 + N_2 w_1' + N_3 w_2 + N_4 w_2' \tag{b}$$

$$w' = N_1' w_1 + N_2' w_1' + N_3' w_2 + N_4' w_2' \tag{c}$$

At node 1, $x = \xi = 0$, and from Eq. (b) we require

$$N_1(0) = 1$$

From Eq. (a) we conclude

$$C_1 = 1 \tag{d}$$

At node 2, $x = L$ and $\xi = 1$. From Eq. (b) we require

$$N_1(1) = 0$$

From Eq. (a)

$$0 = C_1 + C_2(1) + C_3(1)^2 + C_4(1)^3$$

Therefore

$$C_1 + C_2 + C_3 + C_4 = 0 \tag{e}$$

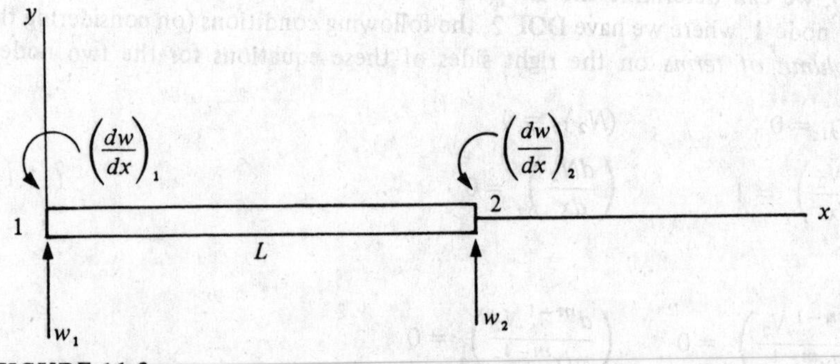

FIGURE 11.3
Flexural element with four degrees of freedom.

At node 1 again, consider Eq. (*c*). We require

$$N_1'(0) = 0$$

From Eq. (*a*)

$$(0 + C_2 + 2C_3\xi + 3C_4\xi^2)_{\xi=0} = 0$$

Therefore

$$C_2 = 0 \tag{f}$$

At node 2 consider Eq. (*c*) again. We require

$$N_1'(1) = 0$$

Therefore

$$(C_2 + 2C_3\xi + 3C_4\xi^2)_{\xi=1} = 0$$

Hence

$$C_2 + 2C_3 + 3C_4 = 0 \tag{g}$$

Solving Eqs. (*d*), (*e*), (*f*), and (*g*) simultaneously, we get

$$C_1 = 1 \qquad C_2 = 0 \qquad C_3 = -3 \qquad C_4 = 2$$

Accordingly,

$$N_1 = 1 - 3\xi^2 + 2\xi^3$$

Now working in a similar manner with N_2, N_3, and N_4, we leave it to you as an assignment to show that

$$\begin{aligned} N_2 &= \xi L(\xi - 1)^2 \\ N_3 &= 3\xi^2 - 2\xi^3 \\ N_4 &= L\xi^2(\xi - 1) \end{aligned} \tag{h}$$

In the example, we calculated the interpolation functions for a Hermite polynomial element. We may use the following notation to describe these interpolation functions:

$$H^n_{mi}(x) \tag{11.12}$$

where

1. n is the order of the highest derivative of the Hermite polynomial. An nth-order Hermite polynomial is a $(2n + 1)$-order polynomial. The Hermite polynomials of the previous example were then first-order Hermite polynomials.

2. m is the order of the derivative for the field variable for which one is computing the interpolation function N_i.
3. i designates the node, in our case 1 or 2.

Thus from the example we have

$$
\begin{aligned}
H_{01}^{1}(\xi) &= 1 - 3\xi^2 + 2\xi^3 = N_1 \\
H_{11}^{1}(\xi) &= L\xi(\xi - 1)^2 = N_2 \\
H_{02}^{1}(\xi) &= \xi^2(3 - 2\xi) = N_3 \\
H_{12}^{1}(\xi) &= L\xi^2(\xi - 1) = N_4
\end{aligned}
\tag{11.13}
$$

Part B
Two-Dimensional Elements

11.5 COMPLETENESS AND GEOMETRIC ISOTROPY

In the case where the field variable is a function of two variables, say x and y, we have for a simple polynomial representation the following:

$$\phi(x, y) = \alpha_1 + \alpha_2 x + \alpha_3 y \tag{11.14}$$

This is a linear variation in two dimensions. For a quadratic variation in two dimensions we would have

$$\phi(x, y) = \alpha_1 + \alpha_2 x + \alpha_3 y + \alpha_4 x^2 + \alpha_5 xy + \alpha_6 y^2 \tag{11.15}$$

In general, we can give polynomials for any degree of variation by using the *Pascal triangle* scheme shown in Fig. 11.4. The procedure for any degree of variation listed on a line to the right is to include all terms on and above this line. Thus the linear variation given by Eq. (11.14) and the quadratic variation given by Eq. (11.15) can readily be found from the scheme.

The various polynomials depicted in the Pascal triangle are called *complete polynomials*, as they contain all the possible two-dimensional coordinates and pro-

α_1

$\alpha_2 x$ $\alpha_3 y$ .... first degree

$\alpha_4 x^2$ $\alpha_5 xy$ $\alpha_6 y^2$ .... second degree

$\alpha_7 x^3$ $\alpha_8 x^2 y$ $\alpha_9 xy^2$ $\alpha_{10} y^3$ .... third degree

$\alpha_{11} x^4$ $\alpha_{12} x^3 y$ $\alpha_{13} x^2 y^2$ $\alpha_{14} xy^3$ $\alpha_{15} y^4$ .... fourth degree

FIGURE 11.4
Pascal triangle. An ith degree polynomial contains all terms at and above the ith level.

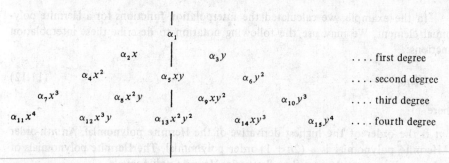

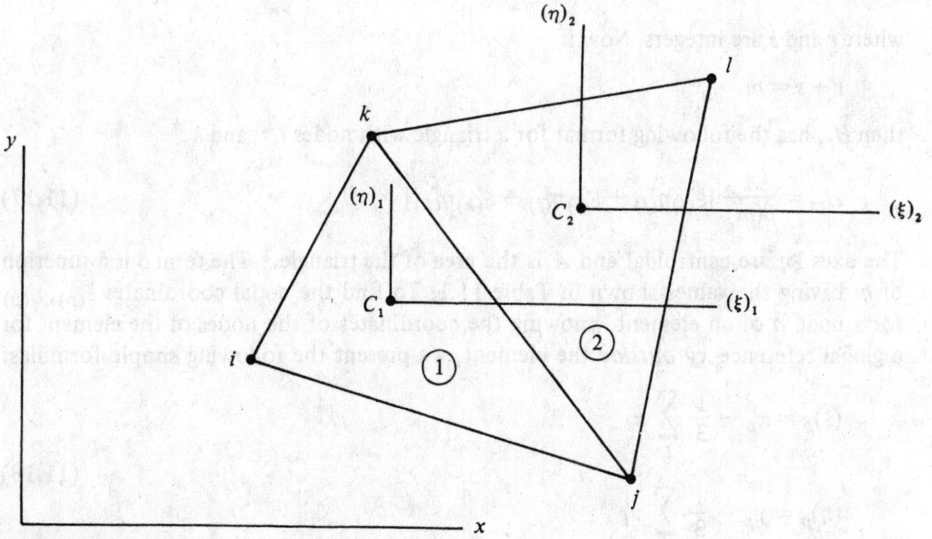

FIGURE 11.5
Triangular elements with centroidal coordinates.

ducts of the coordinates for a specified degree of polynomial. This kind of completeness is important for the polynomial approach in order to achieve good results with the finite element method.

In this regard we introduce the property of *geometric isotropy*, which allows for a linear transformation from one Cartesian coordinate to another without change of form of the field variable function for an element. This implies that the polynomial expression for the field variable is unchanged for such a change of axes. We would expect that an analysis with polynomials should not be changed by simply a change of position of the axes, and so the property of geometric compatibility is important for our formulations. We can achieve geometric isotropy if we use a *complete polynomial*. If we do not use a complete polynomial, but one with fewer terms, we can still achieve geometric isotropy if the terms extracted from the polynomial in the Pascal triangle are such as to leave an array of terms symmetric about the indicated centerline (see Fig. 11.4).

11.6 USE OF CENTROIDAL RECTANGULAR COORDINATES

To illustrate the use of geometric isotropy, we show two adjacent triangular elements in Fig. 11.5, with a global reference xy. Also shown are axes at the centroid of each element. Having geometric isotropy, we can use the *local* references for each element for calculations rather than the single *global* reference xy.

To simplify our work with these elements, we can use this system of centroidal, parallel coordinate axes with one set of axes for each element. For the computation of $[K]^e$, we will have integrations over triangle e as follows:

$$H_{rs} = \iint_{A^e} \xi^r \eta^s \, dA \tag{11.16}$$

where r and s are integers. Now if

$$r+s=m$$

then H_{rs} has the following format for a triangle with nodes i, j, and k.†

$$H_{rs}=\frac{A}{\beta(m)}\left[\xi_{(i)}^{r}\eta_{(i)}^{s}+\xi_{(j)}^{r}\eta_{(j)}^{s}+\xi_{(k)}^{r}\eta_{(k)}^{s}\right] \tag{11.17}$$

The axes $\xi\eta$ are centroidal and A is the area of the triangle.‡ The term β is a function of m having the values shown in Table 11.1. To find the nodal coordinates $\xi_{(p)}, \eta_{(p)}$ for a node p of an element, knowing the coordinates of the nodes of the element for a global reference xy *outside* the element, we present the following simple formulas:

$$\begin{aligned}(\xi)_p &= x_p-\frac{1}{3}\sum_l x_l\\ (\eta)_p &= y_p-\frac{1}{3}\sum_l y_l\end{aligned} \tag{11.18}$$

We can thus carry out our integrations with ease.

We can assemble the global stiffness matrix in the usual way. The use of different axes $\xi\eta$ for the elements in no way changes the usual procedure as long as they are parallel to xy.§ We shall use this approach when we study plates in a later chapter.

†The parentheses around the subscripts indicate no summation.

‡The area A of the triangle is given as

$$A=\tfrac{1}{2}[\xi_i\eta_j+\xi_j\eta_k+\xi_k\eta_i-\eta_i\xi_j-\eta_j\xi_k-\eta_k\xi_i]$$

§To show this, consider element 1 and a function $f(x, y)$ of the global coordinates (see Fig. 11.5). Consider the integral

$$\iint_{A^{(1)}} f(x,y)\,dx\,dy \tag{a}$$

Suppose we wish to change coordinates to the local centroidal coordinates $\xi\eta$ for element 1. From calculus, we know that this is accomplished as follows:

$$\iint_{A^{(1)}} f(x,y)\,dx\,dy=\iint_{A^{(1)}} f[x(\xi,\eta),y(\xi,\eta)]\,|J|\,d\xi\,d\eta \tag{b}$$

where $|J|$ is the determinant of the well-known Jacobian matrix, given as

$$[J]=\begin{bmatrix}\dfrac{\partial x}{\partial \xi} & \dfrac{\partial y}{\partial \xi}\\[2mm] \dfrac{\partial x}{\partial \eta} & \dfrac{\partial y}{\partial \eta}\end{bmatrix} \tag{c}$$

For axes xy and $\xi_1\eta_1$ merely translated relative to each other, $|J|$ must equal unity, as you may readily demonstrate. We can then conclude on considering Eq. (*b*) above that the integral for global coordinates xy of f over area $A^{(1)}$ is the same as the integral of f in local coordinates over the same area. We shall use this fact in Chapter 13 to set up stiffness matrices, where we will use local parallel centroidal axes for each element rather than one global set of rectangular axes, with a resulting reduction of labor.

TABLE 11.1
Values of β for Eq. (11.17)

	m				
	1	2	3	4	5
β	∞	12	30	60	120

11.7 FORMATION OF INTERPOLATION FUNCTIONS FOR TRIANGULAR ELEMENTS BY POLYNOMIALS

We have already shown in Section 10.3 how one proceeds to get the interpolation function after the field variable has been expressed as a parameter-laden polynomial. Recall that

$$[N] = \lfloor C \rfloor [A]^{-1} \tag{11.19}$$

where $\lfloor C \rfloor$ contains the various coordinates and products of coordinates in the polynomial and $[A]$ is the coefficient matrix.

The triangular elements thus far discussed have three nodes and three degrees of freedom. There is linear variation of ϕ and there will be C^0 continuity. This is true, as we shall demonstrate later, because we have two nodes on a side of a triangle. For the linear variation in the element, two values of the field variable along an edge (at the two nodes) uniquely determine the linear variation along a side. And so on a contiguous side between two elements terminating at the same nodes (see Fig. 11.5), we must have the same values of ϕ along this side between the elements and thus we have C^0 continuity. For quadratic variation with C^0 continuity we will need three nodes on each side of the triangular element so that for a quadratic polynomial the value of ϕ will be uniquely determined on a side. For C^0 continuity for cubic variation, we need four nodes on each side, and so on. In each of these cases we can proceed as described at the outset of Section 11.5 by using complete polynomials or polynomials properly reduced from the complete polynomial and computing $[N]$ from $\lfloor C \rfloor [A]^{-1}$.

11.8 TRIANGULAR ELEMENTS WITH NATURAL COORDINATES

Another approach is to use *natural coordinates*. In general, for one-dimensional, two-dimensional, and three-dimensional elements, these coordinates range in value from 0 to 1 within each element.† A natural coordinate will have a value of unity at one node and be zero at the other nodes in the manner discussed earlier for interpolation functions.

†For triangular elements the natural coordinates are also called area coordinates. For one-dimensional linear elements the natural coordinates L_i are noting that $\mathfrak{L}_1(x) = L_1$ and $\mathfrak{L}_2(x) = L_2$.

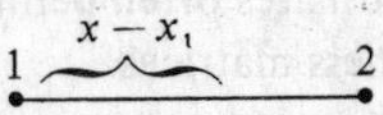

$$L_1 = \frac{x_2 - x}{x_2 - x_1} \qquad L_2 = \frac{x - x_1}{x_2 - x_1}$$

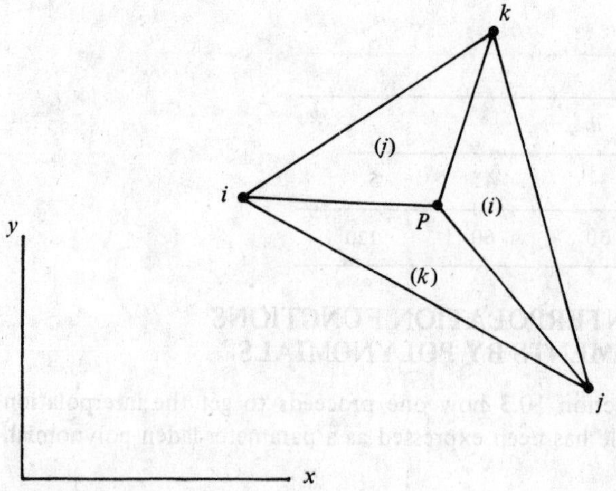

FIGURE 11.6
Triangular coordinates.

For the triangular element presented in Fig. 11.6, we have shown three internal triangles with an apex at P. Each triangle is identified by a letter corresponding to the opposite node. Clearly, if we know the areas of these triangles we can ascertain the position P inside the triangle. Thus, these areas act as coordinates in that they can locate a point P inside the element and on the boundaries of the element. Accordingly, we shall denote A_i/A as the natural or the triangular coordinate L_i. Thus, we can say

$$L_i = \frac{A_i}{A}$$

$$L_j = \frac{A_j}{A} \tag{11.20}$$

$$L_k = \frac{A_k}{A}$$

Clearly, $L_i + L_j + L_k = 1$ and so only two of the coordinates are independent, as can be expected for a two-dimensional space. You should have little difficulty concluding (on considering Fig. 11.6) that

$$\begin{aligned} &\text{for } P \text{ at node } i \quad && L_i = 1 \quad L_j = 0 \quad L_k = 0 \\ &\text{for } P \text{ at node } j \quad && L_i = 0 \quad L_j = 1 \quad L_k = 0 \\ &\text{for } P \text{ at node } k \quad && L_i = 0 \quad L_j = 0 \quad L_k = 1 \end{aligned} \tag{11.21}$$

We show these properties for a triangular element with three nodes in Fig. 11.7. We shall use this kind of approach when we study plane stress and plates and we shall later expand on the details for the use of natural coordinates for triangular elements. We merely wish to point out now that using natural coordinates often permits simple closed-form integration procedures when determining stiffness matrices.

If there is linear variation, as already pointed out, we need two nodes on a side

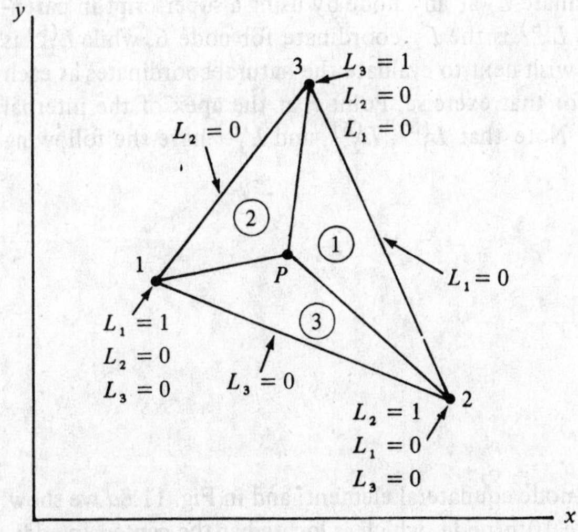

FIGURE 11.7
Specific values of triangular coordinates.

of a triangle in order to achieve C^0 continuity. For higher-order elements, we need more nodes on the element side to achieve C^0 continuity. We have discussed a way of computing the appropriate interpolation functions involving the inverse of the coefficient matrix $[A]$. To avoid taking this inverse when using triangular elements we may use natural coordinates to give the appropriate interpolation functions for C^0 continuity. For this purpose, consider Fig. 11.8*a*, where we show a higher-order element.

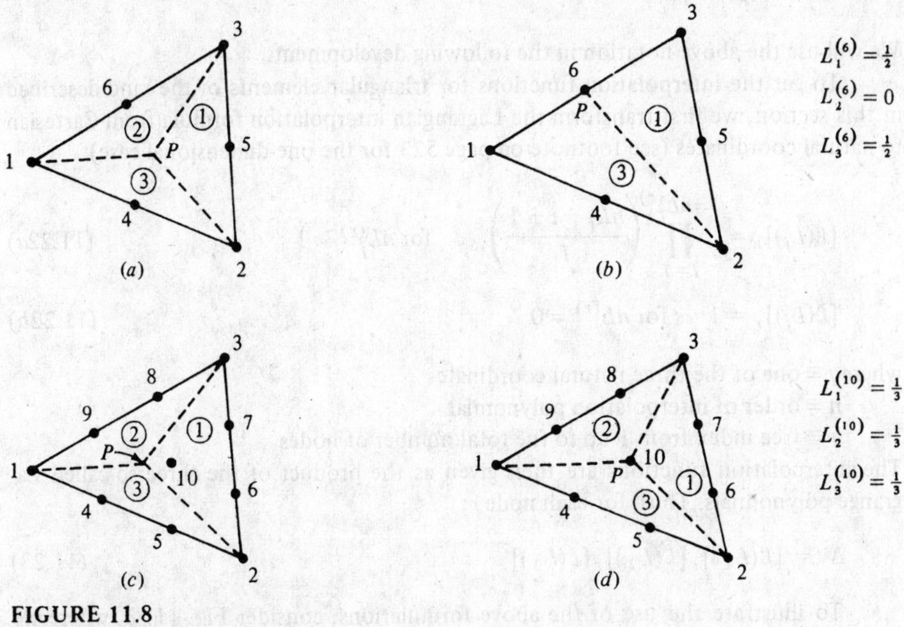

FIGURE 11.8
Higher-order triangular elements.

We will express the natural coordinate L_i at any node by using a superscript in parentheses to identify the node. Thus $L_2^{(6)}$ is the L_2 coordinate for node 6, while $L_1^{(5)}$ is the L_1 coordinate at node 5. We wish next to evaluate the natural coordinates at each node. Let us first take node 6 for that exercise. Point P at the apex of the internal triangles is shown in Fig. 11.8*b*. Note that $L_1^{(6)}$, $L_2^{(6)}$, and $L_3^{(6)}$ have the following values:

$$L_1^{(6)} = \frac{A_1}{A} = \frac{1}{2}$$

$$L_2^{(6)} = \frac{A_2}{A} = \frac{0}{A} = 0$$

$$L_3^{(6)} = \frac{A_3}{A} = \frac{1}{2}$$

In Fig. 11.8*c* we show a 10-node equilateral element, and in Fig. 11.8*d* we show the natural coordinates of P at the 10th node, which is located at the center. In addition, you should be able to say for the fourth node of this element

$$L_1^{(4)} = \frac{A_1}{A} = \frac{2}{3}$$

$$L_2^{(4)} = \frac{A_2}{A} = \frac{1}{3}$$

$$L_3^{(4)} = \frac{A_3}{A} = 0$$

We will use the above notation in the following development.

To get the interpolation functions for triangular elements of the kind described in this section, we first transform the Lagrangian interpolation formula from Cartesian to natural coordinates (see footnote on page 523 for the one-dimensional case):

$$[\mathcal{L}(L_j)]_r = \prod_{i=1}^{i=nL_j^{(r)}} \left(\frac{nL_j - i + 1}{i} \right) \qquad \text{for } nL_j^{(r)} \geqslant 1 \tag{11.22a}$$

$$[\mathcal{L}(L_j)]_r = 1 \qquad \text{for } nL_j^{(r)} = 0 \tag{11.22b}$$

where $j \equiv$ one of the three natural coordinates
$n \equiv$ order of interpolation polynomial
$r \equiv$ free index from 1 up to the total number of nodes

The interpolation functions are then given as the product of the three so-called Lagrange polynomials. Thus, for each node r:

$$N_r = [\mathcal{L}(L_1)]_r [\mathcal{L}(L_2)]_r [\mathcal{L}(L_3)]_r \tag{11.23}$$

To illustrate the use of the above formulations, consider Fig. 11.8*a*, where we have quadratic variation of the field variable. Thus $n = 2$. Let us compute N_1, thus making $r = 1$. Considering $j = 1$ first in Eq. (11.22), it should be easily seen from our earlier discussion in this section that $nL_j^{(r)} = 2L_1^{(1)} = (2)(1) = 2$ and so we must use Eq. (11.22*a*). We can then say

$$[\mathcal{L}(L_1)]_1 = \prod_{i=1}^{i=2L_1^{(1)}} \left(\frac{2L_1 - i + 1}{i}\right) = \prod_{i=1}^{i=2} \left(\frac{2L_1 - i + 1}{i}\right)$$

$$= \left(\frac{2L_1 - 1 + 1}{1}\right)\left(\frac{2L_1 - 2 + 1}{2}\right)$$

$$= 2L_1 \left(\frac{2L_1 - 1}{2}\right) = L_1(2L_1 - 1) \tag{11.24}$$

Considering $[\mathcal{L}(L_2)]_1$ and $[\mathcal{L}(L_3)]_1$ we see that $L_2^{(1)} = L_3^{(1)} = 0$ and so, using Eq. (11.22*b*), we get unity for these expressions. From Eq. (11.23) we then have

$$N_1 = [(L_1)(2L_1 - 1)](1)(1) = L_1(2L_1 - 1) \tag{11.25}$$

We leave it as an exercise to evaluate the other five interpolation functions.

In Chapter 13 we shall see that the stiffness matrix as well as the forcing functions will be integrals of the triangular coordinates, for which, as indicated earlier, simple formulas will be available.

11.9 RECTANGULAR ELEMENTS

We have already described how to get interpolation functions by using a parameter-laden polynomial and inverting the coefficient matrix. In the last section we showed how we could get interpolation functions for triangular elements by using Lagrange polynomials in conjunction with natural coordinates, thus avoiding taking the inverse of the coefficient matrix $[A]$. We will now consider rectangular elements, for which we will first use Lagrange polynomials and then Hermite polynomials. The rectangular elements have sides parallel to the global axes xy.

With this in mind, consider a normalized rectangular element with centroidal coordinates $\xi\eta$ parallel to global reference xy (Fig. 11.9†). There are only four nodes

†In Fig. 11.9 and in subsequent diagrams in this chapter, the rectangle is shown as depicted in reference xy and not as depicted in the nondimensional reference $\xi\eta$, where the rectangle would normally be shown as a square.

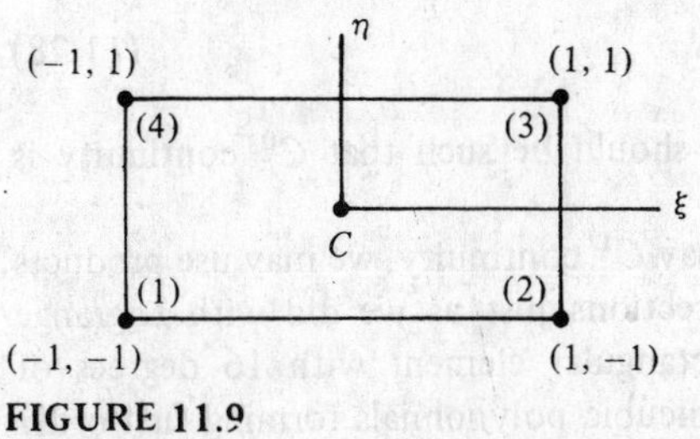

FIGURE 11.9
Rectangular element with a normalized reference $\xi\eta$ at the centroid.

so that we have linear variation of the field variable ϕ. We will propose for this case that

$$\phi = \sum_{i=1}^{4} N_i \phi_i$$

where, using Lagrange polynomials in the ξ and η directions, we have

$$\begin{aligned} N_1 &= [\mathcal{L}_1(\xi)]\,[\mathcal{L}_1(\eta)] \\ N_2 &= [\mathcal{L}_2(\xi)]\,[\mathcal{L}_2(\eta)] \\ N_3 &= [\mathcal{L}_3(\xi)]\,[\mathcal{L}_3(\eta)] \\ N_4 &= [\mathcal{L}_4(\xi)]\,[\mathcal{L}_4(\eta)] \end{aligned} \tag{11.26}$$

We have thus used at each node the product of *Lagrange* interpolation functions in the ξ and η coordinate directions. Why does such a system qualify as proper interpolation function? Note first that $N_1 = \mathcal{L}_1(\xi_1)\mathcal{L}(\eta_1) = 1$ at node 1, and that at all other nodes [i.e., at (ξ_2, η_2), (ξ_3, η_3), and (ξ_4, η_4)] the interpolation function N_1 is zero. We can easily verify this when we spell out the products for this *linear* interpolation case. Thus we have, noting that $\xi_1 = -1$, $\eta_1 = -1$, $\xi_2 = 1$, $\eta_2 = -1$, etc., and observing Fig. 11.9

$$\begin{aligned} N_1(\xi, \eta) &= \frac{\xi - \xi_2}{\xi_1 - \xi_2}\,\frac{\eta - \eta_4}{\eta_1 - \eta_4} = \frac{1}{4}(\xi - 1)(\eta - 1) \\ N_2(\xi, \eta) &= \frac{\xi - \xi_1}{\xi_2 - \xi_1}\,\frac{\eta - \eta_3}{\eta_2 - \eta_3} = -\frac{1}{4}(\xi + 1)(\eta - 1) \\ N_3(\xi, \eta) &= \frac{\xi - \xi_4}{\xi_3 - \xi_4}\,\frac{\eta - \eta_2}{\eta_3 - \eta_2} = \frac{1}{4}(\xi + 1)(\eta + 1) \\ N_4(\xi, \eta) &= \frac{\xi - \xi_3}{\xi_4 - \xi_3}\,\frac{\eta - \eta_1}{\eta_4 - \eta_1} = -\frac{1}{4}(\xi - 1)(\eta + 1) \end{aligned} \tag{11.27}$$

We can thus verify the statement made above. We can also form higher-order rectangular elements in the same way. At each node i we then have

$$N_i = [\mathcal{L}_i(\xi)]\,[\mathcal{L}_i(\eta)] \tag{11.28}$$

The order that one uses for the polynomials should be such that C^0 continuity is achieved.

If we desire greater than C^0 continuity, say C^1 continuity, we may use products of *Hermite* polynomials along coordinate directions just as we did with Lagrange polynomials. Thus, consider a four-node rectangular element with 16 degrees of freedom (see Fig. 11.10). We shall use here bicubic polynomials forming first-order

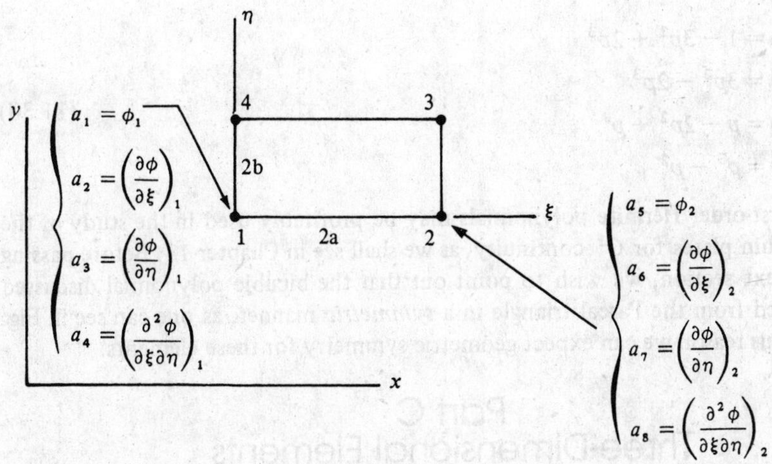

FIGURE 11.10
Hermite rectangle with 16 degrees of freedom.

Hermite polynomials. We have for the four nodes the following interpolation functions, using normalized coordinates $\xi\eta$:

$$\text{Node 1}\begin{cases} N_1(\xi,\eta) = [H_{01}^1(\xi)]\,[H_{01}^1(\eta)] \\ N_2(\xi,\eta) = [H_{11}^1(\xi)]\,[H_{01}^1(\eta)] \\ N_3(\xi,\eta) = [H_{01}^1(\xi)]\,[H_{11}^1(\eta)] \\ N_4(\xi,\eta) = [H_{11}^1(\xi)]\,[H_{11}^1(\eta)] \end{cases} \qquad \text{Node 2}\begin{cases} N_5(\xi,\eta) = [H_{02}^1(\xi)]\,[H_{01}^1(\eta)] \\ N_6(\xi,\eta) = [H_{12}^1(\xi)]\,[H_{01}^1(\eta)] \\ N_7(\xi,\eta) = [H_{02}^1(\xi)]\,[H_{11}^1(\eta)] \\ N_8(\xi,\eta) = [H_{12}^1(\xi)]\,[H_{11}^1(\eta)] \end{cases}$$

$$\text{Node 3}\begin{cases} N_9(\xi,\eta) = [H_{02}^1(\xi)]\,[H_{02}^1(\eta)] \\ N_{10}(\xi,\eta) = [H_{12}^1(\xi)]\,[H_{02}^1(\eta)] \\ N_{11}(\xi,\eta) = [H_{02}^1(\xi)]\,[H_{12}^1(\eta)] \\ N_{12}(\xi,\eta) = [H_{12}^1(\xi)]\,[H_{12}^1(\eta)] \end{cases} \qquad \text{Node 4}\begin{cases} N_{13}(\xi,\eta) = [H_{01}^1(\xi)]\,[H_{02}^1(\eta)] \\ N_{14}(\xi,\eta) = [H_{11}^1(\xi)]\,[H_{02}^1(\eta)] \\ N_{15}(\xi,\eta) = [H_{01}^1(\xi)]\,[H_{12}^1(\eta)] \\ N_{16}(\xi,\eta) = [H_{11}^1(\xi)]\,[H_{12}^1(\eta)] \end{cases} \tag{11.29}$$

The desired first-order Hermite polynomials that go into Eq. (11.29) [see Eqs. (11.13)]† are for a variable p, which could be ξ or η.

†The L's do not appear in Eqs. (11.30) because the degrees of freedom for derivatives here (see Fig. 11.10) are taken with respect to the normalized coordinates $\xi\eta$ rather than xy as before. Furthermore, you may directly demonstrate that for Eq. (11.29), the interpolation functions and their derivatives take on the proper value of zero or unity at the nodes in accordance with the characteristic of interpolation functions. For example, consider DOF 7 where $\xi = 1$, $\eta = 0$ and corresponding to $(\partial\phi/\partial\eta)_2$. We can see from Eq. (11.30) that $(\partial N_7/\partial\eta)_{1,0} = 1$, while $(\partial N_i/\partial\eta)_{1,0} = 0$ for all other 15 DOF.

$$\begin{aligned} H_{01}^1(p) &= 1 - 3p^2 + 2p^3 \\ H_{02}^1(p) &= 3p^2 - 2p^3 \\ H_{11}^1(p) &= p - 2p^2 + p^3 \\ H_{12}^1(p) &= p^3 - p^2 \end{aligned} \tag{11.30}$$

The first-order Hermite polynomials may be profitably used in the study of the bending of thin plates for C^1 continuity, as we shall see in Chapter 13. Before passing on to the next section, we wish to point out that the bicubic polynomial discussed here is formed from the Pascal triangle in a *symmetric* manner, as one can see in Fig. 11.11. For this reason we can expect geometric symmetry for these elements.

Part C Three-Dimensional Elements

11.10 GENERAL COMMENTS

We have already described certain three-dimensional elements. These elements can be constructed by procedures that are direct extensions of the two-dimensional elements. Thus we can form interpolation functions for the field variable by working with polynomials in three variables. In this regard we have an extrapolation of the Pascal triangle into a tetrahedron for discussing completeness of polynomials. This is shown in Fig. 11.12.

Also, we can use natural coordinates in three dimensions in conjunction with tetrahedron elements. In Fig. 11.13 we show a tetrahedron and, inside, four internal tetrahedrons having a common apex at point P. The volume of each internal tetrahedron V_i divided by the total volume of the tetrahedron V gives the natural coordinate L_i. Thus:

$$\begin{aligned} L_1 &= \frac{V_1}{V} \\ L_2 &= \frac{V_2}{V} \\ L_3 &= \frac{V_3}{V} \\ L_4 &= \frac{V_4}{V} \end{aligned} \tag{11.31}$$

$$\begin{array}{ccccccc} & & & 1 & & & \\ & & x & & y & & \\ & x^2 & & xy & & y^2 & \\ x^3 & & x^2y & & xy^2 & & y^3 \\ & x^3y & & x^2y^2 & & xy^3 & \\ & & x^3y^2 & & x^2y^3 & & \\ & & & x^3y^3 & & & \end{array}$$

FIGURE 11.11
Bicubic polynomial terms.

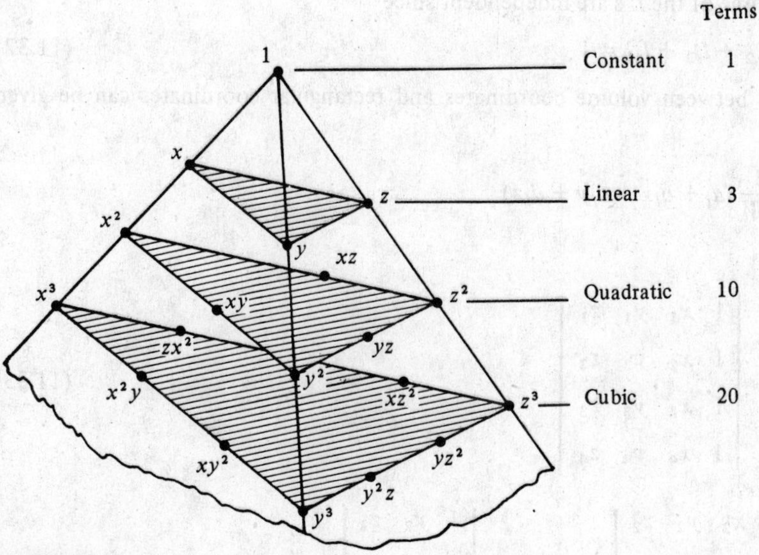

FIGURE 11.12
Pascal tetrahedron.

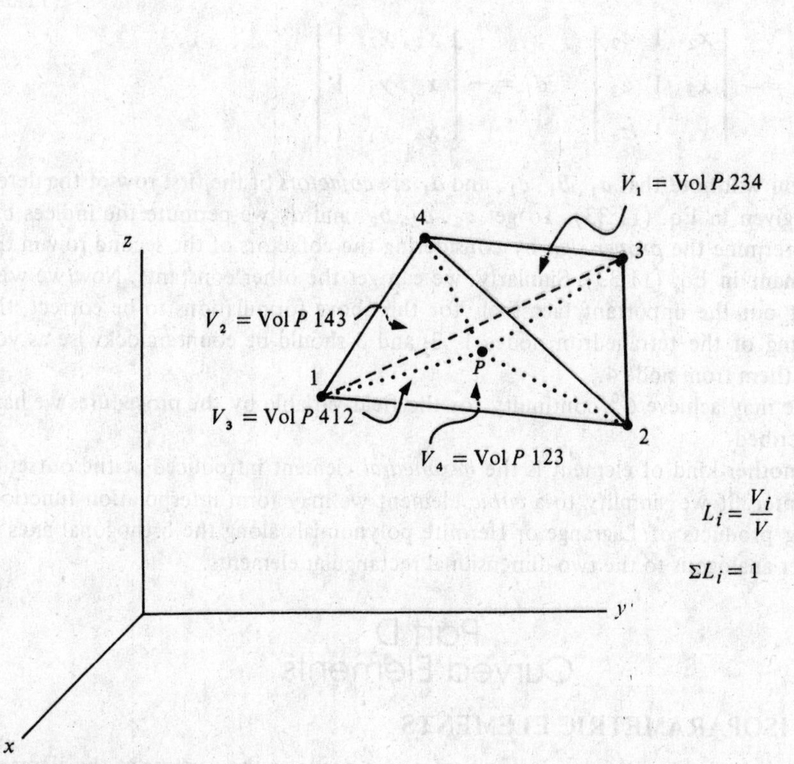

FIGURE 11.13
Natural or volume coordinates.

Only three of the L's are independent since

$$L_1 + L_2 + L_3 + L_4 = 1 \tag{11.32}$$

The relation between volume coordinates and rectangular coordinates can be given as follows:

$$L_i = \frac{1}{6V}(a_i + b_i x + c_i y + d_i z)$$

where

$$V = \frac{1}{6}\begin{vmatrix} 1 & x_1 & y_1 & z_1 \\ 1 & x_2 & y_2 & z_2 \\ 1 & x_3 & y_3 & z_3 \\ 1 & x_4 & y_4 & z_4 \end{vmatrix} \tag{11.33}$$

$$a_1 = \begin{vmatrix} x_2 & y_2 & z_2 \\ x_3 & y_3 & z_3 \\ x_4 & y_4 & z_4 \end{vmatrix} \qquad b_1 = -\begin{vmatrix} 1 & y_2 & z_2 \\ 1 & y_3 & z_3 \\ 1 & y_4 & z_4 \end{vmatrix}$$

$$c_1 = -\begin{vmatrix} x_2 & 1 & z_2 \\ x_3 & 1 & z_3 \\ x_4 & 1 & z_4 \end{vmatrix} \qquad d_1 = -\begin{vmatrix} x_2 & y_2 & 1 \\ x_3 & y_3 & 1 \\ x_4 & y_4 & 1 \end{vmatrix} \tag{11.34}$$

You will note that a_1, b_1, c_1, and d_1 are *cofactors* of the first row of the determinant given in Eq. (11.33). To get a_2, c_2, b_2, and d_2 we permute the indices but must determine the *proper sign* by considering the cofactors of the second row in the determinant in Eq. (11.33). Similarly, we can get the other constants. Now we wish to point out the important fact that, for the above formulations to be correct, the numbering of the tetrahedron nodes 1, 2, and 3 should be counterclockwise as you observe them from node 4.

We may achieve C^0 continuity for the field variable by the procedures we have just described.

Another kind of element is the *hexahedral* element introduced at the outset of the chapter. If we simplify to a *cubic* element we may form interpolation functions by using products of Lagrange or Hermite polynomials along the orthogonal axes in a manner analogous to the two-dimensional rectangular elements.

Part D
Curved Elements

11.11 ISOPARAMETRIC ELEMENTS

We have discussed *simple* elements, where nodes occur at the vertices of the element. For *higher-order* elements the nodes may also occur on the faces of the element as well as inside it. It is to be pointed out that nodes inside the element are not as effective as those on the faces and are not often used. The use of additional nodes increases the accuracy of the process, so that in general fewer elements need be used in such cases than for simple elements.† But using comparatively larger elements then poses a problem on curved and irregular boundaries. For this reason it would appear advisable to be able to form elements with *curved* boundaries from those with straight boundaries such as triangles and rectangles. In this section we will address ourselves to this effort.

What we want to do is map a simple one-, two-, or three-dimensional element into a curved element having a desired curved shape. In Fig. 11.14*a* we show as an example a rectangular element with nondimensional local coordinates $\xi\eta$. This may be mapped into a quadrilateral in the xy plane, as shown in the diagram, with vertices

†The use of higher-order, larger elements does not always decrease the amount of computer time. Generally more computer time is needed. However, the use of larger, higher-order elements does present an economy in data preparation—that is, in the discretization process, numbering of nodes, characterization of element properties, etc.

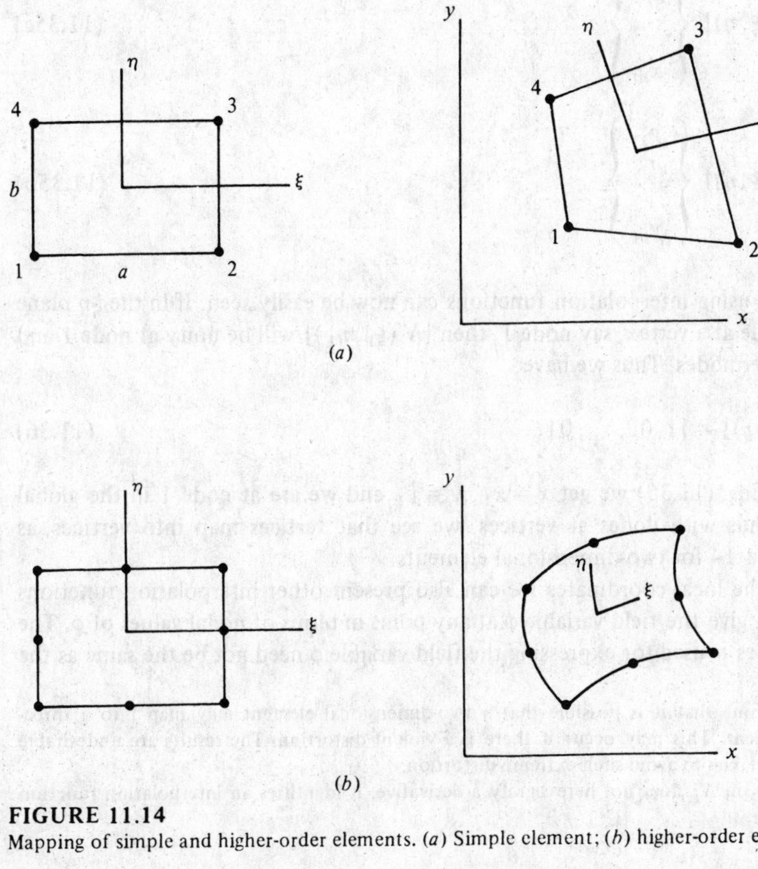

FIGURE 11.14
Mapping of simple and higher-order elements. (*a*) Simple element; (*b*) higher-order element.

mapping into vertices.† In Fig. 11.14*b* we see the mapping of a higher-order element into a shape with curved sides.

How do we find transformations that will map from $\xi\eta\zeta$ coordinates for a simple shape (i.e., using local coordinates) to the desired, more complex shape in the *xyz* reference (i.e., into global coordinates)? We desire a function f that maps local coordinates $\xi\eta\zeta$ into global coordinates so that vertices map into vertices; that is,

$$\begin{Bmatrix} x \\ y \\ z \end{Bmatrix} = f \begin{Bmatrix} \xi \\ \eta \\ \zeta \end{Bmatrix}$$

A way of accomplishing this task it so use *interpolation* functions N_i' expressed in local coordinates $\xi\eta\zeta$ since, as will be explained, such functions have the necessary properties to carry out the desired mapping.‡ Thus we propose that for *m nodes* and for two dimensions:

$$x = \lfloor N'(\xi, \eta) \rfloor \begin{Bmatrix} x_1 \\ \cdot \\ \cdot \\ x_m \end{Bmatrix} \tag{11.35a}$$

$$y = \lfloor N'(\xi, \eta) \rfloor \begin{Bmatrix} y_1 \\ \cdot \\ \cdot \\ y_m \end{Bmatrix} \tag{11.35b}$$

The reason for using interpolation functions can now be easily seen. If in the $\xi\eta$ plane we are at a node at a vertex, say node 1, then $\lfloor N'(\xi_1, \eta_1) \rfloor$ will be unity at node 1 and zero at the other nodes. Thus we have

$$\lfloor N'(\xi_1, \eta_1) \rfloor = \lfloor 1, 0, \ldots, 0 \rfloor \tag{11.36}$$

so that from Eqs. (11.35) we get $x = x_1$, $y = y_1$ and we are at node 1 in the global coordinates. Thus with nodes at vertices, we see that vertices map into vertices, as shown in Fig. 11.14 for two-dimensional elements.

Now in the local coordinates we can also present other interpolation functions $N_i(\xi, \eta)$, which give the field variable ϕ at any point in terms of nodal values of ϕ. The number of nodes n used for expressing the field variable ϕ need not be the same as the

†We point out that it is possible that a two-dimensional element may map into a three-dimensional element. This may occur if there is a violent distortion. The results are undesirable and care must be taken to avoid such extreme distortion.

‡The prime on N_i' does not here signify a derivative, it identifies an interpolation function different from N_i.

number of nodes m used for the mapping involved $N_i'(\xi, \eta)$. Thus, in Fig. 11.15*a* we have fewer nodes for interpolation of the unknown field than we have nodes for mapping. We call this case *superparametric*. In Fig. 11.15*b*, we have the reverse, more nodes for interpolation than for mapping, and we call this case *subparametric*. We thus see in these two cases that the order of the functions N_i and N_i' need be different in the cases where $n \neq m$. Finally, if in Fig. 11.15*c* the nodes for mapping are the same as the nodes for interpolation, then the order of the interpolation functions is the same and $[N] = [N']$. We call such cases *isoparametric*. We shall only consider this last case, although one finds the use of subparametric cases in practice.

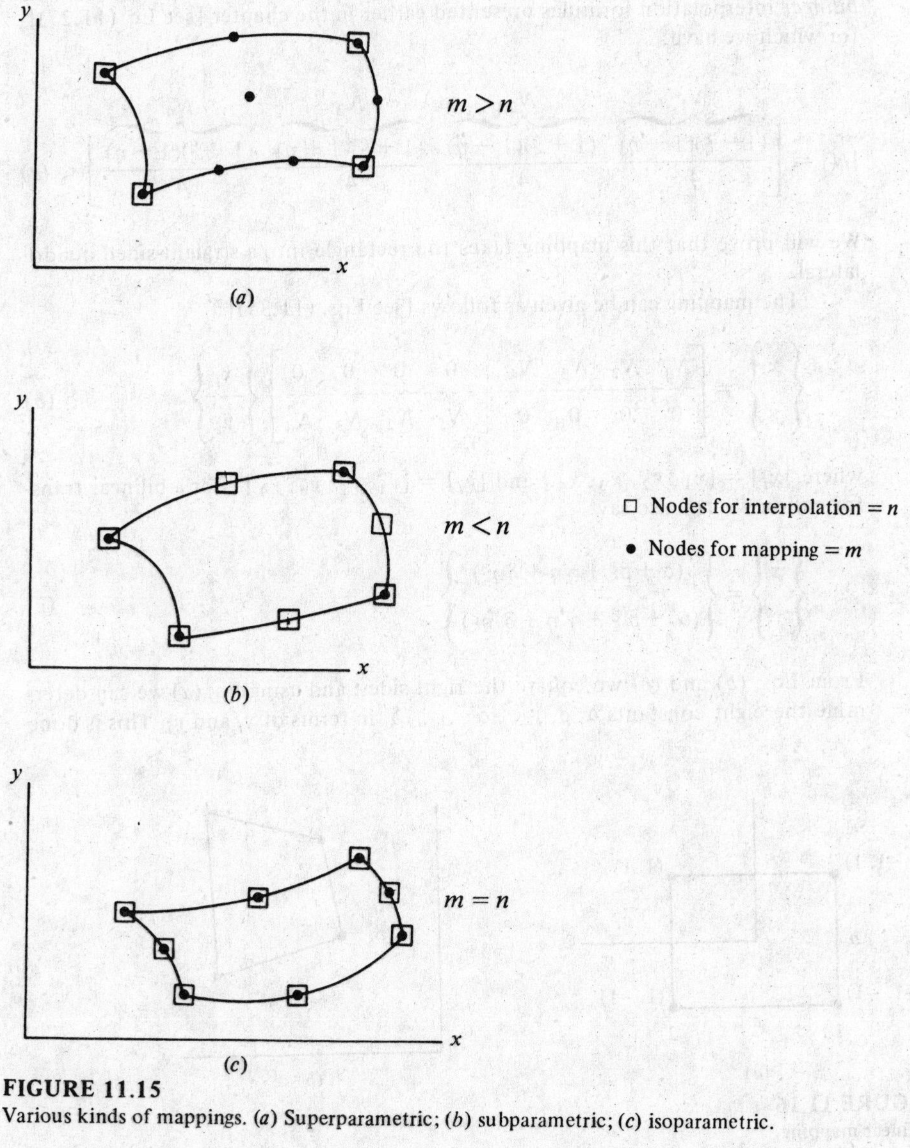

FIGURE 11.15
Various kinds of mappings. (*a*) Superparametric; (*b*) subparametric; (*c*) isoparametric.

Consider next the question of conformability and continuity of adjacent elements that have been mapped and distorted from the simple "parent" elements. First, we point out that for a one-to-one mapping from the $\xi\eta\zeta$ reference to the xyz reference, the *Jacobian* (to be discussed later) must have a constant sign at all points in the domain mapped. Then if for the "parent" elements the interpolation functions satisfy continuity conditions, the distorted elements in the global reference must be contiguous with each other (no gaps) and must also satisfy continuity requirements. We now illustrate some of the above steps in the following example.

EXAMPLE 11.3 We shall map the simple rectangle shown in Fig. 11.16*a*, using *bilinear* interpolation formulas presented earlier in the chapter [see Eq. (11.27)], for which we have:

$$\lfloor N \rfloor = \left\lfloor \overbrace{\frac{(1-\xi)(1-\eta)}{4}}^{N_1}, \overbrace{\frac{(1+\xi)(1-\eta)}{4}}^{N_2}, \overbrace{\frac{(1+\xi)(1+\eta)}{4}}^{N_3}, \overbrace{\frac{(1-\xi)(1+\eta)}{4}}^{N_4} \right\rfloor \qquad (a)$$

We will prove that this mapping takes the rectangle into a straight-sided quadrilateral.

The mapping can be given as follows [see Eqs. (11.35)]:

$$\begin{Bmatrix} x \\ y \end{Bmatrix} = \left[\begin{array}{cccc|cccc} N_1 & N_2 & N_3 & N_4 & 0 & 0 & 0 & 0 \\ \hline 0 & 0 & 0 & 0 & N_1 & N_2 & N_3 & N_4 \end{array}\right] \begin{Bmatrix} x_i \\ \hline y_i \end{Bmatrix} \qquad (b)$$

where $\lfloor x_i \rfloor = \lfloor x_1, x_2, x_3, x_4 \rfloor$ and $\lfloor y_i \rfloor = \lfloor y_1, y_2, y_3, y_4 \rfloor$. For a bilinear transformation we can also say:

$$\begin{Bmatrix} x \\ y \end{Bmatrix} = \begin{Bmatrix} (\alpha + \beta\xi + \gamma\eta + \delta\eta\xi) \\ (\alpha' + \beta'\xi + \gamma'\eta + \delta'\eta\xi) \end{Bmatrix} \qquad (c)$$

From Eqs. (*b*) and (*c*) we equate the right sides, and using Eq. (*a*) we can determine the eight constants $\alpha, \beta, \ldots, \alpha', \ldots, \delta'$ in terms of x_i and y_i. This is done

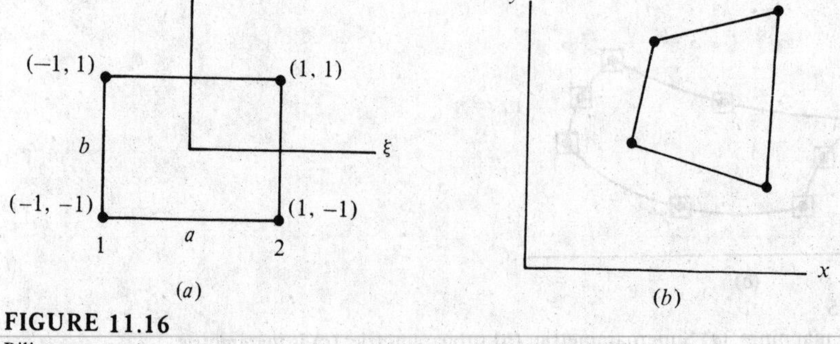

FIGURE 11.16
Bilinear mapping.

by equating the coefficients of ξ, η, $\eta\xi$, and the remaining constants in the two equations to get eight equations to solve for the eight unknown constants. Now along edge 1-2, $\eta = -1$. Then from Eq. (*c*)

$$x = \alpha + \beta\xi + \gamma(-1) + \delta(-1)\xi$$
$$y = \alpha' + \beta'\xi + \gamma'(-1) + \delta'(-1)\xi$$

Eliminating ξ, we get a linear relation between x and y of the form

$$y = C_1 + C_2 x$$

This is a straight line, which for $C_2 \neq 0$ is no longer parallel to the x axis. Looking at the other sides of the parent rectangle in the same way, we see that we get a straight-sided quadrilateral.

11.12 EVALUATION OF MATRICES WITH CURVED ISOPARAMETRIC ELEMENTS

When using the curved elements for finite element calculations we will need to integrate over the domains of the element in the global reference. Thus for $[K]^e$ and $\{q\}^e$ for a three-dimensional problem we will need to evaluate:

$$[K]^e = \iiint_\Omega [B]^T [D] [B] \, dx \, dy \, dz \tag{11.37}$$

$$\{q\} = \iiint_\Omega [N]^T \{b\} \, dx \, dy \, dz \tag{11.38}$$

Also, in computing $[B]$ we will take partial derivatives of N_i with respect to *global coordinates*. But functions N_i are given for local coordinates $\xi\eta\zeta$. Let us consider the latter operation first. Using the chain rule of differentiation, we have for $\partial N_i/\partial\xi$

$$\frac{\partial N_i}{\partial \xi} = \frac{\partial N_i}{\partial x}\frac{\partial x}{\partial \xi} + \frac{\partial N_i}{\partial y}\frac{\partial y}{\partial \xi} + \frac{\partial N_i}{\partial z}\frac{\partial z}{\partial \xi} \tag{11.39}$$

Considering the derivatives $\partial N_i/\partial\eta$ and $\partial N_i/\partial\zeta$ in a similar manner, we can express these three derivatives in the following format:

$$\begin{Bmatrix} \dfrac{\partial N_i}{\partial \xi} \\ \dfrac{\partial N_i}{\partial \eta} \\ \dfrac{\partial N_i}{\partial \zeta} \end{Bmatrix} = \begin{bmatrix} \dfrac{\partial x}{\partial \xi} & \dfrac{\partial y}{\partial \xi} & \dfrac{\partial z}{\partial \xi} \\ \dfrac{\partial x}{\partial \eta} & \dfrac{\partial y}{\partial \eta} & \dfrac{\partial z}{\partial \eta} \\ \dfrac{\partial x}{\partial \zeta} & \dfrac{\partial y}{\partial \zeta} & \dfrac{\partial z}{\partial \zeta} \end{bmatrix} \begin{Bmatrix} \dfrac{\partial N_i}{\partial x} \\ \dfrac{\partial N_i}{\partial y} \\ \dfrac{\partial N_i}{\partial z} \end{Bmatrix} = [J] \begin{Bmatrix} \dfrac{\partial N_i}{\partial x} \\ \dfrac{\partial N_i}{\partial y} \\ \dfrac{\partial N_i}{\partial z} \end{Bmatrix} \tag{11.40}$$

where

$$[J] = \begin{bmatrix} \dfrac{\partial x}{\partial \xi} & \dfrac{\partial y}{\partial \xi} & \dfrac{\partial z}{\partial \xi} \\ \dfrac{\partial x}{\partial \eta} & \dfrac{\partial y}{\partial \eta} & \dfrac{\partial z}{\partial \eta} \\ \dfrac{\partial x}{\partial \zeta} & \dfrac{\partial y}{\partial \zeta} & \dfrac{\partial z}{\partial \zeta} \end{bmatrix} \tag{11.41}$$

is the familiar *Jacobian matrix*. We can solve for global derivatives by using $[J]^{-1}$. Thus we have

$$\begin{Bmatrix} \dfrac{\partial N_i}{\partial x} \\ \dfrac{\partial N_i}{\partial y} \\ \dfrac{\partial N_i}{\partial z} \end{Bmatrix} = [J]^{-1} \begin{Bmatrix} \dfrac{\partial N_i}{\partial \xi} \\ \dfrac{\partial N_i}{\partial \eta} \\ \dfrac{\partial N_i}{\partial \zeta} \end{Bmatrix} \tag{11.42}$$

We now further consider the bilinear element of Example 11.3.

EXAMPLE 11.4 Find the Jacobian matrix for the bilinear, isoparametric element of Example 11.3.

We go back to Eqs. (11.35) with $[N']$ replaced by $[N]$ for the isoparametric case. We see now that on using Eq. (*a*) of Example 11.3 in Eqs. (11.35), we have

$$\frac{\partial x}{\partial \xi} = \frac{\partial}{\partial \xi}[N]\{x_i\} = \frac{1}{4}\lfloor -(1-\eta), (1-\eta), (1+\eta), -(1+\eta) \rfloor \{x_i\}$$

$$\frac{\partial x}{\partial \eta} = \frac{\partial}{\partial \eta}[N]\{x_i\} = \frac{1}{4}\lfloor -(1-\xi), -(1+\xi), (1+\xi), (1-\xi) \rfloor \{x_i\}$$

$$\frac{\partial y}{\partial \xi} = \frac{\partial}{\partial \xi}[N]\{y_i\} = \frac{1}{4}\lfloor -(1-\eta), (1-\eta), (1+\eta), -(1+\eta) \rfloor \{y_i\}$$

$$\frac{\partial y}{\partial \eta} = \frac{\partial}{\partial \eta}[N]\{y_i\} = \frac{1}{4}\lfloor -(1-\xi), -(1+\xi), (1+\xi), (1-\xi) \rfloor \{y_i\}$$

We get, on carrying out the matrix multiplication,

$$\frac{\partial x}{\partial \xi} = J_{11} = \frac{1}{4}[(1-\eta)(x_2 - x_1) + (1+\eta)(x_3 - x_4)]$$

$$\frac{\partial x}{\partial \eta} = J_{21} = \frac{1}{4}[-(1-\xi)(x_1 - x_4) + (1+\xi)(x_3 - x_2)]$$

$$\frac{\partial y}{\partial \xi} = J_{12} = \frac{1}{4}[(1-\eta)(y_2 - y_1) + (1+\eta)(y_3 - y_4)]$$

$$\frac{\partial y}{\partial \eta} = J_{22} = \frac{1}{4}[-(1-\xi)(y_1 - y_4) + (1+\xi)(y_3 - y_2)]$$

We thus have evaluated $[J]$.†

Our next consideration is that of integration over a domain in the distorted geometry in *xyz*. The volume element $dx\,dy\,dz$ needed for this calculation can be given as

$$dx\,dy\,dz = \det[J]\,d\xi\,d\eta\,d\zeta \tag{11.43}$$

so that an integration for $[K]^e$ becomes

$$[K]^e = \int_{-1}^{+1}\int_{-1}^{+1}\int_{-1}^{+1} [B]^T[D][B]\det[J]\,d\xi\,d\eta\,d\zeta \tag{11.44}$$

and for $\{q\}^e$ we have

$$\{q\}^e = \int_{-1}^{+1}\int_{-1}^{+1}\int_{-1}^{+1} [N]^T\{b\}\det[J]\,d\xi\,d\eta\,d\zeta \tag{11.45}$$

Now to get $[K]^e$ we will use Eq. (11.42) for $\partial N_i/\partial x$, and so on, to establish $[B]$, thus making use of the inverse of the Jacobian matrix for the element and getting results in terms of the local reference $\xi\eta\zeta$. Substituting into Eq. (11.44), we can now get $[K]^e$. And for $\{q\}$ we already have $[N]$ in terms of the local reference, so that with $[J]$ also in terms of $\zeta\eta\xi$ we may carry out integrations indicated by Eq. (11.45). The procedure is now the same as other procedures we have presented. We will use the isoparametric element in Chapter 17.

11.13 A COMMENT ON CONVERGENCE FOR CURVED ELEMENTS

We pointed out earlier that the interpolation functions must be such that constant strain is maintained as the elements are made smaller and that the displacement field for an element must reflect rigid-body motion when the nodal displacements are compatible with rigid-body motion. These are requirements for convergence of the process. If these conditions are satisfied in the parent elements, they will prevail in the curved geometry of the elements, as we now show.

Consider a three-dimensional element having n degrees of freedom. For rigid-body movement we require the following:

†The *curved geometry* in the *global reference* has thus been fed into the program for an element in the process of finding the Jacobian matrix $[J]$ for that element.

$$\{\vec{u}(x,y,z)\} = \begin{Bmatrix} u(x,y,z) \\ v(x,y,z) \\ w(x,y,z) \end{Bmatrix} = \begin{Bmatrix} (\alpha_1 + \alpha_2 x + \alpha_3 y + \alpha_4 z) \\ (\bar{\alpha}_1 + \bar{\alpha}_2 x + \bar{\alpha}_3 y + \bar{\alpha}_4 z) \\ (\bar{\bar{\alpha}}_1 + \bar{\bar{\alpha}}_2 x + \bar{\bar{\alpha}}_3 y + \bar{\bar{\alpha}}_4 z) \end{Bmatrix} \tag{11.46}$$

where all the alphas are constants. But for n degrees of freedom we can say, on considering for simplicity the x component of $\vec{u}(x,y,z)$, namely $u(x,y,z)$:

$$u(x,y,z) = \lfloor N(\xi,\eta,\zeta)\rfloor\{a\} = \sum_i^n (N_i)(a_i) \tag{11.47}$$

where $\lfloor N \rfloor$ is the first row of the $[N]$ matrix. At any node i, we then require that

$$u(x_i,y_i,z_i) = a_i = \alpha_1 + \alpha_2 x_i + \alpha_3 y_i + \alpha_4 z_i \tag{11.48}$$

Now replacing the a_i in Eq. (11.47) using Eq. (11.48) and part of the matrix Eq. (11.46), we have

$$\alpha_1 + \alpha_2 x + \alpha_3 y + \alpha_4 z = \sum_i^n (N_i)(\alpha_1 + \alpha_2 x_i + \alpha_3 y_i + \alpha_4 z_i)$$

This becomes

$$\alpha_1 + \alpha_2 x + \alpha_3 y + \alpha_4 z = \left(\sum_i^n N_i\right)\alpha_1 + \left(\sum_i^n N_i x_i\right)\alpha_2 + \left(\sum_i^n N_i y_i\right)\alpha_3 + \left(\sum_i^n N_i z_i\right)\alpha_4$$

Equating coefficients, we come up with necessary conditions for rigid-body movement and constant strain. Thus, for each row of the $[N]$ matrix:

$$\sum_i^n N_i = 1 \tag{11.49a}$$

$$\sum_i^n N_i x_i = x \tag{11.49b}$$

$$\sum_i^n N_i y_i = y \tag{11.49c}$$

$$\sum_{i}^{n} N_i z_i = z \tag{11.49d}$$

The last three conditions are the mapping formulations for isoparametric elements [see Eqs. (11.35)] and hence are satisfied when we use isoparametric elements. We need then to ensure that for the interpolation functions

$$\sum_{i}^{n} N_i = 1 \tag{11.50}$$

We can go to the other components of the displacement field and get similar results for the other two submatrices of $[N]$, namely the other two rows.

In Chapter 17 we shall extend the finite element method to other fields and the conditions for rigid-body movement and constant strain become the constancy of derivatives of the dependent variable of the finite element process.

11.14 CLOSURE

In Chapters 9 and 10 we concentrated on the notation and basic formulations for finite elements while dealing with simple elements. The elements were conformable and we obtained many good results with minimal effort. However, we generally use nonconformable elements that are more complex. Good results for such elements can be achieved only under the right circumstances, requiring careful considerations beyond those employed for the truss element and beam elements. In this chapter we have addressed these considerations. Thus, we have discussed how to choose appropriate elements as well as the number of degrees of freedom for the element. Most important for us are the methods proposed for forming the interpolation functions and for attaining desired characteristics of these functions to achieve good results. Accordingly, when we study plane stress, plates, and torsion in succeeding chapters we shall need to lean heavily on this vital chapter.

PROBLEMS

11.1 Consider a one-dimensional element with three nodes and three degrees of freedom. Using a polynomial representation:

$$\phi = \alpha_0 + \alpha_1 x + \alpha_2 x^2$$

determine the interpolation functions N_1, N_2, and N_3.

11.2 For a Lagrange polynomial element with five nodes, what are the Lagrange interpolation functions $\mathcal{L}_2$ and $\mathcal{L}_4$?

11.3 Determine the interpolation functions for the Hermite polynomial element in Example 11.2.

11.4 Express the field variable w in two dimensions as a complete third-degree polynomial.

11.5 Express a fourth-degree polynomial that achieves *geometric isotropy* with 13 terms.

11.6 Show that at any point in the triangular element (Fig. 11.17),

$$L_i = \frac{s}{h}$$

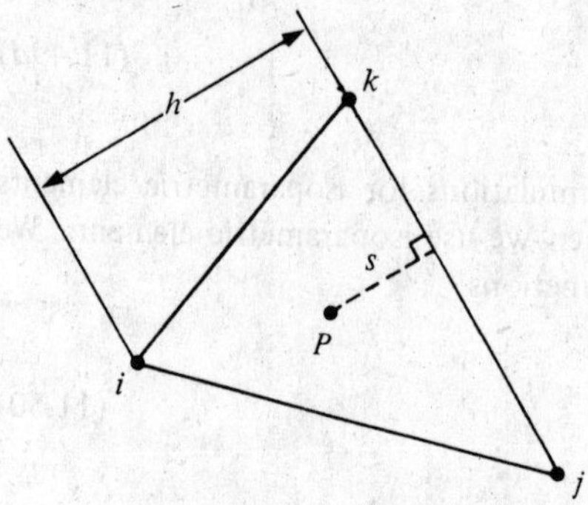

FIGURE 11.17

Also show that $L_i = 0$ at nodes j and k and is unity at node i. Hence for a 3-DOF triangular element, L_i has the properties of N_i. For such cases, we can say $L_i = N_i$, $L_j = N_j$, and $L_k = N_k$. That is, natural coordinates are the same as interpolation functions.

11.7 In Fig. 11.8c determine the natural coordinates for $L^{(7)}$, $L^{(9)}$, and $L^{(4)}$.

11.8 For the six-node quadratic element in Fig. 11.8a, find the six interpolation functions using Lagrange interpolation polynomials.

11.9 Justify the fact that in Eq. (11.27) for the rectangular element in Fig. 11.9,

$$N_1(\xi, \eta) = \frac{\xi - \xi_2}{\xi_1 - \xi_2} \frac{\eta - \eta_4}{\eta_1 - \eta_4}$$

Then show that

$$N_1(\xi, \eta) = \tfrac{1}{4}(\xi - 1)(\eta - 1)$$

11.10 Using Eqs. (11.27), show that the interpolation functions have the required property that field variable ϕ takes on the proper nodal displacement at each of the four nodes of the rectangular element.

11.11 Consider an element with two nodes and four degrees of freedom for which we desire a Hermite polynomial representation. The highest derivative of the field variable taken as a degree of freedom is unity. At node 2, what is the notation given by Eq. (11.12) for the Hermite polynomial for the fourth degree of freedom? Also, express the Hermite polynomial itself.

11.12 For an element with three nodes and nine degrees of freedom with the highest derivative at a node as 2, what is the Hermite polynomial representation as given by Eq. (11.12) for the fifth degree of freedom?

11.13 Formulate N_2 and N_{14} by using the Hermite polynomial given by Eq. (11.30) for a 16-DOF rectangular element of Fig. 11.9.

11.14 For an n-DOF element with n nodes the completeness condition (see Section 11.3, condition 2A) requires for equal values ϕ_i that the field variable ϕ be constant. Show that this results in the following requirement for the interpolation functions:

$$\sum_{1}^{n} N_i = 1$$

12

PLANE STRESS

12.1 INTRODUCTION

We now consider plane stress problems of the type discussed in Part B of Chapter 9. We will use triangular elements with the nodal points at the corners. Such an element is shown in Fig. 12.1. It will be our primary task to determine the nodal displacements for a body in plane stress in terms of the external loads that we shall consider to be applied in some way on the nodes.

12.2 TRIANGULAR COORDINATES

We start by considering a triangular element e in which there are three internal triangles, each having a common apex P, as shown in Fig. 12.2.

As pointed out in Section 11.7, the ratios of the areas of the internal triangles to the outside element triangle are *natural* or *area* coordinates. Recall that these natural coordinates, denoted as L_i, are given as (see Fig. 12.2)†

$$L_i = \frac{\mathcal{A}_i}{\mathcal{A}}$$

$$L_j = \frac{\mathcal{A}_j}{\mathcal{A}}$$

$$L_k = \frac{\mathcal{A}_k}{\mathcal{A}}$$

where $L_i + L_j + L_k = 1$.

†We are using a script $\mathcal{A}$ for area here to avoid confusion with the coefficient matrix $[A]$.

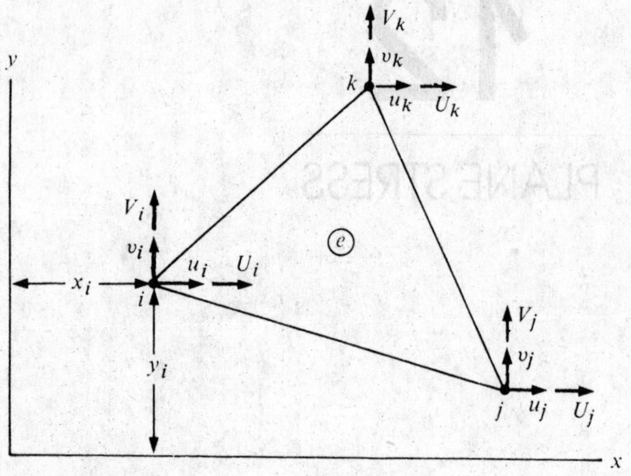

FIGURE 12.1
Triangular element.

Let us now compute the area of the triangular element in terms of its nodal point coordinates. Thus, considering sides $\vec{ij}$ and $\vec{ik}$ of the triangle in Fig. 12.1, we can say

$$A = \tfrac{1}{2}(\vec{ij} \times \vec{ik}) \cdot \mathbf{k}$$

$$A = \tfrac{1}{2}[(x_j - x_i)\mathbf{i} + (y_j - y_i)\mathbf{j}] \times [(x_k - x_i)\mathbf{i} + (y_k - y_i)\mathbf{j}] \cdot \mathbf{k} \qquad (12.1)$$

We have here a triple scalar product, so that we can say

$$A = \frac{1}{2} \begin{vmatrix} (x_j - x_i) & (y_j - y_i) & 0 \\ (x_k - x_i) & (y_k - y_i) & 0 \\ 0 & 0 & 1 \end{vmatrix} \qquad (12.2)$$

$$\begin{cases} L_i = \dfrac{A_i}{A} \\[1em] L_j = \dfrac{A_j}{A} \\[1em] L_k = \dfrac{A_k}{A} \end{cases}$$

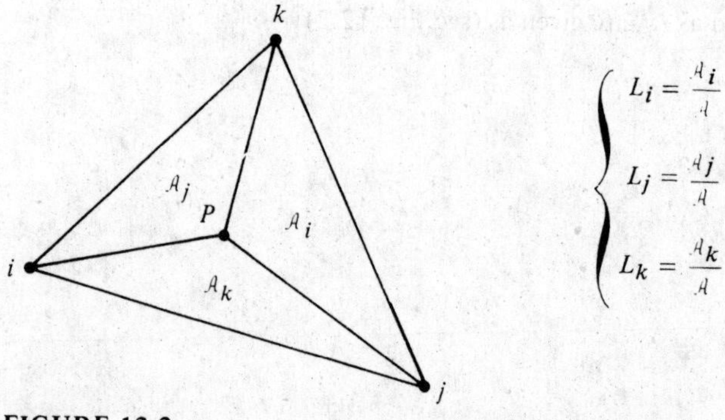

FIGURE 12.2
Triangular coordinates.

We get, on carrying out the determinant,

$$A = \tfrac{1}{2}[(x_j - x_i)(y_k - y_i) - (y_j - y_i)(x_k - x_i)] \tag{12.3}$$

We now introduce the following notation:

$$\begin{aligned} \bar{a}_i &= x_k - x_j & \bar{b}_i &= y_j - y_k \\ \bar{a}_j &= x_i - x_k & \bar{b}_j &= y_k - y_i \\ \bar{a}_k &= x_j - x_i & \bar{b}_k &= y_i - y_j \end{aligned} \tag{12.4}$$

Hence the area of the triangle can be given as follows:

$$A = \tfrac{1}{2}[\bar{a}_k\bar{b}_j - (-\bar{b}_k)(-\bar{a}_j)] = \tfrac{1}{2}(\bar{a}_k\bar{b}_j - \bar{a}_j\bar{b}_k) \tag{12.5a}$$

By permuting the subscripts, we also have

$$A = \tfrac{1}{2}(\bar{a}_i\bar{b}_k - \bar{a}_k\bar{b}_i) \tag{12.5b}$$

$$A = \tfrac{1}{2}(\bar{a}_j\bar{b}_i - \bar{a}_i\bar{b}_j) \tag{12.5c}$$

As a next step we shall compute the area corresponding to one of the triangular coordinates, say A_j. This area is shown shaded in Fig. 12.3 and point P is located by coordinates xy.

Again we form a triple scalar product:

$$A_j = \tfrac{1}{2}[(\vec{iP} \times \vec{ik})\cdot\mathbf{k}]$$

$$A = \tfrac{1}{2}[(x - x_i)\mathbf{i} + (y - y_i)\mathbf{j}] \times [(x_k - x_i)\mathbf{i} + (y_k - y_i)\mathbf{j}]\cdot\mathbf{k}$$

We thus get

$$A_j = \tfrac{1}{2}[(x_k y_i - x_i y_k) + x(y_k - y_i) + y(x_i - x_k)]$$

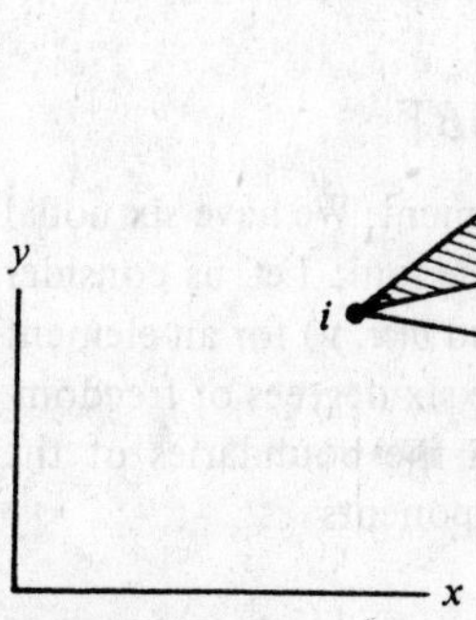

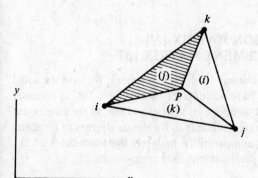

FIGURE 12.3
Compute area of (j).

Using the notation

$$d_j = x_k y_i - x_i y_k \tag{12.6}$$

and the notation presented in Eq. (12.4), we have

$$A_j = \tfrac{1}{2}(d_j + x\bar{b}_j + y\bar{a}_j) \tag{12.7a}$$

We can get the other areas by permuting the subscripts. Thus:

$$A_k = \tfrac{1}{2}(d_k + x\bar{b}_k + y\bar{a}_k) \tag{12.7b}$$

$$A_i = \tfrac{1}{2}(d_i + x\bar{b}_i + y\bar{a}_i) \tag{12.7c}$$

We can now give the natural coordinates as follows:

$$L_i = \frac{1}{2A}(d_i + x\bar{b}_i + y\bar{a}_i) \tag{12.8a}$$

$$L_j = \frac{1}{2A}(d_j + x\bar{b}_j + y\bar{a}_j) \tag{12.8b}$$

$$L_k = \frac{1}{2A}(d_k + x\bar{b}_k + y\bar{a}_k). \tag{12.8c}$$

In matrix form we have

$$\begin{Bmatrix} L_i \\ L_j \\ L_k \end{Bmatrix} = \frac{1}{2A}\begin{bmatrix} d_i & \bar{b}_i & \bar{a}_i \\ d_j & \bar{b}_j & \bar{a}_j \\ d_k & \bar{b}_k & \bar{a}_k \end{bmatrix}\begin{Bmatrix} 1 \\ x \\ y \end{Bmatrix} \tag{12.9}$$

We now have the transformation from rectangular coordinates to triangular coordinates.

In the next section we will determine the interpolation functions for this element as well as the matrix $[B]$.

12.3 INTERPOLATION FUNCTION MATRIX $[N]$ AND STRAIN–NODAL DISPLACEMENT MATRIX $[B]$

As a first step, let us set forth a displacement field for an element. We have six nodal displacements for the element, so we have six degrees of freedom. Let us consider most simply that the displacement field components $u(x, y)$ and $v(x, y)$ for an element are *linear functions* of x and y and have six constants α_i for the six degrees of freedom. In this way we can ensure kinematic compatibility between the boundaries of the elements. Accordingly, we have for the displacement field components:

$$\begin{aligned} u(x, y) &= \alpha_1 + \alpha_2 x + \alpha_3 y \\ v(x, y) &= \alpha_4 + \alpha_5 x + \alpha_6 y \end{aligned} \tag{12.10}$$

In matrix form we can say

$$\{\vec{u}(x,y)\}^e = [C(x,y)]\{\alpha\}^e = [C(x,y)]\begin{Bmatrix}\alpha_1\\ \alpha_2\\ \alpha_3\\ \alpha_4\\ \alpha_5\\ \alpha_6\end{Bmatrix} \tag{12.11}$$

where

$$[C]^e = \begin{bmatrix}1 & x & y & 0 & 0 & 0\\ 0 & 0 & 0 & 1 & x & y\end{bmatrix} \tag{12.12}$$

Next, we wish to relate $\{\alpha\}$ to the nodal displacement vectors $\{a\}$ for the element. At the nodal point i, we can say

$$\{\vec{u}(x_i, y_i)\} = \begin{Bmatrix}u_i\\ v_i\end{Bmatrix} = \{a_i\} \tag{12.13}$$

Using Eq. (12.11) to replace $\{\vec{u}(x_i, y_i)\}$, we have, on rearranging the above equation:

$$\{a_i\} = [C(x_i, y_i)]\{\alpha\}$$

Similarly, at nodal points j and k we have

$$\{a_j\} = [C(x_j, y_j)]\{\alpha\}$$

$$\{a_k\} = [C(x_k, y_k)]\{\alpha\}$$

The last three equations can be incorporated into a single matrix equation involving submatrices:

$$\begin{Bmatrix}\{a_i\}^e\\ \{a_j\}^e\\ \{a_k\}^e\end{Bmatrix} = \begin{bmatrix}[C(x_i, y_i)]\\ [C(x_j, y_j)]\\ [C(x_k, y_k)]\end{bmatrix}\{\alpha\}^e \tag{12.14}$$

Finally, we have

$$\{a\}^e = [A]^e\{\alpha\}^e \tag{12.15}$$

where the coefficient matrix $[A]$ is

$$[A]^e = \begin{bmatrix} 1 & x_i & y_i & 0 & 0 & 0 \\ 0 & 0 & 0 & 1 & x_i & y_i \\ 1 & x_j & y_j & 0 & 0 & 0 \\ 0 & 0 & 0 & 1 & x_j & y_j \\ 1 & x_k & y_k & 0 & 0 & 0 \\ 0 & 0 & 0 & 1 & x_k & y_k \end{bmatrix} \tag{12.16}$$

For convenience, we shall now introduce another form of the nodal displacement matrix. Let

$$\{\tilde{a}\} = \begin{Bmatrix} u_i \\ u_j \\ u_k \\ v_i \\ v_j \\ v_k \end{Bmatrix} \tag{12.17}$$

Then from Eq. (12.14) we have, using the above vector,

$$\{\tilde{a}\}^e = [\tilde{A}]\{\alpha\}^e \tag{12.18}$$

where

$$[\tilde{A}] = \begin{bmatrix} 1 & x_i & y_i & 0 & 0 & 0 \\ 1 & x_j & y_j & 0 & 0 & 0 \\ 1 & x_k & y_k & 0 & 0 & 0 \\ 0 & 0 & 0 & 1 & x_i & y_i \\ 0 & 0 & 0 & 1 & x_j & y_j \\ 0 & 0 & 0 & 1 & x_k & y_k \end{bmatrix} \tag{12.19}$$

With Eq. (12.18) we can solve for the constant $\{\alpha\}$ by taking the inverse

$$\{\alpha\}^e = [\tilde{A}]^{-1}\{\tilde{a}\}^e \tag{12.20}$$

Now go back to Eq. (12.11) and replace $\{\alpha\}^e$ with the above equation. We have

$$\{\vec{u}(x,y)\}^e = [C(x,y)]\,[\tilde{A}]^{-1}\{\tilde{a}\}^e$$

From our basic considerations [see Eq. (9.59*a*)], we can now give the interpolation function matrix $[N]$ for the element as

$$[N] = [C(x,y)]\,[\tilde{A}]^{-1} \tag{12.21}$$

We may in this manner arrive at $[N]$. For this purpose we next compute $[\tilde{A}]^{-1}$ as

$$[\tilde{A}]^{-1} = \frac{1}{2A}\begin{bmatrix} x_jy_k - x_ky_j & x_ky_i - x_iy_k & x_iy_j - x_jy_i & 0 & 0 & 0 \\ y_j - y_k & y_k - y_i & y_i - y_j & 0 & 0 & 0 \\ x_k - x_j & x_i - x_k & x_j - x_i & 0 & 0 & 0 \\ 0 & 0 & 0 & x_jy_k - x_ky_j & x_ky_i - x_iy_k & x_iy_j - x_jy_i \\ 0 & 0 & 0 & y_j - y_k & y_k - y_i & y_i - y_j \\ 0 & 0 & 0 & x_k - x_j & x_i - x_k & x_j - x_i \end{bmatrix}$$

where A is the area of the element. Now noting Eqs. (12.4) and (12.6), we may rewrite the above matrix in a more compact form:

$$[\tilde{A}]^{-1} = \frac{1}{2A}\begin{bmatrix} d_i & d_j & d_k & 0 & 0 & 0 \\ \bar{b}_i & \bar{b}_j & \bar{b}_k & 0 & 0 & 0 \\ \bar{a}_i & \bar{a}_j & \bar{a}_k & 0 & 0 & 0 \\ 0 & 0 & 0 & d_i & d_j & d_k \\ 0 & 0 & 0 & \bar{b}_i & \bar{b}_j & \bar{b}_k \\ 0 & 0 & 0 & \bar{a}_i & \bar{a}_j & \bar{a}_k \end{bmatrix} \tag{12.22}$$

To get $[N]$, we use Eq. (12.21) with Eqs. (12.12) and (12.22), so that

$$[N] = \frac{1}{2A}\begin{bmatrix} 1 & x & y & 0 & 0 & 0 \\ 0 & 0 & 0 & 1 & x & y \end{bmatrix}\begin{bmatrix} d_i & d_j & d_k & 0 & 0 & 0 \\ \bar{b}_i & \bar{b}_j & \bar{b}_k & 0 & 0 & 0 \\ \bar{a}_i & \bar{a}_j & \bar{a}_k & 0 & 0 & 0 \\ 0 & 0 & 0 & d_i & d_j & d_k \\ 0 & 0 & 0 & \bar{b}_i & \bar{b}_j & \bar{b}_k \\ 0 & 0 & 0 & \bar{a}_i & \bar{a}_j & \bar{a}_k \end{bmatrix}$$

Hence:

$$[N] = \frac{1}{2A}\begin{bmatrix} (d_i + x\bar{b}_i + y\bar{a}_i) & (d_j + x\bar{b}_j + y\bar{a}_j) & (d_k + x\bar{b}_k + y\bar{a}_k) & 0 & 0 & 0 \\ 0 & 0 & 0 & (d_i + x\bar{b}_i + y\bar{a}_i) & (d_j + x\bar{b}_j + y\bar{a}_j) & (d_k + x\bar{b}_k + y\bar{a}_k) \end{bmatrix} \tag{12.23}$$

Now go to Eq. (12.8). Note that the nonzero elements of the above matrix are simply triangular coordinates. We then get for the interpolation function matrix:

$$[N] = \begin{bmatrix} L_i & L_j & L_k & 0 & 0 & 0 \\ 0 & 0 & 0 & L_i & L_j & L_k \end{bmatrix} \tag{12.24}$$

Note the simplicity of $[N]$ in terms of triangular coordinates.†

We now embark on the task to find $[B]$. From Eq. (9.59*a*), we have

$$\{\vec{u}(x,y)\}^e = [N]\{\tilde{a}\}^e$$

or

$$\{\vec{u}(x,y)\}^e = \begin{bmatrix} L_i & L_j & L_k & 0 & 0 & 0 \\ 0 & 0 & 0 & L_i & L_j & L_k \end{bmatrix} \begin{Bmatrix} u_i \\ u_j \\ u_k \\ v_i \\ v_j \\ v_k \end{Bmatrix} \tag{12.25}$$

Hence:

$$\{\vec{u}(x,y)\}^e = \begin{bmatrix} (L_i u_i + L_j u_j + L_k u_k) \\ (L_i v_i + L_j v_j + L_k v_k) \end{bmatrix}$$

This in turn can be written as

$$\{\vec{u}(x,y)\}^e = \begin{bmatrix} u_i & u_j & u_k \\ v_i & v_j & v_k \end{bmatrix} \begin{Bmatrix} L_i \\ L_j \\ L_k \end{Bmatrix} \tag{12.26}$$

Now return to Eq. (12.9) to replace the last column vector of Eq. (12.26) so that we shift from triangular to rectangular coordinates:

†For the case of one degree of freedom at a node, we would have the very simple result

$$[N] = \lfloor L_i, L_j, L_k \rfloor$$

Thus $N_i = L_i$. We shall use this result later in the text.

$$\{\vec{u}(x,y)\}^e = \begin{bmatrix} u_i & u_j & u_k \\ v_i & v_j & v_k \end{bmatrix} \left(\frac{1}{2A}\right) \begin{bmatrix} d_i & \bar{b}_i & \bar{a}_i \\ d_j & \bar{b}_j & \bar{a}_j \\ d_k & \bar{b}_k & \bar{a}_k \end{bmatrix} \begin{Bmatrix} 1 \\ x \\ y \end{Bmatrix} \tag{12.27}$$

If we carry out the matrix multiplication on the right side of Eq. (12.27), we get

$$\begin{Bmatrix} u(x,y) \\ v(x,y) \end{Bmatrix}^e = \frac{1}{2A} \begin{Bmatrix} [u_i(d_i + \bar{b}_i x + \bar{a}_i y) + u_j(d_j + \bar{b}_j x + \bar{a}_j y) + u_k(d_k + \bar{b}_k x + \bar{a}_k y)] \\ [v_i(d_i + \bar{b}_i x + \bar{a}_i y) + v_j(d_j + \bar{b}_j x + \bar{a}_j y) + v_k(d_k + \bar{b}_k x + \bar{a}_k y)] \end{Bmatrix} \tag{12.28}$$

We now have the displacement vector at any point in the element in terms of the geometry and the nodal displacements.

As a final step, we can now find the strain-nodal displacement matrix $[B]$. For small strains we have, using the previous equation,

$$\begin{aligned} \epsilon_{xx} &= \frac{\partial u(x,y)}{\partial x} = \frac{1}{2A}(\bar{b}_i u_i + \bar{b}_j u_j + \bar{b}_k u_k) \\ \epsilon_{yy} &= \frac{\partial v(x,y)}{\partial y} = \frac{1}{2A}(\bar{a}_i v_i + \bar{a}_j v_j + \bar{a}_k v_k) \\ \gamma_{xy} &= \frac{\partial u(x,y)}{\partial y} + \frac{\partial v(x,y)}{\partial x} = \frac{1}{2A}(\bar{a}_i u_i + \bar{a}_j u_j + \bar{a}_k u_k + \bar{b}_i v_i + \bar{b}_j v_j + \bar{b}_k v_k) \end{aligned} \tag{12.29}$$

If we put the above into matrix notation, we will involve the strain-nodal displacement matrix $[B]$. Thus:

$$\begin{Bmatrix} \epsilon_{xx} \\ \epsilon_{yy} \\ \gamma_{xy} \end{Bmatrix} = \frac{1}{2A} \begin{bmatrix} \bar{b}_i & \bar{b}_j & \bar{b}_k & 0 & 0 & 0 \\ 0 & 0 & 0 & \bar{a}_i & \bar{a}_j & \bar{a}_k \\ \bar{a}_i & \bar{a}_j & \bar{a}_k & \bar{b}_i & \bar{b}_j & \bar{b}_k \end{bmatrix} \begin{Bmatrix} u_i \\ u_j \\ u_k \\ v_i \\ v_j \\ v_k \end{Bmatrix} \tag{12.30}$$

Hence, from Eq. (9.59*d*) we have

$$[B] = \frac{1}{2A} \begin{bmatrix} \bar{b}_i & \bar{b}_j & \bar{b}_k & 0 & 0 & 0 \\ 0 & 0 & 0 & \bar{a}_i & \bar{a}_j & \bar{a}_k \\ \bar{a}_i & \bar{a}_j & \bar{a}_k & \bar{b}_i & \bar{b}_j & \bar{b}_k \end{bmatrix} \tag{12.31}$$

12.4 CONSTITUTIVE LAW MATRIX [D] AND STIFFNESS MATRIX [K]

We shall now take Hooke's law for plane stress as a constitutive law:

$$\begin{aligned}
\epsilon_{xx} &= \frac{1}{E}(\tau_{xx} - \nu\tau_{yy}) \\
\epsilon_{yy} &= \frac{1}{E}(\tau_{yy} - \nu\tau_{xx}) \\
\gamma_{xy} &= \frac{1}{G}(\tau_{xy}) = \frac{2(1+\nu)}{E}\tau_{xy}
\end{aligned} \tag{12.32}$$

In matrix notation

$$\{\epsilon(x,y)\} = \begin{Bmatrix} \epsilon_{xx} \\ \epsilon_{yy} \\ \gamma_{xy} \end{Bmatrix} = \frac{1}{E}\begin{bmatrix} 1 & -\nu & 0 \\ -\nu & 1 & 0 \\ 0 & 0 & 2(1+\nu) \end{bmatrix}\begin{Bmatrix} \tau_{xx} \\ \tau_{yy} \\ \tau_{xy} \end{Bmatrix} \tag{12.33}$$

Solving for the stresses in terms of strain produces

$$\begin{Bmatrix} \tau_{xx} \\ \tau_{yy} \\ \tau_{xy} \end{Bmatrix} = \{\sigma\} = \frac{E}{1-\nu^2}\begin{bmatrix} 1 & \nu & 0 \\ \nu & 1 & 0 \\ 0 & 0 & \frac{1-\nu}{2} \end{bmatrix}\{\epsilon(x,y)\} \tag{12.34}$$

We then have for $[D]$ the following familiar matrix:

$$[D] = \frac{E}{1-\nu^2}\begin{bmatrix} 1 & \nu & 0 \\ \nu & 1 & 0 \\ 0 & 0 & \frac{1-\nu}{2} \end{bmatrix} \tag{12.35}$$

We are now ready for the element stiffness matrix. From Eq. (9.59f) we have for $[K]^e$:

$$[K]^e = \iiint_V [B]^T[D]\,[B]\,dv$$

Since $[B]$ and $[D]$ are sets of constants, and using t to be the uniform thickness of the element, we have

$$[K]^e = [B]^T [D] [B]\, t\mathcal{A} \tag{12.36}$$

What remains is to carry out the above product of matrices by using Eqs. (12.31) and (12.35):

$$[K]^e = \frac{1}{4\mathcal{A}^2}\, t\mathcal{A}\, \frac{E}{1-\nu^2} \begin{bmatrix} \bar{b}_i & 0 & \bar{a}_i \\ \bar{b}_j & 0 & \bar{a}_j \\ \bar{b}_k & 0 & \bar{a}_k \\ 0 & \bar{a}_i & \bar{b}_i \\ 0 & \bar{a}_j & \bar{b}_j \\ 0 & \bar{a}_k & \bar{b}_k \end{bmatrix} \begin{bmatrix} 1 & \nu & 0 \\ \nu & 1 & 0 \\ 0 & 0 & \dfrac{1-\nu}{2} \end{bmatrix} \begin{bmatrix} \bar{b}_i & \bar{b}_j & \bar{b}_k & 0 & 0 & 0 \\ 0 & 0 & 0 & \bar{a}_i & \bar{a}_j & \bar{a}_k \\ \bar{a}_i & \bar{a}_j & \bar{a}_k & \bar{b}_i & \bar{b}_j & \bar{b}_k \end{bmatrix} \tag{12.37}$$

Using the notation

$$\frac{1}{4\mathcal{A}^2}\, t\mathcal{A}\, \frac{E}{1-\nu^2} = \kappa \tag{12.38}$$

we carry out first the product of the last two matrices to get:

$$[K]^e = \kappa \left[\begin{array}{ccc} \bar{b}_i & 0 & \bar{a}_i \\ \bar{b}_j & 0 & \bar{a}_j \\ \bar{b}_k & 0 & \bar{a}_k \\ \hdashline 0 & \bar{a}_i & \bar{b}_i \\ 0 & \bar{a}_j & \bar{b}_j \\ 0 & \bar{a}_k & \bar{b}_k \end{array}\right] \left[\begin{array}{ccc:ccc} \bar{b}_i & \bar{b}_j & \bar{b}_k & \nu\bar{a}_i & \nu\bar{a}_j & \nu\bar{a}_k \\ \nu\bar{b}_i & \nu\bar{b}_j & \nu\bar{b}_k & \bar{a}_i & \bar{a}_j & \bar{a}_k \\ \dfrac{\bar{a}_i(1-\nu)}{2} & \dfrac{\bar{a}_j(1-\nu)}{2} & \dfrac{\bar{a}_k(1-\nu)}{2} & \dfrac{\bar{b}_i(1-\nu)}{2} & \dfrac{\bar{b}_j(1-\nu)}{2} & \dfrac{\bar{b}_k(1-\nu)}{2} \end{array}\right]$$

Note that we have partitioned the matrices to help manage the calculation. We now get

$$[K]^C = \kappa \left[\begin{array}{ccc|ccc}
\left(\bar{b}_i^2 + \frac{\bar{a}_i^2(1-\nu)}{2}\right) & \left(\bar{b}_i\bar{b}_j + \frac{\bar{a}_i\bar{a}_j(1-\nu)}{2}\right) & \left(\bar{b}_i\bar{b}_k + \frac{\bar{a}_i\bar{a}_k(1-\nu)}{2}\right) & \left(\bar{b}_i\bar{a}_i\nu + \frac{\bar{a}_i\bar{b}_i(1-\nu)}{2}\right) & \left(\bar{b}_i\bar{a}_j\nu + \frac{\bar{a}_i\bar{b}_j(1-\nu)}{2}\right) & \left(\bar{b}_i\bar{a}_k\nu + \frac{\bar{a}_i\bar{b}_k(1-\nu)}{2}\right) \\
\left(\bar{b}_j\bar{b}_i + \frac{\bar{a}_j\bar{a}_i(1-\nu)}{2}\right) & \left(\bar{b}_j^2 + \frac{\bar{a}_j^2(1-\nu)}{2}\right) & \left(\bar{b}_j\bar{b}_k + \frac{\bar{a}_j\bar{a}_k(1-\nu)}{2}\right) & \left(\bar{b}_j\bar{a}_i\nu + \frac{\bar{a}_j\bar{b}_i(1-\nu)}{2}\right) & \left(\bar{b}_j\bar{a}_j\nu + \frac{\bar{a}_j\bar{b}_j(1-\nu)}{2}\right) & \left(\bar{b}_j\bar{a}_k\nu + \frac{\bar{a}_j\bar{b}_k(1-\nu)}{2}\right) \\
\left(\bar{b}_k\bar{b}_i + \frac{\bar{a}_k\bar{a}_i(1-\nu)}{2}\right) & \left(\bar{b}_k\bar{b}_j + \frac{\bar{a}_k\bar{a}_j(1-\nu)}{2}\right) & \left(\bar{b}_k^2 + \frac{\bar{a}_k^2(1-\nu)}{2}\right) & \left(\bar{b}_k\bar{a}_i\nu + \frac{\bar{a}_k\bar{b}_i(1-\nu)}{2}\right) & \left(\bar{b}_k\bar{a}_j\nu + \frac{\bar{a}_k\bar{b}_j(1-\nu)}{2}\right) & \left(\bar{b}_k\bar{a}_k\nu + \frac{\bar{a}_k\bar{b}_k(1-\nu)}{2}\right) \\
\hline
\left(\bar{a}_i\bar{b}_i\nu + \frac{\bar{b}_i\bar{a}_i(1-\nu)}{2}\right) & \left(\bar{a}_i\bar{b}_j\nu + \frac{\bar{b}_i\bar{a}_j(1-\nu)}{2}\right) & \left(\bar{a}_i\bar{b}_k\nu + \frac{\bar{b}_i\bar{a}_k(1-\nu)}{2}\right) & \left(\bar{a}_i^2 + \frac{\bar{b}_i^2(1-\nu)}{2}\right) & \left(\bar{a}_i\bar{a}_j + \frac{\bar{b}_i\bar{b}_j(1-\nu)}{2}\right) & \left(\bar{a}_i\bar{a}_k + \frac{\bar{b}_i\bar{b}_k(1-\nu)}{2}\right) \\
\left(\bar{a}_j\bar{b}_i\nu + \frac{\bar{b}_j\bar{a}_i(1-\nu)}{2}\right) & \left(\bar{a}_j\bar{b}_j\nu + \frac{\bar{b}_j\bar{a}_j(1-\nu)}{2}\right) & \left(\bar{a}_j\bar{b}_k\nu + \frac{\bar{b}_j\bar{a}_k(1-\nu)}{2}\right) & \left(\bar{a}_j\bar{a}_i + \frac{\bar{b}_j\bar{b}_i(1-\nu)}{2}\right) & \left(\bar{a}_j^2 + \frac{\bar{b}_j^2(1-\nu)}{2}\right) & \left(\bar{a}_j\bar{a}_k + \frac{\bar{b}_j\bar{b}_k(1-\nu)}{2}\right) \\
\left(\bar{a}_k\bar{b}_i\nu + \frac{\bar{b}_k\bar{a}_i(1-\nu)}{2}\right) & \left(\bar{a}_k\bar{b}_j\nu + \frac{\bar{b}_k\bar{a}_j(1-\nu)}{2}\right) & \left(\bar{a}_k\bar{b}_k\nu + \frac{\bar{b}_k\bar{a}_k(1-\nu)}{2}\right) & \left(\bar{a}_k\bar{a}_i + \frac{\bar{b}_k\bar{b}_i(1-\nu)}{2}\right) & \left(\bar{a}_k\bar{a}_j + \frac{\bar{b}_k\bar{b}_j(1-\nu)}{2}\right) & \left(\bar{a}_k^2 + \frac{\bar{b}_k^2(1-\nu)}{2}\right)
\end{array}\right] \tag{12.39}$$

To make the arithmetic more tractable, we shall employ simple notation for the four submatrices. Thus for the upper left submatrix, we denote each term as K_{ln}^{11}, where l and n are free indices that cover the entire submatrix. Taking $i = 1, j = 2$, and $k = 3$ for the element, we have

$$[K_{ln}^{11}] = \begin{bmatrix} K_{11}^{11} & K_{12}^{11} & K_{13}^{11} \\ K_{21}^{11} & K_{22}^{11} & K_{23}^{11} \\ K_{31}^{11} & K_{32}^{11} & K_{33}^{11} \end{bmatrix} \tag{12.40}$$

where for each term we have

$$K_{ln}^{11} = \bar{b}_l \bar{b}_n + \frac{\bar{a}_l \bar{a}_n (1-\nu)}{2} \tag{12.41}$$

Similarly, we denote the second upper submatrix as K_{ln}^{12}, so that for each term we have

$$K_{ln}^{12} = \bar{b}_l \bar{a}_n \nu + \frac{\bar{a}_l \bar{b}_n (1-\nu)}{2} \tag{12.42}$$

Now the lower first submatrix is simply the transpose of the second upper submatrix and is denoted as $[K_{ln}^{12}]^T$. Finally, the lower right submatrix we denote as $[K_{ln}^{22}]$, where for each term we have

$$K_{ln}^{22} = \bar{a}_l \bar{a}_n + \frac{\bar{b}_l \bar{b}_n (1-\nu)}{2} \tag{12.43}$$

We can summarize the results as follows:

$$[K]^e = \kappa \left[\begin{array}{c|c} [K_{ln}^{11}] & [K_{ln}^{12}] \\ \hline [K_{ln}^{12}]^T & [K_{ln}^{22}] \end{array} \right] \tag{12.44}$$

where we can fill out the 36 terms by using Eqs. (12.41)-(12.43).

12.5 SURFACE TRACTIONS AND BODY SURFACES

We now return to Eq. (9.59g), which we rewrite here:

$$\{q\} - \{f\} = [K]^e \{\tilde{a}\} \tag{12.45}$$

Recall that $\{q\}$ is the force vector acting *on* the element from nodal points and $\{f\}$ stems from initial strain and residual stress. We shall examine in detail the body force and surface traction contributions to $\{q\}$. Thus, going to Eq. (9.43), we can say

$$\{q\}_{\text{body}} = \iiint_V [N]^T\{b\}\,dv \tag{12.46a}$$

$$\{q\}_{\text{traction}} = \oiint_{\mathcal{A}} [N]^T\{T\}\,d\mathcal{A} \tag{12.46b}$$

For $[N]$ we use Eq. (12.24), so that for Eq. (12.46a) above we have

$$\{q\}_{\text{body}} = \iint_{\mathcal{A}} \begin{bmatrix} L_i & 0 \\ L_j & 0 \\ L_k & 0 \\ 0 & L_i \\ 0 & L_j \\ 0 & L_k \end{bmatrix} \begin{Bmatrix} b_x \\ b_y \end{Bmatrix} t\,d\mathcal{A} \tag{12.47}$$

where t is the thickness of the element. To aid in the integration, we present without proof the following integration formula.

$$\iint_{\mathcal{A}} (L_i^{\alpha} L_j^{\beta} L_k^{\gamma})\,d\mathcal{A} = 2\mathcal{A}\frac{\alpha!\,\beta!\,\gamma!}{(\alpha+\beta+\gamma+2)!} \tag{12.48}$$

where α, β, and γ are constants. Considering the body force distribution b_x and b_y to be uniform and t to be constant, we can use the above formula in Eq. (12.47):

$$\{q\}_{\text{body}} = t\iint_{\mathcal{A}} \begin{Bmatrix} L_i b_x \\ L_j b_x \\ L_k b_x \\ L_i b_y \\ L_j b_y \\ L_k b_y \end{Bmatrix} d\mathcal{A} = t \begin{Bmatrix} b_x \iint L_i\,d\mathcal{A} \\ b_x \iint L_j\,d\mathcal{A} \\ b_x \iint L_k\,d\mathcal{A} \\ b_y \iint L_i\,d\mathcal{A} \\ b_y \iint L_j\,d\mathcal{A} \\ b_y \iint L_k\,d\mathcal{A} \end{Bmatrix} = t \begin{Bmatrix} \frac{\mathcal{A}}{3} b_x \\ \frac{\mathcal{A}}{3} b_x \\ \frac{\mathcal{A}}{3} b_x \\ \frac{\mathcal{A}}{3} b_y \\ \frac{\mathcal{A}}{3} b_y \\ \frac{\mathcal{A}}{3} b_y \end{Bmatrix} \tag{12.49}$$

Hence:

$$\{q\}_{\text{body}} = \frac{At}{3}\begin{Bmatrix} b_x \\ b_x \\ b_x \\ b_y \\ b_y \\ b_y \end{Bmatrix} \tag{12.50}$$

We now replace the body forces acting on the element by equal x and y components of forces on each node.

As for the traction loadings, we shall replace the distributions on the boundaries of the element by the rigid-body static equivalents, using forces acting on the nodal points. Thus consider side i, j of an element having the linear distribution shown in Fig. 12.4. From rigid-body statics, we can find the forces $(r_i)_y$ and $(r_j)_y$ to be

$$\begin{aligned} (r_i)_y &= \left[\frac{T_i}{2} + \frac{1}{6}(T_j - T_i)\right] Lt \\ (r_j)_y &= \left[\frac{T_i}{2} + \frac{1}{3}(T_j - T_i)\right] Lt \end{aligned} \tag{12.51}$$

We thus have the complete means of determining the force vector $\{q\}$ to be used in Eq. (12.45).

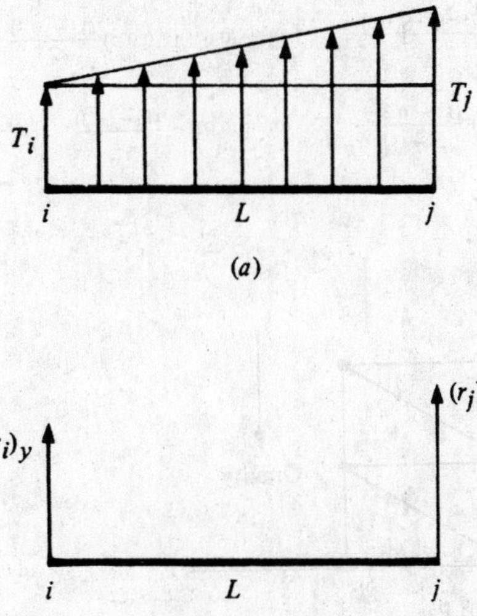

(b)

FIGURE 12.4
Replacement of surface traction distribution.

Finally, remember that $\{f\}$ is determined for an element by consulting Eq. (9.46), where dv becomes $t\,dA$ and the volume integral becomes a surface integral. In this chapter we will not include initial strain or residual stress so we can take $\{f\}$ to be zero.

We shall now consider an example of a finite element computation for a single element.

EXAMPLE 12.1 We now examine in detail a triangular element in plane stress. Accordingly, consider element 1 of the grid shown in Fig. 12.5, having nodal points 1, 2, and 3. The following data apply to the problem:

$$E = 21 \times 10^{10}\ \text{Pa} \qquad \gamma = 1.4 \times 10^5\ \text{N/m}^3$$
$$\nu = 0.3 \qquad T_1 = 7 \times 10^5\ \text{Pa}$$
$$t = 5\ \text{mm} \qquad T_2 = 1.05 \times 10^6\ \text{Pa}$$

We first will calculate the $\bar{a}$'s, $\bar{b}$'s, and $\bar{d}$'s for the element as defined by Eqs. (12.4) and (12.6). Thus:

$$\begin{aligned} \bar{a}_1 &= x_3 - x_2 = 0.075 - 0.075 = 0 & \bar{b}_1 &= y_2 - y_3 = 0 - 0.025 = -0.025 \\ \bar{a}_2 &= x_1 - x_3 = 0 - 0.075 = -0.075 & \bar{b}_2 &= y_3 - y_1 = 0.025 - 0 = 0.025 \\ \bar{a}_3 &= x_2 - x_1 = 0.075 - 0 = 0.075 & \bar{b}_3 &= y_1 - y_2 = 0 - 0 = 0 \end{aligned} \qquad (a)$$

$$\begin{aligned} \bar{d}_1 &= x_2 y_3 - x_3 y_2 = 1.875 \times 10^{-3}\ \text{m}^2 \\ \bar{d}_2 &= x_3 y_1 - x_1 y_3 = 0\ \text{m}^2 \\ \bar{d}_3 &= x_1 y_2 - x_2 y_1 = 0\ \text{m}^2 \end{aligned} \qquad (b)$$

Now go to Eq. (12.39) and substitute to get the stiffness matrix. Thus for $[K^{11}_{ln}]$ we have, on consulting Eq. (12.41),

$$[K^{11}_{ln}] = \begin{bmatrix} (-0.025)(-0.025) + (0) & (-0.025)(0.025) + (0) & (0) + (0) \\ (0.025)(-0.025) + (0) & (0.025)(0.025) + \dfrac{(0.075)^2(1 - 0.3)}{2} & (0) + (0.075)(-0.075)\dfrac{(1 - 0.3)}{2} \\ (0) + (0) & (0) + (-0.075)(0.075)\dfrac{(1 - 0.3)}{2} & (0) + (0.075)^2\dfrac{(1 - 0.3)}{2} \end{bmatrix}$$

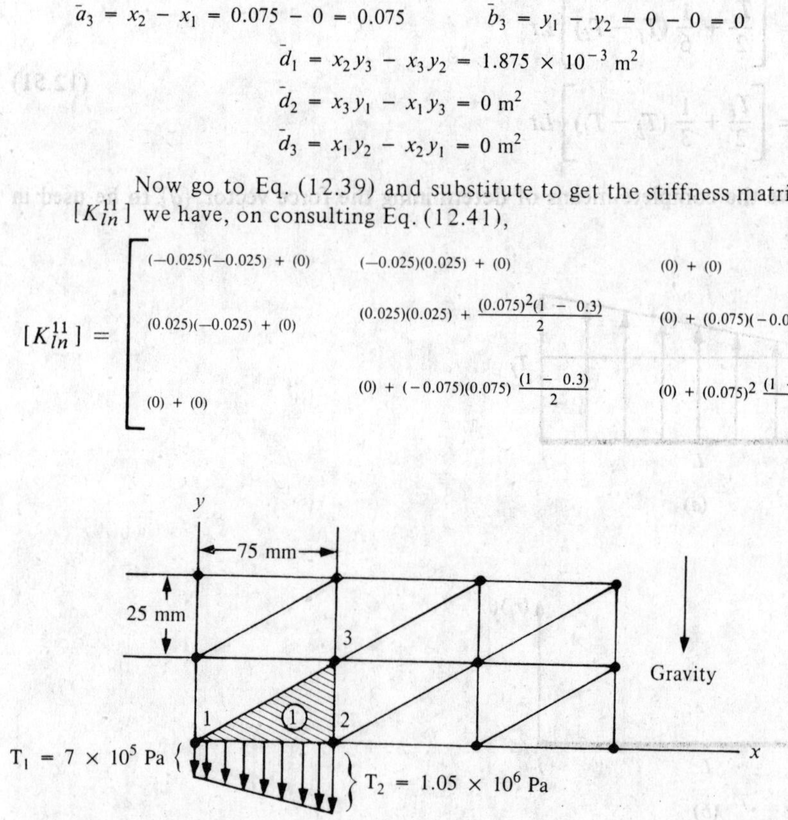

FIGURE 12.5
Element 1 of a grid.

Therefore

$$[K_{ln}^{11}] = \begin{bmatrix} 6.25 & -6.25 & 0 \\ -6.25 & 25.94 & -19.69 \\ 0 & -19.69 & 19.69 \end{bmatrix} \times 10^{-4} \tag{c}$$

Next, for $[K_{ln}^{12}]$ we have, using Eq. (12.42),

$$[K_{ln}^{12}] = \begin{bmatrix} 0+0 & (-0.025)(-0.075)(0.3)+0 & (-0.025)(0.075)(0.3)+0 \\ (0.025)(0)+\dfrac{(-0.075)(0.025)(1-0.3)}{2} & (0.025)(-0.075)(0.3)+\dfrac{(-0.075)(0.025)(1-0.3)}{2} & (0.025)(0.075)(0.3)+0 \\ 0+\dfrac{(0.075)(-0.025)(1-0.3)}{2} & 0+\dfrac{(0.075)(0.025)(1-0.3)}{2} & 0+0 \end{bmatrix}$$

Hence:

$$[K_{ln}^{12}] = \begin{bmatrix} 0 & 5.625 & -5.625 \\ 6.56 & -12.19 & 5.625 \\ -6.56 & 6.56 & 0 \end{bmatrix} \times 10^{-4} \tag{d}$$

Finally, we get $[K_{ln}^{22}]$ by using Eq. (12.43):

$$[K_{ln}^{22}] = \begin{bmatrix} 0+\dfrac{(-0.025)(-0.025)(1-0.3)}{2} & 0+\dfrac{(-0.025)(0.025)(1-0.3)}{2} & 0+0 \\ 0+\dfrac{(0.025)(-0.025)(1-0.3)}{2} & (-0.075)(-0.075)+\dfrac{(0.025)(0.025)(1-0.3)}{2} & (-0.075)(0.075)+0 \\ 0+0 & (0.075)(-0.075)+0 & (0.075)(0.075)+0 \end{bmatrix}$$

Hence:

$$[K_{ln}^{22}] = \begin{bmatrix} 2.188 & -2.188 & 0 \\ -2.188 & 58.44 & -56.25 \\ 0 & -56.25 & 56.25 \end{bmatrix} \times 10^{-4} \tag{e}$$

We can now give the transpose of the matrix $[K_{ln}^{12}]$ to get $[K_{ln}^{21}]$. Thus:

$$[K_{ln}^{21}] = \begin{bmatrix} 0 & 6.56 & -6.56 \\ 5.625 & -12.19 & 6.56 \\ -5.625 & 5.625 & 0 \end{bmatrix} \times 10^{-4} \tag{f}$$

We can now given the stiffness matrix for the element.

$$[K]^{(1)} = \kappa \left[\begin{array}{ccc|ccc} 6.25 & -6.25 & 0 & 0 & 5.625 & -5.625 \\ -6.25 & 25.94 & -19.69 & 6.56 & -12.19 & 5.625 \\ 0 & -19.69 & 19.69 & -6.56 & 6.56 & 0 \\ \hline 0 & 6.56 & -6.56 & 2.188 & -2.188 & 0 \\ 5.625 & -12.19 & 6.56 & -2.188 & 58.44 & -56.25 \\ -5.625 & 5.625 & 0 & 0 & -56.25 & 56.25 \end{array}\right] \tag{g}$$

We have next for κ [see Eq. (12.38)]

$$\kappa = \frac{1}{4(9.375 \times 10^{-4})^2}(0.005)(9.375 \times 10^{-4})\frac{21 \times 10^{10}}{1 - 0.3^2} = 3.077 \times 10^{11} \text{ N/m}^3$$

Now we consider the weight of element as the body force. We get, on using Eq. (12.50),

$$\{q\}^{(1)}_{\text{body}} = \frac{(9.375 \times 10^{-4})(0.005)}{3} \begin{Bmatrix} 0 \\ 0 \\ 0 \\ -1.4 \times 10^5 \\ -1.4 \times 10^5 \\ -1.4 \times 10^5 \end{Bmatrix} = \begin{Bmatrix} 0 \\ 0 \\ 0 \\ -0.2188 \\ -0.2188 \\ -0.2188 \end{Bmatrix} \text{N} \tag{h}$$

Finally, we consider the surface traction on side 1–2. We can use Eqs. (12.51) for this purpose.

$$\{q_i\}_{\text{trac}} = \begin{Bmatrix} 0 \\ -\left[\dfrac{T_i}{2} + \dfrac{1}{6}(T_j - T_i)\right] \end{Bmatrix} Lt = \begin{Bmatrix} 0 \\ -\left[\dfrac{7 \times 10^5}{2} + \dfrac{1}{6}(3.5 \times 10^5)\right] \end{Bmatrix} (0.075)(0.005)$$

$$= \begin{Bmatrix} 0 \\ -153.13 \end{Bmatrix} \text{N}$$

$$\{q_j\}_{\text{trac}} = \begin{Bmatrix} 0 \\ -\left[\dfrac{T_i}{2} + \dfrac{1}{3}(T_j - T_i)\right] \end{Bmatrix} Lt = \begin{Bmatrix} 0 \\ -\left[\dfrac{7 \times 10^5}{2} + \dfrac{1}{3}(3.5 \times 10^5)\right] \end{Bmatrix} (0.075)(0.005)$$

$$= \begin{Bmatrix} 0 \\ -175 \end{Bmatrix} \text{N}$$

Hence we have for the vector $\{q\}$:

$$\{q\}^{(1)} = \begin{Bmatrix} 0 \\ 0 \\ 0 \\ -0.2188 \\ -0.2188 \\ -0.2188 \end{Bmatrix} + \begin{Bmatrix} 0 \\ 0 \\ 0 \\ -153.13 \\ -175 \\ 0 \end{Bmatrix} = \begin{Bmatrix} 0 \\ 0 \\ 0 \\ -153.349 \\ -175.219 \\ -0.2188 \end{Bmatrix} \quad \text{(i)}$$

We have thus determined the key features of a single triangular element.

EXAMPLE 12.2 Consider a cantilever beam with four elements as shown in Fig. 12.6. The following data apply:

$$E = 21 \times 10^{10} \text{ Pa}$$

$$t = 8 \text{ mm}$$

$$\gamma = 72{,}000 \text{ N/m}^3$$

$$\nu = 0.3$$

Determine the displacements.

In Fig. 12.7 we have identified the elements and the nodal points. We will consider the stiffness matrix for one element in detail and then give the results for the other elements.

Element *a*

$$x_1 = 0 \text{ m} \qquad y_1 = -0.05 \text{ m}$$

$$x_2 = 0.1 \text{ m} \qquad y_2 = -0.05 \text{ m}$$

$$x_3 = 0 \text{ m} \qquad y_3 = 0.05 \text{ m}$$

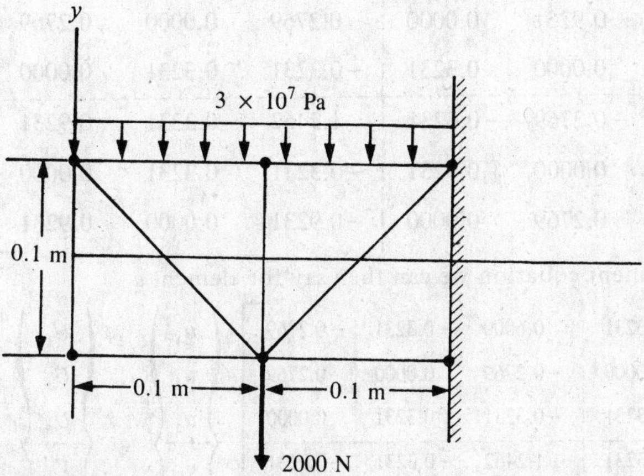

FIGURE 12.6
Cantilever with four elements.

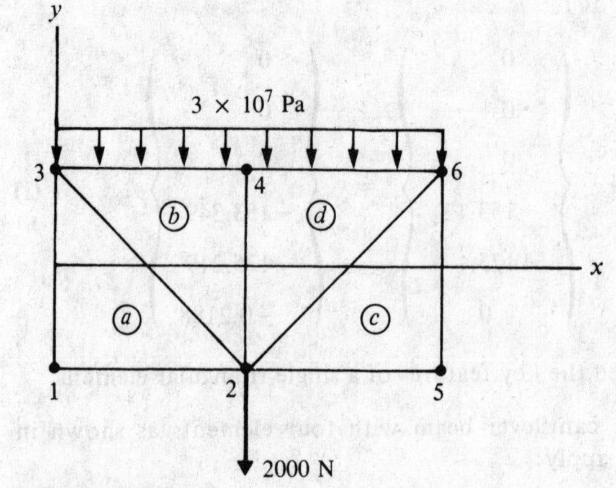

FIGURE 12.7
Elements and nodal points identified for cantilever beam.

$$\bar{a}_1 = x_3 - x_2 = -0.1 \text{ m} \quad \bar{b}_1 = y_2 - y_3 = -0.1 \text{ m}$$
$$\bar{a}_2 = x_1 - x_3 = 0 \text{ m} \quad \bar{b}_2 = y_3 - y_1 = 0.1 \text{ m}$$
$$\bar{a}_3 = x_2 - x_1 = 0.1 \text{ m} \quad \bar{b}_3 = y_1 - y_2 = 0 \text{ m}$$
$$d_1 = x_2 y_3 - x_3 y_2 = 0.0005 \text{ m}^2$$
$$d_2 = x_3 y_1 - x_1 y_3 = 0 \text{ m}^2$$
$$d_3 = x_1 y_2 - x_2 y_1 = 0.0005 \text{ m}^2$$

First we note that for κ we have, using Eq. (12.38),

$$\kappa = \frac{1}{4(0.005)^2}(0.008)(0.005)\frac{2.1 \times 10^{11}}{1 - 0.3^2} = 9.231 \times 10^{10} \text{ N/m}^3$$

For $[K]^a$ we have, upon using Eq. (12.39),

$$[K]^a = 10^9 \left[\begin{array}{ccc|ccc} 1.2462 & -0.9231 & -0.3231 & 0.6000 & -0.3231 & -0.2769 \\ -0.9231 & 0.9231 & 0.0000 & -0.2769 & 0.0000 & 0.2769 \\ -0.3231 & 0.0000 & 0.3231 & -0.3231 & 0.3231 & 0.0000 \\ \hline 0.6000 & -0.2769 & -0.3231 & 1.2462 & -0.3231 & -0.9231 \\ -0.3231 & 0.0000 & 0.3231 & -0.3231 & 0.3231 & 0.0000 \\ -0.2769 & 0.2769 & 0.0000 & -0.9231 & 0.0000 & 0.9231 \end{array}\right]$$

Using the basic finite element equation we can then say for element a

$$10^9 \left[\begin{array}{ccc|ccc} 1.2462 & -0.9231 & -0.3231 & 0.6000 & -0.3231 & -0.2769 \\ -0.9231 & 0.9231 & 0.0000 & -0.2769 & 0.0000 & 0.2769 \\ -0.3231 & 0.0000 & 0.3231 & -0.3231 & 0.3231 & 0.0000 \\ \hline 0.6000 & -0.2769 & -0.3231 & 1.2462 & -0.3231 & -0.9231 \\ -0.3231 & 0.0000 & 0.3231 & -0.3231 & 0.3231 & 0.0000 \\ -0.2769 & 0.2769 & 0.0000 & -0.9231 & 0.0000 & 0.9231 \end{array}\right] \begin{Bmatrix} u_1 \\ u_2 \\ u_3 \\ \hline v_1 \\ v_2 \\ v_3 \end{Bmatrix} = \begin{Bmatrix} U_1 \\ U_2 \\ U_3 \\ \hline V_1 \\ V_2 \\ V_3 \end{Bmatrix}^a$$

By a similar procedure we can find the corresponding equation for the other three elements. We simply quote the results:

$$
10^6 \left[\begin{array}{ccc|ccc}
0.3231 & -0.3231 & 0.0000 & 0.0000 & -0.3231 & 0.3231 \\
-0.3231 & 1.2462 & -0.9231 & -0.2769 & 0.6000 & -0.3231 \\
0.0000 & -0.9231 & 0.9231 & 0.2769 & -0.2769 & 0.0000 \\
\hline
0.0000 & -0.2769 & 0.2769 & 0.9231 & -0.9231 & 0.0000 \\
-0.3231 & 0.6000 & -0.2769 & -0.9231 & 1.2462 & -0.3231 \\
0.3231 & -0.3231 & 0.0000 & 0.0000 & -0.3231 & 0.3231
\end{array}\right]
\left(\begin{array}{c} u_2 \\ u_3 \\ u_4 \\ \hline v_2 \\ v_3 \\ v_4 \end{array}\right)
=
\left(\begin{array}{c} U_2 \\ U_3 \\ U_4 \\ \hline V_2 \\ V_3 \\ V_4 \end{array}\right) \qquad b
$$

$$
10^6 \left[\begin{array}{ccc|ccc}
0.9231 & -0.9231 & 0.0000 & 0.0000 & 0.2769 & -0.2769 \\
-0.9231 & 1.2462 & -0.3231 & 0.3231 & -0.6000 & 0.2769 \\
0.0000 & -0.3231 & 0.3231 & -0.3231 & 0.3231 & 0.0000 \\
\hline
0.0000 & 0.3231 & -0.3231 & 0.3231 & -0.3231 & 0.0000 \\
0.2769 & -0.6000 & 0.3231 & -0.3231 & 1.2462 & -0.9231 \\
-0.2769 & 0.2769 & 0.0000 & 0.0000 & -0.9231 & 0.9231
\end{array}\right]
\left(\begin{array}{c} u_2 \\ u_5 \\ u_6 \\ \hline v_2 \\ v_5 \\ v_6 \end{array}\right)
=
\left(\begin{array}{c} U_2 \\ U_5 \\ U_6 \\ \hline V_2 \\ V_5 \\ V_6 \end{array}\right) \qquad c
$$

$$
10^6 \left[\begin{array}{ccc|ccc}
0.3231 & 0.0000 & -0.3231 & 0.0000 & -0.3231 & 0.3231 \\
0.0000 & 0.9231 & -0.9231 & -0.2769 & 0.0000 & 0.2769 \\
-0.3231 & -0.9231 & 1.2462 & 0.2769 & 0.3231 & -0.6000 \\
\hline
0.0000 & -0.2769 & 0.2769 & 0.9231 & 0.0000 & -0.9231 \\
-0.3231 & 0.0000 & 0.3231 & 0.0000 & 0.3231 & -0.3231 \\
0.3231 & 0.2769 & -0.6000 & -0.9231 & -0.3231 & 1.2462
\end{array}\right]
\left(\begin{array}{c} u_2 \\ u_4 \\ u_6 \\ \hline v_2 \\ v_4 \\ v_6 \end{array}\right)
=
\left(\begin{array}{c} U_2 \\ U_4 \\ U_6 \\ \hline V_2 \\ V_4 \\ V_6 \end{array}\right) \qquad d
$$

We now wish to assemble the global matrix for the entire beam. To do this in a simple way, we will expand each of the stiffness matrix equations for an element to include all the nodal point displacements of the grid as shown below for element a.

$$
\left[\begin{array}{cccccc|cccccc}
1.2462 & -0.9231 & 0.3231 & 0 & 0 & 0 & 0.6 & -0.3231 & -0.2769 & 0 & 0 & 0 \\
-0.9231 & 0.9231 & 0 & 0 & 0 & 0 & -0.2769 & 0 & 0.2769 & 0 & 0 & 0 \\
-0.3231 & 0 & 0.3231 & 0 & 0 & 0 & -0.3231 & 0 & 0 & 0 & 0 & 0 \\
0 & 0 & 0 & 0 & 0 & 0 & 0 & 0 & 0 & 0 & 0 & 0 \\
0 & 0 & 0 & 0 & 0 & 0 & 0 & 0 & 0 & 0 & 0 & 0 \\
0 & 0 & 0 & 0 & 0 & 0 & 0 & 0 & 0 & 0 & 0 & 0 \\
0 & 0 & 0 & 0 & 0 & 0 & 0 & 0 & 0 & 0 & 0 & 0 \\
\hline
3.1247 & -0.2769 & 0.3231 & 0 & 0 & 0 & 1.2462 & -0.3231 & -0.9231 & 0 & 0 & 0 \\
-0.3231 & 0 & 0.3231 & 0 & 0 & 0 & 0.3231 & 0.3231 & 0 & 0 & 0 & 0 \\
-0.2769 & 0.2769 & 0 & 0 & 0 & 0 & 0.9231 & 0 & 0.9231 & 0 & 0 & 0 \\
0 & 0 & 0 & 0 & 0 & 0 & 0 & 0 & 0 & 0 & 0 & 0 \\
0 & 0 & 0 & 0 & 0 & 0 & 0 & 0 & 0 & 0 & 0 & 0 \\
0 & 0 & 0 & 0 & 0 & 0 & 0 & 0 & 0 & 0 & 0 & 0 \\
0 & 0 & 0 & 0 & 0 & 0 & 0 & 0 & 0 & 0 & 0 & 0
\end{array}\right]
\left(\begin{array}{c} u_1 \\ u_2 \\ u_3 \\ u_4 \\ u_5 \\ u_6 \\ \hline v_1 \\ v_2 \\ v_3 \\ v_4 \\ v_5 \\ v_6 \end{array}\right)
=
\left(\begin{array}{c} U_1 \\ U_2 \\ U_3 \\ U_4 \\ U_5 \\ U_6 \\ \hline V_1 \\ V_2 \\ V_3 \\ V_4 \\ V_5 \\ V_6 \end{array}\right) \qquad a
$$

We do this to other elements and add corresponding terms of each element stiffness matrix to form the global stiffness matrix. Thus, the left-hand side of the global basic equation is readily obtained. The right-hand side of the global basic equation, you will recall from Chapter 9, is simply the *external* force vector acting on the nodes. Thus, we show in Fig. 12.8 replacements at nodes 3, 4, and 6 for the external uniform loading. The supporting forces are denoted as $(R_5)_x$, $(R_5)_y$, $(R_6)_x$, and $(R_6)_y$ at nodes 5 and 6. Let us consider next gravitational body forces on the elements. From Eq. (12.50) we can say:

$$\{q\}^e_{\text{body}} = \frac{At}{3}\begin{Bmatrix} 0 \\ 0 \\ 0 \\ \gamma \\ \gamma \\ \gamma \end{Bmatrix} = \frac{(8)(0.008)}{3}\begin{Bmatrix} 0 \\ 0 \\ 0 \\ -72{,}000 \\ -72{,}000 \\ -72{,}000 \end{Bmatrix} = \begin{Bmatrix} 0 \\ 0 \\ 0 \\ -1{,}536 \\ -1{,}536 \\ -1{,}536 \end{Bmatrix} \text{N}$$

We shall neglect this very small force contribution, so that we have for the forces:

$$\begin{Bmatrix} U_1 \\ U_2 \\ U_3 \\ U_4 \\ U_5 \\ U_6 \\ V_1 \\ V_2 \\ V_3 \\ V_4 \\ V_5 \\ V_6 \end{Bmatrix}_{\text{global}} = \begin{Bmatrix} 0 \\ 0 \\ 0 \\ 0 \\ (R_5)_x \\ (R_6)_x \\ 0 \\ -2{,}000 \\ -12{,}000 \\ -24{,}000 \\ (R_5)_y \\ -12{,}000 + (R_6)_y \end{Bmatrix}$$

Accordingly, we have for the global equation, on noting that $u_5 = v_5 = u_6 = v_6 = 0$,

$$
10^9
\left[\begin{array}{cccccc|cccccc}
1.2462 & -0.9231 & -0.3231 & 0.0000 & 0.0000 & 0.0000 & 0.6000 & -0.3231 & -0.2769 & 0.0000 & 0.0000 & 0.0000 \\
-0.9231 & 2.4923 & 0.0000 & -0.6462 & -0.9231 & 0.0000 & -0.2769 & 0.0000 & 0.6000 & 0.0000 & 0.2769 & -0.6000 \\
-0.3231 & 0.0000 & 1.2462 & -0.9231 & 0.0000 & 0.0000 & -0.3231 & 0.6000 & 0.0000 & -0.2769 & 0.0000 & 0.0000 \\
0.0000 & -0.6462 & -0.9231 & 2.4923 & 0.0000 & -0.9231 & 0.0000 & 0.0000 & -0.3231 & 0.0000 & 0.0000 & -0.3231 \\
0.0000 & -0.9231 & 0.0000 & 0.0000 & 1.2462 & -0.3231 & 0.0000 & 0.3231 & 0.0000 & 0.0000 & -0.6000 & 0.2769 \\
0.0000 & 0.0000 & 0.0000 & -0.9231 & -0.3231 & 1.2462 & 0.0000 & -0.6000 & 0.0000 & 0.2769 & 0.3231 & 0.0000 \\
\hline
0.6000 & -0.2769 & -0.3231 & 0.0000 & 0.0000 & 0.0000 & 1.2462 & -0.3231 & -0.9231 & 0.0000 & 0.0000 & 0.0000 \\
-0.3231 & 0.0000 & 0.6000 & 0.0000 & 0.3231 & -0.6000 & -0.3231 & 2.4923 & 0.0000 & -1.8462 & -0.3231 & 0.0000 \\
-0.2769 & 0.6000 & 0.0000 & -0.3231 & 0.0000 & 0.0000 & -0.9231 & 0.0000 & 1.2462 & -0.3231 & 0.0000 & 0.0000 \\
0.0000 & 0.0000 & -0.2769 & 0.0000 & 0.0000 & 0.2769 & 0.0000 & -1.8462 & -0.3231 & 2.4923 & 0.0000 & -0.3231 \\
0.0000 & 0.2769 & 0.0000 & 0.0000 & -0.6000 & 0.3231 & 0.0000 & -0.3231 & 0.0000 & 0.0000 & 1.2462 & -0.9231 \\
0.0000 & -0.6000 & 0.0000 & 0.3231 & 0.2769 & 0.0000 & 0.0000 & 0.0000 & 0.0000 & -0.3231 & -0.9231 & 1.2462
\end{array}\right]
\left\{\begin{array}{c}
u_1 \\ u_2 \\ u_3 \\ u_4 \\ 0 \\ 0 \\ \hline v_1 \\ v_2 \\ v_3 \\ v_4 \\ 0 \\ 0
\end{array}\right\}
=
\left\{\begin{array}{c}
0 \\ 0 \\ 0 \\ 0 \\ (R_5)_x \\ (R_6)_x \\ \hline 0 \\ -2{,}000 \\ -12{,}000 \\ -24{,}000 \\ (R_5)_y \\ -12{,}000 + (R_6)_y
\end{array}\right\}
$$

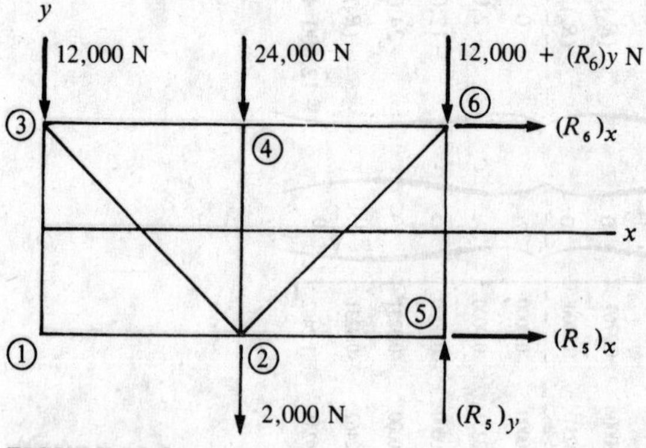

FIGURE 12.8
Joint loadings.

Eliminating the 5th and 6th as well as the 11th and 12th rows and columns, we get the nonsingular reduced stiffness matrix. Thus

$$10^9 \begin{bmatrix} 1.2462 & -0.9231 & -0.3231 & 0.0000 & 0.6000 & -0.3231 & -0.2769 & 0.0000 \\ -0.9232 & 2.4923 & 0.0000 & -0.6462 & -0.2769 & 0.0000 & 0.6000 & 0.0000 \\ -0.3231 & 0.0000 & 1.2462 & -0.9231 & -0.3231 & 0.6000 & 0.0000 & -0.2769 \\ 0.0000 & -0.6462 & -0.9231 & 2.4923 & 0.0000 & 0.0000 & -0.3231 & 0.0000 \\ 0.6000 & -0.2769 & -0.3231 & 0.0000 & 1.2462 & -0.3231 & -0.9231 & 0.0000 \\ -0.3231 & 0.0000 & 0.6000 & 0.0000 & -0.3231 & 2.4923 & 0.0000 & -1.8462 \\ -0.2769 & 0.6000 & 0.0000 & -0.3231 & -0.9231 & 0.0000 & 1.2462 & -0.3231 \\ 0.0000 & 0.0000 & -0.2769 & 0.0000 & 0.0000 & -1.8462 & -0.3231 & 2.4923 \end{bmatrix} \begin{Bmatrix} u_1 \\ u_2 \\ u_3 \\ u_4 \\ v_1 \\ v_2 \\ v_3 \\ v_4 \end{Bmatrix} = \begin{Bmatrix} 0 \\ 0 \\ 0 \\ 0 \\ 0 \\ -2{,}000 \\ -12{,}000 \\ -24{,}000 \end{Bmatrix}$$

To get the solution we find the inverse of $[K]$ and solve for the eight unknown nodal displacements. We get as the prime result of the analysis:

$$\begin{Bmatrix} u_1 \\ u_2 \\ u_3 \\ u_4 \\ v_1 \\ v_2 \\ v_3 \\ v_4 \end{Bmatrix} = \begin{Bmatrix} 0.3087 \\ 0.2635 \\ -0.4193 \\ -0.3121 \\ -1.6910 \\ -0.7951 \\ -1.7362 \\ -0.9569 \end{Bmatrix} \times 10^{-4} \text{ m}$$

12.6 CONVERGENCE OF PROCESS FOR PLANE STRESS

In this section we shall compare results of the finite element process for a tip-loaded cantilever beam for which we have (in Section 1.19) an exact solution from the theory

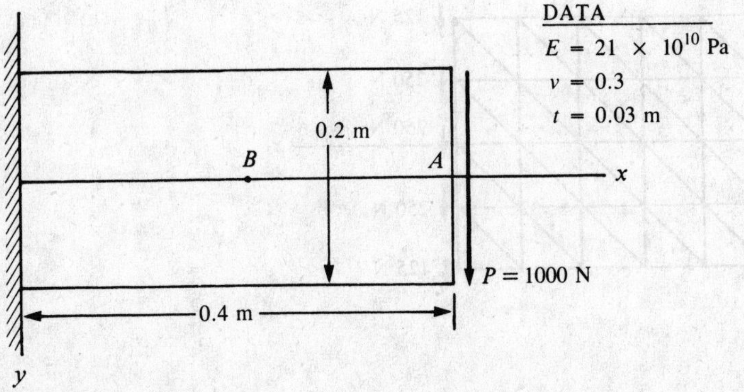

FIGURE 12.9
Tip-loaded cantilever beam.

of elasticity. The cantilever beam is shown in Fig. 12.9. The load P is distributed uniformly over the end of the beam. In Fig. 12.10 we show 16 elements and in Fig. 12.11 we show 64 elements.

We compare results for the vertical deflection at point A on the neutral axis.

16-Element Analysis:

$$[v_{16}]_A = 3.5096 \times 10^{-6} \text{ m}$$

64-Element Analysis:

$$[v_{64}]_A = 4.9841 \times 10^{-6} \text{ m}$$

Exact Analysis [Eq. (1.133)]:

$$[v_{\text{exact}}]_A = 5.08 \times 10^{-6} \text{ m}$$

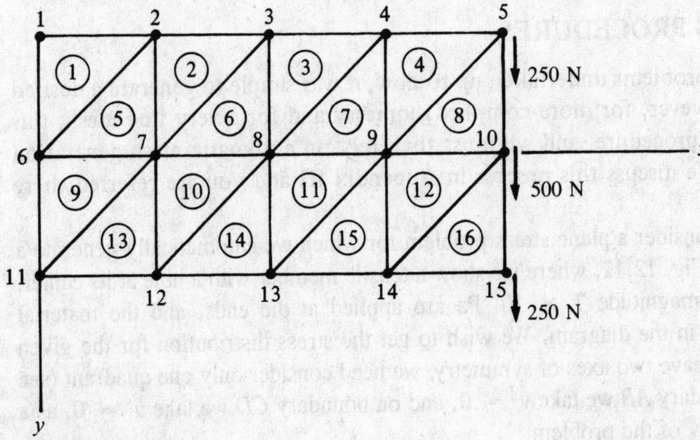

FIGURE 12.10
Cantilever with 16 elements.

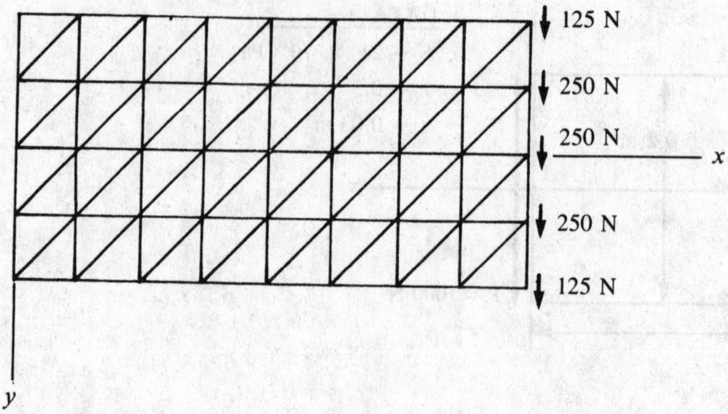

FIGURE 12.11
Cantilever with 64 elements.

Now consider point B along the neutral axis. Here we get:

16-Element Analysis:

$$[v_{16}]_B = 1.2162 \times 10^{-6} \text{ m}$$

64-Element Analysis:

$$[v_{64}]_B = 1.687 \times 10^{-6} \text{ m}$$

Exact Analysis [Eq. (1.133)]:

$$[v_{\text{exact}}]_B = 1.5873 \times 10^{-6} \text{ m}$$

We see that we get satisfactory results even with only 64 elements and that the convergence is obvious.

12.7 GRIDDING PROCEDURES

In the plane stress problems undertaken up to now, it was simple to generate a desired mesh by hand. However, for more complex geometry and for a very fine mesh, this becomes a difficult procedure, and we must then rely on automatic mesh generation via the computer. We discuss this process in Appendix III and you are referred there for details.

We will now consider a plane stress problem for which we automatically generate a grid. Thus, examine Fig. 12.12, where we show a tensile member with a hole at its center. Uniform stresses of magnitude 7×10^7 Pa are applied at the ends, and the material properties are shown in the diagram. We wish to get the stress distribution for the given loading. Because we have two axes of symmetry, we need consider only one quadrant (see Fig. 12.13). On boundary AB we take $v = 0$, and on boundary CD we take $u = 0$, as a result of the symmetry of the problem.

We shall use 48 elements of varying size and orientation (see Fig. 12.14) with smaller elements appearing around the hole, where we can expect large gradients

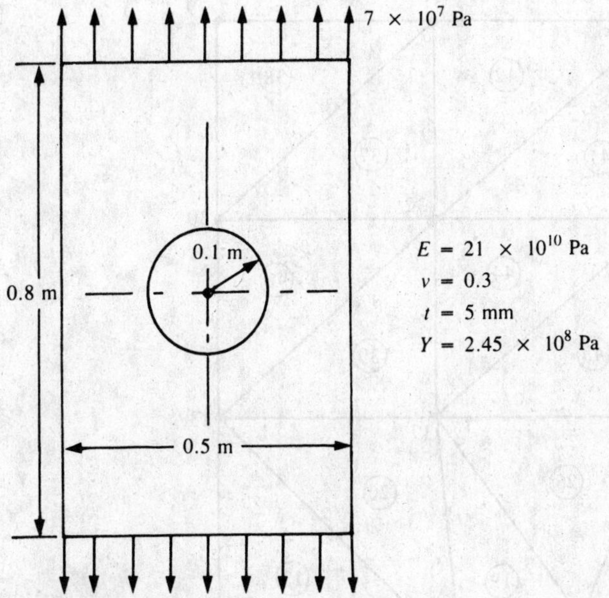

FIGURE 12.12
Plane stress problem.

in the displacement field. This grid was generated after we selected three large subregions, including one having the circular boundary from the hole. We then pick nodes along the boundaries of the trapezoids, putting in some closely spaced nodes on and near the circle. Once this has been done, we can employ a program that will give us the coordinates of a system of nodes, including the ones we chose, to form a mesh, as shown in Fig. 12.14.

In this problem we can make use of the "canned" SAP IV program, which, having the coordinates of the nodes and the loading and boundary conditions along

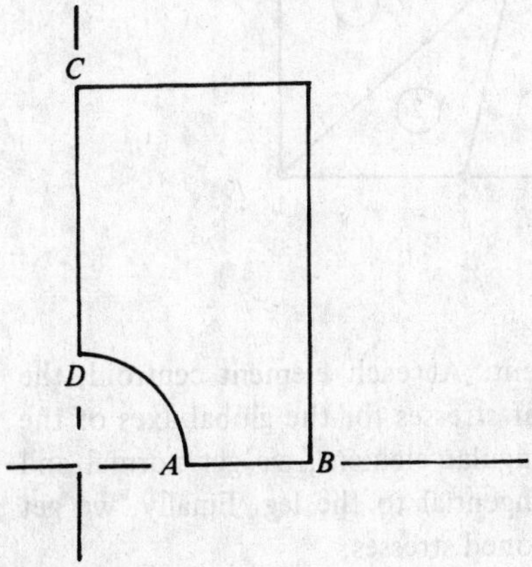

FIGURE 12.13
Consider first quadrant.

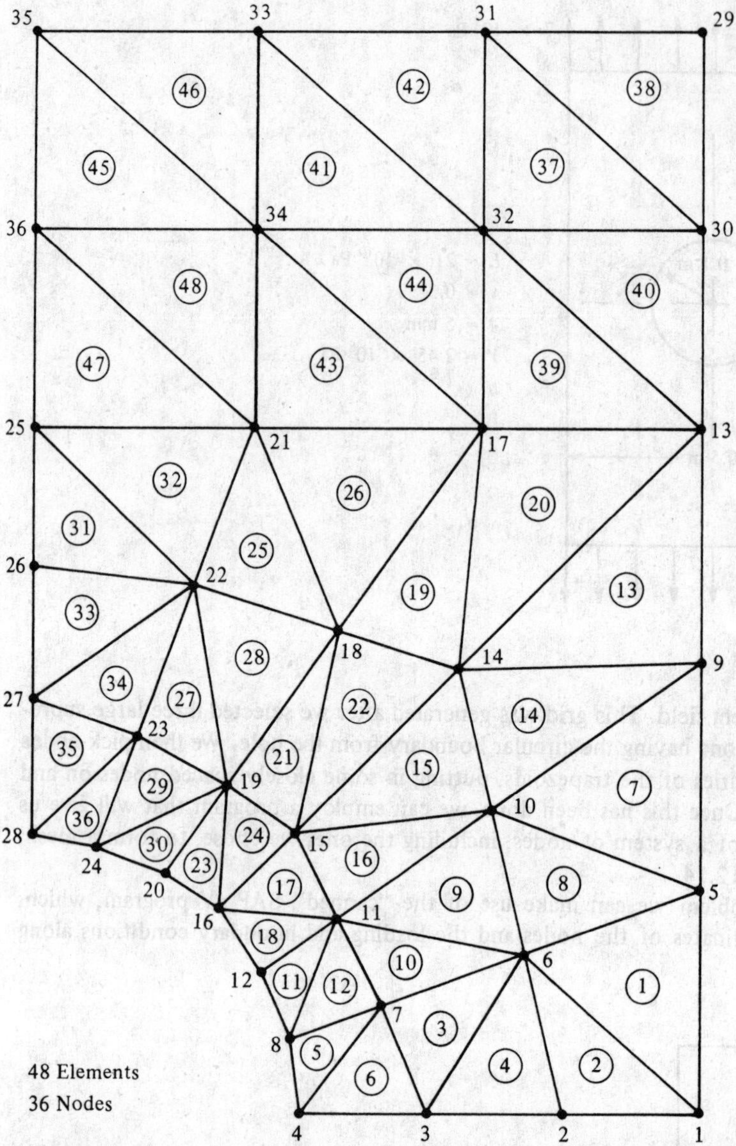

FIGURE 12.14
Automatically generated grid.

the axes of symmetry, we can readily implement. At each element centroid, the SAP IV program will give us the normal and shear stresses for the global axes of the problem. At the center of each leg of each triangular element, we get normal and shear stresses directed along axes normal and tangential to the leg. Finally, we get principal stresses corresponding to the abovementioned stresses.

We will only report that for the centroid and legs of element 5 we get the largest principal stress of 2.177×10^8 Pa. This gives a stress concentration of 3.11, which is a bit higher than 3, which is what the theory of elasticity gives for such a problem with a small hole. Improved results could be obtained by making a finer mesh, particularly near the hole.

12.8 CLOSURE

In this chapter we made use of triangular elements with natural coordinates to study the case of plane stress. We got very good results for a variety of problems. Furthermore, we discussed and used a computer-generated mesh for triangular elements. There are computer-generated mesh programs for other more complex elements but these programs are difficult. You are referred to the literature for such additional undertakings.

In the next chapter we consider the more difficult problem of plates. A number of approaches will be examined and compared. As a consequence, there will be ample opportunity to use some of the key formulations of Chapters 11 and 12.

PROBLEMS

12.1 Find $[K]$ and $\{q\}$ for the element in Fig. 12.15.

12.2 Following Example 12.2, find the nodal displacements for the plate shown in Fig. 12.16. Use the computer to invert a matrix.

12.3 For the body in Fig. 12.17, solve for nodal displacements using 8 elements.

12.4 Solve for the nodal displacements for the cantilever beam in Fig. 12.6 for a single vertical load at the tip of 4,500 N downward.

12.5 A body is lying flat as shown in Fig. 12.18. Find the nodal displacements for the grid shown, using SAP IV or other "canned" program.

*12.6 Find the nodal displacements in Problem 12.5 with your own complete computer program.

12.7 Using the gridding program given in Appendix III, find elements for one quadrant of the plate shown in Fig. 12.19. Have smaller elements near the slot, as done in the example of Section 12.7.

12.8 For the grid developed in Problem 12.7 determine the stresses, using SAP IV or other "canned" program.

*12.9 For the grid formulated in Problem 12.7 find the nodal displacement with your own computer program.

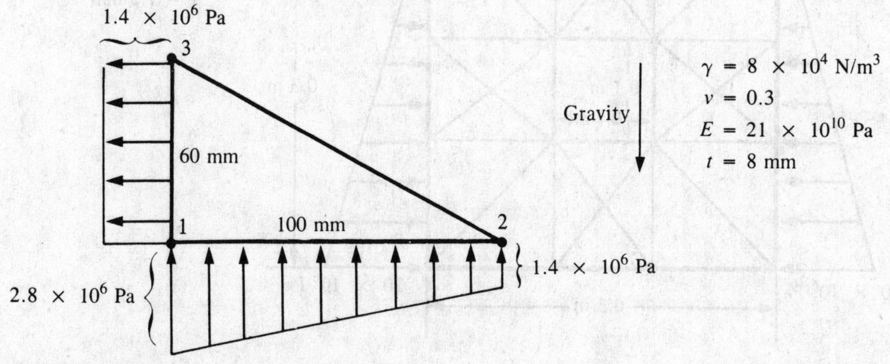

FIGURE 12.15

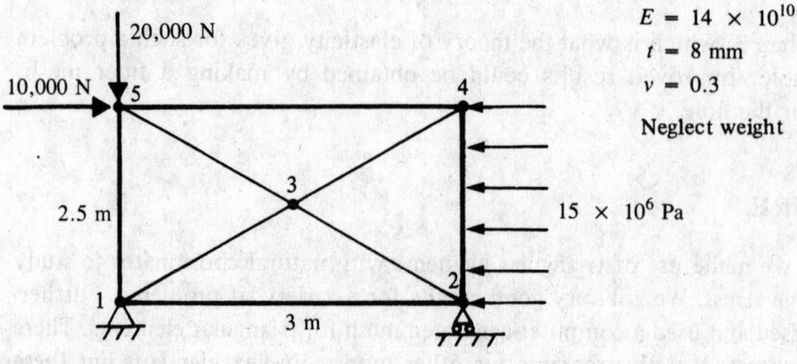

FIGURE 12.16

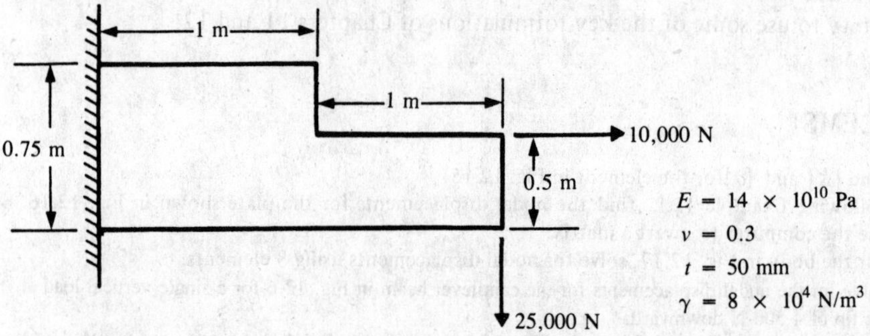

FIGURE 12.17

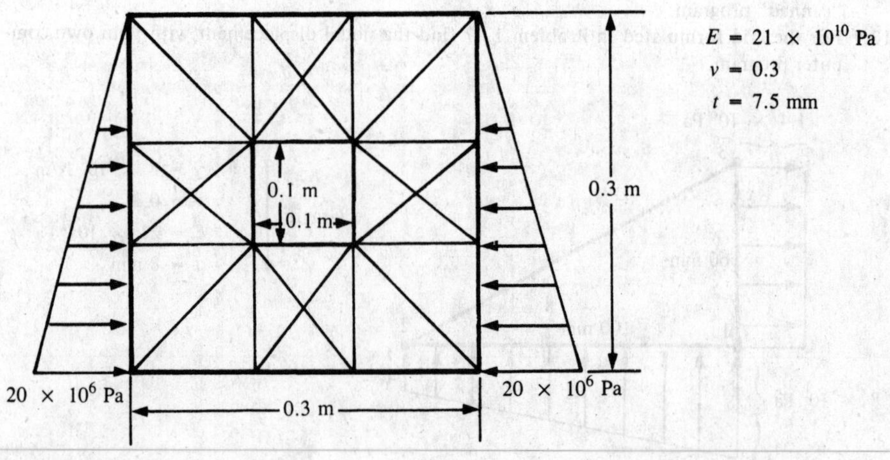

FIGURE 12.18

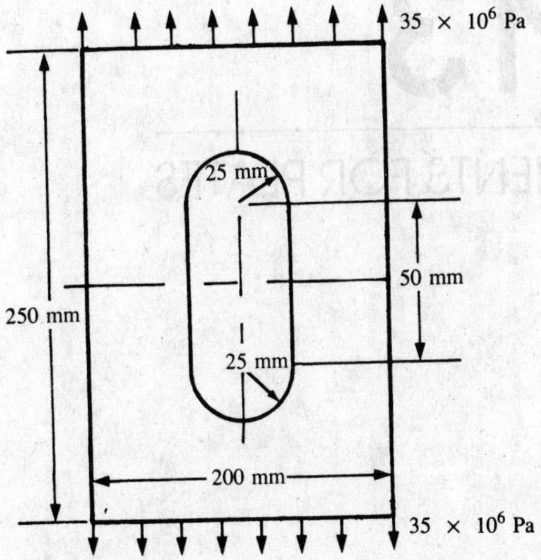

FIGURE 12.19

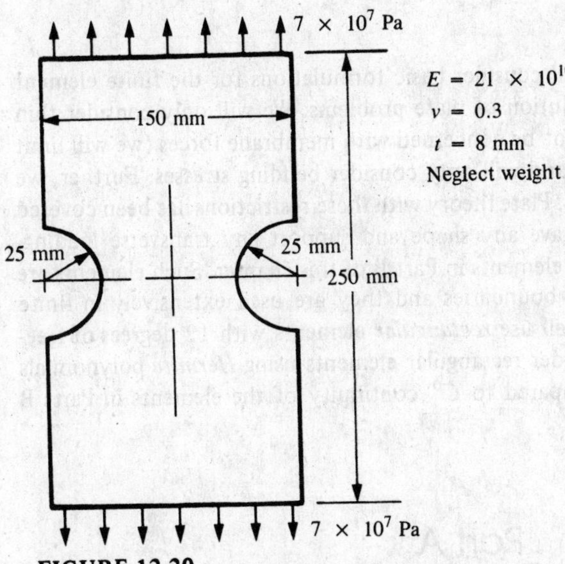

FIGURE 12.20

12.10 Using the gridding program in Appendix III, form a mesh for one quadrant of the plate shown in Fig. 12.20.

12.11 For the grid developed in Problem 12.10, find the stresses with SAP IV or other "canned" program.

*12.12 For the grid developed in Problem 12.10, solve for nodal displacements with your own computer program. Determine the nodal displacements and stresses.

13

FINITE ELEMENTS FOR PLATES

13.1 INTRODUCTION

In Part A of this chapter we shall consider basic formulations for the finite element approach for the approximate solution of plate problems. We will only consider thin plates. For this reason we shall not be concerned with membrane forces (we will limit our selves to small deformation) and will only consider bending stresses. Further, we shall only consider isotropic plates. Plate theory with these restrictions has been covered in Chapter 6. The plates may have any shape and support any transverse loading.

We shall then use *triangular* elements in Part B of this chapter. Such elements are particularly suitable for irregular boundaries and they are used extensively in finite element analysis. In Part C we shall use *rectangular* elements with 12 degrees of freedom. Finally, in Part D will consider rectangular elements using *Hermite* polynomials to achieve C^1 continuity as compared to C^0 continuity of the elements in Parts B and C.

Part A
Basic Considerations

13.2 GENERAL CONSIDERATIONS

In this section we shall apply the general approach given in Part B of Chapter 9 to triangular elements. Much of the results will also apply to the rectangular elements to be studied. We shall box in those results which have general applicability.

We show a triangular element e in Fig. 13.1, with three nodal points i, j, and k at the corners of the triangle. At each node we will have three nodal displacements. They are

w vertical deflection

θ_x slope of plate in y direction

$$\theta_x = \frac{\partial w}{\partial y}$$

θ_y slope of plate in x direction

$$\theta_y = -\frac{\partial w}{\partial x}$$

(We shall use the same nodal displacements as above for each of the four corner nodes of a rectangular element.) We thus have for the triangular element:

$$\{a\}^e = \begin{Bmatrix} \{a_i\} \\ \{a_j\} \\ \{a_k\} \end{Bmatrix} = \begin{Bmatrix} (w)_i \\ (\theta_x)_i \\ (\theta_y)_i \\ \hline (w)_j \\ (\theta_x)_j \\ (\theta_y)_j \\ \hline (w)_k \\ (\theta_x)_k \\ (\theta_y)_k \end{Bmatrix} \tag{13.1}$$

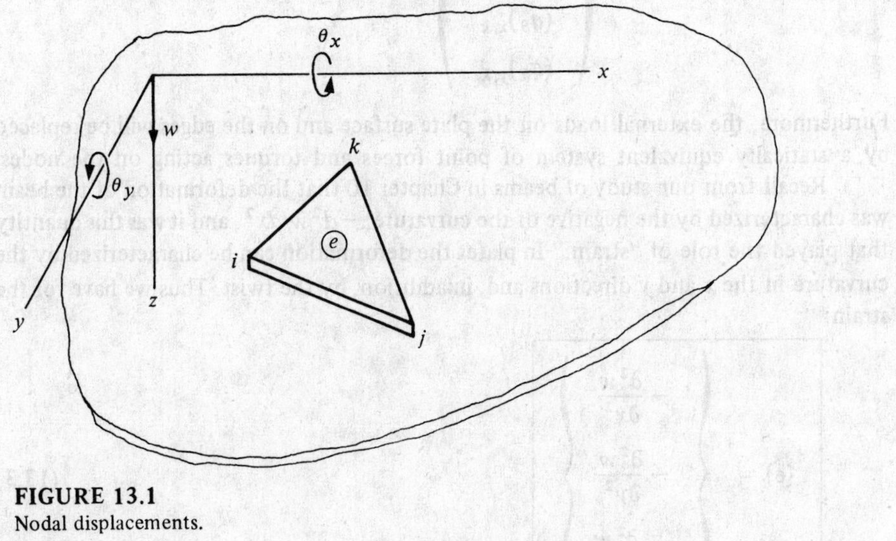

FIGURE 13.1
Nodal displacements.

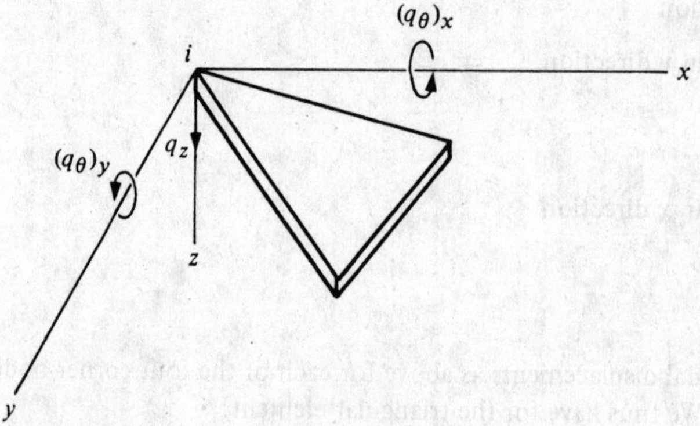

FIGURE 13.2
Nodal forces at node *i*.

The corresponding nodal forces $\{q\}$ consist of two torques and a force. These are shown in Fig. 13.2. These apply *on* the element *from* the nodes. We thus have for $\{q\}$:

$$\{q\}^e = \left\{ \begin{array}{c} \{q_i\} \\ \hdashline \{q_j\} \\ \hdashline \{q_k\} \end{array} \right\} = \left\{ \begin{array}{c} (q_z)_i \\ (q_\theta)_{x,i} \\ (q_\theta)_{y,i} \\ \hdashline (q_z)_j \\ (q_\theta)_{x,j} \\ (q_\theta)_{y,j} \\ \hdashline (q_z)_k \\ (q_\theta)_{x,k} \\ (q_\theta)_{y,k} \end{array} \right\} \tag{13.2}$$

Furthermore, the external loads on the plate surface and on the edges will be replaced by a statically equivalent system of point forces and torques acting on the nodes.

Recall from our study of beams in Chapter 10 that the deformation of the beam was characterized by the negative of the curvature, $-d^2w/dx^2$, and it was this quantity that played the role of "strain." In plates the deformation can be characterized by the curvature in the x and y directions and, in addition, by the twist. Thus we have for the strain:

$$\boxed{\{\epsilon\} = \left\{ \begin{array}{c} -\dfrac{\partial^2 w}{\partial x^2} \\ -\dfrac{\partial^2 w}{\partial y^2} \\ -2\dfrac{\partial^2 w}{\partial x\, \partial y} \end{array} \right\}} \tag{13.3}$$

As for the "stresses" to be associated via a constitutive law with the strains, we have

$$\{\sigma\} = \begin{Bmatrix} M_x \\ M_y \\ M_{xy} \end{Bmatrix} \tag{13.4}$$

We can now determine the matrix $[D]$ for plates by expressing Hooke's law for the plate as presented in Chapter 6. Thus, we express Eqs. (6.12) as follows:

$$\begin{aligned} M_x &= \frac{Eh^3}{12(1-\nu^2)}\left(-\frac{\partial^2 w}{\partial x^2} - \nu\frac{\partial^2 w}{\partial y^2}\right) \\ M_y &= \frac{Eh^3}{12(1-\nu^2)}\left(-\frac{\partial^2 w}{\partial y^2} - \nu\frac{\partial^2 w}{\partial x^2}\right) \\ M_{xy} &= (1-\nu)\frac{Eh^3}{12(1-\nu^2)}\left(-\frac{\partial^2 w}{\partial x\,\partial y}\right) \end{aligned} \tag{13.5}$$

Hence,

$$\{\sigma\} = \begin{Bmatrix} M_x \\ M_y \\ M_{xy} \end{Bmatrix} = [D]\{\epsilon\} = [D]\begin{Bmatrix} -\dfrac{\partial^2 w}{\partial x^2} \\ -\dfrac{\partial^2 w}{\partial y^2} \\ -2\dfrac{\partial^2 w}{\partial x\,\partial y} \end{Bmatrix} \tag{13.6}$$

Comparing Eqs. (13.5) and (13.6), we can see that

$$[D] = \frac{Eh^3}{12(1-\nu^2)}\begin{bmatrix} 1 & \nu & 0 \\ \nu & 1 & 0 \\ 0 & 0 & \dfrac{1-\nu}{2} \end{bmatrix} \tag{13.7}$$

Let us now formulate the stiffness matrix $[K]$ for an element. We use the minimum potential energy principle as we did in Section 9.4. At that time, we used it for any body where we considered actual stresses and strains and not the quantities called stresses and strains in this section. We shall work from first principles here and shall arrive at an equation for $\delta^{(1)}\pi = 0$ in terms of our stresses and strains.

We shall initially consider as indicated above that all the external force systems act in a statically equivalent manner on the *nodes* themselves and not on the element, which we consider separate from the nodes. We reiterate that $\{q\}$ is the force system *from the nodes onto the element.* Thus, from Eq. (6.30) we have for π of the element, wherein we used real stresses and strains,

$$\pi = \frac{D}{2} \iint_R \left\{ (\nabla^2 w)^2 + 2(1-\nu) \left[\left(\frac{\partial^2 w}{\partial x\, \partial y} \right)^2 - \left(\frac{\partial^2 w}{\partial x^2} \right) \left(\frac{\partial^2 w}{\partial y^2} \right) \right] \right\} dx\, dy$$

$$- \{a\}^T \{q\} \tag{13.8}$$

where

$$D = \frac{Eh^3}{12(1-\nu^2)} \tag{13.9}$$

In the next section we proceed with the extremization process in a manner that parallels the general procedure of Section 9.4.

13.3 FINITE ELEMENT EQUATIONS FOR PLATES

By taking the first variation of π, as done in Section 6.4, we arrive at the following equation:

$$\delta^{(1)}\pi = 0 = \frac{Eh^3}{24(1-\nu^2)} \iint_R \left\{ 2\,\nabla^2 w \left[\delta \left(\frac{\partial^2 w}{\partial x^2} \right) + \delta \left(\frac{\partial^2 w}{\partial y^2} \right) \right] \right.$$

$$+ 2(1-\nu) \left[2\, \frac{\partial^2 w}{\partial x\, \partial y}\, \delta \left(\frac{\partial^2 w}{\partial x\, \partial y} \right) - \frac{\partial^2 w}{\partial x^2}\, \delta \left(\frac{\partial^2 w}{\partial y^2} \right) \right.$$

$$\left. \left. - \frac{\partial^2 w}{\partial y^2}\, \delta \left(\frac{\partial^2 w}{\partial x^2} \right) \right] \right\} dx\, dy - \{\delta a\}^T \{q\} \tag{13.10}$$

You may now show that the above equation is identical to the following equation. This can readily be done by expanding out both equations and comparing terms.

$$0 = \frac{Eh^3}{12(1-\nu^2)} \iint_R \left(-\frac{\partial^2 w}{\partial x^2} - \nu\, \frac{\partial^2 w}{\partial y^2} \right) \delta \left(-\frac{\partial^2 w}{\partial x^2} \right) dx\, dy$$

$$+ \frac{Eh^3}{12(1-\nu^2)} \iint_R \left(-\frac{\partial^2 w}{\partial y^2} - \nu\, \frac{\partial^2 w}{\partial x^2} \right) \delta \left(-\frac{\partial^2 w}{\partial y^2} \right) dx\, dy$$

$$+ (1-\nu)\, \frac{Eh^3}{12(1-\nu^2)} \iint_R 2 \left(-\frac{\partial^2 w}{\partial x\, \partial y} \right) \delta \left(-\frac{\partial^2 w}{\partial x\, \partial y} \right) dx\, dy$$

$$- \{\delta a\}^T \{q\} \tag{13.11}$$

Considering Eq. (13.5) and using our "stresses" M_x, M_y, and M_{xy}, we have for the above equation:

$$\iint_R \left[M_x \, \delta \left(-\frac{\partial^2 w}{\partial x^2} \right) + M_y \, \delta \left(-\frac{\partial^2 w}{\partial y^2} \right) + M_{xy} \, \delta \left(-2 \frac{\partial^2 w}{\partial x \, \partial y} \right) \right] dx \, dy$$

$$- \{\delta a\}^T \{q\} = 0 \tag{13.12}$$

Also notice that our "strains" are present. Now the physical meaning of the above formulation should be obvious. That is, the moment M_x, times the variation of the curvature in the x direction, gives an increment of the potential energy π, etc. In matrix notation, using our $\{\sigma\}$ and $\{\epsilon\}$ for plates, we have for the above equation:

$$\iint_R \{\delta \epsilon\}^T \{\sigma\} \, dx \, dy - \{\delta a\}^T \{q\} = 0$$

Now going back to Eq. (9.59*d*), we replace $\{\delta \epsilon\}^T$ and introduce $\{\delta a\}^T [B]^T$. Also, using Eq. (13.6) to replace $\{\sigma\}$, we get

$$\iint_R \{\delta a\}^T [B]^T [D] \{\epsilon\} \, dx \, dy - \{\delta a\}^T \{q\} = 0$$

Now we replace $\{\epsilon\}$ once again using Eq. (9.59*d*) and, extracting the nodal displacement terms from inside the integral, we get the following result:

$$\{\delta a\}^T \left[\left(\iint_R [B]^T [D] [B] \, dx \, dy \right) \{a\} - \{q\} \right] = 0 \tag{13.13}$$

This equation can be written as

$$\{\delta a\}^T \left([K] \{a\} - \{q\} \right) = 0$$

where

$$[K]^e = \iint_R [B]^T [D] [B] \, dx \, dy \tag{13.14}$$

Because $\{\delta a\}^T$ is arbitrary, we finally have

$$[K]^e \{a\}^e = \{q\}^e \tag{13.15}$$

We have arrived at the familiar formulation for the stiffness matrix, only now we integrate over an area R of the plate rather than a volume. This parallels exactly our modeling of a beam in terms of the movement of a line (Chapter 3) and of a plate in terms of the movement of a surface (Chapter 6). The formulation of $[B]$ depends on the element shape.

Let us next evaluate $\{q\}$, which you will recall applies *to* the element *from* the nodes. To do this, we consider the nodes for the element as a system of free bodies. We shall consider the force system $\{q'\}$, which is the *reaction* to $\{q\}$ to act on these nodes. (Thus we have $\{q'\} = -\{q\}$.) Also acting on the nodes is a loading system that is statically equivalent to the external loading on the element. These external loads on the element may consist of a pressure distribution $p(x, y)$, point forces P_i, edge force distributions $r(x, y)$, and edge moment distributions M (see Fig. 13.3). The static equivalent of these external forces on the nodes must be in equilibrium with the force system $\{q'\}$, which is $-\{q\}$. We will use the principle of *virtual work* for the nodal system of the element, using the above system of forces. Thus:

$$\{\delta a\}^T\{-q\} + \iint_R p(x,y)\,\delta w(x,y)\,dx\,dy + \sum_i P_i\,\delta w_i(x_i, y_i)$$

$$+ \int_L r(x,y)\,\delta w(x,y)\,dl + \int_L M\left[\frac{\partial}{\partial n}(\delta w)\right] dl = 0 \qquad (13.16)$$

Note that the virtual work stemming from $p(x,y), P_i, r(x,y)$, and M is the virtual work of these loads directly on the element itself. If P_i is at a point for which there is zero deflection, such as a *fixed* node, then the virtual displacement δw_i there must be zero and we get zero contribution for this P_i in Eq. (13.16). Similarly, if $r(x, y)$ is along an edge of an element for which there is zero displacement, then δw along this edge must be zero, giving a zero contribution to virtual work in Eq. (13.16).† Now we note that

$$w = [N]^{(e)}\{a\}^{(e)} = \{a\}^T[N]^T$$

$$\delta w = \{\delta a\}^T[N]^T$$

†An example would be when the edge is along the boundary, which is simply supported or clamped.

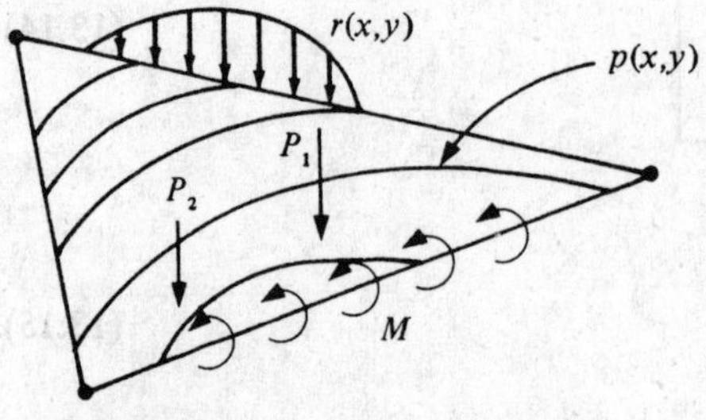

FIGURE 13.3
Element showing various external loadings.

We then have for Eq. (13.16), on using the above relation,

$$\{\delta a\}^T\{-q\} + \{\delta a\}^T \iint_R p(x,y)[N]^T\,dx\,dy + \{\delta a\}^T \sum_i (P_i)[N(x_i,y_i)]^T$$

$$+ \{\delta a\}^T \int_L r(x,y)[N(x,y)]^T\,dl + \{\delta a\}^T \int_L M\,\frac{\partial [N(x,y)]^T}{\partial n}\,dl = 0$$

Hence

$$\{q\} = \iint_R p(x,y)[N]^T\,dx\,dy + \sum_i (P_i)[N(x_i,y_i)]^T + \int_L r(x,y)[N(x,y)]^T\,dl + \int_L M\,\frac{\partial [N(x,y)]^T}{\partial n}\,dl \tag{13.17}$$

We will consider triangular elements in the next section. Rectangular elements will be considered in a later section.

Part B
Triangular Elements

13.4 TRIANGULAR ELEMENTS

The triangular element discussed in the preceding section has nine degrees of freedom. The polynomial representing the deflection will accordingly have nine terms. We then express w as follows:

$$w = \alpha_1 + \alpha_2 x + \alpha_3 y + \alpha_4 x^2 + \alpha_5 xy + \alpha_6 y^2 + \alpha_7 x^3 + \alpha_8 (x^2 y + xy^2) + \alpha_9 y^3 \tag{13.18}$$

Note first that by having the terms $x^2 y$ and xy^2 with a common coefficient α_8, we do not have a complete polynomial (see Fig. 11.4), which would require 10 coefficients. One difficulty arising from using the above polynomial is that the stiffness matrix will be singular if two sides of any element are *parallel to the global xy axes*. Also, while results from Eq. (13.18) are satisfactory for some engineering applications, one does not achieve good accuracy in general. Nevertheless, we shall first work with this element for instructional purposes. In particular, we shall present the details of finite element analysis without giving formulas for the stiffness matrix and the force matrix. And in Example 13.1 we shall solve a plate problem by using the detailed calculations with no shortcuts. Our intention, accordingly, is to work at a very basic level first, before presenting time-saving shortcuts. We then consider the same element by using *natural coordinates* and thereby eliminate the aforementioned difficulties associated with the Cartesian coordinate treatment of the 9-DOF element. Here we do present certain detailed formulas for your use.

The question next arises as to whether the function from Eq. (13.18) affords continuity of displacement and slope along the boundaries between elements. To answer this, consider side 1-2 of a triangular element in Fig. 13.4, oriented for convenience so that side 1-2 is along the y axis. Along this axis $x = 0$. We can express w, θ_x, and θ_y along this axis as follows:

$$\begin{aligned} w_{x=0} &= \alpha_1 + \alpha_3 y + \alpha_6 y^2 + \alpha_9 y^3 \\ (\theta_x)_{x=0} &= \left(\frac{\partial w}{\partial y}\right)_{x=0} = \alpha_3 + 2\alpha_6 y + 3\alpha_9 y^2 \\ (\theta_y)_{x=0} &= -\left(\frac{\partial w}{\partial x}\right)_{x=0} = -\alpha_2 - \alpha_5 y - \alpha_8 y^2 \end{aligned} \tag{13.19}$$

At node 1, $y = 0$ and we have for the nodal displacements

$$\begin{aligned} w_1 &= \alpha_1 \\ (\theta_x)_1 &= \alpha_3 \\ (\theta_y)_1 &= -\alpha_2 \end{aligned} \tag{13.20}$$

At node 2, $y = l$ and we have for the nodal displacements:

$$\begin{aligned} w_2 &= \alpha_1 + \alpha_3 l + \alpha_6 l^2 + \alpha_9 l^3 \\ (\theta_x)_2 &= \alpha_3 + 2\alpha_6 l + 3\alpha_9 l^2 \\ (\theta_y)_2 &= -\alpha_2 - \alpha_5 l - \alpha_8 l^2 \end{aligned} \tag{13.21}$$

We have here only six equations (13.20) and (13.21) to determine the seven constants α_i in terms of the nodal displacements. Notice, however, that the four equations for

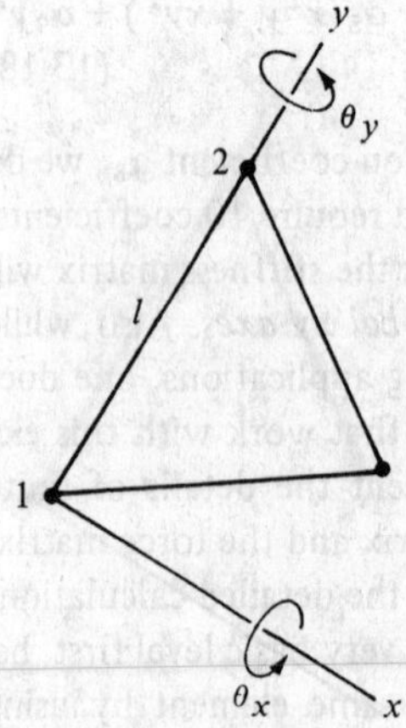

FIGURE 13.4
Element with edge along y axis.

w and θ_x at the two nodes have only four different α's. From these equations we can get $\alpha_1, \alpha_3, \alpha_6$, and α_9 in terms of the nodal displacements. Since the nodal displacements at a node are common for all elements having this node, two contiguous elements having the same border such as 1-2 with the same nodes must have continuity of the displacement w and the slope θ_x along this border. This must be true since w and θ_x along a border 1-2 are completely determined by the values of $\alpha_1, \alpha_3, \alpha_6$, and α_9 [see Eqs. (13.19)]. Further, these α's in turn are completely determined by the nodal displacements w and θ_x at the nodes 1 and 2 [see Eqs. (13.20) and (13.21)]. Hence, w and θ_x are the same along the common border 1-2 of the contiguous elements. Note, however, that the remaining constants are not uniquely determined by Eqs. (13.20) and (13.21), since we have only two equations left to determine three constants. Thus, the slope θ_y is not determined uniquely in terms of the nodal constants at 1 and 2 and hence we will generally not have continuity of the slope *normal* to the boundary line 1-2.† Despite this lack of continuity, the results of this approach are satisfactory for some engineering problems.

Now consider a finite element as shown in Fig. 13.1. We can say at node i:

$$w_i = \alpha_1 + \alpha_2 x_i + \alpha_3 y_i + \alpha_4 x_i^2 + \alpha_5 x_i y_i + \alpha_6 y_i^2 + \alpha_7 x_i^3 + \alpha_8 (x_i^2 y_i + x_i y_i^2) + \alpha_9 y_i^3 \tag{13.22}$$

In matrix form we have

$$(w)_i^e = \lfloor 1, x_i, y_i, x_i^2, (x_i y_i), y_i^2, x_i^3, (x_i^2 y_i + x_i y_i^2), y_i^3 \rfloor \{\alpha\}$$

Similarly, we have for $(\theta_x)_i^e$, noting that it equals $\partial w/\partial y$,

$$(\theta_x)_i^e = \lfloor 0, 0, 1, 0, x_i, 2y_i, 0, (x_i^2 + 2x_i y_i), 3y_i^2 \rfloor \{\alpha\}$$

Finally, we have, noting that $(\theta_y)_i^e = -\partial w/\partial x$,

$$(\theta_y)_i^e = \lfloor 0, -1, 0, -2x_i, -y_i, 0, -3x_i^2, (-2x_i y_i - y_i^2), 0 \rfloor \{\alpha\}$$

In general, we have for all the nodal displacements for node i:

$$\begin{Bmatrix} w_i \\ (\theta_x)_i \\ (\theta_y)_i \end{Bmatrix} = \begin{bmatrix} 1 & x_i & y_i & x_i^2 & x_i y_i & y_i^2 & x_i^3 & (x_i^2 y_i + x_i y_i^2) & y_i^3 \\ 0 & 0 & 1 & 0 & x_i & 2y_i & 0 & (x_i^2 + 2x_i y_i) & 3y_i^2 \\ 0 & -1 & 0 & -2x_i & -y_i & 0 & -3x_i^2 & (-2x_i y_i - y_i^2) & 0 \end{bmatrix} \{\alpha\}$$

Hence, for all nodes we have for element e:

$$\{a\}^e = [A]\{\alpha\}^e \tag{13.23}$$

†For this reason we call this element a *nonconforming* element.

where the matrix $[A]$ is given as follows:

$$[A] = \begin{bmatrix} 1 & x_i & y_i & x_i^2 & x_i y_i & y_i^2 & x_i^3 & (x_i^2 y_i + x_i y_i^2) & y_i^3 \\ 0 & 0 & 1 & 0 & x_i & 2y_i & 0 & (x_i^2 + 2x_i y_i) & 3y_i^2 \\ 0 & -1 & 0 & -2x_i & -y_i & 0 & -3x_i^2 & (-2x_i y_i - y_i^2) & 0 \\ \hline 1 & x_j & y_j & x_j^2 & x_j y_j & y_j^2 & x_j^3 & (x_j^2 y_j + x_j y_j^2) & y_j^3 \\ 0 & 0 & 1 & 0 & x_j & 2y_j & 0 & (x_j^2 + 2x_j y_j) & 3y_j^2 \\ 0 & -1 & 0 & -2x_j & -y_j & 0 & -3x_j^2 & (-2x_j y_j - y_j^2) & 0 \\ \hline 1 & x_k & y_k & x_k^2 & x_k y_k & y_k^2 & x_k^3 & (x_k^2 y_k + x_k y_k^2) & y_k^3 \\ 0 & 0 & 1 & 0 & x_k & 2y_k & 0 & (x_k^2 + 2x_k y_k) & 3y_k^2 \\ 0 & -1 & 0 & -2x_k & -y_k & 0 & -3x_k^2 & (-2x_k y_k - y_k^2) & 0 \end{bmatrix} \tag{13.24}$$

We can solve for the constants α by inverting matrix $[A]$ so that, from Eq. (13.23),

$$\{\alpha\}^e = [A]^{-1}\{a\}^e \tag{13.25}$$

We shall not work out $[A]^{-1}$ here but will leave it for computation for particular cases. The computer will be needed for these computations.

To get the matrix $[B]$, we first relate the strain field to the $\{\alpha\}$ matrix. Thus, using Eqs. (13.3) and (13.18), we have

$$\{\epsilon\}^e = \begin{Bmatrix} -\dfrac{\partial^2 w}{\partial x^2} \\ -\dfrac{\partial^2 w}{\partial y^2} \\ -2\dfrac{\partial^2 w}{\partial x\,\partial y} \end{Bmatrix} = \begin{bmatrix} 0 & 0 & 0 & -2 & 0 & 0 & -6x & -2y & 0 \\ 0 & 0 & 0 & 0 & 0 & -2 & 0 & -2x & -6y \\ 0 & 0 & 0 & 0 & -2 & 0 & 0 & -4(x+y) & 0 \end{bmatrix} \begin{Bmatrix} \alpha_1 \\ \alpha_2 \\ \alpha_3 \\ \alpha_4 \\ \alpha_5 \\ \alpha_6 \\ \alpha_7 \\ \alpha_8 \\ \alpha_9 \end{Bmatrix}$$

Hence

$$\{\epsilon\}^e = [H]\{\alpha\}^e \tag{13.26}$$

where

$$[H] = \begin{bmatrix} 0 & 0 & 0 & -2 & 0 & 0 & -6x & -2y & 0 \\ 0 & 0 & 0 & 0 & 0 & -2 & 0 & -2x & -6y \\ 0 & 0 & 0 & 0 & -2 & 0 & 0 & -4(x+y) & 0 \end{bmatrix} \tag{13.27}$$

Now replace $\{\alpha\}$ in Eq. (13.26), using Eq. (13.25):

$$\{\epsilon\}^e = [H]\,[A]^{-1}\{a\}^e \tag{13.28}$$

We see now [consult Eq. (9.59*d*)] that we have for $[B]$

$$[B] = [H]\,[A]^{-1} \tag{13.29}$$

Knowing $[D]$ and $[B]$ for an element, we can get $[K]^{(e)}$. Thus:

$$[K]^e = \iint_R [B]^T [D]\,[B]\,dx\,dy$$

To get the interpolation function matrix $[N]$ for an element we note that

$$w = \lfloor N \rfloor \{a\} = \lfloor N \rfloor [A]\{\alpha\} \tag{13.30}$$

Going to Eq. (13.18), we also see that

$$w = \lfloor 1, x, y, x^2, xy, y^2, x^3, (x^2y + xy^2), y^3 \rfloor \{\alpha\} \tag{13.31}$$

Recalling from Chapter 10 [see Eq. (10.11)] that the first matrix on the right side of the above equation is denoted as $\lfloor C \rfloor$, we have from Eqs. (13.30) and (13.31):

$$\lfloor C \rfloor = \lfloor N \rfloor [A] \tag{13.32}$$

Hence we have for $[N]$, as in Chapter 10:

$$\lfloor N \rfloor = \lfloor C \rfloor [A]^{-1} \tag{13.33}$$

Knowing $\lfloor N \rfloor$, we can get $\{q\}$ from Eq. (13.17).

In other parts of this text we shall usually set forth, for an element, completed formulations for the stiffness matrix $[K]^e$ and the force vector $\{q\}$. However, for the 9-DOF triangular element of this section, we will not do this so that in the following example we will have the opportunity of working out the details presented in this section. This will afford the reader a chance to actually carry out the steps formulated above.† Accordingly, we now consider a simple example, using a coarse grid.

EXAMPLE 13.1 Consider a square, clamped plate loaded uniformly with $p = 14 \times 10^4$ Pa (Fig. 13.5). The dimensions of the plate are 3 m × 3 m × 3 mm. Take $E = 21 \times 10^{10}$ Pa and $\nu = 0.3$.

We show in Fig. 13.6 a four-element plate. We shall consider element 1 in detail for the element stiffness matrix. We will simply give the results for the other three stiffness matrices and assemble the global stiffness matrix. We will

†There are times when it is more economical to work the problem by inverting the coefficient matrix $[A]$ for each element in the computer than by using evaluated forms for $[K]$, etc.

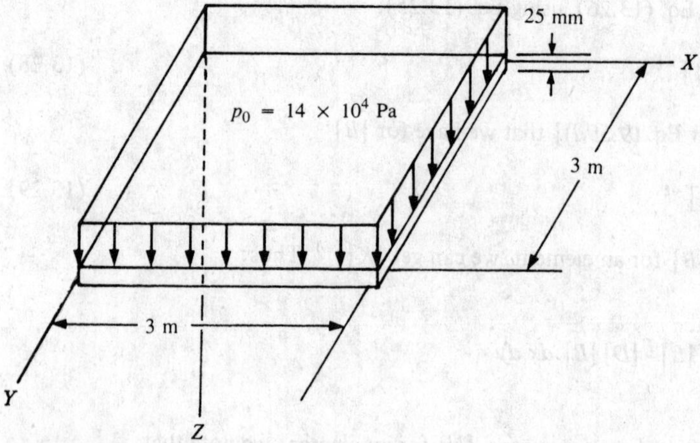

FIGURE 13.5
Clamped plate.

compute the deflection w for the nodal points and compare our result at the center with that of plate theory. We will then give results for smaller grid sizes.

We first go to Eq. (13.24) to evaluate $[A]$. We get from element 1, using local centroidal coordinates xy parallel to XY (see Fig. 13.6),

$$[A]^{(1)} = \begin{bmatrix} 1.0000 & -0.5000 & 1.5000 & 0.2500 & -0.7500 & 2.2500 & -0.1250 & -0.7500 & 3.3750 \\ 0.0000 & 0.0000 & 1.0000 & 0.0000 & -0.5000 & 3.0000 & 0.0000 & -1.2500 & 6.7500 \\ 0.0000 & -1.0000 & 0.0000 & 1.0000 & -1.5000 & 0.0000 & -0.7500 & -0.7500 & 0.0000 \\ 1.0000 & 1.0000 & 0.0000 & 1.0000 & 0.0000 & 0.0000 & 1.0000 & 0.0000 & 0.0000 \\ 0.0000 & 0.0000 & 1.0000 & 0.0000 & 1.0000 & 0.0000 & 0.0000 & 1.0000 & 0.0000 \\ 0.0000 & -1.0000 & 0.0000 & -2.0000 & 0.0000 & 0.0000 & -3.0000 & 0.0000 & 0.0000 \\ 1.0000 & -0.5000 & -1.5000 & 0.2500 & 0.7500 & 2.2500 & -0.1250 & -1.5000 & -3.3750 \\ 0.0000 & 0.0000 & 1.0000 & 0.0000 & -0.5000 & -3.0000 & 0.0000 & 1.7500 & 6.7500 \\ 0.0000 & -1.0000 & 0.0000 & 1.0000 & 1.5000 & 0.0000 & -0.7500 & -3.7500 & 0.0000 \end{bmatrix} \qquad (a)$$

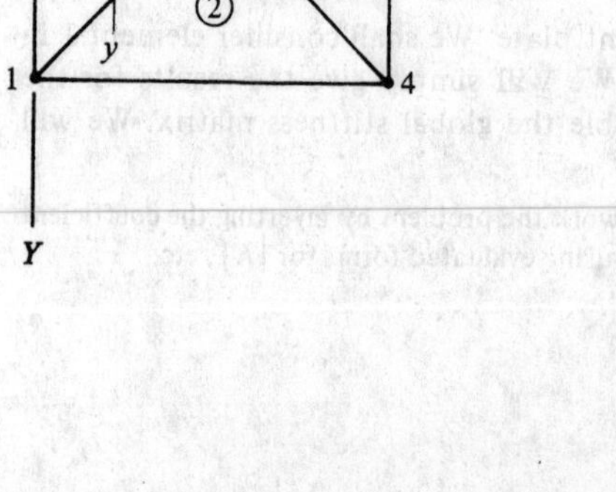

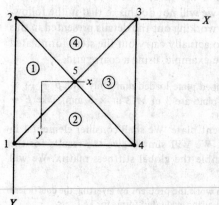

FIGURE 13.6
Four-element plate.

The inverse of $[A]^{(1)}$ may be computed as (using a computer):

$$[A^{(1)}]^{-1} = \begin{bmatrix} 0.4815 & -0.3333 & -0.2222 & 0.2593 & -0.2222 & 0.1111 & 0.2593 & 0.2222 & 0.0000 \\ -0.4444 & 0.3333 & 0.0000 & 0.8889 & 0.0000 & 0.3333 & -0.4444 & -0.3333 & 0.0000 \\ 0.4444 & -0.2222 & -0.1111 & 0.0000 & 0.1111 & 0.0000 & -0.4444 & -0.2222 & 0.1111 \\ -0.5556 & 0.3333 & 0.6667 & 0.4444 & 0.6667 & 0.0000 & 0.1111 & 0.0000 & 0.0000 \\ -0.2222 & 0.1111 & -0.1111 & 0.0000 & 0.4444 & 0.0000 & 0.2222 & 0.1111 & 0.1111 \\ -0.1111 & 0.2222 & 0.1111 & 0.0000 & 0.2222 & 0.0000 & 0.1111 & -0.1111 & -0.1111 \\ 0.5185 & -0.3333 & -0.4444 & -0.5926 & -0.4444 & -0.4444 & 0.0741 & 0.1111 & 0.0000 \\ -0.2222 & 0.1111 & 0.2222 & 0.0000 & 0.4444 & 0.0000 & 0.2222 & 0.1111 & -0.2222 \\ -0.0741 & 0.1111 & 0.0000 & 0.0000 & 0.0000 & 0.0000 & 0.0741 & 0.1111 & 0.0000 \end{bmatrix} \qquad (b)$$

Now we get $[B]$ by using Eqs. (13.29) and (13.27).

$$[B] = \begin{bmatrix} 0 & 0 & 0 & -2 & 0 & 0 & -6x & -2y & 0 \\ 0 & 0 & 0 & 0 & 0 & -2 & 0 & -2x & -6y \\ 0 & 0 & 0 & 0 & -2 & 0 & 0 & -4(x+y) & 0 \end{bmatrix} [A]^{-1}$$

Hence:

$$[B] = \begin{bmatrix} (1.11 - 3.11x + 0.444y) & (-0.667 + 2x - 0.222y) & (-1.33 + 2.67x - 0.444y) \\ (0.222 + 0.444x + 0.444y) & (-0.444 - 0.222x - 0.667y) & (-0.222 - 0.444x) \\ (0.444 + 0.889x + 0.889y) & (-0.222 - 0.444x - 0.444y) & (0.222 - 0.889x - 0.889y) \end{bmatrix.$$
$$\begin{matrix} (-0.889 + 0.356x) & (-1.33 + 2.67x - 0.899y) & 2.67x \\ 0 & (-0.444 - 0.889x) & 0 \\ 0 & (-0.889 - 1.78x - 1.78y) & 0 \end{matrix}$$
$$\left.\begin{matrix} (-0.222 - 0.444x - 0.444y) & (-0.667x - 0.222y) & 0.444y \\ (-0.222 - 0.444x - 0.444y) & (0.222 - 0.222x - 0.667y) & (0.222 + 0.444x) \\ (-0.444 - 0.889x - 0.889y) & (-0.222 - 0.444x - 0.444y) & (-0.222 + 0.889x + 0.889y) \end{matrix}\right] \qquad (c)$$

Next we go to Eq. (13.7) for $[D]$.

$$[D] = \frac{(21 \times 10^{10})(0.025)^3}{12(1 - 0.3^2)} \begin{bmatrix} 1 & 0.3 & 0 \\ 0.3 & 1 & 0 \\ 0 & 0 & \frac{1-0.3}{2} \end{bmatrix}$$

$$= (3.005 \times 10^5) \begin{bmatrix} 1 & 0.3 & 0 \\ 0.3 & 1 & 0 \\ 0 & 0 & 0.35 \end{bmatrix} \qquad (d)$$

We can finally start to get the stiffness matrix for element 1. Thus we go to Eq. (13.14):

$$[K]^e = \iint_A [B]^T[D][B] \, dx \, dy \qquad (e)$$

We now present the integration procedure.

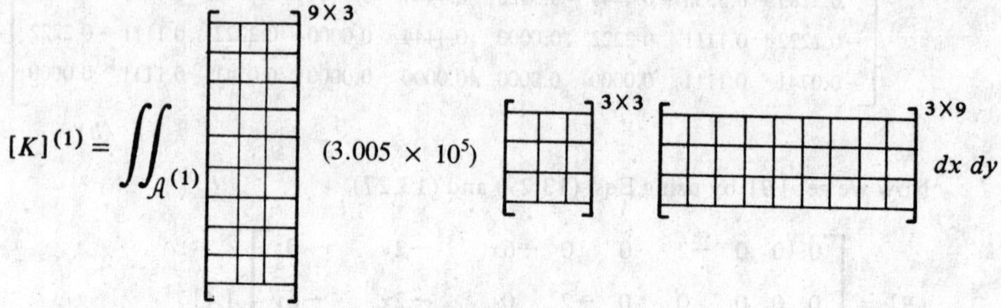

$$[K]^{(1)} = \iint_{A^{(1)}} \quad (3.005 \times 10^5) \qquad \qquad dx \, dy \qquad (f)$$

where matrix superscripts indicate the number of rows and columns. Carrying out multiplication and integration, using Eqs. (11.16) and (11.17) and Table 11.1, we have

$[K]^{(1)} =$

10^7

	w_1	$(\theta_x)_1$	$(\theta_y)_1$	w_5	$(\theta_x)_5$	$(\theta_y)_5$	w_2	$(\theta_x)_2$	$(\theta_y)_2$	
	0.20	−0.139	−0.193	−0.16	−0.227	−0.0671	−0.0401	0.00601	0.02	w_1
	−0.139	0.106	0.132	0.106	0.156	0.0436	0.0331	−0.003	−0.0165	$(\theta_x)_1$
	−0.193	0.132	0.207	0.16	0.223	0.0571	0.0331	−0.016	−0.0235	$(\theta_y)_1$
	−0.16	0.106	0.160	0.16	0.16	0.0801	0	−0.026	0	w_5
	−0.227	0.156	0.223	0.16	0.288	0.0341	0.0671	−0.0025	−0.0371	$(\theta_x)_5$
	−0.0671	0.0436	0.0571	0.0801	0.0541	0.0601	−0.013	−0.0165	0.003	$(\theta_y)_5$
	−0.0401	0.0331	0.0331	0	0.0671	−0.013	0.0401	0.02	−0.02	w_2
	−0.00601	−0.0030	−0.016	−0.026	−0.0025	−0.0165	0.02	0.0265	−0.00651	$(\theta_x)_2$
	0.02	−0.0165	−0.0235	0	−0.0371	0.003	−0.02	−0.00651	0.0205	$(\theta_y)_2$

$$(g)$$

We now get $[K]^{(2)}$, $[K]^{(3)}$, and $[K]^{(4)}$ in the same manner and assemble the global matrix by inserting the values of the element stiffness matrices into the appropriate places of the grid. The final result for the global stiffness matrix is:

$[K] = 10^7$

	w_1	$(\theta x)_1$	$(\theta y)_1$	w_2	$(\theta x)_2$	$(\theta y)_2$	w_3	$(\theta x)_3$	$(\theta y)_3$	w_4	$(\theta x)_4$	$(\theta y)_4$	w_5	$(\theta x)_5$	$(\theta y)_5$
w_1	0.401	−0.333	−0.333	−0.0401	0.00601	0.0200	0	0	0	−0.0401	0.0200	0.0601	−0.321	−0.294	−0.294
$(\theta x)_1$		0.313	0.264	0.0331	−0.003	−0.0165	0	0	0	0.0331	−0.0235	−0.016	0.266	0.213	0.267
$(\theta y)_1$			0.313	0.0331	−0.016	−0.0235	0	0	0	0.0331	−0.0165	−0.003	0.266	0.267	0.213
w_2				0.0801	0.0401	−0.0401	−0.0401	−0.0331	−0.0331	0	0	0	0	0.0801	−0.0801
$(\theta x)_2$					0.0471	−0.013	−0.0200	−0.0235	−0.0165	0	0	0	−0.0260	0.0005	−0.0536
$(\theta y)_2$						0.0471	−0.00601	−0.0160	−0.003	0	0	0	0.0260	−0.0536	0.0005
w_3							0.401	0.333	0.333	−0.0401	−0.0060	−0.0200	−0.321	0.294	0.294
$(\theta x)_3$								0.313	0.264	−0.0331	−0.0030	−0.0165	−0.266	0.213	0.267
$(\theta y)_3$									0.313	−0.0331	−0.0160	−0.0235	−0.266	0.267	0.213
w_4			symmetric							0.0801	−0.0401	0.0401	0	−0.0801	0.0801
$(\theta x)_4$											0.0471	−0.0130	0.0260	0.0005	−0.0536
$(\theta y)_4$												0.0471	−0.0260	−0.0536	0.0005
w_5													0.641	0	0
$(\theta x)_5$														0.697	0.216
$(\theta y)_5$															0.697

(h)

Going to Eq. (13.17) to get $\{q\}$, we have

$$\{q\}^e = \iint_A [N]^T (14 \times 10^4)\, dx\, dy \tag{i}$$

To get $[N]^T$, we go to Eq. (13.33), where, knowing $[A]^{-1}$ and $\lfloor C \rfloor$ [see Eq. (13.31) for $\lfloor C \rfloor$], we have for $[N]$:

$$[N] = \lfloor 1, x, y, x^2, xy, y^2, x^3, (x^2y + xy^2), y^3 \rfloor [A]^{-1}$$

For element 1 we have for $\lfloor N^{(1)} \rfloor^T$:

$$\lfloor N^{(1)} \rfloor^T = \left\{ \begin{array}{l} (0.481 - 0.444x + 0.444y - 0.556x^2 - 0.222xy - 0.111y^2 + 0.519x^3 \\ \quad - 0.222(x^2y + xy^2) - 0.0714y^3) \\ (-0.333 + 0.333x - 0.222y + 0.333x^2 + 0.111xy + 0.222y^2 - 0.333x^3 \\ \quad + 0.111(x^2y + xy^2) + 0.111y^3) \\ (-0.222 - 0.111y + 0.667x^2 - 0.111xy + 0.111y^2 - 0.444x^3 \\ \quad + 0.222(x^2y + xy^2)) \\ (0.259 + 0.889x + 0.444x^2 - 0.593x^3) \\ (-0.222 + 0.111y + 0.667x^2 + 0.444xy + 0.222y^2 - 0.444x^3 \\ \quad + 0.444(x^2y + xy^2)) \\ (0.111 + 0.333x - 0.444x^3) \\ (0.259 - 0.444x - 0.444y + 0.111x^2 + 0.222xy + 0.111y^2 + 0.741x^3 \\ \quad + 0.222(x^2y + xy^2) + 0.741y^3) \\ (0.222 - 0.333x - 0.222y + 0.111xy - 0.111y^2 + 0.111x^3 \\ \quad + 0.111(x^2y + xy^2) + 0.111y^3) \\ (0.111y + 0.111xy - 0.111y^2 - 0.222(x^2y + xy^2)) \end{array} \right\} \tag{j}$$

so that for $\{q\}^{(1)}$ we have, on using Eqs. (*i*), (11.16), and (11.17), and Table 11.1,

$$[q]^{(1)} = 10^5\, [\underbrace{1.26, -0.709, -0.394}_{\text{Node 1}}, \underbrace{0.945, -0.315, 0.315}_{\text{Node 5}}, \underbrace{0.945, 0.551, -0.0788}_{\text{Node 2}}] \tag{k}$$

We get the $\{q\}$ vectors for all four elements and we can then get a vector $\{q\}^T$ having 15 terms. We thus obtain:

$$\{q\}^T = 10^5\, [2.52, -1.10, -1.10, 1.89, 0.63, -0.63, 2.52, 1.10, 1.10, 1.89, -0.63, 0.63, 3.78, 0, 0]$$

The *reduced* matrix equation $[K]\{a\} = \{q\}$ becomes

$$10^6 \begin{bmatrix} 6.41 & 0 & 0 \\ 0 & 6.97 & 2.16 \\ 0 & 2.16 & 6.97 \end{bmatrix} \begin{Bmatrix} w_5 \\ (\theta_x)_5 \\ (\theta_y)_5 \end{Bmatrix} = 10^5 \begin{Bmatrix} 3.78 \\ 0 \\ 0 \end{Bmatrix}$$

Solving for the unknowns yields

$$w_5 = 0.059 \text{ m}$$

$$(\theta_x)_5 = 0.18 \times 10^{-8} \approx 0$$

$$(\theta_y)_5 = 0.121 \times 10^{-8} \approx 0$$

Now compare w_5 with that of the theory of plates, where from Eq. (6.92) we have:

$$w_{\max} = 0.0202 \frac{(p)(1.5^4)}{Eh^3/12(1-\nu^2)}$$

$$= 0.0202 \frac{(14 \times 10^4)(1.5^4)}{(21 \times 10^{10})/12(1 - 0.3^2)} = 0.04765 \text{ m}$$

We get poor results here with only four elements. With 64 triangular elements we get $w_{\max} = 0.0566$ m, showing an improvement.

In the next section we shall present methods of greatly improving results by using natural coordinates for flat plate bending with triangular coordinates.

13.5 IMPROVED METHODS FOR NATURAL COORDINATES

As pointed out earlier, we can avoid the difficulties of the 9-DOF element set forth in the previous section. One procedure is to use natural coordinates for the element. The following parameter-laden polynomial of triangular coordinates has been shown to lead to good results:

$$\begin{aligned} w = {} & \alpha_1 L_1 + \alpha_2 L_2 + \alpha_3 L_3 + \alpha_4 (L_2^2 L_1 + \tfrac{1}{2} L_1 L_2 L_3) + \alpha_5 (L_3^2 L_2 + \tfrac{1}{2} L_1 L_2 L_3) \\ & + \alpha_6 (L_1^2 L_3 + \tfrac{1}{2} L_1 L_2 L_3) + \alpha_7 (L_2 L_1^2 + \tfrac{1}{2} L_1 L_2 L_3) + \alpha_8 (L_3 L_2^2 + \tfrac{1}{2} L_1 L_2 L_3) \\ & + \alpha_9 (L_1 L_3^2 + \tfrac{1}{2} L_1 L_2 L_3) \end{aligned} \tag{13.34}$$

In Chapter 12 we showed that the triangular coordinates are related to x and y by the following equation:

$$\begin{Bmatrix} L_1 \\ L_2 \\ L_3 \end{Bmatrix} = \frac{1}{2A} \begin{bmatrix} d_1 & \bar{b}_1 & \bar{a}_1 \\ d_2 & \bar{b}_2 & \bar{a}_2 \\ d_3 & \bar{b}_3 & \bar{a}_3 \end{bmatrix} \begin{Bmatrix} 1 \\ x \\ y \end{Bmatrix} \tag{13.35}$$

where $\mathcal{A}$ is the area of the triangle and where

$$\begin{aligned} d_1 &= x_2 y_3 - x_3 y_2 \\ \bar{b}_1 &= y_2 - y_3 \\ \bar{a}_1 &= x_3 - x_2 \end{aligned} \tag{13.36}$$

Other terms are formed by permuting the indices. Hence, we can form the following partial derivatives in terms of natural coordinates:

$$\begin{aligned} \frac{\partial}{\partial x} &= \frac{\partial}{\partial L_1}\left(\frac{\partial L_1}{\partial x}\right) + \frac{\partial}{\partial L_2}\left(\frac{\partial L_2}{\partial x}\right) + \frac{\partial}{\partial L_3}\left(\frac{\partial L_3}{\partial x}\right) \\ &= \frac{1}{2\mathcal{A}}\left[\bar{b}_1 \frac{\partial}{\partial L_1} + \bar{b}_2 \frac{\partial}{\partial L_2} + \bar{b}_3 \frac{\partial}{\partial L_3}\right] \\ \frac{\partial}{\partial y} &= \frac{\partial}{\partial L_1}\left(\frac{\partial L_1}{\partial y}\right) + \frac{\partial}{\partial L_2}\left(\frac{\partial L_2}{\partial y}\right) + \frac{\partial}{\partial L_3}\left(\frac{\partial L_3}{\partial y}\right) \\ &= \frac{1}{2\mathcal{A}}\left[\bar{a}_1 \frac{\partial}{\partial L_1} + \bar{a}_2 \frac{\partial}{\partial L_2} + \bar{a}_3 \frac{\partial}{\partial L_3}\right] \end{aligned} \tag{13.37}$$

In order to determine the interpolation functions, we go back to Eq. (13.34) and evaluate the nine nodal values from w and from the first derivatives, namely, $\partial w/\partial y$ and $-\partial w/\partial x$. Thus, noting that $L_1 = 1$ at node 1, while $L_2 = L_3 = 0$ there, etc., we have

$$\begin{aligned} w_1 &= \alpha_1 \\ \left(\frac{\partial w}{\partial y}\right)_1 &= \frac{1}{2\mathcal{A}}(\bar{a}_1\alpha_1 + \bar{a}_2\alpha_2 + \bar{a}_3\alpha_3 + \bar{a}_3\alpha_6 + \bar{a}_2\alpha_7) \\ -\left(\frac{\partial w}{\partial x}\right)_1 &= -\frac{1}{2\mathcal{A}}(\bar{b}_1\alpha_1 + \bar{b}_2\alpha_2 + \bar{b}_3\alpha_3 + \bar{b}_3\alpha_6 + \bar{b}_2\alpha_7) \\ \hline w_2 &= \alpha_2 \\ \left(\frac{\partial w}{\partial y}\right)_2 &= \frac{1}{2\mathcal{A}}(\bar{a}_1\alpha_1 + \bar{a}_2\alpha_2 + \bar{a}_3\alpha_3 + \bar{a}_1\alpha_4 + \bar{a}_3\alpha_8) \\ -\left(\frac{\partial w}{\partial x}\right)_2 &= -\frac{1}{2\mathcal{A}}(\bar{b}_1\alpha_1 + \bar{b}_2\alpha_2 + \bar{b}_3\alpha_3 + \bar{b}_1\alpha_4 + \bar{b}_3\alpha_8) \\ \hline w_3 &= \alpha_3 \\ \left(\frac{\partial w}{\partial y}\right)_3 &= \frac{1}{2\mathcal{A}}(\bar{a}_1\alpha_1 + \bar{a}_2\alpha_2 + \bar{a}_3\alpha_3 + \bar{a}_2\alpha_5 + \bar{a}_1\alpha_9) \\ -\left(\frac{\partial w}{\partial x}\right)_3 &= -\frac{1}{2\mathcal{A}}(\bar{b}_1\alpha_1 + \bar{b}_2\alpha_2 + \bar{b}_3\alpha_3 + \bar{b}_2\alpha_5 + \bar{b}_1\alpha_9) \end{aligned} \tag{13.38}$$

We can express the above equation as

$$\{a\} = [A]\{\alpha\} \tag{13.39}$$

as we have in the past, where the coefficient matrix $[A]$ is

$$[A] = \frac{1}{2\mathcal{A}}\begin{bmatrix} 2\mathcal{A} & 0 & 0 & 0 & 0 & 0 & 0 & 0 & 0 \\ \bar{a}_1 & \bar{a}_2 & \bar{a}_3 & 0 & 0 & \bar{a}_3 & \bar{a}_2 & 0 & 0 \\ -\bar{b}_1 & -\bar{b}_2 & -\bar{b}_3 & 0 & 0 & -\bar{b}_3 & -\bar{b}_2 & 0 & 0 \\ 0 & 2\mathcal{A} & 0 & 0 & 0 & 0 & 0 & 0 & 0 \\ \bar{a}_1 & \bar{a}_2 & \bar{a}_3 & \bar{a}_1 & 0 & 0 & 0 & \bar{a}_3 & 0 \\ -\bar{b}_1 & -\bar{b}_2 & -\bar{b}_3 & -\bar{b}_1 & 0 & 0 & 0 & -\bar{b}_3 & 0 \\ 0 & 0 & 2\mathcal{A} & 0 & 0 & 0 & 0 & 0 & 0 \\ \bar{a}_1 & \bar{a}_2 & \bar{a}_3 & 0 & \bar{a}_2 & 0 & 0 & 0 & \bar{a}_1 \\ -\bar{b}_1 & -\bar{b}_2 & -\bar{b}_3 & 0 & -\bar{b}_2 & 0 & 0 & 0 & -\bar{b}_1 \end{bmatrix} \tag{13.40}$$

Inverting Eq. (13.39), we have

$$\{\alpha\} = [A]^{-1}\{a\} \tag{13.41}$$

To get matrix $[B]$, we first relate the strain field $\{\epsilon\}$ to the matrix $\{\alpha\}$. Thus, noting Eqs. (13.3) and (13.26),

$$\{\epsilon\} = \begin{Bmatrix} -\dfrac{\partial^2 w}{\partial x^2} \\ -\dfrac{\partial^2 w}{\partial y^2} \\ -2\dfrac{\partial^2 w}{\partial x\,\partial y} \end{Bmatrix} = [H]\{\alpha\} \tag{13.42}$$

Using Eq. (13.34) and the differentiation formulas of Eq. (13.37) successively, we get the following result for the matrix $[H]$:

$$
[H] = -\frac{L_1}{(2A)^2}\begin{bmatrix} 0 & 0 & 0 & 2\bar{b}_2^2 + \bar{b}_2\bar{b}_3 & \bar{b}_2\bar{b}_3 & 4\bar{b}_1\bar{b}_3 + \bar{b}_2\bar{b}_3 & 4\bar{b}_1\bar{b}_2 + \bar{b}_2\bar{b}_3 & \bar{b}_2\bar{b}_3 & 2\bar{b}_3^2 + \bar{b}_2\bar{b}_3 \\ 0 & 0 & 0 & 2\bar{a}_2^2 + \bar{a}_2\bar{a}_3 & \bar{a}_2\bar{a}_3 & 4\bar{a}_1\bar{a}_3 + \bar{a}_2\bar{a}_3 & 4\bar{a}_1\bar{a}_2 + \bar{a}_2\bar{a}_3 & \bar{a}_2\bar{a}_3 & 2a_3^2 + \bar{a}_2\bar{a}_3 \\ 0 & 0 & 0 & \bar{b}_2\bar{a}_3 + \bar{b}_3\bar{a}_2 + 4\bar{a}_2\bar{b}_2 & \bar{a}_2\bar{b}_3 + \bar{b}_2\bar{a}_3 & \bar{a}_2\bar{b}_3 + \bar{a}_3\bar{b}_2 + 4(\bar{a}_1\bar{b}_3 + \bar{a}_3\bar{b}_1) & \bar{a}_2\bar{b}_3 + \bar{a}_3\bar{b}_2 + 4(\bar{a}_1\bar{b}_2 + \bar{a}_2\bar{b}_1) & \bar{a}_2\bar{b}_3 + \bar{a}_3\bar{b}_2 & \bar{a}_2\bar{b}_3 + \bar{a}_3\bar{b}_2 + 4\bar{a}_3\bar{b}_3 \end{bmatrix} -
$$

$$
\frac{L_2}{(2A)^2}\begin{bmatrix} 0 & 0 & 0 & 4\bar{b}_1\bar{b}_2 + \bar{b}_1\bar{b}_3 & 2\bar{b}_3^2 + \bar{b}_1\bar{b}_3 & \bar{b}_1\bar{b}_3 & 2\bar{b}_1^2 + \bar{b}_1\bar{b}_3 & \bar{b}_1\bar{b}_3 + 4\bar{b}_2\bar{b}_3 & \bar{b}_1\bar{b}_3 \\ 0 & 0 & 0 & 4\bar{a}_1\bar{a}_2 + \bar{a}_1\bar{a}_3 & 2\bar{a}_3^2 + \bar{a}_1\bar{a}_3 & \bar{a}_1\bar{a}_3 & 2\bar{a}_1^2 + \bar{a}_1\bar{a}_3 & \bar{a}_1\bar{a}_3 + 4\bar{a}_2\bar{a}_3 & \bar{a}_1\bar{a}_3 \\ 0 & 0 & 0 & 4(\bar{a}_1\bar{b}_2 + \bar{a}_2\bar{b}_1) + \bar{b}_1\bar{a}_3 + \bar{b}_3\bar{a}_1 & 4\bar{b}_3\bar{a}_3 + \bar{a}_1\bar{b}_3 + \bar{a}_3\bar{b}_1 & \bar{a}_1\bar{b}_3 + \bar{a}_3\bar{b}_1 & 4\bar{a}_1\bar{b}_1 + \bar{a}_1\bar{b}_3 + \bar{a}_3\bar{b}_1 & \bar{a}_1\bar{b}_3 + \bar{a}_3\bar{b}_1 + 4(\bar{a}_2\bar{b}_3 + \bar{a}_3\bar{b}_2) & \bar{a}_1\bar{b}_3 + \bar{b}_1\bar{a}_3 \end{bmatrix} -
$$

$$
\frac{L_3}{(2A)^2}\begin{bmatrix} 0 & 0 & 0 & \bar{b}_1\bar{b}_2 & \bar{b}_1\bar{b}_2 + 4\bar{b}_2\bar{b}_3 & 2\bar{b}_1^2 + \bar{b}_1\bar{b}_2 & \bar{b}_1\bar{b}_2 & 2\bar{b}_2^2 + \bar{b}_1\bar{b}_2 & \bar{b}_1\bar{b}_2 + 4\bar{b}_1\bar{b}_3 \\ 0 & 0 & 0 & \bar{a}_1\bar{a}_2 & \bar{a}_1\bar{a}_2 + 4\bar{a}_2\bar{a}_3 & 2\bar{a}_1^2 + \bar{a}_1\bar{a}_2 & \bar{a}_1\bar{a}_2 & 2\bar{a}_2^2 + \bar{a}_1\bar{a}_2 & \bar{a}_1\bar{a}_2 + 4\bar{a}_1\bar{a}_3 \\ 0 & 0 & 0 & \bar{a}_1\bar{b}_2 + \bar{a}_2\bar{b}_1 & \bar{a}_1\bar{b}_2 + \bar{a}_2\bar{b}_1 + 4(\bar{a}_2\bar{b}_3 + \bar{b}_2\bar{a}_3) & \bar{a}_1\bar{b}_2 + \bar{a}_2\bar{b}_1 + 4\bar{a}_1\bar{b}_1 & \bar{a}_1\bar{b}_2 + \bar{a}_2\bar{b}_1 & 4\bar{a}_2\bar{b}_2 + \bar{a}_1\bar{b}_2 + \bar{a}_2\bar{b}_1 & \bar{a}_1\bar{b}_2 + \bar{a}_2\bar{b}_1 + 4(\bar{a}_1\bar{b}_3 + \bar{b}_1\bar{a}_3) \end{bmatrix}
\tag{13.43}
$$

We now replace $\{\alpha\}$ in Eq. (13.42) to get

$$\{\epsilon\} = [H][A]^{-1}\{a\}$$

We see that we have for $[B]$

$$[B] = [H][A]^{-1}$$

Knowing $[D]$ and $[B]$ we can get the element stiffness matrix:

$$[K]^e = \iint_{\mathcal{A}} [B]^T [D][B]\, dA \tag{13.44}$$

To aid in the integration we have the following integration formula, first presented in Chapter 12 (Eq. (12.48):

$$\iint_{\mathcal{A}} L_i^{\alpha} L_j^{\beta} L_k^{\gamma}\, dx\, dy = 2\mathcal{A}\, \frac{\alpha!\,\beta!\,\gamma!}{(\alpha+\beta+\gamma+2)!} \tag{13.45}$$

where α, β, and γ are integer constants.

We next express Eq. (13.34) in matrix form as

$$w = \lfloor C \rfloor \{\alpha\}$$

where

$$\begin{aligned} \lfloor C \rfloor = \lfloor & L_1, L_2, L_3, (L_2^2 L_1 + \tfrac{1}{2} L_1 L_2 L_3), (L_3^2 L_2 + \tfrac{1}{2} L_1 L_2 L_3), \\ & (L_1^2 L_3 + \tfrac{1}{2} L_1 L_2 L_3), (L_2 L_1^2 + \tfrac{1}{2} L_1 L_2 L_3), (L_3 L_2^2 + \tfrac{1}{2} L_1 L_2 L_3), \\ & (L_1 L_3^2 + \tfrac{1}{2} L_1 L_2 L_3) \rfloor \end{aligned} \tag{13.46}$$

Now expressing $\{\alpha\}$ above by using Eq. (13.41), we have the familiar result

$$w = \lfloor C \rfloor [A]^{-1}\{a\} = \lfloor N \rfloor \{a\} \tag{13.47}$$

from which we can now determine the interpolation matrix as $\lfloor C \rfloor [A]^{-1}$. Thus, using *i, j, k* notation for subscripts we get after lengthy calculations,

$$\{N\} = \left\{ \begin{array}{l} L_i + L_i^2 L_j + L_i^2 L_k - L_i L_j^2 - L_i L_k^2 \\ -\bar{b}_k (L_i^2 L_j + \tfrac{1}{2} L_i L_j L_k) + \bar{b}_j (L_k L_i^2 + \tfrac{1}{2} L_i L_j L_k) \\ -\bar{a}_k (L_i^2 L_j + \tfrac{1}{2} L_i L_j L_k) + \bar{a}_j (L_k L_i^2 + \tfrac{1}{2} L_i L_j L_k) \\ \hdashline L_j + L_j^2 L_k + L_j^2 L_i - L_j L_k^2 - L_j L_i^2 \\ -\bar{b}_i (L_j^2 L_k + \tfrac{1}{2} L_i L_j L_k) + \bar{b}_k (L_i L_j^2 + \tfrac{1}{2} L_i L_j L_k) \\ -\bar{a}_i (L_j^2 L_k + \tfrac{1}{2} L_i L_j L_k) + \bar{a}_k (L_i L_j^2 + \tfrac{1}{2} L_i L_j L_k) \\ \hdashline L_k + L_k^2 L_i + L_k^2 L_j - L_k L_i^2 - L_k L_j^2 \\ -\bar{b}_j (L_k^2 L_i + \tfrac{1}{2} L_i L_j L_k) + \bar{b}_i (L_j L_k^2 + \tfrac{1}{2} L_i L_j L_k) \\ -\bar{a}_j (L_k^2 L_i + \tfrac{1}{2} L_i L_j L_k) + \bar{a}_i (L_j L_k^2 + \tfrac{1}{2} L_i L_j L_k) \end{array} \right\} \tag{13.48}$$

Using $[N]$ in Eq. (13.17) while using Eqs. (13.35) to get x and y variables in terms of natural coordinates, we can get the force vector $\{q\}$ with the aid of Eq. (13.45). We may then reach the basic element equation

$$[K]^e\{a\} = \{q\}^e$$

We get the following results for the square clamped plate loaded uniformly that we have been considering in Example 13.1 and subsequently:

For 4 elements	$w_{max} = 0.131$ m
For 8 elements	$w_{max} = 0.0717$ m
For 16 elements	$w_{max} = 0.0546$ m
For 32 elements	$w_{max} = 0.0530$ m
For 64 elements	$w_{max} = 0.04912$ m
Exact result	$w_{max} = 0.04765$ m

We see a nice convergence toward the exact result as the number of elements is increased. Also, we avoid the pitfalls described earlier for triangular elements.

Part C
Rectangular Elements

13.6 RECTANGULAR ELEMENTS

We now turn to the rectangular element shown in Fig. 13.7, where we have four nodal points, one at each corner. There will again be three nodal displacements at each node, just as in the case of the triangular element. There will accordingly be 12 degrees of freedom per element.

Our procedure is to present the basic formulations first and then, in Example 13.2, present numerical details of the evaluation of the key matrices as we did for

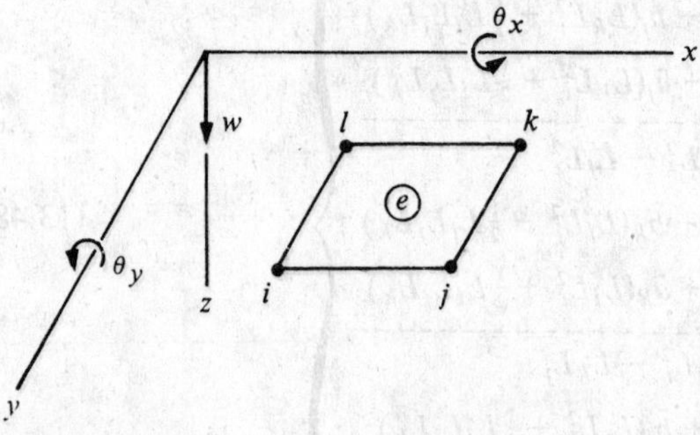

FIGURE 13.7
Rectangular element.

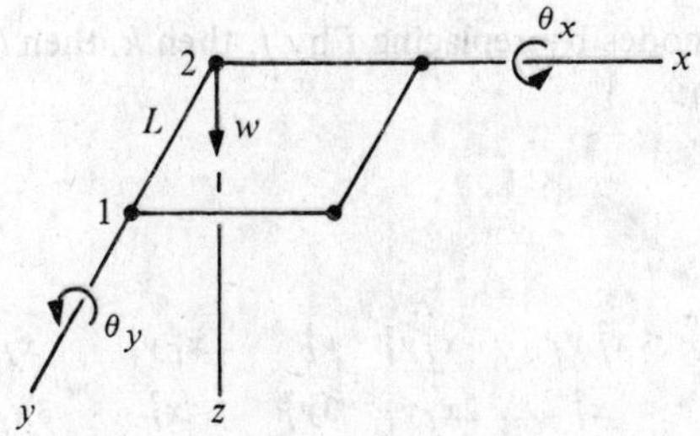

FIGURE 13.8
Element with side 1-2 along y axis.

triangular elements in Example 13.1. This is done to ensure that the reader fully understands the basics. Once this basic exercise is finished, we present a worked-out formula for the stiffness matrix for rectangular elements.

We start by presenting a polynomial with 12 constants α_i. This is an incomplete fourth-degree polynomial (see Fig. 11.4) but has been reduced from 15 terms to 12 terms symmetrically. Thus:

$$w = \alpha_1 + \alpha_2 x + \alpha_3 y + \alpha_4 x^2 + \alpha_5 xy + \alpha_6 y^2 + \alpha_7 x^3 + \alpha_8 x^2 y + \alpha_9 xy^2 + \alpha_{10} y^3 + \alpha_{11} x^3 y + \alpha_{12} xy^3 \tag{13.49}$$

By taking one side of a rectangular element and placing the y axis along this element (see Fig. 13.8) we can compute w, θ_x, and θ_y at the endpoints, just as we did for the triangular element at the outset of the preceding section. The same arguments that were then used to compute the α's in terms of the nodal displacement can again be applied to the rectangular element. We will ask you as an exercise to then show that the polynomial expressed in Eq. (13.49) will permit compatibility of deflection and tangential slope *along* the contiguous boundary between two elements while allowing for a discontinuity in the normal slope along the boundary. Nevertheless, even with this partial satisfaction of compatibility between elements, we can still get good results. Noting again that $\theta_x = \partial w/\partial y$ and $\theta_y = -\partial w/\partial x$, we express the nodal coordinates at joint i as follows:

$$\begin{Bmatrix} w_i \\ (\theta_x)_i \\ (\theta_y)_i \end{Bmatrix} = \begin{bmatrix} 1 & x_i & y_i & x_i^2 & x_i y_i & y_i^2 & x_i^3 & x_i^2 y_i & x_i y_i^2 & y_i^3 & x_i^3 y_i & x_i y_i^3 \\ 0 & 0 & 1 & 0 & x_i & 2y_i & 0 & x_i^2 & 2x_i y_i & 3y_i^2 & x_i^3 & 3x_i y_i^2 \\ 0 & -1 & 0 & -2x_i & -y_i & 0 & -3x_i^2 & -2x_i y_i & -y_i^2 & 0 & -3x_i^2 y_i & -y_i^3 \end{bmatrix} \begin{Bmatrix} \alpha_1 \\ \alpha_2 \\ \alpha_3 \\ \alpha_4 \\ \alpha_5 \\ \alpha_6 \\ \alpha_7 \\ \alpha_8 \\ \alpha_9 \\ \alpha_{10} \\ \alpha_{11} \\ \alpha_{12} \end{Bmatrix} \tag{13.50}$$

We can get similar expressions for the other nodes by replacing i by j, then k, then l. We would then have the following formulation:

$$\{a\} = [A]^e\{\alpha\}$$

where

$$[A]^e = \begin{bmatrix} 1 & x_i & y_i & x_i^2 & x_i y_i & y_i^2 & x_i^3 & x_i^2 y_i & x_i y_i^2 & y_i^3 & x_i^3 y_i & x_i y_i^3 \\ 0 & 0 & 1 & 0 & x_i & 2y_i & 0 & x_i^2 & 2x_i y_i & 3y_i^2 & x_i^3 & 3x_i y_i^2 \\ 0 & -1 & 0 & -2x_i & -y_i & 0 & -3x_i^2 & -2x_i y_i & -y_i^2 & 0 & -3x_i^2 y_i & -y_i^3 \\ \hline 1 & x_j & y_j & x_j^2 & x_j y_j & y_j^2 & x_j^3 & x_j^2 y_j & x_j y_j^2 & y_j^3 & x_j^3 y_j & x_j y_j^3 \\ 0 & 0 & 1 & 0 & x_j & 2y_j & 0 & x_j^2 & 2x_j y_j & 3y_j^2 & x_j^3 & 3x_j y_j^2 \\ 0 & -1 & 0 & -2x_j & -y_j & 0 & -3x_j^2 & -2x_j y_j & -y_j^2 & 0 & -3x_j^2 y_j & -y_j^3 \\ \hline 1 & x_k & y_k & x_k^2 & x_k y_k & y_k^2 & x_k^3 & x_k^2 y_k & x_k y_k^2 & y_k^3 & x_k^3 y_k & x_k y_k^3 \\ 0 & 0 & 1 & 0 & x_k & 2y_k & 0 & x_k^2 & 2x_k y_k & 3y_k^2 & x_k^3 & 3x_k y_k^2 \\ 0 & -1 & 0 & -2x_k & -y_k & 0 & -3x_k^2 & -2x_k y_k & -y_k^2 & 0 & -3x_k^2 y_k & -y_k^3 \\ \hline 1 & x_l & y_l & x_l^2 & x_l y_l & y_l^2 & x_l^3 & x_l^2 y_l & x_l y_l^2 & y_l^3 & x_l^3 y_l & x_l y_l^3 \\ 0 & 0 & 1 & 0 & x_l & 2y_l & 0 & x_l^2 & 2x_l y_l & 3y_l^2 & x_l^3 & 3x_l y_l^2 \\ 0 & -1 & 0 & -2x_l & -y_l & 0 & -3x_l^2 & -2x_l y_l & -y_l^2 & 0 & -3x_l^2 y_l & -y_l^3 \end{bmatrix} \tag{13.51}$$

Next, we express the strain $\{\epsilon\}$ in terms of the vector $\{\alpha\}$. Thus:

$$\{\epsilon\}^e = \begin{Bmatrix} -\dfrac{\partial^2 w}{\partial x^2} \\ -\dfrac{\partial^2 w}{\partial y^2} \\ -2\dfrac{\partial^2 w}{\partial x\,\partial y} \end{Bmatrix} = \begin{bmatrix} 0 & 0 & 0 & -2 & 0 & 0 & -6x & -2y & 0 & 0 & -6xy & 0 \\ 0 & 0 & 0 & 0 & 0 & -2 & 0 & 0 & -2x & -6y & 0 & -6xy \\ 0 & 0 & 0 & 0 & -2 & 0 & 0 & -4x & -4y & 0 & -6x^2 & -6y^2 \end{bmatrix} \begin{Bmatrix} \alpha_1 \\ \alpha_2 \\ \alpha_3 \\ \alpha_4 \\ \alpha_5 \\ \alpha_6 \\ \alpha_7 \\ \alpha_8 \\ \alpha_9 \\ \alpha_{10} \\ \alpha_{11} \\ \alpha_{12} \end{Bmatrix} \tag{13.52}$$

Hence we have the matrix $[H]$ as used in Eq. (13.52):

$$[H] = \begin{bmatrix} 0 & 0 & 0 & -2 & 0 & 0 & -6x & -2y & 0 & 0 & -6xy & 0 \\ 0 & 0 & 0 & 0 & 0 & -2 & 0 & 0 & -2x & -6y & 0 & -6xy \\ 0 & 0 & 0 & 0 & -2 & 0 & 0 & -4x & -4y & 0 & -6x^2 & -6y^2 \end{bmatrix} \tag{13.53}$$

And from Eq. (13.29), we have for the $[B]$ matrix:

$$[B]^e = [H][A]^{-1} \tag{13.54}$$

We can now go to Eq. (13.14) to determine the stiffness matrix for the element. Thus finding $[A]^{-1}$ on the computer for the element and using $[H]$ from Eq. (13.53) (which is valid for all elements) we have

$$[K]^e = \iint_A [B]^T[D][B]\, dx\, dy \tag{13.55}$$

where $[D]$ is given for all elements by Eq. (13.7).

We will need the interpolation function $[N]$. In Eq. (13.33), we have for $[N]^e$

$$[N]^e = \lfloor C \rfloor [A]^{-1} \tag{13.56}$$

where for the rectangular element we get $\lfloor C \rfloor$ from Eq. (13.49). Thus

$$w = \lfloor C \rfloor \{\alpha\}$$

and so

$$\lfloor C \rfloor = \lfloor 1, x, y, x^2, xy, y^2, x^3, x^2y, xy^2, y^3, x^3y, xy^3 \rfloor \tag{13.57}$$

which is clearly valid for all elements.

We thus have $[K]^e$ and can compute $\{q\}^e = \iint p(x, y)[N^T]\, dx\, dy$ for each element. Finally, we assemble the global equation for finite elements and solve for the deflection. We shall illustrate this in the following example, using a coarse grid.

EXAMPLE 13.2 We shall again consider the square clamped plate of Example 13.1, this time using four rectangular elements as a starter. See Fig. 13.5 for the key diagram of the example and Fig. 13.9 for the rectangular elements. You will

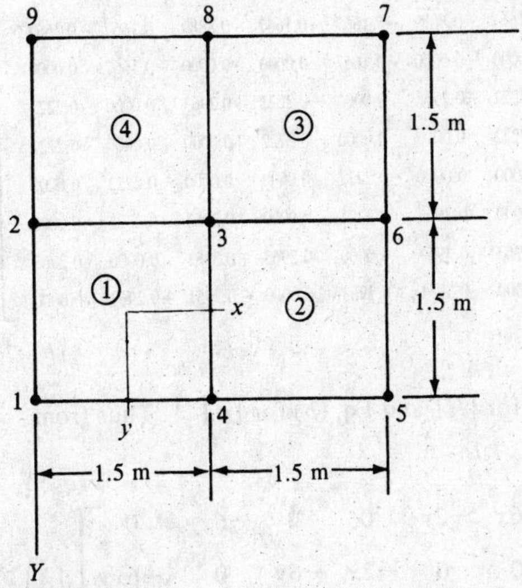

FIGURE 13.9
Four-rectangular-element plate; xy are centroidal axes for each element and XY are global axes.

note that XY are denoted as global axes. It will be simplest to work with local centroidal axes for each element. Each set of axes must be parallel to XY. We get the same results as if we worked with XY. This is because the determinant of the Jacobian matrix is unity under such a change of axes.

As in Example 13.1, we shall first work with element 1 in detail and then give the results for the other elements. Thus, going to Eq. (13.51), we have for $[A]^{(1)}$, using the local centroidal axes xy shown in Fig. 13.9:

$$[A]^{(1)} = \left[\begin{array}{cccccccccccc}
1.0000 & -0.7500 & 0.7500 & 0.5625 & -0.5625 & 0.5625 & -0.4219 & 0.4219 & -0.4219 & 0.4219 & -0.3164 & -0.3164 \\
0.0000 & 0.0000 & 1.0000 & 0.0000 & -0.7500 & 1.5000 & 0.0000 & 0.5625 & -1.1250 & 1.6875 & -0.4219 & -1.2656 \\
0.0000 & -1.0000 & 0.0000 & 1.5000 & -0.7500 & 0.0000 & -1.6875 & 1.1250 & -0.5625 & 0.0000 & -1.2656 & -0.4219 \\
\hline
1.0000 & 0.7500 & 0.7500 & 0.5625 & 0.5625 & 0.5625 & 0.4219 & 0.4219 & 0.4219 & 0.4219 & 0.3164 & 0.3164 \\
0.0000 & 0.0000 & 1.0000 & 0.0000 & 0.7500 & 1.5000 & 0.0000 & 0.5625 & 1.1250 & 1.6875 & 0.4219 & 1.2656 \\
0.0000 & -1.0000 & 0.0000 & -1.5000 & -0.7500 & 0.0000 & -1.6875 & -1.1250 & -0.5625 & 0.0000 & -1.2656 & -0.4219 \\
\hline
1.0000 & 0.7500 & -0.7500 & 0.5625 & -0.5625 & 0.5625 & 0.4219 & -0.4219 & 0.4219 & -0.4219 & -0.3164 & -0.3164 \\
0.0000 & 0.0000 & 1.0000 & 0.0000 & 0.7500 & -1.5000 & 0.0000 & 0.5625 & -1.1250 & 1.6875 & 0.4219 & 1.2656 \\
0.0000 & -1.0000 & 0.0000 & -1.5000 & 0.7500 & 0.0000 & -1.6875 & 1.1250 & -0.5625 & 0.0000 & 1.2656 & 0.4219 \\
\hline
1.0000 & -0.7500 & -0.7500 & 0.5625 & 0.5625 & 0.5625 & -0.4219 & -0.4219 & -0.4219 & -0.4219 & 0.3164 & 0.3164 \\
0.0000 & 0.0000 & 1.0000 & 0.0000 & -0.7500 & -1.5000 & 0.0000 & 0.5625 & 1.1250 & 1.6875 & -0.4219 & -1.2656 \\
0.0000 & -1.0000 & 0.0000 & 1.5000 & 0.7500 & 0.0000 & -1.6875 & -1.1250 & -0.5625 & 0.0000 & 1.2656 & 0.4219
\end{array}\right]
\begin{array}{l}
\\ \text{Node (1)} \\ \\ \\ \text{Node (2)} \\ \\ \\ \text{Node (3)} \\ \\ \\ \text{Node (4)} \\ \\
\end{array} \qquad (a)$$

The inverse of $[A]$ is then:

$$[A]^{-1} = \left[\begin{array}{cccccccccccc}
0.2500 & -0.0938 & -0.0938 & 0.2500 & -0.0938 & 0.0938 & 0.2500 & 0.0938 & 0.0938 & 0.2500 & 0.0938 & -0.0938 \\
-0.5000 & 0.1250 & 0.1250 & 0.5000 & -0.1250 & 0.1250 & 0.5000 & 0.1250 & 0.1250 & -0.5000 & -0.1250 & 0.1250 \\
0.5000 & -0.1250 & -0.1250 & 0.5000 & -0.1250 & 0.1250 & -0.5000 & -0.1250 & -0.1250 & -0.5000 & -0.1250 & 0.1250 \\
0.0000 & 0.0000 & 0.1667 & 0.0000 & 0.0000 & -0.1667 & 0.0000 & 0.0000 & -0.1667 & 0.0000 & 0.0000 & 0.1667 \\
-0.8889 & 0.1667 & 0.1667 & 0.8889 & -0.1667 & 0.1667 & -0.8889 & -0.1667 & -0.1667 & 0.8889 & 0.1667 & -0.1667 \\
0.0000 & 0.1667 & 0.0000 & 0.0000 & 0.1667 & 0.0000 & 0.0000 & -0.1667 & 0.0000 & 0.0000 & -0.1667 & 0.0000 \\
0.2963 & 0.0000 & -0.2222 & -0.2963 & 0.0000 & -0.2222 & -0.2963 & 0.0000 & -0.2222 & 0.2963 & 0.0000 & -0.2222 \\
0.0000 & 0.0000 & 0.2222 & 0.0000 & 0.0000 & -0.2222 & 0.0000 & 0.0000 & 0.2222 & 0.0000 & 0.0000 & -0.2222 \\
0.0000 & -0.2222 & 0.0000 & 0.0000 & 0.2222 & 0.0000 & 0.0000 & -0.2222 & 0.0000 & 0.0000 & 0.2222 & 0.0000 \\
-0.2963 & 0.2222 & 0.0000 & -0.2963 & 0.2222 & 0.0000 & 0.2963 & 0.2222 & 0.0000 & 0.2963 & 0.2222 & 0.0000 \\
0.3951 & 0.0000 & -0.2963 & -0.3951 & 0.0000 & -0.2963 & 0.3951 & 0.0000 & 0.2963 & -0.3951 & 0.0000 & 0.2963 \\
0.3951 & -0.2963 & 0.0000 & -0.3951 & 0.2963 & 0.0000 & 0.3951 & 0.2963 & 0.0000 & -0.3951 & -0.2963 & 0.0000
\end{array}\right] \qquad (b)$$

Now get $[B]^{(1)}$ by using Eq. (13.53) for $[H]$ and Eq. (b) for $[A]^{-1}$. Thus from Eq. (13.54) we have

$$[B]^{(1)} = \begin{bmatrix}
0 & 0 & 0 & -2 & 0 & 0 & -6x & -2y & 0 & 0 & -6xy & 0 \\
0 & 0 & 0 & 0 & 0 & -2 & 0 & 0 & -2x & -6y & 0 & -6xy \\
0 & 0 & 0 & 0 & -2 & 0 & 0 & -4x & -4y & 0 & -6x^2 & -6y^2
\end{bmatrix} [A]^{-1}$$

$$
[B]^{(1)} = \left[\begin{array}{ccc|ccc|ccc|ccc}
-1.78x - 2.37xy & 0 & -0.333 + 1.33x - 0.444y + 1.78xy & 1.78x + 2.37xy & 0 & 0.333 + 1.33x + 0.444y + 1.78xy & 1.78x - 2.37xy & 0 & 0.333 + 1.33x - 0.444y - 1.78xy & -1.78x + 2.37xy & 0 & -0.333 + 1.33x + 0.444y - 1.78xy \\
1.78y - 2.37xy & -0.333 + 0.444x - 1.33y + 1.78xy & 0 & 1.78y + 2.37xy & -0.333 - 0.444x - 1.33y - 1.78xy & 0 & -1.78y - 2.37xy & 0.333 + 0.444x - 1.33y - 1.78xy & 0 & -1.78y + 2.37xy & 0.333 - 0.444x - 1.33y + 1.78xy & 0 \\
1.78 - 2.37x^2 - 2.37y^2 & -0.333 + 0.889y + 1.78y^2 & -0.333 - 0.889x + 1.78x^2 & -1.78 + 2.37x^2 + 2.37y^2 & 0.333 - 1.78y^2 & -0.333 + 0.889x + 1.78x^2 & 1.78 - 2.37x^2 - 2.37y^2 & 0.333 + 0.889y - 1.78y^2 & 0.333 - 0.889x - 1.78x^2 & -1.78 + 2.37x^2 + 2.37y^2 & -0.333 - 0.889y + 1.78y^2 & 0.333 + 0.889x - 1.78x^2
\end{array}\right] \qquad (c)
$$

From Example 13.1, we have for $[D]$ for this problem:

$$[D] = 3.005 \times 10^5 \begin{bmatrix} 1 & 0.3 & 0 \\ 0.3 & 1 & 0 \\ 0 & 0 & 0.35 \end{bmatrix} \tag{d}$$

We next compute the stiffness matrix for element 1 as follows:

$[K]^{(1)} = \iint_A$ 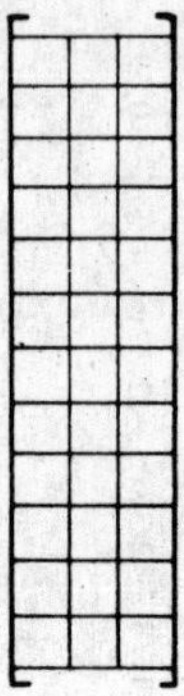 $(3.005 \times 10^5) \begin{bmatrix} 1 & 0.3 & 0 \\ 0.3 & 1 & 0 \\ 0 & 0 & 0.35 \end{bmatrix}$ 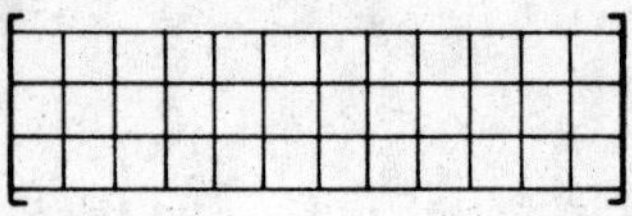 dx

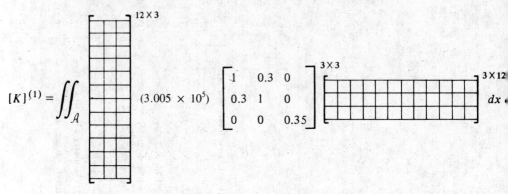

Carrying out the matrix products and integrating, we have

$$[K]^{(1)} = 10^6 \begin{bmatrix}
1.41 & -0.489 & -0.489 & -0.609 & -0.112 & -0.429 & -0.192 & -0.172 & -0.172 & -0.609 & -0.429 & -0.112 \\
-0.489 & 0.457 & 0.901 & -0.112 & 0.144 & 0 & 0.172 & 0.114 & 0 & 0.429 & 0.186 & 0 \\
-0.489 & 0.09 & 0.457 & 0.429 & 0 & 0.186 & 0.172 & 0 & 0.114 & -0.112 & 0 & 0.144 \\
0.609 & -0.112 & 0.429 & 1.41 & -0.489 & 0.489 & -0.609 & -0.429 & 0.112 & -0.192 & -0.172 & 0.172 \\
-0.112 & 0.144 & 0 & -0.489 & 0.457 & -0.09 & 0.429 & 0.186 & 0 & 0.172 & 0.114 & 0 \\
0.429 & 0 & 0.186 & 0.489 & -0.09 & 0.457 & 0.112 & 0 & 0.144 & -0.172 & 0 & 0.114 \\
-0.192 & 0.172 & 0.172 & -0.609 & 0.429 & 0.112 & 1.41 & 0.489 & 0.489 & -0.609 & 0.112 & 0.429 \\
-0.172 & 0.114 & 0 & -0.429 & 0.186 & 0 & 0.489 & 0.457 & 0.09 & 0.112 & 0.144 & 0 \\
-0.172 & 0 & 0.114 & 0.112 & 0 & 0.144 & 0.489 & 0.09 & 0.457 & -0.429 & 0 & 0.186 \\
-0.609 & 0.429 & -0.112 & -0.192 & 0.172 & -0.172 & -0.609 & 0.112 & -0.429 & 1.41 & 0.489 & -0.489 \\
-0.429 & 0.186 & 0 & -0.172 & 0.114 & 0 & 0.112 & 0.144 & 0 & 0.489 & 0.457 & -0.09 \\
-0.112 & 0 & 0.144 & 0.172 & 0 & 0.114 & 0.429 & 0 & 0.186 & -0.489 & -0.09 & 0.457
\end{bmatrix}$$

The global stiffness matrix may then be formed from the four element stiffness matrices to form a 27×27 matrix. We may then form the *reduced* matrix by noting that at all the nodes except node 3, we have homogeneous boundary conditions for w, θ_x, and θ_y. The reduced matrix for K is

$$[K]_{red} = 10^6 \begin{bmatrix} 5.64 & 0 & 0 \\ 0 & 1.83 & 0 \\ 0 & 0 & 1.83 \end{bmatrix} \tag{e}$$

We next need the forcing field vectors $\{q\}$. Considering element 1, for which $[A]^{-1}$ is given by Eq. (*b*), we get for $[N]^{(1)}$, on noting Eqs. (13.56) and (13.57),

$$\lfloor N \rfloor^{(1)} = \lfloor 1, x, y, x^2, xy, y^2, x^3, x^2y, xy^2, y^3, x^3y, xy^3 \rfloor [A^{(1)}]^{-1} \qquad (f)$$

For example, using Eq. (*b*) in the above equation, we have for $N_1^{(1)}$:

$$N_1^{(1)} = (0.250 - 0.5x + 0.5y - 0.889xy + 0.296x^3 - 0.296y^3 + 0.395x^3y + 0.395xy^3)$$

We can now compute the force vector $q_1^{(1)}$ for element 1 as follows:

$$q_1^{(1)} = \int_{-0.75}^{0.75}\int_{-0.75}^{0.75} (N_1^{(1)})(p)\, dx\, dy$$

$$= \int_{-0.75}^{0.75}\int_{-0.75}^{0.75} (0.250 - 0.5x + 0.5y - 0.889xy + 0.296x^3 - 0.296y^3 + 0.395x^3y + 0.395xy^3)(14 \times 10^4)\, dx\, dy$$

$$= 0.788 \times 10^5 \text{ N}$$

The other force vectors can similarly be computed for all the elements. The global force vector for the reduced equations is as follows:

$$\{q\}_{\text{red}} \approx \begin{Bmatrix} 0.315 \times 10^6 \\ 0 \\ 0 \end{Bmatrix} \qquad (g)$$

where we have set the negligibly small terms equal to zero on the right side of the equation.

From Eqs. (*e*) and (*g*), we then have:

$$10^6 \begin{bmatrix} 5.64 & 0 & 0 \\ 0 & 1.83 & 0 \\ 0 & 0 & 1.83 \end{bmatrix} \begin{Bmatrix} w_3 \\ (\theta_x)_3 \\ (\theta_y)_3 \end{Bmatrix} = 10^5 \begin{Bmatrix} 0.315 \times 10^6 \\ 0 \\ 0 \end{Bmatrix} \qquad (h)$$

We get $w_3 = 0.0558$ m. We are 17% from the exact solution of 0.04765 m presented in Example 13.5. Keep in mind, however, that we have used only four nonconforming elements.

13.7 OTHER RESULTS FOR RECTANGULAR ELEMENTS

Now, using the program given in Appendix IV, we can carry out the previous computations for finer grids. We can also make use of the fact that, for the square plate we are dealing with here, we can use one quadrant of this plate (see Fig. 13.10) to give

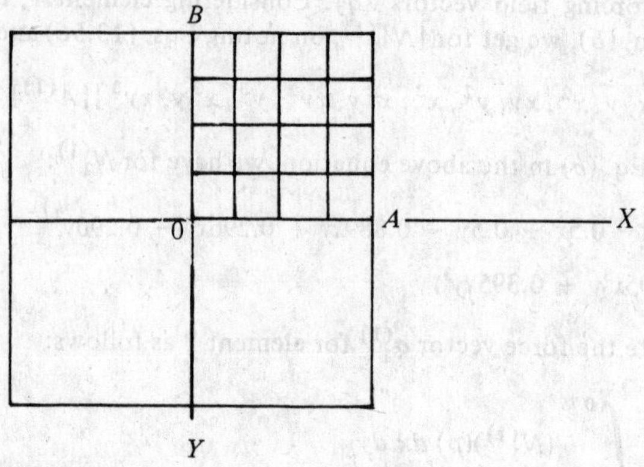

FIGURE 13.10
Rectangular plate, using one quadrant.

complete information for the entire plate. Note that in so doing we must take $\theta_x = 0$ along side *OA* and $\theta_y = 0$ along side *OB*. The following are the results for such computations:

Total number of elements	w_{max}
16	0.053 m
64	0.0492 m

We get satisfactory convergence toward the correct result.

We have also solved the plate problem by using SAP IV with rectangular elements. We get the following results:

Total number of elements	w_{max}
16	0.04177 m
64	0.04594 m

In addition to deflection, the SAP IV program will also give bending and twisting moments.

As a final consideration for rectangular elements, we note from Example 13.2 that the computation of the stiffness matrix as carried out step by step (for instructional purposes) was long as laborious. In Eq. (13.58) we present the stiffness matrix worked out for Eq. (13.55). In future undertakings, now that we have seen the development of solutions in great detail in Examples 13.1 and 13.2, we shall give the resulting formulations as we are now doing to avoid subjecting the reader to all of the laborious details. Thus, for an $a \times b$ rectangular plate element we have

$$[K]^{(e)} = \frac{D}{15ab}[\kappa]^{(e)} \tag{13.58}$$

where, using the notation,

$$\frac{a}{b} = r \tag{13.59}$$

$$D = \frac{Eh^3}{12(1-\nu^2)} \tag{13.60}$$

in Table 13.1, we have the terms for $[\kappa]$.

TABLE 13.1
Stiffness Data for Rectangular Elements

$\kappa_{11} = 60/r^2 + 60r^2 + 30\nu + 42(1-\nu)$	$\kappa_{94} = 15b/r - 15a\nu - 3a(1-\nu)$
$\kappa_{21} = 30ar + 15b\nu + 3b(1-\nu)$	$\kappa_{95} = 0$
$\kappa_{22} = 20a^2 + 4b^2(1-\nu)$	$\kappa_{96} = 10b^2 \; 4a^2(1-\nu)$
$\kappa_{31} = -30b/r - 15a\nu - 3a(1-\nu)$	$\kappa_{97} = 30b/r + 15a\nu + 3a(1-\nu)$
$\kappa_{32} = -15ab\nu$	$\kappa_{98} = -15ab\nu$
$\kappa_{33} = 20b^2 + 4a^2(1-\nu)$	$\kappa_{99} = 20b^2 + 4a^2(1-\nu)$
$\kappa_{41} = -60/r^2 + 30r^2 - 30\nu - 42(1-\nu)$	$\kappa_{10,1} = 30/r^2 - 60r^2 - 30\nu - 42(1-\nu)$
$\kappa_{42} = 15ar - 15b\nu - 3b(1-\nu)$	$\kappa_{10,2} = -30ar - 3b(1-\nu)$
$\kappa_{43} = 30b/r + 3a(1-\nu)$	$\kappa_{10,3} = -15b/r + 15a\nu + 3a(1-\nu)$
$\kappa_{44} = 60/r^2 + 60r^2 + 30\nu + 42(1-\nu)$	$\kappa_{10,4} = -30/r^2 - 30r^2 + 30\nu + 42(1-\nu)$
$\kappa_{51} = 15ar - 15b\nu - 3b(1-\nu)$	$\kappa_{10,5} = -15ar + 3b(1-\nu)$
$\kappa_{52} = 10a^2 - 4b^2(1-\nu)$	$\kappa_{10,6} = -15b/r + 3a(1-\nu)$
$\kappa_{53} = 0$	$\kappa_{10,7} = -60/r^2 + 30r^2 - 30\nu - 42(1-\nu)$
$\kappa_{54} = 30ar + 15b\nu + 3b(1-\nu)$	$\kappa_{10,8} = -15ar + 15b\nu + 3b(1-\nu)$
$\kappa_{55} = 20a^2 + 4b^2(1-\nu)$	$\kappa_{10,9} = -30b/r - 3a(1-\nu)$
$\kappa_{61} = -30b/r - 3a(1-\nu)$	$\kappa_{10,10} = 60/r^2 + 60r^2 + 30\nu + 42(1-\nu)$
$\kappa_{62} = 0$	$\kappa_{11,1} = 30ar + 3b(1-\nu)$
$\kappa_{63} = 10b^2 - a^2(1-\nu)$	$\kappa_{11,2} = 10a^2 - b^2(1-\nu)$
$\kappa_{64} = 30b/r + 15a\nu + 3a(1-\nu)$	$\kappa_{11,3} = 0$
$\kappa_{65} = 15ab\nu$	$\kappa_{11,4} = 15ar - 3b(1-\nu)$
$\kappa_{66} = 20b^2 + 4a^2(1-\nu)$	$\kappa_{11,5} = 5a^2 + b^2(1-\nu)$
$\kappa_{71} = -30/r^2 - 30r^2 + 30\nu + 42(1-\nu)$	$\kappa_{11,6} = 0$
$\kappa_{72} = -15ar + 3b(1-\nu)$	$\kappa_{11,7} = -15ar + 15b\nu + 3b(1-\nu)$
$\kappa_{73} = 15b/r - 3a(1-\nu)$	$\kappa_{11,8} = 10a^2 - 4b^2(1-\nu)$
$\kappa_{74} = 30/r^2 - 60r^2 - 30\nu - 42(1-\nu)$	$\kappa_{11,9} = 0$
$\kappa_{75} = -30ar - 3b(1-\nu)$	$\kappa_{11,10} = -30ar - 15b\nu - 3b(1-\nu)$
$\kappa_{76} = 15b/r - 15a\nu - 3a(1-\nu)$	$\kappa_{11,11} = 20a^2 + 4b^2(1-\nu)$
$\kappa_{77} = 60/r^2 + 60r^2 + 30\nu + 42(1-\nu)$	$\kappa_{12,1} = -15b/r + 15a\nu + 3a(1-\nu)$
$\kappa_{81} = 15ar - 3b(1-\nu)$	$\kappa_{12,2} = 0$
$\kappa_{82} = 5a^2 + b^2(1-\nu)$	$\kappa_{12,3} = 10b^2 - 4a^2(1-\nu)$
$\kappa_{83} = 0$	$\kappa_{12,4} = 15b/r - 3a(1-\nu)$
$\kappa_{84} = 30ar + 3b(1-\nu)$	$\kappa_{12,5} = 0$
$\kappa_{85} = 10a^2 - b^2(1-\nu)$	$\kappa_{12,6} = 5b^2 + a^2(1-\nu)$
$\kappa_{86} = 0$	$\kappa_{12,7} = 30b/r + 3a(1-\nu)$
$\kappa_{87} = -30ar - 15b\nu - 3b(1-\nu)$	$\kappa_{12,8} = 0$
$\kappa_{88} = 20a^2 + 4b^2(1-\nu)$	$\kappa_{12,9} = 10b^2 - a^2(1-\nu)$
$\kappa_{91} = -15b/r + 3a(1-\nu)$	$\kappa_{12,10} = -30b/r - 15a\nu - 3a(1-\nu)$
$\kappa_{92} = 0$	$\kappa_{12,11} = 15ab\nu$
$\kappa_{93} = 5b^2 + a^2(1-\nu)$	$\kappa_{12,12} = 20b^2 + 4a^2(1-\nu)$

Part D
Hermite Polynomial Elements

13.8 INTRODUCTION

In Section 11.9 we explained how we could achieve C^1 continuity for rectangular elements by using Hermite polynomials. We thus arrived at Eqs. (11.29), which, with the aid of Eqs. (11.30), spell out the associated interpolation functions. The rectangular element (see Fig. 13.11) has four nodes, one at each corner, and 16 degrees of freedom. The polynomials are bicubic and the Hermite polynomials are first-order. Thus, for $N_1(\xi, \eta)$ we have

$$N_1(\xi, \eta) = H_{01}^1(\xi) H_{01}^1(\eta) = (1 - 3\xi^2 + 2\xi^3)(1 - 3\eta^2 + 2\eta^3) \tag{13.61}$$

where we are using normalized local coordinates $\xi\eta$ (see Fig. 13.12). We thus have 16 interpolation functions, which we denote in matrix form as $[N]^{1 \times 16}$, where the superscripts indicate that we have a 1 row by 16 column matrix.

13.9 KEY MATRICES

We will now set forth the key matrices for carrying out a finite element procedure. First we consider Eq. (13.3) relating the strains to the field variable w. It is seen by inspection that the matrix operator $\{L\}$ is

$$\{L\} = \begin{Bmatrix} -\dfrac{\partial^2}{a^2\,\partial\xi^2} \\ -\dfrac{\partial^2}{b^2\,\partial\eta^2} \\ \dfrac{-2}{ab}\dfrac{\partial^2}{\partial\xi\,\partial\eta} \end{Bmatrix}$$

where a and b are the side lengths of a rectangular element as shown in Fig. 13.11.

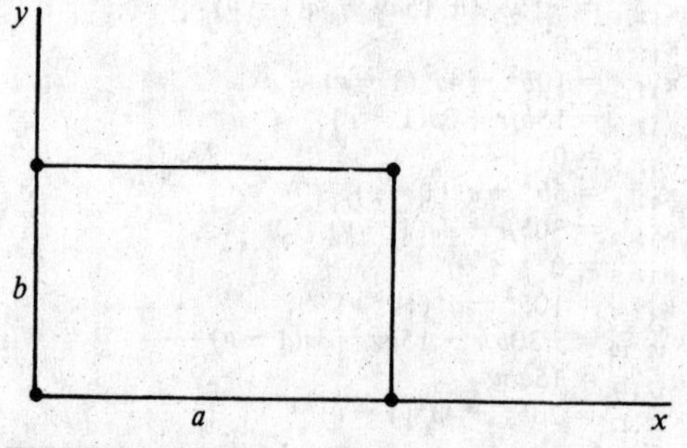

FIGURE 13.11
Rectangular element with sides a and b.

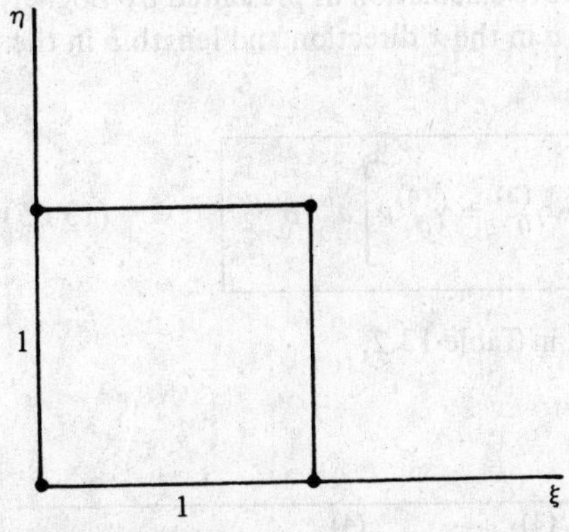

FIGURE 13.12
Normalized coordinates for rectangular element.

Now going to Eq. (9.59*c*), we can get the $[B]$ matrix for an element. We have

$$[B] = \{L\}[N] = \underbrace{\left[\begin{array}{c|c|c|c} \cdots & \cdots & \cdots & \cdots \\ \cdots & \cdots & \cdots & \cdots \\ \cdots & \cdots & \cdots & \cdots \end{array}\right]}_{3\times 16} \tag{13.62}$$

where the dots represent terms of the matrix. The first dot, for instance, is B_{11}:

$$B_{11} = \frac{-1}{a^2}\left(\frac{\partial^2}{\partial \xi^2}\right)[(1-3\xi^2+2\xi^3)(1-3\eta^2+2\eta^3)]$$

$$= \frac{1}{a^2}(6-12\xi)(1-3\eta^2+2\eta^3) \tag{13.63}$$

The matrix $[D]$ for the constitutive law [see Eq. (13.7)] yields a 3×3 matrix, which we now rewrite as follows:

$$[D] = \frac{Eh^3}{12(1-\nu^2)}\begin{bmatrix} 1 & \nu & 0 \\ \nu & 1 & 0 \\ 0 & 0 & \dfrac{1-\nu}{2} \end{bmatrix} \tag{13.64}$$

We get the element stiffness matrix from Eq. (13.14). We have, again using dots as above,

$$[K]^e = \int_0^1 \int_0^1 \underbrace{\begin{bmatrix} \cdots \\ \cdots \\ \cdots \\ \cdots \\ \vdots \\ \cdots \end{bmatrix}}_{[B]^T\ 16\times 3} \left(\frac{Eh^3}{12(1-\nu^2)}\right) \underbrace{\begin{bmatrix} \cdots \\ \cdots \\ \cdots \end{bmatrix}}_{[D]\ 3\times 3} \underbrace{\left[\begin{array}{c|c|c|c} \cdots & \cdots & \cdots & \cdots \\ \cdots & \cdots & \cdots & \cdots \\ \cdots & \cdots & \cdots & \cdots \end{array}\right]}_{[B]\ 3\times 16} ab\, d\xi\, d\eta$$

At this time we shall quote the results of the above calculation as presented by Bogner et al. (1965) for a rectangular element of length a in the x direction and length b in the y direction. We have

$$K_{ij} = \frac{D}{ab}\left[\gamma_{ij}^{(1)}\left(\frac{b}{a}\right)^2 + \gamma_{ij}^{(2)}\left(\frac{a}{b}\right)^2 + \gamma_{ij}^{(3)} + \gamma_{ij}^{(4)}\nu\right] a^{\lambda_{ij}} b^{\mu_{ij}} \tag{13.65}$$

where $\gamma_{ij}^{(1)}, \gamma_{ij}^{(2)}, \gamma_{ij}^{(3)}, \gamma_{ij}^{(4)}, \lambda_{ij}$, and μ_{ij} are given in Table 13.2.

TABLE 13.2
Bogner et al. Data for Stiffness Matrix

i	j	$\gamma_{ij}^{(1)}$	$\gamma_{ij}^{(2)}$	$\gamma_{ij}^{(3)}$	$\gamma_{ij}^{(4)}$	λ_{ij}	μ_{ij}
1	1	156/35	156/35	72/25	0	0	0
2	1	78/35	22/35	6/25	6/5	1	0
2	2	52/35	4/35	8/25	0	2	0
3	1	22/35	78/35	6/25	6/5	0	1
3	2	11/35	11/35	1/50	6/5	1	1
3	3	4/35	52/35	8/25	0	0	2
4	1	11/35	11/35	1/50	1/5	1	1
4	2	22/105	2/35	2/75	2/15	2	1
4	3	2/35	22/105	8/75	2/15	1	2
4	4	4/105	4/105	8/225	0	2	2
5	1	54/35	−156/35	−72/25	0	0	0
5	2	27/35	−22/35	−6/25	−6/5	1	0
5	3	13/35	−78/35	−6/25	0	0	1
5	4	13/70	−11/35	−1/50	−1/10	1	1
5	5	156/35	156/35	72/25	0	0	0
6	1	27/35	−22/35	−6/25	−6/5	1	0
6	2	18/35	−4/35	−8/25	0	2	0
6	3	13/70	−11/35	−1/50	−1/10	1	1
6	4	13/105	−2/35	−2/75	0	2	1
6	5	78/35	22/35	6/25	6/5	1	0
6	6	52/35	4/35	8/25	0	2	0
7	1	−13/35	78/35	6/25	0	0	1
7	2	−13/70	11/35	1/50	1/10	1	1
7	3	−3/35	26/35	−2/25	0	0	2
7	4	−3/70	11/105	−1/150	−1/30	1	2
7	5	−22/35	−78/35	−6/25	−6/5	0	1
7	6	−11/35	−11/35	−1/50	−6/5	1	1
7	7	4/35	52/35	8/25	0	0	2
8	1	−13/70	11/35	1/50	1/10	1	1
8	2	−13/105	2/35	2/75	0	2	1
8	3	−3/70	11/105	−1/150	−1/30	1	2
8	4	−1/35	2/105	−2/225	0	2	2
8	5	−11/35	−11/35	−1/50	−1/5	1	1
8	6	−22/105	−2/35	−2/75	−2/15	2	1
8	7	2/35	22/105	2/75	2/15	1	2
8	8	4/105	4/105	8/225	0	2	2
9	1	−54/35	−54/35	72/25	0	0	0
9	2	−27/35	−13/35	6/25	0	1	0

TABLE 13.2 (*Cont.*)
Bogner et al. Data for Stiffness Matrix

i	j	$\gamma_{ij}^{(1)}$	$\gamma_{ij}^{(2)}$	$\gamma_{ij}^{(3)}$	$\gamma_{ij}^{(4)}$	λ_{ij}	μ_{ij}
9	3	−13/35	−27/35	6/25	0	0	1
9	4	−13/70	−13/70	1/50	0	1	1
9	5	−156/35	54/35	−72/25	0	0	0
9	6	−78/35	13/35	−6/25	0	1	0
9	7	22/35	−27/35	6/25	6/5	0	1
9	8	11/35	−13/70	1/50	1/10	1	1
9	9	156/35	156/35	72/25	0	0	0
10	1	27/35	13/35	−6/25	0	1	0
10	2	9/35	3/35	2/25	0	2	0
10	3	13/70	13/70	−1/50	0	1	1
10	4	13/210	3/70	1/150	0	2	1
10	5	78/35	−13/35	6/25	0	1	0
10	6	26/35	−3/35	−2/25	0	2	0
10	7	−11/35	13/70	−1/50	−1/10	1	1
10	8	−11/105	3/70	1/150	1/30	2	1
10	9	−78/35	−22/35	−6/25	−6/5	1	0
10	10	52/35	4/35	8/25	0	2	0
11	1	13/35	27/35	−6/25	0	0	1
11	2	13/70	13/70	−1/50	0	1	1
11	3	3/35	9/35	2/25	0	0	2
11	4	3/70	13/210	1/150	0	1	2
11	5	22/35	−27/35	6/25	6/5	0	1
11	6	11/35	−13/70	1/50	1/10	1	1
11	7	−4/35	18/35	−8/25	0	0	2
11	8	−2/35	13/105	−2/75	0	1	2
11	9	−22/35	−78/35	−6/25	−6/5	0	1
11	10	11/35	11/35	1/50	6/5	1	1
11	11	4/35	52/35	8/25	0	0	2
12	1	−13/70	−13/70	1/50	0	1	1
12	2	−13/210	−3/70	−1/150	0	2	1
12	3	−3/70	−13/210	−1/150	0	1	2
12	4	−1/70	−1/70	1/450	0	2	2
12	5	−11/35	13/70	−1/50	−1/10	1	1
12	6	−11/105	3/70	1/150	1/30	2	1
12	7	2/35	−13/105	2/75	0	1	2
12	8	2/105	−1/35	−2/225	0	2	2
12	9	11/35	11/35	1/50	1/5	1	1
12	10	−22/105	−2/35	−2/75	−2/15	2	1
12	11	−2/35	−22/105	−2/75	−2/15	1	2
12	12	4/105	4/105	8/225	0	2	2
13	1	−156/35	54/35	−72/25	0	0	0
13	2	−78/35	13/35	−6/25	0	1	0
13	3	−22/35	27/35	−6/25	−6/5	0	1
13	4	−11/35	13/70	−1/50	−1/10	1	1
13	5	−54/35	−54/35	72/25	0	0	0
13	6	−27/35	−13/35	6/25	0	1	0
13	7	13/35	27/35	−6/25	0	0	1
13	8	13/70	13/70	−1/50	0	1	1
13	9	54/35	−156/35	−72/25	0	0	0
13	10	−27/35	22/35	6/25	6/5	1	0

TABLE 13.2 (*Cont.*)
Bogner et al. Data for Stiffness Matrix

i	j	$\gamma_{ij}^{(1)}$	$\gamma_{ij}^{(2)}$	$\gamma_{ij}^{(3)}$	$\gamma_{ij}^{(4)}$	λ_{ij}	μ_{ij}
13	11	−13/35	78/35	6/25	0	0	1
13	12	13/70	−11/35	−1/50	−1/10	1	1
13	13	156/35	156/35	72/25	0	0	0
14	1	78/35	−13/35	6/25	0	1	0
14	2	26/35	−3/35	−2/25	0	2	0
14	3	11/35	−13/70	1/50	1/10	1	1
14	4	11/105	−3/70	−1/150	−1/30	2	1
14	5	27/35	13/35	−6/25	0	1	0
14	6	9/35	3/35	2/25	0	2	0
14	7	−13/70	−13/70	1/50	0	1	1
14	8	−13/210	−3/70	−1/150	0	2	1
14	9	−27/35	22/35	6/25	6/5	1	0
14	10	18/35	−4/35	−8/25	0	2	0
14	11	13/70	−11/35	−1/50	−1/10	1	1
14	12	−13/105	2/35	2/75	0	2	1
14	13	−78/35	−22/35	−6/25	−6/5	1	0
14	14	52/35	4/35	8/25	0	2	0
15	1	−22/35	27/35	−6/25	−6/5	0	1
15	2	−11/35	13/70	−1/50	−1/10	1	1
15	3	−4/35	18/35	−8/25	0	0	2
15	4	−2/35	13/105	−2/75	0	1	2
15	5	−13/35	−27/35	6/25	0	0	1
15	6	−13/70	−13/70	1/50	0	1	1
15	7	3/35	9/35	2/25	0	0	2
15	8	3/70	13/210	1/150	0	1	2
15	9	13/35	−78/35	−6/25	0	0	1
15	10	−13/70	11/35	1/50	1/10	1	1
15	11	−3/35	26/35	−2/25	0	0	2
15	12	3/70	−11/105	1/150	1/30	1	2
15	13	22/35	78/35	6/25	6/5	0	1
15	14	−11/35	−11/35	−1/50	−6/5	1	1
15	15	4/35	52/35	8/25	0	0	2
16	1	11/35	−13/70	1/50	1/10	1	1
16	2	11/105	−3/70	−1/150	−1/30	2	1
16	3	2/35	−13/105	2/75	0	1	2
16	4	2/105	−1/35	−2/225	0	2	2
16	5	13/70	13/70	−1/50	0	1	1
16	6	13/210	3/70	1/150	0	2	1
16	7	−3/70	−13/210	−1/150	0	1	2
16	8	−1/70	−1/70	1/450	0	2	2
16	9	−13/70	11/35	1/50	1/10	1	1
16	10	13/105	−2/35	−2/75	0	2	1
16	11	3/70	−11/105	1/150	1/30	1	2
16	12	−1/35	2/105	−2/225	0	2	2
16	13	−11/35	−11/35	−1/50	−1/5	1	1
16	14	22/105	2/35	2/75	2/15	2	1
16	15	−2/35	−22/105	−2/75	−2/15	1	2
16	16	4/105	4/105	8/225	0	2	2

TABLE 13.3
Bogner et al. Data for Nodal Loads

	Concentrated load P at (a, b)	Uniform load p	Distributed edge moment M				Distributed edge load $r(x, y)$			
			on $x=0$	on $x=a$	on $y=0$	on $y=b$	on $x=0$	on $x=a$	on $y=0$	on $y=b$
w_1	0	$pab/4$	0	0	0	0	$br/2$	0	$ar/2$	0
w_{x1}	0	$pa^2b/24$	$bM/2$	0	0	0	0	0	$a^2r/12$	0
w_{y1}	0	$pab^2/24$	0	0	$aM/2$	0	$b^2r/12$	0	0	0
w_{xy1}	0	$pa^2b^2/144$	$b^2M/12$	0	$a^2M/12$	0	0	0	0	0
w_2	0	$pab/4$	0	0	0	0	$br/2$	0	0	$ar/2$
w_{x2}	0	$pa^2b/24$	$bM/2$	0	0	0	0	0	0	$a^2r/12$
w_{y2}	0	$-pab^2/24$	0	0	0	$aM/2$	$-b^2r/12$	0	0	0
w_{xy2}	0	$-pa^2b^2/144$	$-b^2M/12$	0	0	$a^2M/12$	0	0	0	0
w_3	P	$pab/4$	0	0	0	0	0	$br/2$	0	$ar/2$
w_{x3}	0	$-pa^2b/24$	0	$bM/2$	0	0	0	0	0	$-a^2r/12$
w_{y3}	0	$-pab^2/24$	0	0	0	$aM/2$	0	$-b^2r/12$	0	0
w_{xy3}	0	$pa^2b^2/144$	0	$-b^2M/12$	0	$-a^2M/12$	0	0	0	0
w_4	0	$pab/4$	0	0	0	0	0	$br/2$	$ar/2$	0
w_{x4}	0	$-pa^2b/24$	0	$bM/2$	0	0	0	0	$-a^2r/12$	0
w_{y4}	0	$pab^2/24$	0	0	$aM/2$	0	0	$b^2r/12$	0	0
w_{xy4}	0	$-pa^2b^2/144$	0	$b^2M/12$	$-a^2M/12$	0	0	0	0	0

As for the forcing function vector $\{q\}$, we go back to Eq. (13.17), which we now rewrite:

$$\{q\} = \iint_R p(x, y)[N]^T \, dx \, dy + \sum_i P_i [N(x_i, y_i)]^T + \int_L r(x,y)[N(x,y)]^T \, dl + \int_L M(x,y) \frac{\partial [N]^T}{\partial n} \, dl \tag{13.66}$$

We show the results for $\{q\}$ of Bogner et al. for an $a \times b$ rectangular element in Table 13.3. Note that $w_x \equiv \partial w/\partial x$ in this table.

13.10 RESULTS FOR UNIFORMLY LOADED CLAMPED PLATE

Applying the Hermite polynomial approach to the problem of Parts B and C of this chapter for 16 elements, we get the following for the maximum deflection:

$$(w_{\max})_{16 \text{ elements}} = 0.0468 \text{ m} \qquad (w_{\max})_{\text{ex}} = 0.04765 \text{ m}$$

Thus, for very few elements, we get good results. This shows the value of using a higher-order element. However, the amount of computation for higher-order conforming elements (as you can see from our outlined computations) is considerable.

13.11 CLOSURE

We have thus considered flat plates by using three different elements, two of which were nonconforming. The triangular elements with nine degrees of freedom in rectangular coordinates gave the least satisfactory results. This can be improved by adding a node inside the triangle and giving this node an additional degree of freedom so as to have a 10-DOF triangle. An 18-DOF triangle with nodes at corners and at midpoints along the sides is another possibility. You are urged to refer to more advanced specialized texts on finite elements for further instruction along these lines. It is to be pointed out in closing that one may also use quadrilateral elements to form conformable elements. In the next chapter we shall use triangular elements once again for the study of torsion. In this chapter we begin a departure from the basic approach we have taken up to this point in Part III. Instead of using the total potential energy as the key quadratic functional, we shall turn to the total complementary potential energy principle. And in Chapter 17, we shall use yet other quadratic functionals.

REFERENCES

F. K. Bogner, R. L. Fox, and L. A. Schmit, Jr., "The Generation of Inter-Element Stiffness and Mass Matrices by the Use of Interpolation Formulas," in *Proceedings of the First Conference on Matrix Methods in Structural Engineering*, AFFDL-IR-66-80, 1965. See also K. H. Huebner, "The Finite Element Method for Engineers," Wiley, New York, 1975.

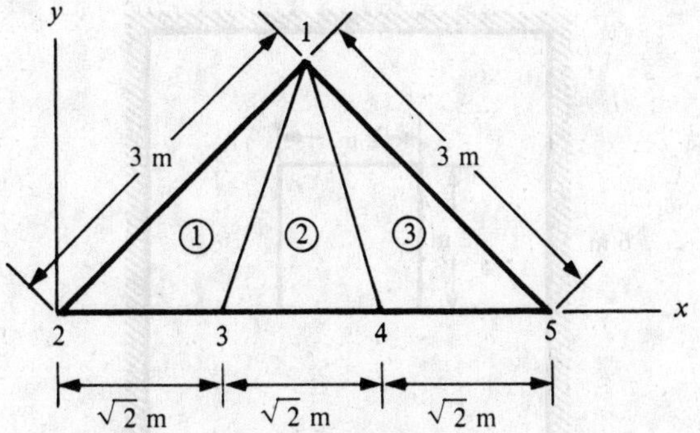

FIGURE 13.13

PROBLEMS

13.1 In Example 13.1 compute the stiffness matrix for element 1 and verify the results given in the example.

13.2 Do Example 13.1 for the case of a simply supported plate. Use whatever data in Example 13.1 apply.

13.3 A triangular plate is shown in Fig. 13.13 for which three elements have been indicated. The plate is clamped along sides 1–2 and 1–5 and is free along side 2–5. The thickness is 30 mm and a uniform loading of 7×10^4 Pa is applied over the surface of the plate. Using the methodology of Example 13.1 determine the deflections at points 3 and 4. Take $E = 21 \times 10^{10}$ Pa and $\nu = 0.3$.

13.4 Do Example 13.1 using triangular (natural) coordinates as set forth in Section 13.5.

13.5 Do Problem 13.3 using natural coordinates as set forth in Section 13.5.

13.6 Using SAP IV or other "canned" program, solve the problem in Example 13.1 for
  (*a*) 4 elements as shown in Fig. 13.6,
  (*b*) 8 elements,
  (*c*) 32 elements.

*13.7 Formulate a complete computer program for natural coordinates and use your program for Problem 13.3.

13.8 In Example 13.2 evaluate the stiffness matrix for element 1. Evaluate 10 terms in Eq. (*b*) and if they check with the example copy the remaining terms from what is given in the example. Continue in this manner to find $[K]^{(1)}$.

13.9 A 6 m × 6 m square plate, simply supported, carries a uniform load of 7×10^4 Pa (Fig. 13.14). Using four rectangular elements for one quadrant determine the deflection at the center. The thickness is 3 mm. Do not consider the weight of the plate. Take $E = 21 \times 10^{10}$ Pa and $\nu = 0.3$.

13.10 A 6 m × 6 m square plate, simply supported, carries a 2×10^4 N force at the center. Using the SAP IV or other "canned" program, solve for the maximum deflection for 16 elements and 64 elements total. Compare your result with the exact result from the theory of elasticity. Take $E = 21 \times 10^{10}$ Pa, $\nu = 0.3$, and the thickness $t = 3$ mm.

13.11 A square plate is clamped along its outer periphery (see Fig. 13.15). It supports a uniform load of 10×10^4 Pa. The inner square boundary is free. Using one quadrant with eight elements, compute the deflections and stresses using SAP IV or other "canned" program. $E = 21 \times 10^{10}$ Pa and $\nu = 0.3$. The thickness is 45 mm.

13.12 Develop a complete computer program for rectangular coordinates and then apply this program to Problem 13.10.

*13.13 A plate is fixed at the ends (see Fig. 13.16) and supports a uniform load of 14×10^4 Pa. The thickness is 3 mm and $E = 21 \times 10^{10}$ Pa. Take $\nu = 0.3$. For a grid approximating the one shown,

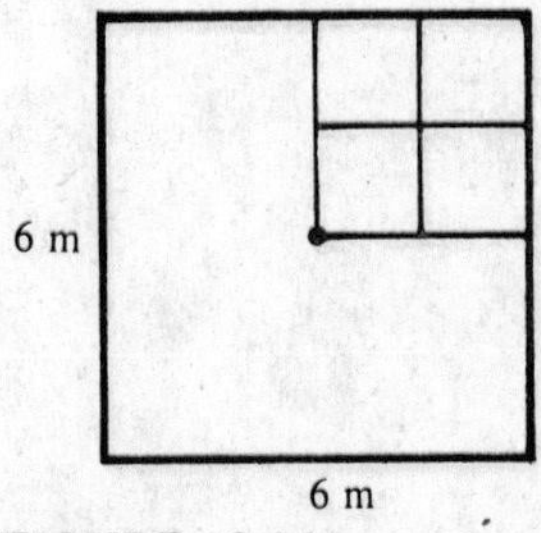

FIGURE 13.14

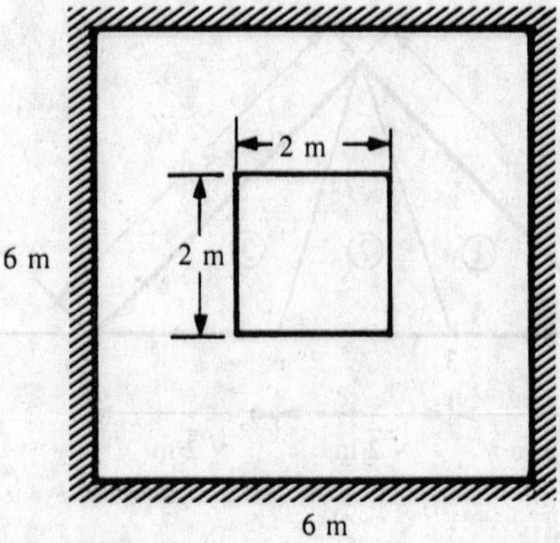

FIGURE 13.15

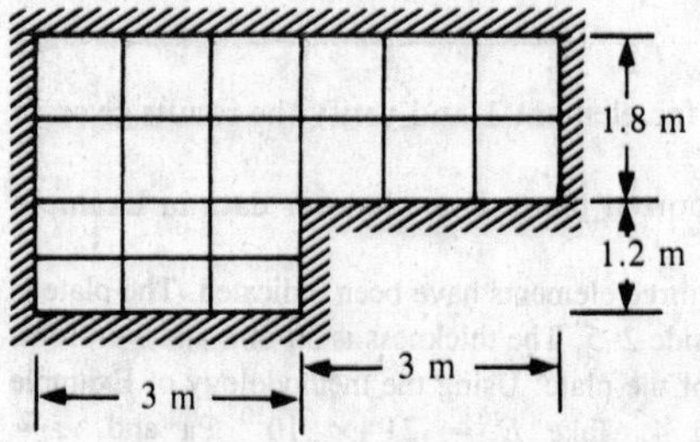

FIGURE 13.16

compute the deflections of the nodes. Use your own computer program or use SAP IV or other "canned" program as directed by your instructor.

*13.14 Develop a program for the computer using Hermite polynomials as set forth in Part D.

*13.15 Use the program developed in Problem 13.14 to solve Problem 13.10.

*13.16 Use the program developed in Problem 13.14 to solve Problem 13.13.

14

TORSION

14.1 INTRODUCTION

Using triangular elements, we shall now examine torsion of noncircular prismatic shafts. One basic difference from our previous undertakings will first be noted. Previously, we used the *total potential energy* as the quadratic functional to develop the key matrix equation for the finite element process. For torsion, we could still do the same by working with the warping function κ described in Section 5.2. Instead, we will find it easier to work with the *stress function* ψ, as discussed in Section 5.3. The *total complementary potential energy* in terms of ψ will then be the quadratic functional, the extremization of which will lead to the basic finite element matrix equation. This step will be the first that we shall take in broadening the finite element process; in Chapter 17 we shall use still other quadratic functionals.

14.2 STIFFNESS AND FORCE MATRICES

You will recall from Section 5.3 that

$$\tau_{xz} = G\alpha \frac{\partial \psi}{\partial y} \qquad \tau_{yz} = -G\alpha \frac{\partial \psi}{\partial x} \tag{14.1}$$

where G is the shear modulus and α is the rate of twist. The boundary-value problem for ψ is as follows:

$$\nabla^2 \psi = -2 \tag{14.2a}$$

$$\psi = 0 \qquad \text{on boundary} \tag{14.2b}$$

$$M = 2G\alpha \iint_A \psi \, dA \tag{14.2c}$$

where the torque transmitted by the shaft is denoted by M. The quadratic functional associated with this boundary-value problem was shown in Section 5.3 to be

$$\pi^* = \frac{G\alpha^2}{2} \iint_A \left[\left(\frac{\partial \psi}{\partial x} \right)^2 + \left(\frac{\partial \psi}{\partial y} \right)^2 - 4\psi \right] dx\, dy \tag{14.3}$$

As indicated at the outset, we shall extremize π^* in small triangular elements (see Fig. 14.1). Taking ψ for element e to be expressed in interpolation functions for three degrees of freedom, we have

$$\psi^e = N_1^e \psi_1^e + N_2^e \psi_2^e + N_3^e \psi_3^e \tag{14.4}$$

where ψ will now be the field variable and the ψ_i are the nodal values of ψ. We then have for π^* for an element:

$$\pi^* = \frac{G\alpha^2}{2} \iint_{A^e} \left\{ \left[\frac{\partial}{\partial x} \left(\sum_{i=1}^{3} N_i \psi_i \right) \right]^2 + \left[\frac{\partial}{\partial y} \left(\sum_{i=1}^{3} N_i \psi_i \right) \right]^2 - 4 \sum_{i=1}^{3} N_i \psi_i \right\} dx\, dy$$

We may rewrite this equation as follows:

$$\pi^* = \frac{G\alpha^2}{2} \iint_{A^e} \left[\left(\sum_{i=1}^{3} \frac{\partial N_i}{\partial x} \psi_i \right)^2 + \left(\sum_{i=1}^{3} \frac{\partial N_i}{\partial y} \psi_i \right)^2 - 4 \sum_{i=1}^{3} N_i \psi_i \right] dx\, dy \tag{14.5}$$

We shall extremize π^* with respect to the nodal values.† As you will recall, this process brings the extremal configuration closer to satisfying the compatibility equations and thus brings us closer to the correct solution for ψ. We thus have, on considering j a free index from 1 to 3 and disregarding $G\alpha^2/2$,

$$\frac{\partial \pi^*}{\partial \psi_j} = 0 = \iint_{A^e} \left[2 \left(\sum_{i=1}^{3} \frac{\partial N_i}{\partial x} \psi_i \right) \frac{\partial N_j}{\partial x} + 2 \left(\sum_{i=1}^{3} \frac{\partial N_i}{\partial y} \psi_i \right) \frac{\partial N_j}{\partial y} - 4N_j \right] dx\, dy$$

$$j = 1, 2, 3$$

†Recall that the use of the stress function in Eq. (14.2) will satisfy the equation of equilibrium automatically.

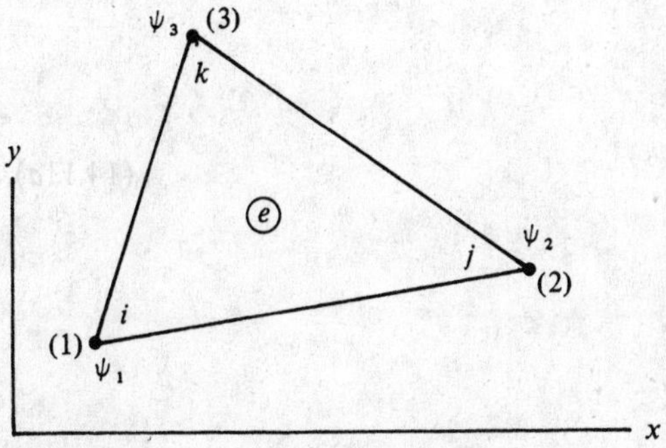

FIGURE 14.1
Triangular element for torsion. Nodes denoted by integers 1, 2, 3 and also by i, j, k.

In matrix notation we have, on canceling a 2,

$$\iint_{A^e} \left\{ \left[\frac{\partial N_i}{\partial x} \frac{\partial N_j}{\partial x} + \frac{\partial N_i}{\partial y} \frac{\partial N_j}{\partial y} \right] \{\psi_i\} - 2\{N_j\} \right\} dx\, dy = 0 \tag{14.6}$$

We can immediately find the terms of the element stiffness matrix:

$$\boxed{K_{ij}^e = \iint_{A^e} \left(\frac{\partial N_i}{\partial x} \frac{\partial N_j}{\partial x} + \frac{\partial N_i}{\partial y} \frac{\partial N_j}{\partial y} \right) dx\, dy} \tag{14.7}$$

The force vector for the element is then:

$$\boxed{\{q\}^e = 2 \iint_{A^e} \{N\}^{(e)}\, dx\, dy} \tag{14.8}$$

14.3 INTERPOLATION MATRICES

The development of the interpolation functions from polynomials of first degree for the triangular element follows very closely what we did in Chapter 12, where we had six degrees of freedom for the element instead of three degrees of freedom as we have here for torsion. Accordingly, we start with a three-term polynomial

$$\psi = \alpha_1 + \alpha_2 x + \alpha_3 y \tag{14.9}$$

for which

$$\lfloor C \rfloor = \lfloor 1, x, y \rfloor \tag{14.10}$$

We can then say

$$\psi_i = \lfloor C(x_i, y_i) \rfloor \{\alpha\}$$
$$\psi_j = \lfloor C(x_j, y_j) \rfloor \{\alpha\} \tag{14.11a}$$
$$\psi_k = \lfloor C(x_k, y_k) \rfloor \{\alpha\}$$

Hence

$$\begin{Bmatrix} \psi_i \\ \psi_j \\ \psi_k \end{Bmatrix} = \begin{bmatrix} 1 & x_i & y_i \\ 1 & x_j & y_j \\ 1 & x_k & y_k \end{bmatrix} \begin{Bmatrix} \alpha_1 \\ \alpha_2 \\ \alpha_3 \end{Bmatrix} \tag{14.11b}$$

Then the coefficient matrix $[A]$ from the above equation is seen to be

$$[A] = \begin{bmatrix} 1 & x_i & y_i \\ 1 & x_j & y_j \\ 1 & x_k & y_k \end{bmatrix} \tag{14.12}$$

The inverse of $[A]$ is

$$[A]^{-1} = \frac{1}{2A} \begin{bmatrix} (x_j y_k - x_k y_j) & (x_k y_i - x_i y_k) & (x_i y_j - x_j y_i) \\ (y_j - y_k) & (y_k - y_i) & (y_i - y_j) \\ (x_k - x_j) & (x_i - x_k) & (x_j - x_i) \end{bmatrix} \tag{14.13}$$

where A is the area of the element. Using Eqs. (12.4) and (12.6), we have for $[A]^{-1}$

$$[A]^{-1} = \frac{1}{2A} \begin{bmatrix} d_i & d_j & d_k \\ \bar{b}_i & \bar{b}_j & \bar{b}_k \\ \bar{a}_i & \bar{a}_j & \bar{a}_k \end{bmatrix} \tag{14.14}$$

We can now get the interpolation functions by using Eq. (12.21). Thus we have

$$[N] = \lfloor C(x, y) \rfloor [A]^{-1} = \frac{1}{2A} \lfloor 1, x, y \rfloor \begin{bmatrix} d_i & d_j & d_k \\ \bar{b}_i & \bar{b}_j & \bar{b}_k \\ \bar{a}_i & \bar{a}_j & \bar{a}_k \end{bmatrix}$$

$$= \frac{1}{2A} \lfloor (d_i + x\bar{b}_i + y\bar{a}_i), (d_j + x\bar{b}_j + y\bar{a}_j), (d_k + x\bar{b}_k + y\bar{a}_k) \rfloor \tag{14.15}$$

and

$$N_i = \frac{1}{2A}(d_i + x\bar{b}_i + y\bar{a}_i)$$

$$N_j = \frac{1}{2A}(d_j + x\bar{b}_j + y\bar{a}_j) \qquad (14.16)$$

$$N_k = \frac{1}{2A}(d_k + x\bar{b}_k + y\bar{a}_k)$$

Now going back to Eq. (14.7) and substituting the above interpolation matrices into the integral, we see that the integrand is a constant after carrying out the partial differentiations. We then have for the terms of $[K]^e$ the following results:

$$K_{ij}^e = \frac{1}{4A}(\bar{b}_i\bar{b}_j + \bar{a}_i\bar{a}_j) \qquad (14.17)$$

so that the stiffness matrix becomes

$$[K]^e = \frac{1}{4A}\begin{bmatrix} (\bar{a}_i\bar{a}_i + \bar{b}_i\bar{b}_i) & (\bar{a}_i\bar{a}_j + \bar{b}_i\bar{b}_j) & (\bar{a}_i\bar{a}_k + \bar{b}_i\bar{b}_k) \\ (\bar{a}_j\bar{a}_i + \bar{b}_j\bar{b}_i) & (\bar{a}_j\bar{a}_j + \bar{b}_j\bar{b}_j) & (\bar{a}_j\bar{a}_k + \bar{b}_j\bar{b}_k) \\ (\bar{a}_k\bar{a}_i + \bar{b}_k\bar{b}_i) & (\bar{a}_k\bar{a}_j + \bar{b}_k\bar{b}_j) & (\bar{a}_k\bar{a}_k + \bar{b}_k\bar{b}_k) \end{bmatrix} \qquad (14.18)$$

Now let us turn to the force vector $\{q\}$ [see Eq. (14.8)]. Note from Eqs. (14.16) and (12.8) that

$$[N] = \lfloor L_i, L_j, L_k \rfloor \qquad (14.19)$$

That is, the interpolation functions equal the triangular coordinates. Hence we can say for $\{q\}^e$ [see Eq. (14.8)]:

$$q_i = 2 \iint L_i \, dA$$

$$q_j = 2 \iint L_j \, dA \qquad (14.20)$$

$$q_k = 2 \iint L_k \, dA$$

Now, using the integration formula given by Eq. (12.48), we can say for $\{q\}^e$

$$\{q\}^e = 2(2A)\frac{1}{3!}\begin{Bmatrix} 1 \\ 1 \\ 1 \end{Bmatrix} = \frac{2A}{3}\begin{Bmatrix} 1 \\ 1 \\ 1 \end{Bmatrix} \qquad (14.21)$$

We thus have all the formulations needed for solving torsion problems with finite elements. The following section illustrates the procedure for a small number of elements.

14.4 SOLUTION OF TORSION PROBLEMS

In this section we shall consider the solution of torsion problems. We first consider a simple cross section for which only four elements are used.

EXAMPLE 14.1 We will consider the torsion of a shaft with a square cross section (Fig. 14.2). Note that we need only work with one-eight of the cross section (tinted) because of the symmetry of the cross section about the diagnosis. The shear modulus G is 21×10^{11} Pa for this case. The rate of twist α for the shaft is 0.003 rad/m. We wish to determine an approximation for the stress distribution function.

A four-element grid is shown in Fig. 14.3. We shall work with one element in detail and then give the results for the remaining elements. Thus, for element 2 we have

$$\begin{cases} x_i = x_2 = 0.025 \text{ m} \\ x_j = x_4 = 0.05 \text{ m} \\ x_k = x_3 = 0.025 \text{ m} \end{cases} \qquad \begin{cases} y_i = y_2 = 0 \text{ m} \\ y_j = y_4 = 0 \text{ m} \\ y_k = y_3 = 0.025 \text{ m} \end{cases}$$

$$\begin{cases} \bar{a}_i = x_k - x_j = 0.025 - 0.05 = -0.025 \text{ m} \\ \bar{a}_j = x_i - x_k = 0.025 - 0.025 = 0 \text{ m} \\ \bar{a}_k = x_j - x_i = 0.05 - 0.025 = 0.025 \text{ m} \end{cases} \qquad \begin{cases} \bar{b}_i = y_j - y_k = 0 - 0.025 = -0.025 \text{ m} \\ \bar{b}_j = y_k - y_i = 0.025 - 0 = 0.025 \text{ m} \\ \bar{b}_k = y_i - y_j = 0 - 0 = 0 \text{ m} \end{cases}$$

We can now get $[K]^2$ by substituting the above results into Eq. (14.18). We thus have

$$[K]^2 = \frac{1}{2}\begin{bmatrix} [(-25)(-25) + (-25)(-25)] & [(-25)(0) + (-25)(25)] & [(-25)(25) + (-25)(0)] \\ [(0)(-25) + (25)(-25)] & [(0)(0) + (25)(25)] & [(0)(25) + (25)(0)] \\ [(25)(-25) + (0)(-25)] & [(25)(0) + (0)(25)] & [(25)(25) + (0)(0)] \end{bmatrix} \times 10^{-3}$$

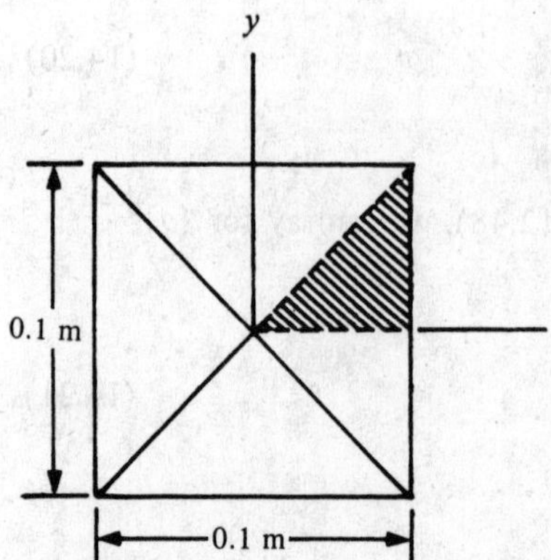

FIGURE 14.2
Square shaft under torsion.

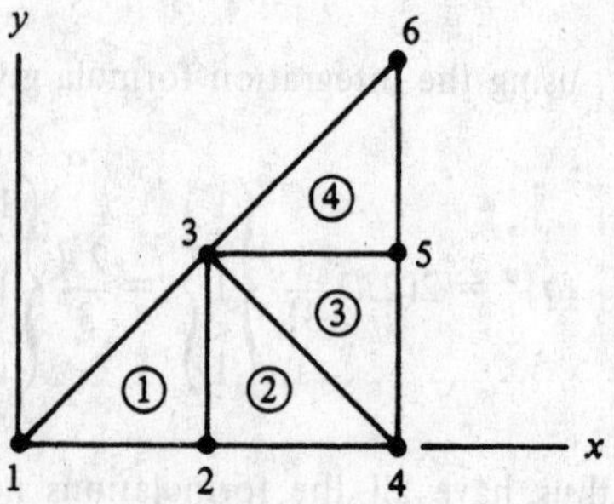

FIGURE 14.3
Four-element grid.

$$[K]^{(2)} = \begin{bmatrix} 1 & -\frac{1}{2} & -\frac{1}{2} \\ -\frac{1}{2} & \frac{1}{2} & 0 \\ -\frac{1}{2} & 0 & \frac{1}{2} \end{bmatrix} \qquad (a)$$

The other three element stiffness matrices are

$$[K]^{(1)} = \begin{bmatrix} \frac{1}{2} & -\frac{1}{2} & 0 \\ -\frac{1}{2} & 1 & -\frac{1}{2} \\ 0 & -\frac{1}{2} & \frac{1}{2} \end{bmatrix} \qquad (b)$$

$$[K]^{(3)} = \begin{bmatrix} \frac{1}{2} & -\frac{1}{2} & 0 \\ -\frac{1}{2} & 1 & -\frac{1}{2} \\ 0 & -\frac{1}{2} & \frac{1}{2} \end{bmatrix} \qquad (c)$$

$$[K]^{(4)} = \begin{bmatrix} \frac{1}{2} & -\frac{1}{2} & 0 \\ -\frac{1}{2} & 1 & -\frac{1}{2} \\ 0 & -\frac{1}{2} & \frac{1}{2} \end{bmatrix} \qquad (d)$$

The global stiffness matrix may now be readily assembled:

$$[K] = \begin{array}{c} \begin{array}{cccccc} 1 & 2 & 3 & 4 & 5 & 6 \end{array} \\ \begin{bmatrix} \frac{1}{2} & -\frac{1}{2} & 0 & 0 & 0 & 0 \\ -\frac{1}{2} & 2 & -1 & -\frac{1}{2} & 0 & 0 \\ 0 & -1 & 2 & 0 & -1 & 0 \\ 0 & -\frac{1}{2} & 0 & 1 & -\frac{1}{2} & 0 \\ 0 & 0 & -1 & -\frac{1}{2} & 2 & -\frac{1}{2} \\ 0 & 0 & 0 & 0 & -\frac{1}{2} & \frac{1}{2} \end{bmatrix} \end{array} \begin{array}{c} \\ 1 \\ 2 \\ 3 \\ 4 \\ 5 \\ 6 \end{array} \qquad (e)$$

Now let us get the force vector for this quadrant. Note that all the areas are equal to each other, having the common value of 3.125×10^{-4} m^2. From Eq. (14.21), we can readily conclude that

$$\{q\} = 2.08 \times 10^{-4} \begin{Bmatrix} 1 \\ 2 \\ 4 \\ 2 \\ 2 \\ 1 \end{Bmatrix} \qquad (f)$$

Next we note that $\psi_4 = \psi_5 = \psi_6 = 0$ because of the boundary conditions. We can then give the reduced matrix equation, using Eqs. (*e*) and (*f*) and deleting the fourth, fifth, and sixth rows and columns. We then get

$$\begin{bmatrix} \frac{1}{2} & -\frac{1}{2} & 0 \\ -\frac{1}{2} & 2 & -1 \\ 0 & -1 & 2 \end{bmatrix} \begin{Bmatrix} \psi_1 \\ \psi_2 \\ \psi_3 \end{Bmatrix} = 2.08 \times 10^{-4} \begin{Bmatrix} 1 \\ 2 \\ 4 \end{Bmatrix} \tag{g}$$

Solving, we have

$$\begin{aligned} \psi_1 &= 1.46 \times 10^{-3}\ \text{m}^2/\text{rad} \\ \psi_2 &= 1.04 \times 10^{-3}\ \text{m}^2/\text{rad} \\ \psi_3 &= 0.938 \times 10^{-3}\ \text{m}^2/\text{rad} \end{aligned} \tag{h}$$

In the example above we determined the stress function ψ at six discrete points. We want the stresses inside the element. Now the stresses are related to the nodal values of ψ via Eqs. (14.1). Thus we have

$$\{\sigma\} = \begin{Bmatrix} \tau_{xz} \\ \tau_{yz} \end{Bmatrix} = G\alpha \begin{Bmatrix} \dfrac{\partial \psi}{\partial y} \\ -\dfrac{\partial \psi}{\partial x} \end{Bmatrix} = G\alpha \begin{Bmatrix} \dfrac{\partial}{\partial y} \\ -\dfrac{\partial}{\partial x} \end{Bmatrix} \psi \tag{14.22}$$

But we can express ψ in terms of its nodal values as follows:

$$\psi = [N]\,\{\psi_i\} \tag{14.23}$$

Inserting into Eq. (14.22), we have

$$\{\sigma\} = G\alpha \begin{Bmatrix} \dfrac{\partial}{\partial y} \\ -\dfrac{\partial}{\partial x} \end{Bmatrix} [N]\,\{\psi_i\} \tag{14.24}$$

We can calculate stresses for element e by using Eq. (14.16):

$$\{\sigma\} = G\alpha \begin{Bmatrix} \dfrac{\partial}{\partial y} \\ -\dfrac{\partial}{\partial x} \end{Bmatrix} \frac{1}{2A} \lfloor (d_i + x\bar{b}_i + y\bar{a}_i), \quad (d_j + x\bar{b}_j + y\bar{a}_j), \quad (d_k + x\bar{b}_k + y\bar{a}_k) \rfloor^e \{\psi_i\} \tag{14.25}$$

$$\{\sigma\} = \frac{G\alpha}{2A^e} \begin{bmatrix} \bar{a}_i & \bar{a}_j & \bar{a}_k \\ -\bar{b}_i & -\bar{b}_j & -\bar{b}_k \end{bmatrix}^e \{\psi_i\} \tag{14.26}$$

EXAMPLE 14.2 Compute the stresses and the torque M in Example 14.1. For element 2 we have, using Eq. (14.26),

$$\begin{Bmatrix} \tau_{xz} \\ \tau_{yz} \end{Bmatrix}^{(2)} = \frac{(21 \times 10^{10})(0.003)}{(2)(3.125 \times 10^{-4})} \begin{bmatrix} -0.025 & 0 & 0.025 \\ 0.025 & -0.025 & 0 \end{bmatrix} \begin{Bmatrix} \psi_2 \\ \psi_4 \\ \psi_3 \end{Bmatrix}$$

We get, after using five-decimal accuracy during computations

$$\begin{Bmatrix} \tau_{xz} \\ \tau_{yz} \end{Bmatrix}^{(2)} = \begin{Bmatrix} -2.625 \times 10^6 \\ 2.625 \times 10^7 \end{Bmatrix} \text{Pa}$$

and for the other elements

$$\begin{Bmatrix} \tau_{xz} \\ \tau_{yz} \end{Bmatrix}^{(1)} = \begin{Bmatrix} -2.625 \times 10^6 \\ 1.05 \times 10^7 \end{Bmatrix} \text{Pa}$$

$$\begin{Bmatrix} \tau_{xz} \\ \tau_{yz} \end{Bmatrix}^{(3)} = \begin{Bmatrix} 0 \\ 2.363 \times 10^7 \end{Bmatrix} \text{Pa}$$

$$\begin{Bmatrix} \tau_{xz} \\ \tau_{yz} \end{Bmatrix}^{(4)} = \begin{Bmatrix} 0 \\ 2.363 \times 10^7 \end{Bmatrix} \text{Pa}$$

Finally, we can easily determine the torque M for this problem by using Eq. (14.2c). Replacing ψ by $[N]$ $\{\psi_i\}$ in the integrand, we can get M for each element. Thus for element 2 we have, using area coordinates [see Eqs. (14.4) and (14.19)],

$$M^{(2)} = 2G\alpha \iint \lfloor L_2, L_4, L_3 \rfloor \begin{Bmatrix} \psi_2 \\ \psi_4 \\ \psi_3 \end{Bmatrix} dA$$

We can now use Eq. (12.48) to carry out the integrations. Thus for $M^{(2)}$ we obtain

$$M^{(2)} = \left[2(21 \times 10^{10})\ (0.003)\right]\left[2(3.125 \times 10^{-4})\,\frac{1}{3!}\right]\ (\psi_2 + \psi_4 + \psi_3)$$

$$= 259.8 \text{ N-m}$$

The total torque M is then 7656.2 N-m, which is eight times the sum of the torques in the half-quadrant where we have been working. The theoretical value for this case is 8911.4 N-m. Using smaller elements would greatly improve the results, as we will see in the next example, where we use a finer mesh for this problem.

EXAMPLE 14.3 We now work the same problem as in Example 14.1, using a fine mesh so as to get more exact computation of the stresses.

Using one-half of a quadrant, we use the gridding program described in Chapter 12 and presented in Appendix III to develop a 64-element mesh, as shown in Fig. 14.4. A full computer program has then been used to get the shear stresses τ_{xz}, τ_{yz} and the torque M for

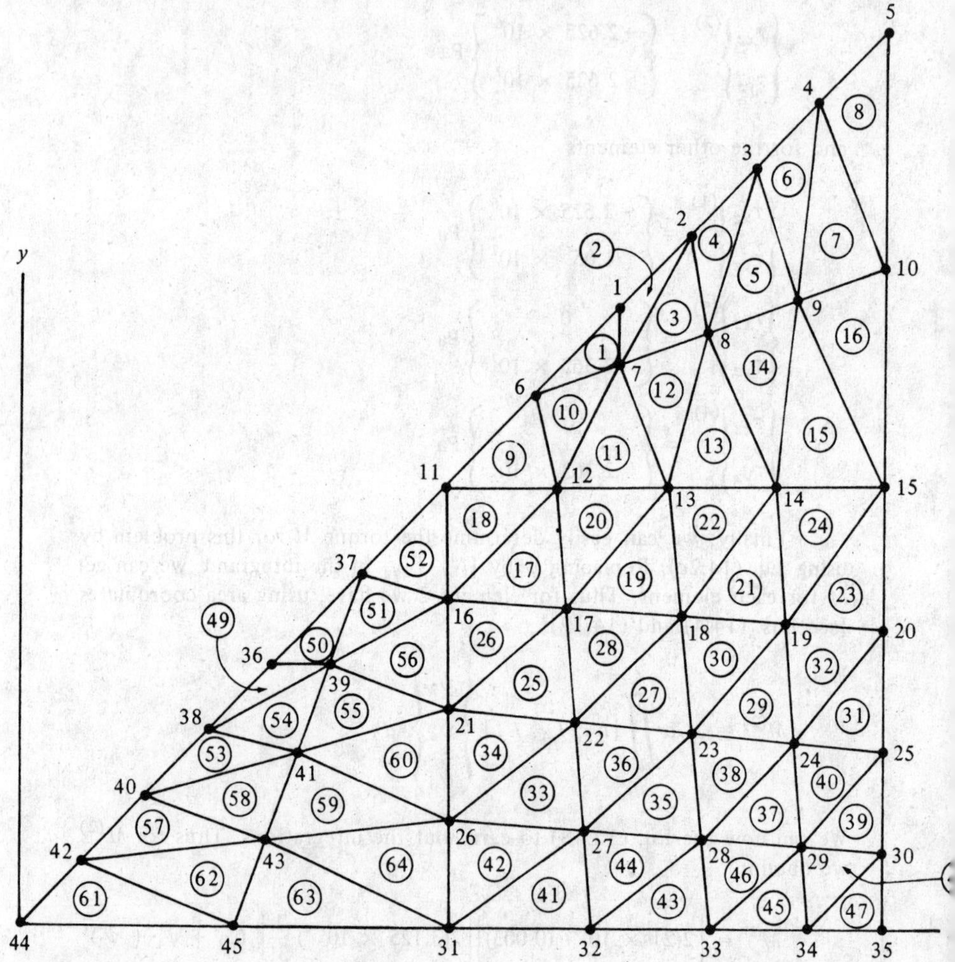

FIGURE 14.4
Gridding for torsion: 64 elements, 45 nodes.

each element. The largest total shear comes at element 47, as is to be expected, with the value:

$$\tau_{max} = 3.855 \times 10^7 \text{ Pa}$$

The total torque for the shaft is

$$M_z = 8\,(1.091 \times 10^3) = 8730.4 \text{ N-m}$$

From Chapter 5, we can give an analytical solution for the maximum shear stress as

$$(\tau_{max})_{analvt} = 3.9375 \times 10^7 \text{ Pa}$$

We thus have a 2% error. The torques check very closely (2.03% error).

In the preceding examples, we knew the angle of twist and we solved for the torque. Suppose that the torque is known; how do we proceed? In that case we set $\alpha = 1$ and solve as we did in the preceding paragraphs for shear stresses and the torque. Now form the ratio β of the given torque M_{given} divided by the computed torque $(M)_{\alpha=1}$; that is

$$\frac{M_{given}}{(M)_{\alpha=1}} = \beta$$

The desired stresses τ for M_{given} is then found from the computed stress for $\alpha = 1$ as follows:

$$\tau = \beta(\tau)_{\alpha=1} \tag{14.27}$$

The desired rate of twist is then simply β.

14.5 CLOSURE

In this chapter we presented a brief treatment of torsion. The matrix working equation was developed by using the total complementary energy functional. We are getting a glimpse here of how the quadratic functional of Chapter 3 plays a role in the finite element process. We shall expand on this theme in Chapter 17, when we study heat transfer and fluid flow applications of the finite element procedure.

The reader will note that the topics covered in Part III of the text parallel those in Part II. Accordingly, we have yet to talk about finite elements for elastic stability and for dynamics. This will be done in the two succeeding chapters, where introductory studies will be carried out.

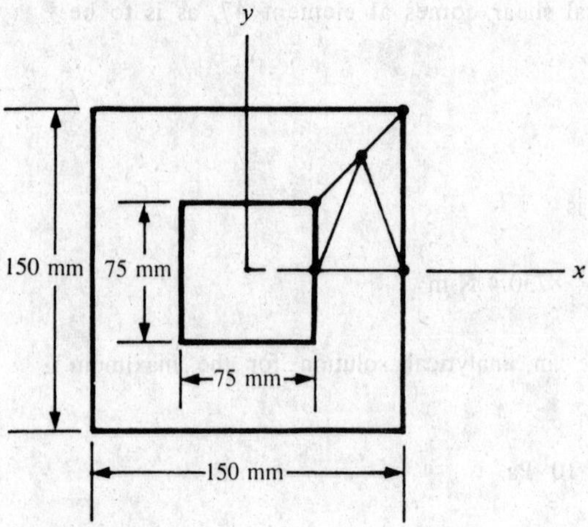

FIGURE 14.5

PROBLEMS

14.1 Using one-eighth of the cross section (see Fig. 14.5), find an approximation of the stress. The shear modulus G is 14×10^{10} Pa. The rate of twist is 0.003 rad/m. Use three elements as shown. Find the torque M for the above rate of twist. Hint: Fill in center square with material having very low modulus G (of order 10^{-3}) and form additional triangular elements there.

14.2 In Problem 14.1 a torque of 600 N-m is applied. What are the stress and rate of twist? Use one-half of a quadrant and the elements shown in Fig. 14.5.

14.3 A shaft with a triangular cross section is shown in Fig. 14.6. Using three elements as shown, what are the stress and the torque? Use $G = 14 \times 10^{10}$ Pa and a rate of twist of 0.006 rad/m.

14.4 In Problem 14.3 a torque of 200 N-m is applied. What are the stresses and the rate of twist if three elements are used, as shown in Fig. 14.6?

14.5 Using the gridding program in Appendix III, form a grid with about 20 elements for the cross section shown in Fig. 14.7.

*14.6 Develop a program for torsion for the computer. The program should give stresses in elements and the rate of twist in terms of the applied moment. Use the program on Examples 14.1 and 14.2 for an applied torque of 1250 N-m.

14.7 Using the grid formed in Problem 14.5 and the program developed in Problem 14.6, find the stresses and rate of twist for a torque of 600 N-m. Take $G = 10.5 \times 10^{10}$ Pa.

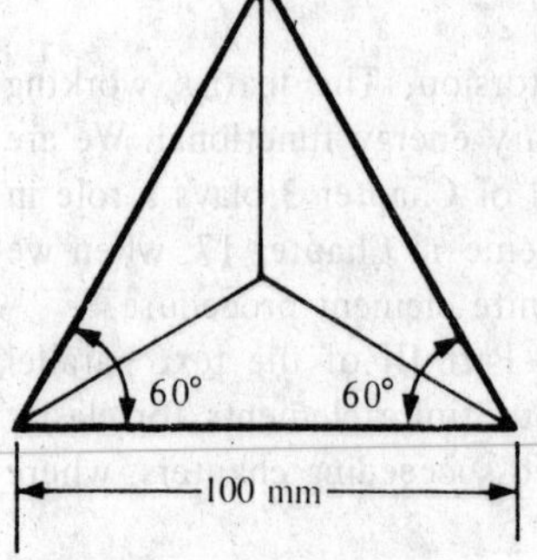

FIGURE 14.6

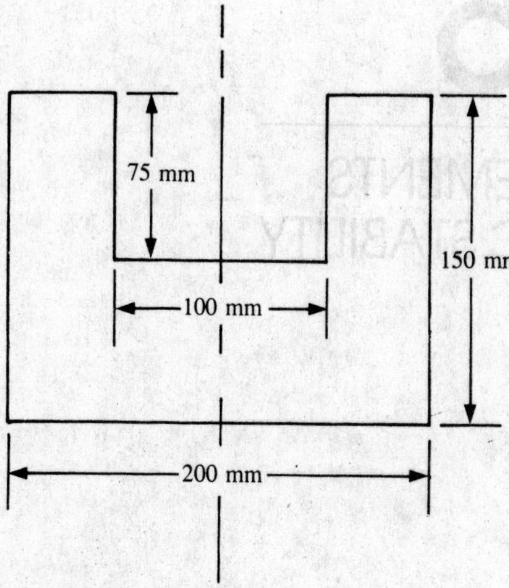

FIGURE 14.7

14.8 Grid the shaft in Problem 14.1 with about 16 elements per half-quadrant, and find the stresses and rate of twist for a torque of 900 N-m. Solve by using the computer program developed in Problem 14.6.

15

FINITE ELEMENTS FOR ELASTIC STABILITY

15.1 INTRODUCTION

We shall briefly examine the use of finite elements for linearized elastic stability analysis of beams and plates. While it is true that buckling of a structure involves nonlinear considerations, a linearized eigenvalue analysis is still very useful to the designer since the results signal an upper bound to structural loading (see Chapter 8). In Part A of this chapter we shall consider beams, and in Part B we shall consider flat plates.

Part A
Beam-Columns† and Columns

15.2 DEVELOPMENT OF THE FINITE ELEMENT EQUATION

As in Chapter 10, we consider an element with four degrees of freedom, as shown in Fig. 15.1. The generalized forces acting from nodes on the element are $\lfloor q_1, q_2, q_3, q_4 \rfloor$. The nodal displacements are $\lfloor w_1, \theta_1, w_2, \theta_2 \rfloor$. Furthermore, the shear deformation is neglected in the ensuing discussion. Hence, only axial strain ϵ_{xx} is present. The total potential energy was presented in Section 8.6 [see Eq. (8.52)] for *columns*, and is restated here as

$$\pi^e = \frac{1}{2}\int_L \left[EI\left(\frac{d^2 w}{dx^2}\right)^2 + P\left(\frac{dw}{dx}\right)^2 \right] dx \qquad (15.1)$$

† A beam-column has transverse loads in addition to an axial load P.

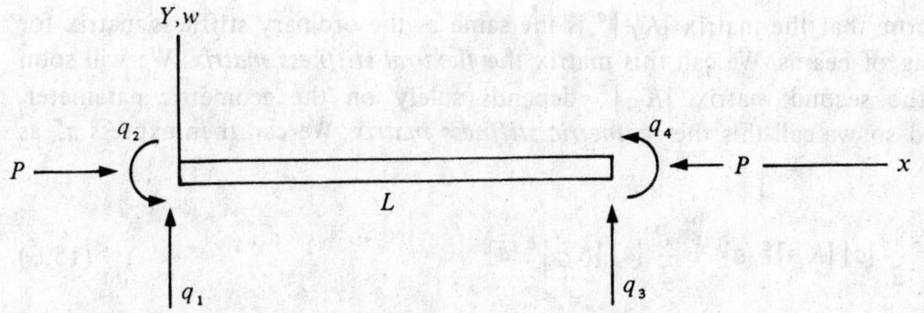

FIGURE 15.1
Element with generalized forces acting from nodes on element.

For a finite element consideration, we express the field variable w in terms of interpolation functions and nodal displacements. Thus:

$$w = [N_1, N_2, N_3, N_4] \begin{Bmatrix} w_1 \\ \theta_1 \\ w_2 \\ \theta_2 \end{Bmatrix} = \lfloor N \rfloor \{a\} \tag{15.2}$$

We now substitute for w in Eq. (15.1), using Eq. (15.2):

$$\pi^e = \frac{1}{2} \int_L \left\{ EI(\lfloor N'' \rfloor \{a\})(\lfloor N'' \rfloor \{a\}) + P(\lfloor N' \rfloor \{a\})(\lfloor N' \rfloor \{a\}) \right\} dx$$

$$= \frac{1}{2} \int_L \left\{ EI(\{a\}^T \{N''\})(\lfloor N'' \rfloor \{a\}) + P(\{a\}^T \{N'\})(\lfloor N' \rfloor \{a\}) \right\} dx \tag{15.3}$$

Extracting the nodal displacement and P from under the integral signs, we have

$$\pi^e = \frac{\{a\}^T}{2} \left(\int_L EI\{N''\} \lfloor N'' \rfloor \, dx \right) \{a\} + \frac{\{a\}^T}{2} \left(P \int_L \{N'\} \lfloor N' \rfloor \, dx \right) \{a\} \tag{15.4}$$

We now set out the following definitions:

$$[K_F]^e = \int_L EI\{N''\} \lfloor N'' \rfloor \, dx \tag{15.5a}$$

$$[K_G]^e = \int_L \{N'\} \lfloor N' \rfloor \, dx \tag{15.5b}$$

You will note that the matrix $[K_F]^e$ is the same as the ordinary stiffness matrix for the bending of beams. We call this matrix the *flexural stiffness matrix*. We will soon see that the second matrix $[K_G]^e$ depends solely on the geometric parameter, length, and so we call this the *geometric stiffness matrix*. We can then express π^e as follows:

$$\pi^e = \frac{1}{2} \lfloor a \rfloor [K_F]^e \{a\} + \frac{P}{2} \lfloor a \rfloor [K_G]^e \{a\} \tag{15.6}$$

Now we turn to the potential of the applied transverse load in order to consider beam-columns. From our work in Chapter 10 on the bending of beams, we see that we can replace the generalized forces $\{q\}$ from the nodes onto the elements by the actual external loads F and actual external torques M acting *on the nodes*. We show these external forces and torques in Fig. 15.2 along with force P. We thus have for the potential of the transverse loads:

$$V^e = -\lfloor w_1, \theta_1, w_2, \theta_2 \rfloor \begin{Bmatrix} F_1 \\ M_1 \\ F_2 \\ M_2 \end{Bmatrix} = -\lfloor a \rfloor \{F\} \tag{15.7}$$

Hence, we have for π^e, the total potential energy for beam-columns, including the transverse loads,

$$\pi^e = \frac{1}{2} \lfloor a \rfloor [K_F]^e \{a\} + \frac{P}{2} \lfloor a \rfloor [K_G]^e \{a\} - \lfloor a \rfloor \{F\} \tag{15.8}$$

We may form the global equation for the beam-column in the usual way. Thus, setting the first variation equal to zero, we have

$$\begin{aligned} \delta^{(1)}\pi = 0 = \frac{1}{2} \lfloor \delta a \rfloor [K_F] \{a\} + \frac{1}{2} \lfloor a \rfloor [K_F] \{\delta a\} \\ + \frac{P}{2} \lfloor \delta a \rfloor [K_G] \{a\} + \frac{P}{2} \lfloor a \rfloor [K_G] \{\delta a\} - \lfloor \delta a \rfloor \{F\} \end{aligned} \tag{15.9}$$

Because the two stiffness matrices are symmetric, we can interchange $\lfloor a \rfloor$ and $\{\delta a\}$ to arrive at the following equation (on canceling out $\lfloor \delta a \rfloor$):

$$[K_F] \{a\} + P[K_G] \{a\} = \{F\} \tag{15.10}$$

This is the beam-column equation for finite elements, and we can now proceed in the usual way.

As for the interpolation functions, we can use the results of Chapter 10 for beams, where we had [see Eq. (10.13)]

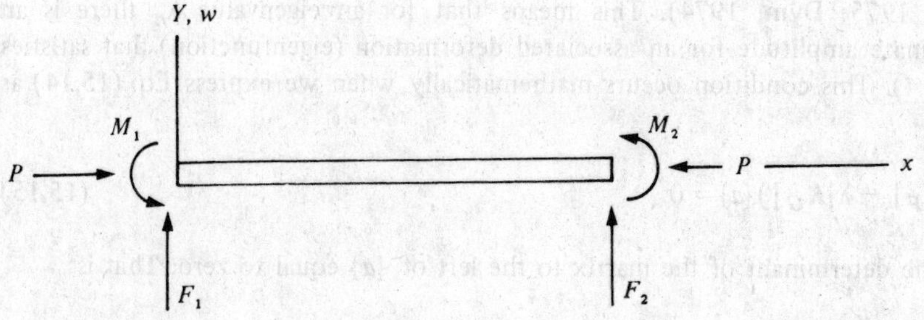

FIGURE 15.2
Element showing external force systems at nodes.

$$\lfloor N \rfloor = \left[\left(1 - 3\frac{x^2}{L^2} + 2\frac{x^3}{L^3}\right), \left(x - 2\frac{x^2}{L} + \frac{x^3}{L^2}\right), \left(3\frac{x^2}{L^2} - 2\frac{x^3}{L^3}\right), \left(-\frac{x^2}{L} + \frac{x^3}{L^2}\right) \right] \tag{15.11}$$

From Eq. (10.22), we then have the flexural stiffness matrix

$$[K_F]^e = \frac{EI}{L^3} \begin{bmatrix} 12 & 6L & -12 & 6L \\ 6L & 4L^2 & -6L & 2L^2 \\ -12 & -6L & 12 & -6L \\ 6L & 2L^2 & -6L & 4L^2 \end{bmatrix} \tag{15.12}$$

Using Eq. (15.11) in Eq. (15.5*b*), we have

$$[K_G]^e = \frac{1}{30L} \begin{bmatrix} 36 & 3L & -36 & 3L \\ 3L & 4L^2 & -3L & -L^2 \\ 36 & -3L & 36 & -3L \\ 3L & -L^2 & -3L & 4L^2 \end{bmatrix} \tag{15.13}$$

Equation (15.10) is the key equation, together with Eqs. (15.12) and (15.13), for solving beam-column problems. If we have no external transverse loads, we have a simple column. Then $\{F\}$ is zero for this case and P is to be considered undetermined. Denoting P as $-\lambda$, we have for Eq. (15.10):

$$[K_F]\{a\} = \lambda[K_G]\{a\} \tag{15.14}$$

We thus have the familiar eigenvalue, eigenfunction problem that we considered in Section 7.13, where ω^2 was the eigenvalue. What we are seeking here are values of λ and the associated nodal displacement vectors that achieve neutral equilibrium

(Shames, 1975; Dym, 1974). This means that for an eigenvalue λ_n there is an indeterminate amplitude for an associated deformation (eigenfunction) that satisfies Eq. (15.14). This condition occurs mathematically when we express Eq. (15.14) as follows:

$$([K_F] - \lambda[K_G])\{a\} = 0 \tag{15.15}$$

and set the determinant of the matrix to the left of $\{a\}$ equal to zero. That is:

$$|[K_F] - \lambda[K_G]| = 0 \tag{15.16}$$

For n elements we will get an $n+1$ degree polynomial in λ. The smallest root of this equation will be the desired approximation of the first buckling load. By using this value of λ in Eq. (15.15), we can get a set of ratios for the nodal displacement components so as to calculate the first buckling mode shape. Higher mode approximations can be found similarly. We illustrate the procedure in the following example.

EXAMPLE 15.1 Figure 15.3 shows a clamped free column with an indeterminate amplitude δ of buckling deformation (dashed lines). The origin of our reference is set at the tip of the column at a distance δ from the undeformed geometry.† We will use only one element, namely the column itself. The boundary conditions for this element are the following homogeneous conditions:

$$\begin{aligned} w_{x=0} &\equiv a_1 = 0 \\ \left(\frac{dw}{dx}\right)_{x=L} &\equiv a_4 = 0 \end{aligned} \tag{a}$$

This means that the first and fourth rows and columns of Eq. (15.14) [and hence of Eq. (15.16)] can be deleted. Hence, going to Eq. (15.16) and using Eqs. (15.12) and (15.13), we have

†In this way we will be getting the buckling load for a simply supported column of length $2L$, as shown below. For this case we have two elements.

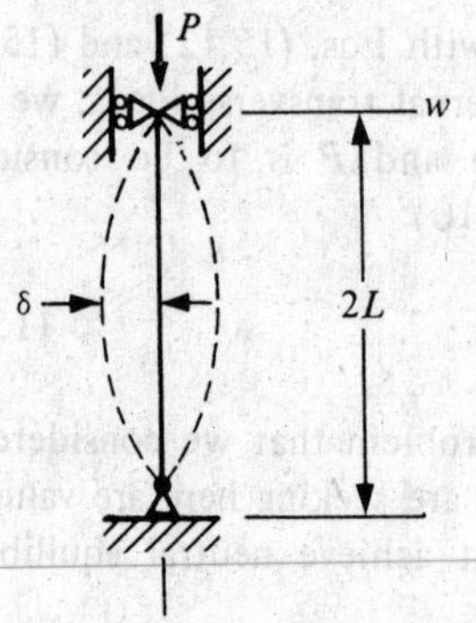

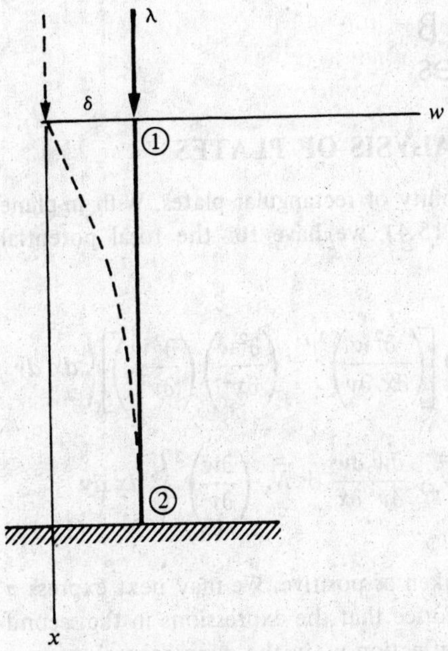

FIGURE 15.3
Clamped free column.

$$\left| \frac{EI}{L^3}\begin{bmatrix} 4L^2 & -6L \\ -6L & 12 \end{bmatrix} - \frac{\lambda}{30L}\begin{bmatrix} 4L^2 & -3L \\ -3L & 36 \end{bmatrix} \right| = 0 \qquad (b)$$

We then get, on combining the matrices and carrying out the evaluation of the determinant,

$$\left(\frac{4EI}{L} - \frac{2\lambda L}{15}\right)\left(\frac{12EI}{L^3} - \frac{6\lambda}{5L}\right) - \left(-\frac{6EI}{L^2} + \frac{\lambda}{10}\right)^2 = 0 \qquad (c)$$

The critical value of λ, and hence the critical force P_{cr}, is then

$$\lambda_{cr} = P_{cr} = 2.486\frac{EI}{L^2} \qquad (d)$$

The exact value for the buckling load for the clamped-free column is

$$P_{cr} = \frac{\pi^2}{4}\frac{EI}{L^2} = 2.467\frac{EI}{L^2} \qquad (e)$$

It is to be pointed out that the procedure presented thus far, namely Eq. (15.16), is not efficient for large-order systems. A better procedure is to rewrite Eq. (15.14) in the following manner (Gallagher, 1975):

$$\frac{1}{\lambda}\{a\} = [K_F]^{-1}[K_G]\{a\} \qquad (15.17)$$

This equation is easily solved by an iterative procedure.

Part B
Plates

15.3 LINEARIZED STABILITY ANALYSIS OF PLATES

We now turn to the case of the elastic stability of rectangular plates. With in-plane edge loads of $\bar{\bar{N}}_x$, $\bar{\bar{N}}_y$, and $\bar{\bar{N}}_{xy}$ (see Fig. 15.4), we have for the total potential energy [see Eq. (8.107)]:†

$$\pi = \frac{D}{2}\int_{-a}^{a}\int_{-b}^{b}\left\{(\nabla^2 w)^2 + 2(1-\nu)\left[\left(\frac{\partial^2 w}{\partial x\,\partial y}\right)^2 - \left(\frac{\partial^2 w}{\partial x^2}\right)\left(\frac{\partial^2 w}{\partial y^2}\right)\right]\right\} dx\,dy$$
$$-\frac{1}{2}\int_{-a}^{a}\int_{-b}^{b}\left\{\bar{\bar{N}}_x\left(\frac{\partial w}{\partial x}\right)^2 + 2\,\bar{\bar{N}}_{xy}\frac{\partial w}{\partial y}\frac{\partial w}{\partial x} + \bar{\bar{N}}_y\left(\frac{\partial w}{\partial y}\right)^2\right\} dx\,dy \tag{15.18}$$

where compressive loadings $\bar{\bar{N}}_x$ and $\bar{\bar{N}}_y$ are taken as positive. We may next express π in terms of nodal displacements.‡ You will notice that the expressions in the second integral in Eq. (15.18) are in terms of the deflection w, in the same general manner as the first integral. Hence, observing Eq. (13.13*b*), the second integral above can be given as three separate expressions, each in the same form as the first integral, namely, (const) $\{a\}^T[\ ]\{a\}$. We thus get for *constant values of* $\bar{\bar{N}}_x$, $\bar{\bar{N}}_y$, and $\bar{\bar{N}}_{xy}$ over each side of the plate element

$$\pi = \frac{1}{2}\{a\}^T[K_F]\{a\} - \frac{\bar{\bar{N}}_x}{2}\{a\}^T[K_{Gx}]\{a\}$$
$$-\frac{\bar{\bar{N}}_y}{2}\{a\}^T[K_{Gy}]\{a\} - \frac{\bar{\bar{N}}_{xy}}{2}\{a\}^T[K_{Gxy}]\{a\} \tag{15.19}$$

where $[K_F]$ is the flexural stiffness matrix used in Chapter 13 on plates. The other stiffness matrices are analogous to the geometric stiffness matrices of the preceding section of this chapter. (This form of π is shown as Eq. (17.17), following which is a short, simple justification.)

These new geometric stiffness matrices become

$$K_{Gx} = \iint \lfloor N_x \rfloor \{N_x\}\,dx\,dy \tag{15.20a}$$

$$K_{Gy} = \iint \lfloor N_y \rfloor \{N_y\}\,dx\,dy \tag{15.20b}$$

$$K_{Gxy} = \iint \lfloor N_x \rfloor \{N_y\}\,dx\,dy \tag{15.20c}$$

†Please do not confuse the interpolation function elements N_i with the in-plane edge loads $\bar{\bar{N}}_x$, etc.

‡As we will soon see, for each natural frequency ω_i the nodal displacements take on a set of values $\{a\}_i$ having a certain set of ratios between the terms. This relative set of values is the ith mode shape or eigenvector. This is the same as the mode shape concept in free vibrations of lumped systems with multi-degrees of freedom.

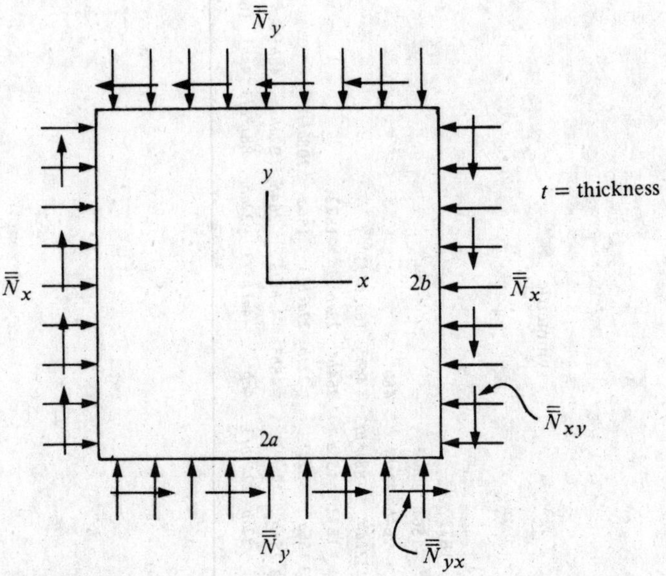

FIGURE 15.4
Rectangular plate with in-plane edge loading.

where $\lfloor N_x \rfloor$ indicates partial derivative of $\lfloor N \rfloor$ with respect to x, etc. The matrix equation of buckling then becomes

$$[K_F]\{a\} - \bar{\bar{N}}_x[K_{Gx}]\{a\} - \bar{\bar{N}}_y[K_{Gy}]\{a\} - \bar{\bar{N}}_{xy}[K_{Gxy}]\{a\} = 0 \tag{15.21}$$

If the in-plane loads have a constant ratio to each other at all times during their buildup, as discussed in Section 8.20, then we can give the above equation as follows:

$$[K_F]\{a\} - P^*(\alpha[K_{Gx}] + \beta[K_{Gy}] + \gamma[K_{Gxy}])\{a\} = 0$$

The term P^* is called the load factor, and α, β, and γ are appropriate constants relating the in-plane loads.

As for interpolation functions, we may use the 16-DOF rectangular element developed from Hermite polynomials in Part D of Chapter 13. The flexural stiffness matrix can be formulated by using Eq. (13.65) and Table 13.2. As for the geometric stiffness matrices, Tables 15.1–15.3 give the desired information for the aforementioned elements.†

We now illustrate the use of these data for the solution of a simple problem.

EXAMPLE 15.2 Consider a square plate, clamped on all sides and subject to uniaxial compression, as shown in Fig. 15.5. Find the

†These tables are taken from Pifko and Isakson (1969).

TABLE 15.1
Data for $[K_{Gx}]$

$$
[K_{Gx}] = \frac{t}{4200}\left(\frac{b}{a}\right)
\begin{bmatrix}
1872 & & & & & & & & & & & & & & & \\
156a & 208a^2 & & & & & & & & & & & \text{symmetric} & & & \\
264b & 22ab & 48b^2 & & & & & & & & & & & & & \\
22ab & 88a^2b/3 & 4ab^2 & 16a^2b^2/3 & & & & & & & & & & & & \\
-1872 & -156a & -264b & -22ab & 1872 & & & & & & & & & & & \\
156a & -52a^2 & 22ab & -22a^2b/3 & -156a & 208a^2 & & & & & & & & & & \\
-264b & 22ab & -48b^2 & -4ab^2 & 264b & -22ab & 48b^2 & & & & & & & & & \\
22ab & -22a^2b/3 & 4ab^2 & -4a^2b^2/3 & -22ab & 88a^2b/3 & -4ab^2 & 16a^2b^2/3 & & & & & & & & \\
-648 & -54a & -156b & -13ab & 648 & -54a & 156b & -13ab & 1872 & & & & & & & \\
54a & -18a^2 & 13ab & -13a^2b/3 & -54a & 72a^2 & -13ab & 52a^2b/3 & -156a & 208a^2 & & & & & & \\
156b & 13ab & 36b^2 & 13ab^2/3 & -156b & 13ab & -36b^2 & 13ab^2/3 & -264b & 22ab & 48b^2 & & & & & \\
-13ab & 13a^2b/3 & -3ab^2 & a^2b^2 & 13ab & -52a^2b/3 & 3ab^2 & -4a^2b^2 & 22ab & -88a^2b/3 & -4ab^2 & 16a^2b^2/3 & & & & \\
648 & 54a & 156b & 13ab & -648 & 54a & -156b & 13ab & -1872 & -156a & 264b & 22ab & 1872 & & & \\
54a & 72a^2 & 13ab & 52a^2b/3 & -54a & -18a^2 & -13ab & -13a^2b/3 & 156a & -52a^2 & -22ab & 22a^2b/3 & 156a & 208a^2 & & \\
-156b & -13ab & -36b^2 & -3ab^2 & 156b & -13ab & 36b^2 & -13ab^2/3 & 264b & 22ab & -48b^2 & -4ab^2/3 & -264b & -22ab & 48b^2 & \\
-13ab & -52a^2b/3 & -3ab^2 & -4a^2b^2 & 13ab & 13a^2b/3 & 3ab^2 & a^2b^2 & -22ab & 22a^2b/3 & 4ab^2 & -4a^2b^2/3 & -22ab & -88a^2b/3 & 4ab^2 & 16a^2b^2/3
\end{bmatrix}
$$

TABLE 15.2
Data for $[K_{Gy}]$

$$
[K_{Gy}] = \frac{t}{4200}\left(\frac{a}{b}\right)
\begin{bmatrix}
1872 &&&&&&&&&&&&&&& \\
264a & 48a^2 &&&&&&&&&&&&&& \\
156b & 22ab & 208b^2 &&&&&&&&&&&&& \\
22ab & 4a^2b & 88ab^2/3 & 16a^2b^2/3 &&&&&&&&&&&& \\
648 & 156a & 54b & 13ab & 1872 &&&&&& & \text{symmetric} &&&& \\
-156a & -36a^2 & -13ab & -3a^2b & -264a & 48a^2 &&&&&&&&&& \\
54b & 13ab & 72b^2 & 52ab^2/3 & 156b & -22ab & 208b^2 &&&&&&&&& \\
-13ab & -3a^2b & -52ab^2/3 & -4a^2b^2 & -22ab & 4a^2b & -88ab^2/3 & 16a^2b^2/3 &&&&&&&& \\
-648 & -156a & -54b & -13ab & -1872 & 264a & -156b & 22ab & 1872 &&&&&&& \\
156a & 36a^2 & 13ab & 3a^2b & 264a & -48a^2 & 22ab & -4a^2b & -264a & 48a^2 &&&&&& \\
54b & 13ab & -18b^2 & -13ab^2/3 & 156b & -22ab & -52b^2 & 22ab^2/3 & -156b & 22ab & 208b^2 &&&&& \\
-13ab & -3a^2b & 13ab^2/3 & a^2b^2 & -22ab & 4a^2b & 22ab^2/3 & -4a^2b^2/3 & 22ab & -4a^2b & -88ab^2/3 & 16a^2b^2/3 &&&& \\
-1872 & -264a & -156b & -22ab & -648 & 156a & -54b & 13ab & 648 & 156a & -54b & -13ab & 1872 &&& \\
-264a & -48a^2 & -22ab & -4a^2b & -156a & 36a^2 & -13ab & 3a^2b & -156a & -36a^2 & 13ab & 3a^2b & 264a & 48a^2 && \\
156b & 22ab & -52b^2 & -22ab^2/3 & 54b & -13ab & -18b^2 & 13ab^2/3 & -54b & -13ab & 72b^2 & 52ab^2/3 & -156b & -22ab & 208b^2 & \\
22ab & 4a^2b & -22ab^2/3 & -4a^2b^2/3 & 13ab & -3a^2b & -13ab^2/3 & a^2b^2 & 13ab & 3a^2b & -52ab^2/3 & -4a^2b^2 & -22ab & -4a^2b & 88ab^2/3 & 16a^2b^2/3
\end{bmatrix}
$$

TABLE 15.3
Data for $[K_{Gxy}]$

$$
[K_{Gxy}] = \frac{t}{300}
\begin{bmatrix}
150 & & & & & & & & & & & & & & & \\
0 & 0 & & & & & & & & & & & & & & \\
0 & 6ab & 0 & & & & & & & & & \text{symmetric} & & & & \\
-6ab & 0 & 0 & 0 & & & & & & & & & & & & \\
0 & 0 & 30b & 6ab & -150 & & & & & & & & & & & \\
0 & 0 & -6ab & -a^2b & 0 & 0 & & & & & & & & & & \\
-30b & -6ab & 0 & 0 & 0 & 6ab & 0 & & & & & & & & & \\
6ab & a^2b^2 & 0 & 0 & -6ab & 0 & 0 & 0 & & & & & & & & \\
-150 & -30a & -30b & -6ab & 0 & 30a & 0 & 6ab & 150 & & & & & & & \\
30a & 5a^2 & 6ab & a^2b & -30a & 0 & -6ab & 0 & 0 & 0 & & & & & & \\
30b & 6ab & 5a^2 & ab^2 & 0 & -6ab & 0 & -ab^2 & 0 & 6ab & 0 & & & & & \\
-6ab & -a^2b & -ab^2 & -a^2b^2/6 & 6ab & 0 & ab^2 & 0 & -6ab & 0 & 0 & 0 & & & & \\
0 & 30a & 0 & 6ab & 150 & -30a & 30b & -6ab & 0 & 0 & 30b & 6ab & -150 & & & \\
-30a & 0 & -6ab & 0 & 30a & -5a^2 & 6ab & -a^2b & 0 & 0 & -6ab & -a^2b & 0 & 0 & & \\
0 & -6ab & 0 & -ab^2 & -30b & 6ab & -5b^2 & ab^2 & -30b & -6ab & 0 & 0 & 0 & 6ab & 0 & \\
6b & 0 & a^2b^2 & 0 & -6ab & a^2b & -ab^2 & a^2b^2/6 & 6ab & a^2b & 0 & 0 & -6ab & 0 & 0 & 0
\end{bmatrix}
$$

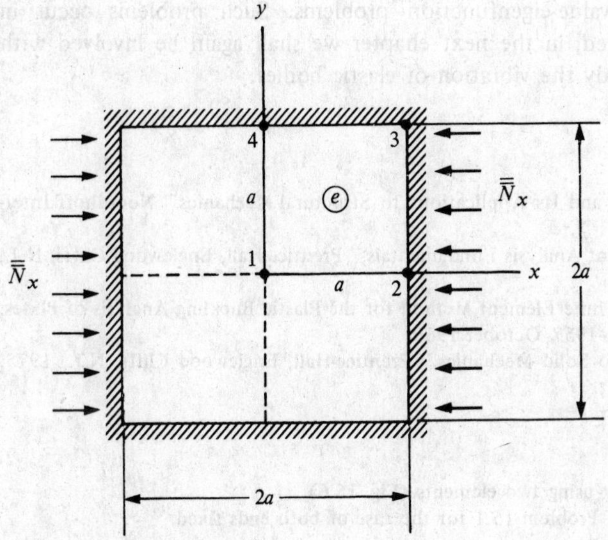

FIGURE 15.5
Clamped plate under uniaxial compression.

buckling load by using four elements as shown, working only with the single element in the first quadrant.

We use a 16-DOF element and note that all the nodal displacements are zero except w_1. This arises from the boundary conditions at the edges of the plate and from symmetry considerations at node 1. Hence, Eq. (15.21) degenerates to a single equation when we reduce the system with $(\bar{\bar{N}}_x)_{cr}$ the desired unknown. We thus have

$$(K_{11})_F - (\bar{\bar{N}}_x)_{cr}(K_{11})_{Gx} = 0 \qquad (a)$$

Going back to Eq. (13.65) and Table 13.2, as well as using Table 15.1, we get

$$\frac{D}{a^2}\left(\frac{156}{35} + \frac{156}{35} + \frac{72}{25}\right) = (\bar{\bar{N}}_x)_{cr}\left(\frac{1}{4200}\right)(1872)$$
$$(\bar{\bar{N}}_x)_{cr} = \frac{D}{a^2}(26.45) \qquad (b)$$

The analytical solution for this problem yields the answer

$$[(\bar{\bar{N}}_x)_{cr}]_{an} = \frac{D}{a^2}(24.8) \qquad (c)$$

We get surprisingly good results for so few elements.

15.4 CLOSURE

In this chapter we have presented a brief introduction to the extensive and important field of elastic stability. We have been primarily concerned with the finite element analysis of eigenvalue-eigenfunction problems. Such problems occur in many areas of study. Indeed, in the next chapter we shall again be involved with such problems when we study the vibration of elastic bodies.

REFERENCES

Dym, C. L., "Stability Theory and Its Applications to Structural Mechanics," Noordhoff International, Leydon, 1974.

Gallagher, R. H., "Finite Element Analysis Fundamentals," Prentice-Hall, Englewood Cliffs, N.J., 1975, Chapter 13.

Pifko, A., and Isakson, G., A Finite Element Method for the Plastic Buckling Analysis of Plates, *AIAA J.*, vol. 7, pp. 1950–1957, October 1969.

Shames, I. H., "Introduction to Solid Mechanics," Prentice-Hall, Englewood Cliffs, N.J., 1975.

PROBLEMS

15.1 Find the buckling load by using two elements (Fig. 15.6).

15.2 Find the buckling load of Problem 15.1 for the case of both ends fixed.

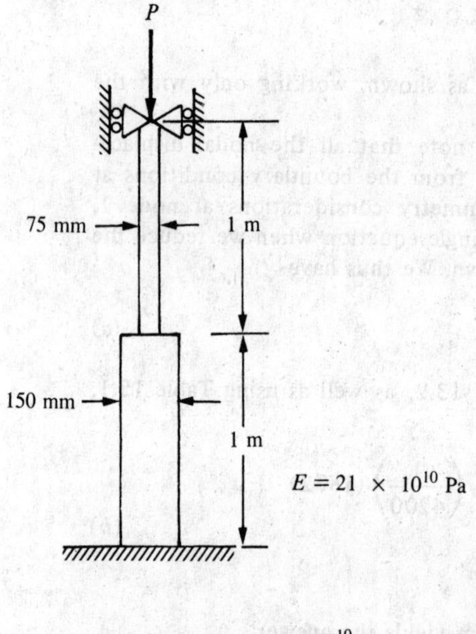

FIGURE 15.6

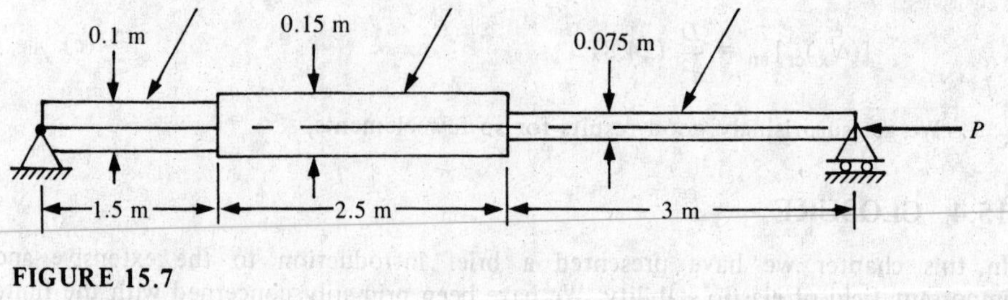

FIGURE 15.7

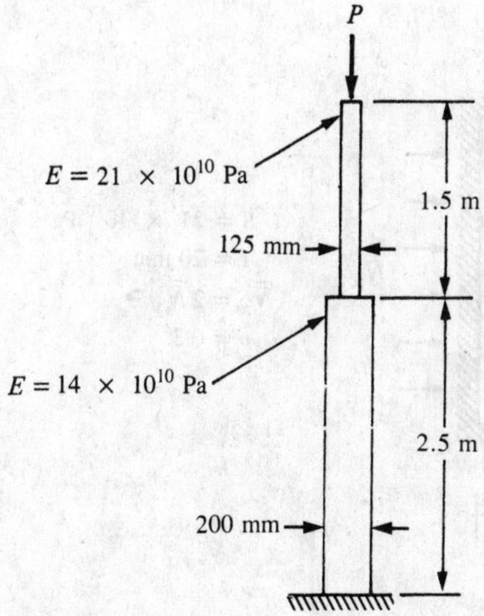

FIGURE 15.8

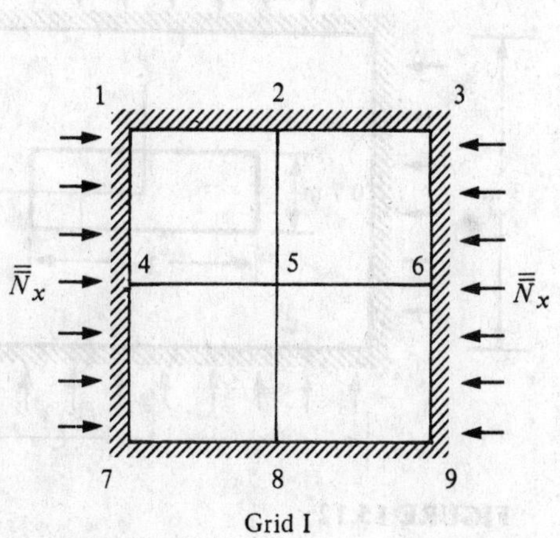

FIGURE 15.9

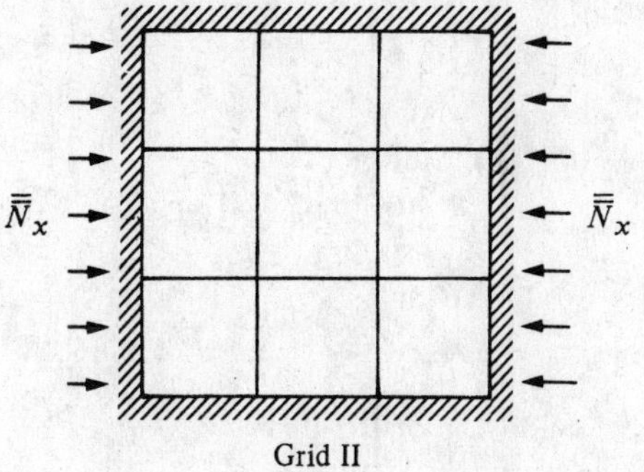

FIGURE 15.10

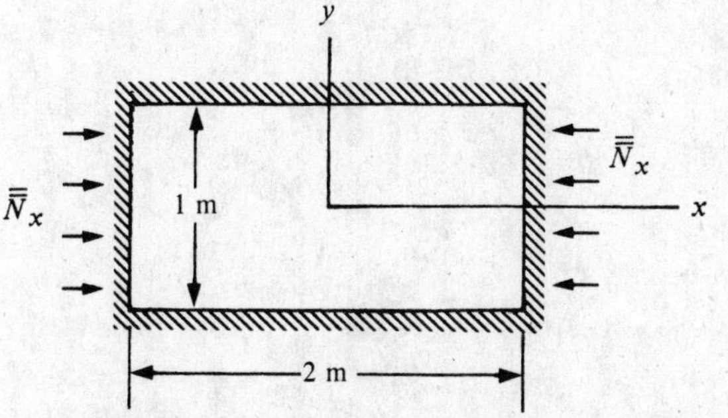

$E = 21 \times 10^{10}$ Pa
$t = 25$ mm
$\nu = 0.25$

FIGURE 15.11

15.3 Find the buckling load by using three elements (Fig. 15.7).

15.4 Find the buckling load for the column in Fig. 15.8. The load P remains vertical during deformation.

15.5 Find the buckling load for GRID I (Fig. 15.9) with all edges clamped.

15.6 Find the buckling load for GRID II (Fig. 15.10) for the case where all edges are clamped.

15.7 Find the buckling load for the plate shown in Fig. 15.11 with all edges clamped.

15.8 Find the buckling load for the clamped plate with a rectangular cutout in the center (Fig. 15.12).

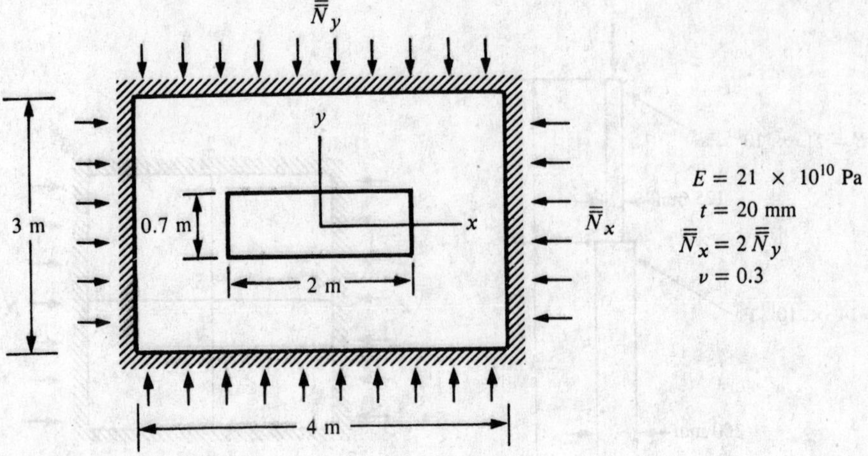
$\bar{\bar{N}}_y$
y
x
3 m
0.7 m
2 m
4 m
$\bar{\bar{N}}_x$
$E = 21 \times 10^{10}$ Pa
$t = 20$ mm
$\bar{\bar{N}}_x = 2\bar{\bar{N}}_y$
$\nu = 0.3$

FIGURE 15.12

16

FINITE ELEMENTS FOR DYNAMICS

16.1 BASIC FINITE ELEMENT EQUATION

We now consider the dynamics of a continuum, using the method of finite elements.† We start by going back to Eq. (7.1), which is a statement of the principle of virtual work that incorporates D'Alembert's principle, with inertia included as a body force distribution. The vector u_i (or $\mathbf{u}$) in Eq. (7.1) has three components u, v, w, so that in matrix notation we can say

$$\{u\} = \begin{Bmatrix} u \\ v \\ w \end{Bmatrix} \tag{16.1}$$

With this in mind, we can express Eq. (7.1) in matrix notation as follows:

$$-\iiint_V \rho\{\delta u\}^T\{\ddot{u}\}\,dv + \iiint_V \{\delta u\}^T\{b\}\,dv + \oiint_S \{\delta u\}^T\{T\}\,dA$$
$$= \iiint_V [\delta\epsilon]^T[\sigma]\,dv \tag{16.2}$$

We next modify this formulation to include *energy dissipation*. We may consider this energy, δW_{diss}, to be expressed as

†The original development of the finite element procedure occurred in the aircraft industry in 1956 in an effort to solve the equations for the dynamic response of complex aircraft structures.

$$\delta W_{\text{diss}} = \delta \left(\iiint_V \{u\}^T \{f\}\, dv \right) \tag{16.3}$$

where $\{f\}$ is the dissipation force (vector) per unit volume. For the case of viscous damping, $\{f\}$ is proportional to the velocity, so that we have

$$\{f\} = c\{\dot{u}\} \tag{16.4}$$

We now discretize the body into small finite elements as we have done earlier, and we consider the displacement field $\{u\}$ to be expressed in terms of nodal displacements $\{a\}$ and interpolation functions N_i. Thus we have

$$\{u\} = [N]\{a\}$$

We will use interpolation functions expressed in terms of spatial coordinates and will consider nodal displacements $\{a\}$ as functions of time. In this discretization of the domain, we use the same interpolation functions as in the static case. We now use in Eq. (16.2) the above equation as well as Eqs. (9.59*d*) and (9.59*e*) written in the forms

$$[\epsilon] = [B]\{a\} \tag{16.5a}$$

$$[\sigma] = [D][\epsilon] \tag{16.5b}$$

We also include viscous damping and so obtain from Eq. (16.2)

$$-\iiint_V \rho\{\delta a\}^T[N]^T[N]\{\ddot{a}\}\, dv + \iiint_V \{\delta a\}^T[N]^T\{b\}\, dv$$

$$+ \oiint_S \{\delta a\}^T[N]^T\{T\}\, dA - \iiint_V \{\delta a\}^T[N]^T c[N]\{\dot{a}\}\, dv$$

$$= \iiint_V \{\delta a\}^T[B]^T[D][B]\{a\}\, dv \tag{16.6}$$

We may now write Eq. (16.6) as follows:

$$\{\delta a\}^T\big([K]\{a\} + [M]\{\ddot{a}\} + [C]\{\dot{a}\}\big) = \{\delta a\}^T\{q\} \tag{16.7}$$

where

$$[K] = \iiint_V [B]^T[D][B]\, dv \equiv \text{stiffness matrix} \tag{16.8a}$$

$$[M] = \iiint_V \rho [N]^T [N] \, dv \equiv \text{mass matrix} \tag{16.8b}$$

$$[C] = \iiint_V c [N]^T [N] \, dv \equiv \text{damping matrix} \tag{16.8c}$$

$$\{q\} = \iiint_V [N]^T \{b\} \, dv + \oiint_S [N]^T \{T\} \, d\mathcal{A} \equiv \text{force vector} \tag{16.8d}$$

We thus end up with the following basic matrix equation for the dynamics of an element:

$$[M]\{\ddot{a}\} + [C]\{\dot{a}\} + [K]\{a\} = \{q\} \tag{16.9}$$

In comparing $[M]$ and $[C]$ in Eqs. (16.8), we see that, for constant ρ and c, they are proportional to each other. That is,

$$[C] = \frac{c}{\rho}[M] = \alpha[M] \tag{16.10}$$

Next, instead of the damping force assumption made in Eq. (16.4), we will consider a viscous damping stress tensor $[\sigma]^d$. We will say that this damping stress is proportional to $[D][\dot{\epsilon}]$. That is,

$$\begin{aligned} &[\sigma]^d \propto [D][\dot{\epsilon}] \\ &[\sigma]^d = \beta[D][\dot{\epsilon}] \end{aligned} \tag{16.11}$$

where β is a constant. The dissipation energy for the element can be given as $-\iiint [\delta\epsilon]^T [\sigma]^d \, dv$. Using Eqs. (9.59$d$), (16.11), and (16.5$a$), we have

$$\begin{aligned} -\iiint_V [\delta\epsilon]^T [\sigma]^d \, dv &= -\iiint_V \{\delta a\}^T [B]^T \big(\beta[D][B]\big)\{\dot{a}\} \, dv \\ &= -\{\delta a\}^T \left[\iiint_V \beta[B]^T [D][B] \, dv\right]\{\dot{a}\} \end{aligned} \tag{16.12}$$

Now comparing the above result with $[K]$ in Eq. (16.8a), we see that

$$\iiint_V \beta[B]^T [D][B] \, dv = \beta[K] \tag{16.13}$$

If we retain *both* dissipative mechanisms, we can generalize $[C]$, the dissipation matrix, to the form

$$[C] = \alpha[M] + \beta[K] \tag{16.14}$$

It is difficult to determine $[C]$ directly because data on the value of $[C]$ are lacking and hence α and β are determined experimentally. The damping matrix $[C]$ as given above is called *Rayleigh* damping.

Returning to Eqs. (16.8), we note that, when computed as indicated, the mass matrix and the damping matrix are called, respectively, the *consistent* mass matrix and the *consistent* damping matrix. This is to distinguish this kind of discretization from that obtained by lumping masses at the nodes, as was done in earlier elementary dynamics problems.

As an example, let us consider plane stress, using triangular elements. The interpolation matrix used in Chapter 12 is given as [see Eq. (12.24)]

$$[N] = \begin{bmatrix} L_i & L_j & L_k & 0 & 0 & 0 \\ 0 & 0 & 0 & L_i & L_j & L_k \end{bmatrix} \tag{16.15}$$

where $L_i = (1/2A)(d_i + x\bar{b}_i + y\bar{a}_i)$, etc. Then, for uniform mass density ρ and thickness t, we have for $[M]$ [see Eq. (16.8b)]:

$$[M] = \rho t \iint_S [N]^T [N] \, dx \, dy \tag{16.16}$$

From this, we see that $[M]$ becomes, on using Eq. (16.15),

$$[M] = \rho t \iint_S \begin{bmatrix} L_iL_i & L_iL_j & L_iL_k & 0 & 0 & 0 \\ L_jL_i & L_jL_j & L_jL_k & 0 & 0 & 0 \\ L_kL_i & L_kL_j & L_kL_k & 0 & 0 & 0 \\ 0 & 0 & 0 & L_iL_i & L_iL_j & L_iL_k \\ 0 & 0 & 0 & L_jL_i & L_jL_j & L_jL_k \\ 0 & 0 & 0 & L_kL_i & L_kL_j & L_kL_k \end{bmatrix} dx \, dy \tag{16.17}$$

Using Eq. (12.48) as an integration formula, we can readily evaluate the terms in $[M]$. Thus we get

$$[M] = \frac{\rho t A}{3}\left[\begin{array}{cccccc} \frac{1}{2} & \frac{1}{4} & \frac{1}{4} & 0 & 0 & 0 \\ \frac{1}{4} & \frac{1}{2} & \frac{1}{4} & 0 & 0 & 0 \\ \frac{1}{4} & \frac{1}{4} & \frac{1}{2} & 0 & 0 & 0 \\ \hdashline 0 & 0 & 0 & \frac{1}{2} & \frac{1}{4} & \frac{1}{4} \\ 0 & 0 & 0 & \frac{1}{4} & \frac{1}{2} & \frac{1}{4} \\ 0 & 0 & 0 & \frac{1}{4} & \frac{1}{4} & \frac{1}{2} \end{array}\right] \tag{16.18}$$

Another approach used for simple systems is, as indicated earlier, the lumping of masses at nodes. This is accomplished by letting $N_i(x, y, z)$ be unity over a specified portion of the element and zero elsewhere. For plane stress, for instance, we can imagine three contributing areas of equal value adjacent to the nodes, as shown in Fig. 16.1. This gives us three equal masses, each one-third of the total mass and each placed at a node. The mass matrix for this case then becomes

$$[M] = \frac{\rho t A}{3}\begin{bmatrix} 1 & 0 & 0 & 0 & 0 & 0 \\ 0 & 1 & 0 & 0 & 0 & 0 \\ 0 & 0 & 1 & 0 & 0 & 0 \\ 0 & 0 & 0 & 1 & 0 & 0 \\ 0 & 0 & 0 & 0 & 1 & 0 \\ 0 & 0 & 0 & 0 & 0 & 1 \end{bmatrix}$$

16.2 FREE VIBRATIONS

We now consider the special case of *undamped* harmonic motion. With no damping and no external forces doing work, Eq. (16.9) becomes

$$[M]\{\ddot{a}\} + [K]\{a\} = 0 \tag{16.19}$$

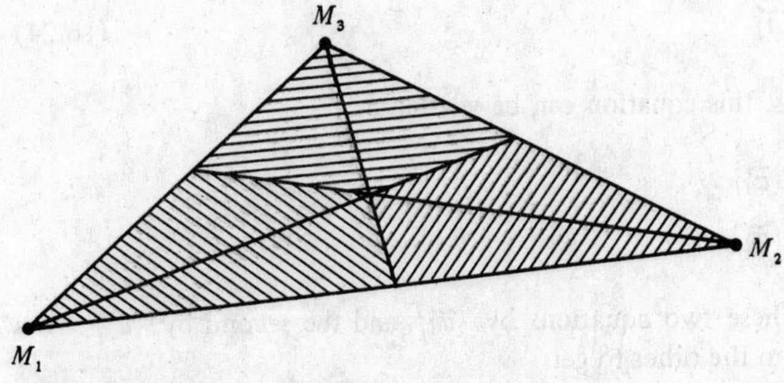

FIGURE 16.1
Lumping of masses for a triangular element.

This is the global equation for free vibration. We now assume a solution for $\{a\}$ of the form:

$$\{a\} = \{\tilde{a}\} e^{i\omega t} \tag{16.20}$$

where $\{\tilde{a}\}$ is a set of constant values at the nodes and is called the *modal* vector, and ω is a *natural frequency*. Substituting for $\{a\}$ in Eq. (16.19), using Eq. (16.20), we get

$$\big([K] - \omega^2 [M]\big)\{\tilde{a}\} = 0 \tag{16.21}$$

This is now an eigenvalue-eigenfunction problem of the same type studied in Chapter 7. For a nontrivial solution of the above equation, it is necessary that the determinant of the coefficients of $\{\tilde{a}\}$ be zero for the *reduced* equation, which includes boundary conditions. That is:

$$|[K] - \omega^2 [M]| = 0 \tag{16.22}$$

[As pointed out in the previous chapter—see Eqs. (15.15)-(15.17)—the above procedure is not efficient for large-order systems. Instead, for such cases we can employ an iterative technique that simultaneously provides eigenvalues ω_i and eigenvectors $\{\tilde{a}\}_i$.] One can show that if the matrices $[K]$ and $[M]$ are positive-definite,† the roots ω_m for Eq. (16.22) will be real. For an nth order determinant in Eq. (16.22), we will get accordingly n natural frequencies, and we can then find, on going back to Eq. (16.21), n natural mode shapes or eigenvectors. These eigenvectors must be orthogonal in the following sense:

$$\{\tilde{a}\}_i^T [M] \{\tilde{a}\}_j = 0 \qquad i \neq j \tag{16.23a}$$

$$\{\tilde{a}\}_i^T [K] \{\tilde{a}\}_j = 0 \tag{16.23b}$$

We can prove this orthogonality in much the same way as we did in Section 7.13. Thus, go back to Eq. (16.21) and express it as follows:

$$[K]\{\tilde{a}\} = \omega^2 [M]\{\tilde{a}\} \tag{16.24}$$

For the ith and jth modes, this equation can be written as

$$[K]\{\tilde{a}\}_i = \omega_i^2 [M]\{\tilde{a}\}_i$$

$$[K]\{\tilde{a}\}_j = \omega_j^2 [M]\{\tilde{a}\}_j$$

Premultiply the first of these two equations by $\{\tilde{a}\}_j^T$ and the second by $\{\tilde{a}\}_i^T$. Now substract one equation from the other to get

†The matrices must be symmetric and the determinants of all principal minors must be positive. This will usually be the case for elastic structures.

$$\{\tilde{a}\}_j^T[K]\{\tilde{a}\}_i - \{\tilde{a}\}_i^T[K]\{\tilde{a}\}_j = \omega_i^2\{\tilde{a}\}_j^T[M]\{\tilde{a}\}_i - \omega_j^2\{\tilde{a}\}_i^T[M]\{\tilde{a}\}_j$$

Since the matrices $[K]$ and $[M]$ are symmetric, we can interchange the position of the $\{\tilde{a}\}$ matrices while keeping the transpose operation in place. Using this once on each side of the above equation, we then get a cancellation of expressions on the left side of the above equation, leaving us with the result

$$(\omega_i^2 - \omega_j^2)\{\tilde{a}\}_j^T[M]\{\tilde{a}\}_i = 0$$

If $i \neq j$ and $\omega_i \neq \omega_j$, then clearly Eq. (16.23*a*) is proved. We leave it to the reader to show that Eq. (16.23*b*) follows directly. If $i = j$, we then get the following results for the expressions in Eq. (16.23), on using E_M and E_K as constants:

$$\begin{aligned} \{\tilde{a}\}_i^T[M]\{\tilde{a}\}_i &= E_M \\ \{\tilde{a}\}_i^T[K]\{\tilde{a}\}_i &= E_K \end{aligned} \tag{16.25}$$

It is usually convenient to *normalize* the eigenvector $E_M = 1$.

EXAMPLE 16.1 We shall find the natural frequencies and the eigenvectors for the tapered cantilever beam shown in Fig. 16.2. Using inches as the length dimension, the mass density ρ is 7845.2 kg-sec^2/m^4 and the modulus of elasticity E is 21×10^{10} Pa.

We have modeled the beam with uniform elements, each of a different diameter, as shown in Fig. 16.3. For the interpolation functions we will use those developed for the static case in Chapter 10, which we now rewrite:

$$\begin{aligned} N_1(x) &= \left(1 - 3\frac{x^2}{L^2} + 2\frac{x^3}{L^3}\right) \\ N_2(x) &= \left(x - 2\frac{x^2}{L} + \frac{x^3}{L^2}\right) \\ N_3(x) &= \left(3\frac{x^2}{L^2} - 2\frac{x^3}{L^3}\right) \end{aligned} \tag{a}$$

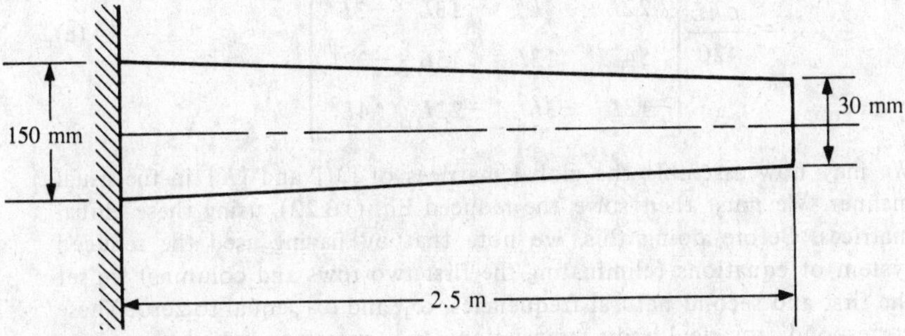

FIGURE 16.2
Circular tapered cantilever beam.

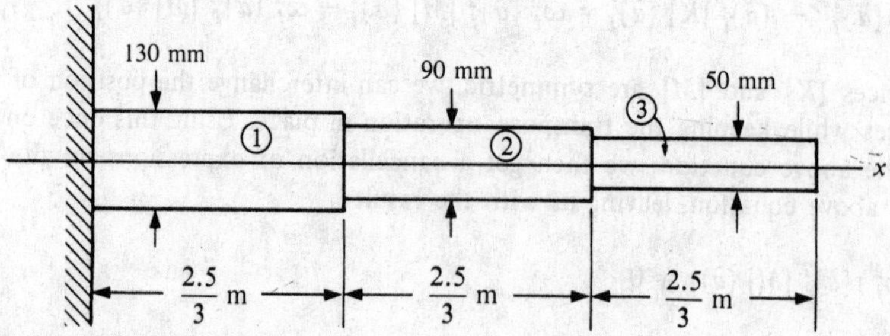

FIGURE 16.3
Cantilever with three elements.

$$N_4(x) = \left(-\frac{x^2}{L} + \frac{x^3}{L^2} \right) \qquad (a) \text{ (Cont.)}$$

The stiffness matrix $[K]$ for an element has already been worked out in Chapter 10 [see Eq. (10.22)] and we rewrite it here:

$$[K]^e = \frac{EI}{L^3} \begin{bmatrix} 12 & 6L & -12 & 6L \\ 6L & 4L^2 & -6L & 2L^2 \\ -12 & -6L & 12 & -6L \\ 6L & 2L^2 & -6L & 4L^2 \end{bmatrix} \qquad (b)$$

Next we shall need the mass matrix. From Eq. (16.8*b*) and Eqs. (*a*), we have

$$[M]^e = \int_0^L \rho A [N]^T [N] \, dx = \rho A \int_0^L \begin{Bmatrix} N_1 \\ N_2 \\ N_3 \\ N_4 \end{Bmatrix} [N_1, N_2, N_3, N_4] \, dx$$

$$= \frac{\rho A L}{420} \begin{bmatrix} 156 & 22L & 54 & -13L \\ 22L & 4L^2 & 13L & -3L^2 \\ 54 & 13L & 156 & -22L \\ -13L & -3L^2 & -22L & 4L^2 \end{bmatrix} \qquad (c)$$

We may now assemble the global matrices of $[M]$ and $[K]$ in the usual manner. We must then solve the reduced Eq. (16.22), using these global matrices. Before doing this, we note that by having used the reduced system of equations (eliminating the first two rows and columns) we set the first and second natural frequencies, ω_1 and ω_2, equal to zero. These correspond to rigid-body translation and rotation, which are zero because of the boundary conditions at node 1 at the base of the cantilever. Using the computer, we can get the remaining natural frequencies in order of magnitude, ω_3 to ω_8, by setting the reduced determinant equal to zero. The first three lowest natural frequencies as well as the exact solutions (Gorman, 1975) are given in Table 16.1.† Although these correspond in this analysis to ω_3, ω_4, and ω_5, they are listed as ω_1, ω_2, and ω_3, signifying the first three nonzero natural frequencies.

Our next task is to get the *eigenvectors*. For this purpose, we go back to Eq. (16.21), which we now express in expanded form. In doing so, set $\omega^2 = \omega_3^2$, the first nonzero eigenvalue, and $\{\tilde{a}\}$ becomes $\{\tilde{a}\}_3$, the first nontrivial eigenvector. Thus we have, on setting $(\tilde{a}_1)_3 = (\tilde{a}_2)_3 = 0$ to satisfy the boundary conditions of the problem, and noting that the system used here is the reduced one,

$$\begin{aligned}(K_{33} - \omega_3^2 M_{33})(\tilde{a}_3)_3 + (K_{34} - \omega_3^2 M_{34})(\tilde{a}_4)_3 + \cdots + (K_{38} - \omega_3^2 M_{38})(\tilde{a}_8)_3 + \cdots &= 0\\ &\vdots\\ (K_{83} - \omega_3^2 M_{83})(\tilde{a}_3)_3 + (K_{84} - \omega_3^2 M_{84})(\tilde{a}_4)_3 + \cdots + (K_{88} - \omega_3^2 M_{88})(\tilde{a}_8)_3 + \cdots &= 0\end{aligned} \qquad (d)$$

We can only get the ratios of the components of the eigenvector. Hence we can choose one of the $\tilde{a}$'s arbitrarily and then solve for the other $\tilde{a}$'s for the third eigenvalue. Accordingly, we choose $(\tilde{a}_7)_3 = 1$ and solve for the remaining $\tilde{a}$'s from Eq. (*d*). We do this for all the remaining eigenvalues ω_i. We then have for the first nontrivial eigenvector:

$$\lfloor \tilde{a} \rfloor_3 = [0.0735, 0.159, 0.346, 0.437, 1.000, 0.909] \qquad (e)$$

The reduced eigenvector matrix $[\tilde{A}]$ can then be given as

$$[\tilde{A}] = [\{\tilde{a}\}_3 \{\tilde{a}\}_4 \{\tilde{a}\}_5 \{\tilde{a}\}_6 \{\tilde{a}\}_7 \{\tilde{a}\}_8] \qquad (f)$$

Hence:

$$[\tilde{A}] = \overset{\begin{matrix}\omega_3 & \omega_4 & \omega_5 & \omega_6 & \omega_7 & \omega_8\end{matrix}}{\begin{bmatrix} 0.073493 & -0.160262 & 0.455896 & -0.113070 & -0.021299 & 0.221331\\ 0.158649 & -0.295477 & 0.464765 & 0.482931 & -1.358051 & 2.837551\\ 0.345560 & -0.330373 & -0.571111 & 0.020785 & 0.296303 & 0.262317\\ 0.436831 & 0.007859 & -1.996993 & -1.911674 & 1.606220 & 6.055418\\ 1.000000 & 1.000000 & 1.000000 & 1.000000 & 1.000000 & 1.000000\\ 0.908598 & 2.241349 & 4.226444 & 7.963498 & 10.871218 & 13.187287 \end{bmatrix}} \qquad (g)$$

where $\omega_3 < \omega_4 \cdots < \omega_8$.

The free vibration can be given as

$$\begin{aligned}\{a\} = {} & C_3\{\tilde{a}\}_3 \sin(\omega_3 t + \alpha_3) + C_4\{\tilde{a}\}_4 \sin(\omega_4 t + \alpha_4)\\ & + \cdots + C_8\{\tilde{a}\}_8 \sin(\omega_8 t + \alpha_8)\end{aligned} \qquad (h)$$

†Because we are using cylinders rather than frustrums as elements, the eigenvalues approach correct results from below.

TABLE 16.1
Natural Frequencies

Frequency	Gorman (1975)	Finite elements (3 cylinders)	Finite elements (6 cylinders)
$\omega_1 (=\omega_3)$	174.7 rad/sec	158.8 rad/sec	183.5 rad/sec
$\omega_2 (=\omega_4)$	545.3 rad/sec	420.6 rad/sec	525.7 rad/sec
$\omega_3 (=\omega_5)$	1242.0 rad/sec	1038.2 rad/sec	1122.3 rad/sec

where the C's and α's are constants to be determined from known initial conditions $\{a\}_{t=0}$ and $\{\dot{a}\}_{t=0}$. Thus, going back to Eq. (h), we have

$$\{a\}_{t=0} = C_3 \{\widetilde{a}\}_3 \sin \alpha_3 + C_4 \{\widetilde{a}\}_4 \sin \alpha_4 + \cdots + C_8 \{\widetilde{a}\}_8 \sin \alpha_8$$

$$\{\dot{a}\}_{t=0} = C_3 \omega_3 \{\widetilde{a}\}_3 \cos \alpha_3 + C_4 \omega_4 \{\widetilde{a}\}_4 \cos \alpha_4 + \cdots + C_8 \omega_8 \{\widetilde{a}\}_8 \cos \alpha_8$$

We have here 12 equations for the 12 unknowns.

16.3 FORCED VIBRATION–MODAL ANALYSIS

In this method, with the eigenvectors of the previous section, we will be able to uncouple the equations so as to have a set of independent (for each mode) differential equations of motion. We start with Eq. (16.9), which we rewrite as

$$[M]\{\ddot{a}\} + [C]\{\dot{a}\} + [K]\{a\} = \{q\} \tag{16.26}$$

We shall assume that the solution to this equation can be written as a linear combination of the n eigenvectors of the free vibration solution of the previous section, wherein each eigenvector $\{\widetilde{a}\}_i$ is multiplied by a function of time, which we denote as $\Lambda_i(t)$. Thus, in matrix notation we can say

$$\begin{aligned}\{a\} &= \lfloor \{\widetilde{a}\}_1, \{\widetilde{a}\}_2, \{\widetilde{a}\}_3, \ldots, \{\widetilde{a}\}_n \rfloor \{\Lambda(t)\} \\ &= [\widetilde{A}]\{\Lambda(t)\}\end{aligned} \tag{16.27}$$

where $[\widetilde{A}]$ is the eigenvector matrix forming an $n \times n$ matrix, which is known from our work in free vibrations.† The idea now is to find the vector $\{\Lambda(t)\}$ for the damped, forced motion equation, Eq. (16.26). For this purpose, we substitute for $\{a\}$ in Eq. (16.26), using Eq. (16.27). Also, we premultiply each expression in Eq. (16.26) by $[\widetilde{A}]^T$. We then have

$$[\widetilde{A}]^T[M][\widetilde{A}]\{\ddot{\Lambda}\} + [\widetilde{A}]^T[C][\widetilde{A}]\{\dot{\Lambda}\} + [\widetilde{A}]^T[K][\widetilde{A}]\{\Lambda\} = [\widetilde{A}]^T\{q\} \tag{16.28}$$

We may express this equation as

†From Eq. (16.27), we see that nodal displacement a_i is given as $[(\widetilde{a}_i)_1 \Lambda_1 + (\widetilde{a}_i)_2 \Lambda_2 + \cdots + (\widetilde{a}_i)_n \Lambda_n]$.

$$[M^*]\{\ddot{\Lambda}\} + [C^*]\{\dot{\Lambda}\} + [K^*]\{\Lambda\} = \{q^*\} \tag{16.29}$$

where

$$[M^*] = [\tilde{A}]^T[M][\tilde{A}] \tag{16.30a}$$

$$[C^*] = [\tilde{A}]^T[C][\tilde{A}] \tag{16.30b}$$

$$[K^*] = [\tilde{A}]^T[K][\tilde{A}] \tag{16.30c}$$

$$\{q\}^* = [\tilde{A}]^T\{q\} \tag{16.30d}$$

The members of each of the starred matrices above are given next:

$$\begin{aligned} M_{ij}^* &= \lfloor \tilde{a} \rfloor_i [M] \{\tilde{a}\}_j \\ C_{ij}^* &= \lfloor \tilde{a} \rfloor_i [C] \{\tilde{a}\}_j \\ K_{ij}^* &= \lfloor \tilde{a} \rfloor_i [K] \{\tilde{a}\}_j \\ q_i^* &= \lfloor \tilde{a} \rfloor_i \{q\} \end{aligned} \tag{16.31}$$

Because the eigenvectors are orthogonal with respect both to $[M]$ and $[K]$ (see Eq. (16.23)], the matrices $[M^*]$ and $[K^*]$ will be diagonal; that is, the terms for $i \neq j$ will be zero. Furthermore, if we employ Rayleigh damping [see Eq. (16.14)], $[C^*]$ will also be diagonal and Eq. (16.29) will represent a system of *uncoupled* equations. Not using dummy index notation here, we can express these equations as

$$m_i\ddot{\Lambda}_i + c_i\dot{\Lambda}_i + k_i\Lambda_i - q_i = 0 \qquad i = 1, \ldots, n \tag{16.32}$$

where

$$m_i = M_{ii}^* = \lfloor \tilde{a} \rfloor_i [M] \{\tilde{a}\}_i \tag{16.33a}$$

$$c_i = C_{ii}^* = \lfloor \tilde{a} \rfloor_i [C] \{\tilde{a}\}_i \tag{16.33b}$$

$$k_i = K_{ii}^* = \lfloor \tilde{a} \rfloor_i [K] \{\tilde{a}\}_i \tag{16.33c}$$

$$q_i = \lfloor \tilde{a} \rfloor_i \{q\} \tag{16.33d}$$

To simplify further, we can *normalize* the modes so that $m_i = 1$. In addition, we express c_i as follows:

$$c_i = 2\omega_i c_i' \tag{16.34}$$

where c_i' is the ratio of damping to its critical value.† Also, noting Eq. (16.21), we can say

$$\omega_i^2 [M] \{\tilde{a}\}_i = [K] \{\tilde{a}\}_i$$

†For information on how to determine c_i' see Clough (1971).

Premultiplying by $\lfloor \tilde{a} \rfloor_i$ and noting Eqs. (16.33*a*) and (16.33*c*), we have

$$\begin{aligned} \omega_i^2 m_i &= k_i \\ \omega_i^2 &= k_i \end{aligned} \tag{16.35}$$

where we have set $m_i = 1$. We can then give Eq. (16.32) as follows:

$$\ddot{\Lambda}_i + 2\omega_i c_i' \dot{\Lambda}_i + \omega_i^2 \Lambda_i - q_i = 0 \tag{16.36}$$

We may solve for Λ_i by the usual methods of differential equations. We point out that the general solution for this equation is given by Duhamel's integral, that is,

$$\Lambda_i = \int_0^t q_i e^{-c_i' \omega_i (t-\tau)} \sin \omega_i (t-\tau)\, d\tau \tag{16.37}$$

We thus have a solution available (see footnote on page 652) for the transient response of a structure under a dynamic loading vector. In practice, we generally estimate only the maximum response for each mode, and a suitable combination of these results is formulated to get the desired results. In closing, we wish to point out that the mode superposition technique described in this section is valid only for linear problems. This means that $[M]$ and $[K]$ must not be functions of the dependent variables a_i.

EXAMPLE 16.2 Consider that on the tip of the tapered cantilever beam in Example 16.1 we apply a vertical sinusoidal force

$$F = 0.001 \sin 20t \text{ N}$$

What is the transient vertical motion of the tip of the cantilever beam, assuming it is at rest initially? Neglect damping. Do not normalize the mass matrix.

The mass matrix $[M^*]$ is a diagonal matrix found by using Eq. (16.30*a*):

$$[M^*] = [\tilde{A}]^T [M] [\tilde{A}]$$

Here $[M]$ is the reduced global mass matrix. Carrying out the computations above by using Eqs. (*c*) and (*g*) from Example 16.1, we get

$$[M^*] = \begin{bmatrix} 7.620 & & & & & \\ & 1.232 & & & & \\ & & 0.807 & & & \\ \text{all others 0} & & & 0.0665 & & \\ & & & & 0.0282 & \\ & & & & & 0.0386 \end{bmatrix} \tag{a}$$

Similarly, for the diagonal stiffness $[K^*]$ we have from Eq. (16.30*c*)

$$[K^*] = [\tilde{A}]^T[K][\tilde{A}]$$

where $[K]$ is the reduced global stiffness matrix. Using Eqs. (*b*) and (*g*) from Example 16.1, we get

$$[K^*] = \begin{bmatrix} 0.1921 \times 10^6 & & & & & \\ & 0.2180 \times 10^6 & & & & \\ & & 0.8701 \times 10^6 & & & \\ \text{all others } 0 & & & 0.4681 \times 10^6 & & \\ & & & & 0.5587 \times 10^6 & \\ & & & & & 0.2049 \times 10^7 \end{bmatrix} \tag{b}$$

Note further that

$$\{q^*\} = [\tilde{A}]^T\{q\}$$

For this problem $\{q\}$ is

$$\{q\} = \lfloor 0, 0, 0, 0, 0.001 \sin 20t, 0 \rfloor^T \tag{c}$$

With the terms of the fifth column of $[\tilde{A}]^T$ of unit value, we get for $\{q^*\}$:

$$\{q^*\} = \lfloor 0.001 \sin 20t, 0.001 \sin 20t, 0.001 \sin 20t, 0.001 \sin 20t, 0.001 \sin 20t, 0.001 \sin 20t \rfloor^T \tag{d}$$

Now substitute into Eq. (16.32), using the results given in Eqs. (*a*), (*b*), and (*d*) of this example. Taking $[C^*]$ equal to zero and noting that we are using the reduced system of equations because of the boundary conditions, we get the following set of six equations for the six Λ's:

$$\begin{aligned}
7.620 \ddot{\Lambda}_3 + 0.1921 \times 10^6 \Lambda_3 &= 0.001 \sin 20t\\
1.232 \ddot{\Lambda}_4 + 0.2180 \times 10^6 \Lambda_4 &= 0.001 \sin 20t\\
0.807 \ddot{\Lambda}_5 + 0.8701 \times 10^6 \Lambda_5 &= 0.001 \sin 20t\\
0.0665 \ddot{\Lambda}_6 + 0.4681 \times 10^6 \Lambda_6 &= 0.001 \sin 20t\\
0.0282 \ddot{\Lambda}_7 + 0.5587 \times 10^6 \Lambda_7 &= 0.001 \sin 20t\\
0.0386 \ddot{\Lambda}_8 + 0.2049 \times 10^7 \Lambda_8 &= 0.001 \sin 20t
\end{aligned} \tag{e}$$

Noting that $\omega_i^2 = k_i/m_i$ and taking ω to be the forcing frequency, we obtain the solution to any of the above equations as

$$\Lambda_i = A_i \cos \omega_i t + B_i \sin \omega_i t + \frac{q_i/m_i}{\omega_i^2 - \omega^2}$$

When $t = 0$, $\Lambda_i = \dot{\Lambda}_i = 0$. Hence we may determine the constants of integration with $q_i = 0.001 \sin 20t$:

$$A_i = 0$$

$$B_i = -\frac{0.001}{m_i} \frac{\omega}{\omega_i} \frac{1}{\omega_i^2 - \omega^2}$$

The general solution is then

$$\Lambda_i(t) = -\frac{0.001}{m_i} \frac{1}{\omega_i^2 - \omega^2} \left(\frac{\omega}{\omega_i} \sin \omega_i t - \sin \omega t \right) \tag{f}$$

Thus we have for the modal displacement vector:

$$\{a\} = [\tilde{A}]\{\Lambda\} \tag{g}$$

Before expanding out the above equation, we note that the first Λ is Λ_3, as indicated in Eqs. (*e*). The "natural frequencies" associated with Λ_1 and Λ_2, being denoted now as ω_1 and ω_2, have zero value. Also, noting that $m_3 = 7.620$, $m_4 = 1.232$, etc., from Eq. (*a*) and that $k_3 = 0.1921 \times 10^6$, $k_4 = 0.2180 \times 10^6$, etc., from Eq. (*b*), we can express Eq. (*g*) as follows:

$$\begin{Bmatrix} a_3 \\ a_4 \\ a_5 \\ a_6 \\ a_7 \\ a_8 \end{Bmatrix} = \begin{bmatrix} \tilde{A}_{33} & \tilde{A}_{34} & \tilde{A}_{35} & \tilde{A}_{36} & \tilde{A}_{37} & \tilde{A}_{38} \\ \tilde{A}_{43} & \tilde{A}_{44} & \tilde{A}_{45} & \tilde{A}_{46} & \tilde{A}_{47} & \tilde{A}_{48} \\ \tilde{A}_{53} & \tilde{A}_{54} & \tilde{A}_{55} & \tilde{A}_{56} & \tilde{A}_{57} & \tilde{A}_{58} \\ \tilde{A}_{63} & \tilde{A}_{64} & \tilde{A}_{65} & \tilde{A}_{66} & \tilde{A}_{67} & \tilde{A}_{68} \\ 1.0 & 1.0 & 1.0 & 1.0 & 1.0 & 1.0 \\ \tilde{A}_{83} & \tilde{A}_{84} & \tilde{A}_{85} & \tilde{A}_{86} & \tilde{A}_{87} & \tilde{A}_{88} \end{bmatrix} \begin{pmatrix} -\frac{0.001}{m_3} \frac{1}{\omega_3^2 - \omega^2} \left(\frac{\omega}{\omega_3} \sin \omega_3 t - \sin \omega t \right) \\ \cdots \\ \cdots \\ \cdots \\ \cdots \\ -\frac{0.001}{m_8} \frac{1}{\omega_8^2 - \omega^2} \left(\frac{\omega}{\omega_8} \sin \omega_8 t - \sin \omega t \right) \end{pmatrix} \tag{h}$$

Since we only want a_7, we have on considering the first three nonzero modes (3–5):

$$a_7 = \sum_{j=3}^{5} -\frac{0.001}{m_j} \frac{1}{\omega_j^2 - \omega^2} \left\{ \frac{\omega}{\omega_j} \sin \omega_j t - \sin \omega t \right\}$$

Inserting numerical data, we have noting that $\omega = 20$ rad/sec

$$a_7 = -6.66 \times 10^{-10} (\sin 158.8t) - 2.1867 \times 10^{-10} (\sin 420.6t)$$
$$-2.2155 \times 10^{-11} (\sin 1038.2t) + 1.1037 \times 10^{-8} (\sin 20t) \text{ m}$$

We thus have the desired information.

16.4 CLOSURE

With this introductory chapter on dynamics, we bring to a close our finite element considerations for structural mechanics. We have covered various aspects of the main topics in structural mechanics considered in Part II of this text. More detailed and more advanced information in this area of study can be readily found in a large and constantly growing number of texts as well as a virtual deluge of research papers in journals.

We will now go back to Chapter 3 on quadratic functionals, and in the next and final chapter we will expand the finite element methodology in conjunction with these quadratic functionals for application to fields of study other than solid mechanics.

REFERENCES

Clough, R. W., Analysis of Structural Vibrations and Dynamic Response, in "Recent Advances in Matrix Methods of Structural Analysis and Design," R. H. Gallagher, Y. Yamada, and J. R. Oden, eds., Univ. of Alabama Press, Huntsville, 1971.

Gorman, Daniel J., "Free Vibration Analysis of Beams and Shafts," Wiley, New York, 1975.

PROBLEMS

16.1 Determine the first four eigenvalues and eigenvectors for free vibration in Fig. 16.4. Take $\rho = 8400$ kg/m^3. Cross-sections are circular.

16.2 Do Problem 16.1 for the case of simple supports at the ends.

16.3 Solve Problem 16.1 by using STRUDL or other "canned" program.

16.4 Find the first three natural frequencies in Fig. 16.5. Take $E_A = 14 \times 10^{10}$ Pa, $E_B = 21 \times 10^{10}$ Pa, $\rho_A = 7500$ kg/m^3, and $\rho_B = 8600$ kg/m^3. Cross-sections are circular. Use three elements.

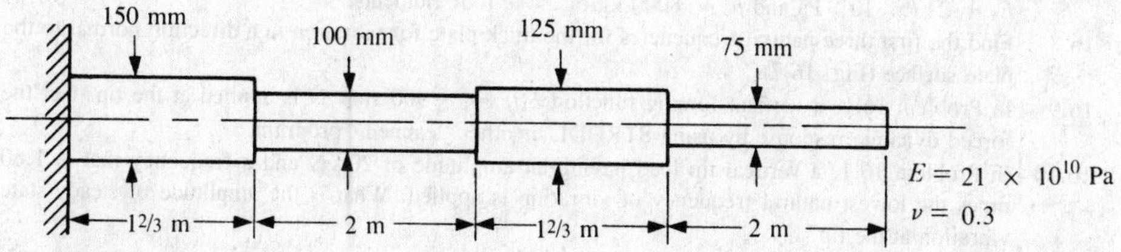

FIGURE 16.4

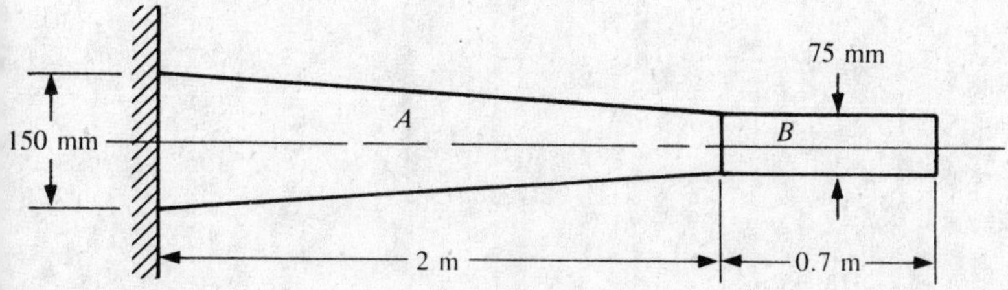

FIGURE 16.5

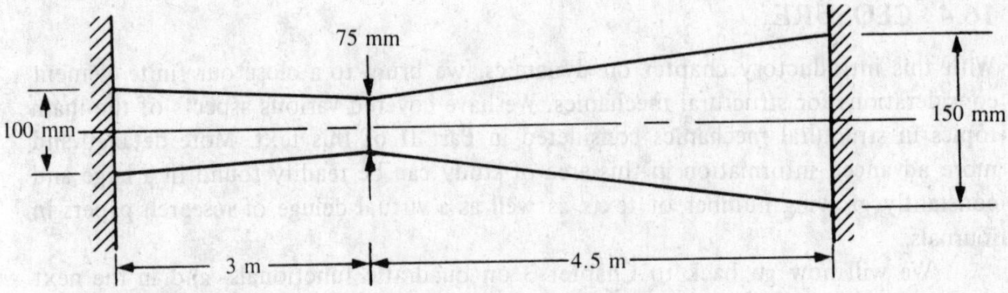

FIGURE 16.6

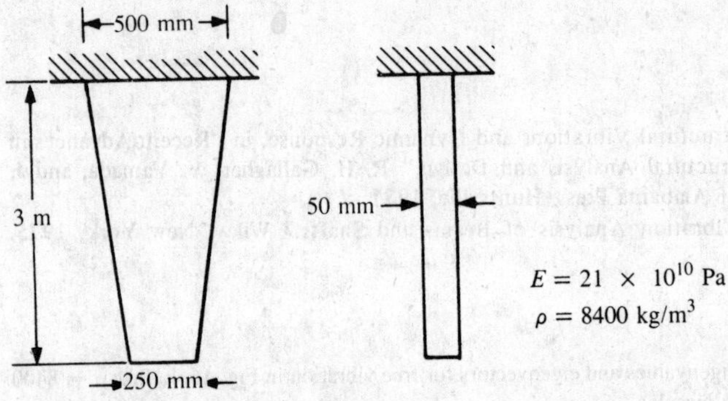

FIGURE 16.7

16.5 Do Problem 16.4 for a simply supported beam.

16.6 Solve Problem 16.4 for a simply supported beam by using STRUDL or other "canned" program.

16.7 Find the first four natural frequencies of vibration for the conical shaft in Fig. 16.6. Use $E = 21 \times 10^{10}$ Pa and $\rho = 8000$ kg/m^3. Use four elements.

16.8 Find the first three natural frequencies for the thick plate for vibration in a direction normal to the plate surface (Fig. 16.7).

16.9 In Problem 16.1, a vertical forcing function $F(t) = -900 \sin t$ N is applied at the tip. Get the forced dynamic response by using STRUDL or other "canned" program.

16.10 In Problem 16.1, a vertical tip load having an amplitude of 200 N and a frequency that is 1.60 times the lowest natural frequency of vibration is applied. What is the amplitude of steady-state vibration at the tip?

17

BROADER ASPECTS OF FINITE ELEMENTS

17.1 INTRODUCTION

We started our studies of finite elements by using a matrix approach for trusses where the members themselves played the role of the finite elements and the movement of the joints played the role of the nodal displacements. With this as a background, we then developed in a general manner the displacement method for the study of finite elements. This study served as the basis for the use of finite elements as applied to the kind of structures considered in Part II of the text. Now we want to make further generalizations which will enable us to better understand the finite element procedure as presented thus far, and which will enable us to consider problems in other fields of study by using the finite element approach. In particular, we shall use aspects of the variational calculus related to quadratic functionals first presented in Part C of Chapter 3.

Part A Variational Considerations—Ordinary Differential Equations

17.2 A SECOND LOOK AT THE GENERAL THEORY OF THE DISPLACEMENT METHOD FOR FINITE ELEMENTS

Recall that in the *Ritz method*, as applied to solid mechanics, we expressed the deformation field in terms of a finite number of carefully chosen coordinate functions ϕ_i, each multiplied by a constant C_i, arranged to form a parameter-laden sum. That is:

$$u(x, y) = \sum_{i=1}^{n} C_i \phi_i$$

where ideally ϕ_i satisfies the boundary conditions. By choosing the C_i so as to extremize π, we are finding the particular deformation whose elastically related stress field satisfies equilibrium. Note that in this process we have extremized π with respect to a *subset* of all possible kinematically compatible deformations for the problem at hand. Hence, we have introduced "hidden constraints" that make the resulting mathematically produced model effectively a stiffer body than the actual case. By increasing the number of coordinate functions and by choosing them carefully, we may formulate a stress field approaching the actual stress field.

We will now show that the finite element approach leading to Eq. (9.47) for an element is the same as the Ritz method when applied to a single element. In the development following Eq. (9.25) we expressed the displacement field $[\vec{u}(x, y)]$ in terms of a finite number of nodal displacement components a_i along with an interpolation function component N_i for each nodal displacement component. Thus [see Eq. (9.27)] we have for the displacement field of an element:

$$\{\vec{u}(x, y)\} = [N]\{a\} \tag{17.1}$$

The interpolation functions N_i are analogous to the coordinate functions ϕ_i of the Ritz method and are chosen ideally to have kinematic compatibility between elements. The nodal displacement components a_i play the same role as the Ritz coefficients C_i. We then substituted $[N]\{a\}$ into Eq. (9.42) so as to reach the familiar finite element equation [Eq. (9.47)]:

$$[K]^e\{a\}^e = \{q\}^e - \{f\}^e \tag{17.2}$$

This equation, when satisfied by properly choosing the nodal displacements a_i, satisfies equilibrium for some elastic element having the same shape and size as the actual element but with "hidden constraints" as discussed above.

We must now point out an essential *difference* between the finite element method as normally used and the Ritz method. In the Ritz method, we considered the *entire domain* to be the element and we increased the number of coordinate functions in a suitable way to decrease the hidden constraints and thus arrive at a deformation approaching the actual case. In the finite element method, the domain is broken down to many small contiguous simple elements connected at the nodal points. The nodal displacements are then adjusted for each element so as to minimize the total potential energy in each small element. The increase in accuracy of the finite element method is achieved by using smaller elements for a given domain and by choosing the appropriate number of nodal points to best describe the deformation of an element. An obvious advantage of the finite element approach for a given domain over the Ritz method for the same domain is that the finite element approach, with its small simple elements, can be made to fit a complex boundary while using simple interpolation functions for a (given) desired accuracy. The Ritz method, on the other hand, requires each of the coordinate functions to fit the entire boundary of the whole domain—a requirement that may be quite difficult to fulfill.

Returning to an element analysis, recall that in Chapter 9 we set the first variation of the quadratic functional equal to zero to extremize the functional, and then we expressed the displacement field in terms of the interpolation functions and nodal displacements to arrive at Eq. (9.47) for the element. We will now show that the resulting stiffness matrix must be symmetric. Thus, expressing the above procedure mathematically, we have

$$\begin{aligned} &\delta^{(1)} I = 0 \\ &\sum_j \frac{\partial I}{\partial a_j}\,\delta a_j = 0 \end{aligned} \tag{17.3}$$

Hence, for arbitrary and independent variations δa_i,

$$\frac{\partial I}{\partial a_j} \equiv \begin{Bmatrix} \dfrac{\partial I}{\partial a_1} \\ \vdots \\ \dfrac{\partial I}{\partial a_n} \end{Bmatrix} = 0$$

This leads to the result:

$$\begin{Bmatrix} \dfrac{\partial I}{\partial a_1} \\ \vdots \\ \dfrac{\partial I}{\partial a_n} \end{Bmatrix} \equiv [K]^e\{a\} + \{f\}^e - \{q\}^e = 0 \tag{17.4}$$

Let us now take the first variation of both sides of the above identity. For the left side we have

$$\begin{aligned} \delta^{(1)} \begin{Bmatrix} \dfrac{\partial I}{\partial a_1} \\ \dfrac{\partial I}{\partial a_2} \\ \vdots \\ \dfrac{\partial I}{\partial a_n} \end{Bmatrix} &= \begin{Bmatrix} \left(\dfrac{\partial}{\partial a_1}\left(\dfrac{\partial I}{\partial a_1}\right)\delta a_1 + \dfrac{\partial}{\partial a_2}\left(\dfrac{\partial I}{\partial a_1}\right)\delta a_2 + \cdots + \dfrac{\partial}{\partial a_n}\left(\dfrac{\partial I}{\partial a_1}\right)\delta a_n\right) \\ \left(\dfrac{\partial}{\partial a_1}\left(\dfrac{\partial I}{\partial a_2}\right)\delta a_1 + \dfrac{\partial}{\partial a_2}\left(\dfrac{\partial I}{\partial a_2}\right)\delta a_2 + \cdots + \dfrac{\partial I}{\partial a_n}\left(\dfrac{\partial I}{\partial a_2}\right)\delta a_n\right) \\ \vdots \\ \left(\dfrac{\partial}{\partial a_1}\left(\dfrac{\partial I}{\partial a_n}\right)\delta a_1 + \dfrac{\partial}{\partial a_2}\left(\dfrac{\partial I}{\partial a_n}\right)\delta a_2 + \cdots + \dfrac{\partial}{\partial a_n}\left(\dfrac{\partial I}{\partial a_n}\right)\delta a_n\right) \end{Bmatrix} \\ &= \begin{bmatrix} \dfrac{\partial}{\partial a_1}\left(\dfrac{\partial I}{\partial a_1}\right) & \dfrac{\partial}{\partial a_2}\left(\dfrac{\partial I}{\partial a_1}\right) \cdots & \dfrac{\partial}{\partial a_n}\left(\dfrac{\partial I}{\partial a_1}\right) \\ \dfrac{\partial}{\partial a_1}\left(\dfrac{\partial I}{\partial a_2}\right) & \dfrac{\partial}{\partial a_2}\dfrac{\partial I}{\partial a_2} \cdots & \dfrac{\partial}{\partial a_n}\left(\dfrac{\partial I}{\partial a_2}\right) \\ \vdots & \vdots & \vdots \\ \dfrac{\partial}{\partial a_1}\left(\dfrac{\partial I}{\partial a_n}\right) & \dfrac{\partial}{\partial a_2}\left(\dfrac{\partial I}{\partial a_n}\right) \cdots & \dfrac{\partial}{\partial a_n}\left(\dfrac{\partial I}{\partial a_n}\right) \end{bmatrix} \begin{Bmatrix} \delta a_1 \\ \delta a_2 \\ \vdots \\ \delta a_n \end{Bmatrix} \end{aligned} \tag{17.5}$$

Hence:

$$\delta^{(1)} \begin{Bmatrix} \dfrac{\partial I}{\partial a_1} \\ \dfrac{\partial I}{\partial a_2} \\ \vdots \\ \dfrac{\partial I}{\partial a_n} \end{Bmatrix} = [K_T]^e \{a\}^e \tag{17.6}$$

The matrix $[K_T]^e$ is called the *tangent* matrix and is important for nonlinear analysis. Note that the terms of $[K_T]$ are

$$(K_T)_{ij} = \frac{\partial^2 I}{\partial a_i \, \partial a_j} \tag{17.7}$$

We see immediately that the tangent matrix $[K]_T$ is *symmetric*. Now go back to Eq. (17.4) and take the first variation of the right side. We get

$$\delta^{(1)}\left([K]^e\{a\} + \{f\}^e - \{q\}^e\right) = [K]^e\{\delta a\} + 0 + 0 \tag{17.8}$$

By comparing the right sides of Eqs. (17.6) and (17.8), we see that for the situation at hand $[K_T]^e = [K]^e$, and so we may conclude that $[K]$ must be symmetric. This is important when we generalize the results of finite elements to fields of study other than linear elasticity.

17.3 GENERALIZATION OF THE FINITE ELEMENT METHOD BEYOND LINEAR ELASTICITY

In Chapter 3 we extremized a specific quadratic functional to solve approximately the corresponding Euler-Lagrange equations. The admissible functions then covered the whole domain of the problem, as indicated earlier. For *homogeneous* boundary conditions the quadratic functional was given as follows, when the operator L of the boundary-value problems was self-adjoint and positive definite:

$$I(u) = \langle Lu, u \rangle - 2\langle f, u \rangle \tag{17.9}$$

where the corresponding Euler-Lagrange equation was of the form:

$$Lu = f \tag{17.10}$$

Furthermore, in Section 3.13 we showed that the following quadratic functional could be used with certain *nonhomogeneous* boundary conditions:

$$J(u) = \langle Lu, u \rangle - 2\langle u, f \rangle + \iint_S P(u) \, dS \tag{17.11}$$

For the Ritz method, in Parts I and II of the text we proceeded to extremize the functional with respect to a subset ϕ_i of all admissible fields by adjusting the values of the constant coefficients of the subset ϕ_i. We used the Ritz method primarily for linear elastic behavior, using various forms of the total potential energy for specific geometries.

For finite elements in Part III we have continued to use the Ritz method, this time for small simple elements of various structures. The functional was either the total potential energy or the total complementary energy. From our studies in Chapter 3, we can now *extend* the finite element approach by using quadratic functionals (17.9) and (17.11) whose Euler-Lagrange equations correspond to fields of study *other* than solid and structural mechanics.

Accordingly, by using appropriate interpolation functions $[N]$ we extremize such functionals with respect to nodal "displacements," which might be temperatures, pressures, or velocities (instead of the actual translations and rotations that we have been using up to this point), to achieve approximate solutions.

We shall now consider an *ordinary differential* equation with the purpose of solving this equation via finite elements. We must first find a quadratic functional for the equation. We will not be able to use Eqs. (17.9) and (17.11) here to get the appropriate quadratic functional. Instead, we will illustrate a way of arriving at I via *manipulation.* The equation we shall study is the same one considered in Part D of Chapter 3.

EXAMPLE 17.1 The differential equation is

$$x^2 \frac{d^2 y}{dx^2} + 2x \frac{dy}{dx} = 6x \qquad (a)$$

where $y(1) = y(2) = 0$ are the boundary conditions. Determine the general form of $[K]^e$ and $\{q\}$.

The operator for the above equation is

$$L = \frac{d}{dx} x^2 \frac{d}{dx} \qquad (b)$$

For finite elements, the boundary conditions will *not be* homogeneous for each element. Homogeneity occurs only at the left of the first element ($x = 1$) and at the right of the last element ($x = 2$). Hence we will form a quadratic functional here by *manipulation.*

Thus we shall multiply $(Ly - f)$ by δy, where we take $\delta y = 0$ at the boundaries, thus requiring the varied functions to have the same boundary conditions as the extremal function. We have

$$\delta y \left(\frac{d}{dx} x^2 \frac{dy}{dx} \right) - \delta y \, 6x = 0 \qquad (c)$$

Integrate over the length of one element with nodes i and j:

$$\int_{L_{i\text{-}j}} \left[\delta y \left(\frac{d}{dx} x^2 \frac{dy}{dx} \right) - \delta y \, 6x \right] dx = 0 \qquad (d)$$

Now integrate the first term in the integral by parts:

$$(\delta y)x^2\left(\frac{dy}{dx}\right)\Bigg|_i^j-\int_{L_{i\text{-}j}}\left\{\left[\delta\left(\frac{dy}{dx}\right)\right]x^2\left(\frac{dy}{dx}\right)+6x\,\delta y\right\}dx=0$$

Noting the requirement $\delta y = 0$ that we have imposed at the ends, and rearranging the above integrand, we have†

$$\int_{L_{i\text{-}j}}\left[\frac{1}{2}\delta\left(\frac{dy}{dx}x^2\frac{dy}{dx}\right)+6x\,\delta y\right]dx=0$$

This may be written as

$$\delta^{(1)}\left\{\int_{L_{i\text{-}j}}\left[\frac{1}{2}x^2\left(\frac{dy}{dx}\right)^2+6xy\right]dx\right\}=0 \qquad (e)$$

Hence, the desired quadratic functional is

$$I_{i\text{-}j}(y)=\int_{L_{i\text{-}j}}\left[\frac{1}{2}x^2\left(\frac{dy}{dx}\right)^2+6xy\right]dx \qquad (f)$$

Consider that we have m elements between $x = 1$ and $x = 2$, each of length L. We then express the function y in terms of the interpolation functions N_i. Note that there will be two interpolation functions for each element because there will be two nodal "displacements" for each element. Thus

$$y=\sum_i N_i a_i=\{N\}^T\{a\} \qquad (g)$$

Going back to Eq. (f) and considering the pth element, we substitute for y from Eq. (g):

$$I=\int_{L_P}\left[\frac{x^2}{2}\left(\sum_i a_i\frac{dN_i}{dx}\right)^2+6x\left(\sum_i a_iN_i\right)\right]dx \qquad (h)$$

We now extremize with respect to a_l, where l is a free index:

$$\delta^{(1)}I=0=\left\{\int_{L_P}\left[x^2\left(\sum_i a_i\frac{dN_i}{dx}\right)\left(\frac{dN_l}{dx}\right)+6xN_l\right]dx\right\}\delta a_l=0 \qquad (i)$$

We may put this equation in matrix form by setting the coefficients of δa_l equal to zero as follows:

†Note that the natural boundary conditions are $x^2\,dy/dx = 0$ at the nodes.

$$[K]^p\{a\}^p = -\{q\}^p$$

If we change the dummy index i to n in Eq. (i), we get for the terms K_{ln} of $[K]^p$:

$$K_{ln} = \int_{L_p} x^2 \left(\frac{dN_l}{dx}\right)\left(\frac{dN_n}{dx}\right) dx \qquad (j)$$

Furthermore, the $\{q\}$ vector is seen to be

$$q_l = \int_{L_p} 6xN_i\, dx \qquad (k)$$

EXAMPLE 17.2 Using four elements each of length $\frac{1}{4}$ unit, solve the boundary-value problem in Example 17.1. Do the same for a smaller mesh and observe the convergence to the exact solution.

We consider now the interpolation functions. Because we have two degrees of freedom for each element, we expect to use a two-term polynomial to express y:

$$y = \alpha_1 + \alpha_2 x \qquad (a)$$

Now going to Eq. (9.59a), we have for an element with nodes i and j:

$$y = [N_i(x)]a_i + [N_j(x)]a_j \qquad (b)$$

In addition to Eq. (a), we require that when $x = x_i$, then $y = a_i$. This in turn means that $N_i(x_i) = 1$ and $N_j(x_i) = 0$. Furthermore, when $x = x_j$, then $y = a_j$, so that $N_i(x_j) = 0$ and $N_j(x_j) = 1$. You can readily demonstrate (see Example 11.1) that the following formulations for the N's satisfy the aforementioned requirements.

$$y^e(x) = \underbrace{\frac{x_j - x}{x_j - x_i}}_{N_i(x)} a_i + \underbrace{\frac{x - x_i}{x_j - x_i}}_{N_j(x)} a_j \qquad (c)$$

We will now work out the details for element 1 from $x = 1$ to $x = 1\frac{1}{4}$. Thus, using Eq. (j) of the previous example, we have for term $K_{11}^{(1)}$:

$$K_{11}^{(1)} = \int_1^{1\frac{1}{4}} x^2 \left(\frac{dN_1}{dx}\right)\left(\frac{dN_1}{dx}\right) dx$$

$$= \int_1^{1\frac{1}{4}} x^2 \left(\frac{-1}{x_2 - x_1}\right)\left(\frac{-1}{x_2 - x_1}\right) dx$$

$$= \int_1^{1\frac{1}{4}} x^2 \left(\frac{-1}{1/4}\right)\left(\frac{-1}{1/4}\right) dx = 5.08 \qquad (d)$$

Finding $K_{12}^{(1)}$, $K_{21}^{(1)}$, and $K_{22}^{(1)}$, we get

$$[K]^{(1)} = \begin{bmatrix} 5.08 & -5.08 \\ -5.08 & 5.08 \end{bmatrix} \qquad (e)$$

We do the same for the other four elements and get the global stiffness matrix:

$$[K] = \begin{bmatrix} \begin{bmatrix} K^1 \end{bmatrix} & & & \\ & \begin{bmatrix} K^2 \end{bmatrix} & & \\ & & \begin{bmatrix} K^3 \end{bmatrix} & \\ & & & \begin{bmatrix} K^4 \end{bmatrix} \end{bmatrix} \qquad (f)$$

where there are zeroes outside the element matrices shown above. Inserting numbers from $[K]^{(1)}$, $[K]^{(2)}$, etc., and adding at the overlaps, we get the global matrix in the form:

$$[K] = \begin{matrix} & 1 & 2 & 3 & 4 & 5 & \\ & \begin{bmatrix} 5.08 & -5.08 & 0 & 0 & 0 \\ -5.08 & 12.66 & -7.58 & 0 & 0 \\ 0 & -7.58 & 18.16 & -10.58 & 0 \\ 0 & 0 & -10.58 & 24.66 & -14.08 \\ 0 & 0 & 0 & -14.08 & 14.08 \end{bmatrix} & \begin{matrix} 1 \\ 2 \\ 3 \\ 4 \\ 5 \end{matrix} \end{matrix} \qquad (g)$$

As for the nodal loads, we have for element 1 from Eq. (k) of the previous example:

$$q_1^{(1)} = \int_1^{1\frac{1}{4}} 6xN_1 \, dx = \int_1^{1\frac{1}{4}} 6x \frac{x_2 - x}{x_2 - x_1} \, dx$$

$$= \int_1^{1\frac{1}{4}} 6x \frac{5/4 - x}{1/4} \, dx = 0.8125$$

$$\qquad (h)$$

$$q_2^{(1)} = \int_1^{1\frac{1}{4}} 6xN_2 \, dx = \int_1^{1\frac{1}{4}} 6x \frac{x - x_1}{x_2 - x_1} \, dx$$

$$= \int_1^{1\frac{1}{4}} 6x \frac{x - 1}{1/4} \, dx = 0.875$$

Similarly, we can find $q_2^{(2)}$ and $q_3^{(2)}$, $q_3^{(3)}$ and $q_4^{(3)}$, etc. Adding $q_2^{(1)}$ and $q_2^{(2)}$, $q_3^{(2)}$ and $q_3^{(3)}$, and so forth, we get

$$\begin{bmatrix} 5.08 & -5.08 & 0 & 0 & 0 \\ -5.08 & 12.66 & -7.58 & 0 & 0 \\ 0 & -7.58 & 18.16 & -10.58 & 0 \\ 0 & 0 & -10.58 & 24.66 & -14.08 \\ 0 & 0 & 0 & -14.08 & 14.08 \end{bmatrix} \begin{Bmatrix} a_1 \\ a_2 \\ a_3 \\ a_4 \\ a_5 \end{Bmatrix} = \begin{Bmatrix} -0.8125 \\ -1.875 \\ -2.247 \\ -2.625 \\ -1.4375 \end{Bmatrix} \tag{i}$$

Now with $a_1 = a_5 = 0$, we can eliminate the first row and column and the last row and column to form the reduced problem by inserting the boundary conditions. Thus we have

$$\begin{bmatrix} 12.66 & -7.58 & 0 \\ -7.58 & 18.16 & -10.58 \\ 0 & -10.58 & 24.66 \end{bmatrix} \begin{Bmatrix} a_2 \\ a_3 \\ a_4 \end{Bmatrix} = \begin{Bmatrix} -1.875 \\ -2.247 \\ -2.625 \end{Bmatrix} \tag{j}$$

We can solve this readily to find

$$\begin{Bmatrix} a_2 \\ a_3 \\ a_4 \end{Bmatrix} = \begin{Bmatrix} -0.449 \\ -0.495 \\ -0.319 \end{Bmatrix}$$

To compare results with the exact solution we have plotted $y(x)_{\text{exact}}$ and the above results in Fig. 17.1. We have also shown in the diagram finite element results for an eight-element calculation.

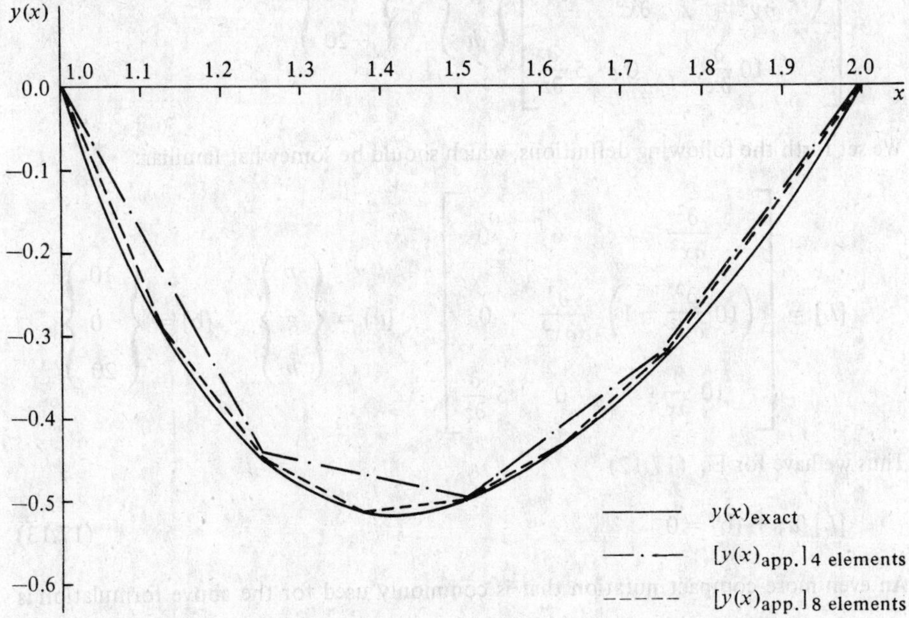

FIGURE 17.1
Exact and approximation solutions for Example 17.2.

*17.4 RESTATEMENT OF KEY VARIATIONAL PRINCIPLES IN MATRIX NOTATION MOST SUITABLE FOR FINITE ELEMENTS

In the previous section we showed the link between variational principles and finite elements and showed briefly also how we could extrapolate the method to fields of study other than solid mechanics. For this reason, we shall go back to certain formulations of Chapter 3 and restate them in a more general manner suitable for the application of finite elements.

First recall that in Part C of Chapter 3 we considered a single differential equation, $Lu = f$. Suppose we have a system of simultaneous equations such as:

$$\begin{aligned}
&\frac{\partial^2 p}{\partial x^2} + 6g + 10 = 0 \\
&\frac{\partial^3 g}{\partial x^3} + 10\,\frac{\partial^2 p}{\partial y^2} + p = 0 \\
&10\,\frac{\partial p}{\partial x} + 5\,\frac{\partial h}{\partial z} + 20 = 0
\end{aligned} \tag{17.12}$$

In matrix form these equations are expressed as:

$$\begin{bmatrix} \dfrac{\partial^2}{\partial x^2} & 6 & 0 \\ \left(10\,\dfrac{\partial^2}{\partial y^2} + 1\right) & \dfrac{\partial^3}{\partial x^3} & 0 \\ 10\,\dfrac{\partial}{\partial x} & 0 & 5\,\dfrac{\partial}{\partial z} \end{bmatrix} \begin{Bmatrix} p \\ g \\ h \end{Bmatrix} = \begin{Bmatrix} -10 \\ 0 \\ -20 \end{Bmatrix}$$

We set forth the following definitions, which should be somewhat familiar:

$$[L] = \begin{bmatrix} \dfrac{\partial^2}{\partial x^2} & 6 & 0 \\ \left(10\,\dfrac{\partial^2}{\partial y^2} + 1\right) & \dfrac{\partial^3}{\partial x^3} & 0 \\ 10\,\dfrac{\partial}{\partial x} & 0 & 5\,\dfrac{\partial}{\partial z} \end{bmatrix} \qquad \{u\} = \begin{Bmatrix} p \\ g \\ h \end{Bmatrix} \qquad \{\mathfrak{b}\} = \begin{Bmatrix} 10 \\ 0 \\ 20 \end{Bmatrix}$$

Thus we have for Eq. (17.12)

$$[L]\{u\} + \{\mathfrak{b}\} = 0 \tag{17.13}$$

An even more compact notation that is commonly used for the above formulation is

$$[A(u)] = 0 \tag{17.14}$$

The self-adjoint property for the system of differential equations now becomes

$$\int_{\Omega} \{v\}^T [L] \{u\}\, d\Omega = \int_{\Omega} \{u\}^T [L] \{v\}\, d\Omega$$

where $\{v\}$ and $\{u\}$ are vectors satisfying appropriate boundary conditions in the field of definition of L. The positive-definite requirement now becomes

$$\int_{\Omega} \{u\}^T [L] \{u\}\, d\Omega \geqslant 0$$

where, for the equality to hold, $\{u\}$ must be a null vector. For self-adjoint, positive definite linear operators, the quadratic functional presented in Chapter 3 for homogeneous boundary conditions is

$$I = \int_{\Omega} (uLu - 2uf)\, d\Omega$$

In the more general context, this now becomes

$$I = \int_{\Omega} \left(\{u\}^T [L] \{u\} + 2\{u\}^T \{\mathfrak{b}\}\right) d\Omega$$

It is customary in finite elements to take for I one half of the above value. Thus we have†

$$\boxed{I = \int_{\Omega} \left(\tfrac{1}{2} \{u\}^T [L] \{u\} + \{u\}^T \{\mathfrak{b}\}\right) d\Omega} \qquad (17.15)$$

We can prove that Eq. (17.15) is valid for the conditions presented. Take the first variation of I and set it equal to zero. We get

$$\delta^{(1)} I = 0 = \int_{\Omega} \left(\tfrac{1}{2} \{\delta u\}^T [L] \{u\} + \tfrac{1}{2} \{u\}^T \delta([L]\{u\}) + \{\delta u\}^T \{\mathfrak{b}\}\right) d\Omega$$

From our work in Chapter 3, we note that in the second expression of the integrand we can interchange the order of the differential operator $[L]$ and the δ operator. Thus we get

†Since we set the first variation equal to zero, the multiplicative factor plays no role in the variational process. Also, the "energy" term $\frac{1}{2}\{u\}^T [L] \{u\}$ now becomes the kinetic energy expression in structural dynamics.

$$\delta^{(1)}I = 0 = \int_{\Omega} \left(\tfrac{1}{2} \{\delta u\}^T [L] \{u\} + \tfrac{1}{2} \{u\}^T [L] \{\delta u\} + \{\delta u\}^T \{\mathfrak{b}\} \right) d\Omega = 0$$

Now use the self-adjoint property of L and note that the first two expressions in the integrand are identical.† Thus,

$$\int_{\Omega} \{\delta u\}^T \left([L] \{u\} + \{\mathfrak{b}\} \right) d\Omega = 0 \tag{17.16}$$

With $\{\delta u\}^T$ arbitrary and with homogeneous boundary conditions, we must conclude from the fundamental lemma that

$$[L] \{u\} + \{\mathfrak{b}\} = 0$$

We have thus demonstrated that the quadratic functional given by Eq. (17.15) is properly stated for Eq. (17.13).

We now present another commonly used representation of the quadratic functional I, this time in terms of the nodal displacements. Accordingly, we first present and then show the validity of the following form for I:

$$\boxed{I = \tfrac{1}{2} \{a\}^T [K] \{a\} + \{a\}^T \{\mathfrak{b}\}} \tag{17.17}$$

Let us now set to zero the first variation of this functional

$$\delta^{(1)}I = 0 = \tfrac{1}{2} \{\delta a\}^T [K] \{a\} + \tfrac{1}{2} \{a\}^T [K] \{\delta a\} + \{\delta a\}^T \{\mathfrak{b}\} \tag{17.18}$$

We showed earlier that the "stiffness" matrix $[K]$ is symmetric. You may then demonstrate yourself that for a matrix $[E]$ of appropriate numbers of rows and columns, the following statement is true:

$$[E]^T [K] [E] = [E] [K] [E]^T$$

By utilizing this fact in Eq. (17.18), we can combine the first two expressions on the right side of the equation so that we have as a result

$$\{\delta a\}^T \left([K] \{a\} + \{\mathfrak{b}\} \right) = 0$$

Since $\{\delta a\}^T$ is arbitrary we conclude

$$[K] \{a\} + \{\mathfrak{b}\} = 0$$

†Note that the variation of u must be consistent with the constraints, and so $\{\delta u\}$ satisfies the same homogeneous boundary conditions as $\{u\}$.

Thus by using Eq. (17.17) for I and setting the first variation equal to zero, we arrive at the proper matrix equation for the finite element process. Hence I may be given as in Eq. (17.17).

*17.5 CONSTRAINED EXTREMIZATION PROCESS FOR FINITE ELEMENTS

First, let us consider the extremization of $I(u_1, \ldots, u_n)$ under a number of constraints that are expressed in the form

$$\begin{aligned} h_1(u_1, \ldots, u_n) &= 0 \\ &\vdots \\ h_m(u_1, \ldots, u_n) &= 0 \end{aligned} \tag{17.19}$$

This set may put into matrix notation as follows:

$$[G]\{u\} + \{Q\} = 0 \tag{17.20}$$

where $[G]$ is an operator. You will recall from Chapter 3 that to extremize a functional while maintaining the function constraints given by Eq. (17.19), we formed I^* with Lagrange multiplier functions. That is,

$$I^* = \int_\Omega F^*\, d\Omega \tag{17.21}$$

where

$$F^* = F + \sum_{i=1}^{m} \lambda_i h_i \tag{17.22}$$

This leads to the following formulation for I:

$$I^* = I + \sum_{i=1}^{m} \int_\Omega \lambda_i h_i\, d\Omega = I + \int_\Omega \{\lambda\}^T \big([G]\{u\} + \{Q\}\big)\, d\Omega \tag{17.23}$$

We now consider the first variation of I^*. Using the delta operator approach, we obtain

$$\delta^{(1)} I^* = 0 = \delta^{(1)} I + \int_\Omega \delta\{\lambda\}^T \big([G]\{u\} + \{Q\}\big)\, d\Omega + \int_\Omega \{\lambda\}^T \delta\big([G]\{u\} + \{Q\}\big)\, d\Omega \tag{17.24}$$

Now to get to finite elements, we shall use interpolation functions N_i for the nodal values a_i and interpolation functions $\bar{N}_i$ for the Lagrange multiplier functions. Thus, for each function u_j and for each Lagrange multiplier function λ_j we have for an element e

$$(u_j)^e = \sum_i (N_i)_j a_i \tag{17.25a}$$

$$\lambda_j = \sum_i (\bar{N}_i)_j e_i \tag{17.25b}$$

where e_i are nodal parameters associated with the λ_j functions. In matrix format we get

$$\{u\}^e = [N]^e \{a\}^e \tag{17.26a}$$

$$\{\lambda\}^e = [\bar{N}]^e \{e\}^e \tag{17.26b}$$

If $[L]\{u\} + \{\mathfrak{b}\} = 0$ forms the Euler-Lagrange equation for the functional $I(u)$, then $\delta^{(1)}I(u)$ can be represented as [see Eq. (17.16)]

$$\delta^{(1)}I(u) = \int_\Omega \{\delta u\}^T\big([L]\{u\} + \{\mathfrak{b}\}\big)\, d\Omega \tag{17.27}$$

which you will recall is one step before using the fundamental lemma. Hence we can express Eq. (17.24) in the following form when we insert the interpolation function as per Eqs. (17.26) and use the above equation:

$$\int_\Omega \{\delta a\}^T [N]^T\big([L]\{u\} + \{\mathfrak{b}\}\big)\, d\Omega + \int_\Omega \{\delta e\}^T [\bar{N}]^T\big([G][N]\{a\} + \{Q\}\big)\, d\Omega$$

$$+ \int_\Omega \{e\}^T [\bar{N}]^T [G][N]\{\delta a\}\, d\Omega = 0$$

where in the last expression $\delta\{Q\} = 0$. Now extract $\{\delta a\}^T$, $\{\delta e\}^T$, $\{e\}^T$, and $\{\delta a\}$ from the integrals. We then have for the above equation

$$\{\delta a\}^T \int_\Omega [N]^T\big([L][N]\{a\} + \{\mathfrak{b}\}\big)\, d\Omega + \{\delta e\}^T \int_\Omega [\bar{N}]^T\big([G][N]\{a\} + \{Q\}\big)\, d\Omega$$

$$+ \{e\}^T \left(\int_\Omega [\bar{N}]^T [G][N]\, d\Omega\right) \delta a = 0 \tag{17.28}$$

We will next rewrite the last expression. You may verify that

$$\{e\}^T \left(\int_\Omega [\bar{N}]^T [G] [N]\, d\Omega \right) \{\delta a\} = \{\delta a\}^T \left(\int_\Omega [N]^T [G]^T [\bar{N}]\, d\Omega \right) \{e\} \qquad (17.29)$$

Inserting this result in Eq. (17.28) and rearranging the integrals, we get

$$\{\delta a\}^T \left[\int_\Omega [N]^T ([L][N]\{a\} + \{\mathfrak{b}\})\, d\Omega + \left(\int_\Omega [N]^T [G]^T [\bar{N}]\, d\Omega \right) \{e\} \right]$$
$$+ \{\delta e\}^T \int_\Omega [\bar{N}]^T ([G][N]\{a\} + \{Q\})\, d\Omega = 0 \qquad (17.30)$$

We use the fundamental lemma next to form the results:

$$\int_\Omega [N]^T ([L][N]\{a\} + \{\mathfrak{b}\})\, d\Omega + \left(\int_\Omega [N]^T [G]^T [\bar{N}]\, d\Omega \right) \{e\} = 0 \qquad (17.31a)$$

$$\left(\int_\Omega [\bar{N}]^T ([G][N])\, d\Omega \right) \{a\} + \int_\Omega [\bar{N}]^T \{Q\}\, d\Omega = 0 \qquad (17.31b)$$

The first integral in Eq. (17.31*a*) can be put in the form

$$[K]\{a\} + \{q\} \qquad (17.32)$$

where

$$[K] = \int_\Omega [N]^T [L][N]\, d\Omega \qquad (17.33a)$$

$$\{q\} = \int_\Omega [N]^T \{\mathfrak{b}\}\, d\Omega \qquad (17.33b)$$

We next set forth the following definition for the second integral in Eq. (17.31*a*):

$$[\bar{K}] = \int_\Omega [N]^T [G]^T [\bar{N}]\, d\Omega \qquad (17.34)$$

Note that the first integral in Eq. (17.31*b*) is the transpose of $[\bar{K}]$. Finally, for the last integral we have

$$\{g\} = \int_{\Omega} [\bar{N}]^T \{Q\}\, d\Omega \tag{17.35}$$

We then have, on using the preceding definitions,

$$[K]\{a\} + [\bar{K}]\{e\} + \{q\} = 0 \tag{17.36a}$$

$$[\bar{K}]^T\{a\} + \{g\} = 0 \tag{17.36b}$$

Thus we can again put this problem in its usual form for finite elements:

$$[K]^e_{\text{total}}\{a\}_{\text{total}} + \{f\}_{\text{total}} = 0 \tag{17.37a}$$

where

$$[K]^e_{\text{total}} = \begin{bmatrix} [K] & [\bar{K}] \\ [\bar{K}]^T & 0 \end{bmatrix} \tag{17.37b}$$

$$\{a\}_{\text{total}} = \begin{Bmatrix} \{a\} \\ \{e\} \end{Bmatrix} \qquad \{f\}_{\text{total}} = \begin{Bmatrix} \{q\} \\ \{g\} \end{Bmatrix} \tag{17.37c}$$

We shall not have space in our limited treatment to employ the above results. It is to be pointed out that some boundary conditions may profitably be considered as constituting constraints, for which the formulations set forth can at times be advantageously employed.

In Part A of this chapter it was shown how we could extend the finite element method to fields of study other than solid mechanics. In Part B we shall consider finite elements for heat conduction.

Part B
Steady-State Heat Conduction

17.6 THE QUADRATIC FUNCTIONAL AND BASIC MATRICES

We can now describe in finite element format steady-state, two-dimensional heat conduction in a solid (see Fig. 17.2). The field variable is the temperature $T(x, y)$ and it satisfies Laplace's equation for an isotropic, homogeneous medium:

$$k\,\nabla^2 T = 0 \tag{17.38}$$

where k is the heat conductivity coefficient and is a constant. On the boundary Γ_I, we consider both heat *conduction*, given as $\tilde{q}(x, y)$ watts per meter squared, and heat

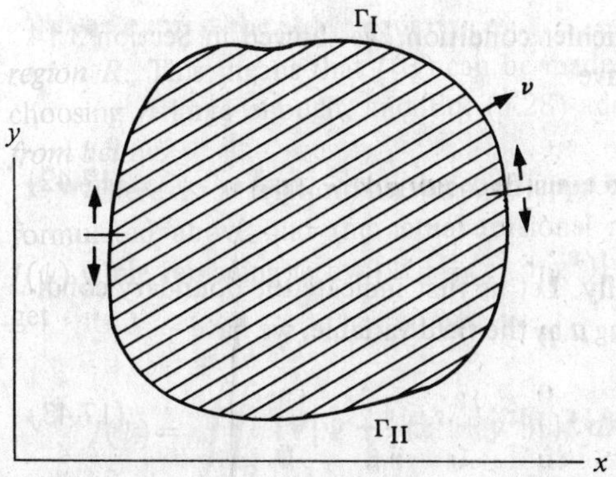

FIGURE 17.2
A unit slice of an infinite prism subject to steady-state heat transfer.

convection, given as $\hbar(T - T_\infty)$ watts per meter squared, where $\hbar$ is the heat transfer convection coefficient and T_∞ is the constant temperature of the surroundings outside Γ_I. A positive $\tilde{q}(x, y)$ will signify heat *leaving* the body at the boundary. On the remaining boundary Γ_{II}, we will specify a constant (in time) temperature distribution $\overline{T}(x, y)$. Thus, on using the well-known *Fourier conduction law* on Γ_I, we have for boundary conditions:

$$k\frac{\partial T}{\partial \nu} + \tilde{q} + \hbar(T - T_\infty) = 0 \tag{17.39}$$

where ν is the outward normal coordinate along the boundary. And on Γ_{II} we have

$$T(x, y) = \overline{T}(x, y) \tag{17.40}$$

You will note that on surface Γ_{II} we have a Dirichlet-type boundary condition, while on surface Γ_I we have what for us is a new, nonhomogeneous boundary condition. To determine the quadratic functional for this boundary-value problem, we must return to Section 3.13. Working in two dimensions rather than three, we first set forth Eq. (3.74) as a starting point for finding the desired quadratic functional. Thus, for $\Gamma_I + \Gamma_{II}$ we have

$$\langle u, L\psi\rangle - \langle \psi, Lu\rangle = \int_{\Gamma_I + \Gamma_{II}} \big(uQ(\psi) - \psi Q(u)\big)\, dl$$

where Q is an operator stemming from the use of Green's formula. For the harmonic operator $(-k\nabla^2)$ as L, we have for the above equation [see Eq. (I.38) in Appendix I]

$$-k\big(\langle u, \nabla^2\psi\rangle - \langle \psi, \nabla^2 u\rangle\big) = -\int_{\Gamma_I + \Gamma_{II}} k\left(u\frac{\partial \psi}{\partial \nu} - \psi\frac{\partial u}{\partial \nu}\right) dl \tag{17.41}$$

For boundary Γ_{II}, where we have a Dirichlet condition, we showed in Section 3.14 that on the right side of Eq. (17.41) we have

$$\int_{\Gamma_{II}} kT \frac{\partial T}{\partial \nu} \, dl \tag{17.42}$$

However, for Γ_I we must proceed carefully. Let us first indicate the boundary condition [Eq. (17.39)] in a simpler form. Using u as the field variable, we have

$$k \frac{\partial u}{\partial \nu} + Cu = h(x, y) \tag{17.43}$$

where C is a constant. Similarly, for a field variable ψ we have

$$k \frac{\partial \psi}{\partial \nu} + C\psi = h(x, y) \tag{17.44}$$

Hence, going back to Eq. (17.41) and using the above two equations, we have for boundary Γ_I on the right side of this equation

$$-\int_{\Gamma_I} [u(h - C\psi) - \psi(h - Cu)] \, dl = \int_{\Gamma_I} (\psi h - uh) \, dl \tag{17.45}$$

Hence, we see that the function $P(u)$ as presented in Section 3.13 [see Eqs. (3.74) and (3.75)] is

$$P(u) = -uh$$

$$\text{Thus } \oint P(u) \, dl = -\oint uh \, dl \tag{17.46}$$

Having determined the right side of Eq. (17.41) for Γ_I and Γ_{II}, we can now go back to Eq. (3.79). In this equation, we have only the boundary condition for Γ_I. Including now the boundary condition for Γ_{II} in addition, we get for Eq. (3.79) for the field variable T the result:

$$I(T) = \langle LT, T \rangle - 2\langle T, 0 \rangle + \int_{\Gamma_{II}} kT \frac{\partial T}{\partial \nu} \, dl - \int_{\Gamma_I} Th(x, y) \, dl + \text{const} \tag{17.47}$$

Comparing Eqs. (17.39) and (17.44), we note that $C = \hbar$, so that for $h(x, y)$ we have:

$$h(x, y) = \hbar T_\infty - \tilde{q}(x, y) \tag{17.48}$$

and so Eq. (17.47) becomes

$$I(T) = \langle LT, T \rangle - 2\langle T, 0 \rangle + \int_{\Gamma_{II}} kT \frac{\partial T}{\partial \nu} \, dl - \int_{\Gamma_I} T[\hbar T_\infty - \tilde{q}(x, y)] \, dl + \text{const} \tag{17.49}$$

Noting Eq. (3.87), we have for $\langle LT, T\rangle$

$$\langle LT, T\rangle = -k \iint_{A} (T\nabla^2 T)\, dA = k \iint_{A} (\nabla T)^2\, dA - k \int_{\Gamma_{\text{I}}+\Gamma_{\text{II}}} T \frac{\partial T}{\partial \nu}\, dl \tag{17.50}$$

We now insert the above result into Eq. (17.49). Then set $\langle T, 0\rangle = 0$ and disregard the constant in Eq. (17.49). Finally, canceling the terms for Γ_{II} we get

$$I(T) = k \iint_{A} (\nabla T)^2\, dA - \int_{\Gamma_{\text{I}}} \left(kT \frac{\partial T}{\partial \nu} + T\hbar T_\infty - T\tilde{q} \right) dl$$

Replacing $k(\partial T/\partial \nu)$ by using Eq. (17.39), we then find

$$I(T) = k \iint_{A} (\nabla T)^2\, dA - \int_{\Gamma_{\text{I}}} T[-\tilde{q} - \hbar(T - T_\infty) + \hbar T_\infty - \tilde{q}]\, dl$$

Combining terms, we reach the desired quadratic functional:

$$\boxed{I = k \iint_{A} (\nabla T)^2\, dA + \int_{\Gamma_{\text{I}}} [2(T\tilde{q}) + \hbar T^2 - 2\hbar T T_\infty]\, dl} \tag{17.51}$$

Our next efforts will be to determine the "stiffness" matrix $[K]^e$ and the "force" vector $\{q\}^e$. The field variable will first be given in terms of interpolation functions as follows:

$$T(x, y)^e = \sum_i N_i a_i = [N]^e \{a\}^e \tag{17.52}$$

Note that a_i are the temperatures at the nodes. Now go to Eq. (17.51) and carry out the first variation, after replacing T as per Eq. (17.52). We have for the free index j

$$\frac{\partial I}{\partial a_j} = 0 = k \iint_{A^e} \left[\frac{\partial}{\partial a_j} \left(\frac{\partial}{\partial x} \sum_i N_i a_i \right)^2 + \frac{\partial}{\partial a_j} \left(\frac{\partial}{\partial y} \sum_i N_i a_i \right)^2 \right] dx\, dy$$

$$+ \int_{\Gamma_{\text{I}}} \left[2\tilde{q} \frac{\partial}{\partial a_j} \sum_i N_i a_i + \hbar \frac{\partial}{\partial a_j} \left(\sum_i N_i a_i \right)^2 - 2\hbar T_\infty \frac{\partial}{\partial a_j} \sum_i N_i a_i \right] dl$$

Carrying out the differentiation,

$$2k \iint_{A^e} \left[\left(\sum_i \frac{\partial N_i}{\partial x} a_i \right) \left(\frac{\partial N_j}{\partial x} \right) + \left(\sum_i \frac{\partial N_i}{\partial y} a_i \right) \left(\frac{\partial N_j}{\partial y} \right) \right] dx\, dy$$

$$+ 2 \int_{\Gamma_{\text{I}}} \left[\tilde{q} N_j + \hbar \left(\sum_i N_i a_i \right) N_j - \hbar T_\infty N_j \right] dl = 0$$

Canceling the factor 2 and rearranging the terms, we get for the above equation

$$\sum_i \left[k \iint_{A^e} \left(\frac{\partial N_i}{\partial x}\frac{\partial N_j}{\partial x} + \frac{\partial N_i}{\partial y}\frac{\partial N_j}{\partial y} \right) a_i \, dx\, dy \right]$$

$$+ \int_{\Gamma_I} \left(\hbar N_i N_j a_i + \tilde{q} N_j - \hbar T_\infty N_j \right) dl = 0 \tag{17.53}$$

The above equation can be put into the following form:

$$[K]^e\{a\} = -\{q\}^e - [\kappa]^e\{a\} \tag{17.54}$$

where the terms of K_{ij}, q_i, and κ_{ij} are

$$K_{ij} = k \iint_{A^e} \left(\frac{\partial N_i}{\partial x}\frac{\partial N_j}{\partial x} + \frac{\partial N_i}{\partial y}\frac{\partial N_j}{\partial y} \right) dx\, dy \tag{17.55}$$

$$q_i = \int_{\Gamma_I} (\tilde{q} - \hbar T_\infty) N_i \, dl \tag{17.56}$$

$$\kappa_{ij} = \int_{\Gamma_I} \hbar N_i N_j \, dl \tag{17.57}$$

The procedure we will follow in evaluating the above integrals is to label the nodes of any element we may be working with as i, j, k in a *counterclockwise* manner, so that in integrating from node i to node j or j to k, etc., we will have the area of the element to our left.Thus in Fig. 17.3 we show an element having node numbers 3, 6, and 8. Note that we can select any of the nodes to be i for our immediate calculations, and then j and k follow as described above. When we have carried out our computations for the element for Eqs. (17.55)-(17.57), we then use $i = 6, j = 8$, and $k = 3$ in assembling the global equations. We now consider the interpolation functions.

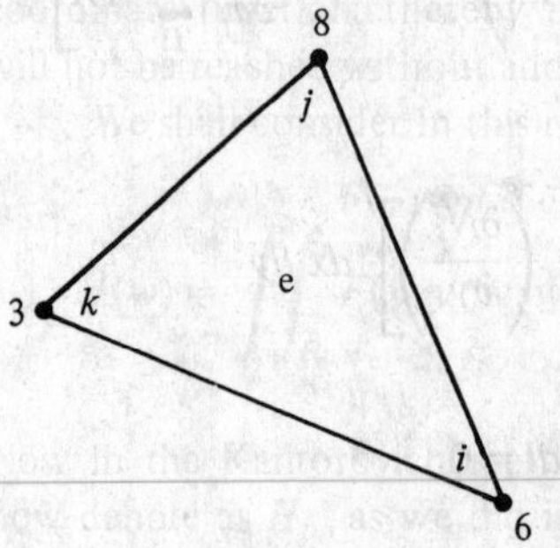

FIGURE 17.3
Element e, using temporary i, j, k identification of nodes.

17.7 INTERPOLATION FUNCTIONS

We already worked out in Section 14.3 interpolation functions for the 3-DOF triangular element that we are using here. We can take the formulations from Eqs. (14.9)-(14.16) to apply here. Substituting Eq. (14.16) into Eq. (17.55), we get for the K_{ij}^e terms of the stiffness matrix:

$$K_{ij}^e = \frac{k}{4A}(\bar{a}_i\bar{a}_j + \bar{b}_i\bar{b}_j) \tag{17.58}$$

The stiffness matrix for an element may now be expressed in detail:

$$[K]^e = \frac{k}{4A}\begin{bmatrix} (\bar{a}_i\bar{a}_i + \bar{b}_i\bar{b}_i) & (\bar{a}_i\bar{a}_j + \bar{b}_i\bar{b}_j) & (\bar{a}_i\bar{a}_k + \bar{b}_i\bar{b}_k) \\ (\bar{a}_j\bar{a}_i + \bar{b}_j\bar{b}_i) & (\bar{a}_j\bar{a}_j + \bar{b}_j\bar{b}_j) & (\bar{a}_j\bar{a}_k + \bar{b}_j\bar{b}_k) \\ (\bar{a}_k\bar{a}_i + \bar{b}_k\bar{b}_i) & (\bar{a}_k\bar{a}_j + \bar{b}_k\bar{b}_j) & (\bar{a}_k\bar{a}_k + \bar{b}_k\bar{b}_k) \end{bmatrix} \tag{17.59}$$

Now consider $\{\tilde{q}\}$ in Eq. (17.56). For global considerations, due to cancellation we will get zero contribution on all *inside* boundaries from the integration. Hence, q_i can be nonzero only at nodes that form the *outer* boundary of the body. We will use an average value of $\tilde{q}$ for a side of a triangular element forming an outer boundary. Note also from Eq. (14.19)

$$\{N\} = \begin{Bmatrix} L_i \\ L_j \\ L_k \end{Bmatrix} \tag{17.60}$$

Hence, we have for side ij of a triangular element e having heat conduction and convection

$$\{q\}_{i\to j} = \int_{i\to j} (\tilde{q}_{\text{av}} - \hbar T_\infty) \begin{Bmatrix} N_i \\ N_j \\ N_k \end{Bmatrix} dl \tag{17.61}$$

This may be written as

$$\{q\}_{i\to j} = (\tilde{q}_{\text{av}} - \hbar T_\infty) \int_{i\to j} \begin{Bmatrix} L_i \\ L_j \\ L_k \end{Bmatrix} dl \tag{17.62}$$

The integration formula along a line segment from i to j of element e in terms of natural coordinates is given as follows, using the notation l_{ij} for length of the line segment:

$$\int_{i\to j} L_i^\alpha L_j^\beta \, dl = \frac{\alpha!\,\beta!}{(\alpha+\beta+1)!}\, l_{ij} \tag{17.63}$$

Thus, going back to Eq. (17.62), we have for L_i

$$\int_{i\to j} L_i\,dl = \frac{(1!)(0!)}{(1+0+1)!}\,l_{ij} = \frac{l_{ij}}{2} \tag{17.64}$$

As for natural coordinate L_j, we get $l_{ij}/2$ also. Since natural coordinate L_k is zero along side $i \to j$, we get zero contribution to nodes i and j from this term. Thus we can say

$$\{q\}_{i\to j} = \int_{i\to j} (\tilde{q}_{\text{av}} - \hbar T_\infty) \begin{Bmatrix} L_i \\ L_j \\ L_k \end{Bmatrix} dl = (\tilde{q}_{\text{av}} - \hbar T_\infty)_{i\to j}\,\frac{l_{ij}}{2} \begin{Bmatrix} 1 \\ 1 \\ 0 \end{Bmatrix} \tag{17.65}$$

for the nodes i, j, k of element e from heat conduction and convection along line $i \to j$. Thus, we are putting heat transfer by conduction and convection for side $i \to j$ half on each node i and j. For a side $j \to k$ of the element we would do similarly, were this side a boundary of the overall domain:

$$\{q\}_{j\to k} = \int_{j\to k} (\tilde{q} - \hbar T_\infty) \begin{Bmatrix} L_i \\ L_j \\ L_k \end{Bmatrix} dl = (\tilde{q}_{\text{av}} - \hbar T_\infty)_{j\to k}\,\frac{l_{jk}}{2} \begin{Bmatrix} 0 \\ 1 \\ 1 \end{Bmatrix} \tag{17.66}$$

And for side $k \to i$ we would have, were it an outer boundary:

$$\{q\}_{k\to i} = \int_{k\to i} (\tilde{q} - \hbar T_\infty) \begin{Bmatrix} L_i \\ L_j \\ L_k \end{Bmatrix} dl = (\tilde{q}_{\text{av}} - \hbar T_\infty)_{k\to i}\,\frac{l_{ki}}{2} \begin{Bmatrix} 1 \\ 0 \\ 1 \end{Bmatrix} \tag{17.67}$$

Finally, we consider Eq. (17.57) for element e. We refer you to Fig. 17.3, where we will assume that side $i \to j$ is part of the overall boundary of the problem domain. The matrix $[\kappa]$ for this element is presented for side $i \to j$ as

$$[\kappa]^e_{i\to j} = \begin{bmatrix} \kappa_{ii} & \kappa_{ij} & \kappa_{ik} \\ \kappa_{ji} & \kappa_{jj} & \kappa_{jk} \\ \kappa_{ki} & \kappa_{kj} & \kappa_{kk} \end{bmatrix}^e_{i\to j} = \begin{bmatrix} \kappa_{66} & \kappa_{68} & \kappa_{63} \\ \kappa_{86} & \kappa_{88} & \kappa_{83} \\ \kappa_{36} & \kappa_{38} & \kappa_{33} \end{bmatrix}^e_{i\to j}$$

Let us consider the term $\kappa_{ii}\ (\equiv \kappa_{66})$. We have

$$(\kappa_{ii})^e_{i\to j} = \hbar \int_{i\to j} L_i L_i\,dl = \hbar \int_{i\to j} L_i^2\,dl \equiv \hbar \int_{i\to j} L_i^2 L_j^0\,dl$$

$$= \hbar\,\frac{2!\,0!}{(2+0+1)!}\,l_{ij} = \frac{\hbar l_{ij}}{3}$$

Next, let us consider the term κ_{ij} ($\equiv \kappa_{68}$). We have

$$(\kappa_{ij})^e_{i\to j} = \hbar \int_{i\to j} L_i L_j \, dl = \hbar \int_{i\to j} L_i^1 L_j^1 \, dl$$

$$= \hbar \frac{1!\,1!}{(1+1+1)!} l_{ij} = \frac{\hbar l_{ij}}{6} \tag{17.68}$$

We continue in this way to get all κ terms for element e stemming from side $i \to j$. Note in this regard that $L_k = 0$ along side $i \to j$ so that $\kappa_{ik} = \kappa_{ki} = \kappa_{jk} = \kappa_{kj} = \kappa_{kk} = 0$. We then have for $[\kappa]^e_{i\to j}$:

$$[\kappa]^e_{i\to j} = \frac{\hbar l_{ij}}{6} \begin{array}{c} \begin{matrix} i & j & k \end{matrix} \\ \begin{bmatrix} 2 & 1 & 0 \\ 1 & 2 & 0 \\ 0 & 0 & 0 \end{bmatrix} \end{array} \begin{matrix} i \\ j \\ k \end{matrix} \tag{17.69}$$

When we assemble the global matrix $[\kappa]$, the result from Eq. (17.69) for the element in Fig. 17.3 becomes

$$[\kappa]^e_{6\to 8} = \frac{\hbar l_{68}}{6} \begin{array}{c} \begin{matrix} 6 & 8 & 3 \end{matrix} \\ \begin{bmatrix} 2 & 1 & 0 \\ 1 & 2 & 0 \\ 0 & 0 & 0 \end{bmatrix} \end{array} \begin{matrix} 6 \\ 8 \\ 3 \end{matrix}$$

Suppose next we have another element e' with a side forming part of the boundary of the domain. We denote the nodes as i, j, and k as indicated earlier and say that side $j \to k$ is on the overall boundary. We can get the matrix $[\kappa]_{j\to k}$ for this element e' as follows, using the same kind of calculation done for the previous element e with side $i \to j$ on the boundary.

$$[\kappa]^{e'}_{j\to k} = \frac{\hbar l_{jk}}{6} \begin{array}{c} \begin{matrix} i & j & k \end{matrix} \\ \begin{bmatrix} 0 & 0 & 0 \\ 0 & 2 & 1 \\ 0 & 1 & 2 \end{bmatrix} \end{array} \begin{matrix} i \\ j \\ k \end{matrix} \tag{17.70}$$

Again, when assembling the global matrix $[\kappa]$ we would replace the i, j, k notation around the matrix above by the proper nodal numbers as we did earlier. Finally, if we have an element e'' where, after inserting i, j, k for the nodes, we have $k \to i$ as a side on the boundary of the domain, then we have for $[\kappa]^{e''}_{k\to i}$ of this element:

$$[\kappa]^{e''}_{k\to i} = \frac{\hbar l_{ki}}{6} \begin{array}{c} \begin{matrix} i & j & k \end{matrix} \\ \begin{bmatrix} 2 & 0 & 1 \\ 0 & 0 & 0 \\ 1 & 0 & 2 \end{bmatrix} \end{array} \begin{matrix} i \\ j \\ k \end{matrix} \tag{17.71}$$

We are now ready to apply this theory to a steady-state, two-dimensional heat conduction problem.

EXAMPLE 17.3 Find the steady-state temperature field for the two-dimensional configuration shown in Fig. 17.4. On the inside surface *BGHC*, we have a heat flow rate $\tilde{q} = -5 \times 10^5\ \text{W/m}^2$. The ambient temperature near this surface is 500 K. The ambient temperature near the outside surface around *AEFD* is 300 K, as shown in the diagram. The heat flow rate for the outside surface is $\tilde{q} = 4 \times 10^5\ \text{W/m}^2$. At surfaces *AB* and *CD* the temperature is maintained at 600 K. The constants for the problem are:

$$\hbar = 2 \times 10^5\ \text{W/m}^2\text{-K}$$

$$k = 200\ \text{W/m-K}$$

Note that there is a plane of symmetry for the problem as shown by the line *M–M*.

We will work the problem first with a rough grid and later with finer grids. We need only work with half of the geometry as indicated in Fig. 17.5, where we show element numbers (circled) and node numbers. Along line 8 → 9 between nodes 8 and 9 we will have no heat transfer because this is a line of symmetry.

Let us proceed by considering in detail element 4, having nodes 1, 3, and 5. Letting i denote node 1, j denote node 3, and k denote node 5 fow now, we have

$$\begin{cases} x_i = 0 \\ y_i = 0 \end{cases} \qquad \begin{aligned} \bar{a}_i &= x_k - x_j = 0 - 0.5 = -0.5\ \text{m} \\ b_i &= y_j - y_k = 0.75 - 1.5 = -0.75\ \text{m} \end{aligned}$$

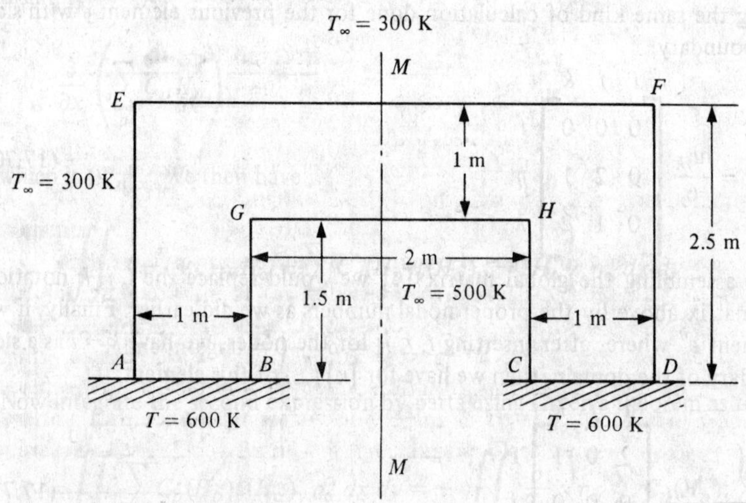

FIGURE 17.4
Steady-state heat conduction.

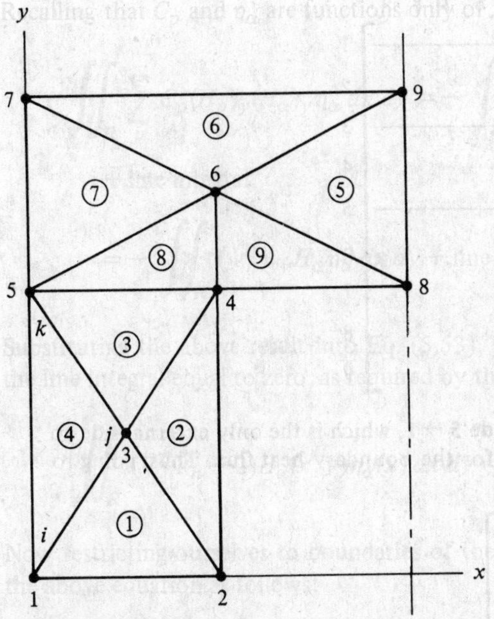

FIGURE 17.5
Rough grid for the heat conduction problem.

$$\begin{cases} x_j = 0.5 \text{ m} & \bar{a}_j = x_i - x_k = 0 - 0 = 0 \\ y_j = 0.75 \text{ m} & \bar{b}_j = y_k - y_i = 1.5 - 0 = 1.5 \text{ m} \end{cases}$$

$$\begin{cases} x_k = 0 \text{ m} & \bar{a}_k = x_j - x_i = 0.5 - 0 = 0.5 \text{ m} \\ y_k = 1.5 \text{ m} & \bar{b}_k = y_i - y_j = 0 - 0.75 = -0.75 \text{ m} \end{cases}$$

The heat transfer stiffness matrix for element 4 is then [see Eq. (17.59)]:

$$[K]^{(4)} = \frac{k}{4\mathcal{A}} \begin{bmatrix} [(-0.5)(-0.5) + (-0.75)(-0.75)] & [(-0.5)(0) + (-0.75)(1.5)] & [(-0.5)(0.5) + (-0.75)(-0.75)] \\ [(0)(-0.5) + (1.5)(-0.75)] & [(0)(0) + (1.5)(1.5)] & [(0)(0.5) + (1.5)(-0.75)] \\ [(0.5)(-0.5) + (-0.75)(-0.75)] & [(0.5)(0) + (-0.75)(1.5)] & [(0.5)(0.5) + (-0.75)(-0.75)] \end{bmatrix}$$

Inserting $k = 200$ W/m-**K**, $\mathcal{A} = (\frac{1}{2})(1.5)(0.5) = 0.375 \text{ m}^2$, and carrying out the arithmetic, we get

$$[K]^{(4)} = 133.3 \begin{bmatrix} 0.8125 & -1.125 & 0.3125 \\ -1.125 & 2.25 & -1.125 \\ 0.3125 & -1.125 & 0.8125 \end{bmatrix} = \begin{bmatrix} 108.3 & -150 & 41.7 \\ -150 & 300 & -150 \\ 41.7 & -150 & 108.3 \end{bmatrix}$$

We may calculate the stiffness matrices easily for the other eight elements. We can thus form the global matrix, remembering now that element nodes i, j, and k correspond to global nodes 1, 3, and 5, respectively. Thus we have for element 4:

$$
\begin{array}{c}
\begin{array}{ccccccccc} 1 & 2 & 3 & 4 & 5 & 6 & 7 & 8 & 9 \end{array} \\
\left[\begin{array}{ccccccccc}
108.3 & & -150 & & 41.7 & & & & \\
 & & & & & & & & \\
-150 & & 300 & & -150 & & & & \\
 & & & & & & & & \\
41.7 & & -150 & & 108.3 & & & & \\
 & & & & & & & & \\
 & & & & & & & & \\
 & & & & & & & & \\
 & & & & & & & &
\end{array}\right]
\begin{array}{c} 1 \\ 2 \\ 3 \\ 4 \\ 5 \\ 6 \\ 7 \\ 8 \\ 9 \end{array}
\end{array}
$$

Let us next get $[\kappa]^{(4)}$ for the side $5 \to 1$, which is the only external side to be considered for this element for the boundary heat flux. Thus, going to Eq. (17.71) we have†

$$
[\kappa]^{(4)}_{5\to 1} = \frac{\hbar l_{51}}{6}\begin{bmatrix} 2 & 0 & 1 \\ 0 & 0 & 0 \\ 1 & 0 & 2 \end{bmatrix}
$$

$$
= \frac{(2 \times 10^5)(1.5)}{6}\begin{bmatrix} 2 & 0 & 1 \\ 0 & 0 & 0 \\ 1 & 0 & 2 \end{bmatrix}
$$

$$
= \frac{10^5}{3}\begin{bmatrix} 3 & 0 & 1.5 \\ 0 & 0 & 0 \\ 1.5 & 0 & 3 \end{bmatrix}.
$$

To insert this into the global matrix for κ we proceed as follows, noting that the first row and column of $[\kappa]^{(4)}$ correspond to global node 1, the second row and column to global node 3, and the third row and column to global node 5. Thus we have

$$
\frac{10^5}{3}
\begin{array}{c}
\begin{array}{ccccccccc} 1 & 2 & 3 & 4 & 5 & 6 & 7 & 8 & 9 \end{array} \\
\left[\begin{array}{ccccccccc}
3 & & 0 & & 1.5 & & & & \\
 & & & & & & & & \\
0 & & 0 & & 0 & & & & \\
 & & & & & & & & \\
1.5 & & 0 & & 3 & & & & \\
 & & & & & & & & \\
 & & & & & & & & \\
 & & & & & & & & \\
 & & & & & & & &
\end{array}\right]
\begin{array}{c} 1 \\ 2 \\ 3 \\ 4 \\ 5 \\ 6 \\ 7 \\ 8 \\ 9 \end{array}
\end{array}
$$

†Note that we are going from 5 to 1, which corresponds to k and i.

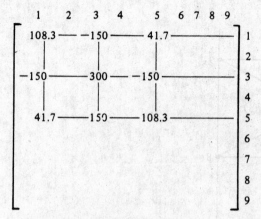

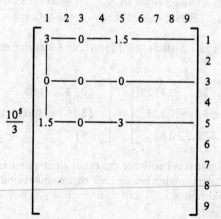

Finally, for the $(\tilde{q}_{av} - \hbar T_\infty)$ contribution for the boundary $5 \rightarrow 1$, we have from Eq. (17.67)

$$\{q\}_{5\rightarrow 1} = (\tilde{q}_{av} - \hbar T_\infty)\left(\frac{l_{51}}{2}\right)\begin{Bmatrix} 1 \\ 0 \\ 1 \end{Bmatrix} \begin{matrix} \text{node 1} \\ \text{node 3} \\ \text{node 5} \end{matrix}$$

or

$$\{q\}_{5\rightarrow 1} = [4 \times 10^5 - (2 \times 10^5)(300)]\left(\frac{1.5}{2}\right)\begin{Bmatrix} 1 \\ 0 \\ 1 \end{Bmatrix}$$

$$= -4.47 \times 10^7 \begin{Bmatrix} 1 \\ 0 \\ 1 \end{Bmatrix} \begin{matrix} \text{node 1} \\ \text{node 3} \\ \text{node 5} \end{matrix}$$

We can get the vector $\{q\}$ in this way for the entire system of elements. Note that on line $1 \rightarrow 2$ of element 1 we have a known temperature $\bar{T}$, so that we know T_1 and T_2. Furthermore, for side $1 \rightarrow 2$ of element 1 we have an unknown heat conduction $\tilde{q}$, but no convection. Then for that element we have for the forcing function [see Eq. (17.65)]:

$$\{q\}^1 = [\tilde{q}_{1\rightarrow 2}]\tfrac{1}{2}\begin{Bmatrix} 1 \\ 1 \\ 0 \end{Bmatrix} \begin{matrix} \text{node 1} \\ \text{node 2} \\ \text{node 3} \end{matrix}$$

Also, on boundary $8 \rightarrow 9$ we have no heat transfer. We thus get no contribution to the forcing function vector from element 5 at nodes 8 and 9.

We are now ready to set up the main matrix equation. We can combine matrices $[K]$ and $[\kappa]$ fo form $[K] + [\kappa] = [\tilde{K}]$. Thus we have

$$[\tilde{K}]\{a\} = -\{q\}$$

Spelled out in detail, we have

$$
\begin{bmatrix}
100{,}217 & -41.667 & -216.67 & 0 & 50{,}042 & 0 & 0 & 0 & 0 \\
-41.667 & 100{,}217 & -216.67 & 50{,}042 & 0 & 0 & 0 & 0 & 0 \\
-216.67 & -216.67 & 866.67 & -216.67 & -216.7 & 0 & 0 & 0 & 0 \\
0 & 50{,}042 & -216.67 & 167{,}383 & -91.67 & -400 & 0 & -33{,}283 & 0 \\
50{,}042 & 0 & -216.67 & -91.667 & 167{,}058 & -50 & -33{,}258 & 0 & 0 \\
0 & 0 & 0 & -400 & -50 & 1{,}000 & -250 & -50 & -250 \\
0 & 0 & 0 & 0 & -33{,}258 & -250 & 200{,}250 & 0 & 66{,}742 \\
0 & 0 & 0 & -33{,}283 & 0 & -50 & 0 & 66{,}842 & -75 \\
0 & 0 & 0 & 0 & 0 & -250 & 66{,}742 & -75 & 13{,}358
\end{bmatrix}
\begin{Bmatrix}
600 \\ 600 \\ T_3 \\ T_4 \\ T_5 \\ T_6 \\ T_7 \\ T_8 \\ T_9
\end{Bmatrix}
=
\begin{Bmatrix}
(\frac{1}{2})(\tilde{q}_{av})_{1-2} + 0.45 \times 10^8 \\
(\frac{1}{2})(\tilde{q}_{av})_{1-2} + 0.75 \times 10^8 \\
0 \\
1.25 \times 10^8 \\
7.50 \times 10^7 \\
0 \\
9.0 \times 10^7 \\
5.0 \times 10^7 \\
6.0 \times 10^7
\end{Bmatrix}
$$

Our next step is to reduce the matrix $[\tilde{K}]$. As described in Section 9.2, we may delete the first two rows and columns because T_1 and T_2 are specified. But in doing so we must simultaneously alter the remaining terms of the $\{q\}$ vector on the right side of the equation. Thus we have

$$\begin{bmatrix} 866.7 & -216.7 & -216.7 & 0 & 0 & 0 & 0 \\ -216.7 & 16{,}738 & -91.67 & -400 & 0 & -33{,}283 & 0 \\ -216.7 & -91.67 & 167{,}058 & -50 & -33{,}258 & 0 & 0 \\ 0 & -400 & -50 & 1{,}000 & -250 & -50 & -250 \\ 0 & 0 & -33{,}258 & -250 & 200{,}250 & 0 & 66{,}742 \\ 0 & -33{,}283 & 0 & -50 & 0 & 66{,}842 & -75 \\ 0 & 0 & 0 & -250 & 66{,}742 & -75 & 133{,}583 \end{bmatrix} \begin{Bmatrix} T_3 \\ T_4 \\ T_5 \\ T_6 \\ T_7 \\ T_8 \\ T_9 \end{Bmatrix}$$

$$= \begin{Bmatrix} 0 - \tilde{K}_{31}(600) - \tilde{K}_{32}(600) \\ 1.25 \times 10^8 - \tilde{K}_{41}(600) - \tilde{K}_{42}(600) \\ 7.50 \times 10^7 - \tilde{K}_{51}(600) - \tilde{K}_{52}(600) \\ 0 - \tilde{K}_{61}(600) - \tilde{K}_{62}(600) \\ 9.0 \times 10^7 - \tilde{K}_{71}(600) - \tilde{K}_{72}(600) \\ 5.0 \times 10^7 - \tilde{K}_{81}(600) - \tilde{K}_{82}(600) \\ 6.0 \times 10^7 - \tilde{K}_{91}(600) - \tilde{K}_{92}(600) \end{Bmatrix} = \begin{Bmatrix} 260{,}000 \\ 9.497 \times 10^7 \\ 4.497 \times 10^7 \\ 0 \\ 9.0 \times 10^7 \\ 5.0 \times 10^7 \\ 6.0 \times 10^7 \end{Bmatrix}$$

Solving for the seven values of T, we get

$$T_3 = 468\text{ K}$$
$$T_4 = 466\text{ K}$$
$$T_5 = 207\text{ K}$$
$$T_6 = 375\text{ K}$$
$$T_7 = 319\text{ K}$$
$$T_8 = 516\text{ K}$$
$$T_9 = 291\text{ K}$$

In Figs. 17.6 and 17.7 we show the preceding problem with a finer grid formed by the mesh-generating computer program. The calculations were carried out completely on the computer.

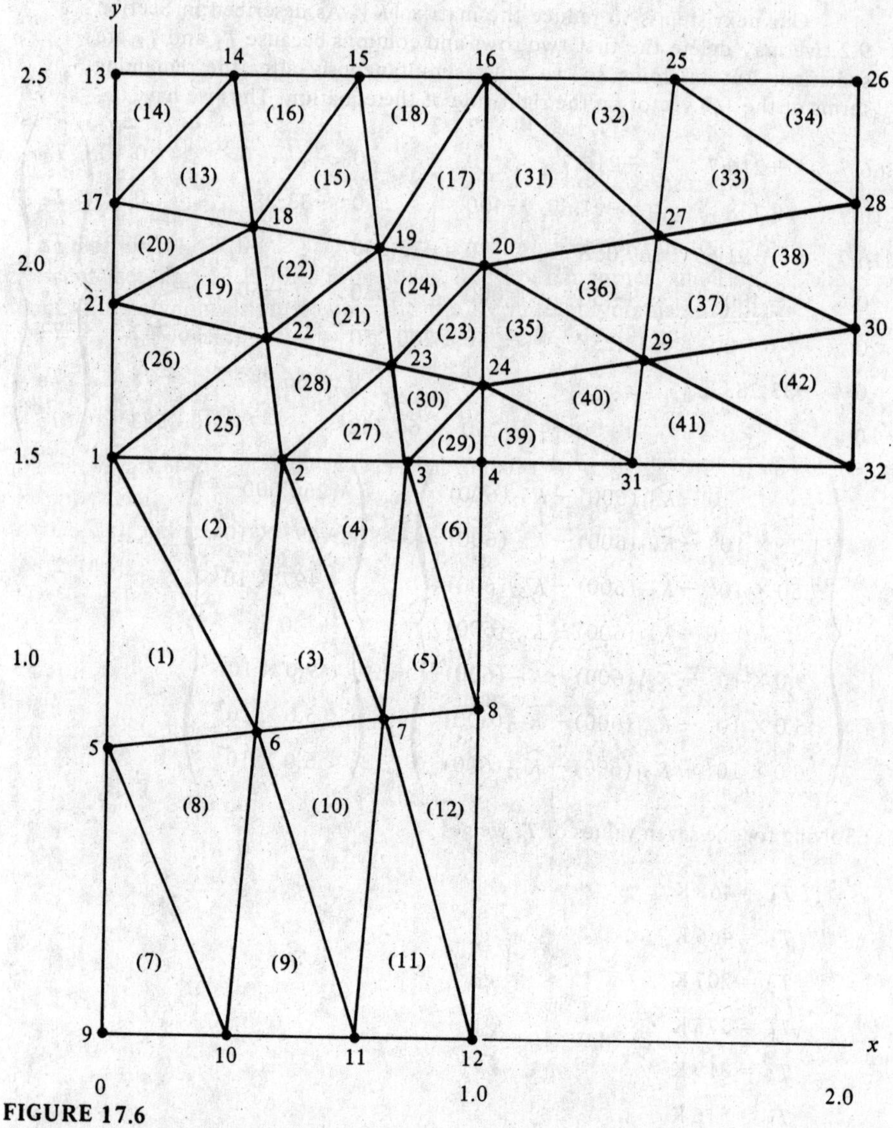

FIGURE 17.6
Gridding for heat conduction problem.

Part C
Method of Weighted Residuals; Galerkin's Method

17.8 INTRODUCTION

In Section 3.15 we introduced the method of *weighted residuals* and the method of *Galerkin* for the entire domain. Recall that in the "virtual work" expressions when the δu_i were taken as the coordinate functions ϕ_i, the equation

$$\iiint_{\Omega} (L\tilde{u} - f)\,\delta u_i \, d\Omega = 0 \tag{17.72a}$$

was transformed into

$$\iiint_{\Omega} (L\tilde{u} - f)\phi_i \, d\Omega = 0 \tag{17.72b}$$

Then we went from the method of weighted residuals to the method of Galerkin by using ϕ_i in $\tilde{u}$. We will use only the Galerkin method here and will apply it to finite elements in a domain.

The main advantage of this approach is that we *do not need* a quadratic functional and so for nonlinear operators we can put this method to good use. For finite

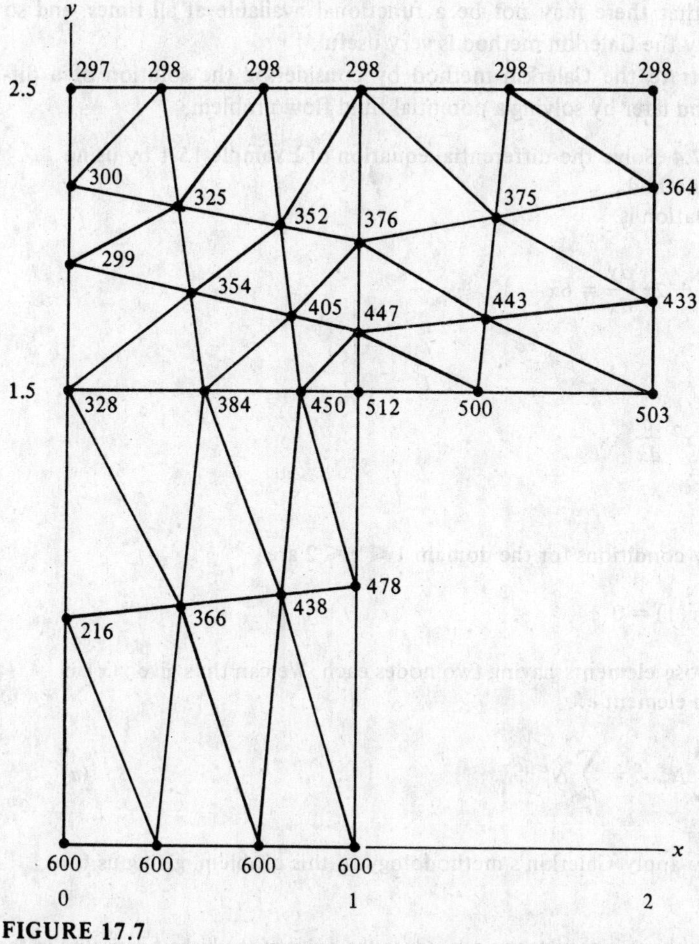

FIGURE 17.7
Results for heat conduction problem.

elements, the ϕ_i become the interpolation functions N_i, and the domain Ω_e for Eqs. (17.72) becomes that of the element e. Hence, for such an element we have

$$\int_{\Omega_e} [L\tilde{u}(x,y,z)-f]N_i\, d\Omega = 0 \tag{17.73}$$

where $\tilde{u}(x, y, z)$ is the "displacement field." The requirement for convergence of this method as the size of the elements decreases is the *same* as for the variational process; namely, the interpolation functions must be chosen so that the dependent variable $\tilde{u}(x, y, z)$ and its derivatives up to one less than the highest order in the integrand are continuous at the boundaries between elements.

We note here that in the quadratic functional the order of derivative for u will be one less than in the corresponding formulation given above for Galerkin's method. Thus, the choice of interpolation functions for the Galerkin method is more difficult.† But keep in mind that there may not be a functional available at all times, and so despite this difficulty the Galerkin method is very useful.

We now illustrate the Galerkin method by considering the solution of a differential equation and later by solving a potential fluid flow problem.

EXAMPLE 17.4 Solve the differential equation of Example 15.1 by using the Galerkin method.

The equation is

$$x^2 \frac{d^2y}{dx^2} + 2x\frac{dy}{dx} = 6x$$

where

$$L \equiv \frac{d}{dx} x^2 \frac{d}{dx}$$

$$f = 6x$$

The boundary conditions for the domain $1 \leqslant x \leqslant 2$ are

$$y(2) = y(1) = 0$$

We choose elements having two nodes each. We can thus give $y(x)$ as follows for an element e:

$$y^e = \sum_r N_r^e a_r^e = \sum_r N_r^e y_r^e \tag{a}$$

We now apply Galerkin's methodology to this problem, giving us for an element:

†If integration by parts is possible, we can reduce the order of the highest derivative in the Galerkin formulation by one and so eliminate this disadvantage.

$$\int_{x_i}^{x_j} \left[\frac{d}{dx}\left(x^2 \frac{dy}{dx}\right) - 6x \right] N_s \, dx = 0 \qquad s = i, j \tag{b}$$

We next integrate the first expression in the integrand by parts:

$$x^2 \frac{dy}{dx} N_s \Bigg|_{x_i}^{x_j} - \int_{x_i}^{x_j} x^2 \frac{dy}{dx} \frac{dN_s}{dx} \, dx - \int_{x_i}^{x_j} 6x N_s \, dx = 0 \qquad s = i, j$$

Now use Eq. (*a*) for *y* in the first integration expression:

$$x^2 \frac{dy}{dx} N_s \Bigg|_{x_i}^{x_j} - \int_{x_i}^{x_j} x^2 \left(\sum_r \frac{dN_r}{dx} a_r \right) \frac{dN_s}{dx} \, dx - \int_{x_i}^{x_j} 6x N_s \, dx = 0$$

$$s = i, j$$

Putting the above equation in matrix form, we have

$$-\int_{x_i}^{x_j} x^2 \left(\left\lfloor \frac{dN}{dx} \right\rfloor \{a\} \right) \left\{ \frac{dN}{dx} \right\} dx - \int_{x_i}^{x_j} 6x \{N\} \, dx + x^2 \frac{dy}{dx} \{N\} \Bigg|_{x_i}^{x_j} = 0 \tag{c}$$

Note that $x^2 \, dy/dx$ is the natural boundary condition. We can now state this equation in the following form for an element:

$$[K]^e \{a\} = -\{q\} + \{Q\} \tag{d}$$

where

$$K_{ij}^e = \int_{x_i}^{x_j} x^2 \left(\frac{dN_i}{dx} \right) \left(\frac{dN_j}{dx} \right) dx \tag{e}$$

$$q_i^e = \int_{x_i}^{x_j} 6x N_i \, dx \tag{f}$$

$$Q_i^e = x^2 \frac{dy}{dx} N_i \Bigg|_{x_i}^{x_j} \tag{g}$$

When we assemble the global system for this problem, the $[K]$ and $\{q\}$ matrices are developed in the usual ways. As far as the Q terms are concerned, we will get a cancellation at all nodes not at the ends of the overall domain. At the ends of the overall domain, the nodal displacements $y_p = a_p$ are prescribed equal to zero and so we delete the first and last rows and columns of Eq. (*d*).

The reduced equations are then identical to those of Examples 17.1 and 17.2 and we arrive, of course, at the same results.

Part D
Two-Dimensional Potential Flow of Fluid

17.9 FLOW CONSIDERATIONS

We will now study incompressible, nonviscous, steady flow in two dimensions. This kind of flow is called *ideal* or *potential* flow. For two-dimensional flow, we use derivatives of the stream function† $\psi(x, y)$ to represent velocity and so satisfy *continuity* requirements. Specifically, the fluid velocity components are given as follows:

$$V_x = \frac{\partial \psi}{\partial y} \qquad V_y = -\frac{\partial \psi}{\partial x} \tag{17.74}$$

Using index notation (as we shall do in this part of the chapter), we have for the above relations:

$$V_i = \epsilon_{ij} \psi_{,j} \tag{17.75}$$

where ϵ_{ij} is the two-dimensional version of the alternating tensor, i.e., $\epsilon_{12} = 1, \epsilon_{11} = \epsilon_{22} = 0$, and $\epsilon_{21} = -1$. Furthermore, Newton's law requires that ψ be *harmonic* so that

$$\nabla^2 \psi = 0$$

or, in index notation,

$$\psi_{,ii} = 0 \tag{17.76}$$

There is also a *velocity potential* ϕ, which must be a harmonic function and which is related to the velocity field $\mathbf{V}$ by

$$\mathbf{V} = -\nabla \phi \tag{17.77}$$

17.10 GALERKIN'S METHOD FOR FINITE ELEMENTS

We will now use Galerkin's method rather than the variational approach since in more complex flows, including compressibility and viscosity, a quadratic functional is not easily established. Accordingly, we express ψ in terms of n interpolation functions $N_i(x, y)$. We thus have for an element:‡

$$\psi(x, y) = N_i \psi_i \tag{17.78}$$

†See Shames, "Mechanics of Fluids," McGraw-Hill Book Co. (1982), chapter 13, for details of two-dimensional potential flow.

‡Note that in the ensuing discussion ψ with one subscript and *nothing* else is a *nodal* value. However, if there are two subscripts or if there is a comma and one or two subscripts, we are dealing with the *field variable* $\psi(x, y)$.

The values ψ_i are the nodal values of the field variable $\psi(x, y)$. Now, orthogonalizing $(\psi_{,ii} - 0)$ with N_j as per Eq. (3.98), we have for n degrees of freedom for an element:

$$\iint_{\mathcal{A}^e} (\psi_{,ii}) N_j \, d\mathcal{A} = 0 \qquad j = 1, 2, \ldots, n \tag{17.79}$$

We shall now call on Green's formula [see Eq. (I.35) in Appendix I]. For two dimensions, this equation states that

$$\iint_{\mathcal{A}} N_{j,i} \psi_{,i} \, d\mathcal{A} = -\iint_{\mathcal{A}} N_j \psi_{,ii} \, d\mathcal{A} + \oint_{\Gamma} N_j \frac{\partial \psi}{\partial \nu} \, ds \tag{17.80}$$

where ν is the outward normal coordinate from the boundary. Consider next an infinitesimal segment dy of this boundary (Fig. 17.8). Establish as the normal coordinate for this element the x axis and as the tangential coordinate the y axis. Then

$$\frac{\partial \psi}{\partial \nu} \equiv \frac{\partial \psi}{\partial x} = -V_y$$

That is, the derivative of ψ in the direction normal to the boundary gives the tangential velocity at this segment. We can generalize to say that $\partial \psi / \partial \nu$ is the tangential velocity V_t along the boundary. That is,

$$\frac{\partial \psi}{\partial \nu} = -V_t \tag{17.81}$$

Note especially that if **V** is *perpendicular* to a portion of the boundary, then $V_t = 0$ and hence $\int N_j(\partial \psi / \partial \nu) \, ds$ must be zero for that portion of the boundary.

Also we note that $\partial \psi / \partial \nu$ is a *directional derivative* and can be written as $(\partial \psi / \partial x_i) \nu_i$, where the ν_i are direction cosines of ν. Thus we may rewrite Eq. (17.80) in the following form:

$$\iint_{\mathcal{A}} N_j \psi_{,ii} \, d\mathcal{A} = -\iint_{\mathcal{A}} N_{j,i} \psi_{,i} \, d\mathcal{A} + \oint_{\Gamma} N_j \frac{\partial \psi}{\partial x_i} \nu_i \, ds \tag{17.82}$$

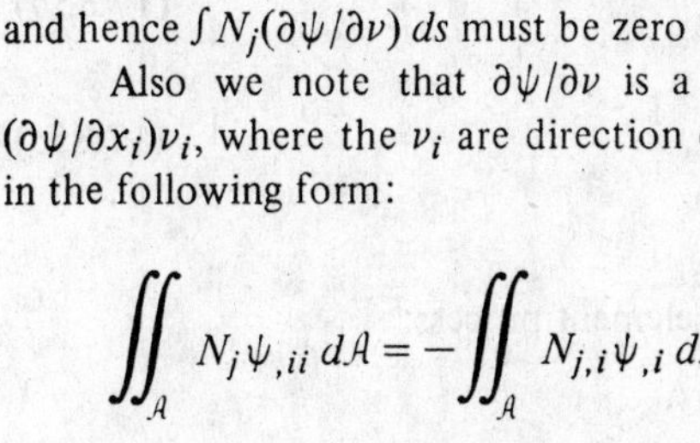

FIGURE 17.8
Computation of V_t.

Now setting the left side of the above equation equal to zero in accordance with Eq. (17.79) and rearranging, we get

$$\iint_{A} N_{j,i}\psi_{,i}\, dA = \oint_{\Gamma} N_j \psi_{,i} \nu_i\, ds \qquad (17.83)$$

Let us next replace $\psi_{,i}$ $(\equiv \partial\psi/\partial x_i)$ using Eq. (17.78). We get, using k as the dummy index,

$$\psi_{,i} = \frac{\partial}{\partial x_i}(N_k \psi_k) = \frac{\partial N_k}{\partial x_i}\psi_k = N_{k,i}\psi_k$$

Substituting this result into the left side of Eq. (17.83) and extracting the nodal values ψ_k from inside the integral, we have for Eq. (17.83):

$$\left[\iint_{A} (N_{j,i} N_{k,i})\, dA\right]\psi_k = \oint_{\Gamma} N_j \psi_{,i}\nu_i\, ds \qquad (17.84)$$

We thus arrive at an equation having the familiar form

$$[K]^e\{a\}^e = \{q\}^e$$

where

$$K^e_{jk} = \iint_{A} N_{j,i} N_{k,i}\, dA \qquad (17.85a)$$

$$a_k = \psi_k \qquad (17.85b)$$

$$q^e_j = \oint_{\Gamma} N_j \psi_{,i}\nu_i\, ds \qquad (17.85c)$$

We thus have the key parts needed for the finite element process.

17.11 INTERPOLATION FUNCTIONS

We have, for triangular elements with three degrees of freedom, the same interpolation functions that were developed in Section 17.7 for steady-state heat conduction. We restate these functions for element e with nodes i, j, and k,

$$[N]^e = \begin{Bmatrix} N_i \\ N_j \\ N_k \end{Bmatrix} = \frac{1}{2A}\begin{bmatrix} d_i & x\bar{b}_i & y\bar{a}_i \\ d_j & x\bar{b}_j & y\bar{a}_j \\ d_k & x\bar{b}_k & y\bar{a}_k \end{bmatrix} \qquad (17.86)$$

where

$$\bar{a}_i = (x_k - x_j) \qquad \bar{b}_i = (y_j - y_k) \qquad d_i = (x_j y_k - x_k y_j)$$

The other constants $\bar{a}_j$, $\bar{b}_j$, etc., can be found by permuting the indices. Note that we are using global coordinates for the above formulations.

We will now concentrate on the vector q_r. Let us go back to Eq. (17.75). You may readily demonstrate that

$$\psi_{,i} = \epsilon_{ki} V_k \tag{17.87}$$

so that Eq. (17.85*c*) becomes, using *r* as a free index,

$$q_r^e = \oint_\Gamma N_r(\epsilon_{ki} V_k)\nu_i \, ds \tag{17.88}$$

We get a nonzero contribution to q_r in the global equation only when we have an *outer* boundary. We will assume for such boundaries that, as an approximation for variable velocity between the two nodes (see Fig. 17.9), the velocity varies *linearly*. Clearly, the shorter the boundary the more correct the results from such an assumption. Using *s* as the distance along the boundary from node 1 to node 2 as indicated in Fig. 17.9, we have

$$[V_p(x,y)]^{(1\to 2)} = V_p^{(1)} + \frac{s}{l_{1\text{-}2}}\left(V_p^{(2)} - V_p^{(1)}\right)$$

$$= \left(1 - \frac{s}{l_{1\text{-}2}}\right) V_p^{(1)} + \frac{s}{l_{1\text{-}2}} V_p^{(2)} \tag{17.89}$$

where superscripts with parentheses represent nodal identification. At any point along the boundary we may also state that

$$[V_p(x,y)]^{(1\to 2)} = \bar{N}_1 V_p^{(1)} + \bar{N}_2 V_p^{(2)} + \bar{N}_3 V_p^{(3)}$$

where the overbars on the interpolation functions N_i are used to indicate values on the boundary. This equation results because of the zero or unity value of N_i at nodes and because of the fact [see Eq. (17.86)] that the interpolation functions N_i give linear variation. Hence, from the previous two equations, we can say

$$\bar{N}_1 = 1 - \frac{s}{l_{1\text{-}2}} \tag{17.90a}$$

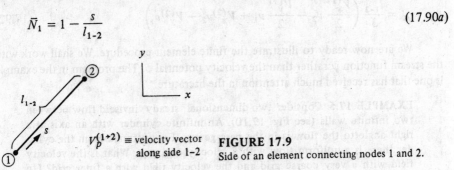

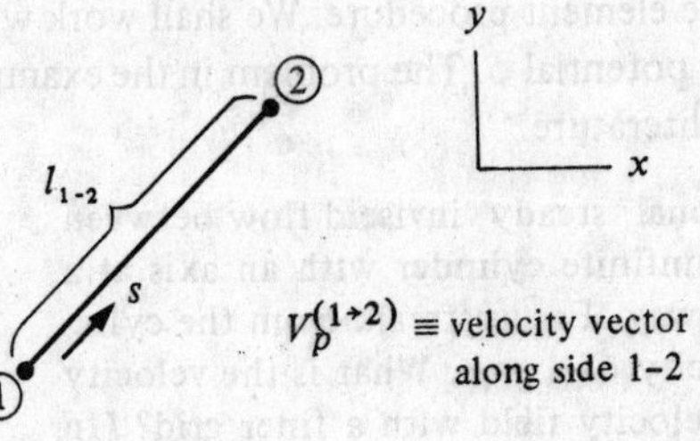

FIGURE 17.9
Side of an element connecting nodes 1 and 2.

$$\bar{N}_2 = \frac{s}{l_{1\text{-}2}} \tag{17.90b}$$

$$\bar{N}_3 = 0 \tag{17.90c}$$

With these results, we can go back to Eq. (17.88) to carry out the integration:

$$q_r^e = \int_{l_{1\text{-}2}} \bar{N}_r \epsilon_{ki} V_k \nu_i \, ds = \int_{l_{1\text{-}2}} \bar{N}_r \epsilon_{ki} (\bar{N}_1 V_k^{(1)} + \bar{N}_2 V_k^{(2)}) \nu_i \, ds$$

$$= \int_{l_{1\text{-}2}} \bar{N}_r \Big\{ \epsilon_{11} (\bar{N}_1 V_1^{(1)} + \bar{N}_2 V_1^{(2)}) \nu_1 + \epsilon_{21} (\bar{N}_1 V_2^{(1)} + \bar{N}_2 V_2^{(2)}) \nu_1$$

$$+ \epsilon_{12} (\bar{N}_1 V_1^{(1)} + \bar{N}_2 V_1^{(2)}) \nu_2 + \epsilon_{22} (\bar{N}_1 V_2^{(1)} + \bar{N}_2 V_2^{(2)}) \nu_2 \Big\} \, ds$$

Noting that $\epsilon_{11} = \epsilon_{22} = 0$ and substituting for $\bar{N}_1$ and $\bar{N}_2$ from Eqs. (17.90*a*) and (17.90*b*), we get

$$q_r^e = \int_{l_{1\text{-}2}} \bar{N}_r \left[-\left(1 - \frac{s}{l_{1\text{-}2}}\right) V_2^{(1)} \nu_1 - \left(\frac{s}{l_{1\text{-}2}}\right) V_2^{(2)} \nu_1 \right.$$

$$\left. + \left(1 - \frac{s}{l_{1\text{-}2}}\right) V_1^{(1)} \nu_2 + \left(\frac{s}{l_{1\text{-}2}}\right) V_1^{(2)} \nu_2 \right] ds$$

For $r = 1$, replace $\bar{N}_1$ from Eq. (17.90*a*) and carry out the integration. We get after collecting terms

$$q_1 = \frac{l_{1\text{-}2}}{3} \left(V_1^{(1)} \nu_2 - V_2^{(1)} \nu_1 + \frac{1}{2} V_1^{(2)} \nu_2 - \frac{1}{2} V_2^{(2)} \nu_1 \right)$$

Replacing the subscripts 1 and 2 for the velocity by x and y, respectively, we have

$$q_1 = \frac{l_{1\text{-}2}}{3} \left(V_x^{(1)} \nu_2 - V_y^{(1)} \nu_1 + \frac{V_x^{(2)}}{2} \nu_2 - \frac{V_y^{(2)}}{2} \nu_1 \right) \tag{17.91}$$

Similarly, we can say

$$q_2 = \frac{l_{1\text{-}2}}{3} \left(\frac{V_x^{(1)}}{2} \nu_2 - \frac{V_y^{(1)}}{2} \nu_1 + V_x^{(2)} \nu_2 - V_y^{(2)} \nu_1 \right) \tag{17.92}$$

We are now ready to illustrate the finite element procedure. We shall work with the stream function ψ rather than the velocity potential ϕ. The problem in the example is one that has received much attention in the literature.

EXAMPLE 17.5 Consider two-dimensional, steady, inviscid flow between two infinite walls (see Fig. 17.10). An infinite cylinder with an axis at a right angle to the flow is in the passageway. Far upstream from the cylinder there is a uniform flow with a velocity of 1 m/s. What is the velocity field with a very coarse grid and the velocity field with a finer grid? (In

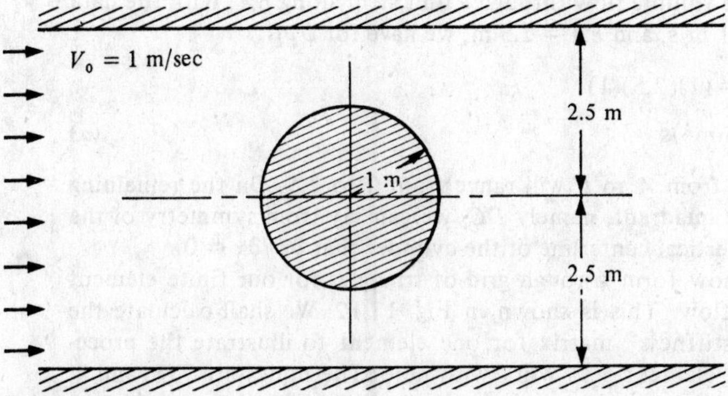

FIGURE 17.10
Two-dimensional inviscid flow.

each case there should be smaller elements near the cylinder since we can expect larger velocity gradients there.)

For this problem we need consider only one quadrant of the flow around the cylinder. We show such a quadrant in Fig. 17.11. The left end of the quadrant is 4 m from the centerline of the cylinder. In this subdomain we can immediately identify two streamlines. Because of the symmetry of the flow about the x axis, we can conclude that along AB in Fig. 17.11 we must have only a horizontal velocity. This makes AB a streamline. We shall denote the value of the streamline as zero. Also, since the flow must be tangent to solid boundaries, we can conclude that $\psi = 0$ continues from B to C along the cylindrical wall. The second streamline in the quadrant diagram is the upper wall ED. The value of this streamline must be such that

$$\psi_{ED} - \psi_{AB} = q_{EA} = (V_0)(\overline{EA})(1)$$

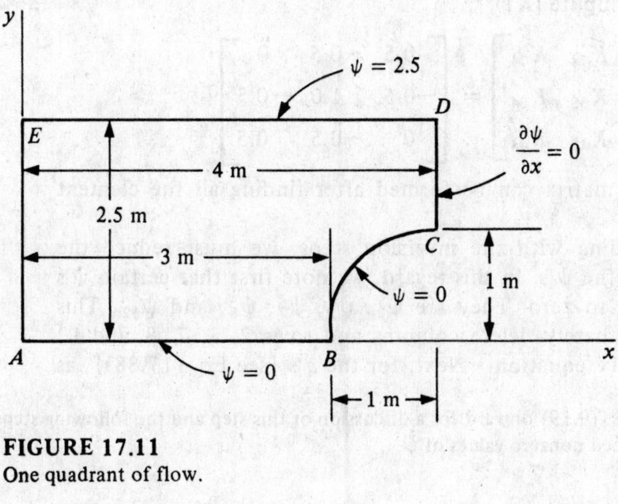

FIGURE 17.11
One quadrant of flow.

where q_{EA} is the volume flow through a unit strip along $\overline{EA}$. With the data $\psi_{AB} = 0$, $V_0 = 1$ m/s, and $\overline{EA} = 2.5$ m, we have for ψ_{ED}:

$$\psi_{ED} - 0 = (1)(2.5)(1)$$

$$\psi_{ED} = 2.5 \text{ m}^2/\text{s} \tag{a}$$

The streamlines from A to E will range from 0 to 2.5. On the remaining boundary of the quadrant, namely DC, we can say from symmetry of the flow about the vertical centerline of the cylinder that $\partial\psi/\partial x = 0$.

We shall now form a rough grid of triangles for our finite element analysis of the flow. This is shown in Fig. 17.12. We shall calculate the element fluid "stiffness" matrix for one element to illustrate the procedure. We choose element 5 to work with here. The nodes are 7, 8, and 4. We denote these nodes for our work with this element as i, j, and k, respectively, as shown in Fig. 17.12.

To compute the K_{pn} terms of the stiffness matrix we go back to Eq. (17.86) and substitute the interpolation functions into Eq. (17.85a). We then get, on carrying out the differentiation and integration operations,

$$K_{pn}^e = \frac{1}{4A}(\bar{a}_p \bar{a}_n + \bar{b}_p \bar{b}_n) \tag{b}$$

Now for element 5 we have

$$\begin{aligned} x_i &= x_7 = 3 \text{ m} & y_i &= y_7 = 0 \\ x_j &= x_8 = 3 + 0.5 = 3.5 \text{ m} & y_j &= y_8 = 0.5 \text{ m} \\ x_k &= x_4 = 3 \text{ m} & y_k &= y_4 = 1 \text{ m} \end{aligned} \tag{c}$$

$$\begin{aligned} \bar{b}_i &= y_j - y_k = -0.5 \text{ m} & \bar{a}_i &= x_k - x_j = -0.5 \text{ m} \\ \bar{b}_j &= y_k - y_i = 1 \text{ m} & \bar{a}_j &= x_i - x_k = 0 \\ \bar{b}_k &= y_i - y_j = -0.5 \text{ m} & \bar{a}_k &= x_j - x_i = 0.5 \text{ m} \end{aligned} \tag{d}$$

We can now easily compute $[K]^{(5)}$:

$$[K]^{(5)} = \begin{bmatrix} K_{77} & K_{78} & K_{74} \\ K_{87} & K_{88} & K_{84} \\ K_{47} & K_{48} & K_{44} \end{bmatrix} = \begin{bmatrix} 0.5 & -0.5 & 0 \\ -0.5 & 1.0 & -0.5 \\ 0 & -0.5 & 0.5 \end{bmatrix}$$

The global stiffness matrix can be formed after finding all the element stiffness matrices.

Before proceeding with the inversion steps, we must reduce the matrix equation for the ψ's. In this regard, we note first that certain ψ's are prescribed equal to zero. They are ψ_2, ψ_3, ψ_7, ψ_8, and ψ_{12}. This means we can immediately delete columns and rows 2, 3, 7, 8, and 12 from the basic matrix equation.† Next, for the q's [see Eq. (17.88)], as

†See Chapter 9 from Eqs. (9.19) onward for a discussion of this step and the following steps applying to nodes with prescribed nonzero values of ψ.

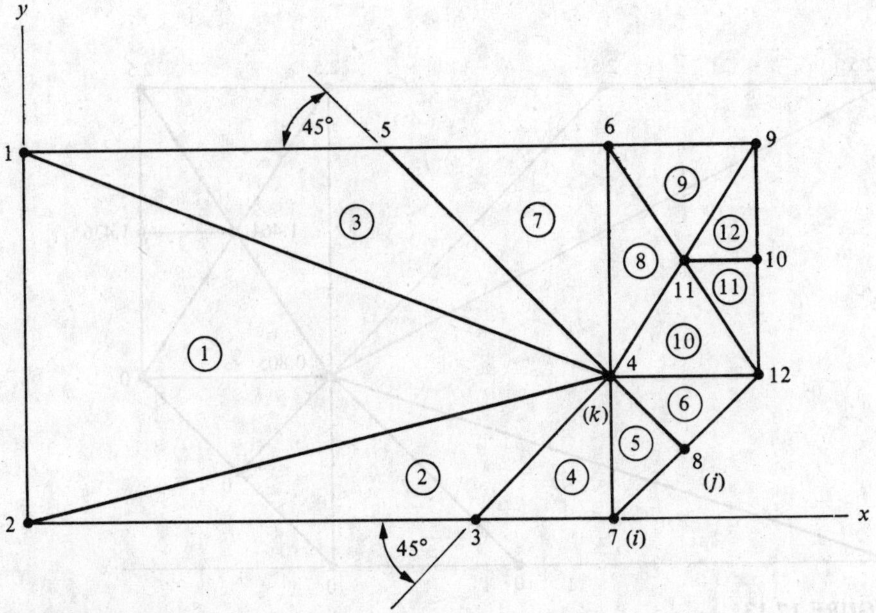

FIGURE 17.12
Coarse grid for potential flow problem.

we pointed out earlier, by integrating in the same direction for all elements (counterclockwise), we get cancellation for all inner legs of the elements so that only q's from the outside boundary need be considered. And on legs 1–2, 9–10, and 10–12 the flow direction is normal to the boundary, and so we get zero contribution to the q's along these legs since $\epsilon_{ki}V_k$ in Eq. (17.88) must be tangent to these boundaries for a nonzero value. Finally, consider the nodes 1, 5, 6, and 9, where the nonzero value of 2.5 has been prescribed. According to our discussion in Chapter 9, we may delete rows and columns 1, 5, 6, and 9 but must properly alter the flux vectors remaining in the equation. We thus get for the reduced finite element equation, on noting that the remaining q's are zero,

$$\begin{bmatrix} K_{4,4} & K_{4,10} & K_{4,11} \\ K_{10,4} & K_{10,10} & K_{10,11} \\ K_{11,4} & K_{11,10} & K_{11,11} \end{bmatrix} \begin{Bmatrix} \psi_4 \\ \psi_{10} \\ \psi_{11} \end{Bmatrix} = \begin{Bmatrix} (0 - K_{4,1}(2.5) - K_{4,5}(2.5) - K_{4,6}(2.5) - K_{4,9}(2.5)) \\ (0 - K_{10,1}(2.5) - K_{10,5}(2.5) - K_{10,6}(2.5) - K_{10,9}(2.5)) \\ (0 - K_{11,1}(2.5) - K_{11,5}(2.5) - K_{11,6}(2.5) - K_{11,9}(2.5)) \end{Bmatrix}$$

Inserting values for the K's, we can solve for the three unknowns to get (see Fig. 17.13):

$$\psi_4 = 0.805 \text{ m}^2/\text{s}$$

$$\psi_{10} = 1.436 \text{ m}^2/\text{s} \qquad (e)$$

$$\psi_{11} = 1.464 \text{ m}^2/\text{s}$$

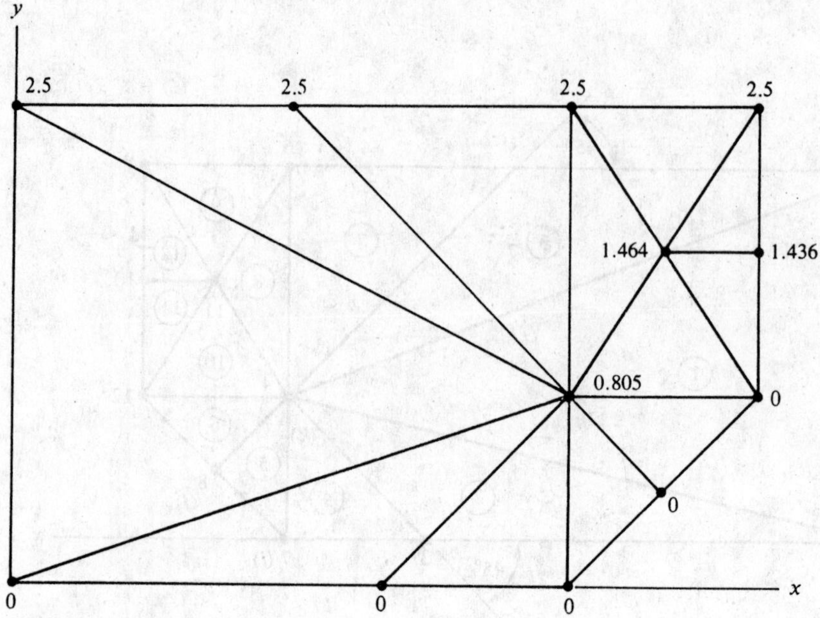

FIGURE 17.13
Values of ψ at nodes.

By taking differences of the ψ's, such as $\psi_9 - \psi_{10}$, and dividing by $y_9 - y_{10}$, we get an average velocity along leg 9–10. Thus:

$$[(V_x)_{9\text{-}10}]_{av} = \frac{\psi_9 - \psi_{10}}{y_9 - y_{10}} = \frac{2.5 - 1.434}{2.5 - 1.75} = 1.421 \text{ m/s} \qquad (f)$$

The approximate velocity field may thus be determined.

17.12 OTHER SOLUTIONS FOR FLOW PROBLEM

We now consider two other solutions to the flow problem of the previous section. First there is a computer solution, for which we automatically grid the flow domain with triangular elements of varying size, using the mesh-generating program presented in Appendix III. The result is shown in Fig. 17.14, where we have plotted the nodes from the node data generated by the mesh-generating program.

In Fig. 17.15 we show the results for ψ using a complete computer program. The velocity over the line segment 1-2 (see Fig. 17.14) can now be computed and related to the results of Example 17.5, wherein we used a rough grid. Thus we have

$$[(V_x)_{1\text{-}2}]_{av} = \frac{\psi_1 - \psi_2}{y_1 - y_2} = \frac{2.5 - 1.756}{2.5 - 2} = 1.488 \text{ m/s}$$

We see from Eq. (*f*) of Example 17.5 that we have rather good agreement.

We now shift to yet a third approach wherein we will use isoparametric elements, which will be described in the following example.

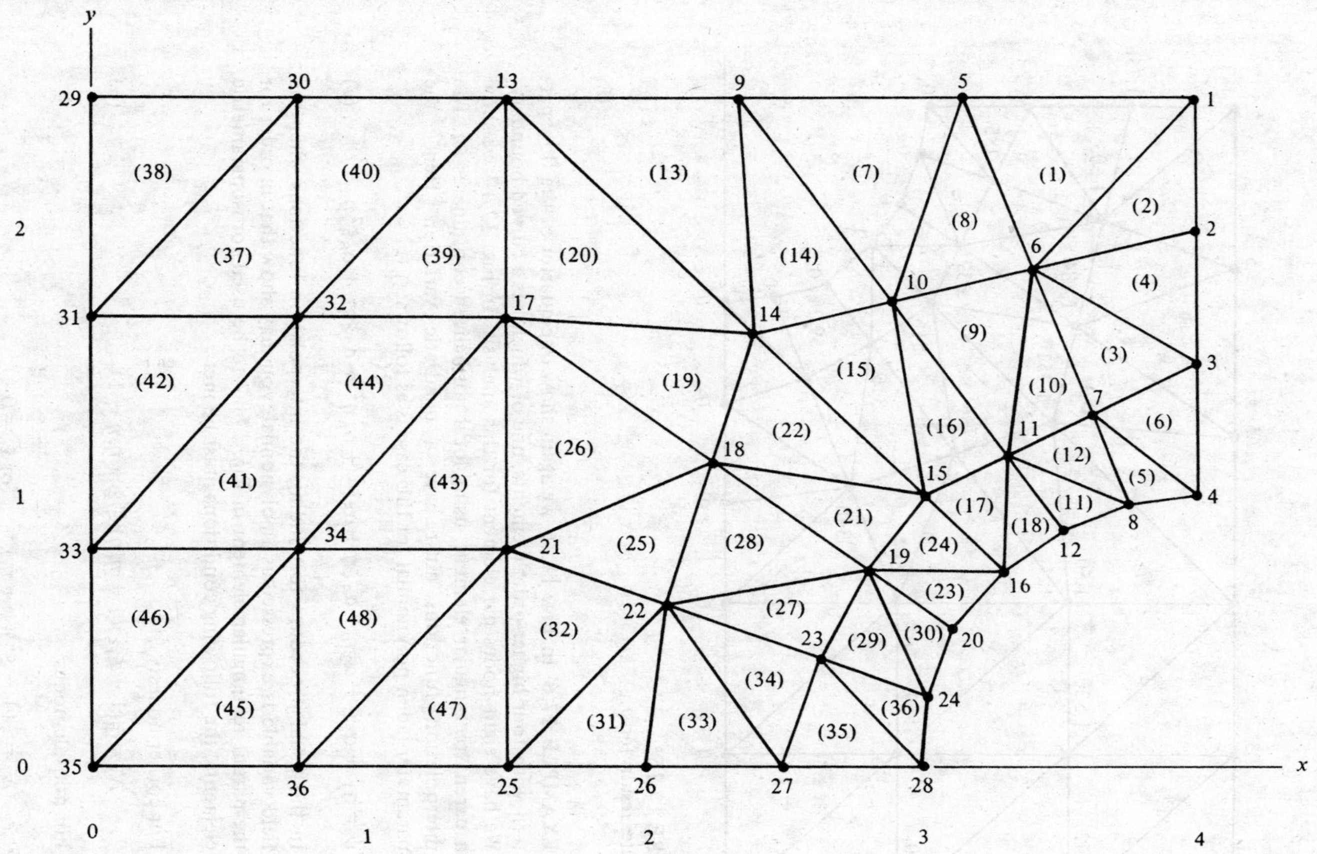

FIGURE 17.14
Computer-generated mesh.

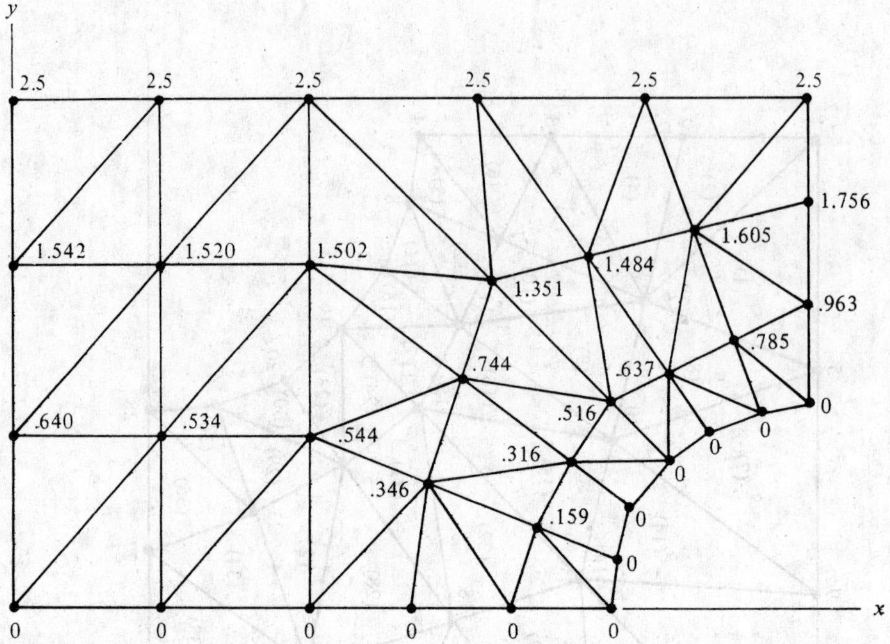

FIGURE 17.15
Computer results for ψ.

EXAMPLE 17.6 In Fig. 17.16 we again show one-quarter of the flow pattern with four higher-order elements, one of which has a curved boundary. We have eight nodes per element for this analysis. In Fig. 17.17 we show a parent rectangular element, using local normalized coordinates $\xi\eta$. The interpolation functions must have a quadratic variation. Hence the parameter-laden polynomial for this case is as follows:

$$\psi = \alpha_1 + \alpha_2 \xi + \alpha_3 \eta + \alpha_4 \xi\eta + \alpha_5 \xi^2 + \alpha_6 \eta^2 + \alpha_7 \xi\eta^2 + \alpha_8 \xi^2 \eta \qquad (a)$$

In the usual procedure for finding the interpolation functions when the field variable is expressed as a polynomial, you may show that in order for the parent rectangular element in Fig. 17.17 to be used for isoparametric elements, the following conditions must be met:

For corner nodes:

$$N_i = \tfrac{1}{4}(1 + \xi_i\xi)(1 + \eta_i\eta)(\xi_i\xi + \eta_i\eta - 1) \qquad (b)$$

For midside nodes

$$\left.\begin{aligned} N_i &= \tfrac{1}{2}(1 - \xi^2)(1 + \eta_i\eta) && \text{for } \xi_i = 0 \\ N_i &= \tfrac{1}{2}(1 + \xi_i\xi)(1 - \eta^2) && \text{for } \eta_i = 0 \end{aligned}\right\} \qquad (c)$$

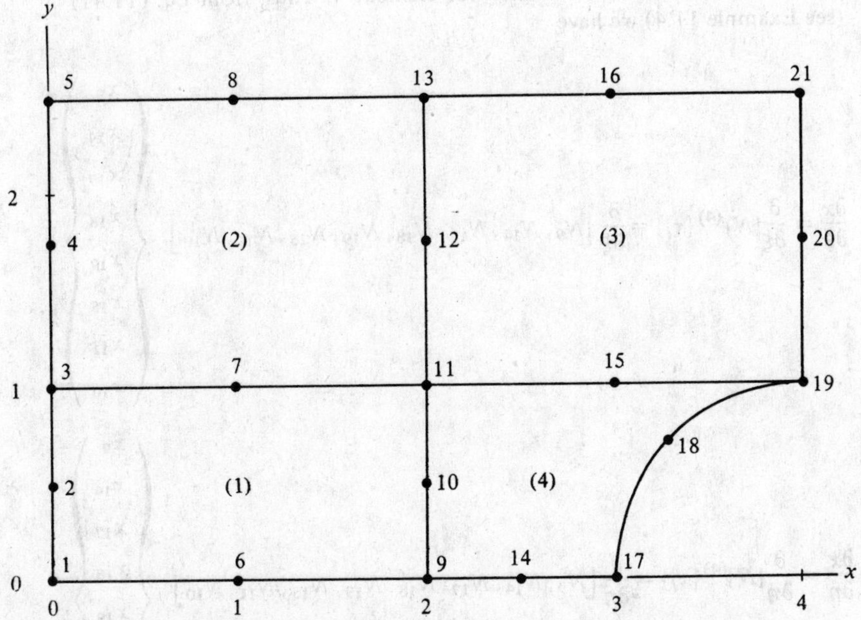

FIGURE 17.16
Flow domain with higher-order elements.

We shall examine element 4 in detail. For this element, using global coordinates, we have

$$\begin{array}{llll} x_9 = 2\text{ m} & x_{11} = 2\text{ m} & x_{15} = 3\text{ m} & x_{18} = 3.3\text{ m} \\ y_9 = 0 & y_{11} = 1\text{ m} & y_{15} = 1\text{ m} & y_{18} = 0.7\text{ m} \\ x_{10} = 2\text{ m} & x_{14} = 2.5\text{ m} & x_{17} = 3\text{ m} & x_{19} = 4\text{ m} \\ y_{10} = 0.5\text{ m} & y_{14} = 0 & y_{17} = 0 & y_{19} = 1\text{ m} \end{array} \qquad (d)$$

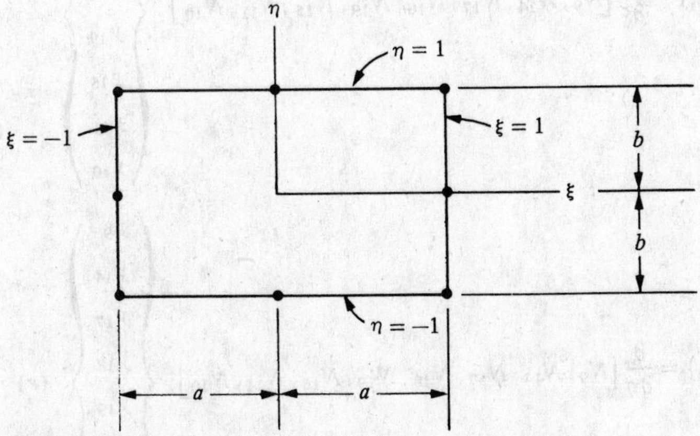

FIGURE 17.17
Parent rectangular element with normalized local coordinates ξη.

We next get the Jacobian matrix for element 4. Thus, from Eq. (11.41) (see Example 11.4) we have

$$\frac{\partial x}{\partial \xi} = \frac{\partial}{\partial \xi}[N]^{(4)}\{x_i\} = \frac{\partial}{\partial \xi}\lfloor N_9, N_{14}, N_{17}, N_{18}, N_{19}, N_{15}, N_{11}, N_{10} \rfloor \begin{Bmatrix} x_9 \\ x_{14} \\ x_{17} \\ x_{18} \\ x_{19} \\ x_{15} \\ x_{11} \\ x_{10} \end{Bmatrix}$$

$$\frac{\partial x}{\partial \eta} = \frac{\partial}{\partial \eta}[N]^{(4)}\{x_i\} = \frac{\partial}{\partial \eta}\lfloor N_9, N_{14}, N_{17}, N_{18}, N_{19}, N_{15}, N_{11}, N_{10} \rfloor \begin{Bmatrix} x_9 \\ x_{14} \\ x_{17} \\ x_{18} \\ x_{19} \\ x_{15} \\ x_{11} \\ x_{10} \end{Bmatrix}$$

$$\frac{\partial y}{\partial \xi} = \frac{\partial}{\partial \xi}[N]^{(4)}\{y_i\} = \frac{\partial}{\partial \xi}\lfloor N_9, N_{14}, N_{17}, N_{18}, N_{19}, N_{15}, N_{11}, N_{10} \rfloor \begin{Bmatrix} y_9 \\ y_{14} \\ y_{17} \\ y_{18} \\ y_{19} \\ y_{15} \\ y_{11} \\ y_{10} \end{Bmatrix}$$

$$\frac{\partial y}{\partial \eta} = \frac{\partial}{\partial \eta}[N]^{(4)}\{y_i\} = \frac{\partial}{\partial \eta}\lfloor N_9, N_{14}, N_{17}, N_{18}, N_{19}, N_{15}, N_{11}, N_{10} \rfloor \begin{Bmatrix} y_9 \\ y_{14} \\ y_{17} \\ y_{18} \\ y_{19} \\ y_{15} \\ y_{11} \\ y_{10} \end{Bmatrix} \qquad (e)$$

By carrying out the differentiations in Eqs. (*e*), we can compute the Jacobian matrix for element 4. This we leave for a computer program.

Going back to Eqs. (17.85*a*) and (11.44), we see that the flow stiffness matrix $[K]$ for element K_{jk} is

$$K_{jk} = \int_{-1}^{+1} \int_{-1}^{+1} (N_{j,i} N_{k,i}) \det [J] \, d\xi \, d\eta$$

where $N_{j,i}$ and $N_{k,i}$ are derivatives with respect to x and y. Using Eq. (11.42), we can obtain those derivatives with respect to local coordinates $\xi\eta$. Thus:

$$\begin{aligned} N_{j,i} N_{k,i} &= \frac{\partial N_j}{\partial x} \frac{\partial N_k}{\partial x} + \frac{\partial N_j}{\partial y} \frac{\partial N_k}{\partial y} \\ &= \left(J'_{11} \frac{\partial N_j}{\partial \xi} + J'_{12} \frac{\partial N_j}{\partial \eta} \right) \left(J'_{11} \frac{\partial N_k}{\partial \xi} + J'_{12} \frac{\partial N_k}{\partial \eta} \right) \\ &\quad + \left(J'_{21} \frac{\partial N_j}{\partial \xi} + J'_{22} \frac{\partial N_j}{\partial \eta} \right) \left(J'_{21} \frac{\partial N_k}{\partial \xi} + J'_{22} \frac{\partial N_k}{\partial \eta} \right) \end{aligned}$$

where

$$[J'] = [J]^{-1}$$

Since the Jacobian of each element has already been determined, the double integral for K_{ij} can then be calculated by means of Gaussian quadrature. Thus, for K_{11} we have

$$\begin{aligned} K_{11} = \int_{-1}^{+1} \int_{-1}^{+1} \Bigg\{ &\left(J'_{11} \frac{\partial N_1}{\partial \xi} + J'_{12} \frac{\partial N_1}{\partial \eta} \right) \left(J'_{11} \frac{\partial N_1}{\partial \xi} + J'_{12} \frac{\partial N_1}{\partial \eta} \right) \\ &+ \left(J'_{21} \frac{\partial N_1}{\partial \xi} + J'_{22} \frac{\partial N_1}{\partial \eta} \right) \left(J'_{21} \frac{\partial N_1}{\partial \xi} + J'_{22} \frac{\partial N_1}{\partial \eta} \right) \Bigg\} \det [J] \, d\xi \, d\eta \end{aligned} \tag{f}$$

For element 4, we have the following correspondence between subscripts in the above equations and nodal identification numbers (see Fig. 17.16):

$$\begin{array}{ll} 1 \to 9 & 5 \to 19 \\ 2 \to 14 & 6 \to 15 \\ 3 \to 17 & 7 \to 11 \\ 4 \to 18 & 8 \to 10 \end{array} \tag{g}$$

To get the first row of $[K]$ for element 4, we compute $K_{9,9}$, $K_{9,14}$, $K_{9,17}$, $K_{9,18}$, $K_{9,19}$, $K_{9,15}$, $K_{9,11}$, and $K_{9,10}$. The second row would consist of $K_{14,9}$, $K_{14,14}$, $K_{14,17}$, $K_{14,18}$, $K_{14,19}$, $K_{14,11}$, and $K_{14,10}$. And so forth for the other rows. We thus get an 8 × 8 matrix. We proceed in a similar manner for the other three elements. We then set up a 21 × 21 global stiffness matrix.

As for the force vectors, we need only worry about outside boundaries. Along *AE* and *DC* (see Fig. 17.11) the flow is perpendiulcar to the boundary and so $q = 0$. On the other hand, the q's on the remaining outside boundary will not be zero. However, the values of ψ at the nodal points at these outside portions of the boundary will be specified, and so for the *reduced* matrix these nonzero values of q will not appear.

Thus, for the specified values of ψ we have the following:

$$\begin{aligned} &\psi_1 = \psi_6 = \psi_9 = \psi_{14} = \psi_{17} = \psi_{18} = \psi_{19} = 0 \\ &\psi_5 = \psi_8 = \psi_{13} = \psi_{16} = \psi_{21} = 2.5 \end{aligned} \tag{h}$$

We reduce the global flow stiffness matrix by deleting the rows and columns corresponding to the subscripts for the specified ψ's, leaving a 9×9 matrix. As indicated, the q's are all zero for the reduced matrix equations. But where the ψ's are specified to be nonzero we must insert appropriate terms to the flux vector, as discussed in Section 9.3. We thus have for the reduced matrix equation

$$\begin{bmatrix} K_{2,2} & K_{2,3} & K_{2,4} & K_{2,7} & K_{2,10} & K_{2,11} & K_{2,12} & K_{2,15} & K_{2,20} \\ K_{3,2} & K_{3,3} & K_{3,4} & K_{3,7} & K_{3,10} & K_{3,11} & K_{3,12} & K_{3,15} & K_{3,20} \\ K_{4,2} & K_{4,3} & K_{4,4} & K_{4,7} & K_{4,10} & K_{4,11} & K_{4,12} & K_{4,15} & K_{4,20} \\ K_{7,2} & K_{7,3} & K_{7,4} & K_{7,7} & K_{7,10} & K_{7,11} & K_{7,12} & K_{7,15} & K_{7,20} \\ K_{10,2} & K_{10,3} & K_{10,4} & K_{10,7} & K_{10,10} & K_{10,11} & K_{10,12} & K_{10,15} & K_{10,20} \\ K_{11,2} & K_{11,3} & K_{11,4} & K_{11,7} & K_{11,10} & K_{11,11} & K_{11,12} & K_{11,15} & K_{11,20} \\ K_{12,2} & K_{12,3} & K_{12,4} & K_{12,7} & K_{12,10} & K_{12,11} & K_{12,12} & K_{12,15} & K_{12,20} \\ K_{15,2} & K_{15,3} & K_{15,4} & K_{15,7} & K_{15,10} & K_{15,11} & K_{15,12} & K_{15,15} & K_{15,20} \\ K_{20,2} & K_{20,3} & K_{20,4} & K_{20,7} & K_{20,10} & K_{20,11} & K_{20,12} & K_{20,15} & K_{20,20} \end{bmatrix} \begin{Bmatrix} \psi_2 \\ \psi_3 \\ \psi_4 \\ \psi_7 \\ \psi_{10} \\ \psi_{11} \\ \psi_{12} \\ \psi_{15} \\ \psi_{20} \end{Bmatrix}$$

$$= \begin{Bmatrix} -K_{2,5}(2.5) - K_{2,8}(2.5) - K_{2,13}(2.5) - K_{2,16}(2.5) - K_{2,21}(2.5) \\ -K_{3,5}(2.5) - K_{3,8}(2.5) - K_{3,13}(2.5) - K_{3,16}(2.5) - K_{3,21}(2.5) \\ -K_{4,5}(2.5) - K_{4,8}(2.5) - K_{4,13}(2.5) - K_{4,16}(2.5) - K_{4,21}(2.5) \\ -K_{7,5}(2.5) - K_{7,8}(2.5) - K_{7,13}(2.5) - K_{7,16}(2.5) - K_{7,21}(2.5) \\ -K_{10,5}(2.5) - K_{10,8}(2.5) - K_{10,13}(2.5) - K_{10,16}(2.5) - K_{10,21}(2.5) \\ -K_{11,5}(2.5) - K_{11,8}(2.5) - K_{11,13}(2.5) - K_{11,16}(2.5) - K_{11,21}(2.5) \\ -K_{12,5}(2.5) - K_{12,8}(2.5) - K_{12,13}(2.5) - K_{12,16}(2.5) - K_{12,21}(2.5) \\ -K_{15,5}(2.5) - K_{15,8}(2.5) - K_{15,13}(2.5) - K_{15,16}(2.5) - K_{15,21}(2.5) \\ -K_{20,5}(2.5) - K_{20,8}(2.5) - K_{20,13}(2.5) - K_{20,16}(2.5) - K_{20,21}(2.5) \end{Bmatrix}$$

By inserting the known values and inverting the K matrix we can solve the six unknown ψ's. The results are shown in Fig. 17.18. We get very good accuracy with only four elements. In this example we have

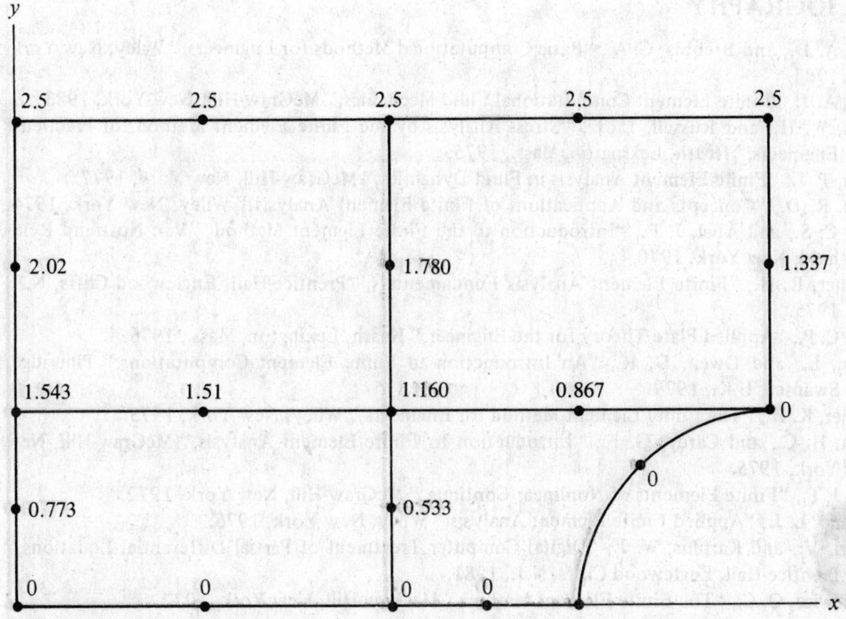

FIGURE 17.18
Values of ψ for flow domain.

indicated how the detailed computations are carried out. However, for most practical situations this should all be done with a computer.

17.13 FINAL CLOSURE

This brings to a close our introductory study of finite elements. You should now be ready to study advanced and specialized texts on finite elements and be able to read the rapidly growing literature in this area.

There are other approximation procedures for the problem we have studied. There are the older well-known finite difference methods, in which the literature abounds. And we wish to point out in closing a recent development called the *boundary integral* method. In this method one studies the boundary integral equations of a boundary-value problem rather than the quadratic functional, as we have been doing in this chapter. The potential benefit of this procedure is that only the boundaries need be discretized, with the result that much smaller systems of algebraic equations are generated. The literature for this method is becoming widespread [see Cruse (1977) and Bannerjee and Butterfield (1981)].

REFERENCES

Bannerjee, P. K., and Butterfield, R., "Boundary Methods in Engineering Science," McGraw-Hill, New York, 1981.

Cruse, T. A., "Mathematical Foundations of the Boundary Integral Equation Method in Solid Mechanics," Pratt and Whitney Aircraft Group, East Hartford, Conn., 1977.

Shames, I. H., "Mechanics of Fluids," 2d ed., McGraw-Hill, New York, 1982.

BIBLIOGRAPHY

Adey, A. D., and Brebbia, C. A., "Basic Computational Methods for Engineers," Wiley, New York, 1983.

Baker, A. J., "Finite Element Computational Fluid Mechanics," McGraw-Hill, New York, 1983.

Bowes, W. H., and Russell, L. T., "Stress Analysis by the Finite Element Method for Practicing Engineers," Heath, Lexington, Mass., 1975.

Chung, T. J., "Finite Element Analysis in Fluid Dynamics," McGraw-Hill, New York, 1977.

Cooke, R. D., "Concepts and Applications of Finite Element Analysis," Wiley, New York, 1974.

Desai, C. S., and Abel, J. F., "Introduction to the Finite Element Method," Van Nostrand Reinhold, New York, 1970.

Gallagher, R. H., "Finite Element Analysis Fundamentals," Prentice-Hall, Englewood Cliffs, N.J., 1975.

Heins, C. P., "Applied Plate Theory for the Engineer," Heath, Lexington, Mass., 1976.

Hinton, E., and Owen, D. R., "An Introduction to Finite Element Computations," Pineridge, Swansea, U.K., 1979.

Huebner, K. H., "The Finite Element Method for Engineers," Wiley, New York, 1975.

Martin, H. C., and Carey, G. F., "Introduction to Finite Element Analysis," McGraw-Hill, New York, 1975.

Oden, J. T., "Finite Elements of Nonlinear Continua," McGraw-Hill, New York, 1972.

Segerlind, L. J., "Applied Finite Element Analysis," Wiley, New York, 1976.

Vemuri, V., and Karplus, W. J., "Digital Computer Treatment of Partial Differential Equations," Prentice-Hall, Englewood Cliffs, N.J., 1981.

Zienkiewicz, O. C., "The Finite Element Method," McGraw-Hill, New York, 1977.

APPENDIX I

CARTESIAN TENSORS

I.1 INTRODUCTION

In this appendix we shall consider elements of Cartesian tensors. We shall employ tensor concepts in much of this text, and the accompanying notation will be used where it is most meaningful. (It will accordingly not be used exclusively.) The material of this appendix has considerably greater application than simply for the study of solid mechanics.

I.2 VECTORS AND TENSORS

In elementary mechanics courses we dealt primarily with two classes of quantities in describing physical phenomena, namely scalars and vectors. The scalar quantity, you will recall, required a single value for its specification and had no direction associated with it. On the other hand, we also worked with quantities that had both magnitude and direction. Accordingly, they could be represented as *directed line segments*. If we could combine these quantities in accordance with the parallelogram law, we call them vectors. Finally, you will recall that we could specify a vector by giving three scalar components of the vector for a given reference frame.

In this appendix we shall view scalars and vectors in a somewhat different manner so as to enable us to properly understand more complex quantities that we shall need for this text, namely tensors. First, note that a scalar quantity does not depend for its value on any particular reference. Stated another way, *the scalar is invariant with respect to a rotation of the coordinate axes that may be associated with the problem.* A vector, on the other hand, does require some reference for purposes of conveying its full meaning. Thus, for a velocity **V** we may refer to reference *xyz* (see Fig. I.1*a*) in conveying its properties in the form:

$$\mathbf{V} = V_x\mathbf{i} + V_y\mathbf{j} + V_z\mathbf{k}$$

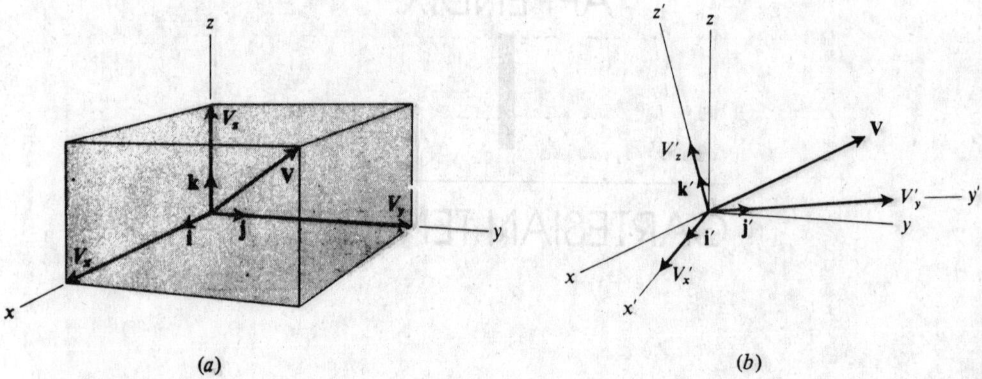

FIGURE I.1
Rectangular components of **V** for two sets of axes.

or we may employ reference $x'y'z'$ as shown in Fig. I.1*b* to convey the same information in the form:

$$\mathbf{V} = V_{x'}\mathbf{i}' + V_{y'}\mathbf{j}' + V_{z'}\mathbf{k}'$$

Despite the need for a reference, *certain aspects* of vectors are independent of the reference. Thus, for example, the magnitude of the vector is the same for any reference that may be used. In other words, we can say that the sum of the squares of the vector components is invariant for a rotation of reference axes. Stated mathematically:

$$V_x^2 + V_y^2 + V_z^2 = V_{x'}^2 + V_{y'}^2 + V_{z'}^2 \equiv \mathbf{V} \cdot \mathbf{V}$$

Also, when considering two vectors it is clear that the relative orientation of the vectors, as given by an angle θ, is independent of the coordinate system. A number of useful operations can accordingly be derived because of these invariances. These are the familiar operations of vector algebra. Vector equations involving such operations then can be stated independent of references.

There are more complex quantities that have certain characteristics independent of coordinates. Furthermore, certain very useful operations can be developed for such quantities which do not depend on a reference and which can convey certain physically important meanings. We will soon define tensor quantities as one such class of quantities. And, just as we have vector equations that represent certain physical laws, so shall we be able to present certain tensor equations that play a similar role. To investigate such quantities we shall first go back to scalars and vectors to establish more useful definitions for these quantities which will permit us to proceed in an orderly way to the more complex quantities. As has been intimated in this discussion, the relation between a quantity and a reference is of critical importance.

I.3 INTRODUCTION TO TENSOR NOTATION

We shall now introduce a notation that will permit us to express the transformation equations of second-order tensors in a very compact form. In doing so we shall be able to better understand the relation between scalars, vectors, and second-order tensors, and we shall at the same time be able to readily extrapolate our thinking to include so-called higher-order tensors.

For this purpose, we introduce the concept of the *free index*. The free index is any letter that appears only once as a subscript in a group of terms. Thus, i is the free index in the following expressions:

$$V_i \quad A_{ij}V_j$$

When we have an expression with a free index such as V_i, we can consider that we are presenting *any one* component of the array of terms formed by having i become an x, a y, or a z. This array accordingly is

$$\begin{pmatrix} V_x \\ V_y \\ V_z \end{pmatrix}$$

Alternatively, V_i can be interpreted to represent the entire *set* of components given above. Clearly, V_i may then be considered a vector since it includes the three rectangular components of the vector. The particular meaning of the free index depends on the *context* of the discussion. Similarly, the expression A_{ij} has two free indices and can represent any one component of the array of terms formed from all possible permutations of the subscripts with i and j taking on the values x, y, and z. This array of terms must then be

$$\begin{pmatrix} A_{xx} & A_{xy} & A_{xz} \\ A_{yx} & A_{yy} & A_{yz} \\ A_{zx} & A_{zy} & A_{zz} \end{pmatrix}$$

Alternatively, the expression A_{ij} can represent the entire set of the above components.

Next we shall introduce the concept of *dummy indices* by prescribing that when letters i, j, k, l, and m are repeated in an expression we *sum* terms formed by letting the repeated indices take on successively the values x, y, and z. Thus, the expression $A_{ij}V_j$ used earlier may represent a set of three terms each of which is a sum of three expressions. Thus, retaining the free index i we have

$$A_{ij}V_j = A_{ix}V_x + A_{iy}V_y + A_{iz}V_z$$

And if we wish to express the set for the free index i we get

$$A_{ij}V_j \equiv \begin{pmatrix} A_{xx}V_x + A_{xy}V_y + A_{xz}V_z \\ A_{yx}V_x + A_{yy}V_y + A_{yz}V_z \\ A_{zx}V_x + A_{zy}V_y + A_{zz}V_z \end{pmatrix}$$

If we have a double set of repeated indices such as $A_{ij}B_{ij}$, we let i take on the value x and sum terms with j ranging over x, y, and z; then we add to this three terms found by letting i become y while letting j range again over x, y, and z; etc. We thus get a *sum* of nine terms. It should be clear that, since a pair of dummy indices always sums out, the particular letters of the set *ijklm* used is immaterial. We can accordingly say that $A_iB_i = A_jB_j = A_mB_m$, etc. These are called inner products.

I.4 SCALARS AND VECTORS

We shall now reconsider scalar and vector quantities in terms of rotation of axes and in terms of the notation that we have just presented. The new perspective that we will achieve will help in setting forth tensor concepts later.

First, note that since a scalar has only a single value we have no need for subscripts. And since there is no change in a scalar when there is a rotation of the reference axes we have

$$M' = M \tag{I.1}$$

as the relation between primed and unprimed axes.

Now consider a vector $\mathbf{V}$ (or V_i) shown in Fig. I.2 as a directed segment. In accord with the parallelogram law, we have for the rectangular components of $\mathbf{V}$:

$$V_x = V\cos(V, x)$$
$$V_y = V\cos(V, y)$$
$$V_z = V\cos(V, z)$$

When the reference is rotated to a primed position as shown in the diagram, we will have a new set of components that again can be computed from the parallelogram law as follows:

$$V'_x = V_x\cos(x', x) + V_y\cos(x', y) + V_z\cos(x', z)$$
$$V'_y = V_x\cos(y', x) + V_y\cos(y', y) + V_z\cos(y', z)$$
$$V'_z = V_x\cos(z', x) + V_y\cos(z', y) + V_z\cos(z', z)$$

We will now introduce for the cosine terms above the letter a with two subscripts. The first subscript identifies a primed axis and the second subscript an umprimed axis. Furthermore, $a_{x'y}$ then represents the cosine of the angle between the x' and y axes and is thus identical in meaning to $\cos(x', y)$. The above equation can then be written as

$$V_{x'} = V_x a_{x'x} + V_y a_{x'y} + V_z a_{x'z}$$
$$V_{y'} = V_x a_{y'x} + V_y a_{y'y} + V_z a_{y'z}$$
$$V_{z'} = V_x a_{z'x} + V_y a_{z'y} + V_z a_{z'z}$$

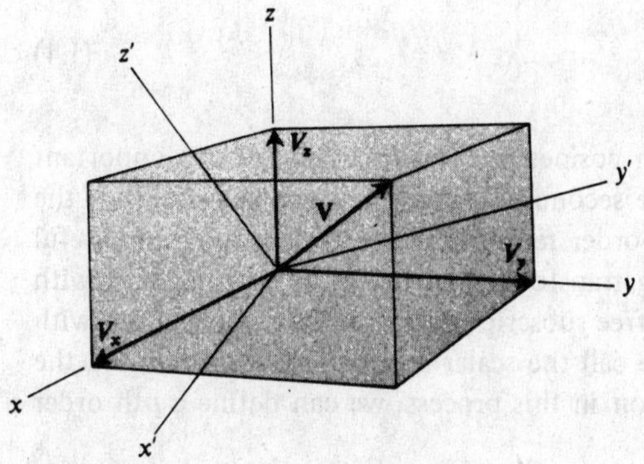

FIGURE I.2
Vector V and its rectangular components.

Using index notation, we can express these equations as

$$V_{i'} = a_{i'j} V_j \tag{I.2}$$

This is a transformation equation for rectangular components of **V** from unprimed to primed axes. This is often written as

$$V'_i = a_{ij} V_j \tag{I.3}$$

where the first subscripts for the a's are understood to represent primed axes without the need for a prime. Also, the prime has been moved to change $V_{i'}$ to V'_i. We can conclude that a *necessary* condition for a quantity to be a vector is that the *components transform according to the above equations under a rotation of axes*. We can also show that this is a *sufficient* condition. Thus, we can define a vector as a set of three components that satisfies the above transformation equations.

I.5 TENSORS: SYMMETRY AND SKEW-SYMMETRY

In the previous section we redefined vectors and scalars in terms of certain transformation equations for the components of these quantities. Thus, considering a primed and an unprimed reference rotated arbitrarily relative to each other about a common origin, we have the following defining transformation equations:

SCALARS: $M = M'$

VECTORS: $A'_i = a_{ij} A_j$

We shall now generalize these transformation equations so as to define other more complex quantities. Thus, we will define a *second-order tensor* as a set of nine components (consequently denoted with two free subscripts, as, for example, A_{ij}) which transforms under a rotation from an unprimed to a primed set of axes according to the following equation:

$$A'_{ij} = a_{ik}a_{jl}A_{kl} \tag{I.4}$$

where the *a*'s are the familiar direction cosines discussed earlier.† Many important quantities in the engineering sciences are second-order tensors. And as a result of the above transformation equations, second-order tensors have certain distinct and useful properties. Notice the way the defining transformation builds up from scalars, with no free subscripts, to vectors, with one free subscript, to the second-order tensor with two free subscripts. For this reason, we call the scalar a *zeroth-order tensor* and the vector a *first-order tensor*. Continuing on in this process, we can define a *p*th order tensor as follows:

$$A'_{q_1 q_2 \cdots q_p} = a_{q_1 j_1} a_{q_2 j_2} \ldots a_{q_p j_p} A_{j_1 j_2 \cdots j_p} \tag{I.5}$$

wherein we have sets of quantities requiring *p* free indices.

We shall say that a tensor such as T_{ijkl} is *symmetric* with respect to any two of its indices–say *jk*–if the values of the tensor components corresponding to these indices are equal to the tensor components corresponding to the reverse of these indices. That is:

$$T_{ijkl} = T_{ikjl} \tag{I.6}$$

In the case of a second-order tensor we have for symmetry the single possibility:

$$A_{ij} = A_{ji} \tag{I.7}$$

In the representation of this set as an array, it means that the terms on one side of the main diagonal may be considered as mirror image values of the terms on the other side of the main diagonal. This is shown in the following representation:

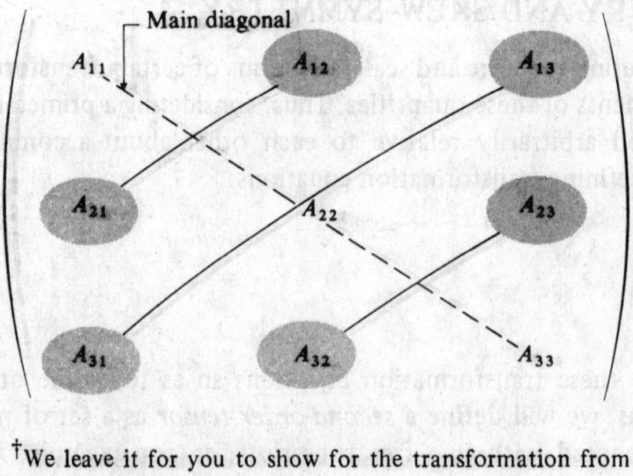

(I.8)

†We leave it for you to show for the transformation from primed to unprimed coordinates that for second-order tensors we have

$$A_{ij} = a_{ki}a_{lj}A'_{kl}$$

Furthermore, a tensor such as T_{ijkl} is said to be *skew-symmetric* in any two of its indices, say jk, if the components of the tensors with indices reversed are negatives of each other. Thus we have as a definition of skew symmetry:

$$T_{ijkl} = -T_{ikjl} \quad \text{skew-symmetric in } jk \tag{I.9a}$$

$$T_{ijkl} = -T_{ilkj} \quad \text{skew-symmetric in } jl \tag{I.9b}$$

Consider next those components of a skew-symmetric tensor wherein the indices involved in the skew symmetry have identical numbers. For T_{ijkl} in Eq. (I.9*a*), for instance, we are referring to T_{i11l}, T_{i22l}, and T_{i33l}, and for T_{ijkl} in Eq. (I.9*b*) we are referring to T_{i1k1}, T_{i2k2}, and T_{i3k3}. It is clear that such terms must be zero. Thus, considering T_{i11l} we see from the skew-symmetry requirement:

$$T_{i11l} = -T_{i11l} \tag{I.10}$$

Clearly for each i and l we must have $T_{i11l} = 0$ in order to satisfy the above condition.

For a second-order tensor we have for skew symmetry the single possibility:

$$B_{ij} = -B_{ji} \tag{I.11}$$

And in the array representation we have

$$\begin{pmatrix} 0 & B_{12} & B_{13} \\ -B_{12} & 0 & B_{23} \\ -B_{13} & -B_{23} & 0 \end{pmatrix}$$

where the main diagonal must be composed only of zeros.

Now suppose we have a tensor T_{ijkl} which is *symmetric* in i, j and a tensor B_{srtp} which is *skew-symmetric* in s, r. It is useful to note that the inner product of these tensors for indices i and s, and j and r, give a set all of whose terms are zero. Thus:

$$T_{ijkl}B_{ijtp} = C_{kltp} = 0$$

Also, if a tensor A_{ijrs} is an arbitrary *skew-symmetric* tensor in indices ij and if $A_{ijrs}B_{ij} = 0$ for a nonzero tensor B_{ij}, then we can consider that B_{ij} must be *symmetric* in ij. It is also true that if A_{ijrs} is an arbitrary *symmetric* tensor in ij for which $A_{ijrs}B_{ij} = 0$ for a nonzero tensor B_{ij}, then B_{ij} must be *skew-symmetric* in ij. Note that ij can be any of the indices of a tensor.

Finally, we will demonstrate that taking the partial derivative of an nth-order tensor $A_{p_1 p_2 \cdots p_n}$ with respect to x_i, where i is a free index, results in a tensor of rank $(n + 1)$. We shall find it convenient here to adopt the convention of denoting partial derivatives of a tensor by making use of commas and indices as follows:

$$\frac{\partial()}{\partial x_i} \equiv ()_{,i} \qquad \frac{\partial^2()}{\partial x_i\, \partial x_j} \equiv ()_{,ij}$$

Accordingly, we wish to show at this time that $A_{j_1 \cdots j_n, i}$ is an $(n + 1)$th-order tensor. Suppose we start with a primed reference x_i'. Then we may say:

$$A'_{j_1 \cdots j_n, i} \equiv \frac{\partial (A'_{j_1 \cdots j_n})}{\partial x_i'}$$

Since $A'_{j_1 \cdots j_n}$ is an nth-order tensor, we use its transformed form in the unprimed reference in the above equation:

$$A'_{j_1 \cdots j_n, i} = \frac{\partial}{\partial x_i'} (a_{j_1 k_1} a_{j_2 k_2} \cdots a_{j_n k_n} A_{k_1 \cdots k_n})$$

Also we note that

$$\frac{\partial}{\partial x_i'} = \frac{\partial}{\partial x_l} \frac{\partial x_l}{\partial x_i'}$$

But $\partial x_l / \partial x_i' = a_{il}$. Hence we have

$$\frac{\partial}{\partial x_i'} = a_{il} \frac{\partial}{\partial x_l}$$

Thus we have

$$A'_{j_1 \cdots j_n, i} = a_{j_1 k_1} \cdots a_{j_n k_n} a_{il} \frac{\partial}{\partial x_l} (A_{k_1 \cdots k_n}) = a_{j_1 k_1} \cdots a_{j_n k_n} a_{il} A_{k_1 \cdots k_n, l}$$

The above statement is the defining one for a tensor of order $(n + 1)$ and so we have demonstrated that $A_{j_1 \cdots j_n, i}$ is an $(n + 1)$th-order tensor.

Since we can perform the operation of taking partials with respect to other coordinates x_j, x_k, etc., we can conclude that

$$A_{p_1 \cdots p_n, jk} = \frac{\partial^2}{\partial x_j \, \partial x_k} (A_{p_1 \cdots p_n})$$

is an $(n + 2)$th-order tensor, while

$$A_{p_1 \cdots p_n, jkl} = \frac{\partial^3}{\partial x_j \, \partial x_k \, \partial x_l} (A_{p_1 \cdots p_n})$$

is an $(n + 3)$th-order tensor.

I.6 VECTOR OPERATIONS USING TENSOR NOTATION: THE ALTERNATING TENSOR

We have already shown that the dot product $\mathbf{A} \cdot \mathbf{B}$ can be expressed as $A_i B_i$ in index notation. Another operation that we have used extensively is the cross product $\mathbf{A} \times \mathbf{B} = \mathbf{C}$, where

$$C_1 = A_2B_3 - A_3B_2$$
$$C_2 = A_3B_1 - A_1B_3 \tag{I.12}$$
$$C_3 = A_1B_2 - A_2B_1$$

In order to be able to get this result by using index notation, we introduce the *alternating tensor* ϵ_{ijk}, defined as follows:

$\epsilon_{ijk} = 0$ for those terms of the set for which i, j, k do not form some permutation of 1, 2, 3. (Example: if any two of the subscripts are equal, then such terms are zero.)

$\epsilon_{ijk} = 1$ for those terms of the set having indices that form the sequence 1, 2, 3 or that can be arranged by an *even* number of permutations to form this sequence.

$\epsilon_{ijk} = -1$ for those terms in the set that require an *odd* number of permutations to reach the sequence 1, 2, 3.

Thus we have

$$\epsilon_{112} = \epsilon_{212} = \epsilon_{331} = \cdots = 0$$

$$\epsilon_{123} = \epsilon_{231} = \epsilon_{312} = 1$$

$$\epsilon_{213} = \epsilon_{321} = \epsilon_{132} = -1$$

Now if we go back to Eq. (I.12), we can readily demonstrate that the following expression represents the terms comprising the set of the cross product:

$$C_i = \epsilon_{ijk}A_jB_k \tag{I.13}$$

Thus, carrying out the double summation over dummy indices j and k, we get for $i = 1$:

$$C_1 = \epsilon_{1jk}A_jB_k = \epsilon_{111}A_1B_1 + \epsilon_{121}A_2B_1 + \epsilon_{131}A_3B_1 + \epsilon_{112}A_1B_2 + \epsilon_{122}A_2B_2 + \epsilon_{132}A_3B_2 + \epsilon_{113}A_1B_3 + \epsilon_{123}A_2B_3 + \epsilon_{133}A_3B_3$$

Now employing the definition of the alternating tensor we find

$$C_1 = A_2B_3 - A_3B_2$$

as you may verify.

Now consider the *triple scalar product* $(\mathbf{A} \times \mathbf{B}) \cdot \mathbf{C}$. Using index notation, we get

$$(\mathbf{A} \times \mathbf{B}) \cdot \mathbf{C} = \epsilon_{ijk}A_jB_kC_i \tag{I.14}$$

Furthermore, you will recall from your mechanics courses that

$$(\mathbf{A} \times \mathbf{B}) \cdot \mathbf{C} = \begin{vmatrix} A_1 & A_2 & A_3 \\ B_1 & B_2 & B_3 \\ C_1 & C_2 & C_3 \end{vmatrix} \tag{I.15}$$

and so comparing Eqs. (I.15) and (I.14) we get the following expression for a determinant:

$$\begin{vmatrix} A_1 & A_2 & A_3 \\ B_1 & B_2 & B_3 \\ C_1 & C_2 & C_3 \end{vmatrix} = \epsilon_{ijk} A_j B_k C_i \tag{I.16}$$

We may put the above relation in a more convenient form from an index point of view by noting that

$$\epsilon_{ijk} A_j B_k C_i = \epsilon_{kij} A_i B_j C_k \tag{I.17}$$

where we have changed the dummy indices. Now ϵ_{kij} can be made into ϵ_{ijk} by two interchanges of indices, and so we have

$$\epsilon_{kij} = \epsilon_{ijk}$$

Accordingly, Eq. (I.17) can be written as

$$\epsilon_{ijk} A_j B_k C_i = \epsilon_{ijk} A_i B_j C_k$$

We can then write Eq. (I.16) in the following manner:

$$\begin{vmatrix} A_1 & A_2 & A_3 \\ B_1 & B_2 & B_3 \\ C_1 & C_2 & C_3 \end{vmatrix} = \epsilon_{ijk} A_i B_j C_k$$

Let us next consider the *field operators* of vector analysis, using index notation. First, we have the *gradient operator*, which, acting on a function ϕ, is given as

$$\nabla \phi = \frac{\partial \phi}{\partial x} \mathbf{i} + \frac{\partial \phi}{\partial y} \mathbf{j} + \frac{\partial \phi}{\partial z} \mathbf{k}$$

To write $\nabla \phi$ in index notation we can express the Cartesian unit vectors as follows:

$$\mathbf{i} = \mathbf{i}_1$$

$$\mathbf{j} = \mathbf{i}_2$$

$$\mathbf{k} = \mathbf{i}_3$$

Accordingly, we can then say

$$\boxed{\nabla \phi = \phi_{,j} \mathbf{i}_j} \tag{I.18}$$

Also, the expression $\phi_{,i}$, with i as a free index, is the index counterpart of $\nabla \phi$.

The *divergence* of a vector, you will recall, is given as follows in vector notation:

$$\text{div } \mathbf{V} = \nabla \cdot \mathbf{V} = \frac{\partial V_x}{\partial x} + \frac{\partial V_y}{\partial y} + \frac{\partial V_z}{\partial z}$$

Using index notation, we get

$$\text{div } \mathbf{V} = \frac{\partial V_i}{\partial x_i} = V_{i,i} \tag{I.19}$$

Next we consider the *curl* operator. Again from vector analysis, we have

$$\text{curl } \mathbf{V} = \nabla \times \mathbf{V} = \begin{vmatrix} \mathbf{i} & \mathbf{j} & \mathbf{k} \\ \dfrac{\partial}{\partial x_1} & \dfrac{\partial}{\partial x_2} & \dfrac{\partial}{\partial x_3} \\ V_x & V_y & V_z \end{vmatrix}$$

$$= \begin{vmatrix} \mathbf{i}_1 & \mathbf{i}_2 & \mathbf{i}_3 \\ \dfrac{\partial}{\partial x_1} & \dfrac{\partial}{\partial x_2} & \dfrac{\partial}{\partial x_3} \\ V_1 & V_2 & V_3 \end{vmatrix}$$

Then by direct analogy to the determinant expression above, we can say

$$\boxed{\nabla \times \mathbf{V} = \epsilon_{ijk}\, \mathbf{i}_i \frac{\partial V_k}{\partial x_j} = \epsilon_{ijk} V_{k,j} \mathbf{i}_i} \tag{I.20}$$

Or considering the index i to be a free index, we can express $\nabla \times \mathbf{V}$ indicially simply as $\epsilon_{ijk} V_{k,j}$.

Finally, we consider the *Laplacian* operator ∇^2. From vector analysis we have

$$\nabla^2 = \frac{\partial^2}{\partial x_1^2} + \frac{\partial^2}{\partial x_2^2} + \frac{\partial^2}{\partial x_3^2}$$

Hence, using index notation, we get

$$\nabla^2 \equiv {}_{,ii}$$

And so

$$\boxed{\nabla^2 \phi \equiv \phi_{,ii}} \tag{I.21}$$

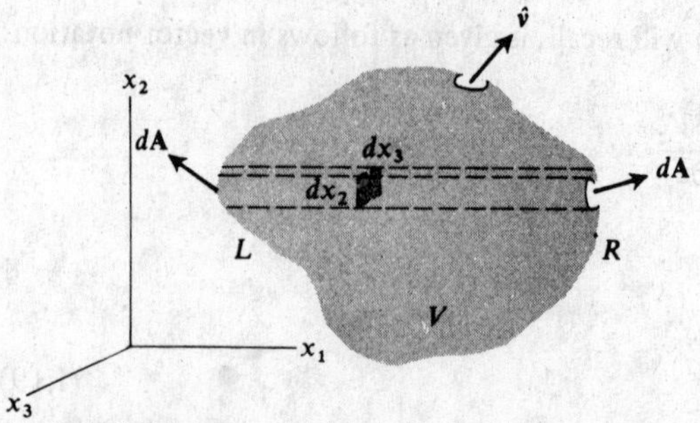

FIGURE I.3
Prismatic element in a body.

I.7 GAUSS'S THEOREM

In the previous section we considered the so-called differential field operators of divergence, curl, etc., in tensor notation. Equally important for us are certain integral theorems. At this time we present Gauss's theorem in a reasonably general form.

Assume we have an nth-order tensor $T_{jk\cdots}$, defined at each point in space (i.e., it is an nth-order tensor *field*). Consider a domain V in space of such a shape that in the direction x_1 lines parallel to the x_1 axis can pierce the boundary only twice (see Fig. I.3). Imagine now that this volume is composed of infinitesimal prisms having sides dx_2 and dx_3, as shown in the diagram. Consider one prism and compute the following integral over the volume δV of this prism:

$$\iiint_{\delta V} \frac{\partial}{\partial x_1}(T_{jk\cdots})\, dx_1\, dx_2\, dx_3$$

Carrying out integration with respect to x_1 we get

$$\iiint_{\delta V} \frac{\partial}{\partial x_1}(T_{jk\cdots})\, dx_1\, dx_2\, dx_3 = \iint (T_{jk\cdots}\, dx_2\, dx_3)_R - \iint (T_{jk\cdots}\, dx_2\, dx_3)_L$$

where the first expression on the right side of the above equation is evaluated at the right end of the prism while the second expression is evaluated at the left end of the prism. Using ν to represent the unit vector normal to the boundary surface of V, and considering ν_1, ν_2, and ν_3 to be the direction cosines of ν,† we can replace dx_2, dx_3 at the right end of the prism by $(+\nu_1\, dA)$ and by $(-\nu_1\, dA)$ on the left end of the

†When we use xyz notation, we generally write ν_1, ν_2, and ν_3 as a_{nx}, a_{ny}, and a_{nz} with n being indicative of the unit normal.

prism.† We thus have

$$\iiint_{\delta V} \frac{\partial}{\partial x_1}(T_{jk\cdots})\,dx_1\,dx_2\,dx_3 = \iint (T_{jk\cdots}\,\nu_1\,dA)_R + \iint (T_{jk\cdots}\,\nu_1\,dA)_L$$

Now, integrating over all the prisms comprising the volume V, we get‡

$$\iiint_V \frac{\partial}{\partial x_1}(T_{jk\cdots})\,dx_1\,dx_2\,dx_3 = \iint_R (T_{jk\cdots})\nu_1\,dA + \iint_L (T_{jk\cdots})\nu_1\,dA$$

The right side of the above equation clearly covers the entire surface of volume V and is accordingly replaceable by a closed-surface integral. We then have the following statement:

$$\iiint_V \frac{\partial}{\partial x_1}(T_{jk\cdots})\,dx_1\,dx_2\,dx_3 = \oiint_S (T_{jk\cdots})\nu_1\,dA \tag{I.22}$$

We developed Eq. (I.22) for the x_1 direction. We could have proceeded in a similar manner in any direction x_i. Accordingly, the generalization of the above statement is given as follows:

$$\boxed{\iiint_V \frac{\partial T_{jk\cdots}}{\partial x_i}\,dv = \oiint_S (T_{jk\cdots})\nu_i\,dA} \tag{I.23}$$

where i is a free index. This is *Gauss's theorem* in a generalized form.§ Using the notation presented for a spatial partial derivative, we can also give this equation as follows:

$$\boxed{\iiint_V (T_{jk\cdots})_{,i}\,dv = \oiint_S (T_{jk\cdots})\nu_i\,dA} \tag{I.24}$$

†The appearance of the minus sign here is a result of the use of the *outward-normal* convention for area vectors. Thus, ν_1 on the left side of the prism is clearly negative for the kind of domain we have chosen to work with and a minus sign must be included so that the product $(-\nu_1\,dA)$ will be the positive number needed to replace $dx_2\,dx_3$ for that side.

‡We are assuming tacitly here, in order to be able to carry out the integration, that the surface of the domain V can be split up into a finite number of parts such that there is a continuously varying tangent plane on each piece. That is, the surface should be *piecewise smooth*.

§This theorem can be extended to bodies whose shape is such that lines parallel to $x_1x_2x_3$ axes cut the surface more than twice, provided such a body can be decomposed into contiguous composite bodies that do have the property specified in the development.

Suppose that $T_{jk\ldots}$ is the zeroth-order tensor ϕ (i.e., a scalar). We then have

$$\iiint_V \phi_{,i}\, dv = \oiint_S \phi \nu_i\, dA$$

If we wish to revert to vector notation, this form of Gauss's law becomes

$$\iiint_V \nabla\phi\, dv = \oiint_S \phi\, d\mathbf{A} \tag{I.25}$$

where $d\mathbf{A} = \boldsymbol{\nu}\, dA$. Suppose next that $T_{jk\ldots}$ is a first-order tensor (i.e., a vector) V_j. We then get

$$\iiint_V V_{j,i}\, dv = \oiint_S V_j \nu_i\, dA \tag{I.26}$$

Next, perform a contraction operation on the free indices i and j (i.e., replace by a set of dummy indices) in the above equation. We get

$$\iiint_V V_{j,j}\, dv = \oiint_S V_j \nu_j\, dA \tag{I.27}$$

You should have no difficulty in writing this equation in terms of vectors to arrive at:

$$\boxed{\iiint_V \operatorname{div} \mathbf{V}\, dv = \oiint_S \mathbf{V} \cdot d\mathbf{A}} \tag{I.28}$$

We will have much use of this form of Gauss's theorem. In this form it is called the *divergence theorem.*†

Let us consider next that the vector field **V** is given as follows:

$$\mathbf{V} = \phi\mathbf{i} + \psi\mathbf{j} + \kappa\mathbf{k}$$

Then Eq. (I.28) can be given in the following way:

†The divergence theorem has a simple physical interpretation. You will recall from your earlier work in mechanics that div **B** represents the net efflux of the flux of vector field **B** per unit volume at a point. Accordingly, the volume integration on the left side of the above equations represents the net efflux of flux of vector field **B** from volume V. This net flux is then equated in this equation to the flux piercing the bounding surface S of the volume V. We have here a statement concerning the conservation of flux for a volume V for the vector **B**.

$$\iiint_V \left(\frac{\partial \phi}{\partial x} + \frac{\partial \psi}{\partial y} + \frac{\partial \kappa}{\partial z}\right) dv = \oiint_S (\phi a_{\nu x} + \psi a_{\nu y} + \kappa a_{\nu z})\, dA \tag{I.29}$$

where $a_{\nu x} = \nu_1$, etc., are the direction cosines of the outer normal to the boundary. In this form the equation is called *Green's theorem.*† If we consider a two-dimensional simplification of Green's theorem, we get the following result:

$$\iint_S \left(\frac{\partial \phi}{\partial x} + \frac{\partial \psi}{\partial y}\right) dA = \oint_\Gamma (\phi a_{\nu x} + \psi a_{\nu y})\, dl \tag{I.30}$$

We now set $\psi = 0$ and $\kappa = 0$ in Eq. (I.29) and let ϕ be the product of two functions u and w. We get on substitution:

$$\iiint_V \frac{\partial (uw)}{\partial x}\, dv = \iiint_V u \frac{\partial w}{\partial x}\, dv + \iiint_V w \frac{\partial u}{\partial x}\, dv = \oiint_S (uw) a_{\nu x}\, dA$$

Rearranging, we reach the following formula representing an *integration by parts*:

$$\iiint_V u \frac{\partial w}{\partial x}\, dv = \oiint_S (uw) a_{\nu x}\, dA - \iiint_V w \frac{\partial u}{\partial x}\, dv$$

Generalizing for any coordinate x_i we get

$$\iiint_V u \frac{\partial w}{\partial x_i}\, dv = \oiint_S (uw) a_{\nu x_i}\, dA - \iiint_V w \frac{\partial u}{\partial x_i}\, dv \tag{I.31}$$

The corresponding integration by parts formula for two dimensions is

$$\iint_S u \frac{\partial w}{\partial x_i}\, dA = \oint_\Gamma (uw) a_{\nu x_i}\, dl - \iint_S w \frac{\partial u}{\partial x_i}\, dA \tag{I.32}$$

We have much occasion in this text to utilize the above formulas.

I.8 GREEN'S FORMULA

Another formula that is useful to us in the text is Green's formula and its various special forms. For this purpose, consider the following integral:

†Green's theorem may be proved without vectorial considerations for any three functions ϕ, ψ, and κ that are piecewise continuous and differentiable. See Kaplan (1951).

$$I = \iiint_V \phi_{,i} \psi_{,i} \, dv \tag{I.33}$$

wherein, you will note, we have as an integrand the inner product of two gradients. The functions ϕ and ψ are assumed to be continuous in the domain V with continuous first and second partial derivatives in this domain. The domain V has all the restriction needed for the development of Gauss's theorem. Note now that:

$$\left(\frac{\partial \phi}{\partial x_1}\right)\left(\frac{\partial \psi}{\partial x_1}\right) = \frac{\partial}{\partial x_1}\left(\phi \frac{\partial \psi}{\partial x_1}\right) - \phi \frac{\partial^2 \psi}{\partial x_1^2}$$

Similar relations may be written for x_2 and x_3 so that Eq. (I.33) may be expressed as follows:

$$\iiint_V \phi_{,i} \psi_{,i} \, dv = -\iiint_V \phi \psi_{,jj} \, dv + \iiint_V \frac{\partial}{\partial x_j}\left(\phi \frac{\partial \psi}{\partial x_j}\right) dv \tag{I.34}$$

We may now use Gauss's theorem for the last integral. That is,

$$\iiint_V \frac{\partial}{\partial x_j}\left(\phi \frac{\partial \psi}{\partial x_j}\right) dv = \oiint_S \left(\phi \frac{\partial \psi}{\partial x_j}\right) \nu_j \, dA$$

But on expanding $(\partial\psi/\partial x_j)\nu_j$ we realize that this is simply the directional derivative of ψ in the direction ν normal to the surface of the domain V. Thus we can say

$$\iiint_V \frac{\partial}{\partial x_j}\left(\phi \frac{\partial \psi}{\partial x_j}\right) dv = \oiint_S \phi \frac{\partial \psi}{\partial \nu} \, dA$$

We can then express Eq. (I.34) in the following manner:

$$\iiint_V \phi_{,i} \psi_{,i} \, dv = -\iiint_V \phi \psi_{,jj} \, dv + \oiint_S \phi \frac{\partial \psi}{\partial \nu} \, dA \tag{I.35}$$

This is a preliminary form of Green's formula which, using vector notation, should be readily recognized to have the following form:

$$\iiint_V \nabla\phi \cdot \nabla\psi \, dv = -\iiint_V \phi \nabla^2 \psi \, dv + \oiint_S \phi \frac{\partial \psi}{\partial \nu} \, dA \tag{I.36}$$

Now write Eq. (I.35) with ϕ and ψ interchanged. Subtracting equations, we see that the left sides cancel, leaving the following result:

$$\iiint_V (\phi\psi_{,jj} - \psi\phi_{,jj})\, dv = \oiint_S \left(\phi \frac{\partial\psi}{\partial\nu} - \psi \frac{\partial\phi}{\partial\nu}\right) dA \tag{I.37}$$

This is the well-known *Green's formula*. In *vector notation* we get

$$\iiint_V (\phi\nabla^2\psi - \psi\nabla^2\phi)\, dv = \oiint_S \left(\phi \frac{\partial\psi}{\partial\nu} - \psi \frac{\partial\phi}{\partial\nu}\right) dA \tag{I.38}$$

In two dimensions the volume integral becomes a surface integral, while the closed integral becomes a line integral.

1.9 KRONECKER DELTA

As a final item we present the useful Kronecker delta, defined as

$$\begin{aligned} \delta_{ij} &= 1 \quad \text{if } i = j \\ \delta_{ij} &= 0 \quad \text{if } i \neq j \end{aligned} \tag{I.39}$$

Thus, $\delta_{12} = \delta_{13} = 0$ and $\delta_{11} = \delta_{33} = 1$.

Consider the expression $\tau_{ij}\delta_{jk}$. For those expressions in the summation where $j \neq k$, we get zero contribution. Only when $j = k$ do we get a nonzero contribution, and for that expression Kronecker delta is unity. We thus get for $\tau_{ij}\delta_{jk}$ the result τ_{ik}. Accordingly, we can say

$$A_{ij}\delta_{jk} = A_{ik} \tag{I.40}$$

and by similar reasoning

$$A_{ij}\delta_{ik} = A_{kj} \tag{I.41}$$

REFERENCE

Kaplan, W., "Advanced Calculus," Addison-Wesley, Reading, Mass., 1951, Chapter 5.

PROBLEMS

I.1 Given the following information:

$$\begin{array}{lll} A_x = 10 & B_x = 3 & C_x = 16 \\ A_y = -15 & B_y = -4 & C_y = 12 \\ A_z = 6 & B_z = 1 & C_z = -3 \end{array}$$

find $A_iC_iB_3$ and $A_iB_iA_2$.

What is the set of terms given by

(*a*) $A_iB_iC_j$ (*b*) $C_jB_kC_j$ (*c*) $A_iB_iC_jA_k$

I.2 Given the following values:

$$\begin{array}{lll} A_1 = 2 & A_2 = 3 & A_3 = 4 \\ B_{11} = 0 & B_{12} = 2 & B_{13} = -2 \\ B_{21} = -3 & B_{22} = 1 & B_{23} = -1 \\ B_{31} = 6 & B_{32} = -3 & B_{33} = 0 \\ C_1 = 1 & C_2 = 3 & C_3 = 8 \end{array}$$

compute or denote the array for the following:

(*a*) A_iB_{2i} (*b*) $B_{j2}C_j$ (*c*) $A_iB_{22}C_j$

I.3 Show that if λ is a constant and A_{ij} is a second-order tensor, then λA_{ij} is a second-order tensor.

I.4 Show that if A_{ij} and B_{ij} are second-order tensors, then $(A_{ij} + B_{ij})$ is a second-order tensor.

I.5 Evaluate the following terms:

(*a*) ϵ_{112} (*b*) ϵ_{132} (*c*) ϵ_{231} (*d*) ϵ_{131}

I.6 Show that $\epsilon_{ijk}\epsilon_{ijp} = 0$ for those terms where the index k does not equal the index p. Also show that $\epsilon_{ijk}\epsilon_{ijp} = 2$ for those terms where the index k equals the index p. Thus, we can say for the representation of $\epsilon_{ijk}\epsilon_{ijp}$

$$\epsilon_{ijk}\epsilon_{ijp} = C_{kp} = \begin{pmatrix} 2 & 0 & 0 \\ 0 & 2 & 0 \\ 0 & 0 & 2 \end{pmatrix}$$

Thus, show that

$$\epsilon_{ijk}\epsilon_{ijp} = 2\delta_{kp}$$

Finally, show that the contraction (make into a pair of dummy indices) of the above over kp gives 6.

I.7 Show that $\delta_{ik}\epsilon_{ikm} = 0$.

I.8 If $\mathbf{V} = 3x\mathbf{i} + 2y^2\mathbf{j} + 10z^3\mathbf{k}$, compute:

(*a*) $V_{i,i}$ (*b*) $V_{i,i2}$ (*c*) $V_{j,3j}$

I.9 If $\phi = x^2 + 2\sin y + xz^3$, compute:

(*a*) $\phi_{,i}$ (*b*) $\phi_{,ii}$ (*c*) $\phi_{,iij}$ (*d*) $\phi_{,ii2}$

I.10 Write indically:

(*a*) $(\nabla^2\phi)\mathbf{A}\cdot\mathbf{B}$ (*b*) $[(\mathbf{A}\times\mathbf{B})\cdot\mathbf{C}]\mathbf{D}$ (*c*) $\nabla^4\phi$

I.11 Given the following second-order tensor:

$$T_{ij} = \begin{bmatrix} x^2 & x^2 + 2yz & 0 \\ x^2 + 2yz & 0 & 0 \\ 0 & 0 & 0 \end{bmatrix}$$

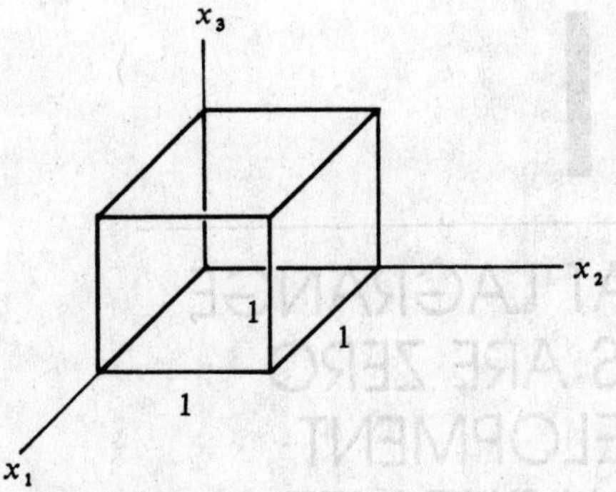

FIGURE I.4

write Gauss's theorem for $T_{12,1}$ for a domain shown in Fig. I.4.

I.12 Check the divergence theorem using

$$\mathbf{V} = x^2 y\,\mathbf{i} + 3zx\,\mathbf{j}$$

for the domain shown in Fig. I.4.

I.13 Integrate by parts the following integral:

$$\int_0^2 x^3 \frac{d}{dx}(\sin x^2)\,dx$$

Do not try to carry out the integration.

I.14 Given the following stress tensor:

$$\tau_{ij} = \begin{pmatrix} 1000 & 0 & -5000 \\ 0 & 3000 & 4000 \\ -5000 & 4000 & -3000 \end{pmatrix}$$

find $\tau_{ij}\delta_{ij}$.

APPENDIX II

TO SHOW THAT LAGRANGE MULTIPLIERS ARE ZERO FOR DEVELOPMENT OF THE RAYLEIGH-RITZ METHOD

We must now verify the assertion in Section 7.16 that the $k-1$ Lagrange multipliers are zero. We note first that for the case where $k = 1$, there is no constraint for the function W as to orthogonality and so the last expression in Eq. (7.167) will not be present. (More formally, for $k = 1$ the summation is empty for this case.) *Then Eq.* (7.167) *can be given as Eq.* (7.168). Now suppose that Eq. (7.168) holds for $S = 1, 2, \ldots, (k-1)$. We will now *prove* that $\lambda^{(1)} = \lambda^{(2)} = \cdots = \lambda^{(k-1)} = 0$ and hence from our discussion of Section (7.16), Eq. (7.168) then holds *also* for $S = k$. On this basis, knowing that Eq. (7.168) holds for $S = 1$, we could then deduce that it holds also for $S = 2$ and so forth for all values of S, thereby justifying the step of setting all the Lagrange multipliers equal to zero.

To show this, we go back to Eqs. (7.167), which we now express using C's associated with the kth approximate eigenfunction as follows:

$$\sum_{j=1}^{n} (G_{ij} - \Lambda^2 E_{ij}) C_j^{(k)} - \tfrac{1}{2}\left(\sum_{S=1}^{k-1} \lambda^{(S)} \sum_{j=1}^{n} C_j^{(S)} E_{ij}\right)\left(\sum_{v=1}^{n}\sum_{p=1}^{n} C_v^{(k)} C_p^{(k)} E_{vp}\right) = 0$$

$$i = 1, 2, \ldots n$$

Now multiply by $C_i^{(f)}$, where f is *any integer less than* k and sum over i from 1 to n. We thus get

$$\sum_{i=1}^{n}\sum_{j=1}^{n} (G_{ij} - \Lambda^2 E_{ij}) C_j^{(k)} C_i^{(f)} - \tfrac{1}{2}\left(\sum_{S=1}^{k-1} \lambda^{(S)} \sum_{j=1}^{n}\sum_{i=1}^{n} C_i^{(f)} C_j^{(S)} E_{ij}\right)$$

$$\times\left(\sum_{v=1}^{n}\sum_{p=1}^{n} C_v^{(k)} C_p^{(k)} E_{vp}\right) = 0$$

But from the orthogonality conditions we have imposed on the approximate eigenfunctions we can conclude [see Eq. (7.166)] that

$$\sum_{j=1}^{n}\sum_{i=1}^{n} C_i^{(k)} C_j^{(f)} E_{ij} = 0$$

in the first expression and that only when $S = f$ do we get other than zero in the second expression. Thus we have

$$\sum_{i=1}^{n}\sum_{j=1}^{n} G_{ij} C_i^{(f)} C_j^{(k)} - \tfrac{1}{2}\lambda^{(f)} \left(\sum_{j=1}^{n}\sum_{i=1}^{n} C_i^{(f)} C_j^{(f)} E_{ij} \right)\left(\sum_{v=1}^{n}\sum_{p=1}^{n} C_v^{(k)} C_p^{(k)} E_{vp} \right) = 0 \tag{II.1}$$

Now let us focus on the first expression in the above equation. It may be rewritten as follows:

$$\sum_{i=1}^{n}\sum_{j=1}^{n} G_{ij} C_i^{(f)} C_j^{(k)} = \sum_{j=1}^{n} C_j^{(k)} \left(\sum_{i=1}^{n} C_i^{(f)} G_{ij} \right) \tag{II.2}$$

We can now go back to Eq. (7.168), which we now assume is valid in this discussion for all values of f. Thus we have

$$\sum_{j=1}^{n} C_j^{(f)} G_{ij} = \Lambda^2 \sum_{j=1}^{n} C_j^{(f)} E_{ij} \qquad i = 1, 2, \ldots, n$$

Using the fact that G_{ij} and E_{ij} are symmetric as a result of the self-adjoint property of the operators L and M [see Eqs. (7.165)] we can change indices in the above equations as follows:

$$\sum_{i=1}^{n} C_i^{(f)} G_{ij} = \Lambda^2 \sum_{i=1}^{n} C_i^{(f)} E_{ij} \qquad j = 1, 2, \ldots, n$$

Now use the above equation to replace the bracketed expression on the right side of Eq. (II.2) so that

$$\sum_{i=1}^{n}\sum_{j=1}^{n} G_{ij} C_i^{(f)} C_j^{(k)} = \sum_{j=1}^{n} C_j^{(k)} \left(\Lambda^2 \sum_{i=1}^{n} C_i^{(f)} E_{ij} \right) = \Lambda^2 \sum_{i=1}^{n}\sum_{j=1}^{n} C_j^{(k)} C_i^{(f)} E_{ij}$$

Again we invoke the orthogonality conditions [Eq. (7.166)] for the approximate eigenfunctions (note $f \neq k$) to show that the last expression above is zero. Thus in returning to Eq. (II.1) we must conclude

$$\tfrac{1}{2}\lambda^{(f)}\left(\sum_{j=1}^{n}\sum_{i=1}^{n} C_i^{(f)}C_j^{(f)}E_{ij}\right)\left(\sum_{v=1}^{n}\sum_{p=1}^{n} C_v^{(k)}C_p^{(k)}E_{vp}\right)=0$$

Since $\iiint WM(W)\,dv$, which is the origin of each of the bracketed expressions above, is positive definite, we can conclude that $\lambda^{(f)}=0$. Now since f can be any integer up to and including $k-1$, we can conclude from above that $\lambda^{(1)}=\lambda^{(2)}\cdots=\lambda^{(k-1)}=0$. Thus having assumed that

$$\sum_{j=1}^{n}(G_{ij}-\Lambda^2 E_{ij})C_j^{(S)}=0 \qquad i=1,2,\ldots,n \tag{II.3}$$

for all values of S up to $k-1$, we see that the Lagrange multipliers up to $\lambda^{(k-1)}$ are zero. But now going to Eq. (7.167) we see that this is the *precise condition for making the above equation valid for* S *equal to* k *also.* We could then similarly prove with Eq. (II.3) valid for S up to k that $\lambda^{(1)}=\lambda^{(2)}=\cdots=\lambda^{(k)}=0$ and thus verify that Eq. (7.168) is valid also for $S=k+1$.

The process can then be continued. It needs only a starting point and, as pointed out at the outset, we do know that [Eq. (II.3)] is valid for $S=1$. With this as the beginning step we can conclude that all the Lagrange multipliers are indeed zero for all values of k.

APPENDIX III

AUTOMATIC MESH GENERATION

We will present a program for decomposing a plane area into a finite element mesh. The area is first broken up into large regions, which may be a rectangle, a triangle, or a general curved-sided quadrilateral. These are shown in Fig. III.1. Each corner has a node and each side has one node. In the triangle in Fig. III.1*b*, we take a point on one of the sides and consider this to be the corner of two sides that would render the triangle as a quadrilateral. In the case shown, node 5 is such a point and is the "corner" of sides 3-5 and 5-7. The numbering is counterclockwise. All previously numbered nodes are skipped.

Thus consider the outside boundary of the area shown in Fig. III.2. It is decomposed into two rectangles denoted as (1) and (2) and a triangle denoted as (3). The nodes are inserted as described above. Note that nodes 4 and 6 are closer to node 5 than to nodes 3 and 7, respectively. This will result in smaller elements from the mesh generation near 5, where we may know ahead of time that there will be large gradients of the dependent variable.

In Fig. III.3, we again show the area and its decomposition—this time identifying in each subarea the four sides of the subarea numbering 1 to 4. Note that inside boundaries have two sides each. Also the direction and positions of the $\xi\eta$ axes for each subarea are shown. We must now "connect" or relate the three regions of the area with each other by using the four sides of each subarea. This is accomplished in Table III.1. Using this table, consider region 1 and side 1. The corresponding number in the array is 2. This means (see Fig. III.3) that side 1 of subarea (1) is a border (and is thus "connected") to subarea (2). Consider as another example region (3) and side 4. From Fig. III.3, we see that this side is not a border of one of our three subareas and so we have the value of 0 in the matrix array of Table III.1. These data are called connectivity data and will be needed for the program.

†Adapted with permission from *Applied Finite Element Analysis* by L. J. Segerlind, John Wiley and Sons, Inc., New York, 1976, Section 18.1.

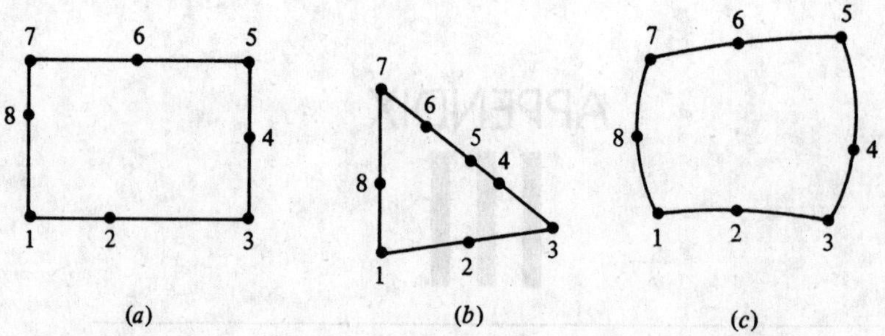

FIGURE III.1
Rectangular, triangular, and general quadrilateral areas.

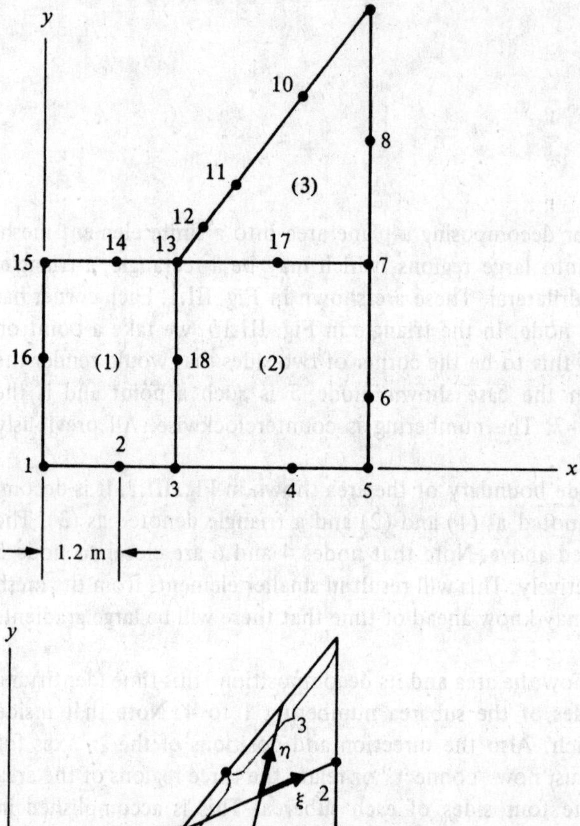

FIGURE III.2
Region decomposed into two rectangles and one triangle, showing nodes.

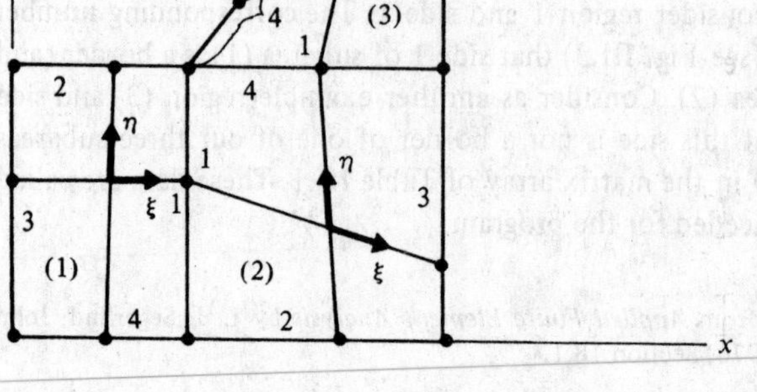

FIGURE III.3
Region showing sides of subareas and local coordinates.

Table III.1

Region	Side 1	2	3	4
1	2	0	0	0
2	1	0	0	3
3	2	0	0	0

We now list the input data for the program of automatic mesh generation. We have a *parameter* card. This card gives two items. They are:

INRG The number of regions (or subareas)
INBP The total number of subregion boundary points for all subregions

For our case, we have three regions (or subareas) and 18 boundary points and we want to punch the element data. Thus, we have on our data card

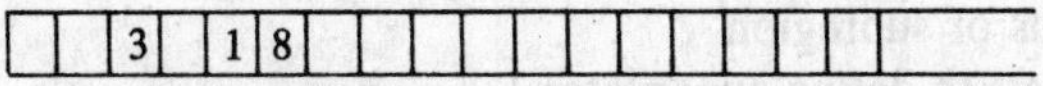

In the next data card set we give the *x coordinate* and *y coordinate* of all 18 boundary points.

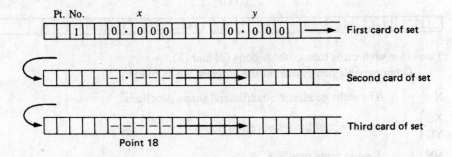

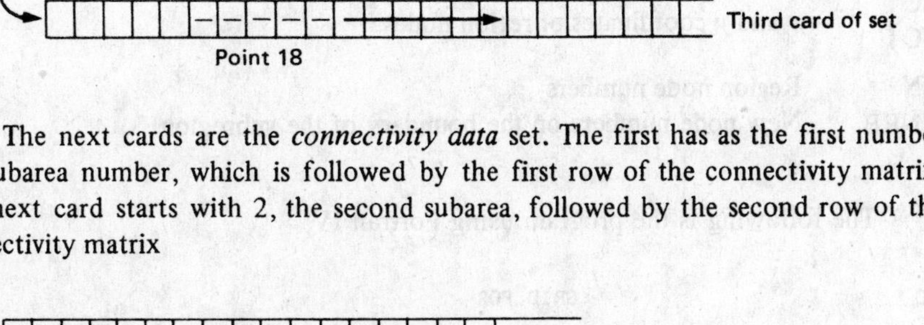

The next cards are the *connectivity data* set. The first has as the first number the subarea number, which is followed by the first row of the connectivity matrix. The next card starts with 2, the second subarea, followed by the second row of the connectivity matrix

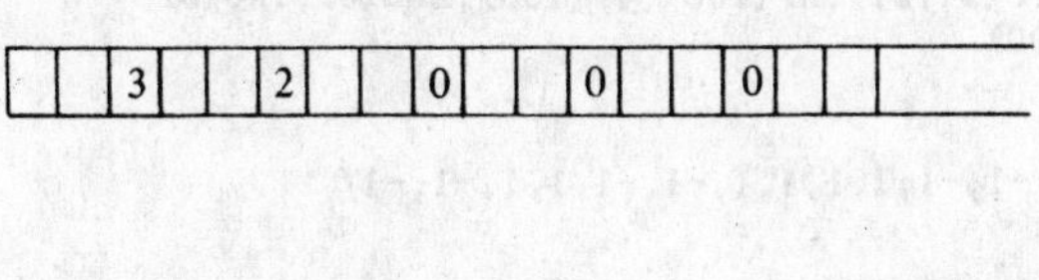

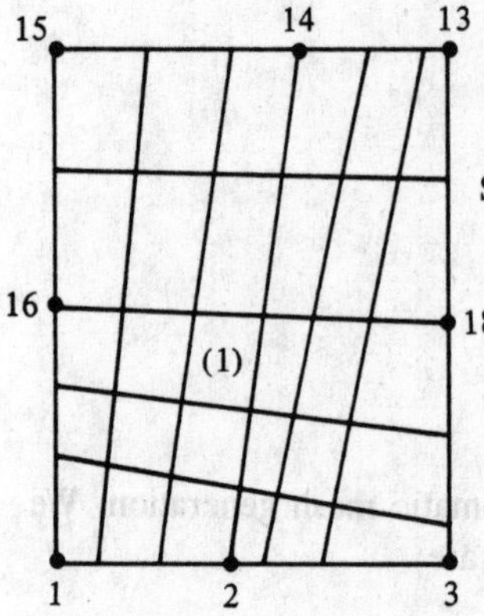

FIGURE III.4
Region (1) showing rows and columns as well as global node numbers. Start at side 1 (see Fig. III.3) and go counter-clockwise.

Finally, we have the *region data*. The numbers on any one card apply to a region with the following notation and data:

NRG	Region number
NROW	Number of rows of subregion
NCOL	Number of columns of subregion
NDN	Global node numbers to define quadrilateral

Thus, for the first card of this set we have NRG = 1. The numbers of rows and columns are the next two numbers on the card. Examining Fig. III.4, we see that NROW = 7 and NCOL = 6. Next inserting NDN, we get for the card for region (1)

1	7	6	3	18	13	14	15	16	1	2

Two other such cards then cover regions (2) and (3).

Other variable names used in the program are as follows:

N	The eight quadratic quadrilateral shape functions
XC, YC	*x* and *y* coordinates of region nodes
NN	Region node numbers
NNRB	New node numbers on the boundary of the subregion

The following is the program using Fortran IV.

```
C                         GRID.FOR
C
C     THIS PROGRAM IS FOR AUTOMATIC MESH GENERATION BY ISOPARAMETRIC
C     INTERPOLATION METHOD.

      DIMENSION ICOMP(4,4),JT(20,4),NDN(8),NN(21,21),NNRB(20,4,21)
      DIMENSION XC(21,21),YC(21,21),XP(100),YP(100),XRG(8),YRG(8)
      DIMENSION X(400),Y(400)
      INTEGER NEL(400,3)
      REAL NP(8)
      DATA ICOMP/-1,-1,1,1,-1,-1,1,1,1,1,-1,-1,1,1,-1,-1/
```

```
      OPEN(4,FILE='IGRID.DAT',STATUS='UNKNOWN')
      OPEN(7,FILE='RGRID.DAT',STATUS='UNKNOWN')
      OPEN(8,FILE='GRIDP.DAT',STATUS='UNKNOWN')
      READ(4,*) INRG,INBP
      READ(4,*)(I,XP(I),YP(I),I=1,INBP)
      DO 5 I=1,INRG
    5 READ(4,*)NRG,(JT(NRG,J),J=1,4)
C*************************************************************************
C     USE ISOPARAMETRIC INTERPOLATION FUNCTION FOR 8 NODES TO GENERATE
C     NODAL COORDINATES FOR EACH SUBREGION.
C*************************************************************************
      DO 60 KK=1,INRG
      READ(4,*)NRG,NROW,NCOL,NDN
      DO 10 I=1,8
      II=NDN(I)
      XRG(I)=XP(II)
   10 YRG(I)=YP(II)
      DETA=2./(NROW-1)
      DSI=2./(NCOL-1)
      DO 15 I=1,NROW
      ETA=-1.+(I-1)*DETA
      DO 15 J=1,NCOL
      SI=-1.+(J-1)*DSI
      NP(1)=-0.25*(1.-SI)*(1.-ETA)*(SI+ETA+1.)
      NP(2)=0.50*(1.-SI**2)*(1.-ETA)
      NP(3)=0.25*(1.+SI)*(1.-ETA)*(SI-ETA-1.)
      NP(4)=0.50*(1.+SI)*(1-ETA**2)
      NP(5)=0.25*(1.+SI)*(1.+ETA)*(SI+ETA-1.)
      NP(6)=0.50*(1.-SI**2)*(1.+ETA)
      NP(7)=0.25*(1.-SI)*(1.+ETA)*(ETA-SI-1.)
      NP(8)=0.50*(1.-SI)*(1.-ETA**2)
      XC(I,J)=0.0
      YC(I,J)=0.0
      DO 15 K=1,8
      XC(I,J)=XC(I,J)+XRG(K)*NP(K)
   15 YC(I,J)=YC(I,J)+YRG(K)*NP(K)
C*************************************************************************
C     GENERATE THE REGION NODE NUMBERS
C*************************************************************************
      KN1=1
      KS1=1
      KN2=NROW
      KS2=NCOL
      DO 35 I=1,4
      NRB=JT(NRG,I)
      IF(NRB.EQ.0 .OR. NRB.GT.NRG) GO TO 35
      DO 20 J=1,4
   20 IF(JT(NRB,J).EQ.NRG) NS=J
      K=NCOL
      IF(I.EQ.2.OR.I.EQ.4) K=NROW
      JL=1
      JK=ICOMP(I,NS)
      IF(JK.EQ.-1) JL=K
      DO 30 J=1,K
      GO TO (22,24,26,28),I
   22 NN(1,J)=NNRB(NRB,NS,JL)
      KN1=2
      GO TO 30
   24 NN(J,NCOL)=NNRB(NRB,NS,JL)
      KS2=NCOL-1
      GO TO 30
```

```
   26 NN(NROW,J)=NNRB(NRB,NS,JL)
      KN2=NROW-1
      GO TO 30
   28 NN(J,1)=NNRB(NRB,NS,JL)
      KS1=2
   30 JL=JL+JK
   35 CONTINUE
      IF(KN1.GT.KN2.OR.KS1.GT.KS2) GO TO 47
      DO 40 I=KN1,KN2
      DO 40 J=KS1,KS2
      ND=ND+1
      NN(I,J)=ND
      X(ND)=XC(I,J)
   40 Y(ND)=YC(I,J)
C***********************************************************************
C     STORE THE BOUNDARY NODE NUMBERS OF THIS SUBREGION FOR REPLACING
C     OF THE BOUNDARY NODE NUMBERS OF THE COMING CONNECTED SUBREGION.
C***********************************************************************
      DO 45 I=1,NCOL
      NNRB(NRG,1,I)=NN(1,I)
   45 NNRB(NRG,3,I)=NN(NROW,I)
      DO 46 I=1,NROW
      NNRB(NRG,2,I)=NN(I,NCOL)
   46 NNRB(NRG,4,I)=NN(I,1)
C***********************************************************************
C     GENERATE THE TRIANGULAR ELEMENTS TOGETHER WITH THEIR NODES.
C***********************************************************************
   47 DO 50 I=1,NROW-1
      DO 50 J=1,NCOL-1
      NR=NR+1
      NE=2*NR
      NEL(NE-1,1)=NN(I,J)
      NEL(NE-1,2)=NN(I,J+1)
      NEL(NE-1,3)=NN(I+1,J)
      NEL(NE,1)=NN(I+1,J)
      NEL(NE,2)=NN(I,J+1)
   50 NEL(NE,3)=NN(I+1,J+1)
   60 CONTINUE
C***********************************************************************
C     PRINT OUT NODE AND ELEMENT NUMBERS AND COORDINATES
C***********************************************************************
      WRITE(7,75)
   75 FORMAT(/,5X,'NODE NUMBER',5X,'X-COORDINATE',5X,'Y-COORDINATE')
      DO 80 I=1,ND
      WRITE(8,82) X(I),Y(I)
   80 WRITE(7,85) I,X(I),Y(I)
   82 FORMAT(2F12.5)
   85 FORMAT(8X,I3,8X,F12.5,5X,F12.5)
      WRITE(7,90)
   90 FORMAT(/,5X,'ELEMENT NUMBER',5X,'NODE(1)',5X,'NODE(2)',5X,
     +'NODE(3)')
      DO 95 I=1,NE
      WRITE(8,96)I,(NEL(I,J),J=1,3),(NEL(I,J),J=1,3)
   95 WRITE(7,99)I,(NEL(I,J),J=1,3)
   96 FORMAT(4I3,1X,3I3)
   99 FORMAT(11X,I3,3X,3(9X,I3))
```

```
C***********************************************************************
C     USE DI3000 GRAPHIC SOFTWARE TO PLOT THE MESH
C***********************************************************************
      WRITE(6,*)' WOULD YOU LIKE TO PLOT THE GRID (Y/N)? IF YOU HAVE
     +GRAPHIC TERMINAL'
      WRITE(6,*)' AND DI3000 SOFTWARE SYSTEM, YOU CAN TRY.'
      READ(5,100) YN
  100 FORMAT(A1)
      IF(YN.EQ.'Y'.OR.YN.EQ.'y') CALL PLOT(ND,NE)
      STOP
      END

      SUBROUTINE PLOT(NN,MM)
      DIMENSION A(3),B(3),NEL(150,3),X(200),Y(200),NOV(200,3),NOE(200)
      INTEGER NN,MM
C***********************************************************************
C     INPUT THE NODE NUMBERS AND COORDINATES FOR EACH ELEMENT.
C***********************************************************************
      REWIND(8)
      READ(8,*) (X(I),Y(I),I=1,NN)
      DO 10 I=1,MM
      READ(8,5)NOE(I),(NOV(I,J),J=1,3),(NEL(I,J),J=1,3)
    5 FORMAT(4A3,1X,3I3)
   10 CONTINUE
C***********************************************************************
C     DETERMINE THE WINDOW OF PLOT
C***********************************************************************
      XMAX=X(1)
      XMIN=X(1)
      YMAX=Y(1)
      YMIN=Y(1)
      DO 20 I=2,MM
      XMAX=AMAX1(XMAX,X(I))
      XMIN=AMIN1(XMIN,X(I))
      YMAX=AMAX1(YMAX,Y(I))
      YMIN=AMIN1(YMIN,Y(I))
   20 CONTINUE
      RATIOX=XMAX-XMIN
      RATIOY=YMAX-YMIN
      XMAX=XMAX+RATIOX/3.
      XMIN=XMIN-RATIOX/3.
      YMAX=YMAX+RATIOY/3.
      YMIN=YMIN-RATIOY/3.
      AR=(YMAX-YMIN)/(XMAX-XMIN)
      XMAX=XMAX*AR
      XMIN=XMIN*AR
      BIG1=RATIOX*0.06/4.
      BIG2=RATIOY*0.08/4.
C***********************************************************************
C     INITIATE THE DEVICE
C***********************************************************************
      CALL JBEGIN
      CALL JDINIT(1)
      CALL JDEVON(1)
      CALL JWINDO(XMIN,XMAX,YMIN,YMAX)
      CALL JVPORT(-1.,1.,-1.,1.)
      CALL JFRAME
      CALL JROPEN(10)
C***********************************************************************
C     PLOT ELEMENTS
C***********************************************************************
      DO 25 I=1,MM
      DO 30 J=1,3
```

```
      A(J)=X(NEL(I,J))
      B(J)=Y(NEL(I,J))
   30 CONTINUE
      CALL JPOLGN(A,B,3)
C***********************************************************************
C     DRAW THE NUMBER OF EACH ELEMENT
C***********************************************************************
      CALL JMOVE((A(1)+A(2)+A(3))/3.,(B(1)+B(2)+B(3))/3.)
      CALL JFONT(6)
      CALL JJUST(2,2)
      CALL JSIZE(BIG1,BIG1)
      CALL JHTEXT(3, NOE(I))
C***********************************************************************
C     DRAW THE NUMBER OF EACH NODE
C***********************************************************************
      DO 25 J=1,3
      CALL JMOVE(X(NEL(I,J)),Y(NEL(I,J)))
      CALL JFONT(1)
      CALL JJUST(2,1)
      CALL JSIZE(BIG2,BIG2)
      CALL JHTEXT(3, NOV(I,J))
   25 CONTINUE
      CALL JRCLOS(10)
      CALL JPAUSE(1)
C***********************************************************************
C     CLOSE THE DEVICE
C***********************************************************************
      CALL JDEVOF(1)
      CALL JDEND(1)
      CALL JEND
      RETURN
      END
```

APPENDIX IV

PLATE PROGRAM FOR TRIANGULAR AND RECTANGULAR ELEMENTS (FORTRAN IV)

```
C                              SI-PLATE.FOR
C
C   THIS PROGRAM IS FOR PLATE UNDER UNIFORM PRESSURE BY USING TRIANGULAR
c   OR RECTANGULAR ELEMENTS. THE NUMBER OF NODES BE LESS THAN 50.
C
      DIMENSION A(12,12),B(12),BDC(12,3),BDX(12,3)
      DIMENSION BDXY(12,3),BDY(12,3),BDY2(12,3),BX2(3,12)
      DIMENSION BXY(3,12),BY2(3,12),CC(12,12),CF(1,12)
      DIMENSION CX(12,12),CX2(12,12),CXY(12,12),CY(12,12),CY2(12,12)
      DIMENSION D(3,3),HCC(3,12),HCX(3,12),HCX2(3,12),HCXY(3,12)
      DIMENSION HCY(3,12),HCY2(3,12),IBC(80,3),QEL(80),QRE(75)
      DIMENSION IDX(80),STIFFRE(75,75),SOL(40,3),S(80,1),BDX2(12,3)
      DIMENSION QMS(150),SHAPE(1,12),STIFFEL(12,12),STIFFMS(150,150)
      DIMENSION WK(500),X2X2(12,12),X2XY(12,12),X2Y2(12,12)
      DIMENSION XP(10),XX(12,12),XX2(12,12),XXY(12,12),XY(12,12)
      DIMENSION XY2(12,12),XYXY(12,12),XYY2(12,12),YY(12,12)
      DIMENSION Y2Y2(12,12),YP(10),YX2(12,12),YXY(12,12),YY2(12,12)
      DIMENSION X(80),Y(80),BC(3,12),BX(3,12),BY(3,12)
      REAL NU
      INTEGER DIMEL,DIMMS,ND(80),IJOB,IA
      COMMON /EQ/XC(10),YC(10),NTYP
      OPEN(16,FILE='RPLATE.DAT',STATUS='UNKNOWN')
      OPEN(15,FILE='IPLATE.DAT',STATUS='UNKNOWN')
      WRITE(16,1)
  1   FORMAT (/////,' COMPUTE THE DEFLECTIONS AND ROTATIONS '
     *,'OF A PLATE UNDER UNIFORM PRESSURE')
C***********************************************************************
C      READ IN THE YOUNG'S MODULUS E(PA.), POISSON'S RATIO NU,
C      THICKNESS H(MM.), AND PRESSURE PR(PA.).
C
C      CONSTRUCT THE MATRIX [D].
C***********************************************************************
      READ (15,*) E,NU,H,PR
      DCOEFF = (E*H**3.0)/(1E9*12.0*(1.0-NU**2.0))
      D(1,1) = DCOEFF
      D(1,2) = DCOEFF*NU
      D(1,3) = 0.0
      D(2,2) = DCOEFF
      D(2,3) = 0.0
      D(3,3) = DCOEFF*(1.0-NU)/2.0
      DO 5 I = 1,3
      DO 5 J = 1,3
  5   D(J,I) = D(I,J)
      WRITE (16,6) ((D(I,J),J=1,3),I=1,3)
  6   FORMAT (/////,' THE MATRIX (D) IS:',///(3E12.4))
```

```
C*********************************************************************
C     READ IN TOTAL NUMBER OF ELEMENTS AND TOTAL NUMBER OF NODES
C     (NELEO,(NNOD) AND THE IDENTIFICATION OF ELEMENT TYPE (NTYP).
C     (NTYP) ALSO REPRESENTS THE NUMBER OF NODES FOR EACH ELEMENT.
C     WE SET
C            NTYP = 3  FOR TRIANGULAR ELEMENT
C                 = 4  FOR RECTANGULAR ELEMENT
C     COMPUTE THE DIMENSIONS OF ELEMENT MATRIX AND MASTER MATRIX
C     (DIMEL), (DIMMS).
C*********************************************************************
      READ (15,*) NTYP,NELE,NNOD
      DIMEL = 3*NTYP
      DIMMS = 3*NNOD
C*********************************************************************
C     READ IN NODAL POINT DATA
C       N        : NODAL NUMBER
C       IBC      : BOUNDARY CONDITION CODE
C      NOTE: EQ.=0: FREE, EQ.=1: FIXED
C       IBC(N,1): Z-DEFLECTION (M)
C       IBC(N,2): X-ROTATION
C       IBC(N,3): Y-ROTATION
C       X(N)     : X-COORDINATE (M)
C       Y(N)     : Y-COORDINATE (M)
C      (ONE DATA CARD FOR EACH NODE)
C*********************************************************************
      WRITE (16,11)
      DO 10 I = 1,NNOD
      READ (15,*) N,(IBC(N,J),J=1,3),X(N),Y(N)
   10 WRITE (16,12) N,(IBC(N,J),J=1,3),X(N),Y(N)
   11 FORMAT (///,'  NODE #     B.C. CODE     X-COORD.     Y-COORD.'//)
   12 FORMAT (I6,7X,3I3,2F12.3)
C*********************************************************************
C     READ IN ELEMENT NUMBER WITH NODE NUMBER SEQUENCE IN C.C.W.
C     DIRECTION.
C          (ONE DATA CARD FOR EACH ELEMENT)
C*********************************************************************
      DO 90 KE = 1,NELE
      READ (15,*) N,(ND(I),I=1,NTYP)
      DO 14 I = 1,NTYP
      I1 = 3*I-2
      XP(I1) = X(ND(I))
   14 YP(I1) = Y(ND(I))
C*********************************************************************
C     CHANGE GLOBAL COORDINATE SYSTEM TO LOCAL CENTROIDAL
C     COORDIANTE SYSTEM.
C*********************************************************************
      SUMX = 0.0
      SUMY = 0.0
      IP = 3*(NTYP-1)+1
      DO 15 I =1,IP,3
      SUMX = SUMX+XP(I)
   15 SUMY = SUMY+YP(I)
      DO 18 I = 1,IP,3
      XC(I) = XP(I)-SUMX/NTYP
   18 YC(I) = YP(I)-SUMY/NTYP
      WRITE(16,20)KE,(XC(I),YC(I),I=1,IP,3)
   20 FORMAT (///,' THE NODAL POINT LOCAL COORD. FOR ELEMENT',I4,
     *' ARE:',//,'      X-COORD.     Y-COORD.',//(2F12.3))
```

```
C***********************************************************************
C     CONSTRUCT MATRIX [A] FOR KE-TH ELEMENT.
C***********************************************************************
      DO 22 I = 1,DIMEL
      DO 22 J = 1,DIMEL
   22 A(I,J) = 0.0
      DO 25 I = 1,IP,3
      A(I  , 1) = 1.0
      A(I  , 2) = XC(I)
      A(I  , 3) = YC(I)
      A(I  , 4) = XC(I)**2
      A(I  , 5) = XC(I)*YC(I)
      A(I  , 6) = YC(I)**2
      A(I  , 7) = XC(I)**3
      A(I+1, 3) = 1.0
      A(I+1, 5) = XC(I)
      A(I+1, 6) = 2.0*YC(I)
      A(I+2, 2) = -1.0
      A(I+2, 4) = -2.0*XC(I)
      A(I+2, 5) = -YC(I)
      A(I+2, 7) = -3.0*XC(I)**2
      IF (NTYP.NE.3) GO TO 23
      A(I  , 8) = XC(I)**2*YC(I)+XC(I)*YC(I)**2
      A(I  , 9) = YC(I)**3
      A(I+1, 8) = XC(I)**2+2.0*XC(I)*YC(I)
      A(I+1, 9) = 3.0*YC(I)**2
      A(I+2, 8) = -2.0*XC(I)*YC(I)-YC(I)**2
      GO TO 25
   23 A(I  , 8) = XC(I)**2*YC(I)
      A(I  , 9) = XC(I)*YC(I)**2
      A(I  ,10) = YC(I)**3
      A(I  ,11) = XC(I)**3*YC(I)
      A(I  ,12) = XC(I)*YC(I)**3
      A(I+1, 8) = XC(I)**2
      A(I+1, 9) = 2.0*XC(I)*YC(I)
      A(I+1,10) = 3.0*YC(I)**2
      A(I+1,11) = XC(I)**3
      A(I+1,12) = 3.0*XC(I)*YC(I)**2
      A(I+2, 8) = -2.0*XC(I)*YC(I)
      A(I+2, 9) = -YC(I)**2
      A(I+2,11) = -3.0*XC(I)**2*YC(I)
      A(I+2,12) = -YC(I)**3
   25 CONTINUE
      WRITE(16,27) KE
   27 FORMAT(///,' THE MATRIX (A) FOR ELEMENT',I4,' IS'//)
      DO 29 I = 1,DIMEL
   29 WRITE (16,30) (A(I,J),J=1,DIMEL)
   30 FORMAT (5X,12F10.4)
C***********************************************************************
C     USE IMSL9.0 SUBROUTINE LINV3F TO INVERT MATRIX [A].
C     NOTE: AFTER LINV3F, [A] CONTAINS THE INVERSION.
C***********************************************************************
      DO 31 I = 1,DIMEL
   31 B(I) = 0.0
      IJOB = 1
      IA=12
      D1 = -1.0
      CALL LINV3F(A,B,IJOB,DIMEL,IA,D1,D2,WK,IER)
      IF (IER.EQ.130) GO TO 999
      WRITE (16,32) KE
   32 FORMAT (///,' THE INVERSION OF MATRIX (A) FOR ELEMENT',I3,' IS')
      DO 33 I = 1,DIMEL
   33 WRITE (16,30) (A(I,J),J=1,DIMEL)
```

```
C*************************************************************************
C      CONSTRUCT MATRIX [H]:
C         [H]=[HCC]+[HCX]*X+[HCY]*Y  FOR TRIANG ELEMENT,
C         [H]=[HCC]+[HCX]*X+[HCY]*Y+[HCX2]*X**2+[HCXY]*X*Y+[HCY2]*Y**2
C      FOR RECTANG ELEMENT.
C*************************************************************************
       DO 35 I = 1,3
       DO 35 J = 1,DIMEL
       HCC(I,J) = 0.0
       HCX(I,J) = 0.0
       HCY(I,J) = 0.0
       HCX2(I,J) = 0.0
       HCXY(I,J) = 0.0
   35  HCY2(I,J) = 0.0
       HCC(1, 4) = -2.0
       HCC(2, 6) = -2.0
       HCC(3, 5) = -2.0
       HCX(1, 7) = -6.0
       HCX(3, 8) = -4.0
       HCY(1 ,8) = -2.0
       IF (NTYP.NE.3) GO TO 36
       HCX(2, 8) = -2.0
       HCY(2, 9) = -6.0
       HCY(3, 8) = -4.0
       GO TO 37
   36  HCX (2, 9) = -2.0
       HCX (3, 8) = -4.0
       HCY (2,10) = -6.0
       HCX2(3,11) = -6.0
       HCXY(1,11) = -6.0
       HCXY(2,12) = -6.0
       HCY2(3,12) = -6.0
       HCY(3,9)=-4.0
C*************************************************************************
C      USE IMSL9.0 SUBROUTINE VMULFF TO COMPUTE [B]:
C          [B] = [H] * (INVERSION OF [A])
C*************************************************************************
   37  N3 = 3
       CALL VMULFF (HCC,A,N3,DIMEL,DIMEL,N3,IA,BC,N3,IER)
       CALL VMULFF (HCX,A,N3,DIMEL,DIMEL,N3,IA,BX,N3,IER)
       CALL VMULFF (HCY,A,N3,DIMEL,DIMEL,N3,IA,BY,N3,IER)
       IF (NTYP.EQ.3) GO TO 38
       CALL VMULFF (HCX2,A,N3,DIMEL,DIMEL,N3,IA,BX2,N3,IER)
       CALL VMULFF (HCXY,A,N3,DIMEL,DIMEL,N3,IA,BXY,N3,IER)
       CALL VMULFF (HCY2,A,N3,DIMEL,DIMEL,N3,IA,BY2,N3,IER)
   38  WRITE (16,39)
   39  FORMAT(///,' THE COEFFICIENT MATRICES OF MATRIX (B) ARE'//)
       WRITE (16,40)
   40  FORMAT(//'  MATRIX (BC) IS:'//)
       DO 41 I = 1,3
   41  WRITE (16,52) (BC(I,J),J=1,DIMEL)
       WRITE (16,42)
   42  FORMAT(//'  MATRIX (BX) IS :'//)
       DO 43 I = 1,3
   43  WRITE (16,52) (BX(I,J),J=1,DIMEL)
       WRITE (16,44)
   44  FORMAT(//'  MATRIX (BY) IS :'//)
       DO 45 I = 1,3
   45  WRITE (16,52) (BY(I,J),J=1,DIMEL)
       IF (NTYP.EQ.3) GO TO 53
       WRITE (16,46)
   46  FORMAT(//'  MATRIX (BX2) IS :'//)
       DO 47 I = 1,3
```

```
 47  WRITE (16,52) (BX2(I,J),J=1,DIMEL)
     WRITE (16,48)
 48  FORMAT(//'  MATRIX (BXY) IS :'//)
     DO 49 I =1,3
 49  WRITE (16,52) (BXY(I,J),J=1,DIMEL)
     WRITE (16,50)
 50  FORMAT(//'  MATRIX (BY2) IS :'//)
     DO 51 I = 1,3
 51  WRITE (16,52) (BY2(I,J),J=1,DIMEL)
 52  FORMAT(1X,12E11.3)
C***********************************************************************
C     CONSTRUCT LOCAL STIFFNESS MATRIX [STIFFEL] FOR KE-TH ELEMENT
C     STEP 1: USE IMSL9.0 SUBROUTINE VMULFM TO COMPUTE
C                  (TRANSPOSE OF [B]) * [D]
C            WHERE
C             [B]=[BC]+[BX]*X+[BY]*Y FOR TRIANGULAR ELEMENT,
C             [B]=[BC]+[BX]*X+[BY]*Y+[BX2]*X**2+[BXY]*X*Y+[BY2]*Y**2
C            FOR RECTANGULAR ELEMENT.
C***********************************************************************
 53  CALL VMULFM(BC ,D,N3,DIMEL,N3,N3,N3,BDC ,IA,IER)
     CALL VMULFM(BX ,D,N3,DIMEL,N3,N3,N3,BDX ,IA,IER)
     CALL VMULFM(BY ,D,N3,DIMEL,N3,N3,N3,BDY ,IA,IER)
     IF (NTYP.EQ.3) GO TO 54
     CALL VMULFM(BX2,D,N3,DIMEL,N3,N3,N3,BDX2,IA,IER)
     CALL VMULFM(BXY,D,N3,DIMEL,N3,N3,N3,BDXY,IA,IER)
     CALL VMULFM(BY2,D,N3,DIMEL,N3,N3,N3,BDY2,IA,IER)
C***********************************************************************
C     STEP 2: COMPUTE (TRANSPOSE OF [B]) * [D] * [B].
C***********************************************************************
 54  CALL VMULFF(BDC ,BC ,DIMEL,N3,DIMEL,IA,N3,CC  ,IA,IER)
     CALL VMULFF(BDC ,BX ,DIMEL,N3,DIMEL,IA,N3,CX  ,IA,IER)
     CALL VMULFF(BDC ,BY ,DIMEL,N3,DIMEL,IA,N3,CY  ,IA,IER)
     CALL VMULFF(BDX ,BX ,DIMEL,N3,DIMEL,IA,N3,XX  ,IA,IER)
     CALL VMULFF(BDX ,BY ,DIMEL,N3,DIMEL,IA,N3,XY  ,IA,IER)
     CALL VMULFF(BDY ,BY ,DIMEL,N3,DIMEL,IA,N3,YY  ,IA,IER)
     IF (NTYP.EQ.3) GO TO 55
     CALL VMULFF(BDC ,BX2,DIMEL,N3,DIMEL,IA,N3,CX2 ,IA,IER)
     CALL VMULFF(BDC ,BXY,DIMEL,N3,DIMEL,IA,N3,CXY ,IA,IER)
     CALL VMULFF(BDC ,BY2,DIMEL,N3,DIMEL,IA,N3,CY2 ,IA,IER)
     CALL VMULFF(BDX ,BX2,DIMEL,N3,DIMEL,IA,N3,XX2 ,IA,IER)
     CALL VMULFF(BDX ,BXY,DIMEL,N3,DIMEL,IA,N3,XXY ,IA,IER)
     CALL VMULFF(BDX ,BY2,DIMEL,N3,DIMEL,IA,N3,XY2 ,IA,IER)
     CALL VMULFF(BDY ,BX2,DIMEL,N3,DIMEL,IA,N3,YX2 ,IA,IER)
     CALL VMULFF(BDY ,BXY,DIMEL,N3,DIMEL,IA,N3,YXY ,IA,IER)
     CALL VMULFF(BDY ,BY2,DIMEL,N3,DIMEL,IA,N3,YY2 ,IA,IER)
     CALL VMULFF(BDX2,BX2,DIMEL,N3,DIMEL,IA,N3,X2X2,IA,IER)
     CALL VMULFF(BDX2,BXY,DIMEL,N3,DIMEL,IA,N3,X2XY,IA,IER)
     CALL VMULFF(BDX2,BY2,DIMEL,N3,DIMEL,IA,N3,X2Y2,IA,IER)
     CALL VMULFF(BDXY,BXY,DIMEL,N3,DIMEL,IA,N3,XYXY,IA,IER)
     CALL VMULFF(BDXY,BY2,DIMEL,N3,DIMEL,IA,N3,XYY2,IA,IER)
     CALL VMULFF(BDY2,BY2,DIMEL,N3,DIMEL,IA,N3,Y2Y2,IA,IER)
C***********************************************************************
C     STEP 3: COMPUTE THE INTEGRAL OF (X**IR)*(Y**IS) OVER
C             THE REGION OF EACH ELEMENT. FOR THE SAKE OF
C             CONVENIENCE, WE CONSTRUCT A EXTERNAL FUNCTION
C             HRS(IR,IS) TO CALCULATE THIS INTEGRAL.
C
C     STEP 4: COMPUTE THE STIFFNESS MATRIX FOR KE-TH ELEMENT.
C***********************************************************************
 55  K0 = 0
     K1 = 1
     K2 = 2
```

```
      K3 = 3
      K4 = 4
      DO 60 I = 1,DIMEL
      DO 60 J = 1,DIMEL
      IF (NTYP.NE.3) GO TO 58
      STIFFEL(I,J) = CC(I,J)*HRS(K0,K0)+(CX(I,J)+CX(J,I))*HRS(K1,K0)
     *+(CY(I,J)+CY(J,I))*HRS(K0,K1)+XX(I,J)*HRS(K2,K0)+(XY(I,J)+
     *XY(J,I))*HRS(K1,K1)+YY(I,J)*HRS(K0,K2)
      GO TO 60
   58 STIFFEL(I,J) = CC(I,J)*HRS(K0,K0)+(CX(I,J)+CX(J,I))*HRS(K1,K0)
     *+(CY(I,J)+CY(J,I))*HRS(K0,K1)+(CX2(I,J)+CX2(J,I)+XX(I,J))*
     *HRS(K2,K0)+(CXY(I,J)+CXY(J,I)+XY(I,J)+XY(J,I))*HRS(K1,K1)+
     *(CY2(I,J)+CY2(J,I)+YY(I,J))*HRS(K0,K2)+(XX2(I,J)+XX2(J,I))*
     *HRS(K3,K0)+(XXY(I,J)+XXY(J,I)+YX2(I,J)+YX2(J,I))*HRS(K2,K1)+
     *(XY2(I,J)+XY2(J,I)+YXY(I,J)+YXY(J,I))*HRS(K1,K2)+(YY2(I,J)+
     *YY2(J,I))*HRS(K0,K3)+X2X2(I,J)*HRS(K4,K0)+(X2XY(I,J)+
     *X2XY(J,I))*HRS(K3,K1)+Y2Y2(I,J)*HRS(K0,K4)
     *+(XYY2(I,J)+XYY2(J,I))*HRS(K1,K3)
     *+(X2Y2(I,J)+X2Y2(J,I)+XYXY(I,J))*HRS(K2,K2)
   60 CONTINUE
C*********************************************************************
C     STEP 5: PRINT THE LOCAL STIFFNESS MATRIX FOR KE-TH ELEMENT.
C*********************************************************************
      WRITE (16,65) KE
   65 FORMAT(///,' THE LOCAL STIFFNESS MATRIX FOR ELEMENT'
     *,I4,' IS'//)
C     DO 67 I = 1,DIMEL
C67   WRITE (16,68) (STIFFEL(I,J),J=1,DIMEL)
C68   FORMAT(1X,12E11.3)
C*********************************************************************
C     CONSTRUCT THE LOCAL FORCE VECTOR FOR KE-TH ELEMENT (QEL).
C*********************************************************************
      CF(1, 1) = HRS(K0,K0)
      CF(1, 2) = HRS(K1,K0)
      CF(1, 3) = HRS(K0,K1)
      CF(1, 4) = HRS(K2,K0)
      CF(1, 5) = HRS(K1,K1)
      CF(1, 6) = HRS(K0,K2)
      CF(1, 7) = HRS(K3,K0)
      IF (NTYP.NE.3) GO TO 72
      CF(1, 8) = HRS(K2,K1)+HRS(K1,K2)
      CF(1, 9) = HRS(K0,K3)
      GO TO 73
   72 CF(1, 8) = HRS(K2,K1)
      CF(1, 9) = HRS(K1,K2)
      CF(1,10) = HRS(K0,K3)
      CF(1,11) = HRS(K3,K1)
      CF(1,12) = HRS(K1,K3)
C*********************************************************************
C     COMPUTE SHAPE FUNCTION FOR KE-TH ELEMENT:
C         (SHAPE) = (CF) * (INVERSION OF [A])
C*********************************************************************
   73 N1 = 1
      CALL VMULFF(CF,A,N1,DIMEL,DIMEL,N1,IA,SHAPE,N1,IER)
      DO 75 J = 1,DIMEL
   75 QEL(J) = PR*SHAPE(1,J)
      WRITE (16,76) KE, (QEL(I),I=1,DIMEL)
   76 FORMAT(///,' THE LOCAL FORCE VECTOR FOR ELEMENT ',I4,' IS',
     *//,1X,12E11.3)
```

```
C***********************************************************************
C     ASSEMBLE THE MASTER STIFFNESS MATRIX [STIFFMS] AND THE
C     MASTER FORCE VECTOR (QMS).
C***********************************************************************
      DO 80 MA = 1,NTYP
      DO 80 MB = 1,NTYP
      DO 80 MC = 1,3
      DO 80 MD = 1,3
      ME = (ND(MA)-1)*3+MC
      MF = (ND(MB)-1)*3+MD
      MG = (MA-1)*3+MC
      MH = (MB-1)*3+MD
   80 STIFFMS(ME,MF) = STIFFMS(ME,MF)+STIFFEL(MG,MH)
      DO 85 MI = 1,NTYP
      DO 85 MJ = 1,3
      MK = (ND(MI)-1)*3+MJ
      ML = (MI-1)*3+MJ
   85 QMS(MK) = QMS(MK)+QEL(ML)
   90 CONTINUE
      WRITE (16,91)
   91 FORMAT(///,' THE MASTER STIFFNESS MATRIX IS')
      DO 92 I = 1,DIMMS
   92 WRITE (16,93) I,(STIFFMS(I,J),J=1,DIMMS)
   93 FORMAT(//,'   ROW',I3,//(2X,9E10.3))
      WRITE(16,94) (QMS(I),I=1,DIMMS)
   94 FORMAT(///,' THE MASTER FORCE VECTOR IS',//
     *,(2X,9E10.3))
C***********************************************************************
C     USE BOUNDARY CONDITION FOR EACH NODE TO REDUCE THE ORIGINAL
C     MASTER STIFFNESS MATRIX AND MASTER FORCE VECTOR INTO
C     REDUCED FORMS [STIFFRE] AND (QRE).
C***********************************************************************
      NDOF = 0
      DO 98 I = 1,NNOD
      DO 98 J = 1,3
      K = (I-1)*3+J
      IF (IBC(I,J).EQ.1) GO TO 98
      NDOF = NDOF+1
      IDX(NDOF) = K
   98 CONTINUE
      DO 100 I = 1,NDOF
      QRE(I) = QMS(IDX(I))
      DO 100 J = 1,NDOF
  100 STIFFRE(I,J) = STIFFMS(IDX(I),IDX(J))
      WRITE (16,101)
  101 FORMAT(///,' THE REDUCED MASTER STIFFNESS MATRIX IS'//)
      DO 102 I = 1,NDOF
  102 WRITE (16,93) I,(STIFFRE(I,J),J=1,NDOF)
      WRITE (16,104) (QRE(I),I=1,NDOF)
  104 FORMAT(///,' THE REDUCED MASTER FORCE VECTOR IS',
     *//,(2X,9E10.3))
C***********************************************************************
C      SOLVE THE EQUATION:
C        [STIFFRE] * (DISPLACEMENT) = (QRE)
C      WHERE (DISPLACEMENT) IS AN UNKNOWN VECTOR WHICH CONTAINS
C      Z-DEFLECTIONS, X-ROTATIONS AND Y-ROTATIONS FOR THE NODES
C      WITH FREE BOUNDARY CONDITIONS.
C***********************************************************************
      IJOB = 2
      IB = 75
```

```
      CALL LINV3F(STIFFRE,QRE,IJOB,NDOF,IB,D1,D2,WK,IER)
      IF (IER.EQ.130) GO TO 999
      DO 105 I = 1,DIMMS
 105  S(I,1) = 0.0
      DO 108 I = 1,NDOF
 108  S(IDX(I),1) = QRE(I)
      DO 112 I = 1,NNOD
      DO 112 J = 1,3
      K = (I-1) *3+J
 112  SOL(I,J) = S(K,1)
      WRITE (16,118)
      DO 115 I = 1,NNOD
 115  WRITE (16,120) I,(SOL(I,J),J=1,3)
 118  FORMAT(///,' THE DEFLECTIONS AND ROTATIONS AT EACH NODE ARE',
     *//5X,'    NODE #     Z-DEFLECTION     X-ROTATION        Y-ROTATION'//)
 120  FORMAT(5X,I8,2X,3E15.3)
 125  FORMAT(/////////,'  S I N G U L A R     M A T R I X')
      GO TO 1000
 999  WRITE (16,125)
1000  STOP
      END

      FUNCTION HRS(IR,IS)
C*****************************************************************************
C     PURPOSE OF THIS FUNCTION: COMPUTE THE INTEGRAL OF
C     (X**IR)*(Y**IS) OVER THE REGION OF EACH ELEMENT.
C*****************************************************************************
      DIMENSION BETA(5)
      COMMON /EQ/XC(10),YC(10),NTYP
      N = IR+IS
      IF (NTYP.EQ.4) GO TO 20
C*****************************************************************************
C     FOR TRIANGULAR ELEMENT, FIRST SET UP THE TABLE OF VALUE FOR
C     BETA.
C*****************************************************************************
      BETA(1) = 0.0
      BETA(2) = 1.0/12.0
      BETA(3) = 1.0/30.0
      BETA(4) = 1.0/60.0
      BETA(5) = 1.0/120.0
C*****************************************************************************
C     COMPUTE THE AREA OF THE TRIANGULAR ELEMENT. THEN COMPUTE THE
C     INTEGRAL:
C         HRS = (AREA)*(BETA)*(SUMMATION OF (X**IR)*(Y**IS)
C*****************************************************************************
      A = 0.5*ABS(XC(1)*(YC(4)-YC(7))+XC(4)*(YC(7)-YC(1))+XC(7)*
     *(YC(1)-YC(4)))
      IF (N.NE.0) GO TO 5
      HRS = A
      GO TO 30
   5  HRS = 0.0
      DO 15 I = 1,7,3
      R = 1.0
      S = 1.0
      IF ( IR.EQ.0) GO TO 7
      DO 6 J = 1,IR
   6  R = R*XC(I)
   7  IF (IS.EQ.0) GO TO 10
      DO 8 J = 1,IS
   8  S = S*YC(I)
  10  HRS = HRS+BETA(N)*A*R*S
  15  CONTINUE
      GO TO 30
```

```
C*******************************************************************
C      FOR RECTANGULAR ELEMENT, THE INTEGRAL
C           HRS = ((-1.0)**N+(-1.0)**IR+(-1.0)**IS+1.0) *
C                  XC(1)**(IR+1)*YC(1)**(IS+1)/((IR+1)*(IS+1))
C      TO MAKE SURE POSITIVE VALUE, WE TAKE THE ABSOLUTE VALUE OF
C      HRS.
C*******************************************************************
   20  A = (-1.0)**N+(-1.0)**IR+(-1.0)**IS+1.0
       HRS = A*XC(1)**(IR+1)*YC(1)**(IS+1)/((IR+1)*(IS+1))
       HRS = ABS(HRS)
   30  RETURN
       END
```

AUTHOR INDEX

SUBJECT INDEX